DEON STOCKERT

The Holt
MODERN BIOLOGY
Program

Modern Biology   Otto and Towle

SUPPLEMENTS TO MODERN BIOLOGY

Biology Investigations   Otto, Towle, and Otto
Laboratory Investigations: Inquiries for Modern Biology   Oakes and House
Tests in Biology   Otto
Alternate Tests in Biology   Otto
Teacher's Edition to the Modern Biology Program

OTHER SUPPLEMENTARY MATERIALS IN BIOLOGY

Environmental Science Sound Filmstrip Program
Holt Biology Film Loops: Educational Research Group
Biological Science: Patterns and Processes Inquiry Slides   Biological Sciences Curriculum Study
Laboratory and Field Studies in Biology   sponsored by National Academy of Sciences—National Research Council
Techniques and Investigations in the Life Sciences   Feldman

HOLT LIBRARY OF SCIENCE   (*paperbacks*)

Animal Photoperiodism, Beck; Experiments in Psychology, Blough and Blough; Our Animal Resources, Fitzpatrick; Our Plant Resources, Fitzpatrick; Protozoa, Hall; Biomedical Aspects of Space Flight, Henry; A Tracer Experiment, Kamen; Human Evolution, Lasker; Life and the Physical Sciences, Morowitz; Photosynthesis, Rosenberg; Viruses, Cells, and Hosts, Sigel and Beasley; Radiation, Genes, and Man, Wallace and Dobzhansky; Cancer, Woodburn; Extraterrestrial Biology, Young

Modern Sex Education   Julian and Jackson

OTHER BASIC TEXTS IN BIOLOGY

Modern Life Science   Fitzpatrick and Hole
Living Things   Fitzpatrick, Bain, and Teter
Biological Science: Patterns and Processes   Biological Sciences Curriculum Study
Modern Health   Otto, Julian, and Tether
Human Physiology   Morrison, Cornett, Tether, and Gratz
Foundations of Life Science   Trump and Volker

*George Porter from
National Audubon Society*

JAMES H. OTTO
ALBERT TOWLE

MODERN BIOLOGY

HOLT, RINEHART and WINSTON, INC.
NEW YORK • LONDON • TORONTO • SYDNEY

James H. Otto

is a biology teacher and head of the Science Department at George Washington High School, Indianapolis, Indiana.

Albert Towle

is a professor of biology and supervisor of biology student teachers at California State University, San Francisco, California.

Cover photograph *Dan Morrill*

Unit opening photographs

1 *Alvin E. Staffan, National Audubon Society*
2 *Photo courtesy of The American Museum of Natural History*
3 *Eric Gravé, Photo Researchers, Inc.*
4 *E. Ellingsen*
6 *N. Smythe, National Audubon Society*
7 *Russ Kinne, Photo Researchers, Inc.*
8 *Douglas Faulkner*

All other photographs are acknowledged on the page on which they appear.

ISBN: 0–03–091337–3

56 032 98

PREFACE

With the flood of new knowledge produced by greatly accelerated biological inquiry in recent years, our picture of life has vastly greater clarity and detail than that of only one or two decades ago. Through the development of new research tools such as the electron microscope and high-speed centrifuge, and techniques for using these tools, biologists have made great advances in cellular biology. Increased knowledge of the cell and its processes has revolutionized many other areas of biology. Improved techniques in biochemistry, including the use of radioisotopes as tracer elements in biochemical reactions, have led to substantial discoveries and new concepts of life at the molecular level. Extensive research in the areas of genetics and microbiology are having far-reaching effects on the lives of all of us. Advances in these areas are bringing us ever closer to the conquest of cancer and other cellular disorders, genetic diseases and disorders and, perhaps, control of the aging process. While much of this knowledge is still at the level of research, many of the latest discoveries have had such an impact on our understanding of life that they need to be included in the high school course in biology. Ecology has taken on new meaning today. It is far more than a branch of biology. It has become a cause and a program for action. Pollution threatens our very survival in the years ahead. Young people are very much aware of this global threat.

The authors of MODERN BIOLOGY have always believed that the learning process should involve a mastery of certain fundamental concepts at the beginning of the course in biology. From these initial understandings, the progression from the cell to protists, to plants and animals, and, finally, to man will come naturally. In following this systematic approach to the study of biology, the student discovers unity in the organisms he studies. Culminating the course, the student explores the ecological relationships of living things and their environmental problems and adaptations. In the study of ecology, it is important that the student realize his place and his responsibility in the living world.

The authors have preserved the approach and methodology that have evolved successfully in thousands of secondary school classrooms and laboratories These features have been tested, tried, and proved effective by thousands of our nation's science teachers. The many professional biologists making significant discoveries in the research laboratories who have learned from earlier editions of this text are evidence of the value of such teaching methods.

MODERN BIOLOGY begins with a consideration of the living condition and discusses the unique properties of living organisms that set life apart from the nonliving. It continues with molecular and cellular biology, from which it moves logically into reproduction and genetics. An understanding of genetics gives meaning to organic variation and methods of scientific classification. Units dealing with microbiology, multicellular plants, invertebrate animal life, the vertebrate animals, and the biology of man follow in logical sequence. The final unit, dealing with ecological relationships, offers a fitting climax and overview of the entire biology course.

In the preparation of this, the eighth revision of MODERN BIOLOGY, the authors have up-dated all areas in which new knowledge is significant to the high school student. The chemistry chapter was re-written for simplification and clarity. Photosynthesis and respiration are discussed in a single chapter in order to compare the matter and energy relations of the two processes. The composition of nucleic acids and the role of DNA, messenger RNA and transfer RNA in protein synthesis are topics included in another revised chapter. A discussion of interferon, a recently discovered body defense against viruses, is included in the chapter on infectious disease. The final unit, dealing with ecology, contains new materials on soil, water, and air pollution and an expanded discussion of forest conservation measures and methods of protecting wildlife and saving endangered species.

As in previous editions of this book, the language of science has been an important consideration. Key scientific terms are included as needed. However, the general style of writing has been kept as informal as possible. The authors have tried to keep the readability in proportion with the age of the average student. Difficult words are pronounced phonetically, and all new words or terms are displayed in boldface italic type and are defined the first time they are used. Italic type is used for emphasis.

The present authors are indebted to the late Truman J. Moon, whose successful texts BIOLOGY FOR BEGINNERS and BIOLOGY were the predecessors of this book. Mr. W. David Otto, a biology teacher at the John Marshall High School in Indianapolis, revised the bibliographies at the end of chapters. The authors are indebted, also, to Dr. J. Paul Burnett of the Lilly Research Laboratories in Indianapolis for his many helpful suggestions.

In regard to the sections in MODERN BIOLOGY dealing with evolution, we feel that we have used scientific data to present this material as theory rather than fact. The information presented allows for the widest possible interpretation that can be applied to any set of values either religious or scientific. We have made every attempt to present this material in a nondogmatic fashion.

CONTENTS

UNIT **1**

THE NATURE OF LIFE

What is life? What is the elusive condition we refer to as the living state? All living things have chemical similarities. They have structural characteristics that set them apart from nonliving materials. They grow by organizing more of their own substance. They reproduce and perpetuate life, generation after generation. They constantly require energy to maintain their many chemical activities. In your study of biology, you will start with an unknown—the living state—and will explore it in many ways. You may never be able to explain life, but your investigations will bring you to a closer understanding of the marvelous condition that is life.

CHAPTER ONE

THE SCIENCE OF LIFE

Biology in a golden age

What is biology? The word comes from the Greek *bios,* meaning "life," and *logos,* meaning "study of" or "science of." Biology is the knowledge about living things that has come to us from previous generations and to which the biologists of our time are contributing. Like the other sciences, biology is a method of investigating events and problems we need to understand and to solve. Biology arose out of man's curiosity about himself and other living things and out of his need to survive and to improve his condition on the planet Earth.

Biology, again like the other sciences, is pushing forward with unbelievable speed. It is no exaggeration to say that we have gained more biological knowledge in the past twenty years than in the previous twenty centuries! Biology is truly in a golden age. Why this sudden explosion of knowledge in our time? Let us examine the position of biology today and see if we can discover what lies behind its many achievements in recent years.

■ *Science has an international heritage.* With the coming of the Renaissance, science began to develop on a worldwide scale. Slowly at first, but with increasing pace, scientists pushed back the barriers of ignorance, superstition, and prejudiced thinking that during the Middle Ages had stifled man's search for understanding. In the sixteenth century, a Belgian medical student, Andreas Vesalius, rebelled against the methods that characterized medieval medicine and established a scientific study of anatomy (Figure 1-1).

In the seventeenth century, William Harvey, an English physician, questioned the ancient belief that blood ebbed and flowed in the veins of the body like the tides of the sea. He proposed instead that it circulated through both the arteries and the veins. Later in that same century, Marcello Malpighi (mahl-PEEG-ee), an Italian scientist, used a microscope and saw in the lung of a frog the capillaries that completed the path between the arteries and veins. This observation provided vital support for Harvey's

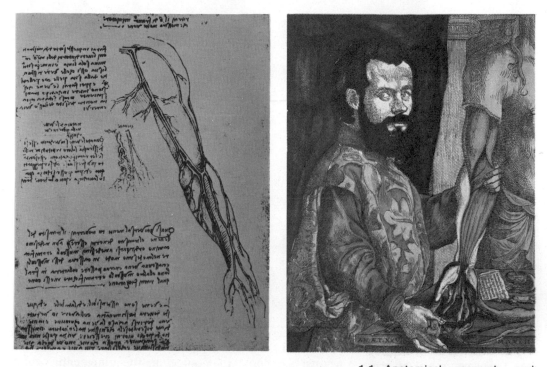

theory of circulation. Harvey's contribution is today regarded as the first major achievement of modern physiology—an important branch of biology.

1-1 Anatomical research and drawings by Leonardo da Vinci (such as the drawing at the left) earned such admiration and respect that they laid the groundwork for the anatomical research of Vesalius (right). (courtesy IBM Corporation; The Granger Collection)

1-2 By the early 1600's, in certain cities of Europe, such as Leyden, it had become fashionable to go to the anatomy theatre to chat. (courtesy WHO)

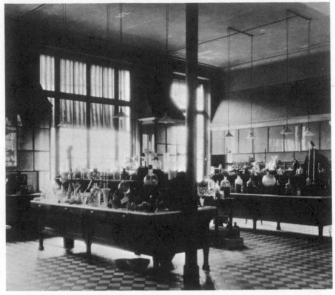

1-3 By the nineteenth century, Louis Pasteur (above left) had founded the science of bacteriology. The photograph of the laboratory for biological chemistry at the Pasteur Institute in Paris (above, right) was taken only a few years after Pasteur's death. (The Granger Collection; The Bettmann Archive)

1-4 Robert Koch discovered how to culture disease-causing bacteria in the laboratory. Identify as many materials and pieces of equipment as you can. (Culver Pictures; The Bettmann Archive)

In the eighteenth century, Edward Jenner, an English doctor, performed the first vaccination when he immunized a small boy against smallpox. In France, in the nineteenth century, Louis Pasteur established the science of bacteriology. A few years later, Robert Koch (KOK) of Germany gave the world the means to investigate infectious diseases by developing techniques for culturing disease-causing bacteria in the laboratory. A century ago, Gregor Mendel (MEND'l), an Austrian monk, conducted

1-5 In the early 1900's, Paul Ehrlich made the first major contribution in the field of chemotherapy. (Culver Pictures)

his famous breeding experiments with peas. Out of his work came several ideas that still serve as the basis for much of the modern science of heredity. In 1909, the German doctor Paul Ehrlich discovered "606," an arsenic compound that would kill the syphilis-causing organism in the human bloodstream. This triumph ushered in the age of chemotherapy in medicine. In 1929, a Scottish physician and bacteriologist, Sir Alexander Fleming, discovered penicillin, the first of a long line of life-saving antibiotics.

■ *Science gradually supplanted superstition and prejudice.* Man is curious by nature. He has always sought explanations for

1-6 Some twenty years later, Sir Alexander Fleming (shown in his laboratory) discovered penicillin, the first modern-day antibiotic.

1-7 Part of a recipe for cough medicine. This page is taken from a thirteenth-century Arabic translation of De Materia Medica, a systematic, 5 volume pharmacopoeia compiled by Dioscorides, a Greek physician and surgeon of the first century. This source, among others, served as an inspiration for later botanical research. (Metropolitan Museum of Art, Rogers Fund, 1913)

the events and phenomena he could not understand. In early times, he was satisfied with answers unsupported by logic or fact, generally arrived at by guesswork or superstitious beliefs. For example, people's lives were supposedly influenced by the position of heavenly bodies at the hour of birth. Did anyone ever find any evidence for this belief? Did anyone even try? As another example, the mud in ponds was supposedly transformed into eels, fish, and frogs. The air from a marsh was thought to cause malaria, which means, literally, "bad air."

It took several centuries to throw off the yoke of superstition and prejudice and establish a new system of objective thinking and investigation. With the new system came scientific freedom: freedom to investigate; freedom to prove and disprove; freedom to base conclusions on observed facts.

■ *Science is a vast body of knowledge.* The fruits of scientific research and investigation are knowledge and understanding. Each generation of scientists receives a heritage of information from preceding generations. This provides the basis for new investigation that will add to the body of knowledge for following generations.

There is no end to scientific exploration. The solving of one problem points to the need for further investigation. You might think of scientific knowledge as a circle of light in a sea of darkness. Through the centuries, the scientists of many nations have expanded this circle enormously. But as the circle has expanded, the size of its perimeter, or the fringe of knowledge, has also increased. In other words, in our circle of biological understanding, the more we learn about life, the more we find remains to be discovered.

■ *Science has progressed at a remarkable rate in areas related to biology.* At one time, definite lines could be drawn between the major areas of science. Today no such separations exist; all scientific knowledge is interrelated. Can we separate chemistry from biology? As you study the organization of the many substances involved in the living condition, you may be surprised to find that everything about life as the biologist has come to understand it can be considered in terms of a vastly complex biochemical system. Growth, response, and heredity all result from chemical activity. In the last analysis, life is chemical in that it involves matter and changes in matter. Life also involves energy changes. The forces that govern the changes in the matter of our earth have a vital influence on all of life.

Can we separate physics, and space science, and earth science, and oceanography from biology? Satellites sent aloft are sending back information that is changing many of our concepts of the earth. The earth is flattened at the poles and bulges at the equator. Its surface has great raised areas and depressions far more extensive than mountain ranges and valleys. Its crust varies in thickness. Oceanographers are exploring the ocean depths

1-8 The moon "rover" was used to help the astronauts explore the surface of the moon. (NASA)

and plotting great currents that circulate water throughout the expanse of the seas. Do all these factors have a profound influence on life and the distribution of living things over the face of the earth? Only in time will the answers become known. These answers will not come easily—they will be found only through the efforts of many generations of biologists and other scientists to come.

■ *The scientist has begun to recognize his own limitations.* In the Middle Ages, "scientific truths" handed down from earlier times were not to be questioned, much less disputed or disproved. Many who dared to disagree with the established authority were ridiculed, persecuted, or forced to flee. Even a generation ago, science was believed to be much more exact than it is known to be today. Fortunately, we no longer consider a scientific explanation to be a final answer. We know that any concept must be held subject to change and revision in the light of new discoveries.

■ *The public has come to accept and support modern science.* Two hundred years ago, townspeople threw rocks at Dr. Edward Jenner when he vaccinated a boy against smallpox, probably saving his life. One hundred years ago, people ridiculed Louis Pasteur when he tried to convince them that invisible microbes were the cause of infectious disease. Today, both these great scientists of the past would undoubtedly receive Nobel prizes for their outstanding contributions to humanity, for our society not only accepts modern science, it also actively supports research programs.

Scientific methods

A scientific method is a logical and orderly procedure of inquiry and investigation. Actually, it is nothing more than the systematic use of common sense. It is this particular method of inquiry that distinguishes scientific study from curious dabbling and hit-or-miss efforts to solve a problem. However, scientific methods are not magic formulas that always lead to solution of a problem. On the contrary, the best-planned, most carefully done scientific experiment can and often does end in failure. But even numerous failures may lead the scientist to final success, since he analyzes each result in the hope of finding a new direction in which to continue his work before he goes on with his investigation. Often this change of direction leads to an even more important discovery than the scientist originally expected.

A number of methods are used in scientific study, depending on the nature of the problem. We shall consider two of these, since both will be used in your biology course.

The research method

In the *research method,* the scientist plans an experiment and outlines the procedure he intends to carry out. It is by this method that new knowledge is gained and new concepts are established. You will have opportunities to use the research method in many phases of your study of biology. A research experiment may be performed by the class, by a group, or by an individual in an area of special interest (Figure 1-9).

The steps a scientist follows in investigating a problem are logical and orderly:

■ *Define the problem.* First of all, scientific research calls for an inquisitive mind and the ability to recognize a problem. For example, how does a root absorb water from the soil? How does light affect the growth of a stem, causing it to bend toward the light? How does a nerve stimulate a muscle and cause it to contract? What controls the rhythmic contractions of the heart? Each of these questions can be answered through well-planned experiments. Problems arise continually in the study of science, and new problems grow out of solutions. In this way successful research leads to new research and to new knowledge.

■ *Collect information relating to the problem.* The scientist does not set out to resolve personally every single aspect of a problem. If he did, science could not progress beyond the limited achievement of a single lifetime. Before beginning an experimentation, the research investigator makes use of all important data and information that relate to the problem. This saves dupli-

1-9 Cecilia Wen-Ya Lo won the first-place scholarship in the Westinghouse Science Talent Search for her research on the process of aging. (courtesy of Westinghouse Corporation)

cation of effort and repetition of work already done. Thus, an extensive library of research papers, scientific journals, and reference books is an important part of a research center. Your textbook and laboratory guide together with supplementary readings will serve as sources of information in solving biological problems in your course.

■ *Formulate a hypothesis.* When available information fails to yield an explanation of the problem, it becomes necessary for the researcher to proceed further by means of experimentation. At this point, he uses his own creative thinking and reasoning to set forth in statement form a tentative explanation or a trial solution for the problem. This statement is called a ***hypothesis.*** It might also be called a scientific hunch or an educated guess. However, while the hypothesis may seem to be a reasonable solution or result in the light of known facts, it cannot be accepted until supported by a large base of evidence. Thus, the research worker must not only be imaginative enough to work out a hypothesis but also open-minded enough to modify or discard it if the evidence fails to support it.

■ *Experiment to test the hypothesis.* The scientist must set up an experiment in which the hypothesis will either be supported or contradicted. While it is often difficult to do, all factors except the one to be tested must be removed or accounted for. We refer to this one factor as the ***single variable,*** or ***experimental factor.*** In other words, the researcher must limit his experiment to the testing of only one condition—the one involved in the hypothesis. Frequently, an experiment is conducted in duplicate, with all factors the same in the second experiment except for the experimental factor. This second, or ***control,*** experiment demonstrates the importance of the missing experimental factor.

■ *Observe the experiment.* What does the experiment prove? What does it disprove? At this point, the scientist must use critical observation. What are the results in relation to the hypothesis? Bear in mind that the experiment that does not work as planned often yields results even more important than those expected.

■ *Organize and record data from an experiment.* Every phase of the experiment—the way it was planned and set up, the conditions under which it was conducted, significant observations made during its progress, and the results—must be recorded accurately. This record may be in the form of notes, drawings, tables, graphs, calculations, or some combination of these. In modern research, data are often processed by means of computers.

■ *Draw conclusions.* Scientific data are valuable only when they are put to use. This is accomplished by drawing valid conclusions from experimental evidence. Such conclusions must be based *entirely* on facts demonstrated in the experiment. If experimental evidence continues to support the hypothesis over a period of time, the hypothesis may come to be called a ***theory.***

■ *Accurately report research methods, results, and conclusions.* Results of scientific research are frequently published in papers and journals, thus becoming valuable contributions to scientific literature. To publish results is a recognized obligation of research scientists. Through the literature, scientists the world over are kept informed of significant developments in their particular fields, and of new research in progress. This cooperative exchange of information saves effort, time, and money and speeds up scientific progress.

Conducting a controlled experiment by the research method

We can illustrate the steps followed in the research method by considering a simple controlled experiment that you can conduct in the laboratory. The experiment will involve the growth and development of bean seedlings. It will relate to one environmental factor—light. We can define the problem in this way: *Is light necessary for the normal growth and development of a bean seedling?*

Having defined the problem, we should next examine various references in the library for information concerning the relation of light to plant growth and nutrition. Much has been written on the subject. However, you may not find a specific answer to your problem about bean seedlings.

At this point, lacking full information, you formulate a hypothesis, or tentative answer to your question. You may assume, for instance, that *light is necessary for the normal growth and development of bean seedlings.* This hypothesis must now be supported or contradicted by experimentation. A logical procedure would involve the germination and growth of two sets of bean seedlings one set in a dark place and the other placed in full light.

Two beans are planted in each of six three-inch pots filled with loose, sandy soil. Three of the pots are marked *experimental* and are placed in a dark cupboard. The other three pots are marked *control* and are set on a window shelf or in some other place where they will receive full light. The temperature should be as nearly uniform as possible in the two locations. The soil in each of the pots must be watered regularly and uniformly throughout the experiment, which lasts about four weeks.

As the experiment progresses, accurate observations must be made each day of the condition of each seedling. The date on which each seed sprouts should be recorded. You should also determine daily and record in a table the length and diameter of the stems, the number and size of the leaves, and the color of the plants.

It is likely that there will be striking differences between the two sets of plants. Those grown in the light probably will have sturdy stems and large, healthy green leaves, whereas those grown in the dark probably will have longer, spindly stems and small yellow leaves. These results provide strong support for the hypothesis: *Light is necessary for the normal growth and development of bean seedlings.*

Recall that in any controlled experiment, while it is necessary to consider all possible factors, there is to be only a single variable. In this experiment, light was the only factor that varied in the experimental and control groups. If any of the other factors had varied, the results would not have been valid. If, for example, the seeds in the dark set had been planted in clay rather than in loam, you would not have known whether the poor growth was caused by lack of light or by poor soil.

While we can conclude from this experiment that light is necessary for the normal growth and development of bean seedlings, this statement immediately raises several more questions. How much light is required? We know that light consists of radiations of various wavelengths, which appear as the colors of the spectrum. Does a plant require red, yellow, green, blue, and violet rays equally, or are certain colors absorbed more than others are? We also know that plants are normally subjected to periods of light and darkness. Is a dark period important? Why is light necessary for the growth of green plants, such as beans? These are but a few of the questions that grow out of a basic experiment involving light and the growth of plants. A hypothesis that has strong support usually leads to many more experiments.

The technical method

While research is a vital part of science, a far greater number of people are engaged in the allied area of technology. The ratio of technicians to research scientists has been estimated to be twenty to one.

The technician is seeking neither support for a hypothesis nor new knowledge. Instead, he uses established procedures and standardized methods to make accurate checks and verify results. The *technical method* involves several steps that can be summarized as follows:

■ *Carry out an outlined procedure without variation.* The results are valid and reliable only if the procedure was followed accurately.

■ *Make accurate observations.* An error or an oversight might make the entire effort worthless.

■ *Record and report all findings.* Again, the technician must be extremely accurate in recording all results of a procedure.

The technical method, often combined with the research method in scientific investigation, is widely used in all branches of science. Technicians follow outlined procedures in checking the bacterial content of water, milk, and other foods; in identifying various bacteria; and in checking the strength of antibiotics and other drugs used in medicine.

You as a student can also learn much by following outlined procedures in your laboratory work. Some of the technical procedures that you will probably use are those followed in preparing materials for microscopic study, in extracting pigments from leaves, and in dissecting specimens in order to study their internal anatomy (Figure 1-10).

Pure and applied science

We often make a distinction based largely on the nature and purpose of the work between *pure science,* or basic research, and *applied science.* In pure science, research is conducted for the sake of knowledge itself. Applied science makes practical use of this knowledge. For example, much basic research has been conducted in recent years on the effects of radiation on living matter. It remained for applied science to make use of this knowledge in destroying cancerous tissue by means of radiation.

1-10 Students in a high school biology laboratory. Try to identify which step of the *technical method* each student is carrying out. (John King)

A biologist looks at life

Earlier generations of biologists made significant progress in developing an understanding of the close interrelationship that

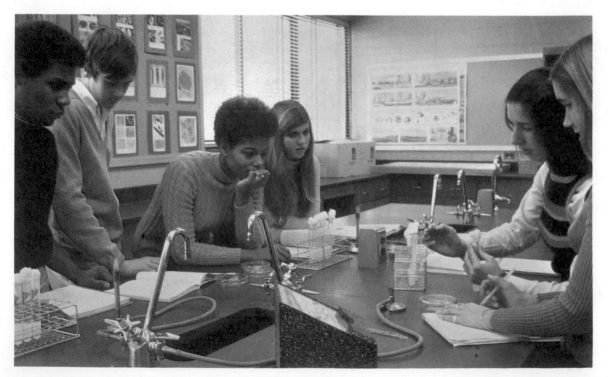

exists between the general structure of living things and how they function. They also reached a basic understanding of the inter-relationships that exist between organisms and their surroundings—both their inanimate physical surroundings and the other plants and animals that share their environment, some supplying needs and others threatening survival.

By using new tools, techniques, and methods, the present-day biologist is making still greater progress in gaining a new, broader understanding of life. Turning to the cell, the biologist investigates the extremely small cell structures—known only through use of the electron microscope—in an attempt to determine how they function. The definite lines that were once thought to exist between biology, chemistry, and physics have become more and more faded. The continuing investigation of such phases of cell activity as growth development, genetic continuity, genetic control, energy relations, and organization of vital chemical substances has been a unifying influence.

As you can see, biology has become a complex science composed of numerous specialized branches, each investigating life in a different way. The names of many of these branches will be used in this book as our study of life continues. Look in a dictionary for a brief description of the areas of study that concern the scientists working in each of these branches.

The microscope—an important tool of the biologist

It is difficult to say who actually invented the microscope. Since a single lens, such as a magnifying glass, is really a *simple microscope,* we would have to go back to the Middle Ages to find its inventor, for lenses capable of magnifying ten to twenty times were ground during that period.

The earliest known *compound microscope*—made in about 1590 by the Janssen brothers, two Dutch spectacles-makers—was a crude instrument with two lenses, one mounted in each end of its tube-within-a-tube barrel (Figure 1-11a). This arrangement permitted one lens to magnify the enlarged image of the other lens, and it also permitted the instrument to be focused by sliding the metal tubes of the barrel together or apart as necessary. Another early compound microscope was used by Galileo, in 1610, to examine biological materials (Figure 1-11b).

Some years later, Anton van Leeuwenhoek (LAVE-in-HOOK), a Dutch merchant, began grinding lenses as a hobby. Altogether, he is said to have used these lenses in making about 250 different simple microscopes, each designed to examine a specific material (Figure 1-11c). One of his early microscopes consisted of a tube for holding a small fish and a frame in which a magnifying lens was mounted. By holding the lens close to his eye, he could see blood surging through vessels in the tail of the fish. With another microscope built to examine pond water, he saw

1-11 Early microscopes: (a) replica of a compound microscope made by the Janssen brothers (courtesy Bausch & Lomb); (b) Galileo's microscope (The Granger Collection); (c) one of van Leeuwenhoek's simple microscopes (Where was the specimen placed? How do you think this microscope was used?) (courtesy Bausch & Lomb); (d) Hooke's microscope. (The Bettmann Archive)

teeming microscopic animals, which he described as "cavorting beasties." With various improvements, van Leeuwenhoek's microscopes could enlarge materials to about three hundred times. Thus, a Dutch merchant pursuing a hobby set the stage for one of the most important fields of today, the field of microbiology.

■ *The modern compound microscope.* The compound microscope has been improved steadily since the time of the Janssen brothers. Improved lenses, which provide greater magnification, and precision mechanical parts are incorporated in the microscopes we use today. An instrument like those in your high school laboratory is shown in Figure 1-12. This light, or optical, microscope contains several sets of magnifying lenses. One set is in the *eyepiece,* or ocular, through which the observer views the magnified material. Other lens systems are contained in *objectives.* Microscopes have from one to four objectives. These, in combination with eyepieces of varying magnifications, give different enlargements. In any case, light, either from a source directly below the lens systems or reflected into the microscope by a mirror, is directed through the lens systems to the observer's eye.

1-12 A modern compound microscope suitable for school laboratory use. (John King)

A standard microscope, used in most high school laboratories, usually provides a low-power magnification of 100 times (100×) and a high-power magnification of 430 times (430×) or 440 times (440×). These magnifications reveal cells and many microscopic plants and animals. Bacteria and other extremely small organisms, as well as the smaller cell structures, require greater magnification. This is provided in a special objective that provides an enlargement of 1,000 to 1,500 times, depending on the eyepiece used with it. Such bacteriological, medical, and research microscopes are high-precision instruments with extremely sensitive optical systems and mechanical parts.

■ *Limitations of the light microscope.* You may wonder why, in this age of precision instruments for nearly every use, it is not possible to build a light microscope with a magnification much greater than 1,500 times. It is true that we could produce the lenses and the mechanical parts for such a microscope. The problem, however, lies in the properties of light itself.

Two factors must be considered in the use of a microscope. One is **magnification,** or the enlargement of an image. The other is **resolution,** or the production of a visible image in which details can be seen. Light rays that have passed through material on the microscope's specimen stage combine with light rays that have been reflected from the surfaces of this material to form an *intermediate image* (Figure 1-13a). Light from a portion of this intermediate image then passes through the lenses of the objective and the eyepiece. Each of these lens systems bends the light rays in a way that results in the spreading apart of the rays. This forms the *magnified image* that reaches the observer's eye. The greater the magnification, the more the rays are bent and spread. This reduces the amount of light that reaches the observer's eye and decreases the resolution. You will notice the difference in the brightness of the field and the resolution under low power (100×) as compared with high power (430 or 440×). There is a limit to which light rays can be spread and still produce an image with sufficient resolution to be visible. This practical upper limit exists at a magnification of about 1,500 to 2,000 times the actual size.

You might compare this problem with the magnification of a newspaper picture. Without magnification, the picture is clear, and the details are sharp. If you view it through a reading glass, you see that the picture is made up of small dots. Now, if you use stronger and stronger lenses, the dots appear farther and farther apart. Finally, so few dots are visible in the field that the picture itself is lost.

■ *The electron microscope.* An entirely new principle of magnification has allowed the biologist to explore a realm of ultramicroscopic particles that the light microscope had never revealed. This principle is reflected in the **electron microscope**. This remarkable instrument substitutes streams of electrons for light.

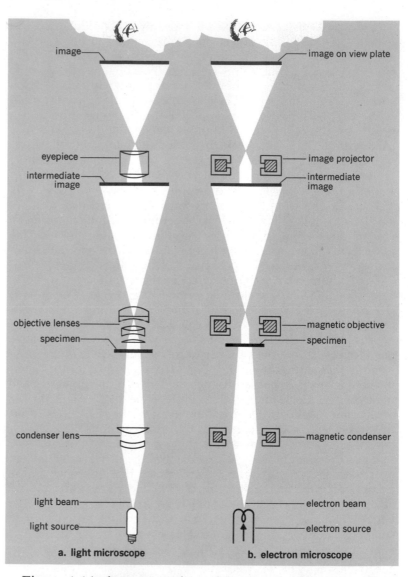

image
image on view plate
eyepiece
image projector
intermediate image
intermediate image
objective lenses
magnetic objective
specimen
specimen
condenser lens
magnetic condenser
light beam
electron beam
light source
electron source

a. light microscope   b. electron microscope

1-13 The optical system of a light microscope compared with the electromagnetic system of an electron microscope.

Figure 1-14 shows a modern electron microscope. We shall follow the electron beam from the energy source to the *view plate* on which the highly magnified image appears (Figure 1-13b). Electrons emitted from a heated, very fine tungsten wire are accelerated as a beam into a vacuum chamber, a chamber nearly free of air molecules with which the electrons would collide and consequently be scattered. An electromagnet, the *magnetic condenser,* focuses the electron beam on the material to be examined. Of the electrons that strike the material, some are scattered as they interact with the material, and others pass through without being deflected greatly; the denser or thicker portions of the ma-

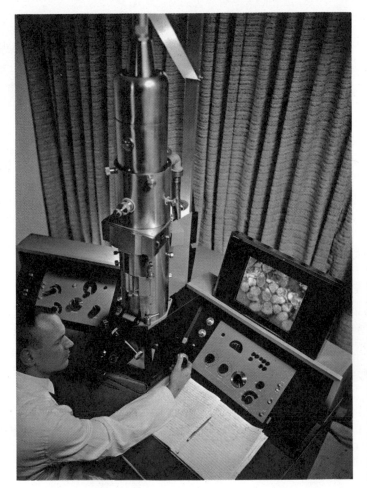

1-14 A modern electron microscope that has been equipped with an auxiliary image intensifier. The image, piped like a television signal, is beamed onto the large fluorescent screen. Because the image is electronically intensified, the electron beam that passes through the specimen is operated at a level lower than would be possible with an electron microscope not so equipped. This results in the specimen lasting longer, since such a beam does damage to the specimen in a given period of time. (courtesy Parke, Davis & Co.)

terial scatter more electrons. The electrons that have not been deflected greatly are then refocused by another electromagnet, the *magnetic objective,* and the image is enlarged by still another electromagnet, the *intermediate image projector.* The electrons then strike a phosphorus-coated view plate (enclosed in leaded glass for safety), producing a visible image. Those areas of the material under study that scattered the electrons rather than permitting them to pass through appear as darker areas in the view plate.

Often, the image on the view plate is photographed for more detailed study. This is done by inserting a photographic plate below the view plate. Prints made from the negatives can be enlarged to make extremely fine details visible to the eye. Of course, only those details that were revealed by the electron microscope appear in the enlargement.

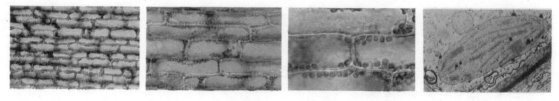

1-15 Cells of the waterweed *Anacharis (Elodea)* at increasing magnification—100X, 1000X, and 13,700X. In the last photograph, a single chloroplast and a small portion of the rest of a single cell fill the image area. (color series: C. Berger, Instrument Division, Nikon Inc., courtesy Myron C. Ledbetter, Brookhaven National Laboratory)

With one set of electromagnets, the instrument magnifies from 1,410 to 32,000 times. Thus, the lowest magnification of the electron microscope overlaps the highest magnification of the light microscope. By changing one of the electromagnets, it is possible to reach a maximum useful magnification of 100,000 to 200,000 times. If this image is enlarged five times photographically, an object in a print appears one million times its actual size.

Now that we have a better understanding of the methods and tools that the biologist uses in his work, we can move on to our consideration of the living condition.

SUMMARY

The knowledge you will acquire and the concepts you will develop during your course in biology represent centuries of contributions of scientists from many lands. Some of this knowledge is centuries old. Much has been acquired in very recent times. On many occasions, your study of biology will lead you to the fringe of biological understanding where hypotheses remain to be formulated and theories developed to extend our knowledge of life still further.

You will have many opportunities to use the methods of inquiry that have led biologists to important discoveries. How curious are you? What do you want to learn about life? How closely do you observe events and occurrences in the living world? What is your place in this vast society of living things? You should find many of these answers in your study of biology.

BIOLOGICALLY SPEAKING

research method	control	applied science
hypothesis	theory	magnification
single variable	technical method	resolution
experimental factor	pure science	electron microscope

QUESTIONS FOR REVIEW

1. Give several examples of scientific achievements that illustrate the international aspects of science.
2. List examples of the influence of progress in areas related to biology on advances in the biological sciences.
3. Describe the steps of the research method.
4. What is a hypothesis?
5. We often think of research as centering in the laboratory. Why is a library equally important?
6. What is the purpose of a control in scientific experimentation?

7. Why is the reporting of methods, results, and conclusions an important part of scientific research?
8. Distinguish between a hypothesis and a theory.
9. Give several examples of application of the technical method.
10. Distinguish between pure and applied science.
11. Name several of the early users of the microscope.
12. Locate the lens systems in a modern compound microscope.
13. Distinguish between magnification and resolution of a light microscope.
14. What energy source is substituted for light in an electron microscope?
15. How is the enlarged image received in an electron microscope?

APPLYING PRINCIPLES AND CONCEPTS

1. Discuss the integral relationship between academic freedom and scientific pursuits.
2. In what ways is science incompatible with superstition?
3. Discuss the ways in which progress in biology has paralleled the perfecting of the microscope.
4. Why is it important that a scientist be willing to recognize his own limitations as well as those of science in general?
5. Outline a controlled experiment designed to test a single experimental factor.
6. Compare the research and technical methods from the standpoint of purpose and procedure.
7. Discuss the principles of an electron microscope that overcome the limitations of the light microscope.

RELATED READING

Books

ANDERSON, M.D., *Through the Microscope.*
Natural History Press (distr., Doubleday & Co., Inc.), Garden City, N.Y. 1965. The use of different kinds of microscopes to probe an unseen world.

ASIMOV, ISAAC, *A Short History of Biology.*
Natural History Press (distr., Doubleday & Co., Inc.), Garden City, N.Y. 1964. Discusses many of the leading biologists (chronologically arranged), explaining their ideas and the fields in which they are important.

JACKER, CORINNE, *Window on the Unknown: A History of the Microscope.*
Charles Scribner's Sons, New York. 1966. Technically sound, yet readable account of the microscope from early times to the field-ion and other highly specialized microscopes of today.

PAYNE, ALMA S., *Discoverer of an Unseen World.*
The World Publishing Co., New York and Cleveland. 1966. The story of Anton van Leeuwenhoek and his microscopes.

CHAPTER TWO

THE LIVING CONDITION

The living and the nonliving

What is life? How did it originate? As a first step in investigating the living condition, we shall make some basic comparisons of living and nonliving things. Consider the natural setting shown in Figure 2-1. It is no problem to separate the living things from the nonliving in this scene. The plants and the mountain lion are

2-1 A mountain lion. Name the living and nonliving things in the photograph. What properties do you use to tell whether something is living or not? (Wilford M. Miller from National Audubon Society)

organisms. That is, they are complete and entire living things, composed of substances that comprise a living system. The water, the rocks, and the fallen log are nonliving. Their substances are different; they do not interact in the way in which the substances interact in the organisms.

It is true, however, that the materials composing the organisms came, directly or indirectly, from the soil, air, and water. It is also true that when the organisms die, the materials of which they are composed will return to the realm of the nonliving. Thus, "living" substances and "nonliving" materials have a close relationship. To distinguish one from the other, we must consider differences in origin, chemical composition, structure, and function. Let us take a closer look at living and nonliving things.

Life is self-perpetuating

Life continues from preexisting life. Biologists call this the principle of *biogenesis* (BY-oe-JENN-uh-siss). The mountain lion shown in Figure 2-1 began its life as a fertilized egg. From this tiny mass of living substance no larger than a pinhead, it grew into a completely new, complex organism. However, the substance from which it grew did not originate in the egg. It was contributed by the mountain lion's parents. In this manner, the life of the parents is perpetuated in the offspring. Similarly, the original substance of the plants came from seeds that were formed from materials of the parent plants. You might compare this perpetuation of life to the lighting of a new fire from one that is already burning.

If a fire goes out, the flame is lost; the fire cannot rekindle itself. This situation can be compared to the death of an organism. But if new supplies of fuel are lighted from burning fires, the flames can be preserved endlessly. Could it be, then, that the original life of the earth still exists in all of the forms in which we find organisms today? Certainly we can find no evidence that living things arise from nonliving materials. Does it surprise you to learn that the concept that *life comes only from life* is a rather recent one?

The myth of spontaneous generation

From ancient times until less than a century ago, people generally believed that certain nonliving materials or dead materials could be transformed directly into living organisms or that one form of life could arise from another entirely different form. That is, a tree could produce a lamb or a goose. We refer to this belief as *abiogenesis* (AY-BY-oe-JENN-uh-siss), or *spontaneous generation.*

The ancients were familiar with the hatching of birds from eggs and with birth in larger animals. However, they knew little or nothing about the growth and development of smaller animals such as insects and worms. Animals of this sort were linked with stories of spontaneous generation.

One of the most astounding accounts of spontaneous generation came from Jean Baptiste van Helmont, a Belgian physician, about three centuries ago. Van Helmont outlined a method, based on a belief dating back to the Greek poet Homer in the ninth century B.C., of producing mice from grains of wheat and human sweat. According to van Helmont, a dirty shirt placed in a container with grains of wheat would produce mice in 21 days. Supposedly the mice would be formed from the fermenting wheat, while the human sweat in the dirty shirt would provide the "active principle" necessary for the process.

Other accounts of spontaneous generation are equally interesting and amazing. Frogs and fish were thought to be generated in clouds during thunderstorms and to fall to the earth with rain. Honeybees supposedly came from the decaying carcasses of animals such as horses. Actually, the insects thought to be honeybees were flies that resemble honeybees. These flies came from maggots that hatched from eggs laid in the carcass. For centuries, however, no one observed this egg-laying, and the belief remained unchallenged.

2-2 Spontaneous generation myths. Geese from trees (The Granger Collection), lambs from melons (The Bettmann Archive), frog and fish from mud, geese from barnacles, mice from wheat.

Redi's blow to spontaneous generation

From ancient times until late in the nineteenth century, the best minds accepted the belief in spontaneous generation without question. No one had ever provided evidence that spontaneous generation could occur, but at that time people drew conclusions without demanding demonstrated facts on which to base them. However, Francesco Redi (RAY-dee), a seventeenth-century Italian scientist, demanded more than unsupported claims and supposition as a basis for his views on spontaneous generation. Can decaying flesh change directly into flies? Redi said No. He claimed that flies came from eggs laid by flies and that the decaying meat provided nothing more than nourishment for the maggots.

In 1668, Redi conducted an experiment in support of his hypothesis that would be considered reliable even by the standards of modern science. According to his own detailed account of the experiment, he placed some pieces of snake, some fish, some "eels of Arno," and a slice of milk-fed veal in each of four clean jars. He then prepared a duplicate set of four jars. One set of jars, which we today would call the *control set,* was left open. The other, the *experimental set,* was covered with parchment and securely sealed with wax. Flies were soon attracted to the open

2-3 Redi's first controlled experiment. This with his second experiment provided evidence that spontaneous generation does not occur.

Experiment 1

open jars

beginning of experiment   end of experiment

sealed jars

beginning of experiment   end of experiment

jars, which they entered to lay eggs. Within a short time, maggots appeared in all of the open jars. Several weeks later, Redi opened the sealed jars and found putrefied meat but no maggots (Figure 2-3). On the basis of evidence from the experiment, Redi concluded that maggots hatch from eggs laid by flies and are not produced by spontaneous generation from decaying animal flesh.

Had he stopped at this point, however, his critics would have argued that air was necessary as an "active principle" in spontaneous generation. It was true that the sealed jars did not admit air. Thus, Redi's first experiment involved not one but two variable factors: air and flies.

In a second experiment, Redi prepared four jars with the same materials as before and covered each with a cloth he called "fine Naples veil." Air passed through the cloth freely, but flies could not. As the meat decayed, flies laid eggs on the cloth, but no maggots appeared in the decaying meat (Figure 2-4). In this second experiment, Redi supplied convincing proof that flies come only from flies. Did his work disprove spontaneous generation? You might think so, but this unscientific, unsupported claim was to remain in general acceptance for more than two centuries after Redi did his work. True, he had proved that flies come only from preexisting flies and that this might also be true of other insects, but he had not demonstrated that worms and other lowly animals do not originate by spontaneous generation.

Microorganisms and abiogenesis

With the development of the microscope, eighteenth-century biologists found various broths and sugar solutions to be swarming with microorganisms. These included bacteria, protozoans, yeasts, molds, and other organisms. This presented a new problem in disproving spontaneous generation: Where had these organisms come from if not from the broths? Two schools of thought developed to explain the origin of microorganisms. One group supported spontaneous generation as the only possible answer. The other insisted that bacteria and other organisms could come

2-4 Redi's second experiment. This experiment was designed to overcome his critics' argument that in his first experiment air was not admitted into the sealed jars.

Experiment 2

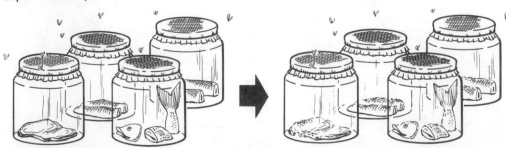

jars covered with fine netting

only from preexisting organisms. (Remember that biologists of that day had no idea that bacteria grow and reproduce and that they are abundant everywhere.)

One of the supporters of the belief in spontaneous generation of microorganisms was the English scientist John Needham. Needham boiled mutton broth in loosely stoppered flasks for a few minutes. After a few days, he examined the broth with a microscope and found it to be teeming with microorganisms. When he repeated the experiment with various meat and vegetable broths, the results were the same. Needham argued that boiling had destroyed all life in the broth and that since the flasks were closed with stoppers, the organisms present after standing must have been formed by spontaneous generation.

Spallanzani's opposition to Needham's conclusions

About twenty-five years after Needham conducted his experiments, Lazzaro Spallanzani, an Italian priest, naturalist, and philosopher, led the opposition in an attack on Needham's experiments as well as on his conclusions supporting the theory of spontaneous generation. Spallanzani was convinced that Needham had not heated his broth sufficiently long to destroy all organisms present and that his flasks were not closed with an airtight seal. He contended that the microorganisms that were abundant in the broth a few days after boiling had grown from surviving organisms or from organisms that entered the flasks from the air. He was convinced they had not come from the broth by spontaneous generation.

In order to demonstrate the flaw in Needham's experiment, Spallanzani conducted a series of experiments of his own. He prepared vegetable infusions by boiling seeds in water for a short time. He placed these infusions in nineteen glass vessels that were then sealed securely by heating the glass and fusing it at the top. All nineteen vessels were placed in boiling water for one hour and then removed and allowed to stand. After several days, Spallanzani opened each flask and examined several drops of the infusion for evidence of life. None of the infusions contained living organisms.

To Spallanzani, this was proof that organisms come only from preexisting organisms present in insufficiently heated infusions or permitted to enter the flasks from the air because the corks are too loose-fitting, not from the infusion itself, as Needham had claimed. But this evidence did not satisfy Needham's supporters. They claimed that boiling for an hour had destroyed the "active principle" or "vegetative power" of the broth or had injured the "elasticity" of the air in the sealed vessels, thus preventing spontaneous generation of organisms.

Spallanzani accepted this challenge and conducted a second series of well-planned experiments designed to prove that heat-

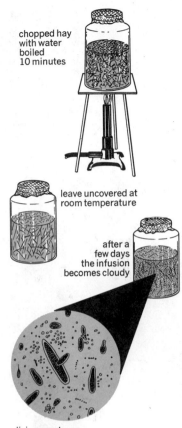

chopped hay with water boiled 10 minutes

leave uncovered at room temperature

after a few days the infusion becomes cloudy

living creatures are discovered by microscopic examination

2-5 Microscopic organisms in a broth infusion.

2-6 Spallanzani conducted many biological experiments unrelated to the work he did to disprove spontaneous generation. Here, he is studying digestion in birds. (The Bettmann Archive)

ing does not destroy an "active principle" in an infusion and thus make it unsuitable for the growth of organisms.

He again made a variety of infusions with the seeds of corn, barley, buckwheat, kidney beans, vetches, beets, and mallow. In addition to the seven kinds of seed infusions, he added an infusion made from the pulverized yolk of a hard-boiled egg. In preparing the infusions, he boiled all of the seeds and the egg yolk in distilled water for the same length of time. In all, he prepared thirty-two infusion flasks and divided them into four sets, each containing identical kinds of infusions. It was Spallanzani's idea that the longer a vessel was heated, the greater would be the amount of "active principle" destroyed. Accordingly, he boiled one set of infusions half an hour; a second set, an hour; a third set, an hour and a half; and a fourth set, two hours.

After boiling, all the vessels were loosely corked, carefully marked, and placed together for several days. Spallanzani could have left all of the flasks uncorked. After all, the object of the experiment was to demonstrate that heating did not destroy the "active principle" necessary for spontaneous generation in an infusion, not to prove that microorganisms entered from the air. If heating *did* destroy the "active principle," the number of microorganisms that appeared should decrease as the length of heating time was increased in the series. Perhaps the "active principle" in those infusions heated one hour and longer would be destroyed to the extent that there would be no organisms at all.

After eight days, Spallanzani examined several drops of each infusion in the four sets. Microorganisms were present in all infusions, but the types as well as the numbers of organisms varied. In all but the corn infusion he found that those boiled the longest contained the greatest number of microorganisms. What had happened? Spallanzani reasoned that in all but the corn infusion, longer boiling dissolved more seed material and enriched the infusion. Thus, rather than destroying the "active principle" in seed infusions, boiling made them more suitable for the growth of microorganisms. He continued to observe his infusions and after a month found all of them to be equally teeming with microorganisms.

Convincing as Spallanzani's experiments should have been, the supporters of the theory of spontaneous generation did not give up. The argument continued for another fifty years, until a young scientist destined to become one of the world's most renowned defeated the theory with evidence that could not be refuted.

Pasteur's decisive defeat of spontaneous generation

It remained for Louis Pasteur, a nineteenth-century French chemist, to deal a final and convincing blow to the theory of spontaneous generation. As a young man, Pasteur was engaged in studies of fermentation and the chemical changes that occur as sugars are converted to alcohol. He had observed the many yeasts and other microorganisms present in fermenting fruit juices and sugar-beet juice and found that the number of organisms increased as fermentation progressed. Pasteur was convinced that the microorganisms associated with fermentation came from the air; hence, dust particles, grape skins, and all other materials exposed to the air harbored great numbers of bacteria, yeasts, and other minute organisms. In testing this hypothesis, Pasteur carried out several experiments that, while simple, were so well planned and convincing that they could not be refuted by even the strongest supporters of the belief in spontaneous generation.

In his first series of experiments, Pasteur used a variety of liquids that would support the growth of bacteria and other microorganisms. These included an infusion of yeast in water; brewer's yeast, sugar and water; and sugar-beet juice. Each of these liquids was sealed in a long-necked flask and boiled for several minutes. Pasteur then took the now sterile flasks to several places in which the air contained different amounts of dust. Flasks opened along dusty roads were quickly contaminated, as evidenced by the abundant growth of microorganisms within a few days. However, flasks opened on hills and mountains showed much less growth of microorganisms. These results bore out his belief that bacteria and other organisms were present in the air with dust particles.

It now remained for Pasteur to find support for his hypothesis under the controlled conditions of his laboratory. The experiment he set up involved his famous swan-necked flasks. He prepared a liquid containing sugar and yeast and poured it into a long-necked flask (Figure 2-7). He then heated the neck of the flask and bent it into an S-shaped curve resembling a swan's neck. After preparing the flask, he boiled the liquid for several minutes. Air forced out of the flask during the boiling returned through the neck as the liquid cooled. Throughout the experiment, air moved through the open neck. Water and dust particles, however, settled in the trap formed by the bent neck. Even though the sugar solution was in direct contact with the outside air, no organisms appeared in the liquid in the flask. Pasteur found that such a flask would remain sterile for more than a year. But whenever the flask was tipped to allow the liquid to flow into the bent neck and contact the trapped dust, microorganisms appeared in the flask in great numbers within a few days.

Thus, Pasteur's simple experiment with his swan-necked flasks refuted all of the arguments his opponents could advance. Boiling had not destroyed the property of the liquid in the flask to support microorganisms, nor had the experiment excluded the "active principle" of the air thought to be necessary for spontaneous generation. Both the liquid and the air in the flask were suitable

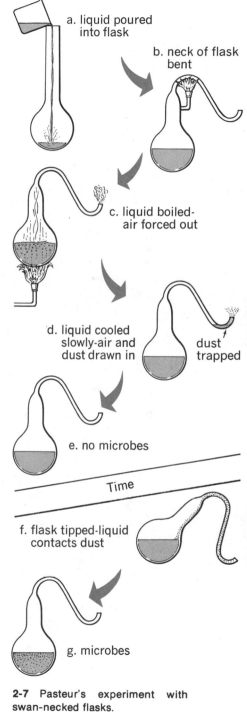

a. liquid poured into flask

b. neck of flask bent

c. liquid boiled-air forced out

d. liquid cooled slowly-air and dust drawn in

dust trapped

e. no microbes

Time

f. flask tipped-liquid contacts dust

g. microbes

2-7 Pasteur's experiment with swan-necked flasks.

START
Tree: 5 pounds
Soil: 200 pounds

END
Tree: 169 pounds, 3 ounces
Soil: 199 pounds, 14 ounces

2-8 Van Helmont's willow tree experiment. From this experiment, he concluded that the living matter of the tree came entirely from water. Of what substance was Van Helmont unaware?

for the growth of bacteria, molds, and other organisms if they were introduced from an outside source.

So it was that Louis Pasteur substituted a proved and valid concept, *biogenesis,* for a theory that had been accepted for centuries without any real evidence: *Life comes only from life*. Experimental evidence has borne this out from the time of Pasteur to the present day. However, we can neither assume that abiogenesis has never occurred nor that it could never happen again. How did life begin in the first place? Will a biologist some day assemble just the right nonliving materials in a test tube and organize living matter? Is a virus particle a form of matter that fluctuates between the nonliving and the living condition? While we accept biogenesis as a fundamental property of the living condition, we must explore this further later in the course.

Van Helmont's tree

Another incorrect conclusion relating to spontaneous generation concerns van Helmont and a willow tree. From what source does a plant derive the substance with which it increases in size and weight year after year? Many people in van Helmont's time assumed that soil changed to the materials of a tree by some form of spontaneous generation. Van Helmont conducted his classic experiment to test this belief.

He filled a large earthenware pot with exactly 200 pounds of previously dried soil. In it, he planted a small willow tree that weighed 5 pounds. He covered the surface of the soil with a metal plate containing holes to admit rainwater and prevent loss of soil from the pot or addition of dust from the air (Figure 2-8). For five years, van Helmont watched the tree grow.

At the close of the experiment, he carefully removed the tree, preserving all of its roots as well as all of the soil. He then weighed the tree and found it to weigh 169 pounds, 3 ounces—a gain in five years of 164 pounds, 3 ounces. Next, he dried the soil and found that it weighed 199 pounds, 14 ounces. Thus, 2 ounces of soil had been used in producing 164 pounds, 3 ounces of tree. At this point, van Helmont arrived at a logical but false conclusion. He concluded that all but 2 ounces of substance in the tree had come from water, the only other source he knew of. What he did not know was that the willow tree had taken in carbon dioxide, an invisible gas, through its leaves and that much of the gain in weight was due to carbon and oxygen from this source. In your study of the chemical processes of plants later in the course, you will discover how this process occurs. You may then think back to van Helmont's excellent experiment but false conclusion.

Organisms have a unique chemical organization

Only in living things do we find the organized activity of a complex system of substances that we refer to as life. We often speak collectively of this complex system as ***protoplasm***. And we would be correct in saying only organisms organize protoplasm.

At one time, biologists thought that protoplasm was actually a living substance. Today we know that it is neither a substance in the sense in which water, salts, sugars, and acids are substances, nor is it living. Furthermore, none of the substances composing protoplasm are living. However, when certain proteins, carbohydrates, fats, and other substances are organized into a system by an organism, a state of chemical activity we call a ***living condition*** is established.

If you analyze the protoplasm of numerous plants and animals, you will find basic similarities in the substances present. In this respect, protoplasm is a unifying characteristic of all organisms. But while there are similarities, there are almost limitless variations in protoplasm. It differs in every kind of organism and between individuals of the same kind. It even varies in different parts of an individual. The materials in your protoplasm are not exactly like those of any other person. Furthermore, these materials are constantly changing. You can see, then, why protoplasm is an indefinite and elusive word. It is still a good biological term, however, so long as you use it properly: not as a definite substance and not as a living material, but as a complex, continually changing system of substances that establishes the living condition.

Organisms have a constant energy requirement

All chemical activities require energy. Since life is basically chemical activity, it requires a constant source of energy. As you will learn, almost all of the energy used by living things comes ultimately from the sun. Both plants and animals, however, obtain energy more directly by the breakdown of complex chemical substances we call *foods*. The energy released in this way is, in turn, used to support other chemical and physical processes. Life continues only as long as these very essential energy transformations occur.

Organisms have a cellular organization

If you conducted a survey of a large number of organisms and parts of organisms, you would discover that regardless of size and complexity, they are all composed of cells—one cell, a few cells, or billions of cells. Because cells enter into the makeup of every organism, they have been referred to as the *common denominator of life*. As you study diverse organisms, from bacteria to seed plants and ameba to man, you will deal continually with cells. Some will be relatively simple in organization, and others will be highly specialized. But the various substances associated with the living condition are always organized in these basic structural units.

Cells may remain for some time after the death of an organism. Thus, we cannot say that the presence of cells is evidence of a living condition. However, we can say that cells are a product of organisms and are never organized in nonliving materials.

If cells are basic units of organisms, then what about the viruses? Are they organisms? Are they living? When you deal with the composition of viruses in Chapter 15, you will discover that virus particles are not cells. They are subcellular, or below the level of cells, in composition. However, they have certain chemical properties that are found only in living cells. Perhaps the viruses are a link between organisms and nonliving substances. We shall explore this possibility more fully in Chapter 15.

Organisms are capable of growth

At least for a time, organisms grow by enlargement. So do crystals and icicles, and they are, of course, nonliving. Is it correct, then, to refer to growth as a characteristic of the living condition? First we must determine just what we mean by growth.

An icicle hanging from a roof may increase in size as water trickles over its surface and freezes without dripping off. This kind of enlargement is growth by external addition. The same would also be true of a crystal growing in a solution.

The growth of living organisms is entirely different. A tree does not grow by taking more of its own substance from the soil, water, and atmosphere and adding it to the material already present. Nor can you say that you grow by adding food to your body or that your body is an accumulation of the foods you have eaten.

There are many complex chemical processes involved in the growth of an organism. During these processes, large and complicated molecules are formed. As a result of these biochemical activities, substances quite unlike those composing nonliving materials are incorporated in the makeup of organisms. Thus, living things do not accumulate their substance; they organize it, or *assimilate* it.

You undoubtedly have recognized that plants and animals often continue to live long after growth has apparently ceased. However, the substances composing organisms are only temporary. Replacement takes place continually. Thus, growth and maintenance without enlargement continue throughout the lives of organisms.

Organisms have a definite form and size range

You can describe a black bear, a jack rabbit, a rainbow trout, a sugar maple tree, or a Douglas fir with reasonable accuracy. With some variation, your description will fit all other animals and plants of the same kind. Furthermore, you can predict the approximate size each will attain at maturity. On what do you base these descriptions and predictions of size? No doubt you

2-9 A zebra with its offspring. Like produces like. (Animals Animals)

would be amazed to see a rainbow trout weighing two hundred pounds or a maple tree towering three hundred feet in a forest. Some fish reach this size, and so do some trees. However, you expect mature plants and animals to resemble their parents in both form and size.

Have you ever wondered why? Perhaps you would say that size and form are determined by heredity. It might be more accurate to say that they are determined by substances called *genes.* Genes cause the rainbow trout to develop a particular form of fins, a certain body form, and a distinctive coloration. All fish have genes, but only rainbow trout have the particular gene structure and combinations that when transmitted to offspring regulate the development of young rainbow trout.

Like produces like. While we are all familiar with this concept, only in recent years have we understood how genes operate in heredity. You will find out much more about the composition and regulatory functions of genes in your study of genetics.

Organisms have a life span

A rock you pick up may be a million years old. And it may remain in its present condition for another million years or more. The substance composing the rock is quite different from the substances composing an organism. Life is activity. When conditions are no longer favorable for this activity, life ceases. Thus, all organisms have a definite period of existence, which we refer to as the *life span.* We may divide this period of existence into periods or stages as follows: (1) beginning, or origin, (2) growth, (3) maturity, (4) decline, and (5) death.

A period of rapid growth follows the formation of an organism. This growth period may last a few minutes, a few weeks, several months, or many years, depending on the organism. As the mass of the plant or animal increases during its period of growth, the rate of growth decreases. Finally maturity is reached. During this period, growth is reduced to repair and replacement of vital substances. Eventually the organism reaches a point at which repair or replacement of all damaged or broken-down materials is impossible. This marks the period of decline, or senility, which is followed inevitably by death.

How long can an organism survive? Here we find great variation. Furthermore, we find that the life spans of organisms are for the most part regulated by factors or conditions that are predetermined. That is, barring disease or accidental death, the life span of any particular plant or animal is about the same as that of all others of its kind. The petunia, marigold, and zinnia plants of your flower garden grow, reproduce, and die in a single season. On the other hand, a white oak tree may live five hundred years. A "big tree" *(Sequoia gigantea)* of California would still be young at that age. These remarkable trees may live thousands of years.

Certain insects live but a few weeks. Five years is old for some fish. The normal life span of a chicken is five to ten years. Horses may reach an age of thirty or more. In the United States, the average life span of man is sixty-eight to seventy years. In a few generations, this may be extended to one hundred years or more.

We can say that a definite life span distinguishes organisms from nonliving materials. Is a life span a positive limitation of the living condition? If an organism can grow and maintain the organization of its substances during most of its life, why can't it live indefinitely? Perhaps it could if certain changes could be avoided.

In your study of biology, you will deal with various one-celled organisms. Among these are bacteria. An individual bacterium may be formed and mature in a half hour or less. After it has matured, it splits into two bacteria, both of which are immature and capable of growth. Thus, a bacterium never dies of old age but continues to live as long as conditions are favorable for growth and reproduction. The same is true of the ameba and other one-celled organisms. Perhaps you have already reasoned that any ameba you see today is part of the first ameba that ever lived.

Continued growth becomes a problem when organisms become larger and many-celled. Despite the fact that our experience has shown us that when a certain size is reached, growth ceases and decline is inevitable, would it be possible to remove a part of a large organism from its own mass, that mass that spells eventual doom, and create an environment in which it can live indefinitely? This was actually done some years ago at the Rockefeller Institute for Medical Research in New York.

On January 17, 1912, Dr. Alexis Carrel removed a small piece of tissue from the heart of a chick just hatching. He placed this mass of throbbing heart tissue in a solution of chicken blood plasma (the fluid part of the blood) and put it in a chamber. The warmth and humidity in the chamber duplicated the conditions in the chick's body. For some time, the heart tissue continued to "beat" normally. Gradually, however, the beat slowed down, and the heart tissue showed evidence of stopping altogether. Apparently, the waste products formed in the active tissue were accumulating gradually and were slowly poisoning it. At this point, the tissue was washed with salt solution, and beating returned to normal. Regular replacement of plasma solved the problem, and the tissue began to grow (Figure 2-10). As the bulk of the tissue increased, a new problem arose. Tissue deep in the mass was no longer receiving nourishment. This problem was corrected by dividing the mass. A routine for care of the tissue culture was established in which new plasma was added and the mass was divided every forty-eight hours. Is the heart tissue still living after more than fifty years? Presumably it could have been had the experiment not ended in the late 1940's. As a part of a normal chick's heart, subject to growth and maturity, it would have died in five to ten years or less. Thus, we can conclude that organisms have definite life spans, even though their substances have the capacity for indefinite life.

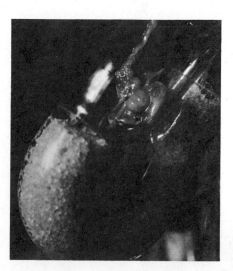

2-10 A piece of the original chick heart tissue continuing to grow in the apparatus developed by Dr. Carrel. (The Rockefeller University)

Organisms have the capacity to reproduce

Since the length of life of a plant or animal is limited, reproduction is necessary for the perpetuation of life. Reproduction takes many forms in the living world. It may be the division of a cell or the formation of a special reproductive cell. It may even consist of the removal from a parent organism of a part capable of independent growth; for example, the cutting of a stem or a root of a plant. The same principle is always involved, however—a mass is divided, or a small portion of a mass is separated from the parent. A seed contains a small amount of the substance of the parent plants. A human being develops from a mass of living material no larger than a pinhead, material contributed by both parents and capable of living and growing for a lifetime.

Organisms are capable of response

Living protoplasm has the capability of responding to external conditions. We refer to this interaction between a living system and its environment as *irritability.* The environmental *stimulus* may be a light factor, temperature, water, sound, pressure, the presence of a chemical substance or a source of food, or a threat to survival. The *response,* or reaction of the organism to the stimulus, varies with the organism's capability. This capability is determined by the structural and physiological organization of the organism. For example, a plant response may be a growth reaction, as when a root pushes through the soil toward a water supply or when a stem grows unevenly and bends toward the light.

Animals are capable of reacting in more complex ways or on higher levels. Sight, hearing, taste, touch, and smell are responses to environmental conditions. More complicated responses include fleeing from an enemy, defending oneself in time of danger, hunting for food, and seeking a place in which to build a nest. Nonliving substances may be influenced by environmental conditions, as happens when water freezes and becomes ice or changes to steam at the boiling temperature. But these changes in form are not responses. Only living organisms are capable of responding to a stimulus.

Organisms have a critical relationship with the environment

All organisms face a constant struggle for life. Part of this struggle centers around requirements for maintaining the living condition. Another part involves a struggle with other forms of life to survive in a highly competitive biological community.

Environmental factors such as light, moisture, oxygen supply, temperature, air currents, soil conditions, and variations in the earth's surface have a direct influence on organisms. Environmental conditions differ in various localities. As these conditions vary, plant and animal life in the region varies. Desert plants and animals cannot survive in moist forests. Nor can prairie life survive in marshes. From the arctic wastelands to the tropics and from mountains to valleys, there are certain kinds of organisms that find each environment ideal.

Its environment must supply plant life with materials that can be organized into the complex chemical substances required as food. Animals in turn eat the plants. Thus, plants become part of the necessary environment of animal life.

Even if the environment is favorable, a plant or animal must compete with other living things. Sometimes the struggle is the competition to acquire the needs of life. At other times, it is a struggle with natural enemies. A small tree growing from the forest floor must compete with many other plants for a place in which to grow. A few survive, while many perish. The robin is a constant threat to the worm and caterpillar. But hawks, crows, and cats, in turn, are a constant threat to robins. The struggle for existence is a problem to all living things.

Variation and adaptation

Conditions in an environment change from time to time. Sometimes these changes are sudden, as they are when there is a severe drought, a destructive storm, or a devastating fire. Other changes may occur much more slowly and over a long period of time. These include climatic changes, changes in soil, and the gradual erosion of hills and mountains. If an organism is not entirely suited to its environment, it can no longer satisfactorily compete with other living things. One of three things must happen: (1) It must *migrate* to more suitable surroundings; (2) *adaptations* must occur; or (3) the organism will *perish* as a species.

Animals capable of movement may leave unfavorable surroundings and seek an environment in which they can meet their needs. These migrations may be seasonal or more permanent, depending on the nature of the environmental changes. Plants lack this motility and must survive or perish in the place in which they are growing. However, even the nonmotile plants produce seeds and fruits or other reproductive structures that may be distributed far from the parent plant. If even a few seeds chance to fall in favorable places, the species survives.

Another characteristic that allows organisms to survive changing conditions is *variation*. This means that no two offspring are exactly alike, nor are they exactly like their parents. While countless variations occur that do not affect survival, some that occur are harmful and may even hasten death. Very occasionally, however, a variation occurs in an organism that improves its chance for survival. If this beneficial variation is passed on to its offspring, it is possible that eventually all organisms of that particular type will have that characteristic. This process, in which the species gradually or rapidly becomes better suited to survive in its environment, is called *adaptation*.

We shall consider one example of adaptation in the deer family. The white-tailed deer ranges over most of North America, Mexico, and Central America. In the north, it is a slender, long-legged animal. It can run at top speed through a dense forest and hurdle logs five feet or more off the ground. In Florida, a tiny deer known as the Key deer lives in the marshes. This "toy" variety of the white-tailed deer weighs at most fifty pounds. It can easily hide in a clump of marsh grass. The Key deer could not survive in the northern forests. Wolves and other flesh-eating animals would have exterminated it years ago. But neither

2-11 These deer illustrate adaptation and natural selection. (Leonard Lee Rue III, James Brogdon from National Audubon Society)

could the northern white-tailed deer find shelter in the marshes of Florida.

In referring to adaptation, we often say that an organism modifies to fit its environment. This does not happen. Plants and animals do not change *in order* to survive. They survive *because of* change. The northern variety of the white-tailed deer did not develop long legs in order to run fast. Rather, those deer that had very long legs and thus were able to run fast survived and produced offspring that also had long legs and could run fast. (Of course, those deer that could not run fast often did not survive long enough to have offspring that would have a similar inability to run fast.) Hence, to have long legs happens to be a favorable variation for the northern white-tailed deer.

Thus, whether variations are favorable or unfavorable, we can add the possibility of variation to our list of characteristics of the living condition.

SUMMARY

The best way to approach the question, What is life? is to compare living and nonliving things. In this chapter, you have considered a few characteristics of the living condition. These properties are common to all living organisms from the most simple and lowly to the most complex and highest in development. They are, however, peculiar to the living condition and do not apply to nonliving things.

Biogenesis is a basic concept accepted by all biologists today. There is no scientific evidence to support the old myth of spontaneous generation. Still, biologists wonder if it did not occur at one time when the first life appeared on the earth. Biologists have proposed several theories to account for the origin of life, but proof and scientific evidence to support these theories are lacking.

It is true that nonliving materials of the earth and atmosphere provide the substances used in the organization of living things. But there is no magic in these transformations. They are part of an orderly, complex, and wonderful chemical activity of organisms, an area of biology into which our discussion of life logically leads.

BIOLOGICALLY SPEAKING

organism	protoplasm	life span
biogenesis	living condition	irritability
abiogenesis	assimilate	variation
spontaneous generation	gene	adaptation

QUESTIONS FOR REVIEW

1. What is the biological concept of an organism?
2. Summarize the meaning of biogenesis.
3. Give examples of myths founded on belief in spontaneous generation.
4. Describe Redi's experiment to disprove spontaneous generation.
5. What evidence in Needham's experiments led him to the false conclusion that microorganisms developed in broth by spontaneous generation?
6. How did Spallanzani refute Needham's conclusion?

7. Describe the flasks Pasteur used in his experiments to disprove spontaneous generation.
8. Explain how van Helmont drew a logical but false conclusion from his experiment with the willow tree.
9. Give a definition of protoplasm in line with modern biological concepts.
10. In what way is growth by assimilation different from growth of nonliving material?
11. Explain how the growth of an organism is regulated internally.
12. List five stages in the life span of an organism.
13. How is the cell the common denominator of life?
14. Define irritability.
15. List several environmental conditions that have a direct influence on living organisms.
16. List three possible consequences of the unsuitability of an organism to its environment.
17. Explain the relationship of variations to adaptations.

APPLYING PRINCIPLES AND CONCEPTS

1. Discuss how Redi's experiments coincide with modern scientific practice. Why was his second experiment of special importance?
2. Describe Spallanzani's method of proving that heat did not destroy the "active principle" in a vegetable infusion and thereby prevent the growth of microorganisms by spontaneous generation.
3. Explain how Pasteur accounted for all variable factors in his experiments disproving spontaneous generation of microorganisms.
4. Outline the biological principles demonstrated in Dr. Alexis Carrel's famous experiment with chick heart tissue.
5. Discuss various ways in which living organisms respond to external stimuli.
6. How is the struggle for existence a problem to all living things?

RELATED READING

Books

ADLER, IRVING, *How Life Began.*
The John Day Company, Inc., New York. 1957. A stimulating discussion of the origin and nature of life and how it differs from the nonliving state.

ASIMOV, ISAAC, *The Wellspring of Life.*
Abelard-Schuman Ltd., New York. 1960. Explains how life came into being and how it reproduces.

BONNER, JOHN T., *Cells and Societies.*
Princeton University Press, Princeton, N.J. 1955. Describes the activities and needs that all living things that are classified together have in common, and explains the variety of ways in which the needs are met.

KEOSIAN, JOHN, *The Origin of Life.*
Reinhold Publishing Corp., New York. 1968. Discusses an answer to the question: "If life arose spontaneously why do we not see evidence of it today?"

OPARIN, ALEXANDER I., *Life: Its Nature, Origin and Development.*
Academic Press Inc., New York. 1962. Traces the development of life from its beginning according to the heterotroph hypothesis formulated by the author.

CHAPTER THREE

THE CHEMICAL BASIS OF LIFE

Biology, matter, and energy

As you continue to explore the living condition and observe life in a great variety of organisms, your need for an understanding of matter and energy will become more and more evident. What chemical substances are removed from the soil and air by the wheat plant when it grows and forms the grain we harvest? How do we transform the wheat and other foods we eat into living body structures?

Every living organism requires a constant supply of energy to maintain the living condition, grow, and reproduce. What is the source of this energy? We know it comes from food. How did it become part of food? In what form is it stored? In what form is it released to support life activities?

Biology includes a study of matter and energy as they relate to living things. The earth and its atmosphere supply the matter used to organize the many substances forming the living world. The sun supplies energy that is stored in chemical compounds formed by living organisms. These chemical compounds serve as the energy sources for maintaining life activities. Scientists understand many of the changes in matter and energy involved in the living condition. Others are still a mystery. Your study of biology will be much more interesting and rewarding if you understand certain of these changes.

What is matter?

You have, no doubt, studied matter and changes in matter in previous science courses. Perhaps you recall that we define matter as *anything that occupies space and has mass*. Anything you could name, except energy, would be included in this definition. Matter includes all solids, liquids, and gases. If we could build a "super" electron microscope and enlarge substances many millions of times, we could see that a solid such as iron, a liquid such as water, or a gas such as oxygen is composed of extremely small particles.

3-1 What chemical substances are removed from the soil and air as the wheat plant grows and forms the grain we harvest? In what form does the plant absorb energy? How is energy stored in the grain?

In the *solid state,* the particles are packed closely together. However, even in a dense solid such as iron or lead, the particles vibrate constantly and have spaces between them (Figure 3-2). In the *liquid state,* the space between particles is much greater. The particles vibrate more actively and migrate freely through the body of the liquid. The space between particles and the movement of particles are greatest in the *gaseous state.* Some gases, such as chlorine, are dense enough to be visible. Often, though, the particles of a gas are so widely separated that the gas is invisible. Air is a mixture of several gases. You cannot see these gases. Does air have mass? Can you weigh it? Does it occupy space? Since the gases composing air are matter, the answer is yes.

Changes in matter

If solid, liquid, and gaseous states of matter are determined by the spacing and activity of particles, a substance could be changed from one state to another by changing the spacing and activity of particles. Such *physical changes* occur regularly.

Water is a good example. At or below the freezing point, water molecules form ice crystals. As heat is applied, the molecules vibrate more rapidly and the ice turns to water. At the boiling point, water changes to steam.

A physical change also occurs as water passes through a plant. It enters the roots and passes through the stem as a liquid. It evaporates from the leaf surfaces and escapes to the atmosphere as vapor. The particles composing ice, water, and vapor are identical in chemical composition. The chemical structure is not altered as it changes from one physical state to another.

When a *chemical change* occurs, matter is changed from one substance to another. A chemical change in water might result in the splitting of water particles and the release of hydrogen and oxygen, the substances that combine to form water. The burning of wood is another example of a chemical change. As the wood burns, substances composing it are changed into other materials. Some are gases that enter the atmosphere. Would it surprise you to learn that one of these gaseous products is water? Residue from the burning remains as ash. Much of this ash is mineral matter. How was the wood formed originally? This required many complex chemical changes in a living tree and materials from the soil and atmosphere. Water was one of these substances used by the tree.

Energy and energy changes

Investigation of the living condition also requires an understanding of energy. Energy supports life and is involved in all life activities. Energy is always associated with matter and with changes in matter. Energy is difficult to define, but is often spoken

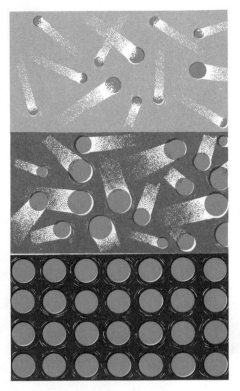

3-2 The solid, liquid, and gaseous states of matter. What characteristics of each state have been illustrated?

of as *the capacity for doing work or causing a change.* When you speak of energy, you may refer to *kinetic energy,* or energy that is actively expressed. This may be *motion* of a mass, such as a moving automobile, or running water. Kinetic energy is also expressed as radiant energy of various wave lengths. These radiations include *electric waves, radio waves, heat waves, visible light, ultra violet rays, X rays, gamma rays,* and *cosmic rays.*

Potential energy is energy that represents a capacity for doing work or causing a change but is not being actively expressed. *Energy of position* is one form of potential energy. *Chemical energy* is another such form.

Potential and kinetic energy are interchangeable. That is, one form can be changed to the other. For example, a boulder lying at the top of a hill has potential energy because of its position. If the boulder is dislodged, its potential energy becomes kinetic energy of motion as it rolls down the hill.

Chemical energy is stored in substances we call compounds. Sugar is such a compound. In fact, a sugar is the basic fuel of organisms. When you think of fuel you probably think of coal, oil, or gasoline. These fuels are also composed of compounds containing chemical energy. Chemical energy in coal remains potential until it is converted to heat and light, forms of kinetic energy, as it burns.

Energy changes also occur in living organisms. Chemical energy in foods is released as heat during chemical changes in the foods. This energy may be trapped and stored in other compounds for later use, converted to mechanical energy, or used to support other chemical reactions involved in maintaining the life of the organism.

Elements—the alphabet of matter

Did it ever occur to you that various numbers and combinations of 26 letters of our alphabet form all the words in a large dictionary? In somewhat the same way, all matter of the earth and atmosphere is composed of 92 natural *chemical elements* and combinations of these elements. Many of these elements are rare, however. Only about 30 are well-known and fewer than half this number form 99 percent of the substances of the earth. We might think of the elements as a chemical alphabet with most of the words formed by 30 of these letters.

Elements are composed of extremely small units of matter called *atoms.* No one has ever seen an atom. Even the largest atoms are less than a fifty millionth of an inch in diameter. The smallest atoms, those of hydrogen, have been estimated to be less than one-fifth this size. Atoms are basic units of matter. This means that *in ordinary chemical reactions* a substance is broken down into its individual atoms. Similarly, you can reduce a word to its letters, but you cannot split the letters.

We designate an atom of an element with a *symbol*. This is chemical shorthand. In most cases, the symbol is the first one or two letters of the name of the element. For example, the abbreviation *C* stands for one atom of the element carbon. *Ca* is calcium, *Cl* is chlorine, *Co* is cobalt, and *Cu* is copper. *H* represents one atom of hydrogen, while *O* is an atom of oxygen. Some chemical symbols are derived from either the Latin or German name for an element. For example, *Na* represents sodium *(natrium)*, *Fe* represents iron *(ferrum)*, and *Ag* is silver *(argentum)*.

The structure of an atom

For several centuries, atoms were thought to be the ultimate, indivisible forms of matter—the smallest particles that could exist. Today, however, we know that atoms are composed of still smaller particles. These same particles make up all atoms. The difference between oxygen, hydrogen, sulfur, iron, gold, and uranium, and all other elements lies not in the kinds of particles composing their atoms, but rather in the number and arrangement of these particles.

Let us explore an atom and begin with the smallest and simplest of all 92 natural elements, the element hydrogen. The central portion of an atom is called its *nucleus*. The nucleus of an ordinary hydrogen atom contains a single particle, known as a *proton*. This proton has a positive charge. Whirling around the nucleus is a second particle. It is extremely small and relatively weightless. This tiny particle, or *electron,* orbiting the nucleus has a negative charge. Thus, the charges of the proton and electron balance each other causing the atom to be neutral.

The second chemical element, helium, has two protons and two electrons. Three protons and three electrons are found in an atom of the element lithium. Carbon, an element you will hear much about in your study of biology, has six protons and six electrons orbiting the nucleus. Oxygen has eight protons and eight electrons. The largest natural atom, uranium, has 92 protons and 92 electrons. We refer to the number of protons in the nucleus as the *atomic number*. There are 92 such numbers in the series of natural elements.

Can you imagine six electrons orbiting the carbon nucleus, eight orbiting the oxygen nucleus, or 92 electrons speeding around the nucleus of uranium? The paths of these electrons have been called an electron cloud. The electrons have also been compared to bees swarming around a hive. Scientists have found that electrons, while orbiting in random and irregular paths, move around the nucleus at given distances. We speak of these as *energy levels*. One or two electrons move at the first energy level, closest to the nucleus. There is a gap to the second energy level in which as many as eight electrons may orbit. Up to eight electrons may also occupy a third energy level. As atoms become larger and larger, the electron arrangement becomes more and more complex.

Let us return to the nucleus of an atom and explore it further. All but the ordinary hydrogen nuclei have one or more additional atomic particles, known as *neutrons*. These particles have approximately the same mass as protons but have no electric charge. Neutrons do not alter the chemical activity of an atom but are involved in its mass. We determine the **atomic mass** of an atom by adding its protons and neutrons. For example, ordinary carbon atoms have 6 protons and 6 neutrons. Thus, we consider carbon to have an atomic number 6 and an atomic mass of 12. Atomic masses are based on this form of carbon. An ordinary hydrogen atom has one-twelfth the mass of the common carbon atoms and thus, is considered to have an atomic mass of 1. An ordinary oxygen atom has 8 protons and 8 neutrons, giving it the atomic number 8 and an atomic mass of 16. If a uranium atom has an atomic mass of 238 and has 92 protons, how many neutrons are present?

What are isotopes?

Several times, in referring to elements such as hydrogen, carbon, and oxygen, we have used the word, "ordinary." Are there other forms of the same element? The answer is yes, but they are still the same element. The number of protons and electrons in a specific element never varies. However, the number of neutrons in the nucleus may vary. For this reason, various atoms of the same element may have different atomic masses. We refer to the different forms of an element as *isotopes.*

Hydrogen is an example of an element with three naturally occurring isotopes. You are familiar with ordinary hydrogen, or protium, with a single proton and an atomic mass of 1. A much rarer form of hydrogen (0.015% in nature) is known as deuterium. This hydrogen isotope has one proton and one neutron in its nucleus. Thus, deuterium atoms have an atomic mass of 2, or twice that of ordinary hydrogen. A third isotope of hydrogen, known as tritium, has one proton and two neutrons in its nucleus. Atoms of this isotope weigh three times as much as ordinary hydrogen.

Various isotopes of carbon may have 5, 6, 7, or 8 neutrons. Ordinary carbon, with 6 neutrons is designated as carbon-12. Carbon-14, an isotope with 8 neutrons has been very important in biological research and will be referred to later. Similarly, there are several isotopes of oxygen. Most oxygen atoms, designated as oxygen-16, have 8 neutrons. Oxygen-18, an isotope of great importance in biology, has 10 neutrons.

Radioisotopes

Certain isotopes of some of the elements have atomic nuclei that can undergo spontaneous change in which charged particles and radiant energy are released. This phenomenon is called

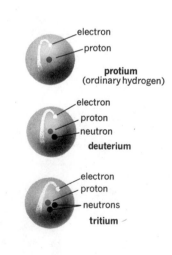

electron
proton

protium
(ordinary hydrogen)

electron
proton
neutron

deuterium

electron
proton
neutrons

tritium

3-3 Representations of the three isotopes of the element hydrogen. While the numbers of protons and electrons are equal and remain unchanged, the number of neutrons varies.

radioactivity. Elements with naturally occurring radioactive isotopes, called *radioisotopes,* are uranium, radium, and thorium. The most common radiations are *alpha particles,* which are helium nuclei (two protons and two neutrons) that travel at a speed of from ten to twenty thousand miles per second, and *beta particles,* which are high speed electrons. *Gamma rays,* which are identical with high energy X rays, are the most penetrating radiations emitted from the nuclei of radioisotopes.

Several radioisotopes used today do not occur naturally. They are produced by exposing nonradioactive isotopes to neutron irradiation in nuclear reactors. Many of these radioisotopes are used in biological and medical research. Others are used in treating thyroid disorders, cancer, and other diseases. You will become familiar with other uses of radioisotopes as you continue your study of biology.

The formation of compounds

Early chemists found, through experimentation, that while some substances consist of only one element, others are composed of a combination of two or more elements. We refer to these chemical combinations of elements as *compounds.* If you think of elements as letters of a chemical alphabet, then compounds are the words they form. We use symbols to designate atoms of the elements. A *formula* is used to indicate a compound. You are familiar with the formula for water, H_2O. This formula shows that a molecule of water is composed of two atoms of hydrogen joined to one atom of oxygen. NaCl is the formula for sodium chloride, or table salt. Cane sugar, $C_{12}H_{22}O_{11}$, contains the elements carbon, hydrogen, and oxygen with 45 atoms in chemical combination. As you become more familiar with compounds, several concepts may be developed.

1. Under certain conditions, *most* elements will combine with one or more other elements. In other words, most elements exhibit the property of *chemical activity.* Certain elements, including helium, neon, argon, krypton, xenon, and radon are almost completely inactive chemically and are spoken of as *inert.*

2. Each element has a characteristic *combining capacity* for joining other elements. That is, a specific atom, such as hydrogen, has chemical properties that cause it to combine with certain other atoms, such as oxygen, but not with all other atoms.

3. In forming compounds, elements combine in *definite proportions*. Thus, the formula for the compound water is always written as H_2O.

4. Compounds exhibit their *own properties* that are unlike the elements composing them. Water has its own properties that are not like those of the hydrogen and oxygen forming it. Similarly, sodium chloride has none of the properties of either the elements sodium or chlorine.

What are chemical bonds?

Chemical bonds are forces that link two or more atoms together in a compound. To understand how bonds are formed, we must examine electron activity more closely.

The chemical activity of atoms involves the number and arrangement of electrons and, more specifically, those at the highest energy level, or farthest from the nucleus. These electrons behave in interesting ways. In some cases, atoms *share* electrons. The shared electrons orbit both atomic nuclei. This sharing of electrons forms a linkage known as a **covalent bond.** Certain atoms share one pair of electrons, each supplying one electron. This linkage is referred to as a single bond. Double or triple bonds are formed when atoms share two or three pairs of electrons. The energy required to break covalent bonds and separate the atoms is known as **bond energy.**

Certain atoms form bonds in an entirely different way. These atoms form compounds by a *transfer* of electrons. Some atoms have a tendency to give up electrons. They are electron *lenders*. Other atoms have a tendency to gain electrons. These are electron *borrowers*. The transfer of one or more electrons from one atom to another establishes a linkage known as an **ionic bond.** Ionic bonds form compounds quite different from those formed by covalent bonds, as you will see.

Molecular and ionic compounds

Atoms that share electrons and form covalent bonds compose units of matter known as **molecules.** These are the smallest units of molecular compounds.

Atoms of gaseous elements, including hydrogen, oxygen, nitrogen, and chlorine form molecules consisting of two atoms of the same element by covalent bonding. We call these **diatomic molecules.** Hydrogen atoms are very unstable and combine, readily, with other hydrogen atoms. When this occurs, the two atoms share their two electrons that orbit both nuclei in the molecule.

We indicate a molecule of hydrogen as H_2. Chlorine atoms, like hydrogen, share one pair of electrons in forming molecular chlorine, Cl_2. Oxygen atoms share two pairs of electrons in forming the molecule, O_2. Three pairs of electrons are shared in molecular nitrogen, N_2. You will find diatomic molecules of gaseous elements involved in many chemical reactions carried on by living organisms.

Atoms of different elements also combine and form molecules by covalent bonding. In forming a molecule of water, two hydrogen atoms form bonds with an oxygen atom. These are single bonds, indicating that the oxygen atom shares a single electron with each hydrogen atom. These bonds are shown in the model of a water molecule illustrated in Fig. 3-5. The two straight lines indicate single bonds. Energy is necessary to break these bonds, split the water molecule, and release hydrogen and oxygen atoms.

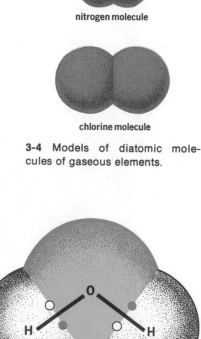

hydrogen molecule

oxygen molecule

nitrogen molecule

chlorine molecule

3-4 Models of diatomic molecules of gaseous elements.

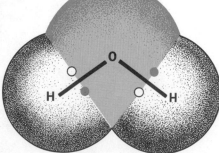

3-5 A representation of the electrons being shared in the covalent bonds of a water molecule.

Now, let us examine a different kind of compound, an ionic compound, formed by electron *transfer,* rather than sharing. We will use common table salt, sodium chloride (NaCl), as our example. You know that common table salt is in the form of crystals. Under a hand lens, salt crystals appear as small cubes. Are these crystals composed of molecules? No, each crystal is formed by an enormous number of positively charged sodium atoms and negatively charged chlorine atoms. Atoms are usually neutral. When they bear positive or negative charges, we call them *ions.* Sodium and chlorine are held together in the salt crystals by the electrical attraction of opposite charges. This linkage is an *ionic bond.* We refer to the compound produced, in this case sodium chloride, as an ionic compound.

How do atoms that are usually neutral change to ions, with positive or negative electrical charges? Examine Figure 3-7 and you will see how this occurs.

A sodium atom has 11 protons and 11 electrons and is electrically neutral. Similarly, a chlorine atom has 17 protons and 17 electrons and is also neutral. However, a sodium atom has a tendency to lose an electron and a chlorine atom has a tendency to gain an electron. The transfer of an electron from sodium to chlorine changes the electrical charge on both atoms. With 11 protons and only 10 electrons, the sodium atom becomes an ion with a charge of $+1$. An additional electron in a chlorine ion gives it 18 electrons with only 17 protons. Thus, a chlorine ion has an electrical charge of -1. Ions are extremely important in biology. Plant roots absorb them, we have them in our blood and tissue fluids, they are essential in muscle contractions and nerve impulses and, in general, are vital in the life of every organism.

Properties of mixtures

No chemical change is involved in forming a *mixture.* No bonding of atoms occurs; no new molecular or ionic substance results. We can best describe a mixture by contrasting its properties with those of a compound:

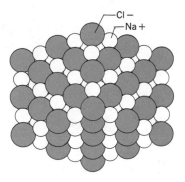

3-6 A model of a portion of a crystal of sodium chloride. The sodium and chloride ions are held together in the crystal by ionic bonds.

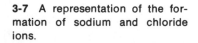

3-7 A representation of the formation of sodium and chloride ions.

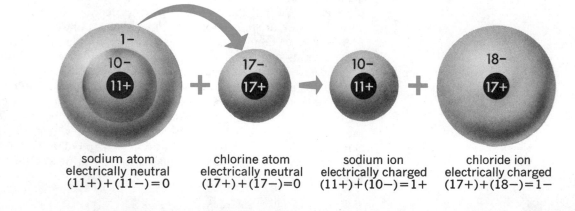

| sodium atom
electrically neutral
$(11+)+(11-)=0$ | chlorine atom
electrically neutral
$(17+)+(17-)=0$ | sodium ion
electrically charged
$(11+)+(10-)=1+$ | chloride ion
electrically charged
$(17+)+(18-)=1-$ |

1. The substances forming a mixture are associated physically or mechanically, rather than chemically, as in a compound.

2. The substances in a mixture can be separated by physical means.

3. The proportions of substances in a mixture are variable, rather than definite.

4. The mixture has the properties of the various substances forming it.

5. Energy changes are not involved in a chemical action in forming a mixture.

Solutions and suspensions

When a mixture of two or more substances is homogeneous, we call it a *solution*. The dissolving medium of a solution is the *solvent;* the dissolved substance is the *solute*. A solution is formed when molecules or ions of a solute are evenly dispersed through molecules of a solvent.

When you add sugar to water, the crystals of sugar dissolve in the water. Molecules of sugar are dispersed evenly among the molecules of water. You can hasten this process by stirring. In this solution, water is the solvent, and sugar is the solute. Sugar water may contain varying amounts of sugar. When at a given temperature a given volume of water is holding all the sugar it can —that is, when no more sugar will dissolve—we say the solution is *saturated*. Molecular solutions may involve a liquid and a gas, two liquids, or a liquid and a solid, as in the example of sugar and water. In all solutions, solute molecules are evenly dispersed through solvent molecules.

Ionic compounds such as sodium chloride are generally very soluble in water. The ion pairs separate, leaving the rigid crystal structure, and spread at random throughout the solution. This process is called *dissociation*. It can be represented by the following equation:

$$Na^+Cl^- \xrightarrow{\text{water}} Na^+ + Cl^-$$

Ionic solutions, such as sodium chloride in water, conduct electricity. We call them *electrolytes*. Molecular solutions, such as sugar water, are *nonelectrolytes;* they do not conduct electricity.

A substance composed of particles that are larger than ions or molecules may form a mixture we call a *suspension*. We can form a suspension by stirring starch into water. The particles may remain dispersed through the water for a time, but they will eventually settle to the bottom. They do this because the force of gravity is greater than the force that holds them in suspension. Thus, the size of the dispersed particles determines whether a substance will be dissolved in a solution or suspended. Generally, particles large enough to form suspensions can be seen with an optical microscope.

Many substances are composed of particles that are larger than the small molecules that form solutions and smaller than the larger particles that settle out in suspensions. The particles of these substances may be very large molecules or groups of smaller molecules. These substances mix with water in *colloidal suspensions*. In describing *colloids* (KAHL-oYDz), we speak of *dispersed particles* rather than solute and of *dispersing medium* rather than solvent. However, particles in a colloid remain suspended, as do molecules or ions in a solution.

Colloidal systems

In the cell, large molecules or groups of molecules are dispersed in water, forming a colloidal suspension. The dispersed molecules have an interesting relationship to water molecules in colloidal systems. Without a change in the water content, a colloid may change from a fluid, or *sol,* to a semisolid, or *gel.* In the sol phase, the dispersed molecules are distributed uniformly through the water molecules. The gel phase results when the dispersed molecules join and produce a spongy network. This physical change traps the water in pools in the molecular net and changes the colloid from a sol to a gel. When the molecules separate and return to a dispersed state, the colloid changes from a gel to a sol. We refer to such a change from sol to gel or gel to sol as a *phase reversal*. This reversal is a common occurrence in living cells.

External conditions such as temperature may cause phase reversals in colloids. Gelatin, for example, forms a gel when it is cool but changes to a sol when it is heated. As the temperature is lowered after heating, it changes back to the gel state. Gelatin is an example of a *reversible colloid*. On the other hand, egg albumen, which is a sol at room temperature, changes to a semisolid when it is heated. A fried egg cannot be changed back to a sol by cooling it. Egg albumen is thus an *irreversible colloid*.

Elements, compounds, and organisms

Earlier in this chapter, we noted that there are 92 natural elements. Now that we are ready to limit our discussion of chemistry to that related to biology, we can narrow our consideration to about eighteen of those elements. These are the elements most common in compounds that form living organisms and supply the substances necessary for their vital chemical activities. The names of these elements and their approximate quantities in the body of a 154-pound man are listed in Table 3-1. While there may be some small variations with regard to amount, a very similar distribution would be found in any organism. Note that those elements present in the greatest amounts are among the most abundant elements of the earth.

Table 3-1 ELEMENTS ESSENTIAL TO MAN

ELEMENT	AMOUNT IN BODY (154-pound man)
oxygen	100.1 pounds
carbon	27.72
hydrogen	15.4
nitrogen	4.62
calcium	2.31
phosphorus	1.54
potassium	0.54
sulfur	0.35
sodium	0.23
chlorine	0.23
magnesium	0.077
iron	0.006
manganese	0.0045
iodine	0.00006
silicon fluorine copper zinc	minute traces

Inorganic compounds—sources of essential elements

If no life were present on the earth, there would still be oxygen, carbon dioxide, nitrogen, and other gases in the atmosphere. There would be water in the air and in rivers, ponds, lakes, and oceans. Mineral compounds would be present in the soil and in the salt water of oceans and seas. These are but a few of the *inorganic compounds* of the earth. We classify them as inorganic because they are not formed by the remarkable processes of organisms. These inorganic compounds of the earth become the chemical sources to build the living world.

Oxygen, as molecular oxygen, composes nearly 21 percent of the mixture of gases making up the atmosphere. This oxygen is essential in supporting respiration in most living organisms. You will discover how oxygen is used in this process in Chapter 6.

Water is the most abundant inorganic compound. It is also the most abundant compound in organisms. Water is so important to organisms that it is a primary factor in determining where living things can survive. We will refer to the role water plays in the living condition again and again during our study of biology. Here are a few ways in which organisms depend on water:

1. Water normally composes from about 65 percent to as much as 95 percent of the total substance of living things. It is the medium in which materials are dispersed in protoplasm.

2. It is the chemical source of hydrogen and some of the oxygen required by living organisms.

3. It is the medium in which soluble materials are absorbed from the environment.

4. It is the medium of transport of foods, minerals, and other vital substances in living systems.

5. Water provides the environment for aquatic organisms.

6. Water pressure provides the firmness that supports soft plant tissues.

Carbon dioxide is an inorganic compound that is a source of carbon as well as of oxygen. The chemical products of organisms contain carbon. Thus, we can say that carbon dioxide is, directly or indirectly, essential for life.

The other essential elements come from *mineral compounds.* These may be in the form of soil minerals, minerals dissolved in water, or salts present in sea water.

Perhaps it has occurred to you that we, as living organisms, do not directly utilize carbon dioxide and many of the essential minerals. We could not produce our bodies from water, carbon dioxide, and minerals. There must be a link between inorganic compounds and the complex carbon-containing substances we require as sources of energy and building materials. The green plants of the earth provide this vital chemical link. You will deal with many phases of this biochemical relationship as you continue your study of biology.

Organic compounds—products of living organisms

With the formation of *organic compounds,* living organisms transform matter into forms unlike any of the inorganic substances of the earth. Early chemists thought that all organic compounds were products of the chemical processes of organisms. This accounts for the name, *organic.* Today, however, we know that this is not the case. In fact, many of the organic compounds we use today are synthetic products of chemical industries.

Whether organic compounds are formed by living organisms or by industrial processes, however, they all have one thing in common. They all contain carbon. Carbon has several properties that make it the key element in organic compounds. Carbon atoms are able to form four covalent bonds with other atoms because of their electron arrangement. These may be single bonds, double bonds, or triple bonds, depending on the number of electrons shared with other atoms. Carbon atoms also have an unusual property of linking with each other. This results in chains or rings of carbon atoms in organic compounds. These carbon groups form the "backbone" or framework to which atoms of other elements attach to form large and complex molecules. Since all organic molecules are organized around carbon atoms, we might think of organic chemistry as *carbon chemistry,* as it is sometimes called.

Organic chemists have been able to determine the exact arrangement of atoms and chemical bonds in the molecules of thousands of organic compounds. These are shown in *structural formulas.* The structural formula for *cholesterol,* a waxy organic substance present in the blood and deposited in the arteries of some people, is shown in Figure 3-8. *There is no need in trying to memorize this structure, or any other we shall show.* However, you can make several interesting observations by examining the structure of cholesterol.

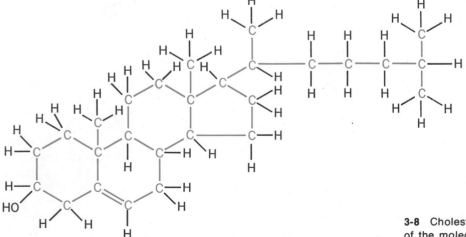

3-8 Cholesterol—the "backbone" of the molecule is shown in blue.

Notice that the carbon atoms form rings in one part of the molecule and a branching chain in another part. The lines indicate bonds. One double bond is shown. The rest are single. Notice that the carbon atoms form bonds with each other as well as with hydrogen atoms and one OH. Notice, also, that each carbon atom forms four single bonds, or two single and one double bond in this molecule. There is another way of designating a molecule of cholesterol. This is by the *empirical formula,* $C_{27}H_{45}OH$. Compare this formula with the structural formula in Figure 3-8.

The organization of organic molecules by living organisms is known as *biosynthesis.* Just how atoms and smaller molecules are assembled in the intricate patterns of organic molecules is a matter of great interest to biochemists today. Somewhere in this maze of cell chemistry must lie the key to life. We shall continue this search in the exploration of organic compounds.

The nature of carbohydrates

Organic compounds that consist of carbon, hydrogen, and oxygen in which the proportion of hydrogen atoms to oxygen atoms is two to one, as in water, are called *carbohydrates.* Examples of carbohydrates are *sugars, starches,* and *celluloses.*

Sugars are vital fuel nutrients, not only in the plants which organize them, but in animals as well. They are of several types. Simple sugars, or *monosaccharides* (MON-uh-SACK-uh-RIDEZ), may contain five or six carbon atoms. Those with five carbon atoms are known as *pentoses. Hexoses* contain six carbon atoms.

Three simple sugars are well-known and important in biology. All are hexoses, with the chemical formula $C_6H_{12}O_6$. *Glucose* is abundant in both plants and animals and is the primary cell

3-9 Structural formulas of three monosaccharides. Note that all three molecules have the same numbers of carbon, hydrogen, and oxygen atoms. Their properties are different because in each molecule the arrangement of the atoms is different. Both chain and ring forms are shown.

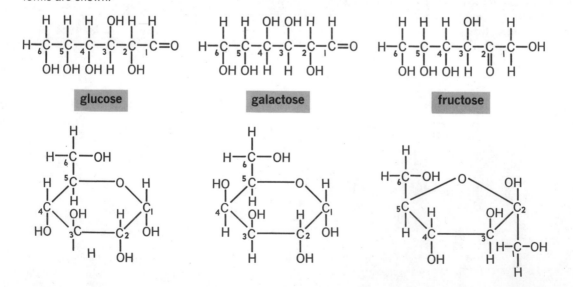

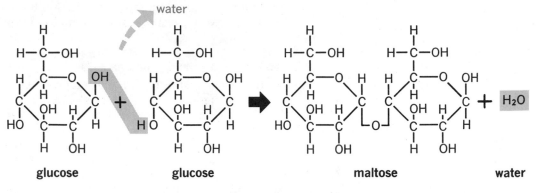

glucose   glucose   maltose   water

3-10 The *dehydration synthesis* of maltose from two molecules of glucose.

fuel. It is also called dextrose, grape sugar, and blood sugar. *Fructose* and *galactose* are other simple sugars with the same empirical formula as glucose. However, if you examine the structural formulas of these sugars in Figure 3-9, you will find differences in the organization of their molecules.

Certain plants combine two molecules of simple sugar (hexose) and form one molecule of a double sugar, or *disaccharide*. One molecule of water is removed in the reaction, called *dehydration synthesis*. Disaccharides have the chemical formula $C_{12}H_{22}O_{11}$. Two glucose molecules are joined in forming maltose, or malt sugar, as shown in Figure 3-10. This process can also be shown by the chemical equation:

$$C_6H_{12}O_6 + C_6H_{12}O_6 \;\rightarrow\; C_{12}H_{22}O_{11} + H_2O$$

In a similar process, a molecule of glucose may combine with a molecule of fructose to form *sucrose,* or cane sugar. This is our common table sugar, produced by the sugar cane and sugar beet. A molecule of glucose joined to a molecule of galactose produces *lactose,* or milk sugar.

Starches are complex carbohydrates composed of glucose units in chains. They are classed as *polysaccharides* (PAH-lee-SACK-uh-RIDEZ). Each glucose unit in a polysaccharide consists of a six-carbon sugar molecule from which a molecule of water has been removed. The chemical formula for a polysaccharide is frequently shown as $(C_6H_{10}O_5)n$. The letter *n* indicates a large number—from dozens to many thousands—of glucose units.

Plant starches include cornstarch, potato starch, tapioca starch, and the starches in rice, wheat, and other grains. Animal starch, or *glycogen,* is produced and stored in the liver as well as in the muscles. When the need arises, the liver converts glycogen to glucose for delivery to the tissues as blood sugar.

Cellulose molecules are larger and more complex than those of starches. They consist of long chains of glucose units that are bonded side by side. This molecular arrangement gives cellulose fibers great strength. Cellulose is formed in the cell walls of plants, where it serves as support. You are familiar with cellulose in the form of wood, paper, cotton, hemp, linen, and other plant products.

Lipids—fats, oils, and related substances

The *lipids* (LIP-idz) constitute a second major class of organic compounds originating in living organisms. They are not soluble in water, but dissolve in such solvents as ether, chloroform, and benzene.

The most abundant lipids are *fats, oils,* and *waxes.* Like the carbohydrates, molecules of these substances contain carbon, hydrogen, and oxygen atoms. In these, however, the ratio of hydrogen atoms to oxygen atoms is much greater than two to one. The body can release much more energy from a given weight of fat than it can from the same amount of carbohydrate.

Fat molecules are composed of one molecule of *glycerol* (glycerin) in combination with three molecules of *fatty acid.* In the process of forming a fat molecule, three molecules of water are released, one from each fatty acid molecule joined to the glycerol molecule, as shown in Figure 3-11. This is another example of dehydration synthesis. During fat digestion, the process is more or less reversed. Water molecules are combined with fat molecules in a process known as *hydrolysis.* This breaks the fat molecule into three molecules of fatty acid and one of glycerol.

Fats occur chiefly in animal tissues and as butterfat in milk and other dairy products. Lipids that are liquid at room temperature are known as *oils.* Vegetable oils include corn oil, cottonseed oil, flaxseed (linseed) oil, peanut oil, and soybean oil.

3-11 The dehydration synthesis of a molecule of fat. Three fatty acid molecules combine with a molecule of glycerol, and yield the fat molecule and three molecules of water.

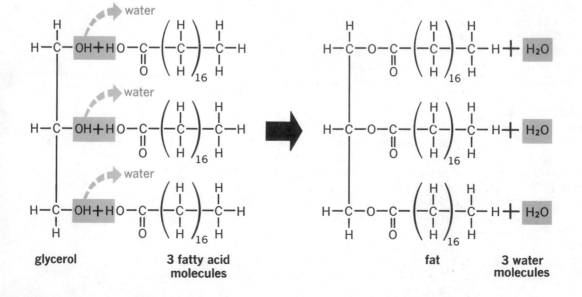

glycerol   3 fatty acid
                  molecules

fat   3 water
             molecules

Waxes, such as beeswax and the waxes found on fruits and leaves, are related to fats. However, waxes consist of fatty acids joined to a long-chain alcohol rather than to glycerol.

Another important group of lipids are the *sterols.* Among these are *ergesterol,* a substance in foods and in the skin from which vitamin D is produced, and *cholesterol,* to which we referred earlier in this chapter.

Proteins and amino acids

Proteins are the most abundant organic compounds in living animal cells. They are very complex in structure and in the enormous number of forms in which they exist.

Structural proteins form the various cell structures you will study in Chapter 4. *Enzymes* are also proteins. These substances, present in cells, control all of the chemical reactions involved in the living condition.

How are these marvelous and complex organic compounds constructed? Protein molecules are made up primarily of carbon, hydrogen, oxygen, and nitrogen atoms. Sulfur is often present and phosphorous and iron are sometimes included.

When protein molecules are carefully dismantled chemically, it is found that they consist of chains of much smaller units called *amino acids,* of which there are about twenty. These are the building blocks of proteins. All amino acids are composed of two groups of atoms. The *amino group* consists of two hydrogen atoms bonded to a nitrogen atom ($-NH_2$) and an *organic acid group* (COOH). The generalized formula for an amino acid is shown in Figure 3-12. *R* in this diagram represents other atoms that vary in different amino acids.

Amino acids are linked together by peptide bonds. The manner in which these bonds are formed by dehydration synthesis is shown in Figure 3-13. Two linked amino acids form a *dipeptide.* These form chains of amino acids, or *polypeptides* that constitute proteins. Many protein molecules consist only of these long-chain polypeptides. Others consist of parallel polypeptide chains joined by cross links.

The number of amino acid units found in proteins may be as few as fifty and as many as three thousand. Not all amino acids are included in any one protein molecule. When you consider that

3-12 The generalized formula for an amino acid.

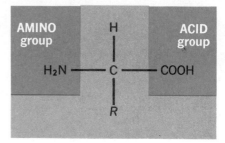

3-13 The dehydration synthesis of a dipeptide. As the water molecule is removed, the two amino acid residues become linked by a *peptide bond.* Long chains of peptides form polypeptides, that compose protein molecules.

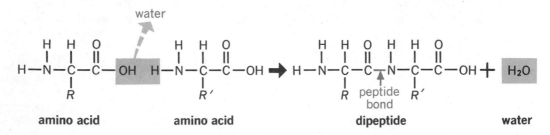

proteins differ not only in the kind of amino acids present but also in their number and arrangement, you can see that the number of different kinds of proteins is almost limitless.

The protein content of different kinds of plants and animals varies greatly. Members of a species have many kinds of proteins in common. For example, certain kinds of proteins are found in all human beings, and they are different from those of other animals or of plants. However, while all human beings are alike in certain parts of their protein makeup, each of us has his own individual protein makeup. Each of our cells is said to contain as many as two thousand different kinds of protein. Could it be, then, that certain proteins make us human, while others make us individuals? This is exactly the case, and later you will discover how polypeptides and, in turn, these proteins are organized and why they make you a human being and an individual human being.

When we consume plant and animal proteins as food, we take in substances that are foreign to our makeup. In other words, the proteins we consume cannot be organized into our own substance. During digestion, however, we break them down into amino acids. Later, these amino acids are recombined to form new polypeptide chains according to our own individual formulas.

The vital role of enzymes

An organism is a living chemical system in which substances are continually changing. Molecules react with other molecules in the chemical phases of the living condition. Some substances are organized, while others are broken down. How are these reactions started? How are they controlled?

Have you ever visited a chemistry laboratory? Here, chemical changes occurring in test tubes and flasks are often rapid and extreme. In some cases, a gas burner is used to start a reaction. In others, considerable heat is given off. But what about the chemical changes that occur in living organisms? These must take place at normal temperatures, under normal conditions, and within the living system of a cell. Furthermore, large numbers of chemical changes must occur simultaneously. One reaction must not interfere with others.

The many chemical activities involved in the living condition could not occur without the action of a large group of proteins known as *enzymes.* These biochemical substances cause chemical reactions to occur and affect their rate but do not enter into the products of the reactions. Biologists refer to enzymes as organic *catalysts.*

Enzymes have another interesting property. They are *specific* in their catalytic action. For example, an enzyme in your saliva (salivary amylase) acts on the starches you eat, changing them to maltose sugar. If you chew starchy foods, such as potatoes, you swallow sugar. This enzyme acts only on starch. Other enzymes

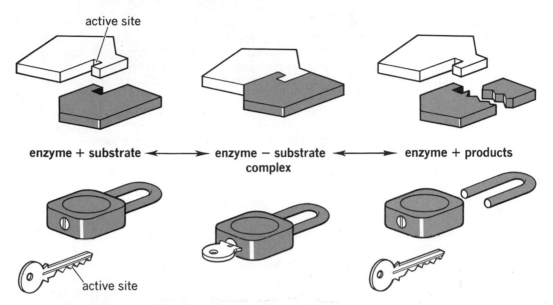

enzyme + substrate ⟷ enzyme – substrate ⟷ enzyme + products
complex

3-14 Models illustrating enzyme specificity.

cause chemical changes in fats, proteins, and other nutrients. Enzymes are associated with growth processes, respiration, and all other chemical activities in organisms.

How are enzymes related to chemical changes? To say that an enzyme is a catalyst merely indicates that it causes a chemical change and affects its rate. Scientists have been searching for these answers for many years and have proposed several models as possible explanations.

Two models of possible enzyme action are shown in Figure 3-14. In one model, an enzyme is visualized as a structure somewhat like a piece of a jig saw puzzle. At one place on the enzyme molecule is an *active site*. A molecule of a particular *substrate*, or substance the enzyme causes to react, can fit this active site and form a temporary bond with the enzyme. This forms an *enzyme-substrate complex*. The complex reacts quickly and forms *products*. The enzyme remains unaltered.

The "lock and key" model is similar. In this case, the enzyme is the key with an active site that fits a particular lock, or substrate. The key in the lock is the enzyme-substrate complex. The key opens the lock, forming products without altering the key. Whether enzymes act in this way is still questionable. However, the general idea of enzymes, active sites, bonding to form the enzyme-substrate complex, and products is logical.

Often, a whole series of chemical changes is involved in a chemical reaction in an organism. The group, or "team," of enzymes in such a series of changes is called an *enzyme system*. Nonprotein molecules that associate with enzymes in certain chemical reactions are known as *coenzymes*. Certain of these reactions will be discussed later.

SUMMARY

What is an organism? From the chemical point of view, any living thing is an intricate combination of proteins, lipids, carbohydrates, and other organic substances associated with water, minerals, and other inorganic materials. An organism is a product of a maze of chemical processes unlike any others that occur as matter changes form.

What is life? We really cannot define it, but we know it is chemical activity supported by energy. Potential energy becomes kinetic energy in supporting the living condition. Living things also change kinetic energy to potential energy and store it for future use.

An organism, then, is a unique chemical system. It is unique, also, in its physical make-up, as you will see in the next chapter.

BIOLOGICALLY SPEAKING

physical change	compound	organic compound
chemical change	formula	biosynthesis
kinetic energy	covalent bond	carbohydrate
potential energy	bond energy	monosaccharide
chemical element	ionic bond	disaccharide
atom	molecule	polysaccharide
symbol	diatomic molecule	lipid
nucleus	ion	protein
atomic number	mixture	enzyme
atomic mass	solution	amino acid
isotope	suspension	polypeptide
radioactivity	colloid	enzyme system
radioisotope	inorganic compound	coenzyme

QUESTIONS FOR REVIEW

1. In what three physical states may matter exist?
2. Distinguish between a physical change and a chemical change.
3. Give several examples of kinetic energy and potential energy.
4. Give an example of the conversion of potential energy to kinetic energy.
5. In what respect are elements the "alphabet" of matter?
6. Locate the protons, electrons, and neutrons in the organization of an atom.
7. Distinguish between the atomic number and the atomic mass of an element.
8. What are isotopes?
9. List three forms of radiation emitted from radioisotopes.
10. Distinguish between covalent and ionic bonding in the formation of compounds.
11. Give an example of a diatomic molecule.
12. Explain why sodium and chlorine ions in sodium chloride have electric charges.
13. Explain the relation of a solvent and a solute in a solution.
14. What factor determines whether a substance will form a solution, a suspension, or a colloid when it is mixed with water.
15. Describe phase reversal in a colloid.
16. List the organic compounds that supply the essential elements to organisms.

17. List at least six ways in which organisms depend on water.
18. Which chemical element forms the "backbone" of all organic compounds?
19. What is biosynthesis?
20. Give several examples of monosaccharides, disaccharides, and polysaccharides.
21. What is the ratio of glycerol to fatty acids in the composition of a fat molecule.
22. Name four elements always found in a protein.
23. Explain the relation of amino acids to protein molecules.
24. What is an enzyme?
25. What is an enzyme system?

APPLYING PRINCIPLES AND CONCEPTS

1. The living world requires oxygen, carbon dioxide, water, and mineral compounds for its existence. Explain why.
2. Discuss the key position of carbon in the organization of organic compounds.
3. From what you have learned about the structure of proteins, explain why the proteins in your body would be "foreign" to another person.
4. Discuss the formation of molecular and ionic compounds in terms of electron activity and chemical bonds formed.
5. Discuss the vital role of enzymes in the chemical reactions occurring in organisms.

RELATED READING

Books

ASIMOV, ISAAC, *The Chemicals of Life.*
   Abelard-Schuman Ltd., New York. 1954. Names of the substances that make life possible, with an explanation of their various uses in nature.
BAKER, JEFFREY J. W., and GARLAND E. ALLEN, *Matter, Energy, and Life.*
   Addison-Wesley Publishing Co., Reading, Mass. 1965. A text for the advanced student providing excellent information on biochemistry and energetics.
DRUMMOND, AINSLIE H., *Atoms, Crystals, and Molecules.*
   American Education Publication, Columbus, Ohio. 1962. An advanced, but understandable, discussion of chemical structure.
LEY, WILLY, *The Discovery of the Elements.*
   Delacorte Press (Dell Publishing Co., Inc.), New York. 1968. A historical approach that demonstrates the development of scientific methods.
READ, JOHN, *A Direct Entry to Organic Chemistry.*
   Harper and Row, Publishers, New York. 1959. Introduction to organic chemistry for the student with limited background in chemistry.
WILSON, MITCHELL, and the Editors of *Life Magazine, Energy.*
   Time-Life Books (Time, Inc.), New York. 1963. Explains in text and fine illustrations the role of energy from photosynthesis to starmaking.

CHAPTER FOUR

THE STRUCTURAL BASIS OF LIFE

Having become familiar with the various elements and compounds, both inorganic and organic, found in organisms, we shall now consider how these materials actually compose a plant or an animal. The *cell* is the structural unit of life. All of the materials in an organism are organized into specific cell structures. In this organization, these materials cease being mere chemical substances and become living matter.

A real knowledge of cellular biology is essential in any area of biological study. Nearly all of the specialized branches of biology relate in some way to cells. Physical scientists ushered in the atomic age as a result of exhaustive studies of molecules, atoms, alpha particles, beta particles, and other invisible forms of matter. Equally important biological discoveries have come from our nation's laboratories, where biologists are finding out more and more about the basic unit of life, the cell.

Three hundred years of cell exploration

More than three hundred years ago, in the year 1665, the British scientist Robert Hooke published a report entitled *Micrographia,* in which he described the observations he had made of slices of cork thin enough to allow microscopic examination. He reported that, to his surprise, the cork consisted of a mass of tiny cavities (Figure 4-1). Logically, since each cavity was enclosed by walls and reminded him of cells in a honeycomb, he called the structures *cells*.

Hooke did not realize, however, that the most important parts of these cells were lacking. He saw only the empty shells of cells that had once contained active, living materials. It is surprising that Hooke did not follow up the discovery of cell walls with an examination of many other parts of plants or animals in which he might have seen cell content. But 170 years passed before another scientist reported a significant discovery relating to cells.

In 1835, the French biologist Dujardin (DOO-ZHAR-DAN) viewed some living cells with a microscope and found that they had content. Dujardin named this material *sarcode,* a term that was later to be changed to *protoplasm.* Three years later, the German botanist Matthias Schleiden proposed, as a result of extensive studies, that all plants are composed of cells. The following year, Theodor Schwann, a German zoologist, made a similar statement regarding animal structure. The work of these men, together with the contributions of other nineteenth-century biologists, established the **cell theory,** which states that:

■ The cell is the unit of structure and function of all living things.
■ Cells come from preexisting cells, by cell reproduction.

For the next hundred years, biologists added greatly to our knowledge of cells and their activities. Their work, however, was restricted by the magnification limits of the light microscope. About thirty years ago, a new era of understanding of cell structure and physiology was ushered in with the development of the electron microscope. New techniques provided additional tools for analyzing the cell structures that the electron microscope revealed. High-speed centrifuges separated cell materials, and advanced methods in biochemistry determined their chemical makeup and molecular structure. By means of new techniques in microsurgery, biologists actually removed cell structures for isolated study. Radioisotopes, serving as *tracer elements,* allowed biologists and biochemists to follow the course of chemical reactions through the machinery of the cell and to relate them to specific cell structures. As a result, biologists have learned more about cells in the past 15 years or so than their predecessors were able to learn in the preceding 300 years!

4-1 Robert Hooke's microscope, and drawings of cork cells as Hooke saw them with that instrument. Two drawings taken from *Micrographia* show (at left) a longitudinal section with the cells split vertically; and (at right) a cross section with the cells split horizontally. (courtesy Bausch & Lomb; inset: The Bettmann Archive)

Today we are discussing cells in an entirely new light. Life has taken on new meaning. We believe that we finally have the answers to many questions about the organization and activity of living cells. However, as reliable as our evidence may seem, we must remember that many of these concepts are still based on hypotheses. We may revise our knowledge of cells many times in the years to come, but each new discovery will bring us nearer to an understanding of the marvels of a living cell.

Cells—the common denominator of life

All the substances composing an organism are contained in its cells. Thus, each cell is a unit mass of protoplasm, or the individual part of which the whole organism is composed. The simplest organisms consist of but one cell. Organisms above this level of organization may be made up of thousands, millions, or even billions of cells. It has been estimated that the human body contains more than 50 thousand billion cells.

The size of the organism is determined not by the *size* of its cells but by their *number*. For the most part, elephant cells are no larger than ant cells. There are just more of them. Large or small, simple or complex, the cell is the unit of structure and function of all living organisms.

Processes of a cell

All of the processes of a living cell involve energy transformations. The source of all of this life energy is chemical activity within the cell. It is difficult to separate one cell activity from all the others that are closely related and occur simultaneously. Similarly, there are specialized cell parts, each of which is concerned with one or more of the vital chemical reactions. Thus, in discussing the cell, we cannot separate structure from function. However, even an artificial separation of the total chemical activity into phases or processes simplifies a discussion of cell structures and their associated functions. The following processes characterize the living condition:

■ *Nutrition.* Food molecules are necessary both to support the processes of the cell and to form its substances. Some cells manufacture their own food molecules. Others ingest them from the environment.

■ *Digestion.* Certain enzymes synthesized in the cell accelerate chemical reactions in which complex food particles are broken down into smaller, soluble units suitable for cell use.

■ *Absorption.* Water, food molecules, ions, and other essential materials are transported into the cell from its environment.

■ *Synthesis.* Cells organize many organic substances including carbohydrates and fats. They also form their own specific proteins. Synthesis of these complex molecules from amino acids is controlled by DNA and RNA and requires the expenditure of energy. The result of these activities is growth and regulation of all chemical activity of the cell by enzyme action.

■ *Respiration*. Chemical energy is released in the cell when certain organic molecules, especially glucose, are split, or degraded. This energy is essential in maintaining life.

■ *Excretion*. Various waste materials are formed as by-products of cell activities. These soluble substances pass from the cell to its environment.

■ *Egestion*. Insoluble, nondigested particles are eliminated by a cell.

■ *Secretion*. Certain cells synthesize molecules that are discharged from the cell and influence extracellular activities. Such cell secretions include vitamins and hormones.

■ *Movement*. Some of the energy released in a cell is used in movement. This may involve the flowing of the cell content, locomotion of the cell by means of special structures, or cellular contractions, as in muscle cells.

■ *Response*. External stimuli such as heat, light, and physical contact alter the activities of a cell and cause a response. This phenomenon is also referred to as *irritability*.

■ *Reproduction*. A cell divides its mass periodically. This results either in an increase in the number of cells within an organism or in an increase in the number of organisms.

Parts of a cell

In classifying the parts of a cell, we can divide them into specific structures (Figure 4-2):

■ The *nucleus,* including the (a) nuclear membrane, (b) nucleoplasm, (c) nucleolus, and (d) chromatin material (chromosomes).

■ The *cytoplasm* (SITE-uh-PLAZZ'm), including the (a) plasma membrane (cell membrane), (b) vacuolar membrane, (c) endoplasmic reticulum, (d) ribosomes, (e) mitochondria, (f) lysosomes, (g) Golgi apparatus, (h) plastids, (i) vacuoles, and (j) nonliving inclusions.

■ The *cell wall,* including the (a) middle lamella (intercellular layer), (b) primary wall, and (c) secondary wall.

Structure of the nucleus

We will begin our discussion of the parts of a cell with the *nucleus,* the control center of all cell activity. Deprived of its nucleus, a cell cannot live very long. When a cell is stained for microscopic examination, the nucleus stands out as the most prominent part. It is usually spherical or oval in shape and often lies near the center of the cell.

The nucleus is bounded by a thin, double *nuclear membrane* that separates the nuclear materials from other parts of the cell. However, the nuclear membrane is not a barrier. What appears

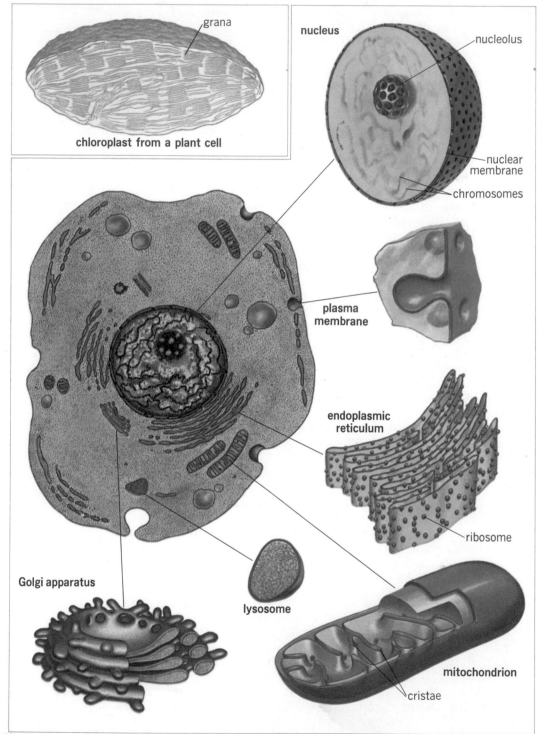

grana

chloroplast from a plant cell

nucleus

nucleolus

nuclear membrane

chromosomes

plasma membrane

endoplasmic reticulum

ribosome

Golgi apparatus

lysosome

mitochondrion

cristae

4-2 Model of an animal cell and organelles.

to be a skinlike covering under the light microscope appears as a porous structure with openings, or gaps, under the electron microscope (Figure 4-3). It is believed that substances from the nucleus pass freely through these pores into the cytoplasm.

The nucleus contains a viscous colloid, rich in protein, which is often referred to as *nucleoplasm*. It also contains one or more spherical bodies called *nucleoli* (singular, *nucleolus*). Distributed in the nucleoplasm are numerous fine strands of *chromatin material*. These are believed to be elongated, active forms of the *chromosomes,* which control the activities of the cell. Chromosomes are composed of DNA joined to protein molecules to form nucleoproteins. DNA molecules are thought to be the key components of chromosomes.

Structure of the cytoplasm

We refer to the cell substances outside the nucleus as *cytoplasm*. Under the light microscope, cytoplasm appears as a semifluid material that fills most of the cell. It often flows through the cell in a manner we call *streaming,* or *cyclosis.* As it changes position in the cell, it may revert from sol to gel and back to sol. The nucleus often flows with the cytoplasm and changes shape as it moves.

The exposed outer edge of the cytoplasm forms a thin molecular layer known as the *plasma membrane,* or *cell membrane.* This thin boundary separates the cell from neighboring cells and from the fluids that bathe it. But like the nuclear membrane, it is not a barrier, for molecules pass through it. However, the plasma membrane is the vital regulator of this molecular traffic. It is selective in allowing certain molecules to pass through and in preventing the passage of others.

The plasma membrane forms where the colloidal cytoplasm borders on another substance, such as a fluid outside the cell. We speak of this boundary as an *interface*. Molecules rearrange along the interface and form a thin, gelatinous layer. You might compare a plasma membrane with the skim that forms on the surface of soup or hot chocolate as it cools. Extremely high magnifications by the electron microscope have revealed that a plasma membrane is not a simple, smooth, skinlike structure, as it appears to be under the light microscope. Rather, it is composed of several layers of molecules. The inner and outer layers of the membrane are composed of protein molecules. Between these is a double layer of fat molecules. The membrane surface is intricately folded, with pouches extending inward and blisters bulging outward. The infolded pouches are tiny gateways in the membrane that allow solid materials to pass through, as you will discover in the next chapter when we discuss the cell and its environment. If a plasma membrane is pierced or cut, cytoplasm may ooze through the opening for a short time. Soon, however, other molecules move into position and plug the opening.

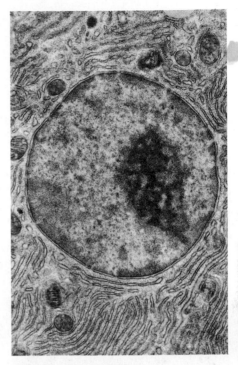

4-3 Electron micrograph of the nucleus. The nucleolus is the dark mass within the nucleus. Note the apparent pores in the nuclear membrane. The darker spheres in the cytoplasm are mitochondria. (courtesy Don W. Fawcett)

A membrane similar to the surface plasma membrane forms along an interface where cytoplasm borders on a central cavity within a cell. Because of its location, we refer to this membrane as a *vacuolar membrane*.

The electron microscope has revealed another characteristic of cytoplasm unknown until recent years. Under the light microscope, cytoplasm appears to be a uniform, semifluid material containing numerous granules. This main body of cytoplasm that looks clear under the light microscope is often referred to as the *cytoplasmic matrix*. It is in this matrix that various visible bodies are suspended. However, a magnification of 40,000 times or more shows a much more intricate structure. Cytoplasm contains a complex system of double membranes that tend to lie parallel to one another. This network, or *endoplasmic reticulum,* fastens to the plasma membrane and nuclear membrane (Figure 4-4). Biologists believe that the endoplasmic reticulum may be a system of canals and parallel double sheets through which materials pass from the plasma membrane to the area of the nucleus. It also provides extensive surfaces on which other cell structures and enzymes are located.

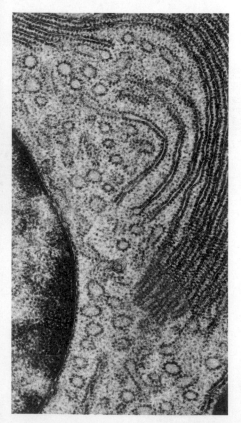

4-4 Electron micrograph of the endoplasmic reticulum and associated ribosomes. A small portion of the nucleus appears at the left. (courtesy Don W. Fawcett)

Cytoplasmic organelles

Various organized bodies are present in the cytoplasm. We refer to these specialized structures as *organelles,* which means "little organs." Certain of these cell organelles are visible under the light microscope. Others are so small that they were not known to exist until we had the electron microscope. Each of these structures is associated with a specific process or activity of the cell. In our present investigation of cells, we are concerned primarily with their structure, composition, and location. In chapters to follow, we will explore their chemical activity in greater detail.

Under extremely high magnification with an electron microscope, you can see tiny, dense granules that are attached to the endoplasmic reticulum and lie between its folds. These bodies are the *ribosomes,* so named because of the large amount of RNA they contain. In addition to RNA received from the nucleus, ribosomes contain protein-synthesizing enzymes. Recent studies indicate that they are the protein factories of the cell. Thus, while they are among the smallest of all cell structures, their function is one of the most vital.

The electron microscope also reveals the detailed structure of rod-shaped bodies in the cytoplasm. These *mitochondria* (MY-tuh-KON-dree-uh) are the centers of cellular respiration, in which energy is released to support cell activities. Hence, mitochondria are the powerhouses of the cell. A *mitochondrion* is enclosed in two membranes, each consisting of a layer of protein molecules and a layer of fat molecules. The inner membrane folds inward at various places, forming partial partitions within the mitochondrion (Figure 4-5). This infolding increases the surface area of the membrane. We refer to these partitions as *cristae*. Mitochondria are known to contain enzymes that split organic molecules and transfer energy to other compounds from which it is released in the cell.

Contained within the cytoplasm of most, if not all, cells are spherical or rounded bodies known as **lysosomes**. They resemble mitochondria but are somewhat smaller and are bounded by a single membrane. Lysosomes are believed to arise from the endoplasmic reticulum, where they enclose newly organized protein-digesting enzymes. Some biologists have referred to lysosomes as "suicide sacs," since release of the enzymes they contain would cause a cell to destroy itself by digesting its own proteins. Normally, lysosomes transfer their enzymes to cavities in the cell in which invading bacteria and other foreign protein substances are digested. Lysosomes also release enzymes in degenerating tissue, causing breakdown of the cells. Such a process is believed to occur when a tadpole gradually resorbs its tail in changing to an adult frog. When a cell dies, rupture of lysosomes results in rapid disintegration of its protein content.

The precise function of certain cell structures remains a mystery. Among these cell unknowns are the *Golgi* (GOWL-jee) *apparatus* (Figure 4-6). They were first seen in 1898 in nerve cells by the Italian neurologist Camillo Golgi. The electron microscope has revealed much more of their structural detail than Golgi was able to determine, but this knowledge has provided little or no clue to their function. Golgi bodies appear as small groups of parallel membranes in the cytoplasm near the nucleus. The membranes form plates with channels along their edges. The fact that they are most numerous in cells composing glands may indicate that they store special substances secreted by other parts of the cell. There is also evidence that they may function in the formation of membranes of the endoplasmic reticulum and in the production of cellulose in plant cells.

Other microscopic bodies present in the cytoplasm are known as **plastids**. Some plastids function as chemical factories, while others serve as storehouses of the cell. Plastids are found most frequently in the cells of plants and in several extremely primitive organisms.

The most familiar plastid is the *chloroplast* (Figure 4-7). These plastids contain green pigments called *chlorophylls*. We often refer to chloroplasts as carbohydrate factories, for they contain the enzymes involved in the organization of carbohydrates. In addition to chlorophylls, chloroplasts may contain pale-yellow pigments called *xanthophylls* (ZANTH-uh-fillz) and deep-yellow or reddish-orange pigments known as *carotenes*.

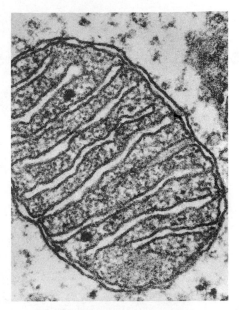

4-5 Electron micrograph of a mitochondrion, showing the outer and inner membranes and cristae. (courtesy George E. Palade, The Rockefeller University)

4-6 Electron micrograph of Golgi apparatus (G). (courtesy Carlo Bruni)

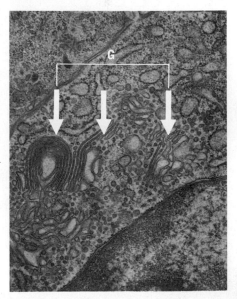

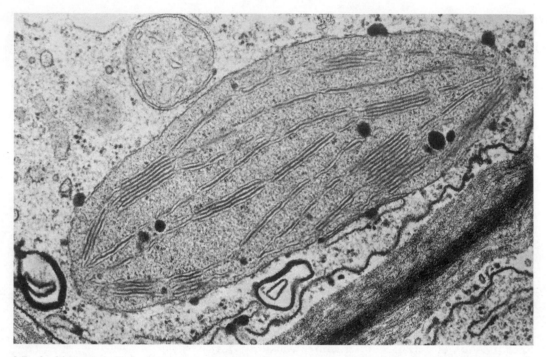

4-7 A chloroplast. The grana are quite apparent in this electron micrograph. (courtesy Myron C. Ledbetter, Brookhaven National Laboratory)

Chloroplasts vary both in size and in shape. In such cells of plants as leaf cells, they are often oval, spherical, or disk-shaped. In many algae cells, they are cuplike, platelike, or ribbonlike. When viewed with a light microscope, chlorophyll appears to be distributed throughout the chloroplast. The electron microscope reveals that the chlorophyll is sandwiched between closely-packed layers of proteins and fatlike lipids in disklike bodies known as *grana.*

Other plastids produce red and blue pigments in addition to yellow and orange. These plastids, often referred to as *chromoplasts,* are found in flower petals and in the skins of fruits such as the tomato, cherry, and pepper. In some cells, the chloroplasts lose their chlorophyll and become chromoplasts. This occurs, for example, in the ripening of a banana or a tomato when the skin changes from green to yellow or red.

Leucoplasts (LOO-kuh-PLASTS) are usually colorless plastids that serve as food storehouses in many plant cells. These plastids contain enzymes necessary to link glucose molecules together and form starch molecules. This may occur temporarily in chloroplasts, but leucoplasts are the principal starch storage centers. Plastids are the only centers of starch storage in a cell. Because starch is insoluble, it never reaches the nucleus or the cytoplasm outside plastids. Leucoplasts can be found in the cells of roots, stems, and in other storage areas of plants. You can see large numbers of them in the cells of the white potato. It is interesting that leucoplasts may develop chlorophyll and change to chloro-plasts when they are exposed to light. This may occur in the cells of the white potato.

Cell vacuoles

Plant cells frequently contain one or more fluid-filled cavities in the cytoplasm. Such cavities are known as *vacuoles*. The bor-dering *vacuolar membrane,* formed by the cytoplasm, regulates the molecular traffic between the vacuole and the cell substances. A young cell often contains several vacuoles that unite as the cell increases in size to form a single, large central vacuole. Pres-sure exerted by the fluids in the central vacuole may force the cytoplasm into a thin layer around the edge of the cell (Figure 4-8).

The fluid in the vacuoles of plant cells is largely water. Ions of mineral compounds, molecules of sugars and other soluble sub-stances, and large molecules of proteins and other organic ma-terials in colloidal suspension in the water form the content of the vacuole that we often refer to as *cell sap.* Other vacuoles may contain food materials and waste products. Still others serve as water-eliminating organelles in many different kinds of one-celled organisms.

Vacuoles of plant cells also contain water-soluble pigments. Common among these are the *anthocyanins* (ANTH-oe-SY-uh-ninz), which appear as shades of scarlet, crimson, blue, purple, and violet. These pigments provide the red shades of autumn leaves, the purple of the turnip top, the red of the radish and beet, and the petal coloration of asters, geraniums, tulips, hyacinths, and many other flowers. In the hydrangea, variations in acidity change the anthocyanin from red to blue. The formation of antho-cyanins is determined by many internal and external factors. The accumulation of sugar seems to stimulate the production of anthocyanins. External factors include both temperature and light—usually low temperature and bright light.

The cell wall

Most plant cells are encased in an outer protective and sup-porting structure, the *cell wall.* At first glance, a cell wall may appear to consist of a single layer, but it is usually composed of several distinct layers. Where two cells lie against each other, each has formed a portion of the wall separating them. The layers of cell walls have been compared to the plastered walls separating two rooms. The center of the wall is like the layer of plasterboard. This portion of a plant cell wall is referred to as the *intercellular layer,* or *middle lamella* (Figure 4-9). It contains various *pectin* (PECK-tin) substances, which are jellylike in tex-ture. Many fruits, including apples, contain a large amount of pectin released during cooking and forms a jelly as it cools.

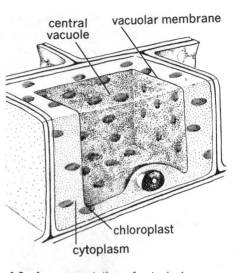

4-8 A representation of a typical plant cell. In a whole cell, the cen-tral vacuole, which is filled with a clear *cell sap,* is completely sur-rounded by cytoplasm. The nucleus can be seen at the bottom of the cell.

4-9 Structure of a plant cell wall. The cell walls are rigid, but they are also porous. This permits mole-cules of all sizes to pass through relatively easily.

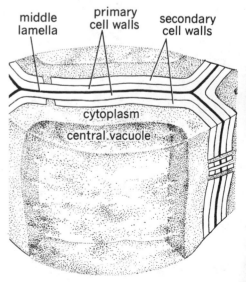

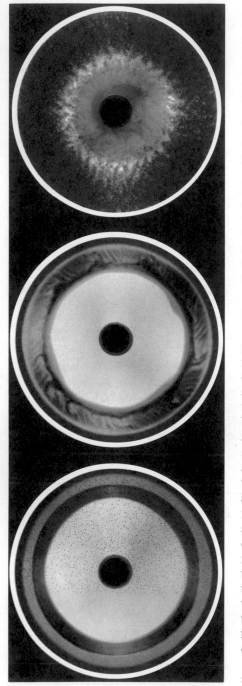

4-10 Centrifuge tubes after spinning and separation of contents into layers. (DeLaval Separator Co.)

Adjacent cells form thin *primary walls* on both sides of the middle lamella. We might compare these to the first coat of plaster on each side of the wall between two rooms. The primary wall is composed of cellulose fibers and pectin. Cellulose, you will recall, is a carbohydrate similar to starch. Soft plant structures such as leaf blades, flower petals, and pulpy fruits have thin primary walls.

Additional cellulose layers are built up in *secondary walls.* These walls are firm and rigid and remain long after the cell is dead. We find thick secondary walls in the cells of wood, plant fibers, and nut shells.

In a primary wall, cellulose fibers are arranged in a network. In secondary walls, they are more or less parallel and crisscrossed in layers. This arrangement of cellulose fibers, somewhat like that of the thin sheets composing plywood, gives these walls great strength. Spaces between the fibers contain pectin and *lignin,* a complex organic compound. Lignin, second only to cellulose in abundance in wood, adds hardness and rigidity to cell walls.

Modern methods of cell investigation

As we have been discussing the structure and chemical composition of extremely small cellular bodies such as ribosomes and mitochondria, haven't you wondered how biologists and biochemists have made these remarkable discoveries? We have already discussed how formerly invisible structures were revealed by the electron microscope. But how could these minute bodies be separated from other cell structures for chemical analysis? In one method, cell materials are pulverized with fine sand as the abrasive. The pulverized material is then placed in an *ultracentrifuge.* As the disrupted cells are spun at speeds up to 100,000 revolutions per minute, the various cell substances and organized bodies settle at different rates and are thus separated into layers. The densest materials, including intact cells and nuclei, occupy the bottom layer in the tube. Next might come a layer of plastids, then a layer of mitochondria. Since the ribosomes are among the smallest and lightest cell structures, they would occupy a layer near the surface, along with pieces of the endoplasmic reticulum. Finally, the layer nearest the top would contain water and various cell fluids. The biologist may then remove material from any layer of the centrifuge tube and find a concentration of the specific cell materials he wishes to examine or analyze.

Although centrifugation separates cell structures, it does not show what is going on in the intact living cell. Recent advances in a technique called *radioautography* have aided the biologist in tracing cell processes. Radioautography is actually a form of photography. In photography, light forms an image on a sensitive substance called a photographic emulsion. You will remember from Chapter 3 that certain isotopes give off radiations spontaneously. These radiations "expose" the photographic emulsion in much the same way that light does.

In radioautography, a substance normally taken in by cells is labeled with a radioactive isotope such as tritium. A slide bearing cells and the labeled substance is then treated with a photographic emulsion. After a time, the emulsion is developed, and the slide is examined under a microscope. The location of the labeled substance is shown by black dots. Thus, if the substance has been taken into the nuclei of the cells, black dots will appear in the nuclei. The number and density of black dots are an indication of the amount taken up. You can see that by this method it is possible to determine what part of the cell takes up how much of the labeled substance. For example, a substance known to be used in the synthesis of DNA may be labeled with tritium. Whenever black dots appear in the nucleus of certain cells on the treated slide, it can be assumed that DNA is being synthesized in these cells (Figure 4-11). In this way, radioautography has been used to investigate the location, timing, and extent of many of the processes in the living cell.

The cellular level of organization

We may consider the cell as the *first* level of the organization of living things. Since the cell is the unit of structure as well as a unit of function of living organisms, a single cell can constitute a complete organism. This is the case in *unicellular* organisms such as bacteria, yeasts, the ameba, and many other forms of life you will study. Some unicellular organisms are grouped together in colonies. In such *colonial* organisms, the cells have no direct functional relationship to one another. More complex plants and animals are composed of many specialized cells working closely together and depending on one another. Such an organization is characteristic of *multicellular* organisms.

Cells and the organization of tissues

When we speak of a multicellular organism, are we referring just to a large number of cells? Is your body a mere mass of billions of cells, all alike and forming a tremendous cell colony? Of course it isn't! If it were, your ability would be limited to the activity of a single kind of cell. Your body and those of complex plants and animals are the results of *cell specialization.* This means that there are many different kinds of cells. Each is developed for a particular kind of activity, which leads us to the subject of tissues.

We define a *tissue* as a group of structurally similar cells performing a similar activity. This is the *second* level of organization of living things. In your body, you have muscle tissue, nerve

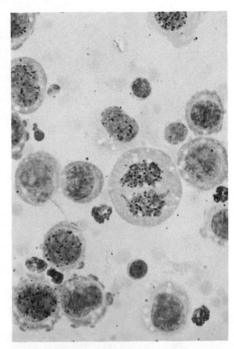

4-11 Radioautograph of cells from a mouse tumor. DNA, labeled with radioactive tritium, appears as black dots. Note that there is a concentration of dots in the nuclear material of the dividing cell (center). (courtesy Renato Baserga)

tissue, bone tissue, cartilage, and many other kinds of tissue. Plants also have tissues. When you name a tissue, you think immediately of a certain kind of activity. Your motion is the best motion skeletal muscles can provide. Your skeleton is the best framework bone cells can comprise. And your ability to hear, see, taste, and control your body is the result of the marvelous activities of your nerve cells.

Each cell in a tissue is a specialist. However, while each cell may be structurally and chemically organized to carry on a specific process with great efficiency, it must also perform all of the processes necessary to maintain its own living condition. That is, it must receive food molecules, respire, synthesize proteins and other essential substances, and generally maintain all basic life activities. In this respect, it is like all other cells. But when a cell is part of a multicellular organism, *division of labor* is possible. A nerve cell can be a specialist in response because other cells supply it with food molecules, furnish it with oxygen, and carry off its waste materials. Such interdependence might be compared to a complex society. The doctor can devote his time to medicine because he can depend on the grocer, the carpenter, the machinist, and other specialists for his nonmedical requirements. These specialists in turn depend on the doctor in matters of health. If each one were required to do everything for himself, he could never limit his attention to carrying out a specialized activity. So it is with cells.

Tissues are grouped to form organs

In higher plants and animals, even a specialized tissue cannot perform a life activity to perfection. This requires several tissues functioning as a unit. We call such a unit of biological organization an *organ*. This brings us to the *third* level of biological organization.

A hand is an organ. It is composed of skin, muscle, bone, tendons, ligaments, blood, and nerves. Your heart, stomach, liver, brain, and kidneys are other organs. A plant stem is also an organ. It has bark tissues, wood, pith, and other tissues, all working together. The stem supports the leaves, flowers, and fruits, and materials move up and down in the stem between the roots and leaves.

Organs may be grouped into systems

In the higher forms of life, especially among animals, several organs may work together as a functioning unit to perform a certain activity. This introduces the *fourth* level of biological organization, the *organ system*. For example, many organs are involved in converting a meal you ate to molecules your cells can use. How many of these organs can you name? To deliver these molecules to cells throughout your body, an efficient transport

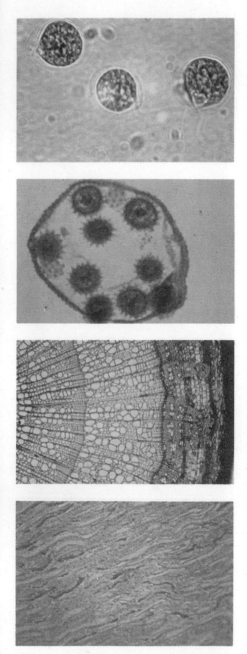

4-12 Levels of organization. A unicellular organism, a colonial organism (The Bergman Associates), seed plant tissue (Photo Researchers, Inc.), striated muscle tissue (William H. Amos).

system is necessary. Your heart, blood vessels, and lymph vessels perform this vital activity. In the higher animals, an organ system is devoted to nearly every one of the activities involved in the living condition.

In the case of higher animals, various systems comprise a *fifth* level of biological organization, the **organism**. However, many organisms exist only at the cellular level. Others exist at the tissue level. Still others have organs but lack well-defined systems. You will become familiar with all levels of organization as you survey the world of plants and animals – in both their primitive and advanced forms.

SUMMARY

If you reduce any organism to its basic components, you will arrive at the cellular level of life. Biology, then, is a study of cells. Cells are the structural units of living matter. As individual units or in enormous numbers, they compose all living things. Cells are, also, the centers of all activities and processes that comprise the living condition.

The cells composing multicellular organisms are amazingly specialized. Because of cells, we can think and remember. We can see, hear, taste, smell, and touch. Cells build our supporting framework and cause it to move. They secrete digestive fluids and pour hormone secretions into the blood.

While all cells contribute to the structure and functions of the total organism, each is an individual living unit, specialized within its own boundary to perform all of the activities essential to the living condition. Each cell has a critical relationship with its environment, as you will see in the next chapter.

BIOLOGICALLY SPEAKING

cell	endoplasmic reticulum	radioautography
cell theory	organelle	unicellular
nucleus	ribosome	colonial
nucleolus	mitochondrion	multicellular
chromatin material	lysosome	cell specialization
chromosome	Golgi apparatus	tissue
cytoplasm	plastid	division of labor
plasma membrane	vacuole	organ
cytoplasmic matrix	cell wall	organ system

QUESTIONS FOR REVIEW

1. In one sense Robert Hooke discovered cells; in another, he did not. Explain.
2. Describe the contributions of Dujardin, Schleiden, and Schwann in early studies of cells.
3. Name the two principles of the cell theory.
4. List the processes of a cell.
5. The nuclear membrane confines certain structures but is not a barrier. Why is this important?
6. Describe the formation and structure of a plasma membrane.

7. Describe the endoplasmic reticulum.
8. Locate the ribosomes in a cell and describe their function.
9. Locate and describe the mitochondria. What is their principal function?
10. In what respect are lysosomes "suicide sacs" in a cell?
11. Why do biologists believe that the Golgi apparatus may function in secretion?
12. Why are chloroplasts sometimes referred to as carbohydrate factories?
13. What pigments other than chlorophyll may be present in chloroplasts?
14. Describe various contents of cell vacuoles.
15. Describe the composition of the wall of a woody plant cell.
16. What two general similarities do cells have in the formation of a tissue?
17. Give several examples of plant and animal organs.
18. Give an example of a human organ system and name as many organs comprising the system as you can.

APPLYING PRINCIPLES AND CONCEPTS

1. In what respect is the cell the basic unit of life?
2. Do you believe that the mere presence in a test tube of all the vital substances composing protoplasm would result in a living condition? Give possible reasons supporting your opinion.
3. Discuss centrifugation as an important research tool in the study of cells.
4. Discuss radioautography as a research tool in the study of cell processes.
5. Discuss the specialization of cell content in various organelles.
6. Discuss various levels of biological organization from cells to complex organisms. In what respect is each level a higher stage of development?

RELATED READING

Books

AFZELIUS, BJORN, *The Anatomy of the Cell.*
The University of Chicago Press, Chicago. 1966. Describes the discoveries made about the structure and ultrastructure of the cell through the use of the electron microscope.
HOFFMAN, JOSEPH G., *The Life and Death of Cells.*
Doubleday and Co. Inc. (Dolphin Books), Garden City, N.Y. 1957. A good book for the student, it describes the living cell in detail and explains why cells die.
HURRY, STEPHEN W., *The Microstructure of Cells.*
Houghton Mifflin, Boston. 1965. A present day account of the evolution of cellular biology into molecular biology. Profusely illustrated with electron micrographs.
LOEWY, ARIEL G., and P. SIEKEVITZ, *Cell Structure and Function.*
Holt, Rinehart and Winston, New York. 1969. A good introduction to the study of the cell and how it functions.
PFEIFFER, JOHN, and the Editors of *Life* Magazine, *The Cell.*
Time-Life Books (Time, Inc.), New York. 1964. Recent discoveries about the cell, with diagrams making up an integral part of the book.
SWANSON, CARL P., *The Cell* (2nd ed.).
Prentice-Hall, Inc., Englewood Cliffs, N.J. 1964. A discussion of the structure of cells, and the tools involved in the investigations of cells.

CHAPTER FIVE

THE CELL AND ITS ENVIRONMENT

Homeostasis—balance on a biological tightrope

Have you ever watched a circus performer walk a tightrope high above the heads of a crowd of tense onlookers? Every step requires precision balance. If you watch the performer closely, you will notice that he leans slightly to the left or to the right to adjust his balance as he takes each precarious step. Balance and adjustment to maintain balance—these are the basic skills in tightrope walking.

In a sense, every organism walks a biological tightrope in remaining alive. Living things maintain an intricate balance in the face of constantly changing conditions, both internal and external. Survival depends on making adjustments to these changing conditions. We refer to the balance that organisms maintain by these self-regulating adjustments as *homeostasis*. Biologists also call this balance a *steady state*.

Homeostasis occurs at all levels in the organization of living things. It involves adjustment of the entire organism. A desert is a place of desolation under the broiling sun but becomes a place of great activity in the evening when insects, birds, reptiles, and mammals come out of their daytime hiding. The walleyed pike seeks the cool water of the depths of a lake when the summer sun heats the surface water. The frog buries itself in mud at the bottom of a pond to escape the cold of winter and the heat of summer; its cells and tissues could not survive either temperature extreme. These adjustments for survival are self-regulating homeostatic activities. Much of your biology course will deal with adjustments of organisms to provide optimum internal environmental conditions for survival.

In the higher animals, we find continuous adjustments at the organ and system level. Organs function in close association with other organs in maintaining an optimum internal environment. All organs depend on heart action to supply their constant environmental requirements with circulating blood. Heart action varies with body needs. During periods of exercise and exertion, the heart speeds up and increases the supply of oxygen and nutrients to the tissues. Increased blood supply also speeds up the removal of cell wastes resulting from increase in cell activity. The kidneys function as blood filters in removing excess mineral ions, water, and cell wastes from the blood. This action is vital in maintaining the proper internal environment in the body. These are but two examples of homeostatic adjustments at the organ and system level. You will find many others as you continue your study of biology.

On the tissue level, we find a precarious homeostatic regulation of the body fluids that bathe cells and provide the immediate internal environment. Fluids that bathe cells must supply food nutrients and oxygen and receive waste products from cell activities. Salts regulate the osmotic concentration of these fluids. A critical acid-base balance must be maintained. In such an optimum environment, the survival problems of our cells are reduced to a minimum. The cells may specialize in such functions as sensitivity, movement, secretion, or excretion. But with this high degree of specialization, they become more and more dependent on ideal environmental conditions.

Our lakes and streams, oceans and seas support an enormous population of one-celled and simple colonial organisms. Here we find single cells or groups of cells adjusting to external environmental changes. Could you expect a nerve cell or a muscle cell removed from your body to survive in a jar of pond water? Certainly not! Removed from the closely regulated internal environment of tissue fluid, such a living cell would quickly perish.

While cells of specialized tissues could not survive in pond water or in other natural aquatic environments, biologists have been successful in growing many such cells in *tissue cultures*. Cells in a tissue culture are grown in nutrient solutions that duplicate the normal cellular environment of an animal body. Oxygen is supplied, and waste products are removed constantly. Only when the need for homeostatic adjustments is reduced to a minimum can specialized cells of higher animals survive in a tissue culture.

The molecular boundary of a cell

To what extent can cells make homeostatic adjustments and maintain their vital balance with the environment? As a first approach to homeostasis, we will consider the way in which the plasma membrane regulates the flow of materials in and out of the cell.

In Chapter 4, we described the molecular composition of the plasma membrane as a double layer of fat molecules lying between an inner and an outer layer of protein molecules. The life of the cell depends on traffic through this "molecular sandwich."

Molecules must enter continuously to supply the chemical activities of the cell. Other molecules, the by-products and waste products of cellular chemical processes, pass outward through the membrane.

Molecules and ions move against the plasma membrane in a steady stream. Some move rapidly; others, more slowly. Some pass through the membrane freely; others in smaller numbers. Some do not penetrate at all. What regulates this movement? Is it the size of the molecules? This is in part the answer, but some large molecules pass through in greater numbers than much smaller ions do. Is it the structure or composition of the membrane? This is also part of the answer. Do conditions outside the membrane or inside the cell regulate the flow? These are also important factors. Each of them has an important influence on the vital relations between the cell and its environment.

The structure of this membrane is of extreme importance in regulating molecular movement through it. While the membrane may appear skinlike under the light microscope, it is apparently porous. Undoubtedly, there are numerous spaces between the various molecules composing it. This would explain why very large molecules such as proteins do not pass through the membrane, while smaller ones penetrate more freely.

Another requirement for passage of most substances through the plasma membrane is that they be soluble in water. A cell is usually bathed in a water solution containing dispersed molecules and ions. A material that forms a colloid in water cannot pass through the membrane, nor can insoluble substances such as starches, which do not disperse among water molecules.

Permeability of membranes

If a substance passes through the membrane, we say that the membrane is *permeable* to that substance. As we have pointed out, however, a plasma membrane allows some substances to penetrate freely, while others pass through more slowly and still others are rejected entirely. A membrane that lets one substance pass through more readily than another substance is said to be a *selectively permeable (semipermeable)* membrane – the type that encloses a cell.

The rate at which various molecules pass through is very important in *absorption,* or the transport of substances through the plasma membrane from the environment into the cell. It is also vital in *excretion,* or the transport of molecules through the membrane from the cell into the surrounding external fluids.

We can classify various substances into groups, based on the rate and degree of passage through a plasma membrane (Table 5-1). What factors determine the rates and degree of passage of these and other substances? Some are purely physical forces over which the cell has no control. Others involve the expendi-

**Table 5-1   THE CAPACITY OF VARIOUS SUBSTANCES TO PENETRATE
THE PLASMA MEMBRANE**

RAPID PENETRATION	SLOWER PENETRATION	VERY SLOW PENETRATION	LITTLE OR NO PENETRATION
gases	monosaccharides	disaccharides	polysaccharides
oxygen	glucose	sucrose	starches
carbon dioxide	fructose	lactose	cellulose
nitrogen	galactose	maltose	proteins
water	amino acids	ions of	lipids (fats)
fat solvents	fatty acids	mineral salts	phospholipids
alcohol	glycerol	acids	
ether		bases	
chloroform			

ture of energy by the cell. In addition, the penetration of some substances is regulated by the structure of the membrane itself.

Diffusion—the spreading of a substance by random molecular motion

Let's visualize a substance as a mass of quivering molecules, bumping into one another like a crowd of people at a bargain counter. Molecules move in straight lines until they collide with other molecules, bounce off, and collide again. The force moving these molecules is internal kinetic energy rather than an outside force. Each molecule thus moves independently of the other molecules. This molecular movement occurs in gases and liquids and to a lesser extent in solids. The collisions cause the molecules to distribute themselves equally in a given volume. We refer to this redistribution of molecules because of their random movements as *diffusion*.

To illustrate diffusion of gaseous molecules, let's imagine that you uncover a jar of ammonia in a closed room. Ammonia molecules are *concentrated* inside the jar. As soon as you open the jar, they start diffusing into the air. Soon you begin to smell ammonia across the room. As diffusion continues, the ammonia odor becomes stronger. More and more ammonia molecules mingle with the molecules of gases in the air. Finally, when the ammonia molecules are distributed equally among the other gas molecules, diffusion (but not the random movement of the molecules) stops. A state of *equilibrium* has been reached.

In accordance with the laws of diffusion, two things happened. Ammonia molecules diffused from the bottle into the air, and air molecules diffused into the ammonia bottle. Both movements were from a region of *greater* molecular concentration to one of *lesser* molecular concentration. This occurred, not in a uniform flow of

5-1 Three stages in the diffusion of a solid and a liquid. Copper sulfate is diffusing into the water as water diffuses into the copper sulfate. Finally, a state of equilibrium is reached when particles of both substances are equally distributed.

molecules, but in a random spreading out resulting from molecular collisions. Diffusion continued until the concentration of air and ammonia was equal in all areas of the room.

Other familiar examples of diffusion

Solids and liquids, or liquids and other liquids, may diffuse as readily as gases *if they normally mix and do not repel each other.* Drop a cube of sugar into a glass of water. As the sugar dissolves, taste the water from time to time. You will be able to detect the increase in the number of sugar molecules as diffusion occurs. Finally, the sugar will dissolve completely without any stirring. Sugar molecules diffused from a region of greater concentration (the lump) into a region of lesser concentration (the water). Meanwhile, water molecules diffused into the sugar. When equilibrium was reached, all parts of the solution tasted equally sweet.

You can watch diffusion occur by dropping some crystals of copper sulfate into a beaker of water. At first, only the solution in the immediate area of the crystals appears blue (Figure 5-1). As the substance continues to dissolve and diffuse, the color spreads, and the water becomes increasingly blue. Finally, the crystals disappear, and the water solution is uniformly a deep-blue color. The direction as well as the rate of diffusion are determined by the *concentration* of the substances involved. The more concentrated the substances, the more rapidly diffusion occurs.

External factors that influence diffusion rates

In addition to molecular concentration, two other factors influence the rate at which diffusion occurs. One of these is *temperature*. The higher the temperature, the greater the speed of molecular movement. Thus, diffusion occurs from an area of higher temperature to one of lower temperature. Similarly, *pressure* accelerates molecular movement, resulting in diffusion from a region of higher pressure to one of lower pressure. Thus, differences in molecular concentration, temperature, and pressure affect diffusion. We refer to the force resulting from these differences as *diffusion pressure*.

Diffusion through a permeable membrane

How strong are the forces of diffusion? To what extent does a membrane interrupt or alter the movement of diffusing molecules? This, of course, depends on the nature of the membrane as well as on the diffusing substances.

First, we shall consider the case of diffusion through a *permeable* membrane. Figure 5-2 shows a simple diffusion apparatus that you can easily set up in the laboratory. The lower bulb of a thistle tube is filled with a sugar solution (commercial syrup can be used). A piece of fine muslin is tied tightly over the open end of the bulb. This thistle tube is then clamped (a ring stand can be used) and submerged in a jar of water.

Since the muslin is permeable both to water molecules and to the sugar molecules, two things will happen. Water molecules will diffuse from the jar into the sugar solution in the thistle tube. At the same time, sugar molecules will diffuse from the thistle tube into the jar. Diffusion will continue until both water molecules and sugar molecules are distributed equally on both sides of the muslin. At this point, a state of equilibrium is reached. Or we can say that the diffusion pressure is now zero because there is equal molecular distribution. What would happen to a cell if its membrane were permeable to this degree? It is true that water and molecules of other substances could enter readily. But wouldn't the cell's own molecules diffuse into the environment just as readily? Logically, a plasma membrane must be selectively permeable, allowing certain materials to enter and leave but retaining the molecules of proteins and other substances that compose the vital structures of the cell.

Diffusion through a selectively permeable membrane

A variation of the preceding experiment is shown in Figure 5-3. The apparatus used is the same except that a selectively permeable membrane (sheep bladder can be used) is substituted for the muslin. Recall from the classification of penetration rates of various molecules that water molecules penetrate a selectively permeable membrane rapidly, while those of sugar pass through more slowly. The sugar solution in the thistle tube has a lower concentration of water molecules than the pure water in the jar. According to the laws of diffusion, the net movement of water molecules through the membrane is from the region of greater concentration in the jar to the region of lower concentration in the thistle tube. This movement occurs freely. However, the membrane permits but little diffusion of the more concentrated sugar molecules in the thistle tube to the jar. Consequently, there is uneven molecular diffusion. Over a period of time (usually several hours), the water level in the tube rises. However, as the solution rises in the tube, another force becomes important —gravity. This force tends to pull the solution in the thistle tube downward so that it exerts a pressure (force) on the inner surface of the membrane. When this force is equal to the diffusion pressure of the water molecules, diffusion ceases.

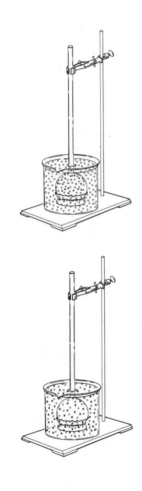

5-2 Diffusion through a permeable membrane: At the beginning of the experiment a sugar solution in the thistle tube is separated from plain water in the jar by a permeable membrane. Water molecules (blue) move through the membrane into the sugar solution as sugar molecules (red) move through the membrane into the water in the jar.

Osmosis—a diffusion of water

The experiment we have just described involved *the diffusion of water through a selectively permeable membrane from a region of greater concentration to a region of lesser concentration*. Biologists refer to this water diffusion as *osmosis*. In the definition of osmosis, *concentration* refers only to water molecules, not to substances dissolved in water. The separation of two different solutions, in this case water and molasses solution, by a selectively permeable membrane creates an *osmotic system*.

As water diffuses into a cell by osmosis, it builds up a pressure known as *turgor* (TUR-gor) *pressure*. Since the plant cell is encased in a wall, it does not stretch or bulge as this internal pressure increases. Internal water pressure forces the plasma membrane and cytoplasm firmly against the wall, causing the cell to become stiff, or *turgid* (TUR-jid). When turgor pressure, which tends to force water molecules out of the cell, becomes as great as the diffusion pressure, which causes them to diffuse into the cell, an equilibrium is reached.

Thin-walled plant tissues such as those composing leaves, flower petals, and soft stems maintain their stiffness by cell turgidity. As long as there is sufficient water supply in the cell environment, this internal turgor pressure is maintained. Otherwise the plant wilts.

We must bear in mind that diffusion and osmosis are purely physical processes over which the cell has no control. No cell energy is involved. Molecules move by their own kinetic (heat) energy. For this reason, we refer to diffusion as *passive transport*. Whether diffusion and osmosis will benefit a cell or destroy it depends on concentration differences inside the cell and in its environment. If the solution outside the cell is *hypotonic*—that is, if it contains a lower concentration of solutes and a higher concentration of water than the cell content—there will be a net movement of water into the cell.

Water problems in animal cells

As you recall, a plant cell has a retaining cell wall that permits the build-up of enough turgor pressure within the cell to stop further diffusion of water into the cell. However, animal cells lack this supporting structure; therefore, in a hypotonic environment they cannot stop this inward diffusion of water. Since this is true, how do animal cells keep from swelling until they burst?

Many one-celled organisms living in a fresh-water environment are equipped with special water-eliminating organelles known as *contractile vacuoles*. These tiny pumps work continuously, eliminating water through the membrane as rapidly as it diffuses into the cell (Figure 5-4). Were it not for these special structures, the cells would soon burst. Fish and other gill-breathing aquatic animals take in large amounts of water with the oxy-

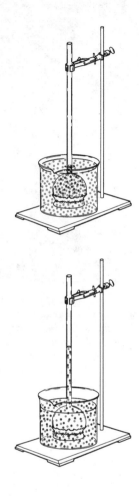

5-3 Diffusion through a selectively permeable membrane: At the beginning of the experiment, a sugar solution in the thistle tube is separated from plain water in the jar by a selectively permeable membrane. Over a period of several hours, water molecules move through the membrane rapidly into the sugar solution.

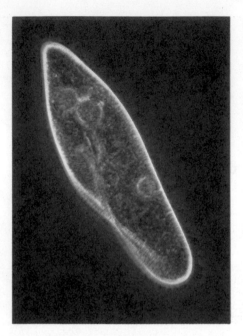

5-4 This contractile vacuole in a one-celled organism eliminates water through the membrane as rapidly as it diffuses into the cell. Otherwise, internal water pressure would burst the cell. (Photo Researchers, Eric Gravé)

5-5 Normal *Anacharis* cells in isotonic solution (left); plasmolyzed *Anacharis* cells in hypertonic solution (right). (Phillip A. Harrington)

gen they absorb into the blood. In these animals, excess water is excreted from the body in urine. Our kidneys, sweat glands, and lungs eliminate excess water and prevent our cells from receiving too much water.

You can observe the effects of excessive water absorption in animal cells by putting animal-like, one-celled organisms into distilled water. The diffusion pressure is greater than their water-eliminating organelles can overcome. The cells soon swell and burst. This bursting of cells from internal pressure is called *cytolysis* (sy-TAHL-uh-siss) or sometimes plasmoptysis. The presence of minerals and other soluble materials in pond water prevents this from happening in nature by reducing the diffusion pressure. Similarly, if you place a drop of blood in distilled water, the cells can be observed to swell and burst almost immediately (Figure 5-6, middle).

Plasmolysis—the loss of cell turgor

Remember that a cell has no control over the movement of water molecules through its plasma membrane. The direction in which water diffuses is determined by differences in the concentration of water inside and outside the cell. If a cell is placed in a *hypertonic* solution—that is, in one in which the water solution outside the cell contains more dissolved substances and therefore has a lower concentration of water molecules than the solution inside the cell—water will diffuse from the cell into its environment. This results in loss of turgor pressure and shrinkage of the cell content. This condition is called *plasmolysis* (plazz-MAHL-uh-siss).

You can observe this water loss in several ways. Slice two pieces of potato; place one in a strong salt solution and the other in tap water. After ten or fifteen minutes, examine both slices. The slice in salt water will be limp, or *flaccid,* indicating that the cells have lost water and therefore turgor pressure. You can watch this process under the microscope as it occurs in a leaf of an aquatic plant like waterweed, *Anacharis (Elodea).* Mount the leaf in water and notice that the cells are turgid. Then add one or

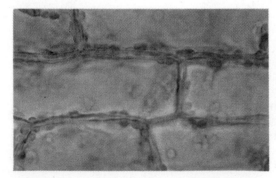

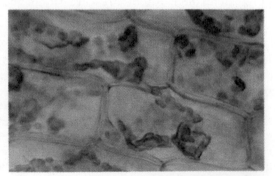

two drops of strong salt solution at the edge of the cover glass. What do you think will happen? The appearance of the cells' contents can be seen in Figure 5-5. In animal cells, such as red blood corpuscles, the entire cell shrinks (Figure 5-6, bottom). Why does the appearance differ from that of the plant cells? Temporary plasmolysis can be corrected by the intake of water. But the cell will die if the condition continues very long. Now you can understand why salt kills grass and why shipwrecked men die from drinking salt water. Plasmolysis explains, also, why too heavy an application of strong fertilizers to the soil may kill plant roots.

Cells in isotonic solutions

Under certain conditions, the concentration of both water and total solute molecules is the same in an external solution as in the cell content. Such solutions are said to be *isotonic*. Water molecules diffuse through the plasma membrane both into and out of the cell. However, the rate of water diffusion is the same in both directions. Thus, the cell neither gains nor loses water.

The water concentration of red blood corpuscles is normally the same as that of the blood plasma in which they are suspended. Thus, corpuscles neither gain nor lose water in the bloodstream (Figure 5-6, top). Isotonic salt solutions are frequently used to bathe living tissues in experimental biology. One such solution, known as Ringer's solution, is a 0.7 to 0.9 percent salt (sodium chloride) solution. This solution is isotonic with the protoplasm of many cells.

Penetration of cell membranes by alcohol and other solvents

Alcohol, ether, and chloroform penetrate the plasma membrane and enter a cell even more rapidly than water. Biologists believe that these solvents dissolve the fat layer of the membrane and pass through quickly. These substances interfere with normal cell activities and have an anesthetic effect on cells. The rapid penetration of the membrane would explain their quick action. The cell membrane is not permanently damaged, because other fat molecules quickly fill the dissolved area.

Diffusion of ions through a plasma membrane

Various mineral salts, acids, and bases form ionic solutions in water. These charged particles penetrate a plasma membrane very slowly. Biologists have made extensive studies of the absorption of ions by a cell, but many questions remain unanswered. One possible explanation suggests that the membrane itself has an electrical charge and that this charge is usually negative. This would repel negatively charged ions, since like charges repel each other. Furthermore, this negative charge would be attractive

5-6 Red blood cells in three solutions. (Phillip A. Harrington)

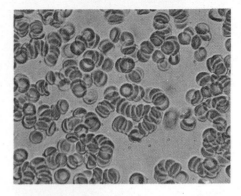

(a) Normal red blood cells in isotonic solution.

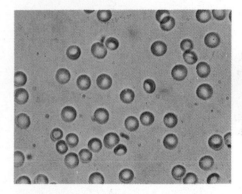

(b) Swollen red blood cells in hypotonic solution. Water pressure may burst the cells.

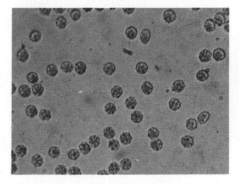

(c) Crenated red blood cells in hypertonic solution. Water loss has caused the cells to shrink.

to any positive ions present in the solution outside the cell. This attraction may be strong enough to prevent many ions from penetrating the membrane. Nevertheless, some ions do penetrate the plasma membrane, and both the movement of the ions through the membrane and the presence of the ions within the protoplasm are important in the chemical activities of the cell.

The principle of active transport

We know that many cells receive or excrete molecules against a diffusion pressure that would normally cause movement in the opposite direction. For example, a root cell may absorb mineral ions from soil solutions that contain fewer ions than the cell already contains. If these ions behaved in accordance with the laws of diffusion, they would move from the cell into the soil water. Similarly, cells of certain algae living in the ocean absorb iodine compounds when the concentration of these substances within the cells is already many thousands of times higher than that of the seawater surrounding them.

It is obvious that some force other than diffusion accounts for this movement through the membrane. Biologists refer to this force as *active transport*. They know that the selectivity of the membrane continually changes with variations in the chemical needs of the cell. They also know that the cell energy is used in transporting these substances through the membrane, although they are not certain just how the energy is used. They have proved that cell energy is involved by measuring the oxygen intake and carbon dioxide release during active transport. If the solute concentration outside of the cell is greater than that of the cell content, solutes enter the cell by diffusion. However, if the solute concentration outside the cell is lower than that of the cell, the oxygen intake and energy release increase sharply. The cell is then using its energy in solute absorption. This is active transport, as opposed to passive transport, or membrane penetration by diffusion pressure alone.

Entry of large molecules into a cell

So far, we have discussed the penetration of a plasma membrane by water molecules, ions, and other materials that could pass through pores in the membrane. But what about larger molecules, such as amino acids, lipids, and even larger masses of material? These particles cannot pass through membrane pores, yet they are known to enter cells.

In the discussion of the structure of a plasma membrane, do you remember the reference to numerous pouches that extend inward into the cell? Biologists believe that large molecules and other particles flow into these pockets and are sealed off as the membrane closes behind them (Figure 5-7). The material is then

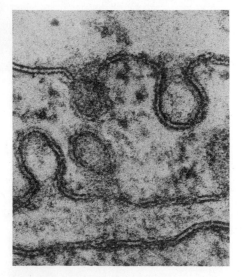

5-7   Electron micrograph showing the pouches that are involved in *pinocytosis*. (courtesy George E. Palade, The Rockefeller University)

enclosed in a membrane within the cell as a vacuole. This engulfing process is referred to as *pinocytosis* (PIN-oe-sy-TOSE-iss). It is another important factor in cell absorption.

SUMMARY

The membrane surrounding a cell lies in the path of heavy molecular traffic. Molecules and ions move against its outer surface from the cell environment and against its inner surface from solutions within the cell.

Because a cell membrane is selectively permeable, various molecules and ions pass through it at different rates. This regulating action tends to allow molecules of water, nutrients, gases, and other substances to pass to and from the cell but retains the molecules composing cell structures. The movement of molecules through a cell membrane is regulated by the forces of diffusion. Diffusion of water to and from the cell is referred to as osmosis. Since the cell has no control over diffusion and osmosis we refer to this molecular movement as passive transport. By expending energy, a cell absorbs mineral ions against the force of diffusion during active transport. Large molecules and other materials that cannot pass through a cell membrane may be engulfed in pouches of the membrane and taken into the cell in vacuoles by the process, pinocytosis.

BIOLOGICALLY SPEAKING

homeostasis	permeable	passive transport
tissue culture	selectively permeable	active transport
diffusion	turgor pressure	hypertonic solution
diffusion pressure	hypotonic solution	plasmolysis
osmosis	contractile vacuole	isotonic solution
osmotic system	cytolysis	pinocytosis

QUESTIONS FOR REVIEW

1. Define homeostasis and explain why homeostatic adjustments are necessary for the survival of an organism.
2. Give examples of homeostatic adjustments at various levels of organization of living things.
3. Distinguish between a permeable and a selectively permeable membrane.
4. Explain diffusion in terms of molecular movement.
5. Under what conditions is a state of equilibrium reached?
6. How are temperature and pressure related to the rate of diffusion?
7. What is diffusion pressure?
8. Define osmosis.
9. What is turgor?
10. Would a cell build up turgor pressure if its membrane were permeable to all of the molecules it contacts? Explain.
11. Explain why animal cells do not build up turgor pressure.
12. Why do blood cells burst quickly if they are put into distilled water?
13. Describe several methods by which animal cells eliminate excess water.

14. Describe the cause of plasmolysis of a plant cell and the physical changes that occur.
15. Why do alcohol, ether, and chloroform penetrate a cell membrane rapidly?
16. What force exceeds diffusion pressure during active transport of minerals through a cell membrane?
17. Describe the entry of substances into a cell by pinocytosis.

APPLYING PRINCIPLES AND CONCEPTS

1. Discuss the homeostatic adjustments you have observed in organisms in your region.
2. It was once thought that the size of the pores in a cell membrane determined molecular penetration. Give evidence to show that this is not always true.
3. Why is it necessary that a cell have a higher solute concentration than the surrounding external solutions?
4. What might happen to the cells of a fresh-water plant if it were placed in salt water? Why would fresh water destroy a salt-water plant?
5. How have biologists demonstrated that active transport involves cell energy?

RELATED READING

Books

GERARD, RALPH W., *Unresting Cells.*
　Harper and Row, Publishers, New York. 1961. A discussion of cells and their functions.
LANGLEY, LEROY L., *Homeostasis.*
　Reinhold Publishing Corp., New York. 1965. The concept of homeostasis, traced through biological control systems in both plants and animals.
SCHMIDT-NIELSEN, KNUT, *Animal Physiology.*
　Prentice-Hall Inc., Englewood Cliffs, N.J. 1960. An interesting presentation by an expert on water balance, salt, and desert animals.
SNIVELY, DR. WILLIAM D., JR., *Sea Within: The Story of Our Body Fluid.*
　J. B. Lippincott Co., Philadelphia. 1960. A serious explanation of the precarious balance of fluid and electrolytes on which our very lives depend.
WINCHESTER, JAMES H., *The Wonders of Water.*
　G. P. Putnam's Sons, New York. 1963. Discusses, in part, those remarkable properties of water that affect the functioning of individual cells.

Articles

HOLTER, HEINZ, "How Things Get Into Cells," *Scientific American,* September, 1961.
　A discussion of the ways materials move into the cell by passive or active transport and by pinocytosis.

CHAPTER SIX

PHOTOSYNTHESIS, RESPIRATION, AND CELL ENERGY

Processes involving glucose and cell energy

This chapter involves two nutritional processes of a cell. Both involve glucose and cell energy, but in different ways. One process is chemically *constructive*. Complex molecules are built up and energy is stored in them. The other process is chemically *destructive*. Complex molecules are broken down and energy is released. In many ways, one process is the complement of the other.

We call the constructive process *photosynthesis*. Glucose is organized in a series of chemical reactions. Light energy is received and converted to chemical energy that is stored in the glucose molecules. Glucose is the basic fuel and energy source for living cells.

Cells must have a continuous supply of energy to support their processes. If a cell is deprived of this energy, even for an instant, activities will cease and the cell will die. How is energy released from glucose? This process involves a series of chemical changes we include in *respiration*.

When you study photosynthesis, you will discover why the process is limited to certain cells. These cells are "self-fueling." Respiration, however, is a vital process in all cells. Thus, you can understand why some cells, and the organisms they compose, must depend on the "self-fueling" cells for their supply of glucose. This nutritional dependence of a large segment of the living world introduces an interesting relationship we shall discuss later. Let us examine this relationship more closely in the imaginary closed environment of a gigantic "super-dome."

Matter and energy relationships of organisms

In our study of cell processes, we must consider not only cells but also the larger structures they form and, of course, entire organisms. Suppose we could completely enclose a whole community, perhaps several square miles in area, and prevent any organisms or materials from entering or leaving the closed environment. We would include wooded areas, farms, open fields, and ponds in the gigantic enclosure. Animal life native to the region and domestic animals would also be included, along with the people who lived there. In other words, life would continue normally except that the whole region would be isolated from the outside world. Could this isolated society continue? Plants and animals would live and die. Their chemical remains would supply the requirements of new generations. Matter would be used and reused, organized in complex substances, and broken down to simpler form, in a continuous cycle, year after year and generation after generation.

Now suppose we covered this giant enclosure and shut out all of the light. Do you think life could continue as before? You have probably concluded that the people, the animals, and even the green plants would be doomed. Life might continue until all the food was consumed; then all would perish. The matter could still cycle, but what about the energy necessary to build food molecules and support life? Can the heat energy released in living cells be trapped and reused? Can the plants and animals capture the energy used in their own activities? No, *energy does not cycle.* When we shut out the light, we cut off the supply of new energy that must enter the living world constantly.

What if we removed all the green plants and left the animals and non-green plants, such as fungi, in full light? Would life continue? No, the animals would be as helpless as if there were no light. They would have energy all around them, but none in the form they could use to supply their cells. Doesn't this place the cells of green plants in a unique position in the world of life? Marvelous, indeed, must be the chemical processes of plant cells that capture sunlight and store chemical energy in the molecules that fuel the cells of living things.

Energy transfer compounds in cells

Light energy is often received in a cell more rapidly than the chemical machinery of photosynthesis can convert it to chemical energy and store it in molecules of glucose. During respiration, energy is released more rapidly than the cell can use it in its processes. Thus, there must be some method of trapping this energy and releasing it in controlled amounts and releasing it as cell processes require it.

Biologists searched for an answer to this question for many years. The discovery of a remarkable energy transfer compound in a cell shed new light on the energy relations of photosynthesis as well as respiration. This substance, *adenosine triphosphate* or *ATP,* is organized in the cell mitochondria and is released to all parts of the cell. The chemical structure of ATP can be changed quickly so as to store energy, or release it. Let us examine the

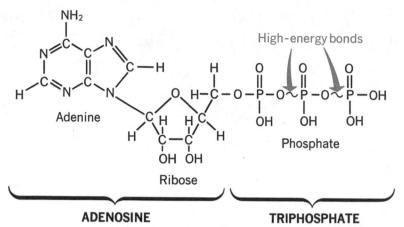

chemical structure of ATP and find out how these important changes occur in the cells of the body.

The structure of a molecule of ATP is shown in Figure 6-1. One part of the molecule is a substance known as *adenine,* which is bonded to a unit of *ribose,* a five-carbon sugar. These two parts of the ATP molecule compose *adenosine.* Smaller *phosphate* molecules form a short chain, or tail, attached to the adenosine. Notice that three phosphates are attached in ATP, hence the name adenosine triphosphate. In the building up of ATP, the first phosphate molecule is firmly bonded to adenosine. This portion of the molecule composes *adenosine monophosphate,* or **AMP.** The second phosphate builds up the molecule to *adenosine diphosphate,* or **ADP.** This phosphate is joined with a high-energy bond. The third phosphate, building up the molecule to ATP, is attached by another higher-energy bond. Thus, you might represent ATP in a simple diagram as A—P—P—P, while ADP would be represented as A—P—P.

The high-energy bond between the second and third phosphate units of ATP is important in the trapping, transfer, and release of energy in a cell. The third phosphate may be rapidly removed, changing ATP to ADP, or attached, changing ADP back to ATP. It is the energy involved in these changes that is important. The bonding of a phosphate to ADP, forming ATP, requires energy. In this reaction, energy is trapped and stored. The removal of one phosphate group from ATP releases energy and reforms ADP, which is then available to form another high-energy phosphate bond. We may show these changes and the storage and release of energy, both of which require enzymes, in a simple equation:

$$ADP + phosphate + energy \rightleftarrows ATP\,(ADP-P)$$

This occurs over and over in a cell in a never ending cycle. A diagram of an ADP—ATP cycle as well as energy sources and uses is shown in Figure 6-2.

6-2 ADP receives energy and attaches a phosphate molecule, changing to ATP. Chemical energy is stored in ATP until it is released with removal of the phosphate molecule, changing it back to ADP to complete the cycle.

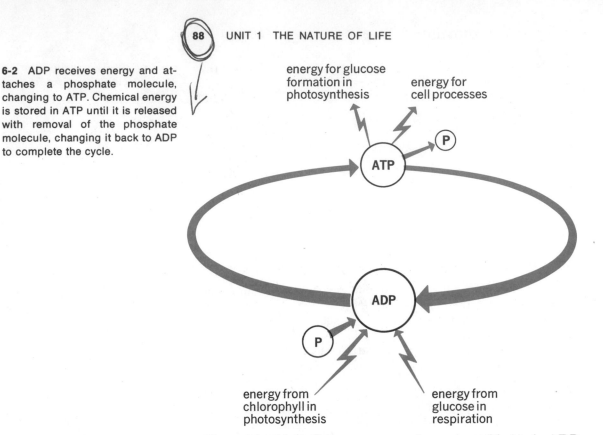

energy for glucose formation in photosynthesis

energy for cell processes

ATP

P

ADP

P

energy from chlorophyll in photosynthesis

energy from glucose in respiration

You might think of these compounds as a portable bank. ADP is the receiving window, while ATP is the paying window. Energy is the money. Energy received from chlorophyll in photosynthesis or released from glucose in respiration is deposited in the bank at the receiving window when a phosphate group is added to ADP. The energy is now in the bank as a chemical bond in the high-energy compound ATP. Securely locked in the ATP vault, the bank may be moved to any part of the cell where energy is needed. As energy is required, it is released from the vault to the paying window as ATP is changed to ADP. Thus, ATP as a portable vault serves to receive energy, store it, transport it, and release it as needed.

With this understanding of ATP, we are now ready to investigate one of the most vital of all chemical processes in which light energy enters a cell and, after a series of chemical changes and energy transfers, is stored in molecules of glucose. Energy from glucose, formed in this process, is supporting your life at this moment.

A marvel of plant life

To explore the machinery of the cell and the process in which fuel molecules are organized, we can select almost any plant part as long as it is green. We might choose a leaf cell, a cell from a green stem, or a hairlike strand of an alga growing in a pond.

Any of the seaweeds could be used, for while they may be brown or red in color, they also contain chlorophylls.

The process we are exploring is known as ***photosynthesis.*** The name nearly defines the process, for *photo* refers to light, while *synthesis* means the building of a complex substance from simpler substances. The simpler substances are *carbon dioxide* and *water,* while the complex substance is *glucose.* You may already have concluded that light is necessary for photosynthesis and that some of the light energy is somehow captured in the sugar molecules. But how is this accomplished? Certainly, carbon dioxide and water will not automatically form glucose merely because they are both in a flask set in sunlight.

Chloroplasts, the machinery of photosynthesis

In our discussion of the parts of a cell, we referred to cytoplasmic inclusions known as *chloroplasts.* Do you remember from our discussion in Chapter 4 that they contain *grana,* the disk-shaped bodies made up of layers of chlorophyll, proteins, and lipids?

While we often speak of chloroplasts as the machinery of photosynthesis, certain cells that carry on the process have no organized chloroplasts. Such primitive cells are found in the blue-green algae. However, the electron microscope reveals that even these cells have grana. They are dispersed through the cytoplasm, not confined to organized chloroplasts. Perhaps we should say, then, that grana rather than chloroplasts constitute the machinery of photosynthesis.

Chlorophyll, the agent of photosynthesis

If bean or corn seedlings are grown in total darkness, the leaves will be yellow rather than green because light is necessary for the production of ***chlorophylls.*** If such a plant is moved to a lighted area, photosynthesis will not occur until chlorophyll has formed in its leaves. Chlorophyll, then, is essential for photosynthesis.

Chlorophyll molecules are embedded partially in the protein disks of the grana and partially in the lipid layers between them. Frequently, molecules of different forms of chlorophyll are mixed in the grana. Four different types of chlorophyll are found in various plant cells. The most abundant form, and the most important in photosynthesis, is chlorophyll *a.* This bright bluish-green pigment has molecules containing carbon, hydrogen, oxygen, nitrogen, and a single atom of magnesium that lies in the center of the molecule (Figure 6-3). The chemical formula for chlorophyll *a* is $C_{55}H_{72}O_5N_4Mg$. A second, yellowish-green pigment is referred to as chlorophyll *b.* Its chemical formula, $C_{55}H_{70}O_6N_4Mg$, shows that it differs from chlorophyll *a* only in the number of hydrogen and oxygen atoms present in the molecule. Small amounts of iron compounds are necessary for the

6-3 The structural formula of the chlorophyll *a* molecule.

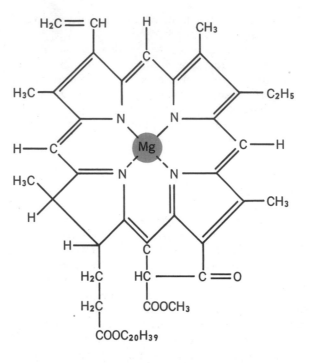

formation of chlorophylls, although iron atoms do not enter into their molecular structure.

Chloroplasts in the cells of seed plants usually contain both of these chlorophylls in the approximate ratio of three parts of chlorophyll *a* to one part of chlorophyll *b*. Certain other plants more primitive than seed plants contain other chlorophylls. Brown algae contain chlorophyll *a* and chlorophyll *c*, while red algae contain chlorophyll *a* and chlorophyll *b*. The purple sulfur bacteria have a unique form of chlorophyll known as **bacteriochlorophyll,** which is contained in granules rather than grana. These bacteria are exceptional in that they are capable of photosynthesis.

Chlorophyll *a* is necessary for photosynthesis, and yet it does not enter into the process chemically. Remember that we call such substances *catalysts*. Other chlorophylls can assist the process by absorbing light energy and transferring it to chlorophyll *a*. Other pigments in the chloroplasts, including xanthophylls and carotenes, can also transfer energy to chlorophyll *a*.

You can remove chlorophyll from cells by heating a plant part, such as a leaf, in alcohol or some other solvent. You might even assume that you could use this catalyst in the laboratory and carry on photosynthesis artificially. Chloroplasts have been removed from cells and under these conditions have retained their photosynthetic abilities. But chlorophyll alone cannot produce a carbohydrate. Thus, photosynthesis seems to be limited to the processes of a living cell or at least to the grana of the chloroplasts.

In addition to chlorophyll, chloroplasts contain *enzymes* that are essential to photosynthesis. Enzymes serve as catalysts and together with chlorophyll cause the chemical reactions involved in the various steps of photosynthesis to occur.

The general nature of photosynthesis

Biologists have had a general understanding of photosynthesis for many years. They assumed, however, that it was a single chemical reaction in which carbon dioxide and water were combined in a cell to form a glucose molecule. Oxygen was known to be a by-product of the process. Chlorophyll has long been recognized as a catalytic agent of the process, although its actual role was not really understood until recent years. Biologists have known, too, that light energy is absorbed during the process and that, somehow, this energy is locked in the bonds of glucose molecules. The overall, simplified chemical equation for photosynthesis is commonly written as follows:

$$6 \ CO_2 + 6 \ H_2O + \text{light energy} \rightarrow C_6H_{12}O_6 + 6 \ O_2$$

carbon water glucose oxygen
dioxide

This equation, while accounting for the materials required and the products formed in the process, fails to summarize the true nature of photosynthesis. Is photosynthesis a single reaction, as the equation indicates? Does carbon dioxide react with water directly to form glucose? Does the oxygen released as a by-product come from the carbon dioxide, the water, or both? How is light energy transformed into chemical bond energy in glucose molecules? How is chlorophyll related to the process? Biologists asked these questions for years. Only recently have they begun to solve the mystery of the most important chemical process in the world.

Tools with which to explore the nature of photosynthesis

By 1941, biologists and chemists had been provided with a new and effective tool for chemical research. Radioisotopes were available from atomic reactors in research centers. Among the available radioisotopes were oxygen-18 and carbon-14. These isotopes could be traced through chemical reactions as tagged atoms, provided the reactions could be stopped at a given instant so that products formed could be isolated and analyzed. The problem now was to find a suitable plant that could be killed and analyzed quickly. It would be difficult to kill leaf cells instantly and remove the products of photosynthesis at various stages. But a tiny, fast-growing alga known as *Chlorella* (Kluh-RELL-uh) proved to be a perfect subject (Figure 6-4). This microscopic one-celled alga can be killed quickly for extraction of the intermediate products that form and change almost in-

6-4 *Chlorella*, a tiny alga, has contributed vastly to our knowledge of photosynthesis. (Walter Dawn)

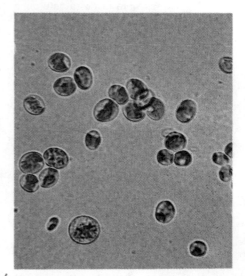

6-5 Calvin's "lollipop," used by Dr. Melvin Calvin to investigate photosynthesis. Radioactively labeled carbon dioxide was bubbled through suspensions of *Chlorella* in the "lollipop" for varying amounts of time, after which they were released into a flask of boiling alcohol, which killed them. Analyses were made to determine what substances had been synthesized in the differing periods of exposure to the $C^{14}O_2$. (Courtesy Melvin Calvin, Lawrence Radiation Laboratory)

stantly during photosynthesis (Figure 6-5). We owe much to this tiny plant, for it yielded one of the most significant biological discoveries of our age—the answer to the riddle of photosynthesis. It is interesting, too, that *Chlorella* is being cultured experimentally in large quantities, with a view to its possible use as food for livestock or even for man.

The secret of photosynthesis unlocked

The first major discovery in the recent investigations of photosynthesis concerned the fate of water molecules in the process. Remember that biologists had not been able to determine whether oxygen came from water or from carbon dioxide. Oxygen-18 provided the first answer. When *Chlorella* cells were grown in water containing oxygen-18 (H_2O^{18}), a remarkable discovery was made. The oxygen streaming from the *Chorella* cells contained the radioisotope! Furthermore, glucose extracted from the cells contained no oxygen-18. It was evident that somewhere in the process, water molecules were split. This process requires a great amount of energy. It can be done electrically by means of an electrolysis apparatus. But here were tiny algae splitting water molecules. Where could this energy be coming from? Since light is necessary for the process, it was the logical energy source. Biologists had known that light provides the energy stored in glucose molecules. But the idea that light energy splits water molecules was an entirely new concept. Chlorophyll plays an important part in this energy transfer. Thus, water molecules, light energy, and chlorophyll molecules are involved in the first, or *light reactions,* phase of photosynthesis.

The splitting of water molecules in the light reactions is only the first phase in glucose production. What happens to the carbon dioxide? Carbon-14 provided the answer, at least in part. The use of $C^{14}O_2$ in controlled photosynthesis led to the discovery of the second, or *dark reactions,* phase of the process. During this phase, carbon is fixed in a series of chemical reactions, none of which requires light (Figure 6-6). In the path of carbon from inorganic carbon dioxide to organic glucose, many steps are involved, and several intermediate products are formed. During these changes, carbon atoms are bonded to form chains. Many other atoms can be joined to these chains to form an almost endless number of organic compounds.

We can summarize the overall chemical changes that occur during both the light and dark reactions in photosynthesis in the following equation:

$$6\ CO_2 + 12\ H_2O^* + \text{light energy} \rightarrow C_6H_{12}O_6 + 6\ H_2O + 6\ O_2^*$$

carbon   water   glucose   water oxygen
dioxide

Oxygen-18 is denoted by an asterisk. Notice that the six molecules of oxygen ($6\ O_2^*$) released as a by-product result from the

6-6 Photosynthesis: an energy source, raw materials, products. The light, or *photo,* reactions occur on the left; the dark, or *synthesis,* reactions occur on the right.

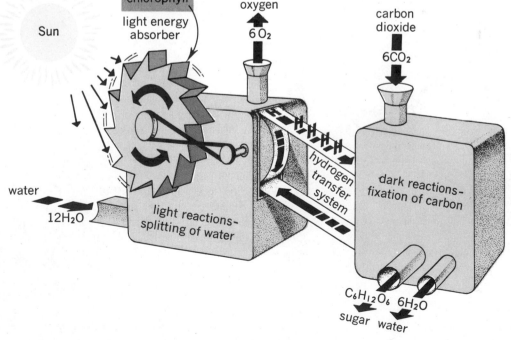
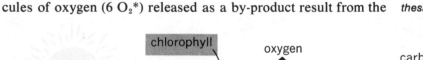

THE LIGHT REACTIONS

The trapping of energy and the splitting of water

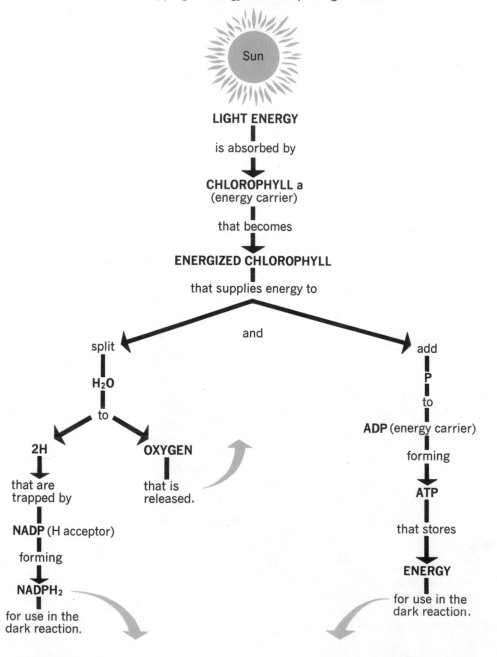

Sun

LIGHT ENERGY

is absorbed by

CHLOROPHYLL a
(energy carrier)

that becomes

ENERGIZED CHLOROPHYLL

that supplies energy to

and

split

H₂O

to

add

P

to

2H

that are
trapped by

OXYGEN

that is
released.

ADP (energy carrier)

forming

NADP (H acceptor)

forming

NADPH₂

for use in the
dark reaction.

ATP

that stores

ENERGY

for use in the
dark reaction.

6-7 The light reactions.

splitting of water molecules. The oxygen in the glucose comes from carbon dioxide. Notice, too, that water enters into the reaction and that it is a product as well. However, the water molecules used in the process are not the same molecules that appear as a product. While this equation represents the overall photosynthetic reaction as we understand it today, we must examine both the light and dark reactions more thoroughly to appreciate what occurs.

The light reactions, or photo phase

The first phase of photosynthesis, sometimes referred to as the *photo phase,* occurs only in the light and requires chlorophyll *a.* Thus, it is obvious that light energy enters into reaction during this phase. As we describe the various activities of the light reactions, it might seem that they form a lengthy series of processes. Actually, this phase of photosynthesis occurs in a split second. Nevertheless, we can best describe the various reactions that take place during the splitting of water molecules by light energy, or *photolysis,* as if they were a series of steps, even though they occur almost simultaneously. As you read the following description, refer often to Figure 6-7.

■ *Chlorophyll is energized.* As chlorophyll molecules lying along the grana of chloroplasts receive light energy, it is believed that a sudden change occurs in the molecules themselves. Electrons move from lower to higher energy levels. The chlorophyll molecules are now said to be *energized,* or excited. Absorption of energy by chlorophyll molecules transforms kinetic light energy into chemical potential energy. Thus, chlorophyll functions as an *energy carrier* in the photo phase. Energy remains in the chlorophyll as long as the molecules remain in this excited condition. As this energy is released, the electrons drop back to a lower energy level. The chlorophyll is no longer energized but may be illuminated again and trap more light energy.

■ *Water molecules are split.* Energy released from energized chlorophyll supplies the force necessary to pry apart the atoms in water molecules. As we stated earlier, hydrogen atoms are joined to oxygen by very strong bonds in a water molecule. Just how these bonds are broken and stable water molecules are split is still an unanswered question. However, we do know that the process requires a great amount of energy, which is supplied by energized chlorophyll.

■ *Additional energy is trapped in ATP.* The light phase of photosynthesis involves another important energy transfer necessary to support later chemical reactions. Chloroplasts contain ADP. Energy not involved in splitting water molecules is released from the energized chlorophyll and is used in converting ADP to ATP. Thus, ATP is a second energy carrier in photosynthesis. This energy transfer frees the chlorophyll molecules to receive additional light energy. Chemical energy stored in ATP will be released as a phosphate unit is removed later in the process.

■ *Hydrogen is trapped by NADP.* Hydrogen is released during the splitting of water molecules in the light reactions; it is captured immediately, thereby preventing its escape from the cell or its recombination with oyxgen to form water. This is accomplished by a coenzyme called NADP *(nicotinamide-adenine dinucleotide phosphate).* We speak of NADP as a **hydrogen acceptor** because it combines readily with hydrogen to form $NADPH_2$. This acceptance, however, is on a "loan basis," for the hydrogen is soon passed to another compound.

The oxygen released when water molecules are split during the photo phase escapes from the cell as a by-product.

The dark reactions, or synthetic phase

The transfer of energy from chlorophyll to ATP during the light reactions "charges" the chloroplast for reactions to follow. With chemical energy available, these activities do not require light and are thus referred to as the *dark reactions.* But this does not imply that the dark reactions must occur in darkness. Actually, they usually occur in the light and accompany the light reactions.

We can summarize the most important result of the dark reactions as *the fixing of carbon in a carbohydrate.* This process occurs in several steps that comprise a cycle. Compounds are formed, broken down, and formed again. In this chemical activity, carbon atoms form chains to which atoms of hydrogen and oxygen can be joined by chemical bonds to form carbohydrate molecules. In this way, the organic chemical world originates. We can summarize the chemical reactions in this stage of photosynthesis as follows (Figure 6-8).

■ *Carbon dioxide is fixed by RDP.* Chloroplasts contain a 5-carbon sugar phosphate known as RDP *(ribulose diphosphate).* The compound is composed of a 5-carbon sugar molecule to which two phosphate groups are attached. Within a fraction of a second after carbon dioxide reaches a chloroplast, it is fixed in a chemical compound by combining with RDP. Thus, RDP serves as the highly important **carbon dioxide acceptor.** The immediate product of this reaction is a 6-carbon sugar that is very unstable. This molecule splits quickly into two molecules of PGA *(phosphoglyceric* [FAHSS-foe-gli-SEHR-ick] *acid).* PGA is thus the first stable product of photosynthesis.

■ *PGA is converted to PGAL.* Again within a fraction of a second, PGA combines with hydrogen supplied by $NADPH_2$. The products of this reaction are PGAL *(phosphoglyceraldehyde* [FAHSS-foe-GLISS-uh-RAL-duh-HIDE]*),* also known as triose phosphate, and water. This reaction requires a large amount of energy, since PGA is a low-energy compound while PGAL has

THE DARK REACTIONS

The fixing of carbon in a carbohydrate

CO₂

combines with

RDP , a 5-carbon sugar in the chloroplast
(CO₂ acceptor),

to form

a very unstable **6-CARBON SUGAR**

that splits quickly and forms

2 molecules of **PGA**, a 3-carbon compound

that combines with

2H supplied by NADPH₂ from the light reaction
(energy supplied by conversion of ATP to ADP)

and forms

PGAL   and   **H₂O**

that can be used as a   that is released
nutrient or converted to   as a by-product.

RDP   and   **GLUCOSE**

that is again used   by combining 2
to combine with CO₂.   molecules of PGAL
                                and substituting
                                H for a phosphate.

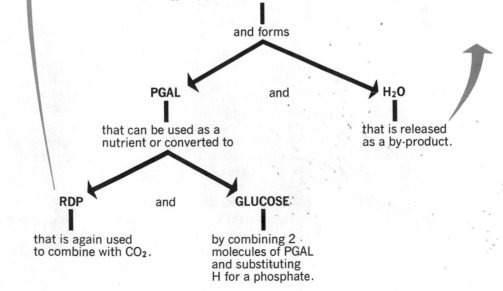

6-8 The dark reactions.

a high energy level. This energy is supplied from ATP by the removal of a phosphate group, converting it to ADP. PGAL can be used directly by a cell and for this reason might be considered to be the principal product of photosynthesis. While a cell does make immediate use of PGAL, much of it is converted to other products for transport from the cell or for storage. We will summarize several of these changes involved in the destiny of PGAL.

■ *The destiny of PGAL.* Since molecules of PGAL contain 3-carbon chains, PGAL can be used directly by a plant cell as a nutrient. In fact, plants nourished artifically with PGAL can survive without photosynthesis or any outside source of organic nutrients. However, a cell produces far more PGAL by photosynthesis than it needs, and much of the PGAL is changed into other products in further chemical reactions.

One essential product of PGAL is additional RDP to refuel the cell for further photosynthesis. You will recall that this 5-carbon sugar phosphate was present in the chloroplast and served as the carbon dioxide acceptor. You can see now how chemical reactions of the dark phase of photosynthesis constitute cycles. RDP was necessary to form PGAL. Part of the PGAL is converted back to RDP in the preparation of the chloroplasts for the next round.

Other molecules of PGAL are converted to glucose. This conversion results from the combination of two molecules of PGAL, the removal of the phosphate group from each molecule, and the substitution of a hydrogen atom for each phosphate group. The formula for PGAL is $C_3H_5O_3 \sim (P)$. The attached phosphate is indicated in the parentheses. You can see how the combination of two PGAL molecules and the substitution of hydrogen for each phosphate would produce a glucose molecule ($C_6H_{12}O_6$).

Fructose, another monosaccharide with the same chemical formula as glucose, can also be produced from PGAL. As you learned in Chapter 3, simple sugars, or saccharide units, may be combined to form double sugars such as sucrose ($C_{12}H_{22}O_{11}$), or may be assembled in long chains to form starches and celluloses. Some plants build PGAL molecules into various oils, such as corn oil, linseed oil, castor oil, peanut oil, soybean oil, and olive oil. Thus, you can see that this important product of photosynthesis is the basic ingredient needed for the synthesis of a wide variety of products.

Conditions for photosynthesis

Since light is the energy source for photosynthesis, you would expect light conditions to have a vital relationship to the process. As you know, sunlight is composed of light rays of varying wavelengths and energy. The various rays that comprise sunlight appear as red, orange, yellow, green, blue, and violet in a rainbow. Together, they are called the *visible spectrum.* Red rays have

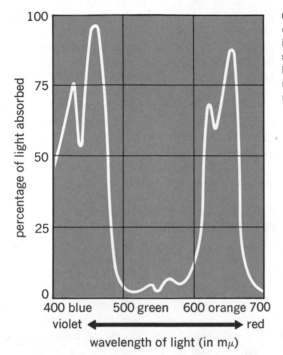

6-9 The absorption spectrum of chlorophyll in alcohol. This curve is similar to the curve for light absorption by chlorophyll in leaves. Note that most of the green light is not absorbed. Does this show why most plants appear green?

6-10 A simple technique for finding the relationship between light intensity and the rate of photosynthesis: Determine the rate at which bubbles of oxygen are released under conditions of different light intensities. (Sol Mednick)

the longest wavelength and the least energy of the visible radiations. Violet rays, at the opposite end of the spectrum, have the shortest wavelength and the highest energy. Chloroplasts absorb these rays in varying amounts. Furthermore, plants vary in the rays they absorb. Most land plants absorb the greatest amount of energy in the form of violet and blue rays and a somewhat smaller amount of energy as red and orange rays. While some of the green and yellow light is absorbed, much is reflected or transmitted. Since only those rays that are not absorbed are visible, you can now account for the green or greenish-yellow color of chloroplasts (Figure 6-9).

Plants such as algae growing in the ocean have different light problems. Seawater absorbs most of the red and violet rays and, in addition, reduces the total intensity of light (Figure 6-10). Much of the energy for photosynthesis in plants of the shallower waters comes from the blue, green, and yellow portions of the spectrum. Deep-water algae, living at depths of fifty to two hundred feet or more, receive most of their light energy from green and blue rays of the spectrum.

Temperature also influences the rate of photosynthesis, but not as much as you might expect. There is, however, a relationship between temperature and the carbon dioxide supply, and the rate of photosynthesis. Plants growing in the normal atmosphere, with a carbon dioxide content of about 0.04 percent by volume, carry on photosynthesis most rapidly at a temperature of about 21°C (70°F). But if the carbon dioxide content is raised to 1.25 percent, photosynthesis occurs most rapidly at a temperature of about 30°C (86°F). Temperature probably causes variation in the activity of enzymes involved in various steps of photosynthesis. However, temperature variations within the normal range during a plant's growing season seem to have little effect on enzyme action until temperatures exceed about 32°C (90°F). From this point on, increase in temperature seems to reduce the rate of photosynthesis.

Water supply is another factor that influences the rate of photosynthesis. Water shortage affects the entire physiology of the cell and therefore reduces the rate of photosynthesis.

The biological significance of photosynthesis

Were it not for photosynthesis, life on our earth would probably be limited to one or two groups of bacteria. There would be no forests or grasslands. Certainly there would be no animal populations. What makes the process so vital to life? It provides the chemical link between the inorganic and the organic chemical worlds. You might liken photosynthesis to the crossing of a bridge, with carbon dioxide and water on one side and PGAL, glucose, and other organic compounds on the other side. Chlorophyll, coenzymes, and energy form the bridge (Figure 6-11).

6-11 Photosynthesis can be likened to a chemical bridge between the inorganic world and the organic world.

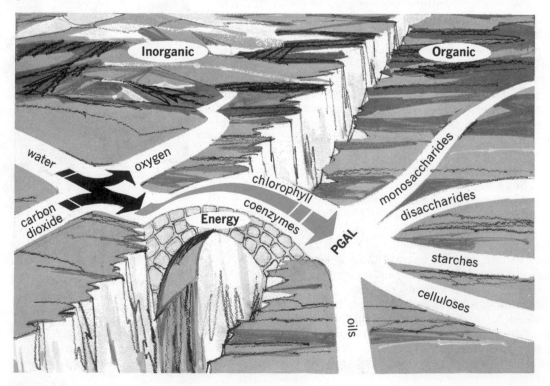

In discussing the food relations of organisms, we refer to green plants as *autotrophs* (AW-tuh-TROFFS). This term means "self-feeders" and refers to the capability of these living things to synthesize organic molecules from inorganic materials. Organisms lacking this capacity are classed as *heterotrophs* (HET-uh-roe-TROFFS), or "other feeders." Animals and nongreen plants such as fungi are among these nutritionally dependent organisms. While many heterotrophs have remarkable chemical abilities, none can produce a molecule of glucose from inorganic materials. In other words, they lack the capacity for photosynthesis and must rely on autotrophs for their basic chemical requirements.

Chemosynthesis

Our discussion of carbohydrate synthesis would not be complete without a brief discussion of a small but very important group of organisms that do not rely on photosynthesis. Certain bacteria organize carbohydrates without using light energy by a process known as *chemosynthesis*. These bacteria have enzyme systems that are capable of trapping energy released during inorganic chemical reactions. Certain of these bacteria are able to add oxygen to hydrogen to form water. Others change ammonia to nitrites, and nitrites to nitrates. Still others receive energy from reactions involving iron compounds and sulfur compounds. Energy from these reactions is used in synthesizing carbohydrates from carbon dioxide. From carbohydrate molecules, chemosynthetic bacteria can form all of their fats, proteins, and nucleic acids. We consider bacteria lowly organisms. But are those organisms lowly? From the standpoint of cell structure and specialization, they are. But from the standpoint of cell chemistry, they are probably as well equipped as any organisms in existence. We, who rank among the most nutritionally dependent of all forms of life, would find life much simpler if we shared the remarkable chemical capabilities of these chemosynthetic bacteria.

Respiration and energy release

All of the chemical reactions in a cell that break down glucose and other food molecules and transfer energy to ATP are included in the process, *respiration*. While the fuel molecules, reactions involved, products formed, and the amount of energy released may vary, respiration in some form is vital to the life of every cell.

You are familiar with the degrading of organic fuel molecules and the transformation of chemical bond energy into heat, light, and mechanical energy. This happens when you burn gasoline in your automobile engine, heat your house with fuel oil, gas, or coal, or burn a log in your fireplace. The energy released during combustion of these organic fuels is released as the fuel molecules are broken down through the addition of oxygen. As you know, this occurs only at a high temperature. Furthermore, the burning that occurs is uncontrolled in that almost all of the molecules are broken down.

While there is some similarity between the breaking down of fuels during combustion and the degrading of fuel molecules in a cell, there are also many differences. The *mitochondria,* in which much of cellular respiration occurs, could not withstand the high temperatures of fuel combustion. Moreover, because the degrading of fuel molecules in a cell is controlled, it occurs in small steps, each of which releases small quantities of energy. How can organic fuel molecules be degraded at the normal temperature of an organism? *Respiratory enzymes* accomplish this and control the process as well.

When you think of respiration in your own body, you probably think of gaseous exchanges during breathing. While we are most aware of breathing and the intake and exhalation of gases, remember that this phase of the process is a gaseous exchange between the blood and the atmosphere. The body cells and, more specifically, the mitochondria of these cells are the seat of respiration. It is here that the cellular "fires" burn constantly.

The fuel for respiration

Any organic molecules present in a cell can be fuel for respiration. These molecules include glucose molecules, fatty acids and glycerol, amino acids, and even vitamins and enzymes.

In Chapter 3, we discussed the chemical bonds that hold atoms together in a molecule. As long as the bonds remain unbroken, the molecule contains stored chemical bond energy. When the molecule is decomposed to atoms or simpler molecules, this energy is set free. Organic fuels used in respiration contain energy that was once solar energy, locked in molecules since they were organized from simpler molecules during photosynthesis. Thus, the energy liberated in respiration is chemical bond energy that was once light energy.

The energy released in a cell during respiration results from the *oxidation* of glucose. Oxidation involves either the addition of oxygen or the removal of hydrogen from a molecule, with an accompanying release of energy. Most of the oxidation in a cell is brought about by the removal of hydrogen.

Stages of cellular respiration

The breakdown of glucose during *cellular respiration* occurs in two major stages involving a series of chemical reactions and many respiratory enzymes. These stages can be summarized as follows:

■ The first stage is referred to as *anaerobic* since it does not require molecular oxygen. This stage occurs outside the mitochondria. In a complex series of chemical reactions involving as

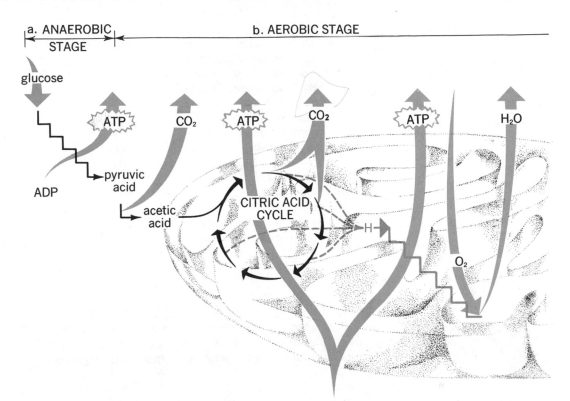

a. ANAEROBIC STAGE

b. AEROBIC STAGE

glucose

ATP

CO₂

ATP

CO₂

ATP

H₂O

pyruvic acid

ADP

acetic acid

CITRIC ACID CYCLE

H

O₂

ADP

6-12 A diagramatic summary of the chemical changes that occur in cellular respiration.

many as 12 enzymes, a molecule of 6-carbon glucose is broken down to two 3-carbon molecules of *pyruvic acid,* as shown in Figure 6-12a. Energy from two molecules of ATP is used in this reaction. However, the splitting of a high-energy glucose molecule to two lower energy pyruvic acid molecules releases energy to form four ATP molecules. Thus, the net gain in this stage of the process is two ATP molecules. This represents approximately 7 percent of the energy in a molecule of glucose.

■ The second stage involves two main series of steps. Since molecular oxygen is required in this stage, we refer to it as *aerobic.* In a sequence of chemical changes, pyruvic acid is broken down and carbon dioxide and water are released, along with considerable energy. As we describe the various steps, follow them in Figure 6-12b.

Two hydrogen atoms and one carbon dioxide molecule are removed from each pyruvic acid molecule to form two *acetic acid* molecules. Carbon dioxide is released. Acetic acid enters a mitochondrion, where the process continues. A molecule of acetic acid is joined to a 4-carbon acid present in the mitochondrion to form *citric acid,* a 6-carbon molecule. This is one of a series of reactions commonly referred to as the citric acid cycle, or

Krebs cycle. As the cycle continues, one carbon atom is re-moved from the citric acid to form a 5-carbon acid and a mole-cule of carbon dioxide. In another step, a second carbon atom is removed, leaving the 4-carbon acid to combine with additional acetic acid and repeat the cycle. Carbon removed in the process is released as additional carbon dioxide.

Hydrogen is also released in reactions of the citric acid cycle. In a second series of steps of the second stage of cellular respira-tion, this hydrogen is involved in a series of reactions called *hydrogen transport.* During these reactions, molecules of ATP are formed from ADP. Energy released in various stages of the pro-cess is trapped in these molecules. Finally, the hydrogen atoms combine with molecular oxygen to form water.

Cellular respiration, including both the anaerobic and aerobic stages, is often represented by the following chemical equation:

$$C_6H_{12}O_6 + 6\,O_2 \;\rightarrow\; 6\,CO_2 + 6\,H_2O + \text{energy (38 ATP)}$$
glucose   oxygen   carbon   water
                              dioxide

Notice that the overall capture of useful energy in the anaerobic and aerobic stages of cellular respiration is represented in 38 molecules of ATP. This is between 55 and 60 percent of the total chemical bond energy in a molecule of glucose.

Fermentation

6-13 A diagramatic summary of the chemical changes that occur in alcoholic fermentation.

Another kind of respiration occurs in which glucose is broken down to simpler compounds by enzyme action in the complete

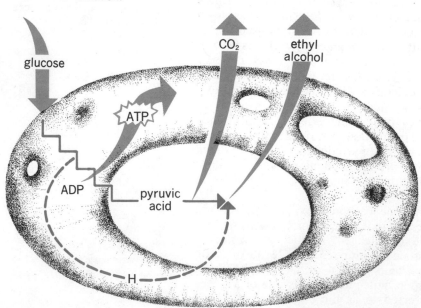

absence of molecular oxygen. As in the anerobic stage of cellular respiration, glucose is broken down to two pyruvic acid molecules, with the liberation of a small amount of energy. The pathways for the further breakdown of pyruvic acid under anaerobic conditions lead in two directions:

■ In yeasts and certain other microorganisms, a carbon dioxide molecule is stripped from pyruvic acid, giving an intermediate product that is reduced to ethyl alcohol as an end product (Figure 6-13). This process, known as *alcoholic fermentation,* can be represented in the following equation:

$$C_6H_{12}O_6 \rightarrow 2\ C_2H_5OH + 2\ CO_2 + \text{energy (2 ATP)}$$
glucose   ethyl alcohol   carbon dioxide

■ In animal tissues, such as muscle, under anaerobic conditions the pyruvic acid is converted to lactic acid as an end product. This process, known as *lactic acid fermentation,* or *glycolysis,* is represented by the following equation:

$$C_6H_{12}O_6 \rightarrow 2\ C_3H_6O_3 + \text{energy (2 ATP)}$$
glucose   lactic acid

When you compare the energy released from glucose in both alcoholic and lactic acid fermentation with cellular respiration, you will see that the former are far less efficient processes (Figure 6-14). Most of the energy in the bonds of glucose remains in the chemical bonds of the fermentation end products. Only a fraction of the energy is made available to the organism.

Uses of cell energy

What happens to the energy released from glucose during respiration? Some of it is given off as heat. In a warm-blooded animal, this heat maintains a constant body temperature. The remaining energy, which was trapped in molecules of ATP, is used to support the many cell activities. These include the synthesis of polysaccharides, fats and oils, as well as nucleic acids and proteins (to be discussed in Chapter 7), active transport, cell division, muscular contractions, and the transmission of nerve impulses.

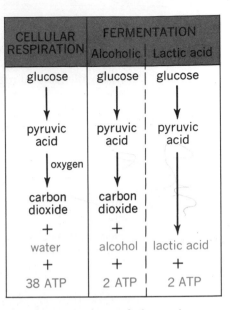

CELLULAR RESPIRATION	FERMENTATION	
	Alcoholic	Lactic acid
glucose	glucose	glucose
↓	↓	↓
pyruvic acid	pyruvic acid	pyruvic acid
↓ oxygen	↓	
carbon dioxide	carbon dioxide	↓
+	+	
water	alcohol	lactic acid
+	+	+
38 ATP	2 ATP	2 ATP

6-14 A comparison of the end products and energy released in cellular respiration and alcoholic and lactic acid fermentation.

SUMMARY

The next time you look at a leaf, just think what a marvelous chemical factory it is! Chlorophyll in the grana of chloroplasts is absorbing light energy that is used to split water molecules. Hydrogen, destined to become part of organic molecules, is being trapped. Energy to support the dark reactions is being transferred from chlorophyll to ATP. Carbon atoms from carbon dioxide are being arranged to form the centers of organic molecules to which hydrogen and oxygen atoms will be added. PGAL is formed, then glucose and other organic compounds.

It takes cells with chloroplasts and the marvel of photosynthesis to bridge the gap between the inorganic chemical world where carbon dioxide and water are abundant and the world of organic compounds. Could we survive without green plants? How important is photosynthesis?

However, the energy stored in glucose is of no value to an organism unless it is released in its cells. This is accomplished in respiration. Now, you can see how photosynthesis and respiration are, in many ways, complementary processes.

BIOLOGICALLY SPEAKING

photosynthesis	dark reactions	heterotroph
respiration	photo phase	chemosynthesis
ATP	photolysis	oxidation
AMP	energy carrier	cellular respiration
ADP	hydrogen acceptor	hydrogen transport
chlorophyll	carbon dioxide acceptor	alcoholic fermentation
bacteriochlorophyll	visible spectrum	lactic acid fermentation
light reactions	autotroph	glycolysis

QUESTIONS FOR REVIEW

1. Explain why living things require a continuous supply of new energy.
2. What two parts compose the adenosine portion of a molecule of ATP?
3. Explain how energy is stored in the conversion of ADP to ATP.
4. Describe the composition of the grana in chloroplasts.
5. List various forms of chlorophyll found in different cells. Which form is necessary for photosynthesis?
6. Why is *Chlorella* an ideal organism to use in research in photosynthesis?
7. What is energized chlorophyll? What part does it play in photosynthesis?
8. Explain two uses of the energy released from energized chlorophyll in photosynthesis.
9. Explain the relationship of NADP and hydrogen in the light reactions.
10. Outline the steps in the fixation of carbon dioxide in the dark reactions.
11. Describe the chemical reactions involved in the conversion of PGA to PGAL.
12. List several organic compounds that can be formed from PGAL.
13. Explain why chloroplasts appear green.
14. Distinguish between an autotroph and a heterotroph.
15. How is chemosynthesis fundamentally different from photosynthesis?
16. What is the biological importance of respiration?
17. Explain why cellular respiration must be controlled and occur slowly.
18. Name two stages in the cellular respiration of glucose.
19. Which of the two stages in cellular respiration yields more energy?
20. What is the total output of energy in cellular respiration in terms of molecules of ATP formed?
21. What product is formed when oxygen is the hydrogen acceptor in aerobic respiration?
22. Name two products of fermentation, other than carbon dioxide.
23. Much of the energy released from glucose in cellular respiration is stored in ATP. In what form does some of it escape?
24. List several processes requiring cell energy.

APPLYING PRINCIPLES AND CONCEPTS

1. Discuss the matter and energy relationships in ATP as an energy carrier.
2. Discuss the role of chlorophyll in photosynthesis.
3. Account for the fact that water is both a requirement and a product of photosynthesis.
4. Discuss the use of radioisotopes as research tools in exploring the nature of photosynthesis.
5. Explain how RDP is involved in a cycle in the dark reactions of photosynthesis.
6. Discuss the nutritional relationships between autotrophs and heterotrophs.
7. Discuss the products formed during the aerobic and anaerobic stages of cellular respiration.
8. Compare the efficiency of cellular respiration and fermentation in supplying the energy needs of cells.
9. Discuss the relation of ATP to respiration.
10. Why might we consider photosynthesis and respiration complementary processes?

RELATED READING

Books

ASIMOV, ISAAC, *Life and Energy.*
Doubleday and Co., Inc., Garden City, N.Y. 1962. A well known author treats a difficult subject thoroughly and gives a good explanation of energy and its role in organisms.

GALSTON, ARTHUR W., *The Life of the Green Plant.*
Prentice-Hall, Inc., Englewood Cliffs, N.J. 1961. An authoritative, detailed work that emphasizes scientific principles.

HAFFNER, RUDOLPH, *The Vital Wheel: Metabolism.*
American Education Publications Inc., New York. 1963. The importance of metabolism is presented in useful pamphlet form.

JAMES, W. O., *An Introduction to Plant Physiology* (6th edition).
Oxford University Press, Inc., New York. 1963. A balanced account of the more elementary aspects of plant physiology.

LEHNINGER, ALBERT L., *Bioenergetics.*
W. A. Benjamin Inc., Menlo Park, California. 1965. A somewhat advanced reference including up-to-date material on photosynthesis and respiration.

ROSENBERG, JEROME, *Photosynthesis.*
Holt, Rinehart and Winston, Inc., New York. 1965. Discusses the basic food making processes in plants.

VanOVERBECK, JOHANNES, *The Lore Of Living Plants.*
McGraw-Hill Book Co., New York. 1964. A coherent story in the realm of plant physiology.

Articles

ARNON, DANIEL I., "The Role of Light in Photosynthesis," *Scientific American,* November, 1960.
An excellent presentation on the manner by which light energy is changed into chemical energy, a form useful to the cell.

GREEN, DAVID E., "The Mitochondrion," *Scientific American,* January, 1964.
Describes the structure of the energy producer of the cell as revealed by the electron microscope.

CHAPTER SEVEN

NUCLEIC ACIDS AND PROTEIN SYNTHESIS

Protein synthesis

Photosynthesis is limited to the cells of autotrophs; however, this cannot be said of *protein synthesis*. No living matter has ever been found that does not contain protein. Every cell organizes its own protein molecules. Thus, protein synthesis is a universal phase of cell anabolism.

There is another striking difference between protein synthesis and carbohydrate synthesis. All cells capable of photosynthesis organize the same PGAL and glucose molecules in the same series of chemical reactions. This is not true in protein synthesis. A cell builds specific proteins that vary from species to species, individual to individual, and, to some extent, within various kinds of cells in the same organism. You may have read accounts of the transplanting of tissues or organs from one individual to another. These transplants are often unsuccessful except between identical twins, because of the recipient's reaction to the proteins of the donor.

A cell thus expresses individuality in protein synthesis. How is the synthesis regulated? What sort of cellular code determines exactly what proteins will be formed and how these molecules are to be constructed? This is a fascinating story. To find the answers to these questions, we must explore a cell more thoroughly. The coding substance lies in the nucleus. The protein "factories" are the ribosomes attached to the endoplasmic reticulum and lying between its folds. How is a code delivered from the nucleus to the ribosomes? What substance in the nucleus provides this code? This was a puzzle to biologists until a few years ago, when two scientists made a discovery that is among the most significant biological advances of all time. What had been isolated pieces of a puzzle suddenly formed a revealing and thrilling picture of life when the missing piece was finally supplied.

What is DNA?

For many years, biologists have looked to the cell nucleus as the center of control of all cell activities. They have been familiar with dark, rod-shaped bodies known as chromosomes which lie in the nucleus. Studies before World War II had established the fact that chromosomes are composed of two substances, *protein* and *nucleic acid*. The question was whether control of the organization of proteins and other cell activities centered in the protein or in the nucleic acid.

Studies at the Rockefeller Institute during World War II provided evidence that the nucleic acid and not the protein controlled all chemical activities of an organism. What were these nucleic acid molecules that possessed such remarkable properties? Their chemical composition was known, but their physical structure—the way their atoms are actually arranged—was not.

In 1953, two young scientists working in the Cavendish Laboratory at Cambridge University made a significant breakthrough which has proved to be one of the most important discoveries of recent years. One member of the team was an American biologist, James D. Watson. The other was a British biophysicist, F. H. C. Crick. Together, they proposed a model of probably the most complex of all organic molecules, *deoxyribonucleic* (dee-OCK-see-ry-boe-NOO-klee-ick) *acid*, or *DNA*.

The structure of DNA

Watson and Crick described the DNA molecule as a double helix, or spiral, consisting of two strands wound around each other like a flexible ladder twisted into a corkscrew (Figure 7-2). Actually, the ladder is in two halves, each providing a side piece, or upright, and half of a rung, or step. The two parts of the ladder are joined by bonds uniting the rungs.

7-1 James D. Watson and F. H. C. Crick, co-discoverers of DNA. (UPI)

7-2 The formation of a double helix can be compared to the twisting of a flexible ladder.

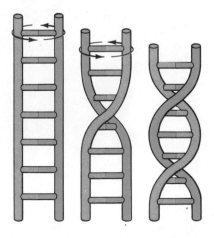

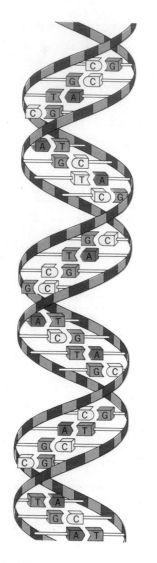

7-3 Model of the double helix of DNA.

Now, we shall examine the DNA ladder more closely and determine the chemical make-up of its parts. Figure 7-3 shows a double helix of DNA. The lower portion is uncoiled to show the parts of the two strands more closely.

The side piece, or backbone, of each strand is composed of *phosphate* portions joined to molecules of *deoxyribose* (dee-ock-see-ry-bose), a five-carbon sugar. The molecular structure of these portions and the manner in which they are joined is shown in Figure 7-4. A nitrogen-containing *base* is attached to each deoxyribose unit, forming part of the rung, or step, of the ladder. Each base is joined to a base of the other DNA strand by weak hydrogen bonds. Each three-part unit of a strand of DNA, consisting of a phosphate, a sugar, and an attached base makes up a *nucleotide*. Thus, you can think of a DNA molecule as a double strand of nucleotides bonded by their bases. The nucleotides in each strand may number several thousand.

Notice in Figure 7-3 that the bases in DNA are of four types. Two are organic molecules known as *purines* (pure-EENS), either *adenine* (AD-uh-NEEN) or *guanine* (GWAH-NEEN). The other two are *pyrimidines* (pie-RIM-uh-DEENS), either *thymine* (THY-MEEN) or *cytosine* (sy-tuh-SEEN). These four bases are designated by the DNA code letters, A, G, T, and C.

The chemical structure of these bases is such that they bond only in certain combinations. Adenine (a purine) always joins thymine (a pyrimidine). Similarly, guanine (a purine) always joins cytosine (a pyrimidine). Figure 7-5 shows four DNA code letters and the manner in which four nucleotides in each strand are joined by hydrogen bonds uniting their bases.

The molecular structure of the four bases of DNA and the hydrogen bonds joining them are shown in Figure 7-6. How many hydrogen bonds join adenine and thymine? How many join guanine and cytosine? Now you can see why they must pair in this manner to form a rung of the DNA ladder.

The replication of RNA

One of the most remarkable properties of DNA is its ability to duplicate its own structure. We refer to this process as *replication*. DNA also directs the synthesis of a near duplicate of its structure, known as *ribonucleic* (RY-boe-noo-KLEE-ick) *acid*, or *RNA*. This process is called *transcription*. The sugar in RNA is *ribose* rather than deoxyribose (it contains one more oxygen atom). A comparison of the molecular structure of these two sugars is shown in Figure 7-7. RNA also differs from DNA in the composition of one of its bases. *Uracil* (YUR-uh-sill), another pyrimidine, is substituted for thymine. Thus, the code letters for bases in RNA are A, G, U, and C. A comparison of the molecular structure of thymine and uracil is shown in Figure 7-8. RNA is usually a single strand and, as a product of DNA, serves as its agent in controlling certain cell activities, as you will soon learn.

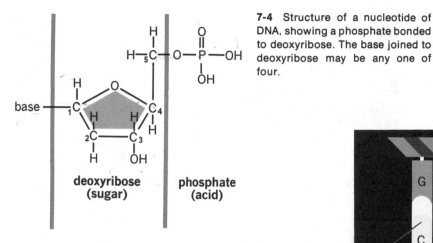

7-4 Structure of a nucleotide of DNA, showing a phosphate bonded to deoxyribose. The base joined to deoxyribose may be any one of four.

deoxyribose (sugar)   phosphate (acid)

7-5 The four bases representing the four code letters in DNA.

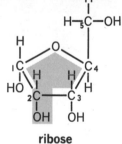

thymine

hydrogen bonds

cytosine   guanine

phosphate unit   sugar unit

7-6 The strands of the DNA molecule are held together by hydrogen bonds (dashed lines) that exist between the purine-pyrimidine base pairs. The arrows indicate the site at which each base is bonded to a DNA strand.

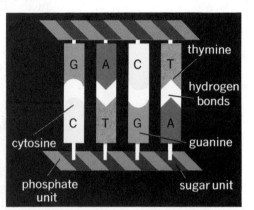

deoxyribose   ribose

7-7 Deoxyribose and ribose compared.

7-8 Thymine and uracil compared. Thymine is found only in DNA; uracil is found only in RNA.

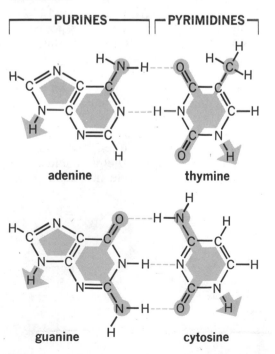

————— PURINES ————— ┌ PYRIMIDINES ┐

adenine   thymine

guanine   cytosine

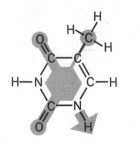

thymine

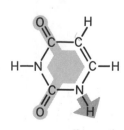

uracil

Three-letter codes in DNA

The nucleotide bases we designated as A, G, T, and C can occur in any linear sequence in a series of one thousand or more in a strand of DNA. However, biologists have found that it is not single bases but groups of three, or *triplets,* that bear an important code. These three-letter code words might be designated as CGT, ACG, or AAA, depending on the order of nucleotides in the DNA strand. How many three-letter code words can be formed from various combinations of the four letters representing the four kinds of nucleotide bases? There are 64 possibilities. What do the triplet codes determine? You will find that each codes an amino acid and that a series of these three-letter codes determines the exact order in which amino acids will be arranged in a specific protein molecule.

Transcription of the DNA code to RNA

DNA lies in the nucleus. Ribosomes are in the cytoplasm. These are the protein "factories" that join amino acids and form polypeptide chains. If DNA controls the synthesis of proteins, how is the pattern, or "blueprint" delivered from "headquarters" in the nucleus to the ribosome "factories?" The answer is RNA, the agent of DNA. This involves a remarkable sequence of events.

We shall begin in the nucleus where a strand of DNA is directing the synthesis of a strand of RNA. We will assume that, among the several thousand nucleotide base triplets in the DNA, three are coded CGT, ACG, and AAA. The corresponding nucleotide triplets in the strand of RNA are coded GCA, UGC, and UUU. Notice that the RNA triplet codes are complementary to, or the opposite of, those of DNA and that uracil (U) is substituted for thymine (T). The transcription of this portion of DNA to RNA is shown in Figure 7-9.

The strand of RNA detaches and may be stored a short time in the nucleolus, or may move immediately through the nuclear membrane into the cytoplasm bearing codes for specific amino acids in its nucleotide bases. We refer to this RNA as *messenger RNA.* As in DNA, there are 64 possible three-letter codes, or *codons* in messenger RNA. There are only 20 amino acids. This would indicate that several codons may code the same amino

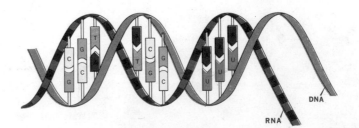

7-9 DNA replicating messenger RNA.

acid. This is true, although biologists have found that certain letter combinations are not involved in coding amino acids. They refer to these as *"nonsense codons."* These are known to include UAA, UAG, and UGA. These codons appear to be termination points in the messenger, to signal the end of a protein chain.

Table 7-1 shows the known messenger RNA triplet codes, or codons, for the 20 amino acids. The three codons included in the messenger RNA we described are coded GCA, UGC, and UUU. By consulting Table 7-1, you can determine that they are coding the amino acids alanine, cysteine, and phenylalanine.

Table 7-1   MESSENGER RNA CODONS FOR THE AMINO ACIDS

AMINO ACID	TRIPLET CODE
alanine	GCA, GCG, GCC, GCU
arginine	CGA, CGG, CGC, CGU, AGA, AGG
asparagine	AAC, AAU
aspartic acid	GAC, GAU
cysteine	UGC, UGU
glutamic acid	GAA, GAG
glutamine	CAA, CAG
glycine	GGC, GGU, GGA, GGG
histidine	CAC, CAU
isoleucine	AUC, AUU, AUA
leucine	CUC, CUU, CUA, CUG, UUA, UUG
lysine	AAA, AAG
methionine	AUG
phenylalanine	UUU, UUC
proline	CCA, CCG, CCC, CCU
serine	UCA, UCG, UCC, UCU, AGU, AGC
threonine	ACA, ACG, ACC, ACU
tryptophan	UGG
tyrosine	UAC, UAU
valine	GUA, GUG, GUC, GUU

The assembly of amino acids and the synthesis of proteins

When a strand of messenger RNA reaches the cytoplasm, it acts as a pattern, or template, for the assembly of amino acids and synthesis of a protein molecule. This process is referred to as *translation*. Several protein molecules will be organized simultaneously along the length of messenger RNA. It is believed that a ribosome moves to the template and attaches at a place where a chain of amino acids will be started. These positions on the template are indicated by certain codons, known as *initiator codons*.

Ribosomes contain another form of RNA, known as *ribosomal RNA*. There is no evidence that this RNA functions directly in protein synthesis. However, it is believed to be synthesized under control of DNA in the nucleus and is combined with proteins in the structure of the ribosomes. There is evidence, also, that ribosomal RNA may function in the attachment of ribosomes to messenger RNA and that it may regulate the enzymes involved in activities of the ribosome.

Dispersed through the cytoplasm are various amino acids that will be joined together in long chains during protein synthesis. The exact kind of amino acids as well as the sequence in which they are linked in the protein chain is determined by the codons in the messenger RNA. These amino acids must be picked up individually and brought to the template in proper order during protein synthesis.

This involves a third type of RNA, known as *transfer RNA*. These strands are shorter than those of messenger RNA, usually consisting of 70-80 nucleotides. A strand of transfer RNA doubles back on itself and forms loops. At one end of the strand is a triplet of nucleotides that provide a specific base code. This code is the opposite of a corresponding codon in the messenger RNA forming the template and is, therefore, referred to as an *anti-codon*. Thus, the anti-codon CGU in transfer RNA is the opposite and complement of GCA in messenger RNA.

The last three nucleotide bases in all strands of transfer RNA are CCA. An amino acid always attaches to the adenosine of this last nucleotide triplet. We know that a strand of transfer RNA, depending on its anti-codon, will pick up only one of the 20 kinds of amino acids. How does the transfer RNA pick up the correct amino acid for delivery during protein synthesis?

It is believed that an activating enzyme in the cytoplasm, using energy supplied from ATP, "recognizes" a specific amino acid and bonds it to the adenosine of the correct transfer RNA. Thus, there must be an enzyme for each of the 20 amino acids.

Protein synthesis involves the assembly of amino acids delivered to a ribosome by transfer RNA in proper order, as determined by the codons in messenger RNA. During formation of a protein chain, a ribosome moves along the template formed by the messenger RNA from one end to the other and "reads" each codon in succession. As the ribosome "reads" a specific codon a transfer RNA unit with the proper anti-codon moves to a site on the ribosome. This places the attached amino acid in correct position to be added to a protein chain.

After delivery of an amino acid to the ribosome, the transfer RNA unit moves to the cytoplasm. Other strands of transfer RNA with attached amino acids move to the ribosome in proper order, as determined by the codons of the messenger RNA until a protein chain is completed. The protein is then released from

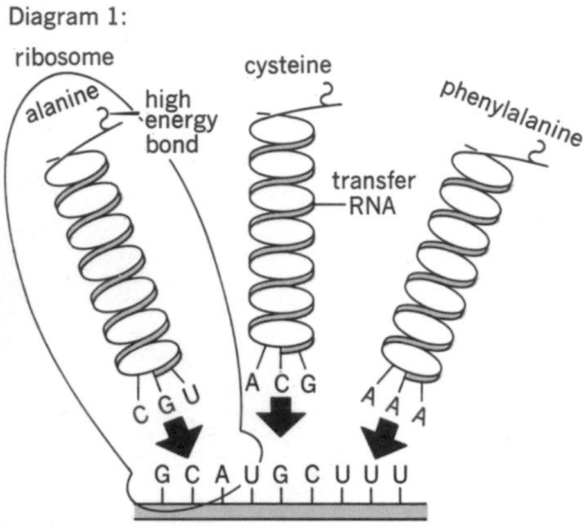

Diagram 1:

ribosome

alanine   high energy bond   cysteine

phenylalanine

transfer RNA

C G U   A C G   A A A

G C A U G C U U U

messenger RNA

Diagram 2:

alanine (protein polypeptide)—cysteine—phenylalanine

ribosome

C G U   A C G A A A

G C A U G C U U U

7-10 Diagram I: Transfer RNA units with attached amino acids are moving to correct positions on the messenger RNA template. Diagram II: Amino acids are being bonded by a ribosome to form a growing protein chain.

the ribosome into the cytoplasm and the ribosome leaves the messenger RNA. Stages in the assembly of a protein chain are shown in Figure 7-10.

While a single ribosome "manufactures" only one protein molecule, electron micrographs have revealed that several ribosomes, each synthesizing a protein molecule may move along the messenger RNA template at the same time. There is evidence, also, that transfer RNA units may attach amino acids more than one time. In some cells, however, they seem to function only once. Similarly, it is believed that a strand of messenger RNA may function as a template in protein synthesis more than once.

Uses of proteins in organisms

Proteins serve many functions in living cells. Some proteins are *structural*. These molecules remain in the cell and form its various parts—its cytoplasm, membranes, and the various cytoplasmic inclusions. *Enzymes* are protein molecules that are essential to all of the cell's chemical activities. We classify these as *intracellular* when they remain in the cell. Other enzymes are *extracellular;* these are secreted from the cell and act as catalysts in chemical reactions that occur outside the cell. Digestive enzymes of most organisms are extracellular enzymes. Still other proteins are *hormones,* which regulate specific activities often far removed from the cells that produce them. Other proteins form pigments in plant cells, the hemoglobin in red blood corpuscles, and the proteins composing blood serum. These are but a few of the protein substances synthesized in cells. We can summarize the importance of proteins by stating that they are *indispensable for life*.

SUMMARY

DNA, by transcribing a nucleotide base code to messenger RNA, determines the kind and the sequence of the amino acids ribosomes will join in chains to form proteins. There are 64 possible triplet codes. These are universal in that each codes the same amino acid in all organisms. Since the DNA code differs among kinds of organisms as well as individuals, the proteins each organism forms are specific and individual.

By determining the composition of structural proteins as well as enzymes and other proteins, DNA controls the physical make-up of cells as well as the many chemical processes involved in their functions. In fact, DNA controls the entire organism. Is it any wonder that DNA is considered the key to life?

As you continue your study of biology, you will find that DNA comes into the discussion again and again. How do organisms grow? How does a particular kind of organism maintain its identity? Why do organisms resemble their parents in some respects and differ in others? DNA provides the answers to all of these questions.

BIOLOGICALLY SPEAKING

DNA	messenger RNA	initiator codon
nucleotide	codon	ribosomal RNA
replication	"nonsense codon"	transfer RNA
triplet	template	anti-codon

QUESTIONS FOR REVIEW

1. Describe the composition of a nucleotide of DNA.
2. Name four nitrogen-containing bases in the nucleotides of DNA.
3. What is replication as it applies to DNA?
4. In what two ways are the nucleotides of RNA different in structure from those of DNA?
5. Explain the origin and destination of messenger RNA.
6. Explain the function of transfer RNA in protein synthesis.
7. Why must a cell contain many different kinds of transfer RNA?
8. Would a group of nucleotide codes in strands of transfer RNA be like those of the messenger RNA or like DNA in the nucleus?
9. How do cell ribosomes function in protein synthesis?
10. List several different kinds of proteins synthesized in cells.

APPLYING PRINCIPLES AND CONCEPTS

1. Discuss the chemical mechanisms by which DNA controls protein synthesis.
2. Explain why proteins are specific in individual organisms.
3. There are four kinds of bases in the nucleotides of messenger RNA. There are 64 possible combinations of these bases in three-base codes. How many single base codes would be possible? How many two-base codes are possible?
4. Why is replication an extremely important property of DNA?
5. Discuss DNA as the "key to life."

RELATED READING

Books

ASIMOV, ISAAC, *The Genetic Code.*
  New American Library, New York. 1963. Written in a popular manner, this book describes the activity involved in the unraveling of the code of DNA.
FRANKEL, EDWARD, *DNA—Ladder of Life.*
  McGraw-Hill Book Co., New York. 1964. How scientists worked out the structure of the DNA molecule and are now deciphering some of its messages.
WATSON, JAMES D., *The Double Helix.*
  New American Library, New York. 1969. A fascinating and lively account of the events leading to the discovery of DNA.

Articles

CRICK, FRANCIS H. C., "The Genetic Code," *Scientific American,* October, 1962. The order in which the bases are attached to the chain of nucleotides is responsible for the genetic code contained therein.
YANOFSKEY, CHARLES, "Gene Structure and Protein Structure," *Scientific American,* May, 1967. A good presentation on the relationship between genes and their control over protein structure.

CHAPTER EIGHT

CELL GROWTH AND REPRODUCTION

Growth of cells

One of the results of protein synthesis in a cell is replacement of worn-out structures. The rate of synthesis, however, normally exceeds the requirements of materials for repair and replacement. Accumulation of these additional materials results in growth of the cell.

Is there a limit to the size a cell can reach? Writers of science fiction have constructed weird tales of cells that do not stop growing. Giant blobs of protoplasm move into cities and flow down streets, engulfing terrorized people who cannot escape. While stories such as these make fascinating reading, you know that they could never happen. Do you know why?

Limitations on cell size

All of the materials necessary to support the life-sustaining processes of a cell must enter the cell through its enveloping membrane. Furthermore, all of the waste products resulting from chemical reactions within the cell must pass through this membrane. Thus, there is a critical relationship between the volume of cell content and the surface exposure of the membrane. As a cell grows, its protoplasmic volume increases, but its membrane surface does not increase proportionally. Supplying the protoplasmic content becomes an increasing problem.

What is a logical solution? Division of the cell into two smaller cells results in the containment of the cell substance within an enlarged membrane surface. This allows more materials to pass into and out of the original cell mass. Because the original mass, now two cells, can grow even more, we might think of cell division as a rejuvenating step in the life of cells.

The nature of cell division

In most kinds of cells, cell growth, or increase in size, is normally followed by cell division, or increase in number. Since cell division involves the splitting of a cell, biologists refer to the process as *fission*. Most cells divide into two approximately equal parts, or undergo **binary fission**.

We call a cell that has undergone growth, and is ready to divide, a *mother cell*. The division of a mother cell results in two approximately equal *daughter cells*. Fission generally involves two distinct phases. One is the duplication and distribution of nuclear materials, as a result of which each daughter cell receives nuclear materials that are identical with those of the mother cell. This duplication of nuclear materials maintains the characteristics of a cell in all of its descendents. We refer to the process in which nuclear materials are duplicated as **mitosis** (my-TOSE-iss).

The second phase of cell fission is division of the cytoplasm into two approximately equal parts. This phase of the process involves a mechanical separation of cytoplasmic structures of the mother cell by a membrane or a wall. Obviously, mitosis must precede cytoplasmic division.

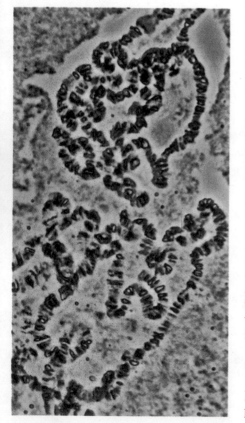

8-1 The giant chromosomes in the nucleus of a cell from the salivary glands of the common fruit fly, *Drosophila melanogaster*. (Walter Dawn)

Mitosis

Before we discuss the sequence of events occurring in mitosis, it would be well to review certain of the nuclear structures discussed in Chapter 4. The key chemical components of the nucleus are strands of DNA wound around a protein core to form *chromosomes* (Figure 8-1). Active groups of nucleotide base triplets that code proteins are believed to constitute *genes*. These are the determiners of all genetic traits, as you will learn in Unit 2.

Before a cell divides, all of the DNA in its nucleus replicates, forming two identical strands in duplicate chromosomes. Distribution of identical chromosomes to daughter nuclei during mitosis preserves the genetic continuity of the organism in all of its cells.

For convenience, the sequence of events preceding and during mitosis is divided into the following stages, or phases: (1) interphase, (2) prophase, (3) metaphase, (4) anaphase, and (5) telophase. These stages are not identified by abrupt changes. However, certain significant events occur that will make it possible for you to distinguish each stage. These stages and the sequence in which they occur are shown in Figure 8-2. Mitosis in an animal cell is illustrated. The process is nearly the same in the cells of higher plants, although there are certain differences we shall describe later. Refer to each stage in Figure 8-2 as it is discussed.

Interphase

The *interphase* period in a nucleus is not a part of mitosis. Rather it is the stage between nuclear divisions. During interphase, the nucleus is controlling cell activities. DNA is replicating

RNA that determines the structure of proteins being formed. The cell is in a period of growth and other activities.

During interphase, the chromosomes are greatly elongated and are diffused in the nucleus as a network of fine threads. One or more nucleoli are clearly visible. The nuclear materials are surrounded by a nuclear membrane (Figure 8-2a).

Near the close of interphase, several events prepare the nucleus for a coming mitotic division. The DNA duplicates its structure and the chromosomes become double threads.

In most animal cells, a small, dense area of cytoplasm can be seen just outside the nuclear membrane. This is a *centrosome.* One or two small granules, the *centrioles,* lie in the center of the centrosome. The electron microscope shows a centriole to consist of a cluster of tiny tubules. The function of these tubules is unknown. It is interesting that two centrioles always lie at right angles to each other in a centrosome. Soon after the chromosomes become double, the centrioles divide in preparation for the coming nuclear division, as shown in Figure 8-2a.

Prophase

Early stages of *prophase* mark the first visible signs that a mitotic division is beginning. The centrioles move apart and, with their centrosomes, migrate to opposite sides of the nucleus (Figure 8-2b). As the centrioles move to their respective locations, cytoplasm surrounding them changes from sol to gel forming fibrils that radiate from the centrioles like rays of a star. These are appropriately called *astral rays.* The astral rays, together with the centrioles, form *asters.*

During prophase, the chromosomes shorten and thicken and become clearly visible with the light microscope. Close examination will reveal that they are double throughout their length except for a small, dark area of attachment. This point of attachment is the *centromere.* We refer to each part of the double chromosome as a *chromatid.* While the position of the centromere may vary, it is usually about midway between the ends of the chromatids (Figure 8-2c).

As prophase continues, additional fibrils develop between the centrioles on opposite sides of the nucleus. These fibrils bow outward in the center to form a three-dimensional structure of threads resembling a football (Figure 8-2d). This is the *spindle.* In describing the spindle, we often refer to each end as a pole and to the point midway between the poles as the equator. Fibers composing the spindle are of two types. Those forming the *central spindle* extend from pole to pole. Other fibers extend only from one pole or the other to the equator. These are known as *traction fibers.*

Late in prophase, the nuclear membrane dissolves, allowing the nuclear substances to mix freely with the cytoplasm. The

a. interphase

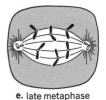

e. late metaphase

b. early prophase

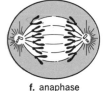

f. anaphase

c. prophase

g. telophase

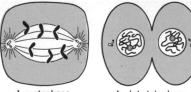

d. metaphase

h. late telophase

8-2 Stages of mitosis in an animal cell.

nucleolus disintegrates and disappears into the cytoplasm. The chromosomes, still consisting of paired chromatids, move to the equator, pulled by their centromeres. What acts on the centromeres to cause this movement is still unknown.

Metaphase

With the random massing of paired chromatids at the cell equator, mitosis enters a second stage, known as *metaphase* (Figure 8-2e). The centromere of each pair pulls apart, resulting in the separation of the chromatids into identical chromosomes. The centromere of each new chromosome becomes attached to a traction fiber. Chromosomes of a pair join to fibers leading to opposite poles.

Anaphase

Immediately after separation of their centromeres, pairs of chromosomes seem to repel each other. Chromosome migration from the equator to the poles constitutes the *anaphase* of mitosis. Each chromosome moves along a fiber of the central spindle, those of a pair moving toward opposite poles (Figure 8-2f). The mechanisms involved in chromosome movement during anaphase are not clearly understood. They appear to be pulled by shortening of the traction fibers. It is possible that this shortening results from the removal of protein molecules. Shortening of the traction fibers appears to be accompanied by lengthening of the central spindle fibers as the chromosomes move along them. During movement toward the poles the centromeres lead and the chromosomes trail behind. Anaphase ends with arrival of the chromosomes at the poles and the formation of clusters.

Telophase

The final stage of mitosis, or *telophase,* is marked by the reorganization of daughter nuclei and the division of the cytoplasm to form two daughter cells (Figure 8-2g). Soon after reaching the respective poles, the chromosomes lengthen and gradually disappear, forming the network of chromatin material characteristic of an interphase nucleus. The spindle fibers and asters of animal cells disappear as their substance reverts from gel to sol. New nuclei are organized, and a membrane forms around each nucleus.

Reorganization of the daughter nuclei is accompanied by division of the cell into two daughter cells of approximately equal size. In animal cells, division usually begins with the appearance of an indentation, or *cleavage furrow,* in the region of the equator (Figure 8-2h). The cleavage furrow deepens and divides the cell into two parts. We mark the end of telophase with the division of the cell and reorganization of daughter nuclei. Both cells then enter a new interphase.

Mitosis in plant cells

The stages of mitosis in cells of higher plants are, basically, like those of the animal cell we described. However, there are certain differences in plant cell mitosis. Cells of an onion root tip, where mitosis is occurring frequently, are often used to illustrate the process.

Plant cells lack centrosomes and centromeres and, therefore, do not form asters. Spindle fibers extend from pole to pole and from the equator to the poles, as in animal cells. However, the form of the spindle is usually more "barrel-shaped" than that of an animal cell.

The most noticeable difference is the manner in which the cell is divided. During telophase, a *division plate* (cell plate) is formed, first in the center of the spindle, then across the cell, dividing it into two daughter cells. Pectin is added to the division plate to form the middle lamella of a cell wall. Daughter cells add cellulose on either side of the middle lamella to form a primary wall.

The time and intervals of mitotic divisions

The time required for a mitotic division to occur varies with the kind of cell as well as with environmental conditions. The entire process may be completed in thirty minutes or less, or it may require several hours (Figure 8-4). Usually, prophase is the longest stage, while metaphase is the shortest.

The frequency of cell division varies in different plant and animal tissues. Generally, it is most frequent in the least specialized tissues. For example, embryonic tissue is in a continuous state of cell division, while nerve tissue divides very infrequently, if at all. For this reason, embryonic tissues are frequently used in the study of the various stages of mitosis.

Cell division is stimulated by injury to a tissue. For example, cells in your skin undergo division slowly under normal conditions. A wound stimulates very rapid cell division and regeneration of the damaged tissues.

Cell division is also stimulated by separation of cells. Do you remember the chicken heart muscle we described in Chapter 2? This tissue lived for more than thirty years and remained active as long as its mass was divided at regular intervals. The same principle applies to tissue cultures. When animal tissue, such as that composing a kidney, is ground or placed in a nutrient solution, the cells undergo rapid division that would not have occurred in the intact kidney. It would seem that the rate of division is held in check by the presence of other cells in a tissue.

Significance of mitosis

Did it ever occur to you that you are an individual unlike any other and that the characteristics that make you an individual are

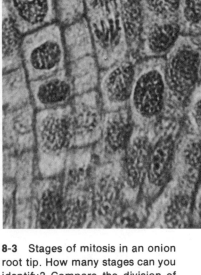

8-3 Stages of mitosis in an onion root tip. How many stages can you identify? Compare the division of plant cells with that of animal cells. (Photo Researchers)

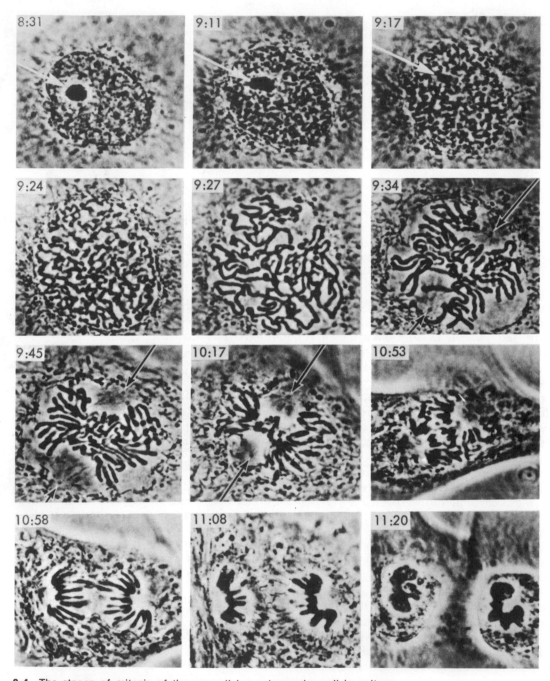

8-4 The stages of mitosis of the same living salamander cell in culture. Prophase: 8:31–9:27; metaphase: 9:34–10:17; anaphase: 10:53–11:08; telophase: 11:20. (courtesy William Bloom, from Bloom & Fawcett, *A Textbook of Histology* (9th ed.), W. B. Saunders Co.)

present in every cell of your body? You began life as a single cell. The billions of cells that compose your body today are descendants of this original cell and the products of countless cell divisions. Each time a cell division occurred, the daughter cells received identical genetic materials from the mother cell. This was accomplished by replication of all genetic materials and equal division of chromosomes each time a division occurred. Mitosis preserves the number as well as the exact genetic make-up of chromosomes. In so doing, it maintains the genetic unity of a cell of an organism.

The division of cell structures other than nuclear materials during cytoplasmic division is not so exact. Distribution of the endoplasmic reticulum, ribosomes, mitochondria, plastids, and other cytoplasmic structures in daughter cells is not exactly equal. This inequality of distribution is not important, however, since the daughter cells soon assemble the missing or deficient cytoplasmic structures.

Asexual reproduction

Reproduction of an organism may be sexual or asexual. Both forms involve cell division. *Sexual reproduction* involves the union of reproductive cells. However, in *asexual reproduction,* cells do not unite.

Binary fission is one form of asexual reproduction by cell division. In a multicellular organism, binary fission merely multiplies the number of cells, resulting in growth or replacement of tissue. In one-celled organisms, however, it constitutes reproduction of the entire organism. Under ideal growth conditions, one-celled organisms can reproduce at surprising rates. Protozoans may divide as often as twice a day. But this rate is far exceeded by certain bacteria that can divide as often as every twenty or thirty minutes.

Budding is another type of asexual reproduction by cell division. It differs from binary fission in that the resulting cells are of unequal size. A bud develops as a knob or bulge on a mother cell. The nucleus divides equally by mitosis, one portion remaining in the mother cell and the other entering the bud. However, the distribution of cytoplasmic materials is unequal, the bud receiving only a small portion. With its full complement of nuclear materials, the bud will grow and soon reach the size of the mother cell and will, in turn, form additional buds. Buds may remain attached or may separate and live as new, single-celled organisms. Although we often associate budding with yeast cells, this type of reproduction also occurs in many other microorganisms.

Spore production is one of the most widespread methods of asexual reproduction. Spores are reproductive cells formed by divisions of special spore mother cells. They are shed by the

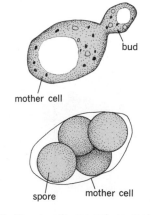

8-5 Yeast cells reproduce rapidly by budding, a form of fission. Under certain conditions, they also form spores.

parent organism when they mature. The spores of some organisms are carried by the wind. If they are produced by aquatic organisms, they usually have whiplike structures with which they swim from the parent to a new location. Spores are often protected by a resistant wall and may survive for long periods under such severe environmental conditions as drying or freezing. Under suitable conditions, a spore may germinate and form an active cell directly. That is, a spore does not fuse with another spore in the production of a new organism. While we usually associate spores with plants and plantlike organisms, you will discover in your study of microbiology certain animal-like cells that produce spores.

All forms of asexual reproduction involve the direct production of a new organism, either from a single cell or from a group of cells. Thus, it is a highly efficient process. Organisms capable of asexual reproduction usually have the capability of increasing their numbers rapidly if conditions for growth are favorable.

Sexual reproduction

Sexual reproduction involves the fusion of two special cells. We refer to these sex cells as *gametes* (GAM-eets). The fusion of gametes is called *fertilization,* or *syngamy* (SING-guh-mee). We refer to the fusion body formed by the union of two gametes as a *zygote* (ZY-GOTE).

In some organisms, all gametes formed are alike. These are called *isogametes*. We cannot designate them as male or female gametes because of their structural similarity. Nevertheless, we know that there are functional differences because only certain isogametes attract each other and fuse during fertilization. *Heterogametes,* however, have structural differences that enable us to designate them as male or female sex cells. We refer to a male gamete as a *sperm* and a female gamete as an *egg,* or *ovum*.

In your study of sexual reproduction in various organisms, you will find interesting differences in gamete production. You will find isogametes being produced by different individuals of a species, indicating maleness and femaleness even though the gametes are identical in form. You will also find both male and female heterogametes being produced by a single individual. More frequently, however, heterogametes are formed by different individuals which we designate as male and female parents.

Cell reproduction and chromosome numbers

Each time a cell nucleus undergoes a mitotic division, pairs of chromosomes, each composed of joined chromatids, appear during prophase and move to the equatorial plate during metaphase. We refer to the chromosomes of these pairs as *homologous* (huh-MAHL-uh-gus) *chromosomes* because they are identical in form and linear arrangement of genes. A *homologue* is a single chromosome of a homologous pair. Is the number of homologous chromosome pairs constant in all cells of an individual organism? Not only is it constant for all cells of the organism, but it is also *constant for all normal individuals of a particular species.* Normally, human body cells contain twenty-three pairs of homologous chromosomes, or forty-six in all. This number is constant in all normal human beings the world over. Furthermore, the distribution and arrangement of genes on these chromosomes is the same. A cell that contains a full set of homologous pairs of chromosomes is said to have the *diploid,* or 2*n,* chromosome number. Mitotic division maintains this chromosome number in all cells, since chromosomes replicate and form chromatids that are equally distributed in the process. Your life began as a single diploid cell containing homologous gene and chromosome pairs. Perhaps you have already figured out that one full set of chromosomes came from one parent and one set from the other.

Chromosome number in gametes

If all human body cells contain twenty-three pairs of chromosomes, or forty-six in all, would this same number be found in an egg or a sperm? If this were true, a fertilized egg would contain ninety-two chromosomes, and all cells produced from it by mitotic division would contain this abnormal chromosome number. This, of course, could not occur. At some time in the formation of eggs and sperms, the diploid chromosome number must be reduced to half. Wouldn't the logical biological answer be separation of the chromosome pairs? This does occur in egg and sperm formation. Each gamete contains but one chromosome of a homologous pair. We refer to this chromosome content as the *haploid,* or *n,* number. (The haploid number is sometimes called the *monoploid* number.) It is important that you understand that the haploid number is not merely half of the diploid chromosome number. All human eggs and sperms contain only twenty-three chromosomes, but not just any twenty-three. These gametes contain one full set of homologues—one of every pair of chromosomes. You might think of an egg, then, as "half a person" even though it contains one of every kind of gene necessary to produce the new individual.

This raises an interesting biological question. If an unfertilized egg were to develop without fertilization by a sperm, would not a haploid organism develop? Did you know that this actually occurs in certain organisms? We refer to it as *parthenogenesis.* One of the best examples is the male, or drone, bee. Drones are produced from unfertilized eggs laid by the queen at certain times of the year. In most organisms, however, an egg does not develop until fertilization has occurred. Thus, each gamete contributes one full set of chromosomes. This results in a zygote with the diploid chromosome number.

(a.)

spermatogonial
cell

beginning of
first meiotic division

(b.)

1a–1a
2b–2b
2a–2a
1b–1b

primary
spermatocyte

1a-1a,1b-1b

2a–2a,
2b–2b

(c.)
synapsis

1a-1a,1b-1b

2a–
2a
2b

(d.)

1a–1a, 1b–1b

2b–2b 2a–2a

(e.)

1a–1a, 1b–1b

2b–2b, 2a–2a

(f.)

1a–1a
2b–2b

1b–1b
2a–2a

secondary
spermatocytes

(g.)

**second meiotic
division**

1a
2b

1a
2b

(h.)

1b
2a

1b
2a

spermatids

1a
2b

1a
2b

1b
2a

1b
2a

(i.)

1a
2b

1a
2b

1b
2a

1b
2a

sperms
(j.)

8-6 Spermatogenesis. Meiotic divisions of a diploid primary spermatocyte give rise to four haploid sperms.

Cells of the *ovaries,* which produce eggs, and those of the *testes,* in which sperms are formed, have the diploid chromosome number, like all other body cells. The question, then, is how the chromosome content is reduced to the haploid number (n) in gamete formation.

Meiosis

To understand how the chromosome number is reduced from diploid to haploid in the production of eggs and sperms, we must explore another type of cell division, known as *meiosis* (my-OSE-iss). In describing mitosis, we started with a mother cell and followed the process in which all chromosomes replicated and were evenly distributed in the nuclei of two daughter cells. The daughter cells were identical to the mother cell because each received a complete set of chromosomes containing the same genetic materials. This occurred in a single division in which one cell formed two daughter cells. In meiosis, however, one cell with the diploid chromosome number gives rise to four haploid cells. This process involves *two consecutive nuclear divisions.* To illustrate meiosis, we shall follow the chromosome changes that occur in *spermatogenesis,* or the formation of sperms. If this were occurring in a human, 23 pairs of chromosomes would be involved. However, we shall reduce this number to two pairs of chromosomes and designate the homologues as 1^a and 1^b and 2^a and 2^b. Follow these chromosomes through the various stages of spermatogenesis shown in Figure 8-6. We shall refer to each stage *by letter* as we discuss the process.

Spermatogenesis — the formation of sperms

Cells in the male reproductive organs, or testes, form *spermatogonial cells* by repeated mitotic divisions (a). These cells, like all others of the body, are diploid and contain the homologous chromosomes 1^a and 1^b and 2^a and 2^b. Spermatogonial cells enlarge and prepare for the *first of two meiotic divisions.* We now refer to these cells as *primary spermatocytes* (b).

During an elongated prophase, sometimes lasting several days, chromosomes in the primary spermatocyte shorten and thicken and become clearly visible with the light microscope, as in mitosis. Each chromosome consists of two chromatids joined by a centromere (b). We may now designate the chromosomes as $1^a - 1^a$, $1^b - 1^b$ and $2^a - 2^a$, $2^b - 2^b$.

An interesting and unexplained event soon occurs. Homologous chromosome pairs, each consisting of joined chromatids, apparently attract each other and come together. We refer to the event as *synapsis.* A synapsed pair of chromosomes consists of four chromatids and is, therefore, called a *tetrad* (c). Near the close of prophase, the tetrads move to the equator. During a brief metaphase, the centromere of each chromosome, still consisting of joined chromatids, attaches to a traction fiber of the spindle (d). These fibers lead to opposite poles.

During the anaphase that follows, one chromosome of a pair in a tetrad, with its chromatids still joined, moves toward one pole. Its mate, with its joined chromatids, moves toward the opposite pole (e). It is important to remember that it is pairs of homologous chromosomes, not chromatids, that separate in the first meiotic division. It should be pointed out, also, that separation of one pair of homologous chromosomes has no relation to the separation of other pairs. We show chromosome $1^a - 1^a$ moving to a pole with $2^b - 2^b$ and $1^b - 1^b$ moving with $2^a - 2^a$. This is only one of several combinations that could be formed at either pole.

Following chromosome migration to the poles, the primary spermatocyte forms a cleavage furrow near the equator (f) and divides, forming two *secondary spermatocytes* (g). Notice that these cells contain only one of each pair of homologous chromosomes and that the chromatids are still joined by the centromere. Since we consider each pair of joined chromatids a single chromosome, the first meiotic division reduces the diploid number in the primary spermatocyte to the haploid number in the secondary spermatocytes.

Both secondary spermatocytes now undergo a *second meiotic division*. In this division, the prophase is brief. The single homologues with their paired chromatids move to the equator. During metaphase in this division, the centromeres divide and pull apart and the chromatids, now chromosomes, are attached to traction fibers leading to opposite poles. During anaphase, the chromosomes move to their respective poles (h). Division of the two secondary spermatocytes produces four *spermatids* (i). Soon, the spermatids mature, develop a "tail" and become functioning sperms (j). Each contains a full haploid set of chromosomes characteristic of the organism.

Oogenesis—the formation of eggs

A similar process occurs in the formation of an egg during *oogenesis* (Figure 8-7). An *oogonial cell* with the diploid chromosome number develops in an ovary (a). This cell matures into a *primary oocyte* (b). As in spermatogenesis, pairs of homologous chromosomes, each consisting of paired chromatids, synapse in the primary oocyte (c) and move as a tetrad to the equator (d). During anaphase, homologous pairs of chromosomes, each with paired chromatids, separate and move to opposite poles (e).

Division of the primary oocyte (f) is unequal, resulting in a *secondary oocyte* and a smaller *first polar body* (g). Both of these cells contain the haploid chromosome number. The secondary oocyte undergoes a *second meiotic division* in which the chromatids separate and move as chromosomes to opposite poles

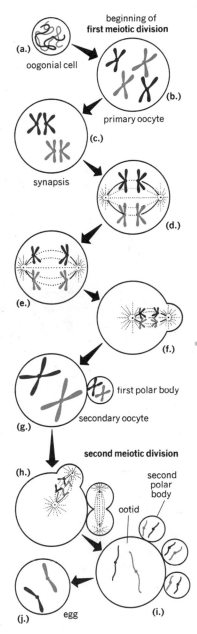

8-7 Oogenesis. Meiotic divisions of a diploid primary oocyte give rise to a haploid egg.

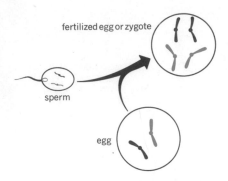

fertilized egg or zygote

sperm

egg

8-8 Fertilization. The combining of the chromosomes of two haploid gametes restores the diploid condition.

(h). Division of the secondary oocyte forms an *ootid* and a *second polar body* (i). In many organisms, the first polar body divides and forms two additional polar bodies. The ootid, with its haploid number of chromosomes, matures and forms the *egg* (j). The three polar bodies formed in the second division have no function and disappear.

Fertilization and the restoration of the diploid chromosome number

Meiosis results in the formation of gametes with the haploid chromosome number. During *fertilization,* an egg and a sperm unite. Each contributes one full set of chromosomes. This restores the diploid number of chromosomes in the fertilized egg, or *zygote.* Soon after fertilization, the zygote divides by mitosis. Thus, the diploid chromosome number is maintained in the daughter cells. This process is continued many millions of times in the formation of cells in a body such as ours.

SUMMARY

Mitosis preserves the genetic characteristics of an organism by transmitting identical chromosomes and genes from a mother cell to daughter cells. Thus, the chromosomes and genes are alike in every body cell.

Normally, meiosis reduces the chromosomes in gametes to the haploid number, thus allowing a union of gametes from different parents without altering the normal chromosome number in the offspring. Each parent contributes a full set of chromosomes to the fertilized egg. This combination of chromosomes is of great biological importance because it introduces genetic characteristics of two parents. Resulting variations in offspring are the basis for genetic change, a mechanism involved in adaptations and the improvement of life, as you will discover in your study of genetics in Unit 2.

BIOLOGICALLY SPEAKING

binary fission	sexual reproduction	diploid
mitosis	asexual reproduction	haploid
interphase	gamete	parthenogenesis
prophase	fertilization	meiosis
metaphase	sperm	spermatogenesis
anaphase	egg	oogenesis
telophase	homologous chromosomes	

QUESTIONS FOR REVIEW

1. What is binary fission?
2. What two phases of cell activity constitute mitotic cell division?
3. Explain the relationship between genes and chromosomes.
4. What changes occur in the DNA of a nucleus during interphase?
5. Describe the formation of asters during prophase in a dividing animal cell.
6. Distinguish between a chromosome and a chromatid.

7. Distinguish between central spindle fibers and traction fibers, which form in the nucleus during prophase.
8. Describe the position of chromosomes during metaphase.
9. Explain how chromatids are attached to spindle fibers.
10. Describe the behavior of chromosomes during anaphase.
11. Summarize chromosome changes during telophase.
12. How does division of the cytoplasm differ in plant and animal cells?
13. Distinguish between sexual and asexual reproduction.
14. Describe several forms of asexual reproduction.
15. Distinguish between a gamete and a spore.
16. How are isogametes different from heterogametes?
17. In what ways are homologous chromosomes alike?
18. Distinguish between the haploid and diploid chromosome numbers.
19. What is the basic difference between mitosis and meiosis in respect to the number of divisions involved and the number of cells formed?
20. Describe the formation of tetrads during prophase in meiosis.
21. Describe chromosome changes that occur in the first meiotic division.
22. What chromosome changes occur in the second meiotic division?
23. Explain why four sperms develop from a spermatogonial cell while an oogonial cell gives rise to a single egg cell.
24. Explain how fertilization restores the diploid chromosome number.

APPLYING PRINCIPLES AND CONCEPTS

1. Discuss the necessity of cell division in perpetuating the life of a cell.
2. Compare the chromosome changes that occur in mitosis and meiosis.
3. Compare division in an animal cell and a higher plant cell following nuclear division.
4. Discuss the importance of exact duplication of chromosomes in daughter cells formed by mitosis in multicellular organisms.
5. Discuss sexual reproduction as a mechanism of change and variation.

RELATED READING

Books

ANDERSON, M.D., *Through the Microscope.*
Natural History Press (distr., Doubleday and Co., Inc.), Garden City, N.Y. 1965. Describes the stages of cell reproduction.

BUTLER, JOHN A. V., *Inside the Living Cell.*
Basic Books, Inc., Publishers, New York. 1959. A very readable, fascinating account of the cell, its structure, and functions, written by a noted biophysicist and chemist.

DEROBERTIS, E. D. P., and others, *General Cytology.*
W. B. Saunders Co., Philadelphia. 1970. Stresses the morphological, physiological, and genetic aspects of modern cytology.

SWANSON, CARL P., *The Cell.*
Prentice-Hall, Inc., Englewood Cliffs, N.J. 1960. A cytology book for the better student, dealing with all phases of the activity of cells.

Articles

MAZIA, D., "Cell Division," *Scientific American,* August, 1953. The remarkable feat of isolating the mitotic apparatus from the remaining cellular material.

UNIT **2**

THE CONTINUITY OF LIFE

When sexual reproduction occurs, both parents transmit chemical instructions that control the development of the offspring. Each new organism resembles its parents in some ways yet differs in others. What chemical controls exert this influence? The answer lies in a search of cells, their nuclei and their chromosomes, and finally the nucleoproteins of which they are composed. Here we find the substance DNA and a genetic code that exerts a control over the development of every inherited trait in an organism.

CHAPTER NINE

PRINCIPLES OF HEREDITY

Heredity and environment

Since the day you were born, two kinds of influences have been interacting to determine your individual makeup. The first of these is *heredity*. Heredity is the transmission of characteristics from parents to offspring. These characteristics include the color of your hair and eyes, your body build, facial features, and many others. The development of these traits is controlled by a chemical code transmitted to you through the reproductive cells of your parents. This code is contained in the genes of which chromosomes are composed. The branch of biology that is concerned with the mechanisms and substance of heredity is, therefore, appropriately called *genetics*.

The second factor involved in your development is *environment*. This includes all the external forces that influence the expression of your heredity. It is sometimes difficult to determine where hereditary influences end and environmental influences begin. For example, body size is controlled by heredity. But it is also determined partially by diet and by the types of activity in which you participate. Similarly, the tanning of your skin is the result of the interaction of sunlight and an inherited ability to produce additional pigment. If you lack this pigment, you will sunburn rather than tan, but you will do neither if you are not exposed to the sun. Your heredity thus determines what you *can* become, but what kind of individual you *do* become depends on the interaction of your heredity and your environment.

What kinds of characteristics are inherited?

In certain respects, all members of a species are alike. For example, man normally inherits those characteristics of the human race that make him like other human beings. These *species characteristics* include the ability to walk erect, grasping fingers, and a highly developed nervous system with a brain superior to that of all other organisms.

In addition to species characteristics, you have inherited certain *individual characteristics* that make you different from all other people. Many of these characteristics are passed on from parent to offspring. The result is that you may resemble your parents to a certain degree but differ from each because you have inherited characteristics from both.

Mendel's work with garden peas

In 1865, Gregor Mendel, an Austrian monk, published the results of a masterful piece of work on the principles of heredity. He was not the first to experiment in the field of inheritance, but his findings were the first of any scientific consequence. His paper, representing years of work with garden peas, was published by the Natural History Society of Brünn, Austria. Mendel had been dead for sixteen years when three other scientists discovered his work and began to make use of his findings. It is, however, a great tribute to Mendel that the conclusions he arrived at from his experiments with garden peas stand today, practically unchanged, as the basis of the science of genetics. It is also very remarkable that his conception of inheritance was developed without a knowledge of chromosomes and their behavior.

During his years as a teacher in a high school in Brünn, Mendel kept a small garden plot at the monastery where he lived. He used several kinds of plants for his experiments, but the work for which he is remembered are the experiments he conducted with garden peas (Figure 9-1). Why did Mendel choose garden peas for his experiment? First, he had observed that they differed in certain characteristics. Some plants were short and bushy, while others were tall and climbing. Some produced yellow seeds, some green seeds; some had colored seed coats, and some white. Mendel identified seven different pairs of contrasting traits in which the plants differed consistently (Figure 9-2).

In order to understand the second reason that Mendel found garden peas ideal for his experiments, you will need a brief introduction to reproduction in the seed plants. You are probably familiar with the fact that the flower is the reproductive structure in seed plants. Every flower of a pea plant bears structures called *stamens,* which produce pollen grains that form sperm nuclei, and a structure called a *pistil,* which contains egg cells at its base. The transfer of pollen from stamens to pistil results in fertilization and is called *pollination.* Pea plants normally carry on *self-pollination,* which means that pollen is transferred from stamens to pistil on the same flower or another flower of the same plant. (*Cross-pollination* involves flowers on two different plants.)

Mendel found that cross-pollination could be assured in pea plants by removing the stamens from a flower and then, when the pistil of this flower matured, transferring flower pollen from another plant to it. The hand-pollinated flower was then carefully

9-1 Gregor Mendel in his garden experimenting with garden peas. (The Bettmann Archive)

gms—DNA

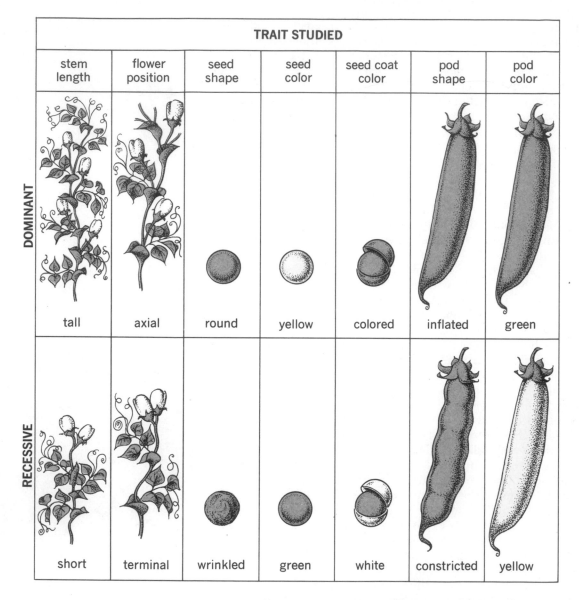

TRAIT STUDIED						
stem length	flower position	seed shape	seed color	seed coat color	pod shape	pod color
DOMINANT						
tall	axial	round	yellow	colored	inflated	green
RECESSIVE						
short	terminal	wrinkled	green	white	constricted	yellow

9-2 Mendel's seven pairs of contrasting traits in garden peas.

protected from any pollen grains that might be transferred to it by wind or insects. Because of the pea's seven pairs of contrasting characteristics, and because cross-pollination was easy to perform, Mendel had selected an ideal subject for breeding experiments.

Mendel discovers the principle of dominance

Mendel allowed several generations of peas to self-pollinate. He found that the seven pairs of contrasting characteristics he had identified were always handed down from parent to offspring. Seeds from tall plants produced other tall plants, and yellow seeds produced plants that always yielded yellow seeds. Mendel's next step was to see what would happen if he crossed two plants with contrasting traits. Selecting tall and short plants, he made hundreds of crosses by transferring the pollen from the tall plants to the pistils of the short ones. When the seeds matured on the short plants, he sowed them to find out the results of his cross. Would the offspring be short like one parent, tall like the other, or of medium height with characteristics of both? He discovered that all the plants were tall, like the plant from which he had taken the pollen in making the cross.

His next step was to determine if it made any difference which plant he used for pollen and which he used to produce the seed. Accordingly, he reversed the process of pollination, using a short plant for pollen and a tall one for seed production. Mendel found that the results were as before—all the offspring were tall.

Mendel then experimented with other characteristics. He limited his study of each cross to a single characteristic involving only *one* trait at a time. For example, he crossed plants contrasting in just one trait, such as seed color—yellow seeds and green seeds. He found that all of the seeds that resulted from this cross were yellow. He further discovered that when he crossed a round-seeded variety with a wrinkled-seeded variety, all the seeds produced were round. He repeated these crosses until he had tested the seven different characteristics. Mendel was surprised to find that in all seven crosses, one of the characteristics present in the parent plant seemed to be lost in the next generation. What would happen if he permitted these offspring to self-pollinate? This step in his experiment was destined to make history, because it led to the discovery of two principles of heredity.

Mendel's conclusions relating to inheritance of traits were based on data accumulated from the study of a large number of offspring. He kept accurate records of all the crosses he made. In recording his generations of crosses, Mendel designated the *parent plants* used in the first cross as P_1. He referred to the generation resulting from this cross as the *first filial,* or F_1, generation. By allowing the tall plants of the F_1 generation to self-pollinate, he produced a *second filial,* or F_2, generation. The results of this self-pollination were quite striking. Some of the plants were tall, while others were short. None were in between. Furthermore, three fourths of the plants were tall, while one fourth were short. The reappearance of short plants in this generation was of great significance. The F_1 plants had possessed a characteristic for shortness without showing it.

When he permitted other F_1-generation plants to self-pollinate, he had the same results. When allowed to self-pollinate, the plants that grew from the yellow peas that had been produced by cross-

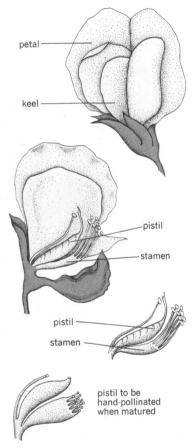

petal

keel

pistil

stamen

pistil

stamen

pistil to be hand-pollinated when matured

9-3 Mendel made an ideal choice for hand-pollination when he selected the garden pea. The keel of a pea blossom surrounds the stamens and pistil. If the keel is spread and the stamens are removed before natural pollination occurs, the flower may be hand-pollinated when the pistil is mature.

ing yellow-seed-producing plants with green-seed-producing plants in turn produced peas of which three fourths were yellow, and one fourth, green.

Mendel's first hypothesis

The fact that tall peas crossed with short peas produced an F_1 generation of tall peas and that short peas reappeared in the F_2 generation led Mendel to reason that something within the plant controlled a characteristic such as height. He called these unknown influences *factors*. Today we call them *genes*. He reasoned further that height in peas was controlled by a pair of factors, since some peas were short, while others were tall. On this basis, he formulated his first hypothesis regarding heredity, the **concept of unit characters**. This states that *the various hereditary characteristics are controlled by factors (genes) and that these factors occur in pairs.* Of course, Mendel did not know about genes or chromosomes, which makes even more remarkable the fact that his concept of unit characters is now a basic principle in genetics.

Mendel reasoned further that the tall plants of his F_1 generation were not like the pure tall parent plants. These peas were carrying a concealed factor for shortness that would reappear in the next generation. This reasoning led to the formulation of his second hypothesis, the **principle of dominance and recessiveness**. This states that *one factor (gene) in a pair may mask or prevent expression of the other.* Mendel gave the name *dominant* to the characteristic, such as tallness, that always appeared in the offspring of a cross between parents with contrasting characters. He called *recessive* the characteristic, such as shortness, that did not appear in the F_1 generation but appeared in the F_2 generation.

Today we know that in Mendel's crosses, one parent was pure tall, having both genes for tallness. The other was pure short, having both genes for shortness. The members of the F_1 generation were all tall but were **hybrid**, a term we use to designate the offspring of a cross between two parents that differ in one or more traits. The members of this generation had one gene for tallness and one for shortness but appeared tall because the gene for tallness was dominant over the one for shortness.

If we let the letter T stand for tall, a pure tall plant would be written TT, indicating that both of its genes for this characteristic were for tallness. The capital T indicates that tallness is dominant over the contrasting characteristic, shortness. Similarly, the small letter t stands for short, and a pure short individual would be designated as tt.

You learned in Chapter 8 that all body cells contain the diploid number of chromosomes. That is, chromosomes are present in *pairs* in these cells. Sex cells, on the other hand, contain the haploid number of chromosomes, having *only one member of each pair.* Remember that genes are located on chromosomes. Consequently, the egg cell in the female organ of the pea plant and the sperms formed by the pollen grain had only one gene for each character. When eggs or sperms are formed by a pure tall pea plant, one sex cell receives one *T,* and the other receives the other *T.* In like manner, the *tt* genes present in all body cells of a pure short plant are separated in meiosis, and each egg or sperm receives one *t* gene. During fertilization, each parent thus contributes one member of each pair of genes, and the diploid number is restored.

Mendel's law of segregation

Mendel based his third hypothesis regarding heredity—now called *Mendel's first law,* the **law of segregation**—on this reasoning: *A pair of factors (genes) is segregated, or separated, during the formation of gametes (spores in lower plants).* That is, a gamete contains only one gene of a pair, the other having gone to another gamete. Furthermore, the composition of one gene is not altered by the presence of another gene in a pair. For example, a recessive gene in a hybrid is not altered by the presence of a dominant gene. If in an offspring of the hybrid the recessive gene is paired with another recessive gene, the recessive character will reappear (Figure 9-4).

Some genetic terms

The genes of any organism can be designated by paired symbols for any characteristic you are studying. These symbols indicate the *genotype* of the organism. The effect of genes in an individual is described as its *phenotype.* It refers to the organism's size, color, structure, and other characteristics. For example, in hybrid tall peas, the genotype is *Tt;* its phenotype is tall. If the paired genes for a particular trait are identical, we call the organism *homozygous* (HOME-oe-ZY-gus) for that trait. An organism having different gene pairs is called *heterozygous* (HET-uh-roe-ZY-gus). The different forms of genes associated with the same trait and giving contrasting effects are called *alleles* (uh-LEELZ). Some genes have three or more alternative alleles. In the gene pair *Tt, T* is an allele of *t; t* is an allele of *T.* In the gene pair *Yy, Y* is an allele of *y; y* is an allele of *Y.* But *T* is *not* an allele of *Y.* The separation of alleles during meiosis demonstrates Mendel's law of segregation.

Method of diagraming Mendel's crosses

In the study of genetics, we use special charts resembling checkerboards to determine the possible results of various crosses. This grid system is called a *Punnett square* after R. C. Punnett, who devised it. The alleles present in gametes from the female are shown across the top of the grid. Alleles in gametes

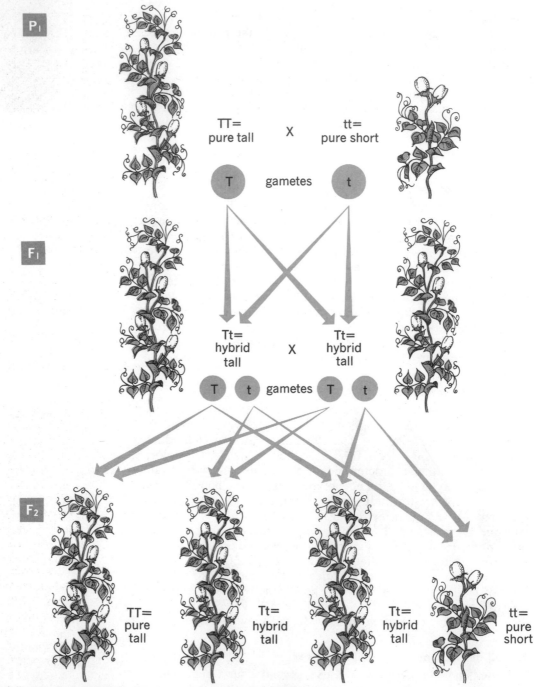

P₁

TT=
pure tall   X   tt=
pure short

T   gametes   t

F₁

Tt=
hybrid
tall   X   Tt=
hybrid
tall

T   t   gametes   T   t

F₂

TT=
pure
tall

Tt=
hybrid
tall

Tt=
hybrid
tall

tt=
pure
short

9-4 Mendel's law of segregation. While the F₁ generation consists only of tall plants, the recessive gene for shortness is again expressed in the F₂ generation. What kinds of plants would result from a cross between the two F₂ plants shown at the bottom left?

from the male parent are shown along the side. You can determine all possible combinations of the gametes by filling in the squares of the grid. Mendel's work with peas can be shown more clearly by diagraming his crosses on grids such as those shown in the right column. In working the grids, just combine the genes from both parents and then proceed to fill in the correct genotypes of the offspring.

A properly filled-out Punnett square not only shows the possible combinations of genes; it also indicates the probability of the occurrence of these gene combinations. However, while we cannot assume that a ratio determined in a square will necessarily apply to a small sample of offspring, it is likely that the ratio will apply to a sample in which a large number of random crosses have occurred.

The monohybrid cross

When one pair of characteristics in an individual is considered in a cross, the individuals possessing mixed genes are called *monohybrids*. A cross between a homozygous tall *(TT)* and a homozygous short *(tt)* pea plant is diagramed as shown in grid A (right column). All of the offspring are monohybrids with the genotype *Tt*. However, they appear equally as tall as the tall parent plant because the gene for tallness is dominant over the gene for shortness.

If the heterozygous *Tt* plants are permitted to self-pollinate, it is easy to see that four combinations can occur (grid B, right). The grid also shows how the genes *T* and *t* from the heterozygous parents, though they combine by chance, result in offspring that are one fourth pure dominant *(TT)*, one half hybrid *(Tt)*, and one fourth pure recessive *(tt)*. This can be expressed as a ratio of 1:2:1, which states that the expected ratio of the genotypes is one fourth pure tall, one-half hybrid tall, and one fourth pure short. Another way to express the result is in phenotypes, which would be expected to occur in a ratio of 3:1—three tall plants to every short plant.

The same scheme explains the ratios resulting from other crosses. The cross between a heterozygous tall *(Tt)* and a homozygous tall *(TT)* pea plant will give the ratios shown in grid C. The phenotype of all the plants is tall.

We can also determine the expected results from the cross between a homozygous short *(tt)* and a heterozygous tall *(Tt)* pea plant. The ratio of the phenotype is one half tall to one half short plants; the ratio of the genotypes is one half *Tt* to one half *tt,* as shown in grid D.

The results that Mendel obtained with two generations of garden peas are listed in Table 9-1. Both Mendel's actual ratio of results and the theoretical probability ratio are shown.

A. RESULTS OF CROSSING TT AND tt

Female → GENES Male ↓	t	t
T	Tt	Tt
T	Tt	Tt

B. RESULTS OF CROSSING Tt AND Tt

Female → GENES Male ↓	T	t
T	TT	Tt
t	Tt	tt

C. RESULTS OF CROSSING TT AND Tt

Female → GENES Male ↓	T	t
T	TT	Tt
T	TT	Tt

D. RESULTS OF CROSSING Tt AND tt

Female → GENES Male ↓	t	t
T	Tt	Tt
t	tt	tt

	P₁ Cross		F₁ Generation	F₂ Generation	Actual Ratio	Probability Ratio
 round	X	wrinkled	 round	5,474 round 1,850 wrinkled	2.96:1	3:1
 yellow	X	green	 yellow	6,022 yellow 2,001 green	3.01:1	3:1
 colored	X	white	 colored	705 colored 224 white	3.15:1	3:1
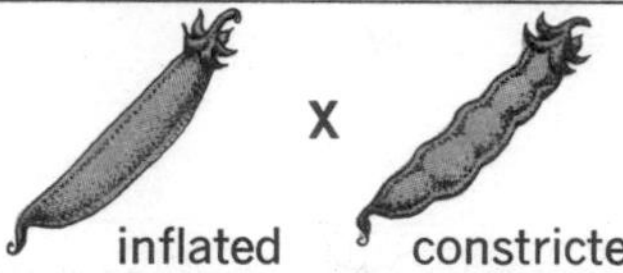 inflated	X	constricted	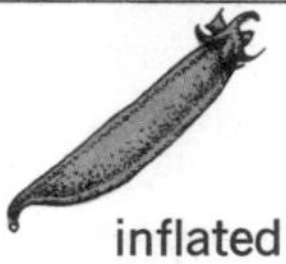 inflated	882 inflated 229 constricted	2.95:1	3:1
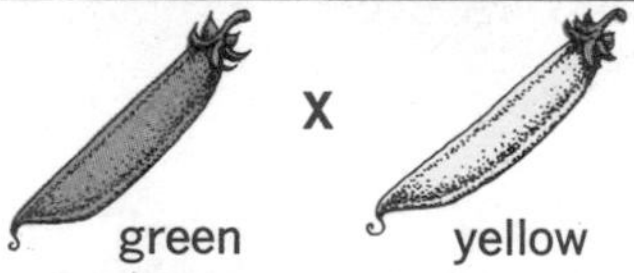 green	X	yellow	 green	428 green 152 yellow	2.82:1	3:1
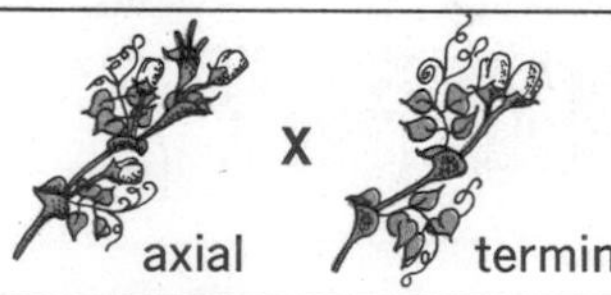 axial	X	terminal	 axial	651 axial 207 terminal	3.14:1	3:1
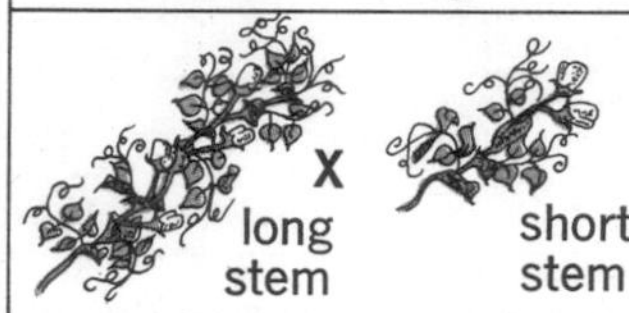 long stem	X	short stem	 long stem	787 long 277 short	2.84:1	3:1

Table 9-1 RESULTS OF MENDEL'S MONOHYBRID CROSSES

Dominant and recessive genes in guinea pigs

The same results Mendel obtained in crossing tall and short peas are shown in the inheritance of coat color in guinea pigs. In these animals, the color black is dominant over white.

Let's see what happens when we cross a homozygous black guinea pig with a homozygous white one. All of the F₁ generation will be heterozygous black. In order to determine whether the animal is carrying a recessive gene for white, we can cross two of the offspring. The expected ratio of the offspring of this cross is one fourth homozygous black, one half heterozygous black, and one fourth homozygous white. If only one animal shows the recessive trait, we have demonstrated that the F_1 generation was heterozygous for white. The expected phenotype of the cross is three fourths black and one fourth white. The grid for the cross between the two heterozygous (*Bb*) offspring of the F_1 generation is diagramed as follows (also see Figure 9-5):

The same ratios would be expected to occur after the crossing of rough-coated and smooth-coated guinea pigs. In the pair of alleles governing coat texture, the gene for rough coat is dominant over that for smooth coat.

RESULTS OF CROSSING Bb AND Bb

Female → GENES Male ↓	B	b
B	BB	Bb
b	Bb	bb

9-5 A cross between a homozygous black guinea pig and a homozygous white guinea pig.

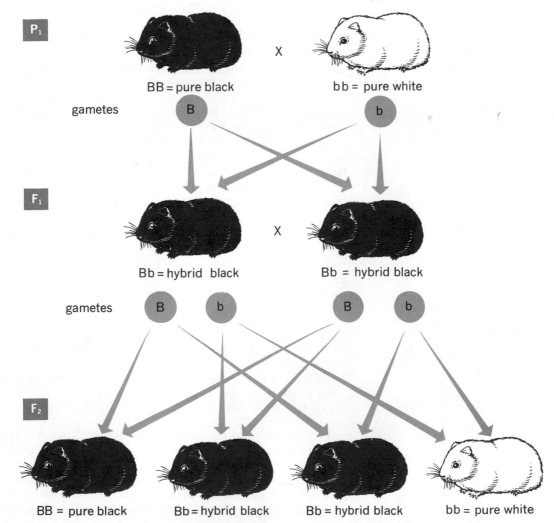

P₁   BB = pure black   X   bb = pure white

gametes   B   b

F₁   Bb = hybrid black   X   Bb = hybrid black

gametes   B   b   B   b

F₂   BB = pure black   Bb = hybrid black   Bb = hybrid black   bb = pure white

Crosses involving two traits

Crosses involving two traits become more complicated than simple crosses, in which only one pair of contrasting traits is considered. The same principles apply, but the possible gene combinations are increased. When two pairs of traits are involved, the individuals possessing mixed genes for both characteristics are called *dihybrids*.

If a pea plant pure for round, green seeds (two traits) is crossed with one pure for wrinkled, yellow seeds, all the seeds produced are round and yellow. The recessive traits of green color and wrinkled seed coat are masked by the two dominant traits. In this cross, R will represent a gene for round seed coat, r for wrinkled, Y for yellow color, and y for green. The F_1 dihybrid seeds all have the genotype *RrYy,* a gene for round seed coat *(R)* having come from one parent and a gene for wrinkled seed coat *(r)* having come from the other. In like manner, one parent supplied a gene for yellow color *(Y),* while the other supplied a gene for green *(y).*

When two plants grown from the dihybrid seeds are crossed, the situation becomes more complicated. Each dihybrid with the genotype *RrYy* can produce four kinds of eggs or sperms. During meiosis, the pairs R and r as well as Y and y must separate and go into different cells. R may pair with Y to form *RY,* or R may pair with y, resulting in *Ry.* Similarly, r may pair with Y to form *rY* or with y to form *ry.* The nature of the offspring in such a cross depends on which eggs and sperms happen to unite during fertilization.

The possible offspring that can result from such a cross and the ratio of their occurrence can be diagramed as in the crossing of a single contrasting trait, except that space must be provided for more possible crosses. Figure 9-6 shows the result of such a cross. One of the parents was pure round, green *(RRyy),* while the other was pure wrinkled, yellow *(rrYY).* You will note that all of the F_1 generation are alike, being dihybrid round, yellow *(RrYy).* In the F_2 generation, however, four different phenotypes have been produced, as follows:

$9/16$ of the seeds are round and yellow (both dominant traits).
$3/16$ of the seeds are round and green (one dominant and one recessive trait).
$3/16$ of the seeds are wrinkled and yellow (the other dominant and the other recessive trait).
$1/16$ of the seeds are wrinkled and green (both recessive traits).

Notice that these four phenotypes occur in an expected ratio of $9:3:3:1$. How many different genotypes can you find? You will note, also, that the genotypes show that yellow seeds may be either pure yellow or hybrid yellow; also, that round seeds may be either pure round or hybrid round. A recessive trait only shows when both genes for the recessive trait are present. Both

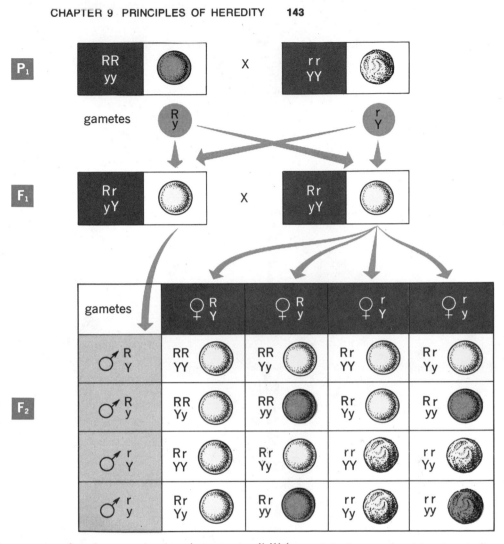

recessive traits appeared only once in the sixteen possibilities.

When heterozygous black, rough-coated guinea pigs are crossed (genes for black and rough are dominant), similar results can be expected: $9/16$ of the offspring are black and rough, $3/16$ are black and smooth, $3/16$ are white and rough, and $1/16$ are white and smooth.

9-6 A cross involving two traits. A pea plant that produces round (R), green (y) seeds is crossed with a pea plant that produces wrinkled (r), yellow (Y) seeds.

The law of independent assortment

The dihybrid crosses you have studied illustrate another of Mendel's hypotheses, the *law of independent assortment*. According to this law, *the separation of gene pairs on a given pair of chromosomes and the distribution of the genes to gametes during meiosis are entirely independent of the distribution of other gene pairs on other pairs of chromosomes.* This law applies only when genes are on different chromosome pairs, since chromosomes, not genes, assort independently.

9-7 Incomplete dominance in four-o'clocks.

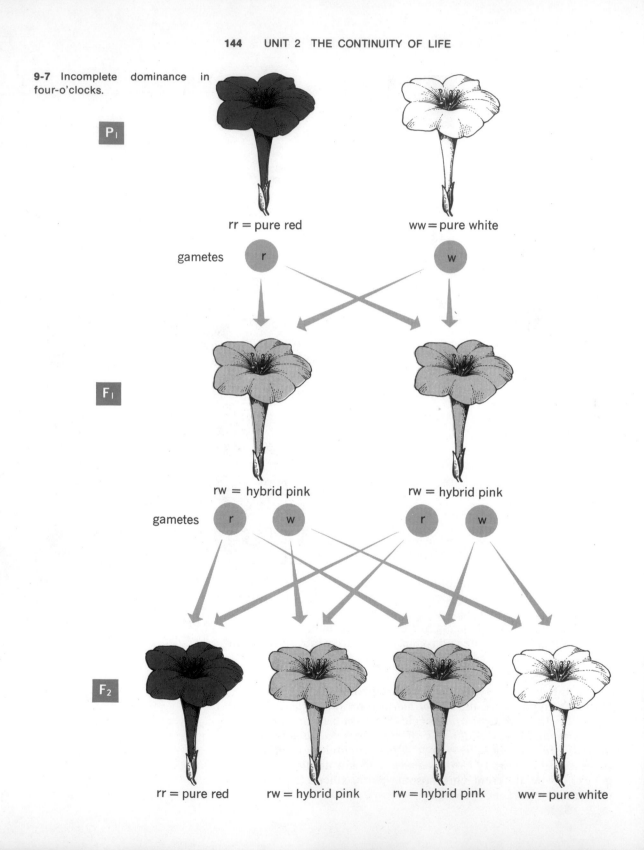

Incomplete dominance

Genes are not always dominant or recessive. In some characteristics, both alleles of a pair can be expressed. This *incomplete dominance,* as it is called, can be illustrated in crossing the flowers of four-o'clocks and snapdragons. When pure red four-o'clocks *(rr)* are crossed with pure white *(ww)* varieties, all of the first generation are pink *(rw)*. Neither red nor white is completely dominant, so both colors are expressed in the heterozygous F_1 offspring as pink. However, when two of these heterozygous pink *(rw)* flowers are crossed, the F_2 generation includes one fourth red, one half pink, and one fourth white individuals. The fact that genes for red and white did not mix in the pink offspring is indicated by the reappearance of both pure characteristics in the F_2 generation (Figure 9-7).

Similarly, the color of Shorthorn cattle illustrates incomplete dominance. A homozygous red animal mated with a homozygous white animal produces a blend of red and white called *roan.* When two roan animals are mated, the expected ratio of the offspring would be one fourth red, one half roan, and one fourth white, illustrating again the $1:2:1$ ratio (Figure 9-8).

Ratios are based on averages

Recall that the ratios obtained in breeding experiments represent *averages* and not definite numbers that will always appear. These ratios are accurate *only when large numbers of individuals are considered.* For example, two roan shorthorns bred four times will not necessarily produce one red calf, two roan ones, and a white one. This is true, of course, because any egg produced by a female heterozygous for a particular trait for which there are only two alleles can *by chance* contain either allele. Further, it is also a matter of chance which of the millions of sperms that are usually available to fertilize any one egg will actually unite with that egg.

Chance ratios can be shown with two coins. When you flip them, they will alight in these possible combinations: two heads, one head and one tail, or two tails. The chance of the appearance of one head and one tail is twice as great as that of two heads or two tails. One of the reasons for the great accuracy of Mendel's work with the peas is the fact that he experimented with such large numbers of plants.

Mendel's hypotheses are widely applicable

Scientists have worked with numerous traits of plants and animals and have shown that Mendel's hypotheses regarding inheritance of traits are remarkably valid. In the chapters to follow, we shall discuss several examples of the workings of Mendelian genetics in man.

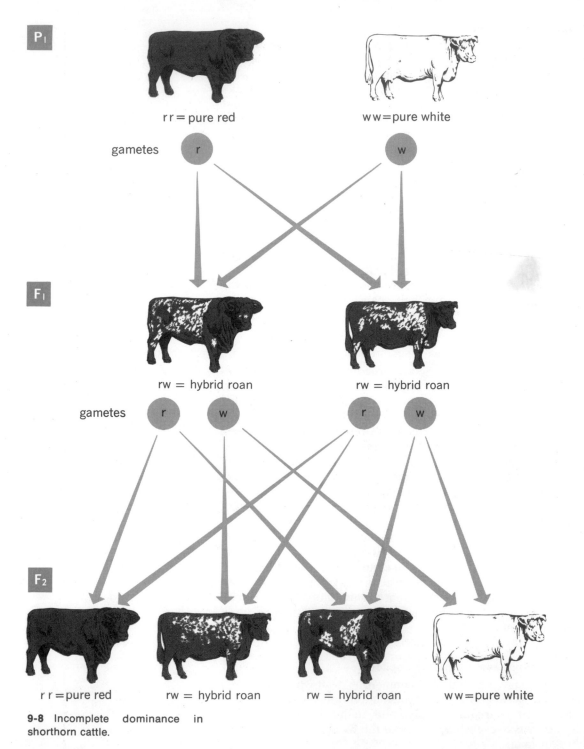

9-8 Incomplete dominance in shorthorn cattle.

SUMMARY

When you consider that Gregor Mendel never saw a chromosome and never heard of a gene, his contribution to science becomes even more remarkable. Within the walls of a monastery garden and with nothing more than several strains of garden peas, he laid the groundwork for a new branch of science. His observations and deductions were so accurate that they remain, today, as a basis for modern genetics. This is a tribute to a man who never realized the importance of his contribution.

Modern genetics has gone far beyond Mendel. Other scientists, building on his laws and principles and with more sophisticated tools for research, have made significant discoveries. Mendel spoke of factors. Today, we call them genes. We even know what genes are and how they produce genetic traits. As you investigate modern genetics in the next chapter, remember that it was Gregor Mendel who established the new and exciting branch of biology we call genetics.

BIOLOGICALLY SPEAKING

heredity	principle of dominance	heterozygous
genetics	and recessiveness	allele
environment	hybrid	monohybrid
species characteristics	law of segregation	dihybrid
individual characteristics	genotype	law of independent
concept of unit	phenotype	assortment
characters	homozygous	incomplete dominance

QUESTIONS FOR REVIEW

1. What two general sets of factors interact in determining the characteristics of an organism? *Heredity + Environment*
2. In what respect did Mendel make a good choice in selecting garden peas for his work? *They carry on self pollination*
3. What observation led Mendel to formulate his hypothesis of unit characters?
4. Mendel found that certain tall peas would produce short plants in a second generation. Which of his hypotheses did he formulate from this observation?
5. List seven pairs of contrasting traits Mendel found in garden peas.
6. Mendel reasoned that a pair of genes was separated in different gametes during meiosis. Which of his laws is based on this reasoning?
7. The gene for black coat color is dominant in guinea pigs. How is homozygous black different from heterozygous black, even though the guinea pigs look alike?
8. When two hybrid animals are crossed, there appear among the offspring homozygous dominant, heterozygous dominant, and homozygous recessive individuals. Explain.
9. When two parents that are heterozygous for one trait are crossed, what ratio of offspring (F_1 generation) is expected to show the dominant trait and what ratio the recessive trait?
10. Explain the law of independent assortment.
11. Based on the law of independent assortment and the number of contrasting traits Mendel investigated, what would you believe to be the minimum number of chromosomes in the garden pea?

12. In what way is incomplete dominance an exception to the principle of dominance and recessiveness?

13. Describe three colors found in Shorthorn cattle and indicate which color is heterozygous.

14. In breeding experiments, why do ratios obtained represent averages rather than definite numbers?

APPLYING PRINCIPLES AND CONCEPTS

1. Outline a possible cross to determine whether a black guinea pig is homozygous or heterozygous for the coat-color trait.

2. In guinea pigs, black coat color is due to a dominant gene, *B,* and white is due to its recessive allele, *b;* short hair to a dominant gene, *S,* and long hair to its recessive allele, *s.* The gene for rough coat, *R,* is dominant over that for smooth, *r.* Cross a homozygous rough, short-haired, black guinea pig with a smooth, long-haired, white one. What are the phenotypes of the F$_1$ and F$_2$ generations?

3. In snapdragons, the inheritance of flower color and size of leaves are examples of incomplete dominance. When red-flowered plants are crossed with white ones, all the flowers are pink. Similarly, when plants with broad leaves are crossed with plants having narrow leaves, the offspring have intermediate leaves. Cross a homozygous red-flowered, broad-leaved plant with a homozygous white-flowered, narrow-leaved plant. What kind of offspring are produced in the F$_1$ generation? Now cross two of these plants and find the phenotype ratio of the offspring. Explain the relationship of the 9:3:3:1 ratio to the one you obtained.

4. Why should a hybridizer know which traits in plants or animals with which he is working are dominant or recessive?

5. Why does the law of independent assortment apply only to certain pairs of genes?

RELATED READING

Books

FAST, JULIUS, *Blueprint for Life.*
  St. Martin's Press, Inc., New York. 1964. The story of modern genetics.
GOLDSTEIN, PHILLIP, *Genetics is Easy* (2nd ed.).
  Lantern Press Inc., New York. 1961. Introduction to the study of genetics for the average biology students.
LUDOVICI, L. J., *Links of Life: The Story of Heredity.*
  G. P. Putnam's Sons Inc., New York. 1962. Helpful to the slower reader in understanding how heredity works.
PETERS, JAMES A., *Classic Papers in Genetics.*
  Prentice-Hall, Englewood Cliffs, N.J. 1959. Several original papers from the outstanding men of genetics including Mendel up to those of the present day.
WEBB, ROBERT N., *Gregor Mendel and Heredity.*
  Franklin Watts Inc., New York. 1963. Tells of the patient research of the Austrian monk whose pea-plant experiments established the basic laws of heredity.
WEBSTER, GARY, *The Man Who Found Out Why.*
  Hawthorn Books Inc., New York. 1963. Describes Gregor Mendel's pioneering work.

CHAPTER TEN

THE GENETIC MATERIAL

The chromosome theory of inheritance

The work of Gregor Mendel seems even more remarkable when you consider that he made his brilliant observations, drew valid conclusions, and formulated hypotheses without any knowledge of genetic materials. It was twenty years after the publication of his paper that the cell nucleus was recognized as the center of hereditary materials. Mendel knew nothing about the segregation of chromosomes during meiosis. In fact, he had never heard of a chromosome. Yet he formulated the law of segregation on the basis of what he observed in crossing garden peas. Mendel described *what* happened in various genetic crosses without having any idea of *the biological mechanisms by which they occurred*.

A number of years after Mendel published his studies of garden peas, biologists began searching for evidence of hereditary information in a cell. The nature of sexual reproduction, in which an egg and a sperm unite during fertilization, was known to biologists at this time. Furthermore, it was evident that each new individual produced by sexual reproduction bore characteristics of both parents. It was clear that these characteristics were transmitted to the offspring by means of certain factors contained in the egg and sperm. Are the entire contents of both egg and sperm involved in this hereditary influence? If the answer is yes, the egg should exert the greater influence, since it is often much larger than the sperm. Do the egg and sperm contribute equally to the genetic makeup of the new individual? What properties do an egg and a sperm have in common that might account for this equal influence? The most obvious similarity between the egg and the sperm is the nucleus in each. These cell structures are of approximately equal size in both egg and sperm. Could the nucleus, then, contain the genetic information? If so, an egg or a sperm must contain only half of the genetic material. Otherwise, this material would be doubled in fertilization.

Questions such as these were being asked by biologists in the 1880's. It was during this time that August Weismann, a German biologist, suggested that in the formation of eggs and sperms, the amount of genetic material was halved; and he predicted that biologists would soon discover how the process occurred. His prediction was soon fulfilled. About one year later, Theodor Boveri, another German biologist, actually observed meiosis in the cells of *Ascaris* (ASK-uh-riss), a common parasitic round-worm.

It is ironic that during these years of searching for proof of reduction division, Mendel's paper containing the evidence for his law of segregation lay forgotten in the library of the Natural History Society of Brünn, Austria. Then, in 1900, three biologists working independently in the area of cell reproduction and heredity were searching scientific libraries for information related to their problem. One was an Austrian biologist, von Tschermak; another, De Vries, a Dutch botanist; and the third, Correns, a German botanist. Each of these three men found copies of Mendel's paper in the library in Brünn. After thirty-five years of neglect, his work was finally rediscovered and put to use. By this time, biologists were familiar with chromosomes and believed that they carried hereditary information. Mendel's paper supplied evidence of their function. This was the information needed to start a landslide of investigation and discovery in genetics.

The gene hypothesis

In 1903, just three years after the rediscovery of Mendel's paper, Walter S. Sutton, a young graduate student at Columbia University in New York, presented a hypothesis of great significance. In a research paper published in the *Biological Bulletin,* he theorized that hereditary particles, or *genes,* are component parts of chromosomes. This was the first reference to the gene as the determiner of a genetic characteristic.

In examining Mendel's work, Sutton found a striking similarity between the behavior of genetic traits of the garden peas he had described and the behavior of chromosomes during meiosis. Mendel had speculated on the segregation of genetic traits in the formation of gametes. Sutton was familiar with the segregation of chromosomes during meiosis. This parallel led Sutton to the conclusion that chromosomes contained the material of heredity.

Furthermore, Sutton was familiar with Mendel's reciprocal crosses in garden peas. You will recall that Mendel reversed many of his crosses to see whether it made a difference in the off-spring if a parent plant was used to bear seeds in one cross and to supply pollen in another. In all of these crosses, the results were the same. Both parents contributed equally to the offspring. Sutton knew that eggs and sperms are not alike in size or in struc-

10-1 Walter S. Sutton, the first to propose that hereditary particles are component parts of chromosomes. (V.A. McKusick)

ture, but he found likeness in their chromosomes. This provided further evidence that chromosomes bear the particles that determine hereditary characteristics. Furthermore, he reasoned that these determiners, or genes, must be situated in identical positions on corresponding chromosomes, a belief later found to be true.

Sutton reasoned further that chromosome pairs maintain their identity with each division of a somatic, or body, cell of an organism. In these divisions, segregation does not occur. All chromosomes split lengthwise, thus providing each daughter cell with identical genetic material. This occurs in mitosis, a process of cell reproduction that you have already studied.

Generally, Sutton found three parallels between Mendel's hereditary factors and the behavior of chromosomes and genes:
- Chromosomes and genes occur in pairs in the zygote and in all somatic (body) cells.
- Chromosomes and genes segregate during meiosis, and only one member of each pair normally enters a gamete.
- Chromosomes and genes maintain their individuality during segregation. Each pair segregates independently of every other pair. This confirmed Mendel's law of independent assortment of the characteristics he had observed in garden peas.

How many genes on a chromosome?

Having established the gene hypothesis, Sutton next speculated about the number of genes on an individual chromosome. If, for example, an organism had ten pairs of chromosomes, would it not have more than ten pairs of genes? Organisms certainly had many more inheritable characteristics than pairs of chromosomes. Undoubtedly, many genes were located on each chromosome. It seemed unlikely to Sutton that genes segregated independently during meiosis. Rather, they must move in sets on a chromosome. In this explanation, Sutton proposed *gene linkage*. What, then, of Mendel's seven pairs of contrasting traits such as tall stem and short stem, round seeds and wrinkled seeds, yellow seeds and green seeds? Mendel based the law of independent assortment on the fact that all of the pairs of traits he studied segregated and recombined independently. Was a gene for each of these traits situated on a different chromosome, or were certain genes linked on the same chromosome? Sutton believed, although he could not prove it, that the genes involved in Mendel's studies must have been on different chromosomes, since they segregated independently. Today we know that he was right in his assumption. The peas Mendel used had seven pairs of chromosomes. By coincidence, each of the contrasting traits he studied was determined by genes on different chromosome pairs.

While Sutton believed that gene linkage must occur, he was never able to establish proof. A few years later, however, another great contributor to genetics found experimental evidence to support Sutton's brilliant deduction.

Discovery of sex chromosomes

Soon after Sutton established the fact that chromosomes and genes are genetic materials, Thomas Hunt Morgan and a group of associates working at Columbia University made a discovery in one tiny fruit fly, a discovery that was destined to make genetic history. Thomas Hunt Morgan was one of the pillars of the study of genetics. His genius was recognized when he was awarded the Nobel Prize in medicine and physiology in 1933 for his research in genetics. Among his associates, all considerably younger, were three men who also made outstanding contributions in genetic research: Calvin Bridges, A. H. Sturtevant, and H. J. Muller. Dr. Muller was also awarded the Nobel Prize in 1946 for outstanding work that we shall discuss later in this chapter.

Dr. Morgan and his associates were growing large numbers of fruit flies in their studies at Columbia University of genetic traits. This small fly is common around overripe fruit. You probably think of it as a pest rather than as a valuable subject for research in genetics. Biologists refer to the fruit fly as *Drosophila melanogaster* (droe-SOFF-uh-luh MELL-uh-noe-GASS-ter), but we shall shorten its name to *Drosophila* here. Several characteristics of *Drosophila* make it an ideal subject for research in genetics. It is easily raised in jars containing mashed bananas and other specially prepared diets. Its life cycle is short, varying from about ten to fifteen days, depending on the environmental temperature. Thus, many generations can be observed in a short time. Furthermore, the sexes of *Drosophila* are easily distinguished. The male is usually smaller than the female, has characteristic combs of dark bristles on its first pair of legs, and has a black-tipped, blunt posterior end (Figure 10-2). Eye color, body color, and wing structure are but a few of the thousands of genetic variations in *Drosophila* with which geneticists are familiar today.

Now, let us return to the one fruit fly that made genetic history. One day Dr. Morgan and his associates were examining a large number of *Drosophila* and, to their surprise, found one fly that had white eyes instead of the normal red eyes. The fly was a male. This unusual fly was mated with a normal red-eyed female. All of the F_1 generation resulting from this mating had normal red eyes. By applying Mendel's principle of dominance, Morgan and his associates concluded that red eyes are dominant over white eyes in *Drosophila*. The investigation was continued with the mating of flies of the F_1 generation to produce an F_2 generation. About three fourths of these flies had red eyes, and about one fourth had white eyes. This, again, conformed with the results Mendel had obtained in garden peas that were hybrid, or heterozygous, for one trait. At this point, Morgan made a significant dis-

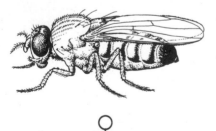

10-2 The common fruit fly, *Drosophila melanogaster.* Note that the male above is smaller than the female, and that the posterior end of the male is darker and blunter than that of the female.

covery. All of the white-eyed flies were males! This could not have been the result of chance. There was a definite association of eye color and sex. Morgan had discovered a *sex-linked trait in Drosophila*. Two problems had now arisen: (1) sex determination in *Drosophila* and (2) the relationship of the gene for white eyes to sex. We shall now consider both of these problems.

Sex determination

At the time when Dr. Morgan and his associates were experimenting with *Drosophila*, no one had examined its chromosomes. This was the next step for Morgan and his associates in the investigation of the white-eyed male. Examination of the nuclei of somatic cells of *Drosophila* revealed four pairs of chromosomes. This is the diploid (2*n*) chromosome number. The cells of female flies contained four kinds of chromosomes with identical mates (Figure 10-3). But there was a difference in the chromosomes of the male. Three pairs were like those of the female, but one chromosome was different. It did not match its mate. Instead of being rod-shaped, it was bent like a hook. The rod-shaped chromosome is called an *X chromosome*, while the hook-shaped member of the pair is named the *Y chromosome*. These are the *sex chromosomes*. The remaining three pairs in *Drosophila*, identical in both males and females, are called *autosomes*.

Following Morgan's fine work on sex chromosomes in *Drosophila*, studies were made of many other animals. Similar chromosomes were found in most animals and in many plants in which there is a sexual difference. The presence of two X chromosomes (XX) produces a female organism, while a single X chromosome paired with a Y chromosome (XY) results in a male.

Further study of sex chromosomes in many organisms has revealed a variation in what we might call XY sex determination. Some organisms have no Y chromosome in males. In these organisms, XX produces a female, while a single X results in a male.

It is easy to diagram sex determination by means of a grid such as you used in showing Mendelian crosses involving a single trait. As we describe the formation of gametes and the segregation of sex chromosomes, and then the fertilization and recombination of chromosome pairs, follow the changes shown in Figure 10-4. In the segregation of sex chromosomes in egg formation, the diploid number XX is reduced to the haploid number X. Thus, all eggs contain a single X chromosome. However, in the formation of sperms, the diploid number XY is reduced to X and Y. In other words, half of the sperms contain the X sex chromosome, and half, the Y chromosome. The sex of the offspring is determined by the chance union of an egg and a sperm. If the X-containing sperm fertilizes the egg, the offspring is female. Union with a Y-containing sperm results in a male.

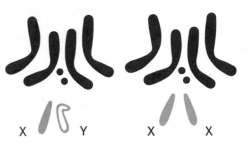

10-3 The sex chromosomes of *Drosophila* are represented in color. The male has one X chromosome and one hook-shaped Y chromosome (left), whereas the female has two X chromosomes.

SEX DETERMINATION

Female CHROMOSOMES Male	X	X
X	XX	XX
Y	XY	XY

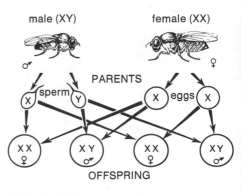

10-4 The inheritance of sex in *Drosophila*.

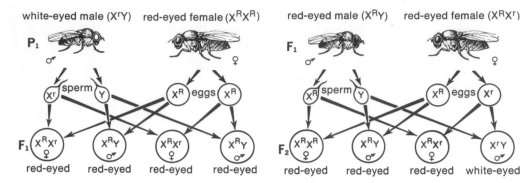

10-5 Sex linkage in *Drosophila.*

INHERITANCE OF SEX-LINKED CHARACTERISTICS

Female SEX CHROMOSOMES Male	X^R	X^R
X^r	$X^R X^r$	$X^R X^r$
Y	$X^R Y$	$X^R Y$

RESULTS OF CROSSING F₁ HYBRIDS RED-EYED MALE AND RED-EYED FEMALE

Female SEX CHROMOSOMES Male	X^R	X^r
X^R	$X^R X^R$	$X^R X^r$
Y	$X^R Y$	$X^r Y$

You may be interested in knowing that sex determination in human beings occurs in the same way. Each of your somatic cells contains twenty-three pairs of chromosomes. Twenty-two of these pairs are autosomes. The remaining pair are sex chromosomes.

Sex linkage

Now that you are familiar with sex chromosomes and their function in the determination of sex, we shall return to a discussion of Morgan's white-eyed male flies. The crosses he made are diagramed at left. As each step is described, follow it both in the diagram and in Figure 10-5. We shall designate the gene for normal red eyes as *R* and use *r* to indicate white eye color. Both of these genes are situated on X chromosomes and are therefore sex-linked. The Y chromosome contains no allele for eye color and therefore plays no part in determining eye color in *Drosophila.*

The original white-eyed male that Morgan discovered had a gene for white eyes on its X chromosome. We can represent this individual as $X^r Y$. The normal red-eyed female, with a gene for normal red eyes, would be indicated as $X^R X^R$. As a result of segregation, the sperms formed by the white-eyed male would be of two types, X^r and Y. All of the eggs produced by the normal red-eyed female would contain X^R. All members of the F₁ generation would be red-eyed, although they would be of two genetic types. Eggs fertilized by half of the sperms would contain $X^R X^r$ chromosomes and would be females, heterozygous for eye color. The other half would contain $X^R Y$. These would be males with a single gene for red eyes. All members of this generation would be red-eyed, however, because of dominance of the red eye gene.

In the mating of two flies of this generation, the offspring shown in the grid at left would occur in the F₂ generation. Half of the eggs produced by a female of this generation would contain X^R, and half would contain X^r. Half of the sperms would contain X^R, and half, Y. Thus, you can see that some of the offspring would be $X^R X^R$ and others would be $X^R X^r$. All of these flies would be red-eyed females. On the other hand, some of the males produced would be $X^R Y$, with red eyes, while others would be $X^r Y$, with white eyes. Actually, in accordance with the law of averages, probably about one fourth of the offspring would be of each type. This would result in an eye-color ratio of three fourths red-eyed and one fourth white-eyed. Furthermore, all of the white-eyed offspring would be males, as Morgan determined.

Will a white-eyed female appear if mating is continued for another generation? This could happen if a heterozygous red-eyed female, $X^R X^r$, were mated with a white-eyed male, $X^r Y$.

The discovery of *sex linkage* in *Drosophila* by Morgan and his associates introduced a new and important principle in genetics. Sex-linked traits are by no means limited to *Drosophila*. Did you know that color-blindness occurs as much as ten times more frequently in men than in women and that a similar ratio is found among victims of hemophilia, or "bleeder's disease"? Does this ratio of occurrence give you a clue? We shall reserve the discussion of the inheritance of these conditions for Chapter 11.

Other characteristics associated with sex are not sex-linked. For example, baldness is far more common in men than in women. Yet it is not a sex-linked trait. Conditions such as this will also be discussed in Chapter 11.

Nondisjunction — abnormal segregation of sex chromosomes

About ten years after Morgan discovered sex-linked traits in *Drosophila*, C. B. Bridges, one of his former graduate students, made another startling discovery. Bridges was working with sex-linked genes that determine eye color in *Drosophila*, much as he had in the earlier studies in which he assisted Morgan. In these genes, however, the alleles were red eyes and vermilion eyes. Red eyes in *Drosophila* are dark red, while vermilion eyes are much brighter red. The gene for normal red eyes is dominant over that for vermilion. These genes normally behave like other sex-linked traits. When a vermilion-eyed female is mated with a red-eyed male, half the offspring are red-eyed females and half are vermilion-eyed males, as you can see in the following grid and in Figure 10-6.

In about one individual in two thousand, however, a striking thing occurred. A vermilion-eyed female mated with a red-eyed male produced a vermilion-eyed female. If you examine the results of such a cross as shown, you will discover that this is impossible under normal conditions. In the grid and in Figure 10-6, the dominant gene for red eye color is represented as R and the recessive gene for vermilion eye color as r. The male fly with red eyes is represented as $X^R Y$, while the vermilion-eyed female is $X^r X^r$. Females in the F_1 generation must receive an X^R chromosome from the male parent and an X^r chromosome from

RESULTS OF CROSSING VERMILION-EYED FEMALE AND RED-EYED MALE

Female SEX CHROMOSOMES Male	X^r	X^r
X^R	$X^R X^r$	$X^R X^r$
Y	$X^r Y$	$X^r Y$

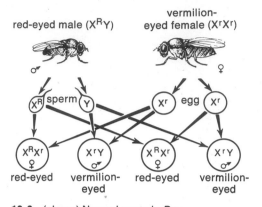

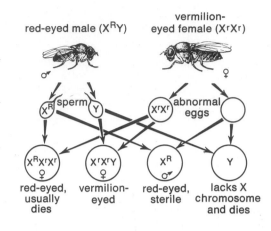

10-6 (above) Normal cross in *Drosophila* between a red-eyed male and a vermilion-eyed female.

10-7 (right) Nondisjunction in *Drosophila.*

the female parent. How, then, could a vermilion-eyed female result? This would require two X^r chromosomes, both of which are in the female parent. Yet it did happen and threatened to upset the whole theory of sex-linked traits. To Bridges, this one-in-two-thousand female fly with vermilion eyes had great genetic significance. Surely it could be accounted for in terms of sex chromosomes and sex-linked traits. The answer was to be found in the cells of the fly. Bridges examined cells from the fly's body and found a remarkable condition. This vermilion-eyed female had two X chromosomes and a Y chromosome. We would represent it genetically as $X^r X^r Y$. The extra X chromosome produced a female, even in the presence of the Y chromosome. The two recessive genes located on these X chromosomes produced vermilion eyes—an impossibility in this mating under normal conditions.

Figure 10-7 shows how abnormal chromosome segregation occurred to produce this unusual situation. We refer to the phenomenon as *nondisjunction.* The red-eyed male parent produced normal sperms, half containing the X^R chromosome and half containing the Y chromosome. However, segregation of the $X^r X^r$ of the female fly did not occur when eggs were formed. Thus, half of the eggs contained $X^r X^r$ chromosomes, as in the somatic cells, while half contained *no sex chromosomes.* Notice in Figure 10-7 that four kinds of offspring can be produced by various egg and sperm combinations:

¼ red-eyed females ($X^R X^r X^r$), which usually die
¼ vermilion-eyed females ($X^r X^r Y$)
¼ red-eyed males (X^R), which lack a chromosome and are sterile
¼ flies (Y), with no X chromosomes and therefore no eye-color genes. (These always die.)

In discovering nondisjunction, Bridges, far from disproving the chromosome theory of heredity, substantiated it further. His work left no doubt that genes are located on chromosomes and that the genes for eye color in *Drosophila* are on the X chromosomes.

Nondisjunction may occur in various autosomal chromosomes as well as in sex chromosomes. Recent studies of this phenomenon in human beings have led to an understanding of how various tragic conditions, such as Down's syndrome (mongolism), are produced. We shall discuss this condition further in Chapter 11.

Gene linkage and crossing over

Keep in mind that it is chromosomes and not individual genes that segregate during meiosis. Furthermore, a single chromosome contains a large number of genes joined together in a linear arrangement. We refer to this condition as *gene linkage*. If a chromosome has fifty genes linked together, its mate will have the alleles, or the fifty corresponding genes. Thus, an organism can have no more pairs of genes sorting out independently than it has pairs of chromosomes.

However, this linkage is not perfect. Segments of chromosomes, bearing many genes, may separate and exchange with a corresponding segment of the other member of a chromosome pair. We call this phenomenon **crossing over**. When do you think a homologous pair of chromosomes would be most likely to exchange segments? When are they in closest contact? Recall the stage in meiosis when the genes have replicated and each chromosome forms two chromatids. The joined chromatids of a homologous pair come together and form a tetrad, or group of four. It is here that segments of two chromatids (one of each homologue) may exchange segments containing varying numbers of genes. For example, let us represent three of the many genes on a chromatid as A, B, and C (Figure 10-8). Corresponding genes on a chromatid of the other homologue are a, b, and c. Now, we will assume that a segment of one chromatid separates between genes A and B. A similar separation occurs between genes a and b of another chromatid. These two chromatids exchange segments and produce a new gene linkage on two of the four chromatids in the tetrad. These new linkages are now a, B, and C and A, b, and c. Following separation of the chromatids in meiosis, eggs or sperms will receive these new gene combinations—combinations that were not present in either parent. We have described a single crossover. Double and even triple crossovers are known to occur.

The genes shown on the chromosome diagrams in Figure 10-8 are widely separated. Biologists have reasoned that the greater the distance is between two genes on a chromosome, the more likely it is that a separation will occur between them. The rate at which these separations occur in specific chromosomes has been used as an index in determining the distance between two

10-8 Crossing over. Following tetrad formation (left), crossing over occurs between adjacent chromatids (center). Note the final result: a new grouping of genes on the chromosomes (right).

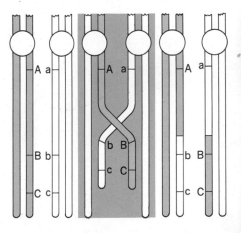

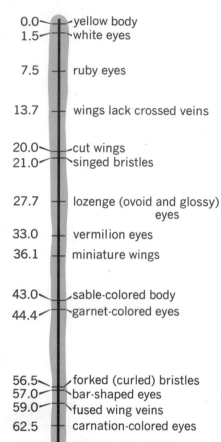

0.0	yellow body
1.5	white eyes
7.5	ruby eyes
13.7	wings lack crossed veins
20.0	cut wings
21.0	singed bristles
27.7	lozenge (ovoid and glossy) eyes
33.0	vermilion eyes
36.1	miniature wings
43.0	sable-colored body
44.4	garnet-colored eyes
56.5	forked (curled) bristles
57.0	bar-shaped eyes
59.0	fused wing veins
62.5	carnation-colored eyes
66.0	bobbed (short) bristles

CENTROMERE

10-9 Chromosome map of an X chromosome of *Drosophila.* The numbers in the left column indicate the order in which the genes are arranged, but do not reflect the actual positions of the genes.

genes. This is one of the ways in which geneticists have been able to construct *chromosome maps* locating the genes in relation to each other (Figure 10-9). Such chromosome mapping has been done for both *Drosophila* and corn. Similar methods are being used to locate specific genes on human chromosomes.

What is a gene?

Our discussion so far has concerned chromosomes and the genetic traits that genes produce. But what is a gene, and how does it function? Why does one gene produce a short pea plant, while another produces a tall plant? To find a possible answer to these questions, we must return to the DNA molecule.

In Chapter 7, you learned that DNA transfers its genetic code to messenger RNA by replication. Messenger RNA, in turn, provides a template for assembling amino acids in polypeptide chains that comprise enzymes and other proteins. Thus, DNA, in controlling protein synthesis, determines the nature of the enzymes, structural proteins, and other protein substances synthesized in an organism. These materials, in turn, are involved in the expression of genetic traits. It is as though DNA were the architect and RNA were the contractor using the blueprint, or genetic code, for assembling proteins in the cell.

Let us now examine the structure of DNA more closely for a possible definition of a gene. In Chapter 7, we referred to current evidence that nucleotide bases in RNA function in groups of three, or triplets, in coding amino acids. We refer to these base triplets as *codons*. Biologists have found that each of the twenty amino acids may be coded by from one to six different codons. In Chapter 3, you learned that a single protein molecule may be composed of as many as from three hundred to three thousand amino acids bonded together in polypeptide chains. The coding of such proteins would involve from three hundred to three thousand codons, or from nine hundred to nine thousand nucleotide bases. Then, consider that a single strand of DNA contains a genetic code for hundreds and perhaps thousands of different proteins.

The action of RNA codons in determining the amino acid composition of a protein leads to a possible definition of a gene. The simplest concept might be that a gene is a sequence of base triplets in a segment of DNA that functions to code a polypeptide chain, that in turn becomes a part of an enzyme or other protein.

DNA is believed to be spiraled around a protein backbone or core in the composition of a chromosome. The nature of bonding of DNA to the protein core seems to determine which segments of the strand are active genetically. It is believed that the portions that are genetically active and that function as genes lie near the surface. There are probably thousands of such genetically active portions along a strand of DNA. The ends of the strand seem to be genetically inactive, as are portions of the segment thought of as "nonsense codons."

Gene action

The action of genes in controlling protein synthesis through RNA codons is reasonably well understood by biologists. However, the mechanisms involved in the expression of a genetic trait, possibly by enzyme action, remain a biological mystery. This is a subject of extensive investigation in molecular genetics today.

As you learned earlier, your genes are present in homologous pairs on the forty-six chromosomes that are normally found in every somatic cell of your body. This genetic makeup was established at the beginning of your life, when a haploid egg and a haploid sperm combined to produce a diploid zygote, your first somatic cell. From that moment on, every gene has replicated prior to a mitotic division. Thus, every cell in your body contains your complete genetic code. If you have blue eyes, every cell contains genes for blue eyes. Yet only in the cells of the iris of your eyes is this trait expressed. Blue-eyed people never have blue skin. The cells of the iris also contain genes that determine the color and texture of your hair. But these genes act only on the cells of hair roots. Genes also determined your blood type, but these genes have had no influence on the shape of your ears or the length of your fingers.

Could it be that genes act only in a particular cell environment? Perhaps in the cytoplasm of various cells there are chemical differences that influence gene action. Gene action is thought to be related to specific enzymes. Perhaps there are chemical substances in cells that influence protein synthesis, stimulating the synthesis of specific enzymes in some kinds of cells and repressing their synthesis in others. Recent research findings in this area suggest that *hormones* may be associated with the control of gene action. Since the characteristics of a cell seem to be determined by the kind as well as by the number of proteins it synthesizes, the action of powerful hormones in altering this process would logically influence gene action.

Environmental influences on genetic traits

Can the action of a gene be altered by conditions external to the cell? If so, a genetic trait may be altered by an environmental condition.

This phenomenon has been demonstrated in variations in hair color of the Himalayan rabbit. The normal color pattern of this breed is white except for black nose, ears, tail, and feet (Figure 10-10). However, biologists have found that temperature variations affect the color pattern. If hair is pulled from any of the black areas, for example, the ears, and the area is wrapped to

10-10 A genetic trait altered by environmental conditions. (a) The normal color pattern of the hair of the Himalayan rabbit. (b) After the hair was removed from a portion of the rabbit's back, an ice pack was kept in position while the hair grew in. (c) New hair in region kept at lower-than-normal temperature is black.

raise the temperature to that of the back or belly, the new hair will be white. Moreover, if hair is pulled from the back and the patch of skin is kept cooler than normal, the new hair will be black. Thus, temperature seems to influence hair color in this rabbit—hair is white in warmer body areas and black in cooler extremities of the body. Temperature differences may influence gene action or enzyme action. In either case, however, a genetic trait is altered by environmental conditions.

A similar temperature influence has been found in *Drosophila*. A gene for curly wings will produce this trait if the flies are raised at 25°C. However, when the flies are raised at a temperature of 16°C, the wings are straight and normal. We know that the gene for curly wings has not disappeared, since the next generation will develop curly wings at 25°C.

Transformation in pneumococcus—proof of DNA

Perhaps you have wondered how biologists have been able to show conclusively that DNA is the genetic material. Some of the most convincing support has been found in studies of pneumococcus, the bacterium that causes one kind of pneumonia. It is interesting that these studies began over forty years ago with the work of an investigator who had never heard of DNA.

In 1928, Frederick Griffith, a British bacteriologist, was working with two strains of pneumococcus. The cells of the two strains were similar—tiny spheres usually joined in pairs or short filaments. However, there was a sharp distinction between the two strains. One formed a slimy capsule that surrounded the cells. The other did not (Figure 10-11). Griffith found that the noncapsulated organisms *did not* cause pneumonia. Apparently the cells were destroyed by white corpuscles when they were injected into a mouse or other test animal (Figure 10-12a). However, mice that received injections of the capsule-forming strain were dead or dying of pneumonia within a short time (Figure 10-12b). Apparently the capsule prevented white corpuscles from destroying cells of the strain.

In one phase of his investigation, Griffith inoculated mice with both strains of pneumococcus. In one injection, he used living, noncapsulated organisms. In the other, he used capsulated organisms that had been killed with heat. Much to his surprise, the mice soon died of pneumonia. When he examined the blood of these mice, he found pneumonia organisms *with capsules!* None of the capsulated organisms he had injected were living. Yet here were capsulated bacteria that when placed in cultures continued to grow and were, therefore, alive. Apparently, the living, noncapsulated cells had been transformed into a capsule-producing strain (Figure 10-12c).

About fifteen years later, Oswald T. Avery and his associates, C. M. MacLeod and M. McCarty, began a search for an explana-

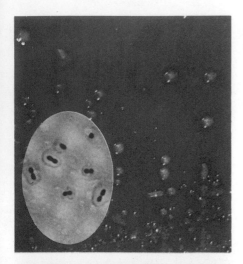
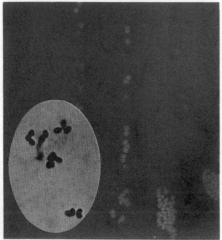

10-11 Pneumococcus. Cells and "smooth" colonies of the capsulated, infectious strain (above), and cells and "rough" colonies of the noncapsulated, noninfectious strain. (Louis Koster; insets: Robert Austrian and *Journal of Experimental Medicine*)

Griffith's investigation with *Pneumococcus*

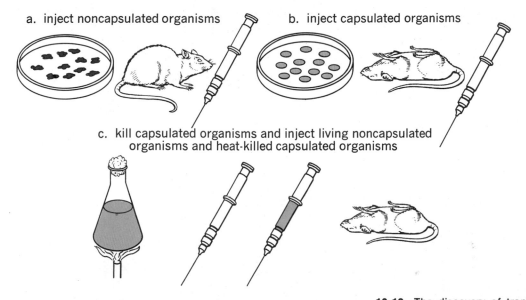

a. inject noncapsulated organisms

b. inject capsulated organisms

c. kill capsulated organisms and inject living noncapsulated organisms and heat-killed capsulated organisms

tion of Griffith's results at the Rockefeller Institute for Medical Research in New York City. In one set of laboratory cultures, they grew noncapsulated Type II pneumonia organisms, whose colonies appear rough (R); in another set, they grew capsulated Type III pneumonia organisms, whose colonies appear smooth (S). Each strain continued to produce its own characteristic cells, generation after generation (Figure 10-13). They next prepared a cell-free extract of heat-killed capsulated Type III-S cells and mixed it with living noncapsulated Type II-R cells. When this mixture was cultured in a sterile medium, it was found that most of the colonies were still of the noncapsulated II-R type. Significantly, however, a few capsulated Type III-S colonies were found. These newly transformed, capsulated cells were then isolated and cultured. Now, when injected into mice, these formerly noninfectious Type II-R cells caused pneumonia. Some factor in the extract made from dead, capsulated organisms had entered these cells and produced a change that was passed on to offspring. Avery and his associates were able to isolate and purify this transforming substance. At first, it was thought to be a protein. However, it proved to be a nucleic acid or, more specifically, DNA.

The discovery of the phenomenon of *transformation* in pneumococcus is of great genetic importance for at least two reasons. It adds evidence that DNA is the gene. Furthermore, it demonstrates that DNA can be transferred from one organism to another and continue to express its genetic qualities.

10-12 The discovery of transformation in pneumococcus-Griffith's investigation. (a) Griffith found that living noncapsulated pneumococci did not cause pneumonia when injected into mice. (b) However, when living capsulated pneumococci were similarly injected, the mice contracted pneumonia, and died. (c) When both living noncapsulated organisms and heat-killed capsulated organisms were injected, the mice died. Living capsulated organisms were recovered from these mice. Apparently some of the noncapsulated pneumococci had become *transformed* into the capsulated type.

Avery and associates' investigation with *Pneumococcus*

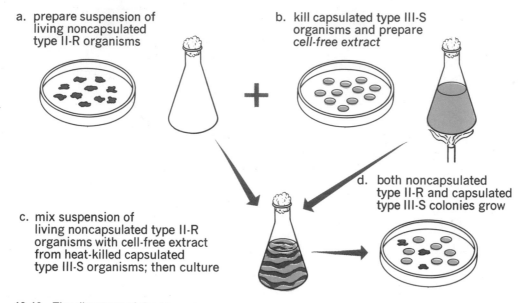

a. prepare suspension of living noncapsulated type II-R organisms

b. kill capsulated type III-S organisms and prepare *cell-free extract*

c. mix suspension of living noncapsulated type II-R organisms with cell-free extract from heat-killed capsulated type III-S organisms; then culture

d. both noncapsulated type II-R and capsulated type III-S colonies grow

10-13 The discovery of the transforming substance-Avery and associates' investigation. (a) & (b) Noncapsulated Type II-R and capsulated Type III-S pneumococci were again cultured, and the capsulated Type III-S organisms were heat-killed. (c) Suspension of living noncapsulated Type II-R cells was mixed with extract from heat-killed capsulated Type III-S cells. (d) When this was cultured on sterile medium, some capsulated Type III-S colonies appeared. It was later determined that DNA from the heat-killed capsulated Type III-S organisms had entered some of the noncapsulated organisms and transformed them into capsulated cells.

Mutations—errors in the genetic code

Normally, genes replicate and chromosomes segregate during meiosis with remarkable accuracy. This may occur millions of times without an error. Variations occur in offspring as new gene combinations are established in sexual reproduction. But the individual genes remain unchanged.

From time to time, however, an organism may appear with a characteristic totally unlike any of those of either parent. This new characteristic is transmitted to offspring. Thus, we know that it represents a sudden genetic change rather than an environmental influence. We refer to such a change as a *mutation* and to the organism possessing it as a *mutant.* Any change in the base coding of a DNA molecule would alter or destroy the trait associated with the gene. This produces a *gene mutation,* also referred to as a *point mutation.* While less common than gene mutations, other mutations may occur as *chromosomal aberrations,* or *chromosome mutations.* These may be the result of nondisjunction during segregation in meiosis, loss of an entire chromosome or a piece of a chromosome, or, possibly, the recombination of chromosome segments by crossing over.

If a mutation occurs in a body cell of a plant or animal, the variation appears in all of the tissue that descends from the original mutant cell. This is known as a *somatic mutation.* It is not transmitted to offspring, since reproductive cells are not involved. Somatic mutations can be preserved in plants, however, by vegetative propagation. Somatic mutations have been the source of many new plant varieties, as we shall discuss in Chapter 12. If a mutation occurs in a reproductive cell, it may be transmitted to offspring. Such *germ mutations* are of great genetic importance.

The nature of gene mutations

Minor mutations may occur from time to time with little or no visible effect on the organism. These are the most common of all mutations. However, a major mutation results in a drastic change in a characteristic. Minor mutations may be either beneficial or harmful. Major mutations are nearly always harmful. Many are *lethal* and result in death of the organism. Mutation of a gene that controls chlorophyll production in a plant is one example of a lethal mutation, since a plant that lacks chlorophyll cannot carry on photosynthesis.

Mutations occur in all levels of life. They are known to occur in viruses and are frequent among bacteria. They have been found in all plants and animals that have been studied genetically. Human beings are also subject to mutations. We shall discuss some of these in Chapter 11.

A mutant gene is nearly always recessive to a normal allele. For this reason, a mutation usually is not expressed in a heterozygous individual. Nevertheless, if the gene is passed to offspring and is paired in that offspring with a similarly mutated gene from the other parent, it is likely that the characteristic controlled by that gene will be altered.

The rate at which different mutations occur varies considerably, because some genes seem to be chemically more stable than others. Some mutate as often as once in two thousand cell divisions, while others mutate only once in millions of cell divisions. Sometimes genes mutate several times in rapid succession. They often mutate and then mutate again back to the original form. Of course, the rate at which many genes mutate is difficult to determine because the mutation has no apparent or obvious effect on the organism itself.

Causes of mutations

One of the most frequent causes of mutations is exposure to high-energy radiation. Cosmic rays from outer space and radiation from radioactive elements may cause natural mutations. They can also be produced experimentally by exposure to X rays, gamma rays, beta particles, and ultraviolet light. Biologists have used artificial sources of radiation to increase both the number and the rate of mutations in organisms under experimental conditions. Temperature increase has also been used to increase the rate at which mutations occur. Certain chemicals have been used to produce mutations; among these are formaldehyde, nitrous acid, peroxide, and mustard gas.

10-14 Dr. Hermann J. Muller, the first to show that radiations cause gene mutations. Identify as many of the items on his laboratory bench as you can. (courtesy Indiana University)

Radiation as a cause of mutations

The first proof that radiation causes mutations came from research conducted by Hermann J. Muller in 1927 (Figure 10-14). Muller was one of the graduate students who worked with Morgan and Bridges at Columbia University in the early studies of inheritance in *Drosophila*. While at the University of Texas, Muller conducted a series of experiments to establish the fact that radiation can cause mutations and that their rate can be increased with artificial radiation. As he had in his earlier work, Muller used *Drosophila* in his investigations. His work, for which he received the Nobel Prize in medicine and physiology in 1946, represents one of the most significant advances in genetics.

Prior to Muller's work, geneticists had unsuccessfully tried many methods of producing mutations artificially. They had experimented with temperature changes, variation in light conditions, different diets, and other factors on various animals, including *Drosophila*. One thing these other investigators probably overlooked that Muller considered was a condition under which a gene mutated. Consider that a pair of genes, normal for an organism and lying close to each other in corresponding positions on corresponding chromosomes, would both be equally affected by chemical changes in the cell or by changes in an environmental condition such as temperature. Suddenly, one gene of the pair mutates, while the other remains unchanged. What could change one gene and not the other? The most likely cause of such a pinpoint effect would be high-energy radiation. It was to prove this idea that Muller began his series of experiments. Muller reasoned, further, that lethal mutations would be the most likely to result from radiation.

The following is a greatly simplified summary of the brilliant work Muller conducted. He selected a male *Drosophila* with known sex-linked traits and radiated it with a strong dose of X rays. This fly was mated with a female with known sex-linked genes on the X chromosomes. The female was not radiated. In the F_1 generation, half of the flies were females, all containing a radiated X chromosome from the male parent and a normal X chromosome from the female parent. The other half were males, containing a normal X chromosome from the female parent and a radiated Y chromosome from the male parent. When two of these particular flies were crossed, four kinds of offspring should have been accounted for in the F_2 generation. This mating should have resulted in about one fourth females with two normal X chromosomes, one fourth females with one normal X chromosome and one radiated X chromosome, one fourth males with a normal X chromosome and a radiated Y chromosome, and one

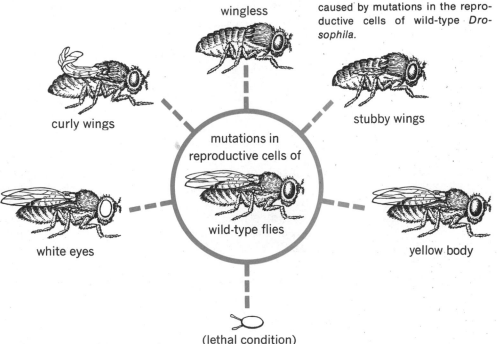

10-15 Some of the changes caused by mutations in the reproductive cells of wild-type *Drosophila*.

wingless

curly wings

stubby wings

mutations in reproductive cells of

wild-type flies

white eyes

yellow body

(lethal condition)

fourth males with both X and Y chromosomes radiated. However, this last group of flies did not appear. The radiated X chromosome bore a lethal gene that prevented them from developing (Figure 10-15). This did not occur in the females that contained a radiated X chromosome because of a dominant (normal) gene on the other X chromosome.

Since Muller's original work, Muller and other geneticists have produced many more mutations in *Drosophila* as well as in other organisms. Artificial radiation has been found to increase the rate at which mutations occur to as much as 150 times above the normal frequency.

A common mold with great genetic significance

Much of the research in genetics today is being conducted with simple organisms, including molds and bacteria. Experimental results can be obtained quickly because of the short life cycles of these organisms. Furthermore, metabolic processes can be studied readily, thus establishing evidence of gene action in cell biochemistry.

One such organism that has been the subject of extensive genetic studies in recent years is the common baker's red mold, *Neurospora crassa*. This common mold, a relative of the yeasts, can be found growing on bread and other food products. As the mold grows, it forms a mass of white threads that penetrate the medium. Within a short time, stalks arise above the medium. The tip of each stalk becomes a spore case in which eight salmon-pink spores are produced. Each spore can produce a new mold plant.

While we would consider *Neurospora* a primitive organism from the standpoint of evolution, it is far from simple in its biochemical capacity. The chemical abilities of this mold far surpass those of more highly evolved organisms, including most animals. *Neurospora* can be grown on a simple medium containing only salts, sugars, and biotin, which is a vitamin.

It is known now that *Neurospora* can synthesize amino acids and produce its own vitamins as well as proteins and can form carbohydrates, fats, nucleic acids, and other essential substances. However, all of these processes require specific enzymes. These, we know today, are related directly to DNA. The fact that many enzymes function in a series of metabolic processes has made *Neurospora* an ideal subject for the study of genetic mutations and the influence of genes on enzyme production.

Altered genes and interrupted metabolic process

Neurospora came into prominence in genetic studies when it was used by two investigators at Stanford University in important experiments in the 1940's. George W. Beadle and Edward L. Tatum produced several mutations in *Neurospora* by exposing the spores to X rays (Figure 10-16). After treatment, the spores were placed in the simple medium normally used to culture the mold. Beadle and Tatum found that certain of the treated spores could not grow in the simple medium. Apparently, a mutation had occurred that prevented these spores from synthesizing certain substances essential for growth. The problem now was to find out what substance or substances could no longer be synthesized and whether a single gene or several genes had mutated in causing that particular condition.

Beadle and Tatum next prepared a simple medium to which they added all vitamins and essential amino acids. The spores germinated and produced apparently normal mold plants in this medium. Spores produced by this mutant mold supplied the investigators with material to test the same mold in other nutrient solutions.

At this point in the investigation, Beadle and Tatum devised a method of determining which one of twenty amino acids the mutated *Neurospora* could no longer synthesize. They prepared 20 tube cultures, each containing the simple nutrient solution, vitamins, and a different amino acid. Each culture was inoculated with spores of the mutated mold. Growth occurred in only one of the twenty cultures, in the tube containing the amino acid *arginine*. This indicated that X rays had altered or destroyed the gene or genes controlling the synthesis of this essential amino acid.

Biochemists had earlier established the fact that arginine is synthesized in a series of chemical steps. This chemical sequence begins with a prior substance from which *ornithine,* a nitrogen-containing organic molecule, is formed. In a second step, *citrulline* is produced from ornithine. Arginine is in turn synthesized from citrulline.

The next question concerned the relation of the damaged genes in the mutated *Neurospora* to this series of processes. Were all three steps normally controlled by a single gene, or was a different gene associated with each step in the process? Beadle and Tatum suggested that each step involved a different gene and that any or all of these genes might be involved in the mutation. We can diagram the chemical steps leading to the formation of arginine and the possible relation of genes to each step:

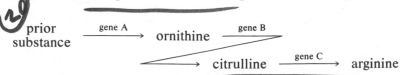

prior
substance $\xrightarrow{\text{gene A}}$ ornithine $\xrightarrow{\text{gene B}}$
$\xrightarrow{\hspace{2cm}}$ citrulline $\xrightarrow{\text{gene C}}$ arginine

If any one or all three of these genes had mutated, the mold could grow if arginine were added to the culture. However, if only gene A had mutated, the mold could grow if ornithine, citrulline, or arginine were added. If gene B or both A and B had mutated, their normal functions could be bypassed by add-

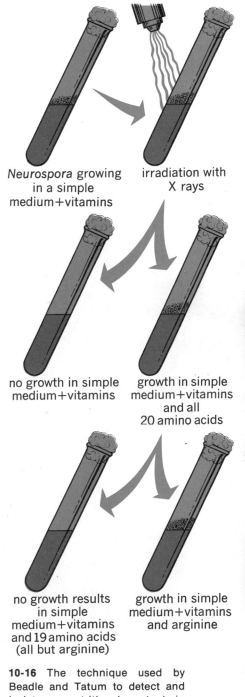

Neurospora growing in a simple medium+vitamins

irradiation with X rays

no growth in simple medium+vitamins

growth in simple medium+vitamins and all 20 amino acids

no growth results in simple medium+vitamins and 19 amino acids (all but arginine)

growth in simple medium+vitamins and arginine

10-16 The technique used by Beadle and Tatum to detect and isolate a nutritional mutant in *Neurospora* and to identify the amino acid it could not synthesize.

ing citrulline or arginine. In the event of a mutation in gene C, arginine would have to be added, even if genes A and B were functioning normally. Beadle and Tatum tested these possibilities by adding the different compounds to the culture media of various mutants. They found that their reasoning that a different gene was in fact involved in each step had been correct.

The work of Beadle and Tatum with various mutated strains of *Neurospora* established the "one gene—one enzyme" hypothesis of gene action. Actually, it was a specific enzyme and not the gene itself that was necessary for each step in the series of reactions leading to arginine. However, a gene was necessary for the synthesis of each enzyme. Mutation destroyed the capability of a gene to produce a polypeptide vital to enzymatic activity, thus blocking the synthesis of arginine. For their brilliant work in investigating gene action, Beadle and Tatum were awarded a share of the Nobel Prize in medicine and physiology in 1958.

Would it surprise you to learn that a similar series of chemical changes occurs in the cells of your liver? In this series of reactions, which biochemists call the *ornithine cycle*, ornithine, in the presence of carbon dioxide and ammonia, is converted into citrulline. In turn, citrulline is condensed with an ammonia molecule to form arginine. However, if this arginine is not used for protein synthesis, it is broken down to form urea, which is excreted, and ornithine, the substance with which we started.

SUMMARY

More than a century has passed since Gregor Mendel conducted his experiments with garden peas and laid the groundwork for the science of genetics. Since the turn of the century scientists have probed deeper and deeper into the wonders of the living cell in search of the genetic material. Sutton, Morgan, Bridges, Griffith, Muller, and many others have made significant contributions to the search. Today with electron microscopes, ultracentrifuges, computers, and other sophisticated research equipment, together with a vast knowledge of biochemistry and cell biology, the science of genetics is moving forward at an amazing rate.

Modern genetics is providing answers to many questions relating to human heredity. Most of the traits we inherit are desirable. Some people are not so fortunate in their inheritance. We are finding out more and more about genetic problems in human inheritance. Often, the discovery of a cause leads to a correction. This is an encouraging application of our knowledge of genetics to the improvement of mankind. This will become more evident to you as you study genes in human populations.

BIOLOGICALLY SPEAKING

gene linkage	crossing over	mutant
sex chromosome	chromosome map	gene mutation
autosome	codon	chromosomal aberration
sex linkage	transformation	somatic mutation
nondisjunction	mutation	germ mutation

QUESTIONS FOR REVIEW

1. What logical conclusion led Weismann to propose that reduction division occurred in the formation of eggs and sperm?
2. List three parallels Sutton found between Mendel's hereditary factors and the behavior of chromosomes and genes.
3. Explain the principle of gene linkage proposed by Sutton.
4. Account for the fact that all of Mendel's traits sorted out independently.
5. Describe the chromosome makeup of *Drosophila.*
6. Explain how sex chromosomes determine the sex of an organism.
7. What is a sex-linked gene?
8. What characteristic did Morgan discover in the white-eyed *Drosophila* of an F_2 generation that led to the discovery of sex linkage?
9. Distinguish between sex chromosomes and autosomes.
10. Explain the mechanism of chromosomal nondisjunction.
11. How does crossing over alter gene linkage?
12. What is a chromosome map?
13. Define a gene in terms of the composition and genetic activity of DNA.
14. Explain the relation of enzymes to gene action.
15. Give an example of an environmental influence on a genetic trait.
16. What observation in Griffith's work led to the discovery of transformation?
17. List several natural causes of mutations.
18. Distinguish between a somatic mutation and a germ mutation.
19. What energy source did Muller use to produce mutations in *Drosophila?*
20. Describe three steps in the synthesis of arginine by *Neurospora.*
21. In what way did X ray-induced mutations alter Tatum's *Neurospora?*

APPLYING PRINCIPLES AND CONCEPTS

1. Discuss the significance of the gene hypothesis proposed by Sutton.
2. Review the work of Morgan and his associates in the discovery of sex chromosomes and sex linkage.
3. Explain why nondisjunction is lethal in many offspring.
4. Discuss possible changes that might occur in a gene when it mutates.
5. Discuss two important genetic principles illustrated in transformation.
6. Discuss the importance of Muller's selection of lethal sex-linked traits.
7. How did the work of Beadle and Tatum with *Neurospora* support the hypothesis of "one gene — one enzyme"?

RELATED READING

Books

ENGEL, LEONARD, *The New Genetics.*
Doubleday and Co., Inc., Garden City, N.Y. 1966. Deals with some of the latest findings in the field of genetics.
GABRIEL, MORDECAI L., (ed.), *Great Experiments in Biology.*
Prentice-Hall, Inc., Englewood Cliffs, N.J. 1955. A presentation of scientific writings in the original including those of the great geneticist, Hermann J. Muller.
SULLIVAN, NAVIN, *The Message of the Genes.*
Basic Books, Inc., Publishers, New York. 1967. A lucid account of the mechanisms of heredity showing that today's molecular biologists have arrived at a clear understanding of life itself.

CHAPTER ELEVEN

GENES IN HUMAN POPULATIONS

The nature of human heredity

Undoubtedly, as you have been reading the many accounts of genetic investigations with garden peas, guinea pigs, *Drosophila*, pneumococcus, *Neurospora,* and other organisms, you have been thinking, "Do these same laws and principles apply to me?" They do, of course, although we probably know more about inheritance in *Drosophila* than we do in human beings. There are several reasons for this. One is the length of the human life span. An investigator can study many generations of *Drosophila* in a few months. Or, in bacteria, he can observe many generations in a week. But we do well to see six human generations in a lifetime. The number of individuals presents another problem in studying human genetics. In many animals and plants, a single mating or a cross may produce hundreds of offspring. When you compare this rate of reproduction with the limited size of human families, you can see that the family represents a very small sampling of genetic possibilities.

Perhaps the greatest problem in studying human heredity is to separate the influence of heredity from that of the environment. Look at a human face. Perhaps this person has his father's ears and his mother's nose. We might account for his eye color and the color of his hair in terms of genes, but don't cares and worries or satisfaction and contentment leave their mark, too? Similarly, genes may influence your height and the general build of your body, but so do your diet, your general health, and the activity of your glands. We know that glandular function is related to emotional states. These, in turn, reflect the conditions and events of your environment. Thus, another set of factors is introduced that complicates investigations in human genetics. This cannot be avoided, since man is a thinking, reasoning, responsive organism.

Another complicating factor in human inheritance is the fact that most people come from mixed ancestry. Few, if any, human genetic traits are pure. Each time a child is born, the genetic backgrounds of two entirely different families are combined in that child.

With all of these complications, however, geneticists have made extensive investigations of human genetics, especially in recent years. We know that the almost limitless number of human hereditary traits is produced by the action of an enormous number of genes on the forty-six chromosomes in each body cell. When you consider the hundreds of genetic traits expressed in *Drosophila,* which has but four pairs of chromosomes, you can get some idea of the great variety of gene combinations we could have with twenty-three chromosome pairs (Figure 11-1). Is it any wonder, then, that among the hundreds of millions of people living in the world today there is no exact genetic duplicate of you? That is, of course, unless you have an identical twin.

Population genetics

You will recall that Mendel established his ideas of heredity by observing several generations of garden peas. No single generation could have supplied the data necessary to establish these principles. When geneticists applied Mendel's hypotheses to human inheritance, the problem of limited numbers arose. Family characteristics could be traced from children to their parents and grandparents. Aunts, uncles, and other close relatives could be used. However, all of these individuals represent a small sampling. A family history might extend a study, but except for certain easily recognized traits, data concerning ancestors are usually vague and unreliable.

Has this thought ever occurred to you? You have two parents, four grandparents, eight great-grandparents, sixteen great-great-grandparents, thirty-two great-great-great-grandparents, and so on, in numbers that double each generation. You may be surprised when you realize what a large segment of the population you may claim as distant relatives!

There is a better method of studying the frequency of genetic traits and predicting how often they will appear than attempting to trace them through families. Rather than extending the time to involve many generations, the scope of the study can be increased to include a much larger sampling of a population at a given time. In other words, it is possible to determine how frequently specific genes appear in a population. From this information, one can predict the probability of appearance of any given genetic trait in offspring. *Population genetics,* as we call this study, is based on *gene frequencies.*

Sampling a population

From time to time, you read the results of an opinion poll in your newspaper. Based on a sampling of the population, the poll will

11-1 Chromosomes of the human female (top) and of the human male. The numbers are those used by geneticists for reference purposes. (Theodore T. Puck)

predict the election of one candidate for political office and the defeat of another. It will even predict the margin of votes for the winning candidate. You may have been amazed at the accuracy of these polls. How can the results of the sampling of a relatively small portion of the population be extended to reflect the opinions of the population of the entire nation accurately? The accuracy of the prediction depends on the selection of a representative group of people in conducting the poll. The national population includes people of all ages, educational backgrounds, professions, trades, businesses, social backgrounds, races, cultural backgrounds, and economic levels. Our sampling, then, must include people from all of these groups and, of course, approximately equal numbers of men and women.

Now, let us apply this method of sampling to a genetic survey that you might conduct in your class. The trait we will investigate will be the ability to taste *phenylthiocarbamide* (FEN'L-THY-oe-KAHR-bah-MIDE) (PTC) on a strip of test paper. This substance is extremely bitter to some individuals, referred to as "tasters," but has no flavor to a "nontaster." The ability to taste PTC is due to a dominant gene T. Thus, a taster may be homozygous *(TT)* or heterozygous *(Tt)*. A nontaster must be a homozygous recessive *(tt)*. Now, let us assume that twenty-four members of your class asked twenty-five people to try to taste PTC on impregnated test papers. This would provide results from a total of six hundred people. The poll should include people of various ages and both sexes in order to provide a representative sample. If possible, include people of different races and geographic and national backgrounds.

Let us assume that 430 people tested were tasters and 170 were nontasters. If we divided the number of tasters, 430, by the total number of people surveyed, 600, we would find the result to be 0.716, or about 72 percent. The number of nontasters would then be about 28 percent of the people tested. This result would be near the national average, which has been determined to be about 65 percent tasters and about 35 percent nontasters.

Having determined the number and the percentage of tasters and nontasters in a sample of a population, it is a simple matter to establish the probable number of each in the total population. Let us assume that the population of your community is about 120,000, so that each person tested in your survey would represent 200 in the total population. Thus, you would expect about 78,000 people in the community to be tasters, while about 42,000 would be nontasters.

Another interesting inherited trait that can easily be sampled in a population is tongue-rolling. Some people can roll their tongues into a U shape, while others cannot. Since the gene for this characteristic is dominant, a tongue-roller may be homozygous *(RR)* or heterozygous *(Rr)*. Those who cannot roll their tongues are homozygous recessives *(rr)*. Since tongue-rolling involves a single pair of genes with dominance, you would expect a survey of a population for this trait to yield results similar to those of the PTC-taster study.

The gene pool

The geneticist refers to all the genes present in a given population as the *gene pool.* We might consider that any individual is a random sample of the genes present in his particular population. How frequently will a particular characteristic appear in a population? This will depend on the frequency in the population of the gene associated with that characteristic.

To return to our survey of tasters and nontasters of PTC, the fact that both of these characteristics appear in a population is evidence that both dominant genes *(T)* and recessive genes *(t)* are present. Furthermore, if you consider the national ratio of about 65 percent tasters to about 35 percent nontasters, you will see that this approximates a three-fourths-to-one-fourth ratio characteristic of a monohybrid cross with dominance. We might assume, then, that about one fourth of our population would be homozygous tasters *(TT)*, one half heterozygous tasters *(Tt)*, and one fourth homozygous nontasters *(tt)*. This would indicate that the genes *T* and *t* are present in about equal numbers.

Might there be populations in some parts of the world where only the dominant gene is present in a population? All members of such a population would be tasters. Similarly, other populations might be entirely recessive for the trait.

As we examine various populations, we can see evidence of a high frequency of certain genes. The American Indian, for example, has characteristic dark, straight hair. His facial features resemble those of the Mongoloid peoples of northeastern Asia, to whom anthropologists believe he is related. Following the migration of the Indians to North America perhaps fifteen to twenty thousand years ago, various culture areas were established from Alaska throughout North America and into Central America and South America. Certain gene pools were established in these culture areas and even in the tribes composing each one. The Eastern Woodland Indians had certain characteristics that distinguished them from Plains Indians or Southwest Farmers and Herders. The Eskimos of the Far North had characteristics that distinguished them from the Northwest Fishermen along the Canadian Pacific Coastal region. Similarly, the African Pygmy of the Belgian Congo area represents a noticeable contrast to the giant Watusi tribesman of the Lake Victoria region.

Although certain genetic traits may predominate in a population at a given time, gene pools change as populations shift. New gene combinations occur. These may appear as variations in

11-2 The ability to roll the tongue is genetically determined. The characteristic is dominant, so that a tongue roller may be either homozygous or heterozygous for the trait. (John King)

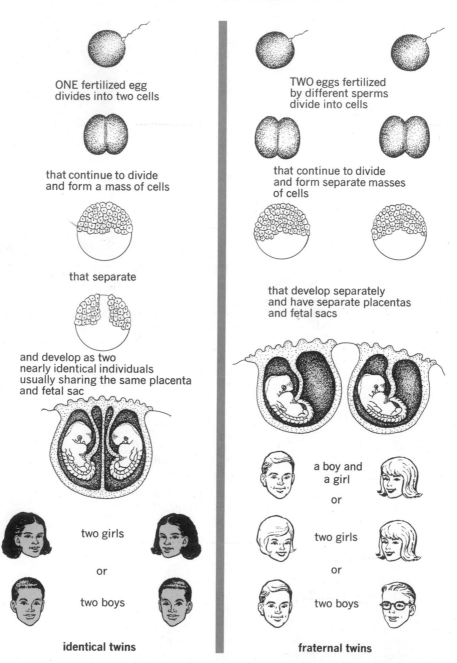

ONE fertilized egg
divides into two cells

that continue to divide
and form a mass of cells

that separate

and develop as two
nearly identical individuals
usually sharing the same placenta
and fetal sac

two girls

or

two boys

identical twins

TWO eggs fertilized
by different sperms
divide into cells

that continue to divide
and form separate masses
of cells

that develop separately
and have separate placentas
and fetal sacs

a boy and
a girl

or

two girls

or

two boys

fraternal twins

11-3 Since identical twins develop from a single egg and sperm, they have
the same genetic makeup. Fraternal twins, developing from different eggs
and sperms, are no more alike than other brothers and sisters.

skin color, hair color and texture, eye color, height and body build, and blood type. The greater the number of people that move from one geographic area to another, the more the gene pool changes.

Genetic studies of twins

Having considered the gene distribution in human populations, let us now consider the expression of genes in individual members of a population. As we mentioned earlier, it is difficult to distinguish between genetic and environmental influence on certain traits in human beings. In an effort to clarify the distinction, geneticists have made studies of twins. Twins are of two types. *Fraternal twins,* the more common type, are two entirely different individuals. Often they are brother and sister. They develop from separate eggs that are fertilized by different sperms, as shown in Figure 11-3, right. They are no more alike genetically and no more closely related than other brothers or sisters in a family. Most fraternal twins live in similar environments. Yet they may be totally different in physical characteristics, personality, emotional makeup, and mental ability. These variations are valuable in helping to determine which characteristics are hereditary and which are environmental.

Identical twins, on the other hand, are nearly the same person in duplicate. Having started life as a single cell, they have the same genetic make-up. Identical twins result when cells that have started development of a single embryo separate and continue as two embryos. This may occur at any stage in early embryonic development. Figure 11-3, left, shows the separation of a mass of cells (blastoderm) into two masses that will continue development of two identical embryos. Notice that the embryos are attached to the same placenta and are sharing the fetal sac.

Since identical twins have the same genetic makeup, their similarities indicate characteristics controlled by genes. They are, of course, the same sex. Usually, one is the mirror image of the other in appearance. They often show a marked likeness in temperament, mental and physical abilities, likes and dislikes, and many other personality traits, although environment may have a strong influence on certain of these characteristics.

Several studies have been made of identical twins who were separated early in life and reared in different environments. In these instances, home and family life, education, day-to-day experiences, friendships, and other environmental influences leave their mark on the personality of each. But the two persons remain amazingly alike in appearance, basic personality traits, and capacity for learning. While they may vary somewhat in intelligence as determined by scores on tests, these differences indicate the influence of different environments rather than changes in basic capacity for learning. Unfortunately, the opportunities for studying identical twins who have been separated are very limited.

11-4 Identical twins have the same genetic makeup, whereas fraternal twins are no more alike than ordinary brothers and sisters. What type of twins are these? (Alpha)

Genes in human populations—inheritance of blood type

Having discussed various methods used in the study of human genetics, we are now ready to consider a few hereditary traits in human beings and the mechanisms involved in their inheritance. One such inherited trait is your blood type.

In 1900, while working in a medical laboratory in Vienna, Dr. Karl Landsteiner made an important discovery in a study of human blood. He found that the mixing of blood from certain people resulted in the clumping, or agglutination, of red corpuscles. In further investigations, he found that the red corpuscles of various people differed in the presence of a protein substance on the corpuscle surface. This substance, which we know as an *agglutinogen,* is one of two types, designated as type A and type B. Some corpuscles have one or the other, while some have both and still others have neither. Thus, we can classify all human blood as type A, B, AB, or O.

In Chapter 42, when we study the composition of blood, we shall deal with the reaction that occurs when certain different blood types are mixed. At present, we are concerned with the inheritance of blood type. Three genes are involved in this inheritance. We call them **multiple alleles,** since more than a single pair of genes is involved in determining the characteristic. However, even though three genes are associated with blood type, only two are present in any single individual. A gene we can designate as *A* produces type-A corpuscles, gene *B* results in type-B corpuscles, and gene *O* does not produce either agglutinogen on the corpuscles. Various combinations of pairs of these three genes result in the four human blood types:

- Genes *AA* or *AO* produce type-A corpuscles.
- Genes *BB* or *BO* produce type-B corpuscles.
- Genes *OO* produce type-O corpuscles (no agglutinogen).
- Genes *AB* produce type -AB corpuscles (both agglutinogens).

In order to identify as alleles the three genes involved in corpuscle types, geneticists write the dominant as *I* and the recessive as *i* and designate the agglutinogens they produce as I^A, I^B, and *i*. Under this system, then, type A may be homozygous $I^A I^A$ or heterozygous $I^A i$. Type B may be homozygous $I^B I^B$ or heterozygous $I^B i$. Type AB is always heterozygous $I^A I^B$, while type O is always homozygous *ii*. It is easy to determine results in crossing individuals with different blood types with a grid such as you used for a monohybrid cross. A cross in which one parent is heterozygous type A ($I^A i$) and the other is heterozygous type B ($I^B i$) is shown at left.

Notice that all four blood types may result in a cross such as this. The ratio of probabilities indicates that one fourth of the offspring should be type A; one fourth, type B; one fourth, type AB; and one fourth, type O. In fact, such an occurrence in a family would be unusual, but it could happen.

RESULTS OF CROSSING $I^A i$ AND $I^B i$

Female GENES Male	I^B	i
I^A	$I^A I^B$	$I^A i$
i	$I^B i$	ii

While biologists have known the frequency with which blood types appear in our own population for many years, only in recent years have surveys been made to determine the distribution of blood types among populations of the entire world. Hence, we have known only for a short time how widely blood-type frequencies differ in world populations. Such variations occur both among racial groups and in populations in different geographic regions. Table 11-1 shows certain of these variations.

Table 11-1 BLOOD TYPE PERCENTAGES IN VARIOUS REGIONS OF THE WORLD

BLOOD TYPE	A	B	AB	O
U.S.A.—white	41.0%	10.0%	4.0%	45.0%
U.S.A.—Negro	26.0%	21.0%	3.7%	49.3%
Swedish	46.7%	10.3%	5.1%	37.9%
Japanese	38.4%	21.8%	8.6%	31.2%
Hawaiian	60.8%	2.2%	0.5%	36.5%
Chinese	25.0%	35.0%	10.0%	30.0%
Australian aborigine	44.7%	2.1%	0.0%	53.1%
North American Indian	7.7%	1.0%	0.0%	91.3%

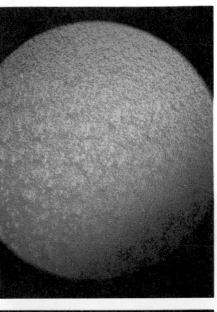

Inheritance of the Rh factor

Many years after the discovery of the A-B-O blood types, Landsteiner and Wiener, an associate, discovered another protein substance in the red corpuscles of a rhesus monkey. The protein was named the *Rh factor,* after the monkey. Later it was found that 85 to 87 percent of the human population in the city of New York had corpuscles containing the Rh factor. Such people are designated as *Rh positive;* people who lack the factor are *Rh negative.* (As you will learn in Chapter 43, serious complications result when Rh-positive blood is mixed with Rh-negative blood that has been sensitized against the factor. However, at this point in our discussion, we will concern ourselves only with the inheritance of the factor.)

At first, it was thought that the Rh factor was a simple trait, present in some people and lacking in others. It was also thought to be controlled by a single pair of alleles in which the gene for the factor was dominant. However, we now know that there are four or more Rh factors.

Sex-linked genes in human beings

Perhaps the best-known sex-linked character in human beings is *red-green color blindness.* While it rarely occurs in females, it appears in approximately 8 percent of the male population. People with this condition cannot easily distinguish red and green,

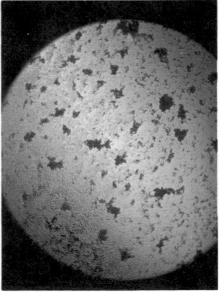

11-5 Rh positive blood corpuscles agglutinated by antibodies in sensitized Rh negative blood. (Lester V. Bergman and Associates)

INHERITANCE OF COLOR BLINDNESS

Female SEX CHROMOSOMES Male	X^C	X^c
X^C	$X^C X^C$	$X^C X^c$
Y	$X^C Y$	$X^c Y$

X   X
normal female

X   X′
carrier female

X′   X′
color-blind female

X   Y
normal male

X′   Y
color-blind male

11-6 The gene for color blindness is indicated in color, and the chromosome that carries it is designated as X′. Notice that two such chromosomes are necessary for a colorblind female while only one is necessary in case of a colorblind male.

both colors appearing as shades of gray. The fact that it appears more frequently in males than in females would indicate that it is sex-linked. The genes associated with red-green color vision are located on the X chromosome. Furthermore, the gene for normal color vision is dominant. Thus, we can indicate various gene combinations as follows, using C for normal red-green color vision and c for color blindness:

$X^C X^C$, a normal female, homozygous for color vision
$X^C X^c$, a carrier female, heterozygous but with normal vision
$X^c X^c$, a colorblind female, homozygous for color blindness
$X^C Y$, a normal male with a single gene for color vision
$X^c Y$, a colorblind male with a single gene for color blindness

Study the grid in the left column to see how a mother who is a carrier of color blindness but who has normal vision might have a son who is colorblind, even though the father has normal vision.

The results of this cross indicate that one fourth of the offspring would be normal females; one fourth, carrier females; one fourth, normal males; and one fourth, colorblind males. Perhaps you can determine a cross in which a colorblind female might result.

Hemophilia (HEE-muh-FILL-ee-uh), or "bleeder's disease," is another sex-linked characteristic similar to color blindness in inheritance. Hemophilia is a condition in which a blood substance necessary for clotting is not produced because of the lack of the gene necessary for its formation. Because of the lack of this clotting substance, victims of hemophilia bleed severely and may even die from loss of blood as a result of wounds that would be slight in a normal person.

Hemophilia tends to run in families and to appear in males. It can appear in a female only when the father has hemophilia and the mother is a carrier. Several genetic studies have been made of families in which hemophilia has occurred frequently. One of the most famous of these is the study of the family of Queen Victoria, shown in Figure 11-7. As you examine this family history, notice how many males were hemophiliacs. Then study the chart and note how many females were carriers.

Other characteristics associated with sex

Baldness is an example of a human *sex-influenced trait.* The gene for baldness is dominant in males but recessive in females. Thus, a mother can transmit baldness to her son without showing it herself. If we represent the gene for baldness as B and the one for normal growth of hair as b, the Bb male would be bald, while a Bb female would not be. However, BB would represent a male or a female with baldness, while bb would produce a normal male and female.

Another type of inheritance, which has been investigated in birds more thoroughly than in human beings, involves *sex-*

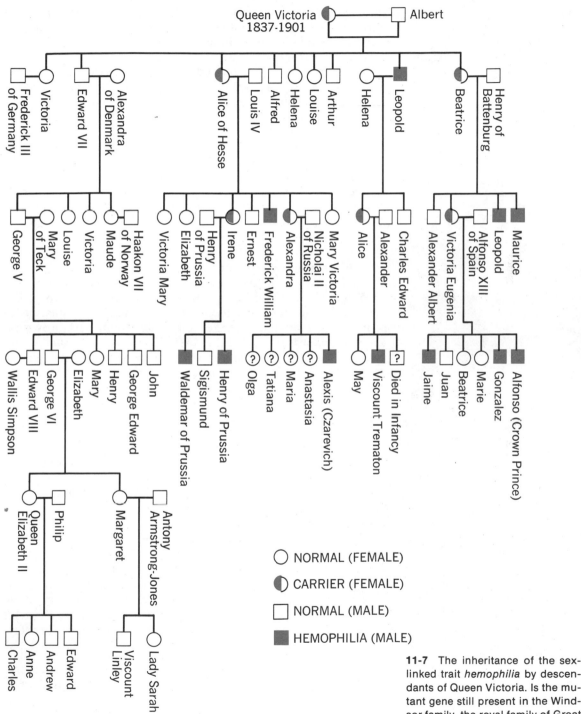

11-7 The inheritance of the sex-linked trait *hemophilia* by descendants of Queen Victoria. Is the mutant gene still present in the Windsor family, the royal family of Great Britain?

○ NORMAL (FEMALE)

◐ CARRIER (FEMALE)

☐ NORMAL (MALE)

■ HEMOPHILIA (MALE)

limited traits. In this type of inheritance, certain genes produce a trait in one sex or the other, but not in both, even though they are carried in both sexes. It appears that sex-limited traits appear only in the presence of sex hormones.

One example of a sex-limited character is the bright plumage of certain male birds, which does not appear in females of the same species. The roosters of most breeds of chickens develop a large comb and wattles and characteristic male plumage, while hens of the same breed develop a different kind of female plumage. Both result from the action of genes in the presence of sex hormones.

In human beings similar sex-limited genes can determine such characteristics as the growth of a beard. A son might inherit a beard characteristic from his mother, even though the characteristic was not expressed in the mother because of the lack of male hormones.

Inheritance of eye color and skin color

The inheritance of eye color is governed by multiple alleles. Blue eyes are produced by a single pair of recessive genes. The genes producing brown pigment are dominant. The various shades of iris pigmentation from hazel to light brown or dark brown are apparently the result of the expression of varying numbers of genes for brown pigmentation. Other genes for pigmentation produce shades of gray ranging to green. Loss of the eye-pigment genes resulting from mutations produces the pink eye of the albino.

Skin color is thought to be determined by several pairs of genes. Some geneticists suggest that two pairs of alleles are involved. Others believe that from four to perhaps as many as eight pairs of genes may function. In any case, variations in skin color—from dark brown to medium brown, light brown, and white—are the result of the expression of varying numbers of genes for skin pigmentation. Other pigment-producing genes are responsible for the characteristic skin color of oriental peoples and the reddish or bronze skin of many Indians.

Are diseases inherited?

The question of whether or not diseases are inherited is of great importance to the geneticist because of its medical aspect. Many geneticists have explored the possibility that certain diseases that appear to run in families may be related to genes. In infections such as tuberculosis, the disease itself cannot be inherited. However, evidence seems to indicate that chemical conditions in the body tissues important in resistance against infections may be inherited. Lack of this resistance would increase the possibility of contracting the disease at some time during one's life. However, proper precautions can prevent tuberculosis.

11-8 What sex-limited traits do you see expressed in this rooster (top) but not in the hen? (Grant Heilman)

On the other hand, diseases resulting from abnormal structure or function of the body organs are more likely to be hereditary. Such a disease is sugar diabetes *(diabetes mellitus),* in which certain cells of the pancreas fail to secrete sufficient insulin. (We shall discuss this condition more fully in Chapter 47.) Diabetes has been found to occur frequently in certain families. It is thought to be caused by a recessive gene. However, the disease is not equally serious in all people. Furthermore, regulation of the diet and body weight may prevent or arrest the condition even when the genes are present. There is also evidence that multiple factors may be involved in the inheritance of diabetes. If such is the case, the seriousness of the disease might be determined by the number of genes present.

Case-history studies of human families indicate that many other human diseases and disorders may be definitely associated with genes. Among these are respiratory allergies, asthma, bronchitis, nearsightedness, farsightedness, and night blindness.

Sickle-cell anemia

This serious blood disease is one of several hereditary anemias. It is most prevalent in the Negro population. Recent figures indicating the frequency of *sickle-cell anemia* among North American Negroes vary considerably. One survey indicates that $8\frac{1}{2}$ percent of the Negro population carry the trait and 0.3 to 1.3 percent have the disease. Sickle-cell anemia is even more widespread among natives of central and western Africa, where it is present in as much as 4 percent of the Negro population.

11-9 Electron micrograph of sickle-cell anemia. (Philips Electronic Instruments)

Sickle-cell anemia was first observed in 1910, when the blood of patients was found to contain abnormal sickle-shaped or wheat-shaped red corpuscles. However, it was not until 1949 that the corpuscles were found to contain an abnormal hemoglobin. The chemical structure of this abnormal hemoglobin was determined in 1957. In the protein chain containing more than 500 amino acids, one amino acid, *valine,* was substituted for *glutamic acid* in normal hemoglobin. This slight variation alters the chemical properties of the hemoglobin and causes the red corpuscles to change from the normal disk shape and become sickle-shaped when the blood circulates through tissues where the oxygen supply is low. The abnormal corpuscles disintegrate in the blood stream. Loss of red corpuscles causes the anemia.

Sickle-cell anemia usually appears during the latter months of the first year of life. The condition usually shortens life considerably by causing increased susceptibility to infections and cellular damage in various body organs.

The gene associated with sickle-cell anemia is thought to be recessive. Therefore, the disease appears only in homozygous individuals. Heterozygous individuals, with a normal allele, may transmit the disease to offspring but do not develop it.

Under proper medical treatment, the life of a person with sickle-cell anemia may be lengthened many years. For this reason, it is important that the disease be discovered early. Today, there are widespread testing programs for babies for early detection of the condition. The disease is also being investigated in extensive research programs.

Is intelligence inherited?

To what extent is intelligence related to genes? This is a very difficult question to answer, since no one has yet given an adequate definition of intelligence. We find many families in which parents and children seem to be highly intelligent. Is this a genetic quality, or is it the product of the home, the school, and other environmental influences?

One of the standard methods of measuring intelligence is a test that involves reasoning, memorizing, calculating, visualizing, word recognition, word usage, and other fundamental thought processes. The score on such an intelligence test is used to determine the *mental age.* Intelligence is then determined by dividing the mental age by the *chronological age,* or age in years. The result multiplied by 100 is the *intelligence quotient,* or I.Q. An average I.Q. is considered to be 100.

There is some evidence that intelligence is related to genes. Geneticists believe that many alleles must be involved. Identical twins provide the best evidence of such a genetic influence. The similarity between their intelligence-test results is striking. There is much less similarity in results between fraternal twins and still less among other brothers and sisters. The similarity continues to decrease as relatives become more distant. Thus, genes and gene combinations yet undiscovered must at least share with environmental influences in determining the level of intelligence.

Inheritance of mental disorders

Extensive genetic studies have supplied evidence that several kinds of mental retardation have a definite genetic basis. Among these disorders is an inherited metabolic disease known as *phenylketonuria* (FEN'L-KEET'N-yoohr-ee-uh), or *PKU*. This disorder is believed to result from the lack of a gene necessary to produce an enzyme that converts the amino acid phenylalanine to a similar amino acid, tyrosine. As a result of this chemical block, phenylalanine is converted to phenylpyruvic acid, which builds up in the body fluids and is excreted in the urine. The accumulation of phenylpyruvic acid in the body invariably causes a form of arrested mental development known as *phenylpyruvic idiocy.*

Studies of pedigrees of families in which phenylketonuria is present indicate that it is a simple recessive trait. However, while individuals who are heterozygous for the trait are apparently normal, they are unable to convert standard doses of phenylalanine to tyrosine as readily as those who are normal and lack the recessive gene entirely.

Clinical tests will reveal the presence of phenylpyruvic acid in the urine of a person with phenylketonuria. If the disease is detected early in life, preferably in the first few weeks, treatment can be given to prevent accumulation of phenylalanine in the body and its conversion to phenylpyruvic acid with resulting brain damage. This treatment consists of restriction of foods containing phenylalanine. Treatment does not alter the genetic defect; environmental control only prevents its tragic effects.

Geneticists have much to learn about the relation of genes to mental deficiencies. Is there some genetic lack in the more than three million mentally retarded people in the United States today? Are certain mental illnesses related to genes? Manic-depressive psychosis is thought to be hereditary, at least to some degree. It seems to be related to a dominant gene, although the presence of the gene may merely produce a tendency that can be prevented from developing by environmental influences. Similarly, a tendency to develop schizophrenia may be associated with genes.

Nondisjunction of human chromosomes

Various chromosomal abnormalities may result from nondisjunction during meiosis in humans much as they occurred in the vermilion-eyed fruit fly you studied in Chapter 10. This may involve various of the somatic chromosomes or the sex chromosomes. Nondisjunction may occur in either the first or the second meiotic division. Often, a chromosome fails to migrate from the equator to a pole during anaphase. This *anaphase lag* results in one daughter cell with an extra chromosome and the other lacking a chromosome.

An egg or a sperm with an extra chromosome involved in fertilization will produce a zygote with three chromosome homologues rather than a normal pair, a condition referred to as *trisomy.* On the other hand, an egg or a sperm lacking a chromosome involved in fertilization will produce a zygote with a single homologue. We call this condition *monosomy.*

Chromosome abnormalities may be detected by examining the nuclei of white blood corpuscles or cells scraped from the lining of the mouth. Since the abnormality was present in the zygote, or fertilized egg, it will be present in every cell of the body.

Down's syndrome

In recent years, scientists have found the genetic basis for *Down's syndrome,* or mongolism. This unfortunate condition results from an extra twenty-first chromosome in all of the body cells, resulting from nondisjunction during meiosis, probably in egg formation. Down's syndrome is often referred to as trisomy-

11-10 Down's syndrome, or mongolism. Almost all individuals with this disorder have forty-seven chromosomes instead of the normal forty-six. This results from the failure of the two chromosomes of pair 21 to separate during oogenesis. (D. H. Carr & M. L. Barr)

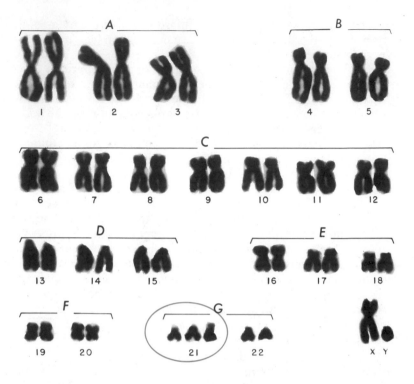

21 because of the presence of three twenty-first chromosomes. All other chromosomes are present in normal numbers, as shown in Figure 11-10.

A syndrome is a group of symptoms. Down's syndrome is marked by severe mental retardation and abnormal physical characteristics, including an enlarged tongue, slanted eyes, and muscle weakness. Palm and footprints are, also, abnormal.

About one in 600 babies are born with Down's syndrome. The number is lower, about one in 1,000, in mothers under 35 years of age, but increases to one in 60 babies in mothers over 45 years of age.

Nondisjunction in human sex chromosomes

Nondisjunction may also occur in human sex chromosomes. This produces a variety of diseases and disorders. Some result from trisomy, or the presence of three sex chromosomes. In monosomy, there is a single sex chromosome.

One such condition is *Turner's syndrome*. This condition, referred to as XO, or monosomy X, results from the absence of a sex chromosome. All body cells have only 45 chromosomes, as shown in Figure 11-11. Turner's syndrome results in a female who is abnormally short, does not develop mature sexual organs, does not menstruate and is, of course, sterile.

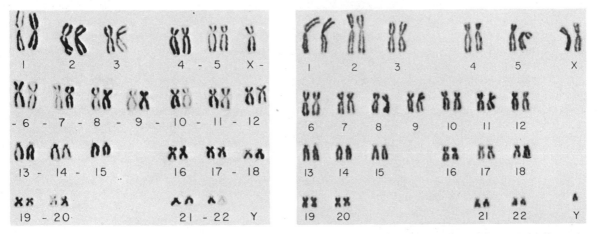

11-11 Chromosomes in Turner's syndrome (XO) above left. (Dr. Lillian Y. Hsu, Mount Sinai School of Medicine, N.Y.)

11-12 Chromosomes in Klinefelter's syndrome (XXY) above right. (Dr. Lillian Y. Hsu, Mount Sinai School of Medicine, N.Y.)

Trisomy X, designated as XXX, produces females with varying characteristics. Some are so-called "super females" with a tendency toward basic male traits. Many are fertile and appear normal but may be retarded mentally. Some females with trisomy X develop symptoms similar to those of Turner's syndrome.

Klinefelter's syndrome results from an XXY combination of sex chromosomes, as shown in Figure 11-12. It is believed that a normal egg bearing an X chromosome fertilized by an XY sperm formed by nondisjunction causes the condition. A person with Klinefelter's syndrome is a male who is sterile and underdeveloped sexually. He is often thin and has a high-pitched voice. In some cases, he is mentally retarded.

Another abnormal condition results when a normal X-bearing egg is fertilized by a YY sperm, formed by nondisjunction during spermatogenesis. This produces an XYY male who is usually over six feet in height and very aggressive. Studies have revealed that a high percentage of inmates for the criminally insane have this chromosome abnormality.

SUMMARY

In this chapter, we have presented but a few applications of the laws and principles of heredity to human genetics. The old saying "like father, like son" should have more meaning to you now. Or, perhaps in your case, it might be more appropriate to say, "like father, like daughter." Based on what you have learned about genetics, wouldn't it be equally true to say, "like mother, like son or daughter?" Your parents contributed equally to your heredity.

Medical genetics is a new and important branch of science. While your doctor can't change your genes, he may be able to help you live with them. Results in controlling many genetic problems are encouraging. They will be even more encouraging in the years ahead.

BIOLOGICALLY SPEAKING

population genetics	red-green color blindness	anaphase lag
gene frequency	hemophilia	trisomy
gene pool	sex-influenced trait	monosomy
fraternal twins	sex-limited trait	Down's syndrome
identical twins	sickle-cell anemia	Turner's syndrome
multiple alleles	intelligence quotient	trisomy X
Rh factor	phenylketonuria	Klinefelter's syndrome

QUESTIONS FOR REVIEW

1. List several physical characteristics of your own that you believe to be produced by gene action.
2. Generally, what is population genetics?
3. How can PTC paper be used as a basis for sampling a population?
4. What is a gene pool?
5. When is a gene pool considered stable?
6. Distinguish between fraternal twins and identical twins.
7. Explain how A and B red-corpuscle agglutinogens establish four basic human blood groups.
8. What are multiple alleles?
9. In what way does the inheritance of blood type involve multiple alleles?
10. Generally, what are the Rh factors?
11. Why is color blindness more frequent in males than in females?
12. In what respect might we consider a gene for hemophilia a lethal gene?
13. Distinguish between sex-influenced and sex-limited traits, giving an example of each.
14. Explain how the inheritance of eye color involves multiple alleles.
15. Give several examples of a possible genetic relation to human diseases.
16. Explain how an intelligence quotient is determined.
17. What specific biochemical deficiency is believed to cause phenylpyruvic idiocy?
18. What chromosome abnormality is believed to be the cause of Down's syndrome (mongolism)?
19. Distinguish between monosomy and trisomy in sex chromosomes.
20. Name four conditions resulting from nondisjunction in the formation of sex chromosomes and show the chromosome make-up of each.

APPLYING PRINCIPLES AND CONCEPTS

1. Discuss environmental influences as complicating factors in the study of human genetics.
2. Discuss the limitations of genetic studies in individual families.
3. How is random sampling applied to studies in population genetics?
4. Discuss various reasons for the shifting of genes in a gene pool.
5. How can identical twins supply valuable data in the study of human genetics?
6. Discuss various problems in establishing proof of gene action in producing intelligence.

7. How does phenylketonuria bear out the "one gene—one enzyme" hypothesis?
8. Determine various egg and sperm chromosome abnormalities that might produce the following: Turner's syndrome, trisomy X, Klinefelter's syndrome, and the XYY condition.

RELATED READING
Books
McKusick, Victor A., *Human Genetics.*
Prentice-Hall, Inc., Englewood Cliffs, N.J. 1964. One of the few authoritative books available to the layman, written in a readable style and incorporating the latest findings.
Papazian, Haig P., *Modern Genetics.*
W. W. Norton and Co., Inc., New York. 1967. Updated and authoritative, a book that will give the better-than-average students a rare insight into genetics.
Pilkington, Roger, *Human Sex and Heredity.*
Franklin Watts, Inc., New York. 1963. An easy to read book telling who's who and why.
Reed, Sheldon C., *Parenthood and Heredity.*
John Wiley and Sons, Inc., New York. 1964. A simplified version of medical genetics that should prove useful and interesting.
Scheinfeld, Amram, *Your Heredity and Environment.*
J. B. Lippincott Co., Philadelphia. 1964. Explains how heredity works to produce a unique individual, and the effect of the environment on the person.
Articles
McKusick, Victor A., "The Royal Hemophilia," *Scientific American,* August, 1965. An account of the gene for hemophilia and how it is found in the royal families of Europe.

CHAPTER TWELVE

APPLIED GENETICS

Plant and animal breeding—a time-honored practice

Selective breeding of plants and animals is an old practice. For centuries, man has made a constant effort to improve the varieties of plants and animals that supply his daily needs. Wheat was grown as a cereal crop by the early Egyptians. Flowers, fruit trees, fowl, sheep, goats, and cattle have been bred for domestication since before recorded history.

The science of genetics originated largely to explain the results of plant and animal breeding. Breeding was a practice of chance selection rather than of scientific application of principles. With the development of genetics as a science, established laws have greatly improved the efficiency of the process. We must remember, however, that the breeding of a plant or animal involves the crossing of not one or two but hundreds of different characteristics. While the laws of heredity apply, the results are not always as predictable as they were for Mendel's seven traits in peas.

Luther Burbank, the genius of plant breeding

Plant breeding will always be associated with the genius of Luther Burbank. He produced many new and different plants on his farm in California.

Burbank's brilliant work began in the summer of 1871 in his native Massachusetts, when he was a young man. He was examining a crop of potatoes one day and happened to notice a fruit maturing on one of the plants. This was an unusual occurrence, because the potato plant flowers regularly but seldom bears fruit; new plants are grown from potatoes rather than from seeds. Burbank saved the fruit. When the seeds ripened, he planted each one in a separate hill. After the plants matured, he dug up the potatoes and discovered that those produced by each plant were different. Some were large; some were small. Some bore many tubers, while others had only a few. One plant yielded better potatoes than any of the others. These were large, smooth, and

12-1 Luther Burbank, the genius of plant breeding. He is probably holding one of the many root crops that his work did so much to improve. (The Bettmann Archive)

numerous. Burbank sold them to a gardener for $150—his first profit from plant breeding. They were named Burbank potatoes in his honor and were the first of a strain that was destined to become popular all over the country. With the profit from his first achievement, Burbank bought a ticket to California, where he established the farm that made him famous.

Objectives in plant breeding

The plant breeder has several purposes in producing new strains or varieties of plants. One of the chief objectives is the production of *more desirable varieties*. Such characteristics as large fruit, large and abundant seeds, vigorous growth, early maturation of fruit, large leaf area in leafy vegetables, and vigorous root growth in root crops are highly profitable. Plant breeders work constantly to improve the quality and quantity of the yield of all crop plants. In addition to the nature of the yield, resistance to disease is highly important. Plant breeders also have been able to produce many varieties of *disease-resistant crops*.

A third objective is an *extension of crop areas* through the production of new varieties. Wheat is an example of this extension through plant breeding. Varieties of spring wheat grow well in the northern sections of the nation, while winter wheat favors the climate of the central states. Wheat growing has been extended even to the Great Plains by the production of varieties of hard wheat. In a similar way, other crops that were once limited to small areas because of climatic requirements or soil conditions can now be grown in many other regions as new varieties.

Plant breeders have even been able to develop entirely *different kinds of plants*. The new plant may be the result of crossing two strains or varieties of two closely related species.

Obtaining desirable varieties by mass selection

As the name implies, *mass selection* consists of the careful selection of parent plants from a great number of individuals. Burbank practiced mass selection when he discovered his famous potato. From all those he grew from seed, he selected the one plant that had the characteristics he considered ideal. Farmers who use seed from their own crops always pick for propagation those from the most desirable plants. Thus, they take advantage of any natural, desirable variations that occur.

Mass selection is important, too, in the production of disease-resistant strains of plants. To show how mass selection operates, let us assume that a cabbage disease has swept into an area, resulting in the destruction of almost the entire crop. As we examine the acres of diseased plants, we find two or three plants that because of some unknown variation have withstood the disease. Seeds from these plants are grown the following season. The same

12-2 Hybridization. The mountain laurel (top) has decorative pink flowers, but grows too tall to be wholly acceptable as an ornamental plant. In contrast, the sandhill, or hairy, laurel (center) has smaller leaves and a lower growth habit— and somewhat less attractive, white flowers. Some of the hybrid offspring of a cross between the two—such as the one shown at the bottom—combine the desirable traits of both. (courtesy Connecticut Agricultural Experiment Station)

disease again strikes the crops, but a few more plants survive than did the year before. These plants have inherited the favorable genetic trait from the parent. Again, seeds from the surviving plants are grown the following season. Each year, more and more plants withstand the disease. The genetic trait for disease resistance, present in the original plants, has become more common in the offspring as they have been selected and allowed to breed. Finally, an entire strain is developed in which this trait exists.

What is outbreeding?

We define *outbreeding* as the crossing of two different varieties to obtain a new one. This is also called *hybridization*. In hybridization, characteristics of two unlike but closely related parents are combined in a new individual by means of artificial crossbreeding. In getting a new hybrid strain, we might choose one parent because of vigorous growth. The other one might be selected because of the fine quality of its fruit or flower (Figure 12-2). Often, because of a new combination of genes, a hybrid possesses qualities not shown in either parent. This *hybrid vigor* may include increased size, fruitfulness, speed of development, and resistance to disease, insect pests, and climatic extremes.

Inbreeding—the opposite of hybridization

After the desired characteristics have been obtained by hybridization and selection, the next step is to propagate these new and different plants. This is a simple matter when vegetative propagation is involved, because such asexual reproduction does not involve the recombination of genes.

After Burbank had selected his potato, propagation was simple. He used pieces containing two or three "eyes" cut from the potato in order to produce more plants exactly like the parent. Had he been forced to grow more potatoes from seed, the situation would have been quite different, because chromosomes from two parents would have been involved. In the same way, the grower can propagate a new variety of apple, peach, or rose by grafting, cutting, or budding without altering the hereditary makeup.

In plants that are propagated by seeds, like corn and wheat, the problem is more difficult. Seed production involves sexual reproduction, in which there is a mixing of numerous traits. Plants produced from seeds are not necessarily like the parent, especially when they are crop plants that have been crossed by man for centuries.

This difficulty can be overcome by generations of *line breeding,* or *inbreeding.* This is the opposite of hybridization, or outbreeding. Self-pollination is carried on to avoid introducing any new characteristics from a new plant. Seeds resulting from self-pollination are planted, and all individuals of the new generation are carefully sorted. Only those with the desired characteristics of the parent are selected as seed plants for the next generation. Again, self-pollination is carried on, after which the resulting plants are sorted with the greatest possible care.

As you repeat this breeding method generation after generation, more and more plants bear the desired characteristics. Eventually, a pure strain is established and is ready for the market. Even then, it is possible that not all plants will produce the pure-strain characteristics, but these individuals can readily be sorted out.

The production of hybrid corn

Years ago, a farmer saved some of the best ears from his corn crop as seed for the next year. By the process of mass selection, he tried to produce more corn like his best plants of the previous season. But the plants that bore these ears were so mixed in their heredity that only some of the kernels bore the genes that had made them productive. And with no control over pollination, the farmer had no idea about the quality of the other parent. The seeds on a single ear might produce many different varieties of corn, some good and some poor. Some of the kernels might have resulted from self-pollination, while others might have been the result of cross-pollination from fields some distance away. It was not unusual to find ears of corn with a mixture of yellow, white, and red kernels. Often, sweet corn and popcorn were mixed with kernels of field corn. Under conditions like these, a yield of twenty to forty bushels per acre was all that could be expected.

Today, hybrid varieties are produced by scientifically controlled artificial pollination (Figure 12-3). When two varieties of corn are cross-pollinated, the offspring are usually more vigorous than either parent. One of the reasons for this seems to be that the desirable dominant characteristics from both parents mask many of the undesirable recessives. The hybrids produced today yield from 100 to 180 bushels of corn per acre. These hybrids have large root systems, sturdy stalks, broad leaves, and large ears. Today, the need is for full, long ears.

Much of hybrid corn planted today is the result of a double-cross in which four pure-line parents are mixed in two crosses. Each pure-line parent has been selected because of its vigor, resistance to disease, or some other desirable trait. Plants resulting from the double-cross, however, are superior to any of the parent strains.

Figure 12-4 shows how hybrid corn is produced by the double-cross method. Four inbred plants, designated as A, B, C, and D, serve as the foundation. These varieties are the result of controlled self-pollination, or line breeding. This is accomplished by covering the developing ears with sacks until the silks are

12-3 Controlled pollination of corn in a research plot. To produce the necessary inbred lines required for hybridizing, care must be taken to insure that natural pollination does not occur. (courtesy W. R. Grace & Co.)

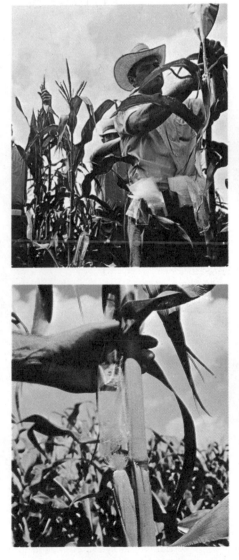

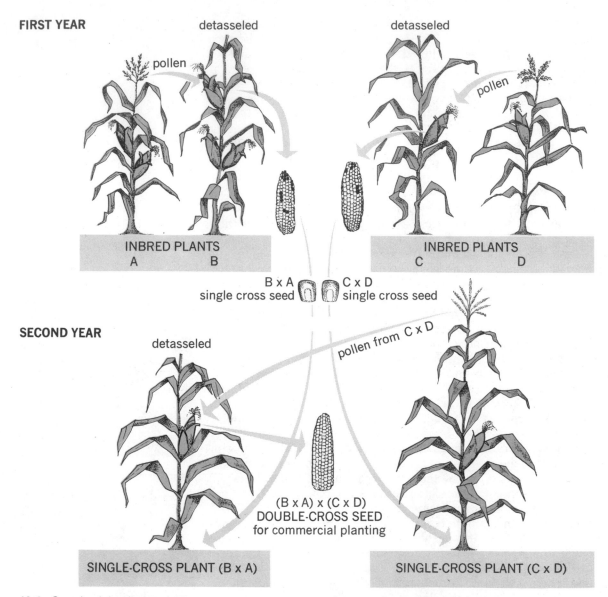

FIRST YEAR

pollen

detasseled

detasseled

pollen

**INBRED PLANTS
A   B**

**INBRED PLANTS
C   D**

B x A
single cross seed

C x D
single cross seed

SECOND YEAR

detasseled

pollen from C x D

(B x A) x (C x D)
DOUBLE-CROSS SEED
for commercial planting

SINGLE-CROSS PLANT (B x A)

SINGLE-CROSS PLANT (C x D)

12-4 Crossing inbred corn plants, then crossing the resulting single crosses to produce double-cross hybrid seed. The four plants—A, B, C, and D—represent the products of four different inbred lines. Strain A is crossed with strain B. Strains C and D are crossed in a similar way. The F_1 generations of these two lines are then crossed to produce the hybrid corn seed that is used by farmers today.

ready to receive pollen. Then pollen collected from the same plant is dusted onto the silks. The plant breeder carefully avoids any contamination of these inbred varieties. During the first cross, plant A is crossed with plant B, which produces a single-cross hybrid, plant AB. In making this cross, the tassel from plant B is removed, and the ear is covered with a bag. When the silks are mature, they are dusted with pollen from plant A. A similar first cross is made between plant C and plant D, resulting in a single-cross hybrid, plant CD. The following season, plant AB is detasseled. The developing ear is covered. This plant is cross-pollinated with plant CD. Kernels from this double-cross are designated as ABCD.

This seed is sold to the farmer for planting. However, he cannot plant the seed produced by this hybrid corn because the genes will sort out in new combinations in the next generation. He might get plants that would have characteristics different from those of the hybrid parents.

Producing new plants by crossing different species

One of the best known of Burbank's plant varieties is the Shasta daisy. He produced this beautiful garden flower by crossing three varieties of daisies—a native oxeye daisy, a European variety, and an Asiatic variety. In a cross between a plum and an apricot, he produced the plumcot. Another hybrid plant was produced by crossing the squash and the pumpkin. He also produced a pitless plum, an improved peach plum, and a thin-shelled walnut. The spineless cactus, used as fodder for cattle, was another of his achievements in plant breeding. One of Burbank's last experiments was an attempt to cross a tomato and a potato to produce a dual-purpose plant that would bear fruit above ground all season and form tubers that could be dug at the end of the season. Unfortunately, such a cross was never perfected.

Plant varieties resulting from mutations

While examining a bed of white tea roses one day, a grower happened to notice a branch that had produced a pink flower. He carefully removed the branch and set it in a cutting bed. The plant that grew from the branch bore all pink flowers. These were budded onto rooted stems and propagated as a *budmutant,* or **sport,** of the white rose. Such somatic gene mutations occur from time to time in plants. In the tea rose, a color gene had mutated in a cell of the stem. All tissues of the branch that developed possessed the mutant trait.

Other varieties that have resulted from a bud mutation include the California navel orange, the Delicious apple, seedless grape, pink grapefruit, and the smooth-skinned peach, or nectarine.

Plant varieties with chromosomal aberrations

Have you ever noticed blueberries on a fruit counter that are twice as large as native blueberries? These blueberries have four sets of chromosomes. This condition, in which there is more than twice the haploid number of chromosomes, is called *polyploidy* (PAHL-i-ploy-dee). Other fruits that have been produced by plants with multiple chromosome numbers include varieties of plums, cherries, grapes, strawberries, and cranberries. A similar increase in the chromosome number occurred in the McIntosh apple. Normal McIntosh apple trees usually have seventeen pairs of, or thirty-four, chromosomes. However, one variety with a fruit more than twice as large as the normal McIntosh has four

Burbanks hybrids

daisies
plumcot
squash + pumkin
pitless plum
thin shelled walnut

sets of chromosomes (the *tetraploid* number 4n), or sixty-eight in all.

A polyploid with an even number of chromosome pairs (4n, 6n, 8n, etc.) is usually fertile. However, the formation of gametes is interfered with in polyploids having an odd number of chromosome pairs (3n, 5n, 7n, and so on). The lack of fertility is not a handicap in obtaining new specimens with the same genetic makeup. The plant breeder can produce these with cuttings and grafting.

Scientists have discovered that polyploidy can be induced artificially through the use of the drug *colchicine* (KAHL-chuh-SEEN). When shoots of plants are put in weak solutions of colchicine, the chromatids are unable to separate during mitosis. This often results in the doubling of the chromosome number. The drug is also available in salve form for use on buds of plants. Blueberries, lilies, and cabbage are a few plants that have been improved through its use. Even interspecies can be produced by using colchicine. A cross between the ordinary cabbage, with twenty chromosomes, and the Chinese cabbage, with eighteen chromosomes, yields a new type with only nineteen chromosomes. This type is sterile. After plant breeders treated it with colchicine, it had thirty-eight chromosomes, but it continued to be sterile. But when plant breeders crossed this plant with a rutabaga having 38 chromosomes, a new type of cabbage was produced. It also had thirty-eight chromosomes, but it was fertile and bred true.

Many years ago, almost all garden iris plants were diploid, while most of the modern ones are tetraploid. The large and beautiful varieties of roses include almost every multiple of chromosomes up to 8n.

Animal breeding

The principles used in plant breeding apply to animal breeding as well. Mass selection has long been a method of producing highly desirable breeds of animals.

The results of years of selective breeding are well illustrated in the modern breeds of poultry. The Leghorn, for example, has been bred for its ability to lay large numbers of eggs. All its energies are directed toward egg production rather than toward the production of body flesh. The Plymouth Rock has been developed as a dual-purpose fowl and is ideal for egg production and meat. Large breeds, like the Brahma, Cochin, and Cornish, are famous for their delicious meat rather than for egg production.

The modern turkey, with massive body and broad breast covered with thick layers of white meat, is quite a contrast to the slender bird the Pilgrims found in the New England forest. The modern turkey has been bred to produce the highest possible amount of meat. It spends its life eating a scientifically prepared diet and building up large, little-used muscles that better suit them to being eaten than to flying and perching high in trees.

Improvement in livestock

Using similar selective breeding methods, domestic cattle have been developed along two entirely different lines. Aberdeen Angus, Hereford, and Shorthorn are breeds of beef cattle. Their low, broad, stocky bodies provide high-quality steaks and roasts for the nation's markets. Dairy breeds, including the Jersey, Guernsey, Ayrshire, Holstein-Friesian, and Brown Swiss have been bred as milk producers. A breed of Shorthorns known as Milking Shorthorns, as well as Red Poll cattle, are classified as dual-purpose breeds because they were developed for milk production as well as beef.

Swine raising is one of the most important divisions of American agriculture, especially in the Corn Belt. Heavy, or lard, type breeds include the Poland China and Berkshire, Hampshire, and Duroc-Jersey. The Yorkshire and Tamworth hogs have long, slender bodies and are classified as lean, or bacon, type hogs.

In livestock breeding, the records of outstanding individuals used in breeding are kept in pedigree and registration papers. Purebred animals can be registered at the headquarters of their respective breeds. Papers must include the names and registration numbers of both sire (male) and dam (female), as well as facts about part of the ancestry. With the use of information provided in these papers, it is possible to cross different strains of the same breed without the danger of introducing undesirable characteristics or losing any good qualities.

Hybrid animals

The mule is an animal that has resulted from the crossing of two entirely different species. This hardy, useful animal is produced by crossing a female horse with a male donkey. The size is inherited from the horse. From the donkey the mule inherits long ears, surefootedness, great endurance, and the ability to live on rough food and to endure hardships. However, with all its hybrid vigor, the mule is usually sterile—that is, it is unable to reproduce.

Several kinds of hybrid cattle have been produced by crossing domestic beef breeds with Brahman cattle. The Brahmans are native to India, where they are known as zebus. The Brahman is a large animal with long, drooping ears and a shoulder hump of fat and cartilage. Colors range from gray to brownish and red. Brahmans have a heritage of resistance to heat, insect attack, and disease. They can endure the humid climate of the Gulf States and the dry summer heat of the Southwest much better than domestic breeds of beef cattle. It was with the idea of crossing Brahman cattle with domestic breeds that Brahmans were first imported from India in 1849 and have been imported in increasing numbers during the past fifty years.

One of the most successful Brahman crosses was conducted at the King Ranch in southern Texas. The original cross was a

12-5 Two breeds of cattle: an example of a milk breed-a Holstein-Friesian cow (top); and an example of a prime beef breed—a Hereford steer. (USDA; American Hereford Association)

12-6 Hybrid animals—Brangus, a result of a cross between a Brahman bull and an Aberdeen-Angus cow. (USDA)

Brahman bull and a Shorthorn cow. Over a period of thirty years, this Brahman-Shorthorn cross was perfected, and a new breed of beef cattle, known as Santa Gertrudis, was established. This breed is now stabilized and carries three eighths Brahman and five eights Shorthorn ancestry. Santa Gertrudis cattle possess the Brahman's skin resistance to hot climate and the Shorthorn's beef qualities.

Brahmans have also been crossed with other breeds of beef cattle. A well-established Brahman-Aberdeen Angus cross is known as the Brangus. Another cross, between Brahman and Hereford cattle, produced the Braford.

Domestic cattle have also been crossed with the bison, or North American buffalo. In this cross, the bison bull is usually mated with the domestic cow. The hybrid offspring is known as a cattalo. The purpose of the cross is to combine the bison's stamina and endurance to weather extremes with the beef quality of the domestic breed. However, this cross has not proved as successful economically as Brahman crosses.

SUMMARY

Plant and animal breeding have come a long way since Luther Burbank made his chance discovery of a new variety of potato a little over 100 years ago. Since that time, genetics has developed as a science and has added a new dimension to plant and animal breeding. The old method of mass selection is still used in choosing the most desirable parent stock for breeding, but varieties and breeds of plants and animals have been improved much more rapidly by applying laws and principles of heredity.

Plant and animal breeding, today, are scientific businesses. Modern technology, together with new and vastly improved strains of agricultural plants and breeds of livestock, has made America the best fed nation on earth. New varieties of hybrid roses, lilies, tulips, daffodils, iris, delphiniums, chrysanthemums, and other garden flowers are produced by growers year after year. New varieties of trees and shrubs add beauty to our yards and parks. We are improving our dogs and cats and even our tropical fish by applying principles of genetics to breeding methods.

BIOLOGICALLY SPEAKING

mass selection	hybrid vigor	polyploidy
outbreeding	inbreeding	colchicine
hybridization	sport	

QUESTIONS FOR REVIEW

1. What are four objectives of plant breeding?
2. Why is inbreeding practiced in plant and animal breeding?
3. Make a comparison of the methods used and the purposes of hybridization and inbreeding.
4. How many pure-line parents are involved in the production of hybrid seed corn by the double-cross method?
5. Name several hybrid plants produced by the genius of the late Luther Burbank.
6. How is natural cross-pollination prevented during the growing of hybrid corn?
7. Name three general types of chickens and a breed representing each type.
8. Name a dual-purpose breed of cattle.
9. In what respect is the mule a true hybrid animal?
10. Why are Brahman cattle good parent stock for breeding purposes?

APPLYING PRINCIPLES AND CONCEPTS

1. If inbreeding is practiced too long, offspring may become weak and inbred. How could this condition be remedied?
2. A farmer does not use seed from his hybrid corn for the next year's crop. Explain why.
3. Outline the method by which poultry breeders have been able to increase egg production by developing three-hundred-egg strains of chickens.
4. What is the importance of pedigrees and registration papers in breeding livestock?

RELATED READING

Books

Burt, Olive W., *Luther Burbank—Boy Wizard.*
The Bobbs-Merrill Co., Inc., New York and Indianapolis. 1962. An easily read book telling about the shy, obscure American boy who eventually became one of the world's great naturalists.

Kraft, Kenneth, and Patricia Kraft, *Luther Burbank: The Wizard and the Man.*
Meredith Press (The Meredith Publishing Co.), Des Moines, Iowa. 1967. A biography of the great experimenter in horticulture.

Articles

Mangelsdorf, Paul C., "Hybrid Corn," *Scientific American,* August, 1951.
Vigorous new crosses of the ancient cultivated plant have transformed agriculture in the Middle West, founded an industry, and taught man a lesson in the efficient utilization of natural processes.

CHAPTER THIRTEEN

ORGANIC VARIATION

Our changing earth

Scientists estimate that the earth is more than five billion years old and that life has been present on it for about two billion years. It is difficult for us to comprehend such long periods of time and to imagine the many changes that must have occurred through the ages.

Has the earth always looked much as we see it today? Have the same kinds of organisms populated its lands and seas? Scientists know that the face of the earth is changing constantly. Through the ages, land masses have risen and settled below the seas several times. Mountains have pushed up, only to be leveled gradually by winds and rain. Rivers have deepened their channels as they carried more and more land to the sea. Climatic conditions have also changed many times. There have been warm periods and cold periods, periods of heavy rain and periods of drought in all regions of the earth. Changes are still taking place, but they happen so slowly that we are not aware of them.

A geological timetable

Through the ages, nature has left records of geological time and change. The layers, or strata, of *sedimentary rock* provide one such record (Figure 13-1). Rock of this type is formed by sediment that settles to the bottom of oceans, seas, lakes, and other bodies of water. Here it is cemented together by chemicals in the water and the pressure of overlying sediment to form rock. New layers settle above older layers, leaving a timetable in rock. The exploration of layers of sedimentary rock from more recent strata near the surface to more ancient strata deep in the earth is a journey through millions of years of geological time.

As ocean bottoms rise and become land masses, these new land masses begin to undergo erosion, and gaps appear in the strata. These gaps are the basis for division of the geological time periods into *eras,* each of which endured for many millions of

years. Eras, in turn, are divided into shorter periods called *periods* and *epochs*. These divisions of geological time are marked by great climatic changes that are known to have occurred.

How are scientists able to determine how long ago these eras, periods, and epochs began? One method is to use the rate of sedimentation. By knowing the rate at which sedimentation occurs today, scientists can make a reasonably accurate estimate of the length of time that is necessary for the formation of rock strata. In a similar way, they can estimate the *relative* duration of geological eras of the past. However, modern science has provided a means of determining the *absolute* duration of geological eras. Strata of sedimentary rock contain elements that undergo radioactive decay and thereby change into other elements. One such element is uranium. A given sample of uranium will change entirely to lead in 4.5 billion years. Thus, if uranium and lead are present in a stratum of sedimentary rock, the proportion of uranium to lead gives an accurate measure of the age of the rock. Interestingly, dating based on recent investigations into rates of sedimentation agrees closely with dating of rock by measurement of the decay of radioactive elements. Thus, the duration of eras, periods, and epochs listed in the geological timetable on page 202 can be considered as relatively accurate.

13-1 Sedimentary rock in the Grand Canyon. In some places in the world, sedimentary rock is as much as ten miles thick. It took many millions of years for such layers to form. (Union Pacific Railroad color photo)

A timetable of life

The geological timetable shown in Figure 13-4 is of great biological significance. To what extent has the living population of the

13-2 A fossil brittle starfish from the Devonian period in a piece of slate. The substance of the starfish was gradually replaced by *pyrite,* which is commonly known as *fool's gold.* (courtesy Ward's)

earth changed during the past two billion years? Physical and climatic changes through the ages must have altered the life on the earth. Strata of rock not only date the geological eras but also contain records of organisms that lived at the time the strata were forming. These records are *fossils.* A fossil may be an imprint in rock, or it may possibly be an entire organism or part of an organism that has been preserved in rock as mineral matter gradually replaced the organic materials composing the plant or animal (Figure 13-2).

In addition, records of the life of past ages, especially of insects, are found in *amber,* a fossil plant resin. The La Brea Tar Pits near Los Angeles have yielded the fossil skeletons and other remains of numerous animals of past ages, including extinct antelope, bison, bears, and saber-toothed cats. These tar pits mark the location of a petroleum spring that formed many thousands of years ago. As the oils in the spring evaporated, a mass of sticky tar and, later, one of thick asphalt was left. When this happened, small rain pools accumulated on the surface. It seems that many mammals and birds were attracted to these pools and became trapped in the sticky tar and asphalt. Large predators attracted by the helpless victims soon found themselves trapped and sinking in the tarry ooze. More complete remains of ancient animals, including the mastodon, now extinct but related to the modern elephant, have been found with flesh intact in the "deep-freeze" of glacial ice.

Several major groups of organisms and the geological times in which they lived are shown in Figure 13-4. Certain of these groups, including bacteria, algae, fungi, and protozoans, are considered primitive organisms. These organisms were present during the most ancient geological eras and periods. More complex forms of life appear in more recent geological times. The biologist interprets this as evidence of change. We refer to the process by which organisms change and new forms appear, while old ones decline or become extinct, as *evolution.* Fossils have provided much of the best evidence to support this concept.

Organic variation and change in living populations

Scientists do not doubt that organisms living today descended from species of previous ages. If you study such modern organisms and then compare them with organisms of ancient geological eras, you can make an interesting observation: Certain forms of life seem to have survived through millions of years with little change. Many others, however, appear to be much more recent in origin and were not represented in earlier populations. These are products of *organic variations.*

In your study of genetics, you found that variations occur constantly. While offspring usually resemble their parents in most respects, they also differ from them slightly. Through the mech-

13-3 Rancho La Brea Tar Pits. (American Museum of Natural History)

anism of organic variation, life is continually striking out in new directions. The changes that species undergo during their descent from their ancestors is the basis of the evolutionary concept. When we examine the organisms living today, we are seeing the products of evolution in different directions. They may have descended from common or similar ancestors of ages ago. The variations that have occurred make them different, yet they have retained many sets of genes that were present in their early ancestors. These characteristics make them similar and indicate a relationship to common ancestors.

Evidences of common ancestry

In the study of classification in Chapter 14, you will find that all organisms fall naturally into groups and subgroups on the basis of their similarities and differences. In establishing any system of classification, as you proceed from large groups to smaller subgroups, the relationship of the members becomes closer and closer. That there are gaps between the smaller subgroups of established systems of classification based on similarities leads scientists to believe that vast numbers of forms intermediate between organisms living today and their remote ancestors have become extinct. If all of these intermediate forms had survived, there would be a gradual transition from common ancestors to every form of life we find today. The paths of evolution would be distinct, but there would be no lines to draw between structural differences. In classification, we follow each slender twig of the tree of life, each of which represents individual organisms living today, to the larger limbs, which indicate common ancestry.

Geological Time Table

Era	Period	Epoch	Years since beginning	Years duration	Conditions and characteristics
Cenozoic	Quaternary	Recent	15,000		moderating climate; glaciers receding
		Pleistocene	1,000,000	1,000,000	warm and cold climates; periodic glaciers
	Tertiary	Pliocene	10,000,000	9,000,000	cold climate; snow building up
		Miocene	25,000,000	15,000,000	temperate climate
		Oligocene	35,000,000	10,000,000	warm climate
		Eocene	55,000,000	20,000,000	very warm climate
		Paleocene	70,000,000	15,000,000	very warm climate
Mesozoic	Cretaceous		120,000,000	50,000,000	warm climate; great swamps dry out; Rocky Mtns. rise
	Jurassic		150,000,000	30,000,000	warm climate; extensive lowlands and continental seas
	Triassic		180,000,000	30,000,000	warm, dry climate; extensive deserts
Paleozoic	Permian		240,000,000	60,000,000	variable climate; increased dryness; mountains rising
	Pennsylvanian (Late Carboniferous)		270,000,000	30,000,000	warm, humid climate; extensive swamps; coal age
	Mississippian (Early Carboniferous)		300,000,000	30,000,000	warm, humid climate; shallow inland seas; early coal age
	Devonian		350,000,000	50,000,000	land rises; shallow seas and marshes; some arid regions
	Silurian		380,000,000	30,000,000	mild climate; great inland seas
	Ordovician		440,000,000	60,000,000	mild climate; warm in Arctic; most land submerged
	Cambrian		500,000,000	60,000,000	mild climate; extensive lowlands and inland seas
(Pre-Cambrian) Proterozoic			1,500,000,000	1,000,000,000	conditions uncertain; first glaciers; first life
(Pre-Cambrian) Archeozoic			3,500,000,000	2,000,000,000	conditions uncertain; earth probably lifeless

13-4 The Earth through the ages. This geological timetable shows the sequence and estimated length of the eras, periods, and epochs.

Time Scale of Life Based On Fossil Records

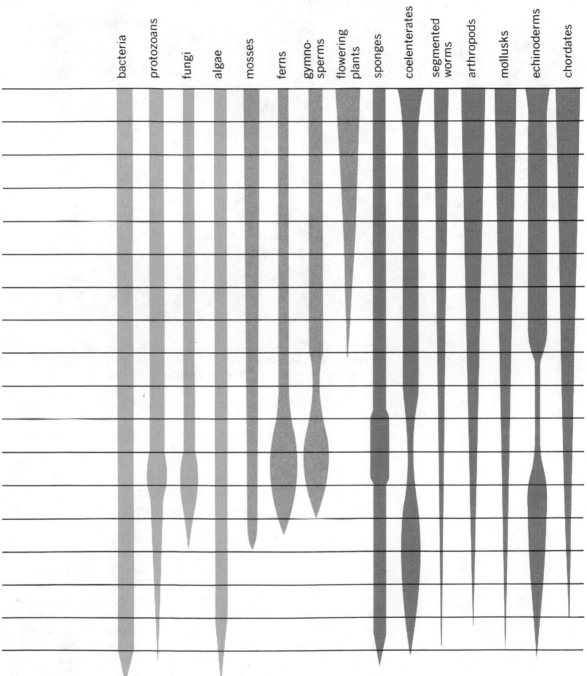

bacteria | protozoans | fungi | algae | mosses | ferns | gymno-sperms | flowering plants | sponges | coelenterates | segmented worms | arthropods | mollusks | echinoderms | chordates

The time scale of major groups of organisms shows the approximate time each group is believed to have originated and, to a limited extent, relative increases and decreases in numbers that have occurred.

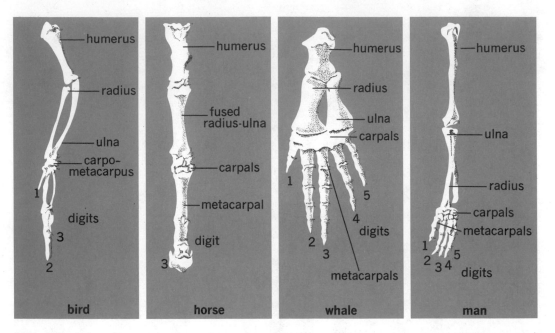

13-5 Homologous bones in the forelimbs of several vertebrate animals.

In both plants and animals, we find body parts that are evidently of similar origin, structure, and development, although they may be adapted for very different functions in various species. We refer to such parts as *homologous organs*. In many seed plants, leaves are found as petals, tendrils, and thorns. Covering tissue of animals may be modified as hoofs, scales, nails, claws, feathers, and hair. The bones of a bird's wing, the front leg of a horse, the paddle of a whale, and the human arm are so similar that, with slight variations, they are even given the same names (Figure 13-5). The *differences* in the structure of homologous organs show the result of variations and adaptation to environmental conditions. However, the *similarities* indicate a genetic relationship and common ancestry.

Most animals have certain structures that are seemingly useless and often degenerate, although similar structures are well developed and functional in related species. We call these structures *vestigial* (vess-TIJ-ee-uhl) *organs*. Vestigial organs are believed to be the remnants of organs and structures that were well developed and functional in an organism's ancestors. Groups of genes inherited from these ancestors still produce these structures, but their importance to survival apparently became ever less significant as the environment and life habits of the organism changed in the course of evolution. The human appendix, for example, is a small structure without any known function. However, in mammals that eat a coarse diet with large amounts of cellulose, this pouch serves as an organ in which mixtures of food and enzymes can remain during digestion. In rabbits, the appendix is large and functional. Other human vestigial structures include the remains of ear muscles, nictitating membranes, which are the remnants of third eyelids, and the coccyx, or tail vertebrae, which is the remains of a tail.

Biologists have found further evidence of common ancestry in *embryology*. The embryos of vertebrate animals, for example, resemble each other closely in early stages of development. In these embryonic stages, it would be difficult to distinguish a developing fish, bird, or mammal. It would seem that all these animals received, from remote common ancestors, sets of genes that control development for a time. Later in the process, however, other genes assume control of the process and cause the fish, bird, and mammal to develop in different ways (Figure 13-6).

Many organisms resemble one another not only in structure and development but in *biochemistry* as well. All evidence that is presently available indicates that all organisms produce nucleic acids, especially DNA, and that all use ATP in energy transfer. Digestive enzymes are similar in many forms of life. There is also similarity in hormone secretions. Many human diabetics owe their staying alive to insulin extracted from the pancreas of hogs and cattle.

Plant and animal breeding provides other evidence that organisms may change. Over twenty-five breeds of dogs have been developed from wild, wolflike ancestors. A dozen kinds of chickens have a common ancestor in the jungle fowl of India. The story of plant and animal breeding, while it does not prove that similar changes have taken place naturally in past ages, does strongly point to that possibility.

Lamarck's theory of evolution

One of the first and most interesting attempts to explain evolution was proposed by the French biologist Jean Baptiste Lamarck. His idea, presented in 1801, placed great stress on the environment as the mechanism responsible for change. Actually, Lamarck's idea of evolution involved three theories:

■ *The theory of need:* The production of a new organ or part of a plant or animal results from a need. For example, the early ancestors of the snake had legs and short bodies. But with the changes in land formations, it became necessary for the snake to walk through narrow places. Then it began to stretch its body and to crawl rather than walk. This assumption formed the basis on which Lamarck formulated his second theory.

■ *The theory of use and disuse:* Organs remain active and strong as long as they are used but disappear gradually with disuse. Lamarck believed that the snakes of each generation continued to stretch and strengthen their bodies. Furthermore, the legs were used less and less because they interfered with crawling, and they finally disappeared.

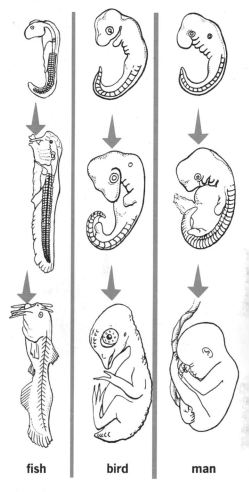

fish   bird   man

13-6 Different stages in the development of three vertebrate animals. Note their similarities in the very early stages and their differences in the later stages.

■ *The theory of inheritance of acquired characteristics:* All that has been acquired or changed in the structure of individuals during their life is transmitted by heredity to the next generation. Lamarck believed that the modern snakes evolved from the forms that had lost their legs through disuse. Thus, they inherited the legless trait from their ancestors.

Lamarck used other examples that he could see in nature to explain his theory of need. He believed that the giraffe evolved from a short-legged, short-necked form. When competition for low-growing grasses became too great, the giraffe began to stretch its neck and forelegs in order to reach the leaves of trees. Each succeeding generation inherited the longer neck resulting from stretching until the trait was inherited by our modern giraffe. Other beliefs held by Lamarck included the idea that frogs and ducks developed webbing in the toes for swimming and that the heron and other wading birds developed their long legs to keep their bodies out of the water. He attempted to explain the evolution of mammals into birds by suggesting that the hair turned into feathers as certain mammals attempted to fly.

Many biologists have tested Lamarck's theory of inheritance, but each experiment presents further evidence that acquired traits are not inherited. One experimenter, August Weismann, tested the theory by cutting off the tails of mice and then mating them. The offspring of the tailess mice had tails. He then cut off the tails of the second generation and mated them. He continued this experiment for twenty generations. The twenty-first generation still had tails of the same length as the first.

The explanation of evolution given by Lamarck may seem very reasonable. Gradual modifications do appear in species from time to time. Parts of plants and animals do change as a result of use and disuse. For example, an athlete develops strong muscles through use, and these muscles become weak from disuse when they are confined in a cast. There is no evidence, however, that these acquired traits are passed on to offspring.

Darwin's theory of natural selection

In 1859, Charles Darwin, an English scientist, published his *On the Origin of Species by Means of Natural Selection.* His theory of **natural selection,** while confined to biology, has also influenced other branches of science. According to Darwin, the chief factors that account for the development of new species from a common ancestry can be summarized as follows:
■ All organisms produce more offspring than can actually survive.
■ Because of overproduction, there is a constant struggle for existence among individuals.
■ The individuals of a given species vary.
■ The fittest, or the best adapted, individuals of a species survive.
■ Surviving organisms transmit variations to offspring.

Overproduction

A fern plant may produce 50 million spores each year. If all the spores resulting from this overproduction matured, in the second year they would nearly cover North America. A mustard plant produces about 730,000 seeds annually. If they all took root and matured, in two years they would occupy an area 2,000 times that of the land surface of the earth. The dandelion would do the same in ten years.

At a single spawning, an oyster may shed 114,000,000 eggs. If all these eggs survived, the ocean would be literally filled with oysters. Within five generations, there would be more oysters than the estimated number of electrons in the visible universe! There is, however, no such actual increase.

The elephant is considered to have a slow rate of reproduction. An average elephant lives to be 100 years old, breeds over a span of from 30 to 90 years, and bears about 6 young. Yet if all the young from one pair of elephants survived, in 750 years the descendants would number 19,000,000.

Struggle for existence

We know that in actuality the number of individuals of a species usually changes little in its native environment. In other words, regardless of the rate of reproduction, only a small minority of the original number of offspring reaches maturity.

Each organism seeks food, water, air, warmth, and space, but only a few can obtain these needs in struggling to survive. This struggle for existence is most intense between members of the same species, because they compete for the same necessities.

Variation among individuals

With the exception of identical twins, each individual varies in some respects from other members of its species. Animal breeders take advantage of this fact when they choose for breeding purposes those individuals with desirable characteristics. Nurserymen are able to produce disease-resistant plants and to control the size and color of the bloom by careful study and cross-fertilization of particular individuals. The variations within a species furnish the material for nature to use in her selection.

Survival of the fittest

If among the thousands of dandelion seeds produced, for example, some have better dispersal devices, these will be carried to a distant and less crowded place where they may survive. Those having poorer adaptations will perish by overcrowding. In so severe a struggle, where only a few out of millions can hope to live, very slight variations in speed, senses, or protection may turn the scale in favor of the better fitted individual. Those with unfavorable variations will sooner or later be wiped out. Darwin called this part of his theory the *survival of the fittest*. In this way, nature selects the characteristics of a certain population by favoring even the slightest variation.

Heredity

In general, offspring resemble their parents. If the parents reach maturity because of special fitness, those of their descendants that inherit most closely the favorable variation will, in turn, be automatically selected by nature to continue their species. Of course, this would be true only if the environment has remained unchanged. While Darwin recognized that genetic variations occur, he had no knowledge of the factors and mechanisms involved.

Mutation theory of evolution

Mutation as a basis for evolution was proposed by Hugo De Vries, a Dutch botanist, when he presented his mutation theory in 1901. De Vries had found two evening primrose plants that were different from their parent stock and that bred true by producing these variations in offspring. He experimented for many years and found that of fifty thousand specimens of evening primrose, at least eight hundred plants showed striking non-inherited variations that were transmitted to offspring. From his study of the evening primrose, De Vries concluded that similar mutations occurred frequently in other organisms and that this was the basis for the evolution of life through the ages.

Mechanisms of evolution

As we review the work of Lamarck, Darwin, and DeVries, we can see that each made a significant contribution to modern concepts of evolution. Lamarck observed a relationship between the organism and its environment. His observation was correct. It was his conclusion—that is, that the environment acted directly on the organism and produced hereditary changes in relation to need—that was in error. Nevertheless, Lamarck did recognize change in living things through the ages and thus provided evidence of evolution.

Darwin also recognized change, but he reversed the hypothesis Lamarck had proposed. He established that the change occurs in the organisms without respect to need. He believed that the environment determines whether or not a variation improves the chance for survival.

De Vries reinforced Darwin's concepts of variation and survival of the fittest. He also explained the mechanism of change in terms of mutations and other alterations in the genetic makeup of organisms.

The modern biologist has been able to go much further in exploring the mechanisms of evolution. He has the advantages of recent advances in genetics to account for variations, biochemistry to explain gene action, anatomy and physiology to establish relationships, and ecology to give evidence of environmental relationships, population movements, and other factors operating on a large scale. Generally, we can relate evolution to the following mechanisms of change: gene mutations, chromosomal mutations, and recombination.

Gene mutations as a cause of change

Mutations have been found in every kind of plant and animal that has been studied genetically — from columbines to fruit flies and from mice to men. In gene mutations, we find the basic source of variations and the material for evolution. In considering gene mutations as a mechanism in evolution, we must examine their nature and frequency more closely to see how they relate to variations.

How frequently do gene mutations occur? Genes tend to be stable. The DNA molecules composing them tend to replicate time after time without chemical alteration. Still, gene mutations do occur and at a rate that geneticists have been able to predict. In studies of mutation in corn, geneticists have found that genes for seed color mutate 492 times per 1,000,000 gametes. This would average about 1 mutation in 2,000 gametes. This may seem to be a very slow rate of mutation. However, when you consider that a single acre of corn contains more than this number of plants, you will realize that the frequency of mutations is greater than you might think.

Many gene mutations are so slight that there is no apparent difference in the organism. The fruit fly is believed to contain about twenty thousand genes in its haploid chromosome number. Thus, the body cells contain as many as forty thousand genes. A mutation in a single gene might not cause a noticeable change. This may account for the fact that change in a population can be detected only over a long period of time, not from one generation to the next.

Mutant genes are nearly always recessive. For this reason, a mutant characteristic does not appear in an offspring in which the other allele is normal. For example, in a pair of genes designated as *AA,* let us assume that one gene mutates to *a.* The mutant characteristic will not appear in a heterozygous *(Aa)* offspring. However, if the frequency of the mutant gene increases in the population, a homozygous offspring with paired mutant genes *(aa)* may appear and the new characteristic will be expressed.

Gene mutations often produce a characteristic that cancels the survival chance of an organism. If such a *lethal gene* is recessive, it is possible for many individuals in a population to possess the gene without the fatal characteristic being expressed. Since individuals homozygous for the lethal trait die without reproducing, there is a tendency toward reduced frequency of a lethal gene in a population. However, additional normal genes may continue to mutate to the lethal gene and perpetuate the problem.

Chromosomal changes

While chromosomal mutations occur less frequently than gene mutations, their effects are usually much more noticeable. In your study of genetics, you found various reasons for changes in chromosomal structure. Generally, such reasons include:

■ *Polyploidy*—gain in the number of sets of chromosomes, resulting in such abnormalities as 3*n* (triploid number) or 4*n* (tetraploid number).

■ *Haploidy*—loss of one entire set of chromosomes.

■ *Nondisjunction*—gain or loss of part of a set of chromosomes.

■ *Crossing over*—exchange of parts of segments of chromosomes between a chromosome pair.

These and other changes in chromosome composition and number produce new characteristics in offspring. Thus, chromosomal mutations are an important mechanism in variation and resulting evolution of organisms.

Recombination as a source of variation

You will recall, from your study of sexual reproduction, that chromosome pairs separate and reduce from the diploid number *(2n)* to the haploid number *(n)* in the formation of gametes. During fertilization, new combinations of chromosomes are formed as the chromosome number is restored to the diploid number *(2n)*. Recombination of chromosomes and genes in sexual reproduction is a mechanism second only to mutation in the production of variation. Furthermore, it is by recombination that mutant characteristics appear as variations in populations.

To show how recombination greatly hastens the appearance of a new characteristic, let us again represent a pair of alleles as *AA*. One of these genes mutates to *a*. The gamete containing this mutant gene unites with a gamete containing a normal allele, producing the gene pair *Aa*. Since the mutant gene is recessive, the characteristic will not appear in this individual. If this organism were to reproduce asexually, all offspring would be identical with the parent. The only chance for the variation (*a*) to appear would be for the second gene of the pair (*A*) to mutate, thus producing the alleles *aa*. The chance that this second mutation might happen in an organism has been calculated to be about one in ten billion!

Now consider what sexual reproduction and recombination could accomplish in this example. When the pair of alleles *A* and *a* separate during meiosis, half of the gametes produced will receive *A*, while the other half will receive the mutant gene *a*.

The new characteristic will not appear in the first generation, because the gene *a* must combine with a normal gene *A*. However, the mutant gene *a* is increasing in frequency. It is now present in half the offspring. As these heterozygous offspring interbreed, they will produce one fourth *AA,* one half *Aa,* and one fourth *aa,* as you determined earlier in a monohybrid cross. Thus, one fourth of the offspring will express the mutation. If this characteristic is favorable to the survival of these organisms, the mutant gene *a* will increase in frequency in the population as a result of selection. In time, all members of the population may possess the mutant characteristic.

Variation, change, and survival

Mutations, chromosomal changes, and recombination provide the genetic basis for variation in organisms. But it is the environment that determines whether or not variations are favorable to survival. Thus, the environment determines the direction as well as the rate of evolution.

Isn't it logical to assume that a population already established in a given environment is well adjusted to the conditions of that environment? Such a population is the result of natural selection and survival of the fittest over a long period of time, during which undesirable variants were eliminated. As long as the environment remains unchanged, it is unlikely that further genetic variation will improve the already adjusted population. But what if the environment were altered? Then genetic variations might improve the chances of survival and lead to the establishment of a new variety within a species or even to a new species. Changes in environmental conditions can occur in two ways:

■ Migration of an organism to another locality.
■ Change in the environmental conditions in a given locality.

Migration, variation, and the benefit of change

When *migration* occurs so that individuals of a population occupy new areas and interbreed with other populations of the same species, new gene combinations are formed. Let us assume that several members of an animal population migrate to a new area. They take with them certain combinations of genes characteristic of the population of which they have been a part. Their arrival in a new area introduces genetic characteristics that have been absent in this population. Because of interbreeding, their offspring are also receiving different genes from the new population. The genetic makeup of this entire population may be altered by the migrating organisms. Thus, one result of migration is the production of *variations through new gene combinations*.

There is another important result. Migration takes plants and animals into new and different environments. This introduces the *benefit of change*. New characteristics that appear in offspring through mutations and recombinations of genes may result in adaptations that are favorable to survival in the new environment.

Migration and selection in animal populations

The migration and redistribution of camels in early times is a good example of the spreading of a population and of selection. At one time, camels in various forms were found throughout Asia, Europe, North America, Central America, and South America. Biologists believe that they had their origin in North America and spread to Asia by traveling across a land bridge that is thought to have existed at that time across the Bering Strait region. A land bridge that acts as a pathway for migrating animals is called a *corridor*. The camel migrations are believed to have extended through North America and Central America to South America. Then, with the coming of the Ice Age, the camel population was eliminated in most areas. This resulted in widely separated camel populations. Today, we have the Asian camel with two humps and the African (Arabian) camel with one hump. Both are well adapted for life in the desert. Relatives in far-off South America evolved into quite a different animal – the South American llama – which lacks a hump but is sure-footed and is protected with a dense coat of hair that adapts it for life in the rocky, mountainous area in which it lives (Figure 13-7).

The individual variations among the camel populations that reached the Andes were probably not very great. But the ones with the slightly heavier coat had a better chance to survive cold winters and to reproduce the following year. Also, the sure-footed animals with shorter legs had a better chance to survive by running away from enemies. The animals that did not possess these favorable variations were eventually eliminated.

The movement of organisms to new environments sometimes involves strange and interesting methods and devices. A mouse may float across a wide body of water on a log and reach a new environment. If other mice are present in the new environment and are capable of interbreeding with the new arrival, new gene combinations may be formed, and mice may strike out in a new evolutionary direction. We refer to such distribution over strong barriers as *sweepstakes dispersal*.

The peppered moth and the impact of environmental change

A moth population in England affords us one of the best examples of the impact on a species of a change in the environment. The peppered moth *(Biston betularia)* is native to the region of Manchester, a British industrial city. Two British scientists, R. A. Fisher and E. B. Ford, conducted the original investigation of a color change that occurred in this moth within a period of just fifty years. More recent studies were conducted by Professor H. B. D. Kettlewell. The result of these investigations has become a classic example of evolution in action.

13-7 Camel ancestor; dromedary or Arabian camel; Bactrian or Asian camel; a South American llama. (The American Museum of Natural History; Rue, National Audubon Society; Jerry Frank; George Holton, Photo Researchers)

Prior to 1845, the peppered moth was light-colored with dark blotches and spots. It could hardly be seen when it rested on the light-gray bark of trees. Then, in 1845, an almost completely black peppered moth was captured in Manchester. A mutation had occurred in a gene determining coloration. Was this a favorable mutation? You would expect a black moth resting on light bark to be an easy prey for a bird (Figure 13-8, left). But something else was happening in Manchester. The city was rapidly becoming an industrial center. Smoke poured from factory chimneys, and soot turned the light-colored bark of the trees nearly black. The black moths could hardly be seen on the dark, sooty bark. With this environmental change, the light moths became easy prey for birds (Figure 13-8, right). In the period from 1845 to 1895, the black peppered moth population increased from one known individual to 99 percent of the population!

This change in the peppered moth population interested Professor Kettlewell, who continued the investigation. He selected two entirely different areas for study. One was a bird reserve in Birmingham, an industrial city similar to Manchester. The other was a countryside area near Dorset, a place where there was no soot on the trees.

In the bird reserve in Birmingham, Kettlewell released 477 black moths and 137 light moths. As he watched through the day, birds fed on the moths as they rested on tree trunks. That night he recovered most of the remaining moths by attracting them to a light. Altogether, he recaptured 40 percent of the black moths but only 19 percent of the light moths.

In the countryside near Dorset, he released 473 black moths and 496 light moths. Again the birds destroyed large numbers as they rested on trees of the region. Only 6 percent of the black moths were recaptured, while 12.5 percent of the light moths survived.

This provides us with an interesting example of a principle biologists now call *industrial melanism* (MELL-uh-NIZZ'm). A mutation, a resulting adaptation to a changed environment, and a predator combined to change an entire population of moths in fifty years. Industrial melanism is a demonstration of the importance of natural selection in the process of evolution.

Isolation as a factor in evolution

Two squirrel populations in the region of the Grand Canyon of the Colorado River illustrate the effect of *isolation* on the evolution of different forms. On the north rim of the Grand Canyon, we find the Kaibab squirrel, with long ears, white tail, and dark underparts. On the south rim we find a similar animal, the Abert squirrel, which has long ears but a gray tail and light underparts (Figure 13-9). Biologists consider these squirrels to be different species. Why have they developed separate characteristics when the two populations are so near? Between the two rims of the can-

13-8 The principle of industrial melanism. In each situation, which peppered moth is likely to survive the longest? (courtesy Dr. H. B. Kettlewell)

13-9 The Kaibab squirrel (top) lives on the north rim of the Grand Canyon, while the Abert Squirrel lives around the south rim. In serving as a barrier, the canyon has had a role in the evolution of these two species. (Sonja Bullaty, Audubon Society; Al Lowry, Photo Researchers)

yon, there is a fast-flowing river. In the depths of the canyon, the temperature may reach 120°F. These factors have acted as barriers to the two groups of squirrels, preventing them from crossing the canyon and interbreeding. Over a period of thousands of years, the two populations have gradually become different from each other, although they probably started out as the same species. In other words, the gene pools of the two groups were separated, so that as mutations occurred in each group, the variations between the two became more pronounced, until they could be considered two species.

A mountain range, a dry plain, a desert, or an ocean may act as a physical *barrier,* preventing plants or animals from interbreeding. The populations of islands give striking examples of isolation. With an ocean as a barrier, there is little or no opportunity for plants and animals to interbreed with others on the mainland. This accounts for the development of totally different populations in the two areas.

Physical barriers are not the only causes of isolation of a species population, however. Any factor that prevents interbreeding may cause isolation. The sockeye salmon of the Fraser River is a good example. This great Canadian river has long been the ancestral spawning area of this salmon. A sockeye salmon hatches far up the river in shallow, cold water. Gradually, it works its way downstream, finally arriving at the mouth of the river, where it disappears in the vast Pacific Ocean. There it lives three or four years until maturity, when a reproductive instinct urges it back to the mouth of the Fraser. Fighting the currents and leaping waterfalls, the salmon finally reaches its ancestral home — perhaps the very pool in which it hatched. Here it reproduces and dies. In the Pacific Ocean, the sockeye mingles with other species of salmon. However, it never interbreeds with other species — only with other sockeyes and always in the Fraser River. Thus, the sockeye salmon has remained an isolated breeding population. As might be expected, this has resulted in differences between the sockeye salmon and the salmon that breed in other streams.

Other barriers to interbreeding include such factors as variation in the mating time of two populations, structural differences that prevent mating, and differences in mating habits. Sometimes interbreeding between two populations does occur, but the resulting hybrids are often sterile.

The development of species

A *species* is a group of organisms distinct from all other organisms and capable of interbreeding with others of its kind. All members of a species have certain genetic likenesses. They have the same number of chromosomes and the same arrangement of genes on the chromosomes. The development of a species is a process called *speciation* (SPEE-shee-AY-shun). All through the ages species have been disappearing, and new species have been developing. New species are forming today, just as they have in past ages.

These mechanisms that we have been discussing—variation, migration, or environmental change, selection, and isolation—result in speciation. To review how speciation may occur, let us condense the changes of perhaps thousands of years into a few sentences. Organism A is a member of a species population adapted to certain environmental conditions. However, the environment is such that migration can occur. Now we will assume that a mutation occurs in a member of the species population A. As the mutant gene increases in frequency in the population, certain organisms express the trait. We will designate these variations as AB. As both A and AB organisms migrate, they occupy a new and different environment. A is not entirely suited to the new conditions and may perish. However, the variation in AB is a favorable adaptation, and AB survives. Additional mutations occur. Those that are favorable to the new environment are preserved in offspring. Finally, a new species that we will call B is produced. Variation and selection have separated A and B to the extent that they no longer interbreed.

13-10 Two species of maples: *Acer saccharum,* the sugar maple; and *Acer rubrum,* the red maple. (Walter Chandoha)

Speciation in maples

Maples are found in many environments of North America. Several kinds have been introduced from Europe and Asia and thrive in our yards and parks and along city streets. Altogether, sixty or seventy maple species are distributed through the Northern Hemisphere. Are all of these maples descendants of the same ancestral stock that lived many ages ago? Similarities in all of these maple species would indicate that they are. How, then, did they become so different?

For example, at some time in the past a variation occurred in a maple that adapted it for life in wetter surroundings than maples had usually occupied. This variation might have been a difference in root structure. This new variety flourished in wet environments. Other variations resulted in further adaptations to wet surroundings. Finally, a new maple species evolved. We call it the silver maple *(Acer saccharinum).* Today the silver maple is found throughout the eastern part of the United States. It towers to a height of sixty to eighty feet in bottomlands, swamps, stream borders, and river floodplains (lowlands that flood seasonally). While it lives best in lowlands, it can also tolerate much drier situations and, for this reason, is widely planted as a fast-growing shade tree. Perhaps you know it as the soft maple. It grows in shade in wet surroundings but requires more light in drier situations. Foresters are familiar with a variation of the silver maple that has deeply cut leaves. The cut-leaf maple grows more slowly

13-11 Adaptive radiation. These North American members of the deer family probably had a common ancestor. However, genetic variations produced adaptations for different environments and modifications in body structure in each species. (Rue from Monkmyer; Harry Engels, National Audubon Society; Pro Pix from Monkmyer)

and seldom reaches the size of a silver maple. Can you think of a reason why? Perhaps this maple is a species in the making.

Maples in various forms occupy a wide range of environments in North America. The sugar maple and black maple thrive in the forests of the eastern United States, while the moosewood maple mingles with the pines and hemlock of the northern forest. The striped maple prefers the elevations of eastern mountain ranges. The bigleaf maple is limited to a narrow belt in the Pacific coastal area from Alaska to California, where it lives in moist situations in foothills and low mountains.

While all of the maple species have distinguishing characteristics, they have certain characteristics common to all maples. All maples in the world produce double-winged fruits known as samaras. In all maples the arrangement of buds on the stem is opposite. Their leaves tend to have a distinctive vein pattern. We refer to these similarities as *genus characteristics.* These represent a stable gene pool that has not changed as maples have evolved during speciation. Characteristics such as these, in addition to indicating evolutionary patterns, are important considerations in classification, a topic we shall discuss in Chapter 14.

Adaptive radiation

The evolutionary pattern we have discussed in speciation is called *adaptive radiation.* This process consists of a branching out of a population to new and different environments. As variations occur, conditions in the environment will determine whether they are favorable or unfavorable. Organisms with favorable variations survive and pass on these characteristics to offspring. Thus, adaptive radiation supports Darwin's concept of natural selection.

The branching out of a population through adaptative radiation is, usually, a slow process. Variations may be slight at first. As the process continues, however, further variations may adapt members of a population to increasingly different environments. Generation after generation, descendants become more and more different from their ancestral type. After many thousands of years, it may be difficult to determine that two different species are related through a distant common ancestor.

To illustrate adaptive radiation over a long period of time, let us consider some North American members of the deer family. Several varieties of mule deer occupy regions of western North America. These deer live in the mountain forests in the summer and move in large herds to the more sheltered valleys in the winter. The elk, a much larger deer, lives in high mountain forests in the summer and, like the mule deer, descends in herds to sheltered valleys in the winter. The moose, the largest member of the deer family, is found in lake and swamp areas in the summer where it feeds on aquatic plants. In the winter, the moose moves to higher ground and finds food and shelter in nearby forests.

Different kinds of caribou live in the northern forests and the barren tundra regions of the far north.

These members of the deer family are not closely related. Yet, if we could trace their ancestries, we might find that they all descended from a common ancestor and that adaptive radiation through the ages has made them different in form and ways of life.

Convergent evolution

In many ways, *convergent evolution,* also called parallel evolution, is the reverse of adaptive radiation. To converge means to come together. In this case, organisms with entirely different ancestries but sharing the same environment become similar in form and way of life through adaptations.

The whale and the seal are good examples. Both of these mammals live in the open sea. They have flippers for locomotion and are powerful swimmers. While both are lung breathers, they can hold their breath and submerge for a considerable time. A thin layer of fat beneath the skin protects both the whale and the seal from loss of body heat in the cold ocean water. These similarities do not indicate relationship. They are the result of adaptations to the same environment.

Even more striking examples of convergent evolution may be found in a comparison of North American and Australian mammals. In your study of mammals in Chapter 38, you will learn about two different groups of mammals. Most North American mammals are *placental mammals.* The pouched mammals, or *marsupials,* well-represented in Australia, are more ancient and are considered more primitive. Placental mammals destroyed most marsupials in North America many ages ago. However, placental mammals did not reach Australia.

13-12 Convergent evolution; the kangaroo, a marsupial, and the deer, a placental mammal are similar in many ways because of adaptations to a similar environment. The same applies to the marsupial mole and the placental mole. Yet the kangaroo is more closely related to the marsupial mole than to the deer. The deer is more closely related to the placental mole than to the kangaroo. (George Leavens, Photo Researchers; Harvey A. Schwartz; F. J. Mitchell; Allan Roberts)

It is interesting that many placental mammals in North America have marsupial counterparts in Australia. Adaptations to similar environments in the two continents have resulted in these similarities through convergent evolution. The deer and the kangaroo have little relationship, yet they have similar head structures and feeding habits. Our common mole is a placental mammal. The counterpart in Australia, leading a similar life, is a marsupial mole. The Tasmanian wolf, a marsupial, resembles the North American wolf, a placental mammal, both in body structure and in feeding habits. The wombat of Australia is much like the North American ground hog. The flying phalanger of Australia is much like our flying squirrel, a placental mammal. The living world is, indeed, full of examples of convergent evolution.

SUMMARY

It is difficult for us to comprehend the changes that have occurred in the earth and its living population through the past ages and changes that are bound to occur in ages yet to come. In the course of a lifetime, each of us has an opportunity for only a brief glimpse at the drama of life that has been going on for many millions of years.

We know that environments are constantly changing. We know, too, that there are continuous variations in organisms. When you put the two together, you have the basis for evolution.

As you look about, you may see a forest, a grassland, a lake, or a bog. Each supports a community of organisms adapted to the conditions of the environment. Will these environments be the same in 100 years, 1,000 years, or 10,000 years? The chances are that the environments will change greatly, and so will the organisms living in them. Perhaps biologists at some time far in the future will search for evidence to determine what life was like in our time.

BIOLOGICALLY SPEAKING

sedimentary rock	homologous organ	isolation
geological era	Lamarckian evolution	barrier
geological period	natural selection	species
geological epoch	survival of the fittest	speciation
fossil	migration	genus characteristic
evolution	corridor	adaptive radiation
organic variation	sweepstakes dispersal	convergent evolution
vestigial organ	industrial melanism	

QUESTIONS FOR REVIEW

1. Explain how sedimentary rock forms a geological time table.
2. Describe two methods by which scientists date sedimentary rock.
3. What is a fossil?
4. What is the biological concept of evolution?
5. Describe various kinds of biological evidence used to establish the common ancestry of organisms.
6. Describe the three theories included in Lamarck's explanation of evolution.

7. What concepts are included in Darwin's theory of evolution?
8. What significant contribution did De Vries make to the evolutionary theory?
9. Explain how gene mutations cause variations in organisms.
10. Explain why a gene mutation seldom appears in the individual in which it occurs.
11. List several causes of chromosomal mutations.
12. Explain how recombination operates as a source of variation.
13. How are mutant traits expressed through recombination?
14. List two direct evolutionary results of migration.
15. Why are corridors important in migration?
16. Explain how a black peppered moth appeared in the light peppered moth population in Manchester, England, in 1845.
17. Describe several barriers that can isolate an environment.
18. How can isolation of a population lead to speciation?
19. Generally, what occurs in adaptive radiation?
20. Give an example of convergent evolution in organisms.

APPLYING PRINCIPLES AND CONCEPTS

1. Discuss organic variation and environmental change as causes of change in living populations.
2. Discuss the importance of structural, physiological, and biochemical similarities in determining the paths of evolution.
3. Explain the development of the long neck of the giraffe according to Lamarck, according to Darwin, and according to modern theory.
4. Discuss the genetic significance of the results of studies of the peppered moth.
5. Discuss mutation, recombination, interbreeding, and selection as mechanisms of evolution.
6. Compare adaptive radiation and convergent evolution.

RELATED READING

Books

DARWIN, CHARLES R., *The Origin of Species.*
Doubleday and Co., Inc., Garden City, N.Y. 1960. A classical reading that explains Darwin's ideas on the way favored organisms are preserved in the struggle for life.

FARRINGTON, BENJAMIN, *What Darwin Really Said.*
Schocken Books Inc., New York. 1967. A comprehensive examination of Darwin's main theories, incorporating quotations from his work.

MOORE, RUTH, *Man, Time, and Fossils.*
Alfred A. Knopf, Inc., New York. 1961. A book for the layman about the gathering of evidence on evolution.

OPARIN, ALEXANDER I., *The Origin of Life.*
(2nd ed.). Dover Publications Inc., New York. 1953. A translation of Oparin's significant contribution relating to the explanation of where life came from.

SHEPPARD, P. M., *Natural Selection and Heredity.*
Harper and Row, Publishers, New York. 1960. A discussion on the relationships existing between evolution, genetics and adaptation.

Articles

Kettlewell, H. B. D., "Darwin's Missing Evidence," *Scientific American,* March, 1959.
An interesting presentation of evolution in action, the story of the peppered moth.

CHAPTER FOURTEEN

THE DIVERSITY OF LIFE

The science of classification

In our study of evolution in the previous chapter, we discussed the various species of maples and the way in which they have probably evolved by adaptive radiation. The science of genetics has given us a foundation for understanding how such evolution occurs. A study of evolution has, in turn, enabled us to find natural relationships between living things. Early biologists, however, attempted to group and name living things without the knowledge of genetics and evolution that we now have. The classification of organisms, a branch of biology known today as *taxonomy,* has been of concern to biologists since ancient times.

The Greek philosopher Aristotle probably devised the first classification system. He divided plants into three groups: the herbs, with soft stems; the shrubs, with several woody stems; and the trees, with a single woody trunk. He divided animals into three groups on the basis of where they lived: the water dwellers, the land dwellers, and the air dwellers.

In the eighteenth century, the great Swedish botanist Carolus Linnaeus (lin-EE-us), despite the fact that he had no knowledge of genetics or evolution, devised a classification system that in many respects is still in use today. Linnaeus believed that the main aim of science was to find order in nature. He recognized the species as the basic natural grouping and thought that species were unchanging. He sought to group all the known plants and animals into a fixed number of species, according to their structural similarities. He disregarded any organism that did not fit into his species categories. Perhaps if Linnaeus, like Darwin, had recognized that the organisms that did not fit into his categories were in the process of change, the theory of evolution would have been developed much earlier.

Linnaeus discarded the common names of plants and gave each one a scientific name made up of Latin words. None of these names was taken from his own language or from any other modern language. There are several reasons for Linnaeus' choice of Latin as the language of classification. First, since it was no longer in use, it was unchanging. Furthermore, many modern languages contain words taken directly from this ancient language. Latin was understood by scientists of all countries. Also, many descriptive Latin words are ideally suited for identifying the characteristics of an organism.

Linnaeus published his list of plant names in 1753 and his list of animal names in 1758. Each scientific name had at least two parts. Usually the name referred either to some characteristic of the organism or to the person who named it. Many of his names are in use today; these can be recognized by the *L.* that appears at the end of each.

How scientific names are written

Linnaeus' system of giving each organism a scientific name of two or more parts is called *binomial nomenclature,* or "two-word naming." The first name refers to the genus and always begins with a capital letter. The species name follows and usually begins with a small letter. The genus name is usually a noun and the species name, an adjective. The placing of the noun before the adjective is not unusual in Latin. We use a similar system in official lists of names where *John Smith* appears as *Smith, John.* In addition, scientific names are usually printed in italic type or underlined.

The species is still the basic group used in classifying organisms. The members of a species are similar in structural characteristics, and they can mate and produce fertile offspring. Thus, all domestic cats are of one species although they may differ in size, color, and shape. All human beings belong to the same species.

Closely related species are placed together in the larger group called the *genus.* For example, *Pinus* is the name of the genus into which all pine trees are grouped. There are many different species of pine trees. *Pinus resinosa,* for example, is the red pine, and *Pinus strobus* is the white pine.

In devising his standardized naming system, Linnaeus enabled biologists to avoid the confusion and the misleading nature of common names. The mountain lion, for example, is also known as the puma, or cougar. It is called the panther, silver lion, American lion, mountain demon, mountain screamer, king cat, sneak cat, varmint, brown tiger, red tiger, and deer killer (see Figure 14-1). Under the Linnaean system, however, all scientists can easily identify this animal as *Felis concolor.* The common house cat is *Felis domesticus* because it belongs to the same genus as the mountain lion but is of a different species. *Felis onca* is the jaguar, while *Felis leo* is the African lion.

14-1 These members of the cat family are related, as indicated by the genus name.

Since organisms are classified according to structural similarity under the Linnaean system, scientific names are not as misleading as most common names are. For example, what is a fish? If you think of a fish as an animal with a backbone, scales, fins, and gills, you are using the name correctly and scientifically. The perch, cod, halibut, bass, and salmon are fishes. But what about the silverfish? It is an insect. We call clams and oysters shellfish. And we call other animals that in no way resemble a true fish crayfish, jellyfish, and starfish (Figure 14-2). We shall use these names, even in the study of biology, because they are familiar. But when you learn more about these animals, you will understand why it is misleading to use the term *fish* in referring to them.

The bases of scientific classification

From the time of Linnaeus until recently, most scientific classification has been based on ***structural similarity***. For example, the cow is structurally similar to the bison and the deer. These are all cud-chewing mammals that have large molar teeth for grinding plant foods and two-toed hoofs adapted for bearing weight and for running over hard ground. Using structural similarity as a basis, biologists have classified organisms into groups, starting with very large divisions and continuing with smaller groups down to a single species. The smaller the classification group, the more similar its members are.

14-2 Common names can be misleading. Although these two animals are commonly known as *jellyfish* and *starfish*, respectively, neither is a true fish. They are unlike structurally, and they are not even closely related to one another. (Systematics—Ecology Program—MBL)

Recently, however, biologists have considered other characteristics in classifying organisms. One such characteristic is *cellular organization,* which includes nuclear structure, plastids, and other cell organelles. Another is *biochemical similarity.* For example, cells of closely related organisms may synthesize the same organic compounds. The most conclusive evidence of common ancestry and relationship between organisms, however, is in their *genetic similarity,* since all of the above characteristics are determined by genes. If two animals have the same number of chromosomes, and if the chromosomes of one animal are identical or very similar in structure to those of the other animal, you can assume that the two animals have many similar or identical genes and are closely related.

Groupings in the Linnaean system

If you had a specimen of each of the more than one million kinds of plants and animals known to exist and started grouping them, where would you begin? Wouldn't it be best to separate them first into plants and animals? This is where the biologist begins his classification. He separates living things first into very large groups called *kingdoms.* Next, he divides each kingdom into smaller groups known as *phyla* (singular, phylum). Each phylum is, in turn, divided into *classes.* A class contains many *orders.* A division of an order is a *family.* A family contains related *genera,* and a genus is composed of more than one *species.* Sometimes individuals of a single species vary slightly, but not enough to be considered separate species. We refer to these as *varieties.* An organism considered a variety has a third part to its scientific name.

As an example of how organisms are classified into these groupings, let us consider man. The kingdom Animalia includes all animals. Most of the animals in the phylum Chordata, including man, have backbones and are therefore included in the subphylum Vertebrata. The class Mammalia includes all animals having mammary glands. The order Primates includes only a certain group of mammals that stand nearly erect. Monkeys, chimpanzees, and gorillas are primates. The family Hominidae (hoe-MIN-uh-DEE) separates early manlike forms from the other primates. The genus *Homo* includes all true men. The species *sapiens* (which means "wise") is the only surviving species of man on earth.

Some biologists further divide man into races. The classification of man and five other organisms can be found in Table 14-1.

Problems in classification

As you examine various biological references, you may wonder at the wide variation in the classification systems. How many kingdoms do we recognize today? How are the phyla of living

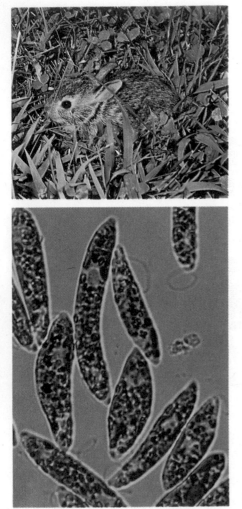

14-3 While it is simple to classify the rabbit as an animal and the grass as a plant, it is quite a different story to classify microscopic organisms, such as the euglena, into their respective groupings. (Allan Roberts; Walter Dawn)

Table 14-1 THE CLASSIFICATION OF SIX DIFFERENT ORGANISMS

	MAN	GRASS-HOPPER	DANDELION	WHITE PINE	AMEBA	TYPHOID BACTERIUM
KINGDOM	Animalia	Animalia	Plantae	Plantae	Protista	Protista
PHYLUM	Chordata	Arthropoda	Tracheophyta	Tracheophyta	Sarcodina	Schizomycophyta
CLASS	Mammalia	Insecta	Angiospermae	Gymnospermae	Rhizopoda	Schizomycetes
ORDER	Primates	Orthoptera	Campanulales	Coniferales	Amoebida	Eubacteriales
FAMILY	Hominidae	Acridiidae	Compositae	Pinaceae	Amoebidae	Bacteriaceae
GENUS	*Homo*	*Schistocerca*	*Taraxacum*	*Pinus*	*Amoeba*	*Eberthella*
SPECIES	*sapiens*	*americana*	*officinale*	*strobus*	*proteus*	*typhosa*

things placed in these kingdoms, and on what basis are these distinctions made? As we attempt to classify the diverse forms of life, from the simplest bacteria and protozoans to seed plants and complex vertebrates, we must remember that any system of grouping is purely man-made. We divide living things into classification groups for our own convenience.

Among the many forms of life that exist today, we recognize that there are gaps that we know were once filled with now-extinct organisms. Biologists use these gaps in establishing classification groups. But this is not an easy task. In many cases, intermediate organisms have survived and form a nearly continuous evolutionary sequence between two groups. This makes it difficult to know where to draw a line in classification. In other cases, the opposite is true. Extinction of large numbers of intermediate organisms leaves a group so isolated that it is difficult to determine its origin or relationship to other forms of life.

What is a plant? What is an animal? When is an organism neither distinctly plant nor animal? These questions did not concern biologists much until recent years. They recognized only two kingdoms: plant and animal. The plant kingdom contained four great phyla, while the animal kingdom included twenty or more, depending on the system used. This traditional classification served very well for many years. Since it is still the basis for many of the books you will use as references, we have included the two-kingdom system in the BIOLOGY INVESTIGATIONS workbook.

In the past few decades, a great amount of research has focused on simpler forms of life. Are these organisms plants? Are they animals? Actually, they are neither. They are more closely related to one another than to organisms we consider definitely plant or animal. Traditional classification has provided no real place for these "in-between" organisms. Thus, biologists have recognized the need for classifying them in a different way. Some systems place bacteria and the blue-green algae in a kingdom called the *Monera* (muh-NEER-uh). These organisms have one striking characteristic in common—they lack an organized nucleus. Some biologists even place the viruses in this kingdom, while others present valid arguments claiming that a virus isn't even a living organism.

A second kingdom, universally recognized in more recent classifications, is the kingdom *Protista* (proh-TIST-uh). There seems to be little doubt that such a kingdom should be established, but there is disagreement as to what phyla should be included. If the kingdom Monera is not recognized, then bacteria should certainly be placed among the protists. Protozoans are also placed in this kingdom. The real problem arises in what to do with the algae and fungi. The simpler members of these groups seem logically to be protists. But these same groups include organisms such as the kelps and mushrooms that are more like true plants. On the basis of size and limited cell specialization, however, it may be more reasonable to place all the various algae and fungi in the same kingdom. For this reason, in this book we shall use a three-kingdom classification system in which the bacteria, fungi, algae, and protozoans constitute the kingdom Protista (see Table 14-2). Remember that this is not the only system, nor is it by any means perfect. We cannot say that one system is right and another wrong. We are simply choosing one system as a convenience in organizing our study of the world of life.

Table 14-2 A MODERN CLASSIFICATION OF ORGANISMS

Kingdom—Protista

Organisms having a simple structure; many unicellular, others colonial or multicellular but lacking in specialized tissue; both heterotrophic and autotrophic; neither distinctly plant nor distinctly animal.

	PHYLUM	ORGANISMS
Algal Protists	Cyanophyta	blue-green algae
	Chlorophyta	green algae
	Chrysophyta	golden-brown algae, or diatoms
	Pyrrophyta	dinoflagellates and cryptomonads
	Phaeophyta	brown algae
	Rhodophyta	red algae
	Schizomycophyta	bacteria
Fungal Protists	Eumycophyta	fungi
	Myxomycophyta	slime fungi
Protozoan Protists	Sarcodina	amoeboid organisms
	Mastigophora	flagellates
	Ciliophora	ciliates
	Sporozoa	*Plasmodium*

Kingdom—Plantae

Multicellular plants having tissues and organs; cell walls containing cellulose; chlorophyll *a* and *b* present and localized in chloroplasts; food stored as starch; sex organs multicellular; autotrophic.

PHYLUM	SUBPHYLUM	CLASS	ORGANISMS
Bryophyta			liverworts, hornworts, and mosses
Tracheophyta	Psilopsida		"whiskferns"
	Lycopsida		club mosses
	Sphenopsida		horsetails (*Equisetum*) and calamites
	Pteropsida	Filicineae	ferns
		Gymno-spermae	seed ferns, cycads, *Ginkgo,* and conifers
		Angio-spermae	flowering plants

Kingdom—Animalia

Multicellular animals having tissues and, in many cases, organs and organ systems; pass through embryonic or larval stages in development; heterotrophic. (Only the major phyla are listed.)

PHYLUM	SUBPHYLUM	ORGANISMS
Porifera		sponges
Coelenterata		jellyfish, sea anemones, corals
Platyhelminthes		flatworms
Nematoda		roundworms
Trochelminthes		rotifers
Bryozoa		bryozoans, sea mosses
Brachiopoda		brachiopods, or lampshells
Mollusca		clams, snails, squids, octopi
Annelida		segmented worms (earthworm)
Arthropoda		insects, spiders, crustaceans
Echinodermata		starfishes, sea urchins
Chordata	Hemichordata	tongue worms, acorn worms
	Tunicata	tunicates
	Cephalochordata	lancelets
	Vertebrata	vertebrates

The definition of a species

An even greater problem in classification is to formulate an exact definition of a species. We have defined a species as an individual or distinct kind of living thing. But this is only a working definition, because biologists cannot always be sure when to classify two similar plants or animals as separate species and when to classify one as a variety of the other. Remember, for example, our discussion of speciation in maples in Chapter 13. It is difficult to determine exactly when to name as two different species two types of maples that are evolving by adaptive radiation. Biologists usually agree, however, on the following two general characteristics of species:

■ A species is a population of organisms whose members, within a considerable range of natural variations, are structurally different from all other organisms.

■ Members of a species can interbreed and produce offspring that are capable of further reproduction. In other words, all members of a species have a close genetic relationship.

SUMMARY

The work of taxonomists is never finished. About 1.5 million species of protists, plants, and animals have been named and classified. However, biologists estimate that there may be as many as 2 to 5 million different kinds of organisms in the total world population. If this is the case, the work of taxonomists is only half completed. In remote regions of the earth and in the depths of the seas, there are, undoubtedly, vast numbers of organisms yet to be discovered. Add to this several times that number of extinct organisms that must be included in a classification of the life of all ages.

Scientists did not devise our classification system. They discovered it. Relationships of organisms are natural and have always existed. We have, merely, found evidence of these relationships in structural, biochemical, and genetic similarities and have established classification groups. In using a modern classification system, we are tracing the branches of a tree of life rooted in the earliest era of life on our earth.

BIOLOGICALLY SPEAKING

taxonomy	genetic similarity	family
binomial nomenclature	kingdom	genus
structural similarity	phylum	species
cellular organization	class	variety
biochemical similarity	order	

QUESTIONS FOR REVIEW

1. What is the science of classification called?
2. On what bases did Aristotle attempt to classify plants and animals?
3. Why is Latin an ideal language for biological classification?
4. Explain the binomial system of naming organisms.
5. Discuss ways in which common names are both confusing and misleading.

6. On what characteristics of an organism is its classification based?
7. Name the classification groups from the largest to the smallest.
8. In what way are evolutionary gaps that represent extinct intermediate organisms important in establishing classification groups?
9. Discuss the problems biologists have encountered in trying to classify organisms into kingdoms.
10. How do most biologists define a species?
11. Distinguish between a natural and an artificial system of classification.
12. How has the Linnaean concept of a species been modified by evolutionary theory?

APPLYING PRINCIPLES AND CONCEPTS

1. Give examples to demonstrate that size, habitat, and diet similarities show no true animal relationships that could be considered in classification.
2. Make a list of plants and animals of your region that have more than one common name.
3. The fox, wolf, and coyote are different species of the dog family. On the other hand, the cocker spaniel, collie, and poodle are breeds of the domestic dog. Distinguish between a species and a breed, or variety.
4. How can the principles of scientific classification be made useful in areas outside biological study?

RELATED READING

Books

DELEVDRYAS, THEODORE, *Plant Diversification*.
  Holt, Rinehart and Winston, Inc., New York. 1965. Shows the ways and why plants have diversified.
DICKINSON, ALICE, *Carl Linnaeus, Pioneer of Modern Botany*.
  Franklin Watts Inc., New York. 1967. The story of the father of classification.
HANSON, EARL D., *Animal Diversity* (2nd ed.).
  Prentice-Hall Inc., Englewood Cliffs, N.J. 1964. An explanation, not just a description, of animal diversity.
RIEDMAN, SARAH LANE, *Naming Living Things*.
  Rand McNally and Company, Chicago. 1963. Explains the way plants and animals are grouped and named.

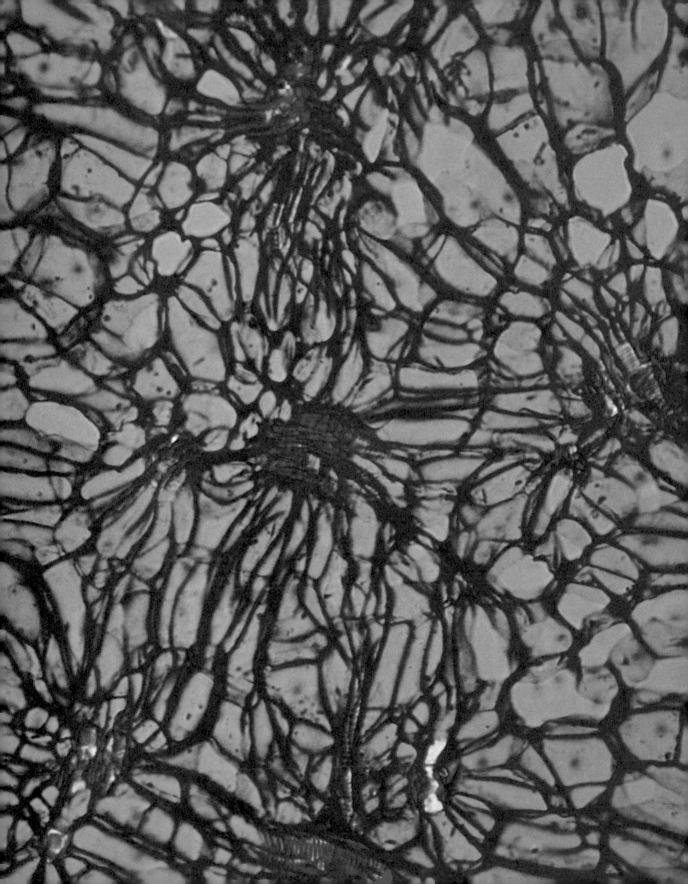

UNIT 3

MICROBIOLOGY

Your study of microbiology will take you to the fringe of life, where systems of substances first show properties that we associate with the living condition. In the composition of a virus, we find crystallizable materials that lack cellular organization but that still reproduce under appropriate conditions. Continuing from this primitive state, we find cells of increasing complexity that constitute entire organisms. We cannot classify them as definitely either plant or animal, and we therefore refer to them as protists. In this vast assemblage of lowly organisms, we find cells grouped in colonies and organisms that approach the multicellular organization of higher forms of life.

CHAPTER FIFTEEN

THE VIRUSES

Viruses—living or nonliving?

In beginning our study of microbiology with the viruses, we are introducing one of the most recent and exciting areas of biology, the science of virology. We are also presenting an entirely new concept of the living and the nonliving, for a virus qualifies for membership in both categories and seemingly shuttles back and forth between the two.

Before biologists had explored viruses and understood their structure and activities, it was easier to draw a line between non-living substances and living organisms. Then our new knowledge of viruses came along and upset what we had thought to be a clear-cut distinction. In a virus, we find a particle definitely linked to biochemical processes in its organization and unlike any non-living material, yet in itself not actually living. But in the presence of a living system within a cell, a virus seems to be very much alive. Could it be that a virus is alive at some times and not at others? Who can say, when there is no definition of life that is satisfactory to all biologists?

Regardless of the classification of viruses as living or non-living, however, it is appropriate to deal with them in a unit on microbiology. If they are nonliving, they are unique, because their influence on cell activities is different from that of any other nonliving material. If they are living, they are certainly the most basic organisms, representing life at the molecular level.

What are viruses?

At the mention of the word *virus,* you probably think of an agent of disease. In many respects, you are right. You are familiar with the polio virus and probably associate viruses with small-pox, chicken pox, influenza, rabies, and the common cold. These are but a few virus diseases. More than three hundred different viruses are known to produce diseases in various organisms. Scientists, too, thought until recent years that all viruses were *pathogenic,* or disease-producing, agents. Does it surprise you to learn that many virus particles are apparently harmless and that few bacteria, plants, and animals escape some kind of virus invasion?

Virus particles are noncellular. Rather, they can be considered to be subcellular for they are below the level of cellular organization. A virus has no nucleus, no cytoplasm, and no surrounding membrane. It is larger than a molecule yet much smaller than the smallest cell. We call viruses *filterable viruses* because they pass through the extremely small pores of the unglazed porcelain filters used in separating bacteria from fluids.

All but the largest virus particles are invisible under even the highest magnification of a light microscope. For this reason, before the invention of the electron microscope, little could be determined about the structure of a virus. With this instrument, even the smallest viruses have been photographed. We are now familiar with the forms of about twenty viruses, as well as with the sizes of their particles.

Viruses occur in a wide variety of shapes. Some have the form of needlelike rods, while others are spherical or cubical or brick-shaped. Still others have an oval or many-sided head and a slender tail.

Viruses are measured in *millimicrons,* for which we use the abbreviation mμ. When you consider that one millimicron is 0.001 micron and that one micron is 0.001 millimeter, you can appreciate the extremely small size of virus particles. Figure 15-1 shows the average diameters of several viruses. The smallest bacteria, which would appear as tiny specks under the high power of your microscope, range from 500 to 750 millimicrons.

The discovery of viruses

Scientists worked with virus diseases long before the existence of viruses was known. Dr. Edward Jenner performed the first vaccination against smallpox in 1796, when he transferred a virus-containing fluid from a cowpox sore on the hand of a dairymaid to a scratch on the arm of an eight-year-old boy. About a century later, Louis Pasteur discovered that the rabies infection centered in the brain and spinal cord. He successfully transmitted the disease to a laboratory animal by injecting it with infected brain and spinal-cord substance. While both of these men made significant medical contributions that we shall discuss later, neither realized that a virus is an agent of infection. Other nineteenth- and twentieth-century investigators working with virus diseases described virus-containing materials as contagious fluids, destructive chemical substances, and destructive enzymes.

One such investigator was Dimitri Iwanowski, a Russian biologist, who was working with diseased tobacco plants in 1892. These plants were infected with a virus disease known as tobacco

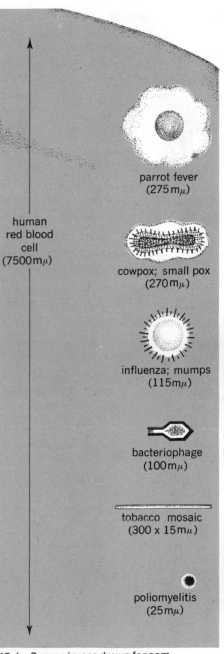

human red blood cell (7500 mμ)

parrot fever (275 mμ)

cowpox; small pox (270 mμ)

influenza; mumps (115 mμ)

bacteriophage (100 mμ)

tobacco mosaic (300 x 15 mμ)

poliomyelitis (25 mμ)

15-1 Some viruses drawn for comparative purposes over a portion of a human red blood cell.

mosaic. The term *mosaic* refers to a curious pattern of light green and yellow areas that appears, especially in the new leaves, as the tissues are destroyed by the virus. As the disease progresses, the leaves become stunted and wrinkled.

Iwanowski squeezed fluid from the infected leaves and rubbed it onto the leaves of healthy plants. Soon the healthy leaves developed the mosaic pattern. Iwanowski repeated his experiment, but this time he passed the fluid through an unglazed porcelain filter with pores small enough to remove all bacteria. Microscopic examination revealed no bacteria or other visible bodies that might cause disease. Nevertheless, healthy plants inoculated with the filtered leaf fluid soon developed mosaic disease.

Unfortunately, Iwanowski did not realize the importance of his work. In his day, bacteria were thought to be the smallest agents of disease. How could a bacteria-free fluid produce an infection? This puzzled Iwanowski. He assumed that the infectious "juice" contained an invisible bacterial poison that had passed through the filter.

Six years later, the Dutch botanist Martinus Beijerinck further investigated the tobacco mosaic disease. In repeating Iwanowski's work, he concluded that an invisible agent, smaller than the smallest bacterium, must be present in the infected "juice." He named this unknown agent *virus,* a Latin word meaning "poison."

Many years later, in 1935, Dr. Wendell Stanley made a significant discovery while working with the tobacco mosaic disease at the Rockefeller Institute. In an effort to isolate the virus, Stanley ground more than a ton of diseased tobacco leaves and extracted the "juice." From this extract, he obtained about one spoonful of needlelike crystals. These crystals could be stored in a bottle in an apparently lifeless condition. Yet when they were suspended in water and rubbed on a tobacco leaf, they produced the mosaic disease. For his outstanding work in isolating and crystallizing the tobacco mosaic virus (TMV), Stanley was awarded the Nobel Prize in chemistry in 1946.

The composition of viruses

In recent years, scientists have determined the chemical composition of certain viruses, including tobacco mosaic and polio. They have found the simplest virus particles to consist of a protein *coat,* or shell, encasing a *core* of nucleic acid. In some viruses, including plant viruses, the core is RNA. In others, it is DNA. Thus, we may consider a simple virus particle to be a giant molecule of nucleoprotein, similar in certain respects to the nucleoproteins that are found in the nuclei of cells. Recent studies have shown that complex viruses may also contain carbohydrates, lipids, metals, and other substances.

Extensive studies have been made of the structure and composition of the tobacco mosaic virus. With an electron microscope

15-2 Tobacco mosaic virus: infected leaves showing mottling and virus particles shown in an electron micrograph. (Marvin Williams, North Carolina State University; courtesy Virus Laboratory, University of California, Berkeley)

magnification of 60,000 times or more, this virus appears to be composed of elongated, rod-shaped bodies (Figure 15-3). The protein coat, composing 95 percent of the virus, consists of more than 2,000 protein subunits. A core of RNA composing 5 percent of the virus is located within the protein coat.

The presence of nucleic acid in a virus tends to place the virus in the category of living organisms. The genetic code in the RNA or DNA gives the virus certain characteristics that are duplicated as new virus particles are formed.

The properties of viruses

It is in the activity of viruses that we find certain characteristics that make them different from cells and of questionable status as living things. A virus particle can be active only in direct association with the content of a specific living host cell. Removed from a cell, a virus ceases all apparent activity, but it still retains its ability to infect a cell.

A virus cannot reproduce independently. That is, it cannot duplicate its own structure in the way in which cells self-reproduce by fission. In order to reproduce, the virus must invade a host cell and assume control of the cell's metabolic processes. Biologists are not sure how this is accomplished. Furthermore, it is probable that various kinds of viruses alter the chemical activities of cells in different ways. One hypothesis suggests that an invading virus alters the enzyme pattern that normally regulates protein synthesis and growth of the cell. In this way, the virus uses the substances and chemical machinery of the cell to form virus particles rather than normal cell constituents. When this happens, the behavior of the virus is similar to that of a gene. Could it be that a virus particle is a gene or a group of genes without a "home" until it invades a cell?

In the discussion of the living condition in Chapter 2, we referred to various processes associated with life. Cells carry on metabolic activities continuously. They use organic molecules for growth and oxidize fuel molecules to supply the energy necessary to support cellular activities. Are metabolic activities essential to a virus? An isolated virus requires no metabolic activity. Within a living cell, it is capable of limited metabolic activity made possible by the machinery of the cell.

The properties of many viruses are altered by the environment in which they are grown. For example, when rabies virus is grown in cells of the brain and spinal cord of dogs, its potency, or *virulence* (VIR-uh-lenss), for man and dogs increases. If the virus is grown in rabbits, however, it becomes less virulent for man and dogs but increases in virulence for rabbits. Hence, one might argue that the structure of a virus may be altered by the chemical nature and activity of the cell in which the new virus particles are produced.

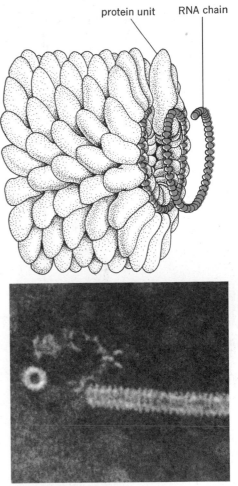

protein unit   RNA chain

15-3 Tobacco mosaic virus. Compare the schematic drawing with the electron micrograph. (courtesy R. W. Horne)

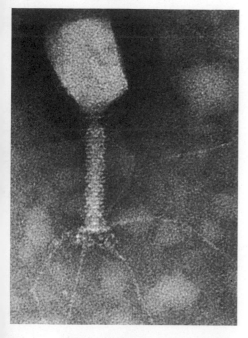

nucleic acid

protein coat

head

collar

sheath

base plate

tail

tail fibers

15-4 A bacterial virus, or phage. Compare the schematic drawing (top) with the electron micrograph. (courtesy T. F. Anderson)

Further variations in the structures of viruses occur as mutations. More than fifty mutant strains of the tobacco mosaic virus have been discovered. These mutant strains differ in virulence and in the symptoms they produce in the host plant.

You have heard of highly virulent strains of influenza virus that strike populations at various intervals, as well as of less virulent strains that cause much less serious infections. Perhaps the virus increases in virulence as it passes from one person to another during an epidemic. Or we may be dealing with various mutant viruses that differ in virulence.

The classification of viruses

If a virus is a product of the disorganized machinery of a host cell, it is logical to assume that a particular kind of cell is necessary to produce a particular kind of virus. The relation of viruses to host cells is, in fact, highly specific. We classify viruses according to the hosts in which they live as follows:

■ *Bacterial viruses,* which invade the cells of bacteria.
■ *Plant viruses,* which live in the cells of seed plants, especially flowering plants.
■ *Human and animal viruses,* which live in human and animal cells.

Actually, the virus-host relationship is even more specific than these large groups would indicate. Bacterial viruses invade only specific kinds of bacteria. Similarly, a plant virus may be specific for the cells of flower petals, leaf tissues, or stem tissues of a particular kind of plant. A specific human or animal virus may require the environment of cells of the skin, the respiratory organs, or the nervous system. It might surprise you to know that some viruses are even more specific than this. Polio viruses attack only the cells of one kind of nerve in the brain and spinal cord. Similarly, mumps is an infection of only one pair of salivary glands; the virus never invades the other pairs.

Bacteriophages

Much of our knowledge of viruses has come from investigations of the bacterial viruses, often referred to as *phages* (FAH-jiz). These are of the "tadpole" form, consisting of a round or many-sided head and a slender tail with attachment fibers (Figure 15-4). The coat of a phage is composed of protein, while the core is usually DNA, although phages that contain RNA have been discovered recently.

The first investigations of phages were conducted by two scientists working independently at about the same time. F. W. Twort, working in England in 1915, made an extensive study of a peculiar phenomenon he observed in the type of spherical bacterium called *staphylococcus* (STAFF-uh-loe-KAH-kus). For some unexplained reason, circular areas of bacterial destruction appeared in masses, or colonies, of these bacteria growing in a culture.

These holes, or *plaques* (placks) as we call them today, grew until the entire colony was destroyed. Twort found further that the unknown agent that caused destruction of the bacteria could be transferred with an inoculating needle to other colonies.

Two years later, in France, F. H. d'Herelle (DAY-RELL) conducted similar studies of a mysterious disease of the dysentery bacillus, a rod-shaped bacterium. He detected plaques that we know today were just like those that Twort had seen. It was d'Herelle who first named this invisible agent of bacterial destruction *bacteriophage,* or "bacteria eater."

The lytic cycle of a virulent phage

While early investigators were able to demonstrate the action of a phage on bacteria, it remained for the electron microscope to reveal what actually happens during the destructive attack. Through the use of the electron microscope, biologists have photographed all stages in the destruction of a bacterial cell.

We refer to the disintegration of a bacterium as a result of invasion by a phage as *lysis* (LICE-is). A phage that produces such a *lytic cycle* of destruction is called a *virulent phage.* As we describe the lytic cycle shown in Figure 15-7, check each numbered stage.

■ We begin with a normal, uninfected bacterium (1).

■ A phage has attached, tail down, to the cell wall of the bacterium by tiny hooks. An enzyme in the tail of the phage is dissolving an opening in the bacterial wall (2).

■ Having formed an opening in the wall, the tail contracts and injects the DNA of the phage core into the bacterial cell. The empty protein coat remains outside (3).

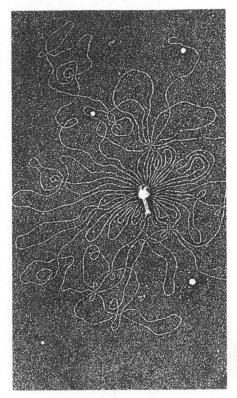

15-5 An "exploded" T₂ bacteriophage. Its central core of DNA is seen to be but a single strand. (courtesy A. K. Kleinschmidt and *Biochimica et Biophysica Acta*)

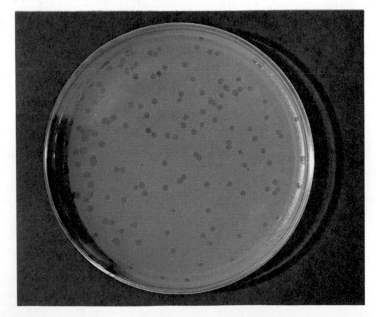

15-6 A bacterial culture spotted with plaques. (Lewis Koster)

bacterial cell

virulent phage

protein coat

phage DNA

15-7   The lytic cycle of destruction caused by a virulent phage.

■ Within a few minutes, the phage DNA appears near the DNA of the host cell, which normally controls the formation of bacterial substances. The phage DNA takes over this control by destroying the bacterial DNA, and the machinery of the bacterium is used to synthesize phage DNA and protein molecules. The bacterium has become a virus factory (4).

■ Soon the bacterium contains three hundred or more phage particles (5).

■ The bacterial cell ruptures, or undergoes lysis, releasing its phage content to attack other bacterial cells (6).

This entire cycle, from the entry of phage DNA to the bursting of the bacterial cell and release of phage particles, requires up to forty-five minutes. If each phage multiplies as much as three hundred times in a lytic cycle, you can see how an entire colony of bacteria composed of many millions of cells can be destroyed within a few hours. The rate of bacterial destruction by a phage can be demonstrated in a broth culture. When a phage in the amount of one billionth the quantity of bacteria is added to such a culture, the bacteria are destroyed in three to four hours.

Biologists have used the following analogy to describe a virulent phage cycle graphically. A tank moves up to the wall of an automobile factory in which an assembly line is in operation. The tank breaks through the wall, and its crew is discharged into the factory to disrupt the assembly line. Machines are reset and, using the same materials that were to have been automobiles, begin to turn out tanks on the assembly line. Soon one hundred or more tanks are rumbling through the plant. A wall is broken through, and the tanks move out to attack other automobile factories, discharge their crews to disrupt the factories' assembly lines, and cause the factories to assemble more tanks!

From a biological point of view, we can draw several important conclusions from an understanding of the lytic cycle of a phage:

■ The free phage cannot reproduce independently of a host. While it contains the DNA "blueprint" for synthesis of enzymes and other proteins, it cannot organize more of its own substance outside of a host cell.

■ A phage particle consists of a protein coat surrounding a DNA core. Both its inner core and its action are similar to genes normally present in bacteria and other cells.

■ Bacterial genes control the synthesis of specific enzymes and other proteins characteristic of the particular organism.

■ Phage DNA within a bacterial cell functions as a gene and apparently substitutes its control for the action of bacterial genes.

■ The phage DNA alters the chemical machinery of the bacterial cell, causing it to synthesize phage particles.

The economic importance of bacterial viruses

At one time, biologists thought virulent phages might offer a new and powerful medical weapon in the treatment of infectious

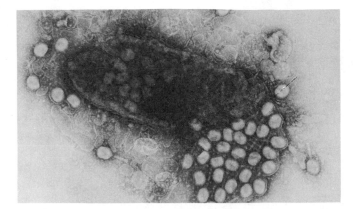

diseases. That is, they thought it might be possible to destroy disease-causing bacteria with a lytic phage – to fight disease with disease. However logical and promising this procedure might have seemed, further investigation has shown that there are many problems and limitations. For one thing, it is difficult to introduce the phage at the site of an infection, often deep in the body tissues. In other diseases, infection is widespread, making it difficult for a phage to contact a sufficient number of bacteria to be effective. Still another problem is the possibility that the human tissue environment may not be suitable for phage action. The conditions in the body are unlike those of a bacterial culture, where phage action may be effective. For these reasons, and perhaps others as well, the possible medical uses of phages are not as promising as we once thought.

Phages may play a harmful role in industries that utilize bacterial action. In the cheese industry, for example, certain bacteria are involved in the manufacturing process. Can you imagine the problem that arises if a phage is accidentally introduced and the cultures of these necessary bacteria are suddenly destroyed? This is a constant danger in industries that use bacterial processes.

Temperate phages – seeds of destruction

In certain cases, a phage may inject its DNA into a bacterial cell without causing production of phage particles. We call this type of phage a *temperate phage.* A chemical property of the bacterial cytoplasm must provide the seeming immunity, or resistance, to the temperate phage, immunity that prevents the phage from becoming virulent.

Biologists believe that the DNA of a temperate phage attaches to the bacterial chromosome and becomes a "foreign" gene that can cause the production of virus particles. In this condition, it is called a *prophage.* When the bacterial genes replicate in preparation for the splitting of chromosomes during cell fission, the phage DNA also replicates. Each resulting daughter cell receives the phage DNA (Figure 15-9).

prophage

bacterial cell
temperate phage
protein coat
bacterial chromosome
phage DNA

15-9 The invasion of a bacterial cell by a temperate phage.

15-10 A Rembrandt tulip, showing the "broken" coloration that results from a virus infection. (Verlin Biggs, Photo Researchers)

In this way, the phage DNA is multiplied generation after generation as a "stowaway" in bacterial cells, causing no immediate damage but being present as a potential "seed of destruction." At any time, a temperate phage may mutate and become insensitive to the inhibitory action of the bacterial cytoplasm. Ultraviolet radiation may also destroy the immunity of the bacterium. In either case, the phage becomes virulent, and the cell undergoes a lytic cycle in which free phage particles are formed.

Plant viruses

Many viruses attack the cells of plants, especially flowering plants, causing serious damage and often killing the plant. The core of many of these viruses is RNA. Plant viruses are often named for the host plant, the specific tissues they attack, or the nature of the symptoms of the infection.

We have already referred to the tobacco mosaic virus, the first discovered. Similar mosaic infections occur in the tomato, potato, bean, and cucumber. One of the most interesting mosaic viruses causes an infection in the cells of flower petals. This results in the light streaks or blotches that contrast with the normal petal coloration. One of the best examples of such a color variation caused by mosaic virus is the "broken" tulip (Figure 15-10). These tulips, which have enjoyed great popularity because of their unusual color pattern, for years were deliberately grown by gardeners, who did not know that the pattern resulted from a virus infection. Other plant virus diseases include the potato leaf roll, curly top of beet, peach rosette, aster yellows, and a serious disease of American elm trees known as phloem necrosis.

Plant viruses may be spread in a number of ways. Insects that suck juices from leaves often act as agents of infection. Among these insects are plant lice, leaf hoppers, mealy bugs, and thrips. Virus infections are sometimes spread when gardeners handle diseased plants.

Human and animal viruses

Many familiar diseases of man and animals are caused by viruses. In most of them, the virus invades only specific primary tissues in which the cell environment is suitable for multiplication of the virus particles. Symptoms of virus infections are as different as the viruses that cause them. Generally, however, a virus infection results in the disrupting of metabolic processes of the cells involved and damage or destruction of tissue. A lasting immunity remains after recovery from many virus infections, the common cold and influenza being notable exceptions. Biologists do not know the basis for this immunity.

Here is a list of the better-known virus diseases: smallpox, cowpox, chicken pox, shingles, cold sores and fever blisters, warts, influenza, measles, German measles (three-day measles), virus

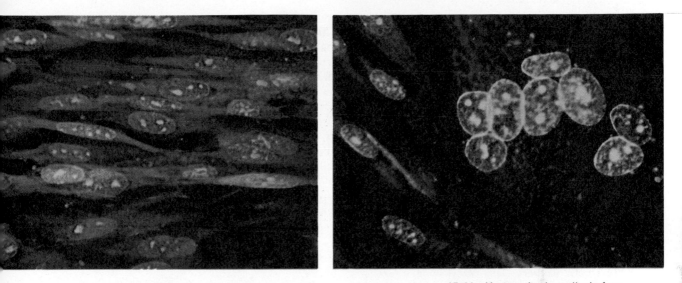

pneumonia, the common cold, parrot fever, yellow fever, infectious hepatitis, infectious mononucleosis (glandular fever), and mumps.

15-11 Human body cells before (left) and three days after infection with measles virus. RNA in the nucleoli and cytoplasm appears yellow-orange in color. (courtesy Chas. Pfizer & Co.)

Are viruses associated with cancer?

One of the most encouraging developments in cancer research in recent years has been the discovery of a possible relationship between viruses and certain forms of cancer. Such an association might provide the basis for long-sought preventive measures or even for cures for certain forms of cancer. To date, no viruses have been definitely linked with human cancers, but results of experiments with animals have been very encouraging. Dr. Ludwik Gross of the Veterans Administration Hospital in New York City has found that a form of leukemia can be induced in mice by injecting them with an extract that contains virus particles. Other investigators, using similar extracts, have induced more than twenty different kinds of malignant tumors in mice, guinea pigs, and hamsters.

Mycoplasmas—a possible link between viruses and cellular organisms

Recent investigation of a group of virus-like organisms of the genus *Mycoplasma* has raised several important biological questions. Could these organisms be a link between viruses and primitive cellular organisms such as bacteria? Mycoplasmas resemble viruses in form and size. However, unlike viruses, which require a cellular environment for activity, mycoplasmas can be grown on artificial culture media in the absence of living cells. Mycoplasmas lack cell walls and, therefore, assume various

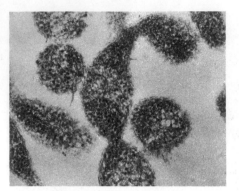

15-12 *Mycoplasma.* The outer membrane has a triple-layered appearance; the nuclear material is unbounded; and there are numerous ribosomes, which appear as black dots. (courtesy Jack Maniloff, University of Rochester)

shapes. The presence of an outer membrane is a definite cellular characteristic. While the method of reproduction of mycoplasmas is not known, the organisms form new bodies and increase in number without requiring the chemical processes of a cell. In this respect, they differ from viruses.

Mycoplasmas were discovered about seventy years ago and have been known to live harmlessly in the human mouth and nasal passages. In recent years, however, they have been found to be associated with several human and animal diseases. Among the human infections possibly caused by mycoplasmas are atypical pneumonia, arthritis, and infections of the urinary tract. Mycoplasmas have also been isolated from bone marrow taken from patients suffering from certain forms of leukemia. Medical discoveries such as these have focused attention on mycoplasmas in recent years.

SUMMARY

A knowledge of viruses is of great biological importance for several reasons. The virus may be an important link between the nonliving and the living. A virus particle consists of a protein coat, or shell, encasing a core of nucleic acid. Both of these substances are associated with living organisms. Yet, a free virus particle removed from a living cell does not perform the processes we associate with the living condition.

The control of cellular metabolic processes by nucleic acids is graphically illustrated in the destruction of a bacterial cell by a virulent phage. Only the DNA or RNA of a phage is injected into the bacterium. Yet, the viral nucleic acid completely disrupts the synthesis of nucleic acid and protein by the bacterium, converting the cell into a virus "factory."

Virus diseases remain a challenge to modern medicine. We are able to prevent certain of these infections, but many of the "weapons" the doctor uses to treat bacterial infections have little action on viruses, as you will learn in Chapter 17.

BIOLOGICALLY SPEAKING

pathogenic	nucleic acid core	lytic cycle
filterable virus	virulence	virulent phage
virus	plaques	temperate phage
protein coat	bacteriophage	prophage
		mycoplasmas

QUESTIONS FOR REVIEW

1. Explain the term *filterable virus.*
2. Describe various shapes of virus particles.
3. From what source did Dr. Wendell Stanley isolate the first virus?
4. What organic substance is found in the coat, or shell, of a simple virus? in the core?

5. What is meant by the virulence of a virus?
6. Classify the viruses into three groups on the basis of the host organism.
7. What contributions in virology were made by F. W. Twort and F. H. d'Herelle?
8. Describe the lytic cycle of a virulent phage virus.
9. List several limitations on using a phage in the treatment of an infectious disease.
10. List several plant virus diseases.
11. Account for the unusual coloration of a "broken" tulip.
12. In what respect is a temperate phage a potential "seed of destruction"?
13. List several well-known human virus diseases.
14. In what respects are mycoplasmas similar to viruses? Why are they considered to be cellular organisms?

APPLYING PRINCIPLES AND CONCEPTS

1. Discuss various factors that may alter the virulence of a virus.
2. Discuss several biological principles illustrated in the lytic cycle of a virulent phage.
3. Compare virus multiplication with the self-reproduction of cells of organisms.
4. Offer a theory to account for the high degree of specificity of viruses.
5. Explain the delayed-action aspect of a temperate phage.

RELATED READING
Books
CURTIS, HELENA, *The Viruses.*
   Natural History Press (distr., Doubleday and Co., Inc.), Garden City, N.Y. 1965. The fascinating history of man's discoveries about viruses.
SIGEL, MOLA M., and ANN R. BEASLEY, *Viruses, Cells and Hosts: An Introduction to Virology.*
   Holt, Rinehart and Winston, Inc., New York. 1965. A well written book introducing virology.
STANLEY, WENDELL, and EVANS G. VALENS, *Viruses and the Nature of Life.*
   E. P. Dutton and Company, Inc., New York. 1961. The essential facts about viruses and the closely related fields of genetics and cancer research.
Articles
ANDREWS, CHRISTOPHER H., "The Viruses and the Common Cold," *Scientific American,* December, 1960. An interesting account of the investigations involved in the tracing and isolation of the viruses that cause our most common illness.
MARAMORSCH, KARL, "Friendly Viruses," *Scientific American,* August, 1960. Attempts to dispel the much held idea that all viruses are harmful and gives insight to those that are beneficial to man.
FRAENKEL-CONRAT, H., "Rebuilding a Virus," *Scientific American,* June, 1960. Shows that a virus is able to tear down and rearrange material in order to duplicate itself.
FRAENKEL-CONRAT, H., "The Genetic Code of a Virus," *Scientific American,* October, 1964. Artificial changes in the hereditary material explain how it directs the synthesis of the three-dimensional molecule of protein.
Life Educational Reprint No. 28, "The Microbe Enemy" (originally published in *Life,* Feb. 18, 1966). Summarizes and updates research in virology and immunology, using color illustrations to present the characteristics of viruses and their interactions with the hosts they invade.

CHAPTER SIXTEEN

BACTERIA AND RELATED ORGANISMS

Bacteria—primitive cellular organisms

As we proceed from the study of viruses to the study of bacteria and their relatives, we shift from a molecular level to a modified cellular level of life. While there is a lack of agreement among biologists, many believe that bacteria were the first forms of life on earth, appearing late in the Archeozoic period (see Figure 13-4). Long before there were green plants capable of photosynthesis, certain primitive bacteria may have utilized energy from iron, sulfur, and nitrogen compounds instead of from the sun. Geologists think that the extensive deposits of iron ore we use today are the result of bacterial action during ancient geologic times. Later, when green plants began building up stores of organic compounds, other kinds of bacteria began using them as a food supply. Still other bacteria invaded the tissues of the plants themselves, as well as the bodies of animals.

Bacteria have survived through the ages and have increased their numbers until today they are the most abundant form of life. Too small to be seen by the naked eye, they live almost everywhere. They thrive in the air, in water, in food, in the soil, and in the bodies of plants and animals. In fact, any environment that can support life in any form will have its population of bacteria.

Louis Pasteur—the father of bacteriology

In the history of biology, there have been a few truly great investigators who have changed the direction of the science and opened a whole new field for experimentation. Louis Pasteur was such a figure. This great scientist of a century ago is no stranger to you. Do you remember his experiment that dealt a crushing blow to the belief in spontaneous generation (see Chapter 2). This experiment was only a small part of this great man's contribution to science. Let's take another look at the man who introduced the world to microorganisms, revolutionized the practice of

244

medicine, and provided much of the basis for the modern science of bacteriology.

Pasteur was born in Dôle, France, in 1822. He graduated from college at an early age with a brilliant record of achievement in chemistry. In 1854, at the age of 32, he was appointed professor of chemistry and dean at the University of Lille. It was here that circumstances combined to alter the course of biological science for generations to come.

In the city of Lille, alcohol was manufactured by fermenting sugar-beet juice in large vats. Chemists at that time thought that alcoholic fermentation was a purely chemical process. They had seen tiny yeast cells growing in fermenting beet and fruit juices but had always considered these cells to be products of spontaneous generation, in no way connected with the fermentation process.

Then, one day, a serious problem arose at the fermentation plant. In several of the vats, the juice was souring rather than becoming alcoholic. What had happened to change the chemical process? Pasteur was called in to try to find the answer.

With his microscope, Pasteur examined the juice that was fermenting properly. He saw numerous oval yeast cells dispersed through the liquid. Over a period of several hours, the cells increased in number, and the alcohol content increased. Were the yeast cells producing the alcohol? Pasteur thought that the answer to this question might be found by studying some sour juice. Accordingly, he examined a drop of liquid from a vat of sour juice. Here he found an entirely different population of microorganisms. Instead of yeast cells, he found quivering rods, smaller than yeasts but apparently alive. He found that this liquid contained lactic acid, not alcohol. Furthermore, the lactic-acid content increased as the microbe population grew.

This discovery led Pasteur to more extensive studies of fermentation. Three years later, he set up a small laboratory in Paris. Here he proved that fermentations are associated with microorganisms and that the products formed depend on the organism involved. Yeasts produce alcohol. Bacteria may form lactic acid in beet juice and in milk.

Fortunately, Pasteur did not end his explorations in microbiology with fermentation studies. Far more important than these was the presentation of his germ theory to the scientific world. If one kind of bacteria could ferment beet juice, might other kinds cause disease? Further discoveries made by this remarkable chemist who became one of the greatest biologists of all time are yet to be discussed.

What are bacteria?

Since the time when Pasteur observed bacteria, they have been considered both animals and plants. Until recently, they were classified with the fungi in the old phylum Thallophyta of the plant kingdom. More recently, biologists, by placing them in the kingdom Protista, have avoided calling them anything but organisms. Together with close relatives, they constitute the protist phylum *Schizomycophyta* (sᴋɪᴢᴢ-uh-ᴍʏ-KAH-fuh-tuh), a name meaning "fission fungi."

Compared with a virus, a bacterium is quite large. You will recall that a single infected bacterial cell may contain as many as 300 phage particles. However, compared with the cells of most other organisms, bacteria are extremely small. We used *milli-microns (mμ)* to measure viruses; bacteria are measured in *microns* (one micron, $1\mu = 0.001$ millimeter). Spherical bacteria range in size from about 0.5μ to 1.5μ in diameter, while the rod-shaped forms average from about 0.2μ to 2μ in width and from 0.5μ to as much as 10μ in length. The average length of bacteria is approximately 1.5μ. When we convert these figures to inches, we see that bacteria range from about 1/50,000 to 1/10,000 of an inch. If these figures mean little to you, consider that several thousand bacteria could be placed on the period at the end of this sentence, or that a small drop of water may contain as many as 50 million bacteria.

You can see some bacteria and even determine their shape if you use the high-power magnification of your microscope (430×). Additional magnification (1,000 to 1,500×), however, is necessary before you can see them clearly. No light microscope can magnify sufficiently to reveal most of the internal structures of bacterial cells. For this we need the 50,000× to 100,000× enlargement of the electron microscope.

Forms of bacteria

While bacteria vary greatly in size, their cells are of three basic shapes. Some bacterial cells tend to exist singly when grown in liquid cultures or broths, while others often remain attached after cell division and form colonies of cells. We can classify the basic cell shapes and groupings as follows (Figure 16-1):

- *coccus* (plural, *cocci*): cells sphere-shaped, or globular
  diplococcus: cells often joined in pairs or short filaments
  staphylococcus: clusters of cells
  streptococcus: filaments, or strings, of cells
  tetrad: groups of four cells arranged as a square
  sarcina: cubes or packets of cells

- *bacillus* (plural, *bacilli*): cells cylindrical, or rod-shaped.
  diplobacillus: cells in pairs
  streptobacillus: cells joined end to end forming a filament or thread

- *spirillum* (plural, *spirilla*): cells in the form of bent rods or cork-screws

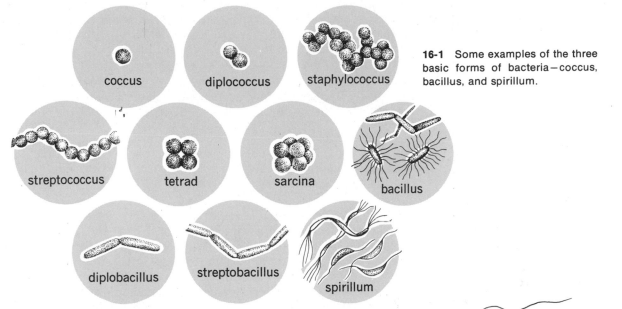

16-1 Some examples of the three basic forms of bacteria—coccus, bacillus, and spirillum.

Structure of a bacterial cell

Bacterial cells are surrounded by a *slime layer* of varying thickness. This gelatinous coat may protect the cell from unfavorable environmental conditions and enable it to stick to the surface of food supply or to a host cell or organism. Some bacteria have a thick slime layer called a *capsule*. Perhaps you remember from our discussion of transformation in Chapter 10 that the pneumonia organism *(Diplococcus pneumoniae)* is an example of such a capsulated organism (see Figure 10-11). It is interesting that the presence of the capsule is an indication of the virulence of the pneumonia organism.

Capsulated forms produce a serious infection. The capsule may protect the capsulated bacteria from the normal body defenses against infectious organisms. Strains of pneumonia organisms lacking capsules are low in virulence or even noninfectious. Perhaps these organisms are easily destroyed by the body defenses.

Some bacterial masses form gelatinous coatings. Have you ever removed the gelatinous "mother of vinegar" from a vinegar jar? This is such a bacterial mass. In other species, including iron bacteria, a *sheath* may surround an entire filament of bacteria. Iron compounds are often deposited in the sheaths of these bacteria, forming the reddish-brown, slimy masses often seen in streams.

Beneath the slime layer or capsule is a *cell wall* that gives the cell its characteristic shape. A thin flexible *plasma membrane* lies just beneath the cell wall and marks the outer edge of the *cytoplasm*. The cytoplasm of bacteria is granular and rich in RNA.

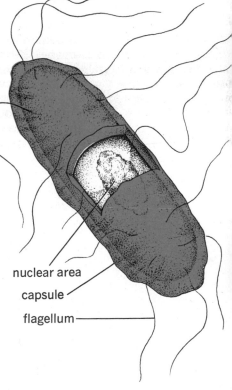

nuclear area

capsule

flagellum

16-2 The general structure of a bacterial cell.

Bacteria lack a nuclear membrane. However, they have been found to contain DNA in genes that are concentrated in chromatin bodies lying in a nuclear area near the center of the cell. Some biologists believe that these bodies constitute a single bacterial chromosome. Mitosis does not occur in bacteria, although there is division of the chromatin bodies each time a cell divides. This results in the equal distribution of DNA in daughter cells.

Many bacteria contain granules of stored food and other materials dispersed through the cytoplasm. A few also contain vacuoles of water and dissolved materials. Mitochondria, the vital centers of respiration in other cells, are lacking in bacteria. Respiratory enzymes, usually found in mitochondria, seem to be concentrated on the cell membrane of a bacterial cell.

Motility in bacteria

Various bacillus and spirillum forms of bacteria are equipped with threadlike whips, or bacterial *flagella* (fluh-JELL-uh) (singular, *flagellum*), that extend from the plasma membrane and propel the cell through water and other fluids. Flagella may be found singly or in tufts at either or both ends of a bacterial cell. In some forms, flagella are found all around the cell. Flagella are visible only when they are treated with special stains. Under the highest magnification of the light microscope, they appear as minute threads. The electron microscope shows them to be cytoplasmic extensions that project through openings in the cell wall (Figure 16-3). Flagella are strands of protein molecules resembling the microscopic fibers composing muscle. Thus, the basis for the muscle contractions of animal organisms may lie in the beating flagella of bacteria.

You can recognize motility in bacteria by observing their random movement in a microscopic field. All cells move independently in a quivering, twisting manner. This *true movement* by means of flagella should not be confused with an oscillating or

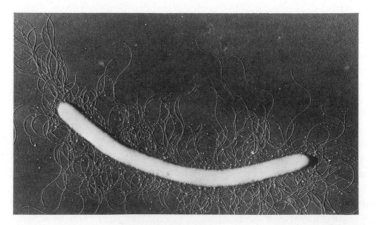

16-3 The flagella of *Proteus vulgaris* show clearly in this electron micrograph. (courtesy Houwink, Van Iterson and *Biochimica et Biophysica Acta*)

bouncing motion known as *Brownian movement.* This motion results from the jarring of very small bacteria, especially coccus forms, caused by collisions of molecules or other moving particles in a fluid with the bacteria.

Conditions for growth of bacteria

In discussing various environmental requirements for the growth of bacteria, we need to make a distinction between activity and survival. Bacteria have the property of remaining dormant when conditions for growth are lacking. For example, they may float high in the air or lie on an object in an inactive condition. Then, when environmental conditions are favorable, they may enter a period of very rapid growth and reproduction.

A *suitable temperature* is essential for bacterial activity, but this temperature varies greatly with different species. The majority of bacteria thrive at moderate temperatures, ranging from 80°F to 100°F. Those that cause human infections grow best at 98.6°F, that is, at normal body temperature. But there are many bacteria that grow best at much lower temperatures, ranging from 32°F to an upper limit of 85°F. These bacteria occur in ocean depths, in the cold soils of the far north, and high in the stratosphere. At the other extreme are bacteria that thrive at temperatures ranging from 110°F to as high as 185°F. These are found in hot springs and in the hot environment of decomposing sewage, silage, and other plant materials.

Moisture is another growth requirement for bacteria. Active bacterial cells are about 90 percent water. In dry surroundings, water loss makes the cells inactive; prolonged dryness will kill bacteria of most species.

Darkness is also a condition for best bacterial activity for most species. Exposure to sunlight may retard growth, and ultraviolet radiation actually kills cells. An application of this can be seen in the partial sterilization of the air in hospital operating rooms by ultraviolet rays emitted from special lamps.

A *suitable food source* is still another requirement for the growth of bacteria. In this respect, bacteria vary greatly. Some are highly specific in their food needs. Most pathogenic bacteria require living tissues or substances similar to their own in chemical composition. Many bacteria are more tolerant and can live on a wide variety of food materials.

Bacterial nutrition

The majority of bacteria lack the ability to synthesize the substances they require and must therefore live in contact with pre-formed organic matter. In this respect, they are *heterotrophic.* The lack of ability to carry on photosynthesis puts them in direct competition with man and other animals for food. Those bacteria that we classify as *saprophytes* (SAP-ruh-FITES) utilize dead organic matter or nonliving organic substances such as food products intended for man's use. *Parasites,* on the other hand, invade the bodies of plants and animals and take their nourishment directly from living tissue. We refer to the organism supporting the parasite as the *host.*

Bacteria secrete powerful enzymes. These organic catalysts are essential in causing chemical changes both inside and outside of the bacterial cell. Many of these are digestive enzymes. An enzyme acts only on one kind of food substance. Thus, the food source or host required by a specific kind of bacterium is determined by the enzymes it produces. Enzyme action simplifies complex organic molecules by converting them to water-soluble substances that can be absorbed through the cell wall and membrane. Enzymes allow saprophytic bacteria to use a variety of organic materials, many of which are useless to most other forms of life. Among these are wood and other cellulose materials. Parasitic bacteria lack many of the enzyme systems found in saprophytic organisms. This explains why they must live in contact with living tissue and utilize enzyme action of the host cells.

A relatively small number of bacteria are *autotrophic.* That is, they synthesize organic compounds from carbon dioxide and other simple inorganic substances. Certain of these organisms utilize inorganic compounds as a source of energy. These are the *chemotrophic,* or *chemosynthetic,* bacteria. Of these, some use iron compounds. Others use sulfur compounds. Still others, including nitrite and nitrate bacteria, oxidize nitrogen compounds. Perhaps the most unusual autotrophic bacteria are those capable of photosynthesis. *Photosynthetic* bacteria contain bacteriochlorophyll, a blue-gray pigment that differs chemically from the chlorophyll of higher plants. The pigments in these bacteria are dispersed in the cytoplasm rather than localized in chloroplasts.

Bacterial respiration

Atmospheric oxygen is an important factor in the growth of bacteria. Some bacteria require oxygen for life; others require its absence. Others have requirements between these two extremes.

Certain bacteria require atmospheric oxygen for respiration, just as most plants and animals do. These *obligate aerobes* (AIR-obez) split glucose molecules and form carbon dioxide and water during respiration. This group includes the bacteria causing diphtheria, tuberculosis, and cholera.

At the other extreme are organisms classed as *obligate anaerobes* (AN-uh-robez). These bacteria cannot grow in the presence of atmospheric oxygen. Among the best-known obligate anerobic bacteria are those that cause tetanus and botulism.

Many bacteria are *facultative anerobes* — those that grow best as aerobes but may grow, at least to some extent, as anaerobes.

Among these organisms, we find *Escherichia coli* (ESH-uh-RICK-ee-uh KOE-LY), the common bacillus of the human intestine. Several disease-producing bacteria, including those causing typhoid, and scarlet-fever, are also facultative anaerobes. Much less common are the **facultative aerobes** – organisms that are primarily anaerobes but are able to maintain limited activity in the presence of atmospheric oxygen.

Recall from Chapter 6 that *fermentation* is an anaerobic process. The products of fermentation vary with the organisms involved. Bacteria that produce alcohol by fermentation split glucose molecules in such a way as to release only a small amount of energy. In this process, carbon dioxide and pyruvic acid are formed. The pyruvic acid is then reduced (hydrogen is added) to form ethyl alcohol. We again can summarize the net fermentation reaction in the following simplified equation:

$$C_6H_{12}O_6 \rightarrow 2\ C_2H_5OH + 2\ CO_2 + \text{energy}$$
$$\text{glucose} \qquad \text{ethyl} \qquad \text{carbon}$$
$$\text{alcohol} \qquad \text{dioxide}$$

Other bacteria produce lactic acid by fermentation. In this reaction, a glucose molecule is split in such a way as to form pyruvic acid, which is then reduced (hydrogen is added) to lactic acid. Again, only a small amount of the energy stored in the glucose molecule is released. We again can summarize this reaction in the following simplified equation:

$$C_6H_{12}O_6 \rightarrow 2\ C_3H_6O_3 + \text{energy}$$
$$\text{glucose} \qquad \text{lactic acid}$$

Another bacterial anaerobic reaction results in the formation of methane. In this reaction, hydrogen is combined with carbon dioxide, forming methane and water and releasing energy. In summary,

$$4\ H_2 + CO_2 \rightarrow CH_4 + 2\ H_2O + \text{energy}$$
$$\text{hydrogen} \quad \text{carbon} \quad \text{methane} \quad \text{water}$$
$$\text{dioxide}$$

Methane is called marsh gas because it may be seen bubbling to the surface of bogs, swamps, and ponds. It is formed from decomposing organic matter accumulated in the bottom of these damp places. Methane is also the major component of natural gas.

Bacterial reproduction

Bacteria multiply by undergoing transverse binary fission (Figure 16-4). When conditions for growth are ideal, they multiply at an amazing rate. Consider that a cell may be formed by a division of a mother cell, grow to maturity, and start dividing again in a period of only twenty minutes! This may not seem alarming until you begin to calculate the results. Start with one cell that divides and becomes two in twenty minutes. These become four

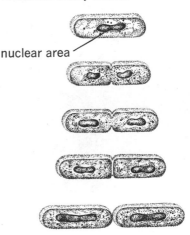

16-4 Bacterial reproduction by transverse binary fission.

nuclear area

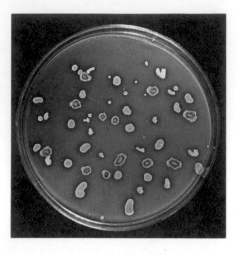

16-5 Colonies of bacteria on a Petri-dish culture. (Walter Dawn)

16-6 Germination of a bacterial endospore of *Bacillus mycoides*. A portion of the spore coat appears at each end of the growing vegetative cell. (courtesy Knaysi, Baker, and Hillier, *Journal of Bacteriology*, **53**: 525, 1947)

in forty minutes and eight at the end of an hour. How many would be formed in twelve hours? in twenty-four hours? If this rate of increase continued for twenty-four hours, a mass of bacteria weighing over 2,000 tons would be formed! If this rate of increase continued through another twenty-four-hour period, the mass would weigh many billion tons! Why, then, don't bacteria cover the earth and crowd out all other living things, as you might guess would happen?

Suppose we start with a single bacterium on a culture medium, such as nutrient agar, within a Petri dish. After a brief period of adjustment to the culture medium, the cell divides. For a few hours, growth and cell divisions occur rapidly. If divisions occur as rapidly as once in twenty minutes, the number of cells will have increased to 512 within three hours. As the mass increases in size, the problem of *mechanical crowding* also increases. With this crowding comes *competition for food*. There is also an *accumulation of metabolic wastes*.

Within a short time, these factors cause a gradual decrease in the rate of growth and reproduction. The mass continues to increase in size, however, and within twenty-four hours, a colony composed of several billion cells may become visible. This colony will continue to increase in size for perhaps another twenty-four hours, after which the colony stabilizes, with some cells dying or becoming inactive.

Spore formation

Many bacillus bacteria form **endospores.** During endospore formation, a new protoplasmic mass develops within a vegetative cell. Then, still within the vegetative cell, this protoplasmic mass, including genetic material, becomes encased in a protective spore coat.

In most species of bacteria, endospore formation is not a form of multiplication. A bacterium forms a single endospore that may germinate after a period and become a single active cell again. Hence, the number of bacteria is not increased by spore formation.

Bacterial endospores, which are usually spherical or oval, are very resistant to the most adverse environmental conditions of heat, cold, and drying, among others. For example, a few years ago, scientists were surprised to find bacterial spores in deep layers of ice in the Antarctic. Apparently these spores had been trapped there for many thousands of years. Yet when they were placed in an environment that was favorable for growth they became active bacillus cells! As another example, some bacteria survive periods of an hour or more in boiling water.

The ability of certain bacteria to survive the most unfavorable environmental conditions while in the endospore stage creates a medical problem. For example, while active tetanus-causing bacteria are restricted to an anaerobic environment, such as that which exists in a closed wound, inactive endospores are abundant in soil, on objects, and even in the air. The endospores are not even affected by being immersed in boiling water or in those cleansing and antiseptic agents that would be effective against active bacterial cells.

Sexual reproduction in bacteria

Bacterial cells can apparently reproduce by fission indefinitely. For many years, however, biologists wondered whether or not sexual reproduction occurred in organisms as low in the scale of development as bacteria. Recent research has established that sexual reproduction does occur, at least in some species of bacteria; perhaps it may even occur in all bacteria.

Extensive studies of sexual reproduction in bacteria have been conducted with *E. coli,* an organism often referred to as the colon bacillus because it normally lives in the human large intestine. Two mating types of this organism have been discovered. One is designated as male, the other as female. In this bacterium, maleness results from the presence of genetic particles that are not present on the chromosome. Femaleness results from the lack of these particles.

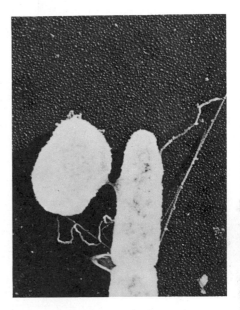

16-7 A photomicrograph of sexual reproduction in bacteria. Note the cytoplasmic bridge joining the two cells. (T. F. Anderson, E. L. Wollman, F. Jacob)

When male and female strains of *E. coli* are mixed in a fluid suspension such as a broth culture, the cells come together. A cytoplasmic bridge joins the two cells. The male cell injects its chromosome along with its sex-determining particles into the female cell. Soon after this process the male cells, lacking genetic materials, die. What do you suppose happens to the female cells? Having received the male genetic particles, they change to male cells.

What is the significance of sexual reproduction in bacteria? Remember that when organisms reproduce sexually, both parents contribute genetic material. The resulting new genetic combinations may produce offspring that differ from both parents. These offspring may be more favorably adapted to their environment and have better chances of survival. In the mating process that occurs in *E. coli,* the male cell that results contains genetic material from both the male and female strains. The offspring arising from fission of this cell will in turn contain newly combined genetic material. It may be that this sexual process has been important in producing the many types of bacteria existing today.

Beneficial activities of soil bacteria

We have heard so much about disease-causing, or *pathogenic* (path-uh-JEN-ick), bacteria that it is easy to get the impression that all bacteria are dangerous to health. This is untrue; the majority of bacteria are entirely harmless, having no direct relation to our lives or to our economy. Many other bacteria are beneficial to man. As a matter of fact, our lives depend on the bacteria that live in the soil. To understand the vital role these bacteria play, we need to consider a much larger segment of the biological society.

Green plants absorb minerals from the soil and use them in synthesizing organic compounds. Animals consume plants and rearrange these compounds to meet their specific needs. In this way, many of the chemical requirements for life come directly or indirectly from the soil. When a plant or animal dies, the chemical substances of which it is composed are left in a complex form—products of life too complex to be used as a source of energy and material by another generation of plants and animals. What if all of the organisms that populated the earth during the past few thousands of years were still lying about in their original chemical condition? Would there be any room on the earth's surface for new generations? Would there still be sufficient chemical supplies in the earth's crust to meet the needs of these new organisms?

You know, of course, that organisms are decomposed after death and that their materials are returned to the earth and atmosphere. Did you know, however, that this results largely from bacterial activity? During decay and putrefaction, bacteria and other soil organisms, including molds, break down complex molecules in dead plant and animal matter and form simpler chemical compounds in meeting their energy and nutritional requirements. In this way, the matter that once composed a living organism is broken down into substances that can be absorbed by plant roots and used in another series of building activities in new organisms.

Build up and break down, synthesize and decompose—matter comes and goes constantly in chemical cycles involving organisms. Bacteria are vital in these chemical cycles. Since chemical cycles involve many kinds of organisms and the interrelation of their lives, we have reserved a full discussion of them for an ecological study in Unit 8.

Bacteria in industrial processes

Bacteria play an important part in the processing of many foods and other products of industry. We will discuss a few of these processes briefly.

Vinegar-making involves two kinds of microorganisms: yeasts and bacteria. In producing vinegar, yeasts ferment sugar in fruit juice and change it into alcohol. The alcohol concentration in the fermented solution reaches 10 to 13 percent by volume before the yeasts are deactivated. Acetic acid bacteria (*Acetobacter*) then oxidize the alcohol, changing it to acetic acid. Commercial vinegars are about 4 percent acetic acid.

Sauerkraut-making is another process involving bacteria. In the preparation of sauerkraut, cabbage leaves are shredded and put in an airtight jar. Anaerobic bacteria *(Lactobacillus)* ferment the sugar in the cabbage leaves and convert it to lactic acid. A small amount of carbon dioxide and alcohol are formed in the process.

Several kinds of bacteria are involved in the *tanning of leather.* After a hide has been preserved by drying or salting, it is soaked and scraped to remove excess flesh. The hair is removed either by bacterial action or by chemical treatment. The final tanning takes place in large tanks or vats in which bacteria and other organisms attack the hides and make them pliable.

The *retting of flax* is a process many centuries old in which the flax stem fibers that will eventually be made into linen are separated from the stem tissues. When cut stems, tied into bundles, are weighted and submerged in tanks of warm water, the water-soluble materials dissolve and form a culture medium suitable for the growth of anaerobic bacteria *(Clostridium).* These organisms slowly ferment the pectin that holds the fibers together. After ten to fourteen days, the intact fibers are removed from the water, thoroughly washed and dried, and processed into linen.

Silage (SY-lij) is a succulent food product that is readily eaten by cattle. In making silage, fodder, such as cornstalks, alfalfa, and clover, is shredded and put into an airtight silo. As the living plant cells respire, they rapidly consume the available oxygen and replace it with carbon dioxide. It is this anaerobic condition that prevents the decay of the plant materials but permits bacteria *(Lactobacillus)* to ferment the sugar in the plants. The lactic acid formed in the fermentation process is of great value in the milk production of dairy cattle.

The *tobacco* industry uses bacterial action in the curing process. Stalks and leaves of the tobacco plant are harvested and hung in special curing barns where "sweating" occurs. Bacteria ferment carbohydrates in the moist leaves, producing special flavors.

Bacteria in the dairy industry

Since milk, in addition to being a nutritious food for human beings, is an ideal culture medium for many bacteria, great care must be exercised in maintaining rigid sanitary conditions throughout the dairy industry. Without this care, milk, ice cream, and other dairy products could help to spread disease until it reached epidemic proportions.

Pasteurization is the destruction by heat of disease-causing and most other organisms in milk, fruit juices, wines, and malt beverages. The principle of pasteurization was discovered by Louis Pasteur and was first used to prevent the souring of wines. Originally, milk was pasteurized to destroy the tuberculosis-causing bacterium. However, since pasteurization destroys milk-souring bacteria as well as pathogenic organisms, pasteurized milk sours more slowly than raw, unpasteurized milk.

Both the temperature and duration of treatment are important in milk pasteurization. The temperature must be high enough to destroy harmful bacteria, yet low enough to avoid overheating, which would reduce the quality of the milk and give it a scorched flavor. The time must be sufficient to allow all of the milk to reach the pasteurization temperature.

Two methods are used in the commercial pasteurization of milk. In the low-temperature, or *holding,* method the milk is heated to 145° F for thirty minutes. In the high-temperature, or *flash,* method the milk is heated to 161°F for fifteen seconds. About 90 to 99 percent of the bacteria present are killed. Rapid cooling retards the growth of surviving organisms. However, pasteurized milk must be kept refrigerated to reduce bacterial growth. It should also be covered to avoid the introduction of additional bacteria from the air.

In making *butter,* cream is first separated from the other milk solids. After separation, the cream is pasteurized to reduce the number of bacteria present. The pasteurized cream is then inoculated with special "starter" bacteria that form lactic acid and act on other ingredients of the cream, producing desirable flavors. After a period of bacterial activity, the cream is cooled and churned. Globules of butterfat stick together and form butter. The liquid that is drained off is *buttermilk.*

Buttermilk is also produced by inoculating pasteurized skim milk with starter bacteria. Certain of the bacteria *(Streptococcus)* form lactic acid; others produce desired flavors and odors. We call this product *cultured buttermilk.*

Most *cheese* is made from whole or skimmed cow's milk, although other kinds of milk, including goat's milk, may be used. Several steps are used in making cheese. Starter bacteria are added to the milk to form lactic acid. This causes curdling, or separation of the milk solids. The *curd* made from whole milk consists of casein (milk protein), fats, and minerals. If skimmed milk is used, the curd is largely casein. After curdling, the curds are separated from the watery, liquid portion known as *whey.* In some types of cheese, including cottage cheese, the curd is eaten while moist and unripened. Butterfat can be added to produce cream cheese. In processing other kinds of cheese, the curd is salted and pressed to remove moisture. A period of "ripening" follows.

In the ripening of soft cheeses, such as Limburger, microorganisms are grown on the surface of the cheese. Enzymes penetrate from the surface and their action causes the distinctive flavor of the cheese. In making hard cheeses, such as Cheddar and Swiss, ripening takes place because of the action of bacteria and molds, which are inoculated throughout the interior of the mass of curd. Both the time and temperature of the ripening process vary with the kind of cheese.

Food spoilage and preservation

Bacteria are among our chief competitors for food. We cannot even estimate the amount of food lost each year because of spoilage resulting from bacterial action. Carbohydrates ferment, protein foods putrefy, and fats and oils become rancid due to bacterial action on them.

Most foods would remain edible for many months or even years if bacteria were not present or if they could not multiply. Hence, we have developed efficient ways of preserving foods. One of the two most important methods of preserving foods is to destroy all bacteria present and seal the food in an appropriate container.

■ *Canning* is an application of this method. Canned foods can be preserved indefinitely if the canning is done properly.

The other method involves environmental control. If foods are held under conditions that will not allow bacterial growth and activity, they will not spoil even though bacteria are present in an inactive or dormant condition. Some examples of applications of this method are:

■ *Salt curing* destroys bacteria by plasmolyzing cells. This method has been used for centuries in preserving fish and pork and other meats.

■ *Refrigeration* not only retards the growth and multiplication of bacteria, but it also reduces the chemical activity of their enzymes.

■ *Quick-freezing* is a far more efficient method of preserving foods that can be frozen solid and thawed without reducing their palatability. At temperatures of 0°F to minus 10°–15°, bacteria cease nearly all activity.

■ *Dehydration* involves the removal of water from a food to the point that bacteria cannot grow or secrete enzymes. Active bacterial cells can be killed by dehydration, although endospores may survive. Dehydrated foods are widely used today. They are easily packaged and shipped and require no refrigeration. They must be kept dry, however, or spoilage will occur rapidly. When wanted for use, they can be rehydrated by adding water.

■ *Chemical preservatives*, once very widely used, are largely discontinued or forbidden by law today, although some foods, including dried fruits, are still preserved with chemicals.

■ *Radiation* may be used much more widely in years to come in the food industries. If meats and other foods are packaged and sealed, then irradiated to kill all organisms present, they will keep indefinitely without the necessity of drying, freezing, or refrigerating them.

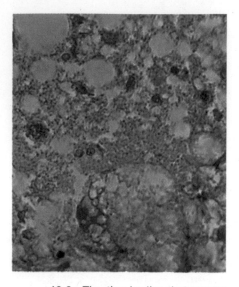

16-8 The tiny bodies that appear blue in the photomicrograph are rickettsiae. They are smaller than bacteria and larger than viruses. (Walter Dawn)

16-9 An electron micrograph of the syphilis organism, *Treponema pallidum.* (Dept. of HEW)

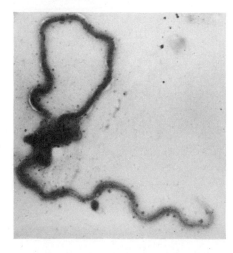

Relatives of bacteria—the rickettsiae

The *rickettsiae* (ri-KET-see-ee) are a group of organisms that seem to be between the bacteria and the viruses. Rickettsiae are tiny rod-shaped or spherical organisms averaging 0.3μ to 0.5μ in size. They are barely visible under the highest magnifications of the light microscope (Figure 16-8).

Rickettsiae resemble bacteria in that they are cellular and reproduce by fission. They are similar to viruses in that they grow and reproduce only in living cells. Thus, they must be cultivated in tissue cultures, chick embryos, or in the tissues of animals.

The rickettsiae were discovered in 1909 by Dr. Howard T. Ricketts, for whom they were named. Dr. Ricketts found the first rickettsial organism in the blood of victims of Rocky Mountain spotted fever. He also demonstrated that cattle ticks transmit this rickettsial infection to human beings. Unfortunately, in 1910 Dr. Ricketts died, a victim of typhus fever, the rickettsial organism he was investigating at the time.

So far as we know, rickettsiae are transmitted only by insects and their relatives, including the human body louse, ticks, and mites. The organisms live in cells of the carrier's intestine but do not cause disease in the carrier. In the human body, however, they invade cells and cause infection. Those who recover from a rickettsial disease have a subsequent immunity to the damaging effects of the organisms, although the organisms may remain in the cells apparently without causing damage.

In addition to Rocky Mountain spotted fever and typhus fever, rickettsiae cause trench fever and the mysterious Q fever. These infections are similar in that they produce fever, skin rashes, and dark blotches beneath the skin. The blotches are caused by hemorrhages resulting from damage to the cells lining small blood vessels.

The spirochetes

A group of microorganisms called *spirochetes* (SPY-ruh-KEETS) seem to lie between the bacteria and the more specialized one-celled protists we call protozoans. Many of the spirochetes fall in the size range of bacteria, with cells 3μ to 15μ in length and 0.5μ in thickness. However, certain spirochetes reach as great a length as 500μ. The cells of spirochetes are long and cylindrical. Some have a spiral shape, while others are tight corkscrews. So far as is known, no spirochete has an organized nucleus. Reproduction is by transverse fission. Endospores are not produced. The cell walls of spirochetes are flexible, permitting the organisms to move through fluids with a characteristic quivering action. All spirochetes were thought to lack flagella until recent studies of certain forms *(Treponema)* with the electron microscope revealed threadlike cytoplasmic projections that look like flagella, but are not true flagella.

We are most familiar with the spirochetes that live in the human body and cause disease. The best-known pathogenic spirochete is the syphilis organism *(Treponema pallidum),* which lives in the bloodstream and may invade the nervous system (Figure 16-9). Spirochetes are spread in the material that is discharged from lesions or eruptions in the skin and mucous membranes during the early stages of the disease.

SUMMARY

In addition to being the most primitive organisms, bacteria are also the most abundant and most widely distributed. Any environment that supports life has its bacterial population.

Bacteria multiply at an amazing rate when conditions for growth are ideal. The usual method of reproduction is fission, a characteristic referred to in their protist phylum name, Schizomycophyta.

Whether we consider bacteria beneficial or harmful depends on the way in which their processes affect our lives. Decay is beneficial when it involves the breakdown of organic matter for reuse by other organisms in a chemical cycle. The same process involving foods is considered harmful spoilage. A relatively small number of pathogenic bacteria cause such infectious diseases as typhoid fever, scarlet fever, diphtheria, and tetanus. Many bacterial diseases are now under the control of modern medicine, as you will discover in the next chapter.

BIOLOGICALLY SPEAKING

coccus	true movement	facultative anaerobes
bacillus	Brownian movement	facultative aerobes
spirillum	saprophytes	endospores
slime layer	parasites	pathogenic
capsule	host	pasteurization
sheath	obligate aerobes	rickettsiae
flagella	obligate anaerobes	spirochetes

QUESTIONS FOR REVIEW

1. What difference did Louis Pasteur find in the population of microorganisms in the two samples of beet-juice extract he examined?
2. Explain why bacteria are classified as protists rather than plants in the recent classification of living things.
3. Expressed in microns, what is the average size of bacteria?
4. Classify bacteria into three groups, based on the shape of their cells.
5. Distinguish between a bacterial slime layer and a capsule.
6. Describe the structure and function of bacterial flagella.
7. Distinguish between true movement and Brownian movement of bacteria.
8. List four environmental requirements for the growth of bacteria.
9. Distinguish between heterotrophic and autotrophic bacteria.
10. Distinguish between aerobic and anaerobic bacteria.
11. Name several products of the fermentation of glucose by microorganisms.
12. Describe the chemical activity of bacteria in forming methane.

13. How rapidly do bacteria multiply under ideal conditions? What stops them from multiplying?
14. Why are certain bacillus bacteria more difficult to destroy than other forms of bacteria?
15. List several industrial uses of bacteria.
16. Describe several uses of bacteria in the dairy industry.
17. List several methods of preserving foods by killing bacterial cells and by controlling the environment and preventing growth and reproduction.
18. How are rickettsiae transmitted to humans?
19. Name several infections caused by rickettsiae.
20. Give an example of an infection caused by a spirochete.

APPLYING PRINCIPLES AND CONCEPTS

1. Discuss several factors that retard the rate of growth and multiplication of bacteria in a culture.
2. Discuss endospore formation in bacilli, including the manner in which spores are formed and their importance in the survival of bacteria.
3. Give evidence to support the recent claim that bacteria reproduce sexually. What may be the significance of this type of reproduction in bacteria?
4. Discuss the position of the rickettsiae in relation to bacteria and viruses.

RELATED READING

Books

BROCK, THOMAS D., *Milestones in Microbiology.*
  Prentice-Hall, Inc., Englewood Cliffs, N.J. 1961. Selections from original papers dealing with spontaneous generation, microbiology, and the germ theory of disease.
BRYAN, ARTHUR, and others, *Bacteriology: Principles and Practice.*
  Barnes and Noble Inc., New York. 1962. Contains newer material on the rickettsiae, revised laboratory techniques, and enlarged tables of antibiotics, with a discussion of virus infections that takes into account the latest findings in that area.
Encyclopedia of Life Sciences (eds.), *The World of Microbes.*
  Doubleday and Co., Inc., Garden City, N.Y. 1965. Beautiful color photographs and a running glossary of scientific and technical terms in margins provide a broad survey of a microscopic world.
KNIGHT, DAVID C., *Robert Koch—Father of Bacteriology.*
  Franklin Watts, Inc., New York. 1961. A simply written biography of the scientist who perfected the technique of culturing disease-causing bacteria in the laboratory.
PELCZAR, MICHAEL J., JR., and ROGER D. REID, *Microbiology.*
  McGraw-Hill Book Company, New York. 1965. A general textbook of microbiology, excellent for reference.
RIEDMAN, SARAH R., *Shots Without Guns, The Story of Vaccination.*
  Rand McNally and Co., Chicago. 1960. Traces the development of vaccines and includes biographies of Jenner, Koch, Pasteur, and Salk.
SCHNEIDER, LEO, *Microbes in Your Life.*
  Harcourt, Brace and World, New York. 1966. A good book for background information includes a discussion of techniques used in microbiology.

CHAPTER SEVENTEEN

INFECTIOUS DISEASE

Pathogenic organisms

In Chapter 16, we learned that some organisms are pathogenic —that they cause a diseased condition when they enter the body and multiply in the tissues. A list of groups of organisms of which some or all members cause infectious diseases would include the viruses, rickettsiae, bacteria, spirochetes, yeasts, molds and moldlike organisms, protozoans (animal-like protist organisms), and parasitic worms. We shall limit our discussion of infectious diseases in this chapter to those caused by bacteria and viruses.

Robert Koch—the father of bacteriological technique

It is appropriate to begin our study of infectious disease with the work of Robert Koch, a German physician who investigated infectious diseases at the same time that Louis Pasteur was conducting his studies in France.

Robert Koch, born the son of a poor miner and one of thirteen children, made his first outstanding contribution to medicine through the study of anthrax, an epidemic disease of animals that is often contracted by man. From early times, anthrax epidemics had spread through sheep, cattle, and other herds. People sometimes contracted anthrax from infected animals or from wool sheared from infected sheep. (This accounts for the name "wool sorter's disease," by which anthrax was known in earlier times.) In examining organs of animals that had died from anthrax, Koch found swarming in the blood vessels numerous rod-shaped bacteria, which he tentatively identified as anthrax germs. His next problem was to find out if these organisms caused the disease. He therefore transferred some of the living bacteria into a cut made at the base of the tail of a healthy mouse. The mouse developed anthrax and died. Koch found large numbers of the same bacteria in the bloodstream of the dead mouse.

Koch was not satisfied until he had actually watched this multiplication. Accordingly, he obtained a drop of sterile fluid

Isolate the organism suspected of causing the disease.

Grow the organisms in laboratory cultures.

Inoculate a healthy animal with the cultured organism and see if it contracts the disease.

If the animal contracts the same disease, examine the diseased animal and re-isolate the organism that caused the disease.

17-1 Koch's Postulates.

from within the eyeball of a freshly killed ox and put a small portion of the spleen of a mouse containing anthrax germs into the drop. Using his microscope, he patiently watched as the bacteria multiplied in the drop. He transferred germs from one drop to another and succeeded in growing them in the complete absence of any mouse spleen or blood. His next step was to inoculate healthy mice with his laboratory-grown organisms. The mice died soon after inoculation, and microscopic examination of the blood disclosed a great abundance of the same rod-shaped organisms. Koch concluded that anthrax was indeed caused by bacteria.

Koch's procedure is summarized in four steps:

1. Isolate the organism suspected of causing the disease. (Koch found anthrax organisms in the bloodstreams of infected animals.)
2. Grow the organism in laboratory cultures. (Koch used sterile fluid from the eyes of oxen.)
3. Inoculate a healthy animal with the cultured organism and see if it contracts the same disease. (Koch inoculated mice with the eye fluid containing germs.)
4. If the animal contracts the same disease, examine the diseased animal and re-isolate the organism that produced the disease. (Koch found that the organisms with which he had inoculated the mouse had increased enormously in numbers in the bloodstream.)

Following his investigation of anthrax, Koch continued to develop laboratory methods for the study of bacteria and infectious diseases. He devised the technique for preparing smears of bacteria on microscope slides and discovered various dyes for staining organisms and making them more visible for microscopic examination. Among his greatest contributions was the development of culture media containing gelatin and agar on which he could grow colonies of bacteria and isolate pure cultures. Many of Koch's stains, culture media, and methods of growing bacteria are still used in laboratories today.

Using the procedure he had developed in the investigation of anthrax, Koch isolated the tuberculosis organism from the lungs of human victims. He succeeded in growing the tuberculosis bacterium on an agar culture medium to which he had added blood.

Following the outstanding work of Pasteur and Koch, scientists in many countries accepted their teachings and adopted their procedures and techniques. The science of bacteriology grew rapidly after this start and entered a "golden age." One by one, infectious diseases were brought under medical control. The battle against infection is still being waged, but more than fifty years of steady progress have provided the doctor with many powerful weapons for preventing infectious diseases.

The spread of infectious organisms

Bacteria, viruses, and other infectious organisms are transmitted and gain entrance into the body in a variety of ways. Biologists often use these characteristics in classifying infectious diseases.

The agents of *food- and water-borne infections* enter the body through the mouth. Many diseases in this group center in the digestive tract, especially in the small and large intestines. During such illnesses, infectious organisms are usually present in the patient's excrement. This waste material is highly infectious. If strict sanitary precautions are not followed, the infection may be spread to other people by direct contact with the patient or by contact with contaminated articles.

More frequently, however, these infections are spread in food and water contaminated with intestinal wastes. We often associate typhoid fever with contaminated food and water. Among other infections of the digestive system that are transmitted through these media are paratyphoid, cholera, and dysentery. Of course, many diseases not of the digestive system can also be transmitted in contaminated food and water. Tuberculosis organisms may travel in milk. Milk is also an agent of transmission of dangerous streptococci from infected cattle. Undulant fever is usually transmitted in milk or dairy products from cows infected with Bang's disease (contagious abortion).

Most *air-borne infections* occur in the respiratory tract. This type of infection is often called a *droplet infection,* since the droplets sprayed into the air by a sneeze or cough may be laden with bacteria and viruses. Air-borne infectious diseases are most common during the fall and winter months. They often strike suddenly and infect large numbers of people. In addition to direct transmission of the agents of these infections by discharges from the nose and throat, these diseases are frequently spread indirectly through the handling of articles contaminated by infected persons.

Among the better-known air-borne respiratory diseases are colds, sinus infections, influenza, measles, German measles (three-day measles), mumps, whooping cough, diphtheria, scarlet fever, tuberculosis, pneumonia, and meningitis.

Certain diseases produce sores, or lesions, on the skin and mucous membranes. Direct contact with material from these lesions spreads infection. Impetigo may be spread by such *contact infection.* Chicken pox and smallpox viruses can be transmitted directly through contact with a lesion, or indirectly through the handling of a contaminated article or through air-borne droplet infection. Diseases spread by contact also include syphilis and gonorrhea.

The unbroken skin is an effective barrier to the entrance of bacteria. However, breaks in the skin may permit *wound in-*

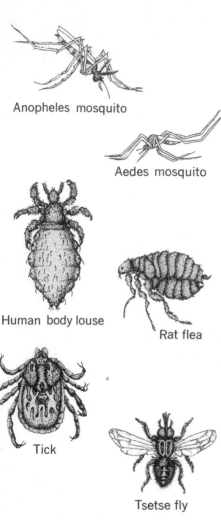

Anopheles mosquito

Aedes mosquito

Human body louse

Rat flea

Tick

Tsetse fly

17-2 Arthropod carriers of infectious diseases. With what disease is each vector associated?

fections unless wounds are cleansed by bleeding and properly treated with antiseptic. Puncture wounds are especially dangerous because of the possibility of tetanus infection. Tetanus bacteria, which are anaerobes, enter the wound as spores clinging to some object. When the wound heals on the surface, it leaves an airtight cavity ideally suited to the tetanus organism. Various staphylococci commonly infect wounds and produce a characteristic yellow pus. Streptococcus wound infections are dangerous and can lead to a fatal general infection.

Human *immune carriers* present a special problem. Immune carriers show no signs of illness, yet they harbor living infectious viruses and bacteria in their bodies and transmit diseases, usually without knowing it. In some diseases, patients who recover may be immune carriers for weeks, months, or even years. The immune-carrier problem is especially great with respect to typhoid fever, diphtheria, scarlet fever, and polio.

Arthropod carriers (insects and their relatives) are associated with certain infectious diseases. Biologists refer to the arthropod carriers as *vectors*. Houseflies and cockroaches carry infectious organisms on their feet and bodies and spread diseases directly or contaminate food and other articles. Other arthropods carry infectious organisms internally and transmit them in bites. Some infectious diseases that are transmitted by carriers are typhus fever, which is transmitted by the human body louse; bubonic plague, which is transmitted by the rat flea; and spotted fever, which is transmitted by the tick. African sleeping sickness is transmitted by the tsetse fly. The *Anopheles* mosquito is involved in the life cycle of the malaria parasite, while the *Aedes* mosquito transmits the virus of yellow fever.

How microorganisms cause disease

Various microorganisms damage the body in several ways during an infection. One such way is through *tissue destruction,* a type of damage illustrated in tuberculosis. The primary infection is usually in the lungs, although other organs may be affected. As tuberculosis organisms multiply in lung tissue, they destroy cells and produce lesions. These allow blood to seep from the capillaries into the air passages, resulting in the hemorrhages characteristic of advanced tuberculosis. Many organisms, including streptococci, destroy blood corpuscles. In the case of typhoid, the bacteria destroy cells of the intestine wall, thereby producing lesions.

Many bacteria form poisonous protein substances that diffuse from the cells and are absorbed by the body tissues. These bacterial products are called *soluble toxins,* or *exotoxins*. Exotoxins may enter the bloodstream and reach body tissues far from the site of the infection. In this way, for example, exotoxin from tetanus-causing bacteria multiplying in a puncture wound in the foot is carried to distant regions of the body, including the jaw muscles, where it causes muscle spasms and rigid paralysis. It is this effect that accounts for the use of the name *"lockjaw"* for a tetanus infection.

Similarly, certain streptococcus bacteria growing in the throat form exotoxins that cause inflammation of tissues in the joints and heart valves—a condition called *rheumatic fever*. The rash of scarlet fever is a reaction of the skin to a powerful exotoxin. Diphtheria is another well-known disease in which exotoxins from an infection in the throat may cause extensive tissue damage throughout the body.

In some cases, exotoxins may be formed in foods that when eaten are absorbed with damaging effects on the body. We refer to these poisonous proteins as **pre-formed toxins**. The resulting illness is *food poisoning*. The most deadly type of food poisoning is botulism, caused by a relative of the tetanus-causing organism. Like tetanus-causing bacteria, botulism-causing bacteria form spores. These spores may reach meat, vegetables, and other foods prior to canning. If the foods are improperly sterilized during canning and are sealed in airtight containers, the spores germinate into vegetative cells that grow in the food and release deadly exotoxins.

If the spoiled food is eaten, the toxins are absorbed and symptoms of botulism appear, usually within twelve to thirty-six hours. These symptoms include double vision, weakness, and paralysis that creeps from the neck region to other parts of the body. Death may result from paralysis of the breathing muscles or from heart failure. This occurs in about 65 percent of the cases of botulism. Since the exotoxins are destroyed by heat, botulism can be prevented by the heating of canned meats and vegetables, especially those that are home-canned, to boiling before they are even tasted. Cans containing spoiled food often bulge because of gas production. The food in these cans should never be used. Other organisms that form toxins in foods include certain *Salmonella* and *Staphylococcus* bacteria. These forms of food poisoning are more common and, fortunately, less deadly than botulism.

Certain bacteria form toxins that remain in the cells and are liberated when the bacteria die and disintegrate. We call these poisonous products **endotoxins**. The release of endotoxins causes severe tissue reactions that may be fatal. Included among the endotoxin diseases are typhoid fever, tuberculosis, cholera, bubonic plague, and bacterial dysentery.

Structural defenses against disease

Even though the body is the normal environment of an enormous number of bacteria and other organisms that cause no harm, there is a constant threat of invasion of the tissues by organisms that are capable of causing disease. To prevent such an invasion, the body is provided with many defenses that function at various stages of the attack. These defenses can be compared with lines of soldiers, each equipped with weapons.

The most effective way of avoiding infectious disease is to prevent the disease-causing organisms from entering the body tissues. This is the function of the structural defenses, composing the *first line. Skin* covers all the external parts of the body and if unbroken is bacteria-proof. In addition, salts and various fatty acids present in perspiration are believed to destroy some bacteria and thereby help to make the skin an even more effective defense against infection. However, natural openings in the skin, such as pores and hair follicles, may allow entry of some microorganisms.

The mouth, digestive tract, respiratory passages, and genital tract are lined with *mucous membranes*. These membranes function much as skin does in preventing microorganisms from invading the body tissues. **Mucus,** secreted by cells of the mucous membranes, is a slimy substance that traps bacteria on the membrane surface. Cells forming the lining of mucous membranes in the respiratory tract are equipped with tiny, hairlike projections that sweep bacteria and other foreign materials upward toward the throat. When such particles irritate the mucous membranes of the throat, a cough results, and the particles, along with droplets of fluid, are blown out into the air. Irritation of the membranes of the nasal passages results in sneezing.

Tears, secreted by tear glands, flow over the eyes continually, not only lubricating them but cleansing them as well. Bacteria and other foreign particles are washed into the tear ducts, which empty into the nasal passages. Recent studies have revealed that both tears and mucus contain enzymes known as **lysozymes** (LICE-uh-ZIEMZ). These enzymes, by destroying the cell walls of many kinds of bacteria, prevent the bacteria from entering the body.

The *acid* secretion of the stomach is another effective structural defense of the body. Great numbers of bacteria enter the stomach in the food we eat. However, few can survive the hydrochloric acid secreted by glands in the stomach wall.

Little is known of the function of *other microorganisms* as a first-line defense. The presence of the normal bacterial population of the digestive system seems to interfere with the growth of invading organisms and thus protects the host. The importance of these intestinal bacteria is demonstrated by what often results from their inadvertent destruction by some medications. Before the bacteria can multiply and reestablish their normal number, other bacteria reproduce in the intestine and cause disease.

Cellular defenses against disease

Once bacteria have passed through the mechanical defenses of the skin, mucous membranes, and stomach, they are met by a *second line* of defense, which operates in the body tissues. The principal defenders of this line are *phagocytic* (FA-guh-SIT-ick) *cells* that engulf bacteria and digest them with enzymes, including lysozymes (Figure 17-3).

Phagocytic cells are either free or fixed. Among the free phagocytic cells are certain blood cells called *leucocytes*. These cells pass through the walls of capillaries and migrate through tissue fluids to the site of a local infection. Here many of them form a wall around the invading organisms and begin to engulf them. Debris of the battle, consisting of blood serum, digested bacteria, and degenerated leucocytes, constitute *pus*.

Often during such an infection, which is still local rather than general, the tissues in the area swell and become inflamed. The redness results from increased flow of blood to the region of the infection. Blood vessels enlarge, and lymph (a clear liquid present in the blood) seeps into the tissue spaces. The lymph aids the struggle by carrying bacteria and leucocytes that have engulfed bacteria to lymph nodes where they are filtered out. All of these reactions are part of the localized inflammation that occurs at the cellular level.

Further destruction of microorganisms occurs in the lining of small blood vessels in the liver, spleen, lungs, bone marrow, and all loose connective tissue. Here, fixed phagocytic cells composing the *reticuloendothelial system* capture and engulf bacteria and leucocytes that contain bacteria.

During the struggle between invading microorganisms and cellular defenses of the body, the body temperature often rises. This temperature elevation, or *fever,* is a host reaction that reduces or inhibits the growth of many bacteria, activates other body defenses, and increases the rate of body metabolism. Thus, fever is beneficial unless it is too high or lasts too long. If this happens, damage or destruction of cells of the host will occur.

The influence of body temperature on the growth of bacteria was demonstrated many years ago by Louis Pasteur. Pasteur was attempting to infect ducks with anthrax-causing bacteria taken from diseased sheep. However, the bacteria would not grow in a duck, in which the body temperature is about 5°F higher than that of a sheep. In other words, the normal body temperature of a duck is the equivalent of a five-degree fever in a sheep. However, when he lowered the body temperature of one of the ducks by immersing its feet in ice water, the anthrax bacteria started growing and produced the infection.

Antibody defenses against disease

The body defenses we have described so far have been nonspecific in that they act on any invading bacteria. We now reach the *third line* of body defenses, which includes various specific *antibodies* that act against specific disease organisms and their products. Bacteria, as well as their toxins, contain protein materials that are foreign to the host organism. We call foreign pro-

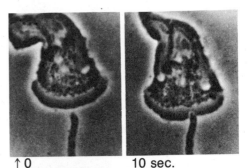
↑ 0   10 sec.

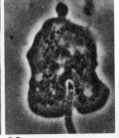
20 sec.   30 sec.

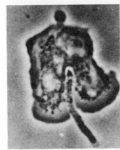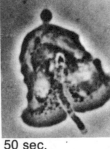
40 sec.   50 sec.

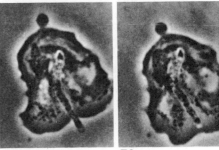
60 sec.   70 sec.

17-3 Phagocytosis. This series of photographs, taken at ten-second intervals, shows a human white blood cell, or leucocyte, engulfing a chain of *Bacillus megaterium* cells. (courtesy James G. Hirsch)

teins that provoke an antagonistic reaction in the host *antigens.* These antigens stimulate the production of specific antibodies in the host. These host antibodies combine with antigens in the blood, and together they eventually remove the foreign material.

Investigations have shown that antibodies are protein molecules found principally in the gamma globulin proteins of blood serum. These antibodies are believed to come from the lymph nodes, spleen, and thymus gland, from which they enter the blood and lymph. Specific antibodies against a specific infection are formed only in the presence of specific bacteria or bacterial products in the body. Antibody production begins within a few hours after the introduction of antigens in the form of viruses or bacteria, or after the release of bacterial products. Within a few days, antibodies enter the bloodstream and continue to increase in quantity for three or four weeks. This period usually marks the highest level of antibody production. Antibodies may remain in the bloodstream for many weeks or even years. The level of a specific antibody in the blood may decline slowly. A second exposure to the antigen involved speeds up production in the lymph nodes and spleen and raises the antibody level rapidly.

Various antibodies act in interesting ways. *Antitoxins* combine with exotoxins formed by such organisms as diphtheria, scarlet fever, and tetanus. These extremely poisonous bacterial secretions are neutralized when they are joined to specific antitoxins. *Agglutinins* are antibodies that cause certain bacteria to clump together, or agglutinate. This process is of value since the masses are more easily destroyed by phagocytic cells. *Cytolysins,* also called *bacteriolysins,* are antibodies that cause certain bacteria to disintegrate. *Precipitins* are little understood antibodies that cause bacteria to settle out, or precipitate, in the blood. This action is an aid in filtering them out in the lymph nodes, spleen, and other organs. *Opsonins* are believed to combine with certain substances in bacterial cell walls and to prepare them for ingestion by phagocytic cells.

Chemical defenses against viruses

Until recently, the nature of the body's chemical defenses against virus diseases was not known. However, an important discovery in 1957 shed some light on this question. Two doctors, Alick Isaacs and Jean Lindenmann, while working at the National Institute for Medical Research in London, England, identified a protein substance produced by cells invaded by a virus. Since this substance was found to interfere with the action of a virus, they named it *interferon.*

Further studies indicate that interferon, itself, may not be the antiviral agent. This action seems to involve a second protein substance synthesized in a cell as a result of stimulation by interferon. Within a few hours after the onset of a virus infection, the interferon system is in operation. As a cell is invaded by a virus, the presence of the virus stimulates the cell to synthesize interferon. This' interferon is released into the intercellular spaces and is taken into other cells. It is believed that these cells are stimulated to produce *interferon messenger RNA* resulting in the synthesis of an *antiviral protein*. This protein protects the cell by preventing the virus from modifying its protein synthesizing machinery to form virus particles. Thus, the virus is prevented from multiplying in the protected cells. The extent of this protection depends on the level of antiviral protein. The action of interferon and antiviral protein is intracellular (within cells) and continues for a period of one to three weeks.

About three days after the onset of a virus infection, specific antiviral antibodies begin to appear in the blood serum. This is a second phase of defense against a virus infection. This defense is extracellular (outside the cell). Antibodies combine with virus particles and make them noninfectious. Phagocytic white corpuscles aid in this defense by engulfing virus particles.

Thus, the interferon system and antiviral antibodies are a two-pronged attack. Interferon acts rapidly and for a short time. It is nonspecific and is, thus, effective against any virus. Antibodies are produced more slowly but are formed for a much longer period of time (in some cases for life). They are specific, in that one kind of antibody acts only on one kind of virus.

Immunity against disease

We refer to resistance of the body against infections as ***immunity***. It may be present at birth, which is the form called *inborn natural immunity,* or it may be *acquired* during the lifetime of the individual.

For the most part, man has a natural immunity to diseases that affect other animals and plants, because conditions in the human body do not support the growth and activity of the infectious organisms that cause these diseases. Such inherited properties as body structure, body temperature, physiological characteristics, and chemical makeup are responsible for this resistance. The general immunity of all people to pathogens of other species is called *species immunity*. There are, however, several notable exceptions to species immunity. Tuberculosis and undulant fever may be transmitted to human beings in milk from diseased cattle. Anthrax may be contracted by contact with lesions in the skin of infected sheep, cattle, horses, and other animals. Tularemia is spread by infected rabbits, while psittacosis, or parrot fever, may be contracted from infected parrots and parakeets.

Acquired immunity may be *active* or *passive,* depending on how it is established. Active immunity may be acquired *naturally* by an individual as he recovers from certain infectious diseases, including diphtheria, scarlet fever, measles, and mumps. During the infection, the body produces specific antibodies against the pathogenic organisms or their products. This antibody production normally continues after recovery, resulting in permanent active immunity. Active immunity may be acquired *artificially* by the introduction into the individual of biological preparations known as *vaccines* that contain dead or weakened pathogenic organisms or their products. The body is thereby stimulated to form its own antibodies without the necessity for the individual actually to suffer the symptoms and dangers of a full-blown case of the disease.

Passive immunity is acquired *artificially* by the transferring into the individual of antibodies that have been produced in other individuals or in animals. We refer to the portion of the blood, or *blood fraction,* that contains antibodies as *serum* (plural, *sera*). While the introduction of serum provides immediate protection against an infection, the protection is temporary. In most cases, the protection is eliminated by the body in from only a few weeks to several months. Temporary passive immunity may also be acquired naturally in an unborn infant through the transfer of antibodies from the mother's blood through the placenta. Other antibodies may be transferred to the baby from an immune mother in the *colostrum,* or the first milk the baby receives during nursing. Immunity transferred from mother to baby usually lasts from six months to a year.

A summary of the types of immunity, how they are established, and their duration is found in Table 17-1.

Our knowledge of immunity and the use of immunity-inducing agents in dealing with infectious diseases is an important part of medicine. The medical achievement we call *immune therapy* represents knowledge contributed by many great scientists in a series of exciting discoveries of the past two centuries. All of us have benefited from these contributions.

Edward Jenner—country doctor

It was during one of the most dreadful smallpox epidemics in England's history that Edward Jenner, a country doctor, made a discovery that was to alter the course of history. Epidemics took their greatest toll in cities. Jenner noticed that the disease seldom struck people who lived in rural areas and worked around cattle. Most farmers and dairy workers had contracted cowpox and had recovered with nothing more serious than a pustule that left a scar. Were they immune to smallpox? If so, why not deliberately infect people with cowpox to protect them from smallpox?

On May 14, 1796, Dr. Jenner had a chance to test his theory. His patient was James Phipps, a healthy boy about eight years old. James' mother, who had great confidence in Dr. Jenner, allowed her son to be used in the test in the hope that he could be spared the dangers of smallpox. Dr. Jenner took his young

Table 17-1 TYPES OF IMMUNITY

TYPE		HOW ESTABLISHED	DURATION
INBORN	species	through inherited anatomical, physiological, and chemical characteristics	permanent
ACQUIRED active	natural	by experiencing an infection during which contact with microorganisms or their products stimulates antibody production	usually lasting or permanent
	artificial	by injecting vaccines, toxoids, or other weakened bacterial products	usually from several years to permanent; booster shots may be necessary
passive	natural	by transfer of antibodies from mother to infant through the placenta prior to birth or in colostrum after birth	from 6 months to 1 year
	artificial	by injecting a serum containing antibodies	usually from 2 or 3 weeks to several months

patient to a dairymaid, Sarah Nelmes, who had on her hand a cowpox pustule resulting from an infection from one of her master's cows. Dr. Jenner made two shallow cuts, each about an inch long, on James Phipps's arm and inoculated the cuts with matter taken from the cowpox sore (Figure 17-4). A pustule developed on the boy's arm, formed a scab, and healed, leaving only a scar. Was James Phipps now immune to smallpox? There was only one way to find out. That was to inoculate him with smallpox.

In July of the same year, Dr. Jenner deliberately inoculated James with matter from a smallpox pustule. During the next two weeks, the doctor watched his patient anxiously for signs of smallpox. They did not appear. Several months later, he repeated the inoculation. The disease did not develop this time, either. James Phipps was definitely immune to smallpox!

Following this famous experiment, Dr. Jenner wrote a paper explaining his method of *vaccination*. At first, the doctors were hostile and would not listen. Many townspeople even organized anti-vaccination campaigns. Gradually, however, the doctors and their patients accepted vaccination, and smallpox epidemics were eliminated.

17-4 The first vaccination. Dr. Edward Jenner inoculates James Phipps with matter taken from a cowpox sore. (© 1960, Parke, Davis & Co.)

Pasteur's famous immunization experiments

About eighty years after Dr. Jenner vaccinated James Phipps, Louis Pasteur conducted his famous immunization experiments. Prior to performing these experiments, Pasteur had made a vaccine containing the weakened bacteria of chicken cholera. He found that he could inject the vaccine into healthy chickens and produce active immunity against the disease. This led him to try a similar procedure against anthrax at about the same time at which Robert Koch was conducting his famous work on the disease.

Pasteur made an anthrax vaccine from bacteria that were weakened after being isolated from the blood of infected animals. He claimed that this vaccine would immunize animals against the disease. Scientists challenged him to prove his theory. This challenge was the opportunity he had waited for. He selected forty-eight healthy animals (mostly sheep) and divided them into two groups: an experimental group and a control group. He gave the animals in one group injections containing five drops of anthrax vaccine. Twelve days later, he gave the same animals a second injection of the vaccine. Fourteen days later, he gave all forty-eight animals injections of living anthrax bacteria. Two days later, the scientists met at the pens where the animals were being kept. Imagine their amazement when they found all of Pasteur's immunized animals alive and healthy and all of the untreated animals dead or dying of anthrax. This famous experiment was an important milestone in the conquest of disease.

Pasteur's experimentation with rabies

During the latter years of his life, Pasteur turned his genius to experimentation with one of the most dreaded of all diseases: rabies, or hydrophobia. We know today that rabies is caused by a virus. Pasteur, of course, did not know of the existence of viruses. But for the sake of clarity, we shall use this term in explaining his work with the disease. During Pasteur's time, this disease was common among dogs, wolves, and other animals. If the virus was transmitted to a human being by an animal bite, the victim was certain to suffer an agonizing death after an incubation period of from a few weeks to six months or more. During this time, the virus slowly destroyed brain and spinal-cord tissue.

The restlessness, convulsions, great thirst, and throat paralysis that climaxed a rabies infection led Pasteur to believe that the infection centered in the brain. However, microscopic examination of the brain tissue of animal victims did not reveal any microorganisms. We can understand why today, for the rabies virus is too small to be seen with an ordinary microscope.

Pasteur found that he could transmit rabies by injecting infected brain tissue from a rabid dog into a healthy one. He repeated the inoculations with rabbits and discovered that the virus gained strength as it was passed from one animal to another. However, if spinal-cord tissue taken from an infected dog or rabbit was dried for fourteen days, the virus lost its strength and could no longer produce the infection. Virus thirteen days old was only slightly stronger. The discovery that the virus weakened with drying and aging led to experiments to find out if rabies immunity could be produced. Pasteur injected fourteen-day-old brain tissue from a rabid animal into a healthy dog, then followed this injection with thirteen-day-old material.

The injections were continued day after day until the dog was given an injection of full-strength virus in the fourteenth injection. The animal suffered no ill effects. The series of injections with material of increasing strength had produced immunity to rabies. The question then was whether or not Pasteur dared to try the series of injections on human victims.

Pasteur's treatment of rabies

On July 6, 1885, a frantic mother took her son to Pasteur's laboratory, begging him to use any method to save her boy's life. The boy had been attacked by a rabid dog two days before he and his mother had reached Pasteur's laboratory. Pasteur had no time to lose and no choice except to give the boy the treatment that had been effective with dogs. The physicians and laboratory assistants he consulted agreed with his decision. On the evening of this important day, the boy was given an injection of rabbit-grown virus that had aged for twelve days. Injections were repeated each day, with successively fresher virus. On the twelfth day, the boy received full-strength virus. After several weeks of observation, he was sent home, the first person to have been immunized against rabies.

This series of injections, known as the Pasteur treatment, is used today to immunize victims of bites by rabid animals. The virus used in making the vaccine was once grown in rabbits. However, there was some danger of serious body reaction to the substances present in rabbit nerve tissue. Hence, a more recent practice is to grow the virus in developing duck embryos, because there is no danger of serious body reaction associated with vaccine made through their use.

The conquest of diphtheria

From the earliest times, diphtheria was one of the worst epidemic killers, especially among children. The bacteria grow in a thick, grayish-white membrane on the back wall of the throat. As the membrane spreads, it may block the opening to the lungs and cause death by strangulation. In addition, the living bacteria give off highly poisonous *exotoxins* that are absorbed through the infected tissues into the blood. These toxins often cause severe damage to the heart, nervous system, and other organs.

The conquest of diphtheria some fifty years ago involved the work of several scientists. One found the rod-shaped bacteria growing in the throats of patients with diphtheria. Another worker cultured the bacteria and developed a stain suitable for use in microscopic studies of the organisms. However, much of the credit for the conquest of this disease belongs to Emil von Behring (vahn BAY-ring), a German bacteriologist.

Von Behring was puzzled by the fact that even though diphtheria organisms remained in the throat, the effects of the disease appeared in organs far removed from the throat. When an extract of the medium on which the bacteria were grown was injected into guinea pigs, the symptoms of diphtheria occurred even though no germs were present. Apparently the extract contained some substance from the bacteria that caused the symptoms. This substance was *toxin.*

It was while he was conducting such experiments that von Behring discovered that guinea pigs and rabbits could be used only once. Apparently they developed immunity to the disease during the first exposure. Could this immunity be transferred to animals that had never been given doses of toxin? In an effort to answer this question, von Behring took a blood fraction from immune animals and injected it into other animals. Exposure to diphtheria toxin now showed that they were immune. Von Behring named the substance that had produced immunity *antitoxin.*

Sheep were used first in the production of diphtheria antitoxin. After extensive testing of the antitoxin in guinea pigs, it was used with great success in the Children's Hospital in Berlin. Von Behring found that immunity resulting from injections of sheep antitoxin lasted only a few weeks. Apparently this kind of antitoxin was slowly destroyed in the human bloodstream. If children could be made to produce their own antitoxin, immunity would be as lasting as it would be if they had had diphtheria and recovered from it. To give diphtheria toxin, however, would be as dangerous as inoculating them with the disease itself. Von Behring reasoned that a mixture of toxin and antitoxin might be safe to use. World War I prevented von Behring from finishing his work on such a *toxin-antitoxin.* However, the work was completed in the United States by Dr. William H. Park and other workers.

Toxin-antitoxin, until it was discovered that some people have a serious reaction to it, was used widely in producing immunity to diphtheria. This problem has been solved by using a *toxoid,* a substance very similar to the toxin and usually derived from it, but lacking the poisonous quality of the toxin. The toxoid is made by treating the toxin with heat or with chemicals, such as formaldehyde. Apparently the toxoid is similar enough to the toxin so that in response to the introduction of the toxoid, the host manufactures an antitoxin that is effective against the toxin itself.

Immune therapy

The principles of immune therapy discovered by Jenner, Pasteur, and von Behring are applied extensively today in the production of vaccines and sera for use in the prevention and treatment of infectious diseases.

The name *vaccine,* from the Latin word *vacca* meaning "cow," is taken from the cowpox virus Jenner used in the first vaccination against smallpox. Cowpox *(vaccinia)* virus is closely related to smallpox *(variola)* virus but is not highly infectious in human beings. However, it is sufficiently similar in its chemical structure to stimulate smallpox antibody production.

This principle is also illustrated in the Sabin polio vaccine, in which a strain of weakened, noninfectious living virus is given by mouth to stimulate the production of antibodies against polio. Other vaccines contain killed or weakened bacteria or viruses that, when injected, act as antigens and stimulate antibody production. Typhoid vaccine is a suspension of killed typhoid organisms, often combined with two other organisms closely related to typhoid in a triple vaccine. Salk polio vaccine contains killed polio virus of all three types. This vaccine is injected into the body to stimulate antibody production and establish immunity against the three types.

In the preparation of influenza vaccine, the virus is first grown in chick embryos in eggs, after which it is made harmless by chemical treatment. Similar methods are used in the production of yellow fever vaccine.

Von Behring's work with diphtheria established another phase of immune therapy, the transfer of serum, the blood fraction that contains antibodies, to establish passive immunity. Sera are used to provide immediate protection to a person who has been exposed to an infectious disease and to bolster natural antibody production during an infection. Gamma globulin, administered in a serum, is used to protect against measles and infectious hepatitis.

Chemotherapy

The conquest of disease is not a war of biology alone, for chemistry plays a very significant part. In *chemotherapy,* specific chemical compounds are used to destroy pathogens without causing injury to the patient. This type of treatment is of great importance, because it bolsters the natural body defenses.

The early development of chemotherapy is associated with the work of a brilliant German chemist, Paul Ehrlich (AYR-lick), in connection with his long search for a cure for syphilis. Ehrlich spent many years attempting to find a drug that would kill the organisms in the bloodstream without damaging the blood or other parts of the body. After 605 unsuccessful attempts, he finally succeeded. His 606th drug was an arsenic compound called *salvarsan*. It was used in treating syphilis many years before the discovery of penicillin.

Other scientists began experimenting with chemicals in the treatment of disease. In 1932, Dr. Gerhard Domagk (DOE-MAHK), another German scientist, discovered that a red dye called *prontosil* had remarkable germ-killing powers. Soon after his discovery of prontosil, he tried it on his own daughter, who was dying of a streptococcic infection that had progressed beyond medical control. It proved to be effective in halting the infection and saved the child's life.

Further investigations on prontosil proved that its germ-killing powers were concentrated in only one part of the drug. This part was isolated and called *sulfanilamide* (SULL-fuh-NILL-uh-MIDE). It was the first of an important family known as the *sulfa drugs*. There are many different ones now; they are used in the treatment of certain infectious diseases. These drugs should be taken only on the advice and recommendation of a physician. They are not cure-alls and may be dangerous.

Antibiotic therapy

The chemical substances called **antibiotics** are now commonly used in the treatment of disease. Antibiotics are products of living organisms. In this respect, they are different from the drugs used in chemotherapy. We can sum up the use of antibiotics by saying that "bugs produce drugs that kill bugs," because our supply of these substances comes from bacteria, molds, and moldlike organisms.

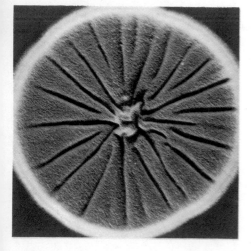

17-5 The mold *Penicillium notatum* growing on a Petri-dish culture. The antibiotic penicillin is obtained from this mold as well as from other species of the same genus. (Walter Dawn)

The wonder drug of World War II, *penicillin*, was the first of the antibiotics. It was discovered accidentally in 1929 by Sir Alexander Fleming, a Scottish physician and bacteriologist. Fleming was working with staphylococcus bacteria in a London hospital. While examining plate cultures of staphylococci, he noticed that several of them contained fluffy masses of mold and that the staphylococcus bacteria had not grown in the immediate vicinity of the mold. Later, the mold colonies turned dark green and were identified as *Penicillium notatum*, a relative of the green-and-white mold we often find on oranges. Its antibacterial secretion was called penicillin. Since the mold and the staphylococci were competing for the same food supply, it seemed that the mold's secretion was an adaptation that destroyed the competitor.

In the opening days of World War II, Dr. Howard Florey and a group of Oxford workers began a search for antibacterial substances that would be useful in combating wound infections. Their attention turned to Fleming's work and, in cooperation with him, they developed penicillin and thoroughly tested it. The result of this work is history.

Today, a penicillin ten times as powerful as Fleming's is available in unlimited quantity and at low cost. Mutant strains of

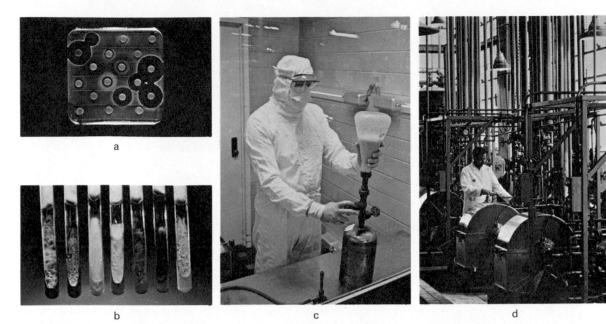

a

b

c

d

Penicillium notatum, produced by exposure to X rays, yield far more penicillin than earlier strains. Biological companies have even produced penicillin synthetically. It is given effectively in large doses by injection with a slowly absorbed procaine salt. It can also be taken by mouth in tablet form and inhaled into the nasal passages in powder form. Ointments are available for use locally and in the eyes. However, *it should never be used in any form unless recommended by a physician.*

17-6 Various steps in antibiotic production. (A) Testing activity against specific bacteria. (B) Culturing antibiotic organisms on agar. (C) A transfer device used to inoculate large vats with antibiotic organisms. (D) Extraction and purification of antibiotics from the producing organisms. (courtesy Eli Lilly and Co.)

Streptomycin

Dr. Selman Waksman became interested in the soil and its relation to life when he was a boy in Europe. Later, he came to the United States and enrolled in Rutgers University. His interest in soil led him to the New Jersey Agricultural Experiment Station at Rutgers. While he was still a student, Waksman discovered a soil organism that he named *Streptomyces griseus.*

After graduate study, Waksman returned to Rutgers as a member of the faculty. With the aid of students, he continued the investigation of soil organisms. Together, Waksman and his students studied the problem of the disappearance of disease organisms when the body of a diseased animal is buried, and they found that products of soil organisms destroyed the pathogens. After years of testing the effect of soil organisms on various pathogens, Waksman discovered *streptomycin,* an antibiotic substance produced by *Streptomyces griseus.*

Streptomycin proved to be an effective drug against tuberculosis. It was also found to be partially effective against whooping cough, some forms of pneumonia, dysentery, gonorrhea, and syphilis. The streptomycin industry grew rapidly, and this antibiotic took its place beside penicillin. Newer antibiotics have largely replaced streptomycin today, although it is still a valuable antibiotic in the treatment of tuberculosis.

Other antibiotics

Today, large numbers of antibiotics have been introduced under a variety of names. Many are of a type known as *tetracycline.* An ideal antibiotic has a *broad spectrum,* or range, of pathogenic organisms against which it is effective. It must also destroy pathogenic organisms without damaging the body tissues or disturbing body functions.

As effective as antibiotics are in treating certain diseases, however, we must not regard them as "cure alls." There are definite limitations in their use. For this reason, *an antibiotic should never be used unless prescribed by a physician.*

One reason for limiting the use of antibiotics is that they sometimes produce side effects. A severe allergic reaction to penicillin is not uncommon. Other antibiotics when taken by mouth destroy the normal bacterial population of the intestine. These bacteria seem to aid normal activity and keep other bacteria, including protein-splitting organisms, in check. Diarrhea and other gastrointestinal disturbances usually follow the destruction of the normal population. (Interestingly, living lactobacilli, like those used in making fermented milk products, taken orally in capsules or tablets help to restore the normal bacterial population and correct the digestive disturbances.)

A second, and perhaps the most important, reason for caution involves the process of adaptation and natural selection. The bacteria causing a disease vary somewhat in their characteristics, as members of any species do. When an antibiotic is used, some of the population of pathogenic bacteria may have adaptations that allow them to resist its effects. Repeated use of the antibiotic may kill off the other, more susceptible bacteria, while allowing the resistant ones to multiply and produce new strains in a short time. This is becoming more and more of a medical problem. For example, penicillin was very effective against staphylococci for many years. The resistant strains became more and more abundant, until today they are a great problem.

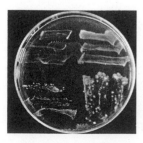

17-7   Determining the sensitivity of a microorganism to various antibiotics. Each of the four paper disks has been soaked in a different antibiotic and placed in a bacterial culture. The size of the *zone of inhibition* reflects the relative effectiveness of each antibiotic. (Walter Dawn)

17-8   Determining the effectivness of different strengths of the same antibiotic against four microorganisms. Each culture medium contains an increasingly larger dose of the same antibiotic. The bacteria in the upper right hand corner of each dish are least inhibited by the antibiotic. (Society of American Bacteriologists)

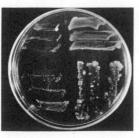
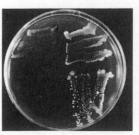
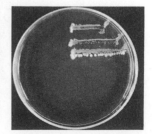

SUMMARY

There was a time, less than a century ago, when whooping cough took its toll of infants and diphtheria was a dreaded disease of childhood. Sanitariums were filled with victims of tuberculosis, many of them young and in the prime years of life. Pneumonia struck down people of all ages, especially during the winter and early spring months. The doctor had but a few drugs in his bag, and most of these were of only limited effectiveness. Recovery from a serious infectious disease was, largely, a matter of bed rest and nursing care.

Then, one of the most dramatic chapters in medical history began to unfold. Scientists discovered antibodies and prepared sera to transfer these vital body defenses to patients struggling against infection. They discovered vaccines to establish active immunity and remove the threat of one infectious disease after another.

With the discovery of various germ-killing drugs, chemotherapy became an important part of the treatment of infectious diseases. The discovery of antibiotics brought still more diseases under chemical control.

All of these advances have added many years to the average life expectancy. We will add still more as new advances are made.

BIOLOGICALLY SPEAKING

immune carriers	mucus	antigens
arthropod carriers	lysozymes	immunity
vectors	phagocytic cells	vaccines
exotoxins	leucocytes	serum
pre-formed toxins	pus	toxin-antitoxin
botulism	antibodies	toxoid
endotoxins	interferon	antibiotics

QUESTIONS FOR REVIEW

1. List the Koch postulates and describe each as a step in the investigation of an unknown disease.
2. Name several procedures and techniques Koch devised for the laboratory study of bacteria.
3. How can food and water become media in which infectious diseases are spread?
4. Name several diseases spread by droplet infection.
5. List several air-borne infections.
6. Name a disease associated with human carriers.
7. Name several arthropod vectors of infection and the diseases with which they are associated.
8. What are pre-formed toxins? Give an example.
9. Distinguish between an exotoxin and an endotoxin.
10. List the principal structural defenses of the body.
11. In what way do lysozymes function in body defenses?
12. Give two examples of phagocytic cells that function at the level of cellular defense.
13. In what way is fever an important body defense?
14. What is an antigen?

15. Distinguish between natural and artificial immunity.
16. From the standpoint of origin and action, distinguish among toxins, anti-toxins, toxin-antitoxins, and toxoids.
17. Distinguish between a vaccine and a serum. Explain how each is used in establishing immunity.
18. What medical contributions were made by Ehrlich and Domagk?
19. How are antibiotics different from other drugs used in chemotherapy?

APPLYING PRINCIPLES AND CONCEPTS

1. Discuss the scientific contribution of Robert Koch in his investigation of anthrax.
2. Discuss the specific and nonspecific body defenses.
3. Compare interferon and antiviral protein with antiviral antibodies from the standpoint of time of production, specificity, place of action, nature of action, and duration in the body.
4. Discuss the contribution of Edward Jenner to the rise of immune therapy.
5. What significant contribution did Emil von Behring make to our knowledge of immunity?
6. Discuss the principle involved in the Pasteur treatment for rabies.
7. Discuss the way in which natural selection produces bacteria resistant to antibiotics.

RELATED READING
Books
BOETTCHER, HELMUTH, *Wonder Drugs: A History of Antibiotics.*
  J. B. Lippincott Co., Philadelphia. 1963. An entertaining history of the "wonder drugs" and the medical discoveries of the last century.
FRAZIER, W. C., *Food Microbiology.*
  McGraw-Hill Book Company, New York. 1958. A textbook, but also useful as a reference book, especially on illnesses caused by improper food handling or microorganisms.
POOLE, LYNN, and GRAY POOLE, *Doctors Who Saved Lives.*
  Dodd, Mead and Co., New York. 1966. The dramatic discoveries by sixteen men of medicine and their associates in the battle against disease.
WALKER, NONA, *Medicine Makers.*
  Hastings House, Publishers, Inc., New York. 1966. A carefully researched history of medicine from the days of witchcraft to the beginning of the growth of the drug companies.
WHEELER, MARGARET F., and WESLEY A. VOLK, *Basic Microbiology.*
  J. B. Lippincott Co., Philadelphia. 1964. Microbiology treated not only in terms of the categories of microorganisms but also in terms of the application of this knowledge to everyday life.

CHAPTER EIGHTEEN

THE PROTOZOANS

The protozoans—four phyla of related protists

Before biologists began to recognize the need to make further distinctions in their system of classification, they classified in a single phylum of the animal kingdom a large number of related one-celled organisms having specialized cell organelles. *Protozoa,* the name given to this phylum, means, literally, "first animals." The protozoans were, in turn, divided into four classes, according to their means of locomotion. Since many of these organisms are not distinctly animal, most biologists today place them in the kingdom *Protista* and consider the classes to be four separate phyla of this kingdom. (See the classification in Chapter 14 for the names of these phyla and some examples.) While some biologists estimate that there are 15,000 protozoan species, others believe that there are as many as 100,000.

The protozoans are fascinating to study. Some have mouths, while others merely engulf their food by surrounding it with their bodies. Some dart about by means of tiny, hairlike projections that act like oars. Others wave flagella in front of them and thus project themselves forward or backward. The highly specialized organelles of protozoans enable them to live efficiently as single cells, carrying on all the life processes in a microscopic universe.

A mass of living jelly

The genus *Amoeba* includes several species of interesting protozoans. An ameba might be described as "animated jelly." On first seeing it, you might mistake it for a nonliving particle. But this tiny blob of grayish jelly moves of its own accord, feeds, reproduces, and performs all the other life processes.

These protozoans can be collected by taking slime from the bottoms of streams and ponds and from the surface of leaves of aquatic plants. Collecting samples and searching through the microscopic world for an ameba is a rewarding experience.

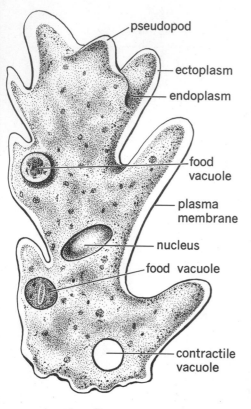

pseudopod

ectoplasm

endoplasm

food
vacuole

plasma
membrane

nucleus

food vacuole

contractile
vacuole

18-1 The structure of an ameba. This protist assumes many different shapes as it moves.

Under the microscope, the ameba looks like an irregular mass of jellylike protoplasm. It is surrounded by a thin *plasma membrane* (Figure 18-1). If you find an active animal, you will notice that the cytoplasm has a constant flowing motion. This streaming cytoplasm presses against the plasma membrane and produces projections called *false feet*, or *pseudopodia* (soo-doe-POE-dee-uh) (singular, *pseudopod*). A pseudopod starts as a bulge at any point on the surface of the organism and enlarges as part of the mass of the ameba flows into it. Soon, another pseudopod may form, and the flow will change to a new direction. The old pseudopod gradually disappears as the cytoplasm flows back to the cell mass. This type of locomotion is called *ameboid movement.* It is responsible for the classification of amebas in the protist phylum *Sarcodina.*

A closer look at amebas, for example, at *Amoeba proteus,* will show that the cytoplasm is of two different types: A clear, watery *ectoplasm* is found next to the cell membrane; the *endoplasm,* found inside the ectoplasm, is more dense and granular and resembles gray jelly with pepper sprinkled through it. If you examine the endoplasm in a moving ameba closely, you will notice that the inner region is more fluid than the outer region. The endoplasm flows more rapidly in the inner region. The nucleus can be seen as a disk-shaped mass that changes its position as the cytoplasm flows.

The oxygen necessary to maintain the life of these protists diffuses through the cell membrane from the surrounding water. Most of the carbon dioxide and soluble wastes of protein metabolism, such as ammonia, pass out through the cell membrane in the same way. However, much useless water comes into the ameba cell, mainly by osmosis but to a lesser extent with the intake of food. If the ameba did not have a method of ridding itself of this water, it would swell up like a balloon and burst. The method that has developed is that the excess water accumulates to form a *contractile vacuole* that, when it reaches maximum size, contracts and discharges the water through a temporary break in the cell membrane. Hence, by forming contractile vacuoles, the ameba maintains a constant or nearly constant internal water concentration.

How the ameba gets food

When an ameba comes into contact with an algal cell or another protozoan that is acceptable as a food source, it engulfs the cell by surrounding it with pseudopodia. Part of the membrane of the ameba now becomes the lining of a *food vacuole* inside the cytoplasm. A new membrane forms quickly on the surface at the point at which the food particle entered the cell. Digestion is accomplished by enzymes, formed by the cytoplasm, that pass into the vacuole and act on the food substances. Digested food is absorbed by the cytoplasm and can then be used as a source of energy and of materials needed to form additional protoplasm. Undigested particles remain in the vacuole and pass out of the cell at any point in the membrane.

Sensitivity in the ameba

Amebas respond to conditions around them. Although they lack eyes, they are sensitive to light and seek areas of darkness or dim light. While they lack nerve endings that we associate with the sense of touch, they react to movement around them. They also move away from objects they touch.

You can see how an ameba responds to food by putting small drops of food organisms in an ameba culture. Under the microscope, you may see amebas move over to the food. The presence of food, perhaps because of chemicals the food gives off, acts as a stimulus to the cells. Unfavorable conditions, such as dryness, lack of food, or cold, cause some species of ameba to become inactive and to withdraw into a rounded mass called a *cyst*. When favorable conditions return, the organisms resume activity.

Reproduction in the ameba

In the presence of abundant food and good environmental conditions, an ameba such as *Amoeba proteus* doubles its volume in a day or two. When the ameba has reached this size, its membrane surface is not large enough to supply such a volume of cytoplasm with adequate food and oxygen or to remove waste. At this point, the ameba reproduces by simple cell division. The ameba ready to divide rounds out into a spherical mass (Figure 18-2). The nucleus divides. The two "daughter" nuclei move to opposite ends of the cell. The cytoplasm then constricts near the center, and the two halves pull apart, forming two distinct cell masses. Each new daughter cell has a nucleus and is capable of independent life and growth. The division process itself takes about an hour.

Paramecium, a complex protozoan

The ameba is an example of a protist cell with few specialized structures, living an independent existence. We shall now observe cell specialization carried to a higher degree.

Various species of the genus *Paramecium* live in quiet or stagnant ponds where scums form. If you want to culture paramecia in the laboratory, collect submerged pond weeds, put them in a jar with pond water, and set the jar aside in a warm place for a few days. As the weeds decay, a scum forms on the surface. Large numbers of paramecia may be found in this scum.

When a drop of water containing paramecia is placed on a slide under the microscope, the most striking characteristic of this unicellular protist is its movement. It appears to swim rapidly

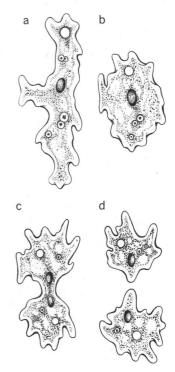

18-2 The ameba reproduces by simple cell division.

anterior end

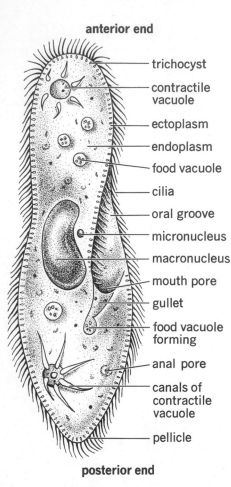

trichocyst

contractile
vacuole

ectoplasm

endoplasm

food vacuole

cilia

oral groove

micronucleus

macronucleus

mouth pore

gullet

food vacuole
forming

anal pore

canals of
contractile
vacuole

pellicle

posterior end

18-3 The structure of the paramecium. In what ways are the paramecium and the ameba alike? In
what ways are they different?

through the thin film of water between the slide and cover glass.
Actually, its rate of movement is quite slow—about three inches
per minute. The microscope, however, magnifies the speed to the
same extent to which it enlarges the object. A few strands of cotton
or filaments of algae added to the drop effectively reduce this
rapid motion. In addition, methyl cellulose (now used as a wall-
paper paste) can be used to increase the viscosity of the solution.
This slows down the paramecia and allows a detailed study of
their structure.

The paramecium is shaped like a slipper (Figure 18-3). Although it does not change its shape as an ameba does, it is by no
means rigid. It often bends around an object it happens to meet
when swimming. The definite shape of the cell is maintained by a
thickened outer membrane, the *pellicle,* that surrounds the plasma
membrane.

Paramecia move by means of hairlike cytoplasmic threads
called *cilia* that project through the cell membrane and pellicle.
These cilia are arranged in rows and lash back and forth like tiny
oars. They cover the entire cell, but are most easily seen along
the edges. Cilia contain protein filaments, which biologists believe
contract alternately to produce the beating. Cilia can beat either
forward or backward, thus moving the cell in either direction and
allowing it to turn. This form of movement places paramecia in
the protist phylum *Ciliophora.*

Another striking feature of the paramecium is a deep *oral groove*
along one side of the cell. This depressed area is lined with long
cilia that cause the animal to rotate around its long axis as it
swims through the water. The paramecium has a definite front
end, or *anterior* part, which is rounded, and a more pointed rear
end, or *posterior* part—a perfect design in streamlining. The oral
groove runs from the anterior end toward the posterior part. The
action of the cilia lining the oral groove and the forward movement of the animal force food particles into the *mouth pore.* The
mouth pore opens into a funnel-like tube, the *gullet,* which extends
into the cytoplasm. Bacteria and other food particles forced into
the gullet enter the cytoplasm within a *food vacuole.* When the
food vacuole reaches a certain size, it breaks away from the
gullet, and a new one begins to form.

The movement of the cytoplasm carries the vacuole in a circular course around the cell. During this circulation, digestion and
absorption occur as they do in the ameba. Undigested food passes
through a special opening in the pellicle called the *anal pore.* This
tiny opening is located near the posterior part of the cell. It is
completely closed except when in use and is quite difficult to see.

A *contractile vacuole* is found near each end of the cell. Surrounding each vacuole are numerous canals that radiate from the
central cavity into the cytoplasm. The canals enlarge as they fill
with water, after which their content is passed to the central
cavity and emptied out at the surface through an opening.

As they do in the ameba, the contractile vacuoles of the paramecium serve primarily to remove excess water that has entered by osmosis and with food. By determining the rate at which the contractile vacuoles fill and contract, biologists have estimated that in about thirty minutes the paramecium discharges a volume of water equal to the cell content. Thus, contractile vacuoles are highly efficient "water pumps." Some soluble waste may be eliminated by the contractile vacuoles, although most of the waste products appear to diffuse through the plasma membrane. Respiration in the paramecium is accomplished by diffusion of oxygen and carbon dioxide through the pellicle.

Sensitivity in the paramecium

The reactions of the paramecium to conditions around it are remarkable, considering that this protist, like the ameba, has no specialized sense organs. Except when they are feeding, the cells swim constantly, bumping into objects, reversing, and moving around them in a trial-and-error fashion. This response to the stimulation caused by bumping into objects is called the *avoiding reaction.* The reaction also occurs in response to such stimuli as excessive heat and cold, presence of chemicals, and lack of oxygen. Paramecia tend to move into regions of low acidity, a response of value to the organisms because bacteria, an important source of food, are found in water that is slightly acid.

The *trichocysts,* which normally appear as minute lines just inside the pellicle, are used as a means of defense. When a larger protozoan approaches, the trichocysts of the paramecium explode special protoplasmic threads into the water through tiny pores. These threads are quite long and give the organisms a bristly appearance. A bit of acetic acid or iodine added to some water that contains paramecia will often cause the trichocysts to discharge.

Reproduction in the paramecium

The paramecium has two different kinds of nuclei, both of which are located near the center of the cell. A large nucleus, or *macronucleus,* regulates the normal activity of the cell. Near the large nucleus is a small nucleus, or *micronucleus,* that functions during reproduction. Some species of paramecia have more than one micronucleus.

Reproduction may involve two distinct processes: *fission* and *conjugation.* Fission in paramecia involves the division of both the macronucleus (amitotically) and the micronucleus (mitotically), after which two daughter cells form (Figure 18-5a). Under ideal conditions, fission may occur two or three times a day. This is equivalent to more than seven hundred generations per year. If all the daughter cells were to live and ideal conditions could be maintained, the total mass of paramecia in five years would be many times greater than the mass of the earth.

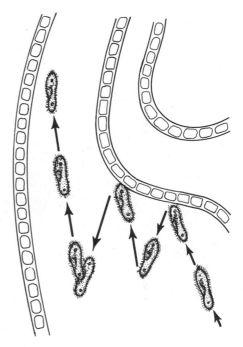

18-4 Avoiding reaction in paramecium.

a. fission

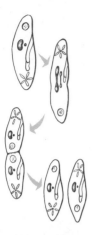

b. conjugation

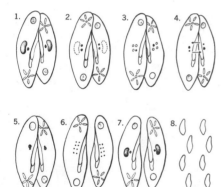

18-5 The two types of reproduction characteristic of paramecia: (a) fission, and (b) conjugation.

After several months of cell divisions, especially in the same environment, paramecia lose vitality and die, unless they undergo conjugation. This process requires the mixing of two mating types, or sexes, which may be designated as + and − or as I and II. Exchange of nuclear materials during conjugation, resulting in revitalizing of the cells, can be summarized as follows (Figure 18-5b):

■ Two cells unite at the oral grooves (1).
■ The micronucleus in each cell divides. The macronucleus degenerates (2).
■ The two micronuclei in each cell divide, forming four micronuclei, three of which degenerate (3).
■ The remaining micronucleus divides unequally, forming a large and a small micronucleus. The cells exchange smaller micronuclei (4).
■ The large (stationary) and small (migrating) micronuclei fuse in each cell (5).
■ The cells separate. In each, the fused micronucleus undergoes three consecutive divisions, forming eight nuclei (6).
■ Of the eight nuclei, four fuse and form a macronucleus, three degenerate, and one remains (7).
■ Two consecutive cell divisions occur, resulting in the formation of four small paramecia from each of the two original conjugants (8).

A flagellated protozoan

Several species of the genus *Euglena* live in fresh-water ponds and streams. Under the microscope, a euglena looks like an oval or a pear-shaped cell. The anterior end is rounded, while the posterior end is usually pointed (Figure 18-6).

Since euglenas possess some characteristics of plants and some of animals, they have been claimed by both botanists and zoologists. Because of one of their methods of locomotion, they are often placed in the protist phylum *Mastigophora* (MASS-ti-GAH-fuh-ruh), although some biologists put them in a phylum of their own. The organism swims by means of a flagellum attached to the anterior end. The flagellum—nearly as long as the one-celled body—rotates, thus pulling the organism rapidly through the water.

Unlike any other member of the phylum *Mastigophora*, the euglena has a second method of locomotion. This type of movement is so characteristic of the euglena that we call it *euglenoid* (yoo-GLEE-NOID) *movement.* It is accomplished by a gradual change in the shape of the entire cell. The posterior portion of the body is drawn forward, causing the cell to assume a rounded form, after which the anterior portion is extended, thus pushing the cell forward. Contraction and elongation of the cell is accomplished by *contractile fibers* composed of specialized protoplasm.

The internal features of a euglena such as *Euglena gracilis* show an interesting combination of plant and animal characteristics. The outer covering is a thin, flexible outer membrane, the *pellicle*. A gullet opening at the anterior end of the cell leads to an enlarged *reservoir*. Since the euglena has never been seen to ingest food, the gullet probably serves only as an attachment of the flagellum. A *contractile vacuole* close to the reservoir discharges water at regular intervals. Near the gullet is a very noticeable red *eyespot*. This tiny bit of specialized protoplasm is especially sensitive to light and serves to direct the organism to bright areas in its habitat. Near the center of the mass of cytoplasm is a large *nucleus* containing a nucleolus.

Perhaps the most striking characteristic of the euglena is the presence of numerous oval chloroplasts scattered through the cytoplasm of the cell. Most species of euglena carry on photosynthesis and, in addition, absorb organic matter from the environment through the cell membrane. Some species lose their chlorophyll in periods of prolonged darkness and live entirely on dissolved organic matter from the water in which they live. This organic matter passes through the cell membrane and into the cell, where it is digested. Thus, these organisms live an autotrophic life but can revert to a heterotrophic type of nutrition during unfavorable environmental conditions.

Euglenas multiply rapidly by regular cell division under ideal conditions. In about a day, a mature organism splits lengthwise, forming two new cells. In less than one month, one euglena may divide to give rise to several million if conditions for growth are favorable. Euglenas are often so numerous in ponds and streams that they cause the water to look bright green.

Table 18-1, which compares the ameba, euglena, and paramecium, may help you to review the degree of specialization of these three common protozoans.

The spore-forming protozoans

The spore-forming protozoans have no method of locomotion. They are all parasitic, absorbing their nourishment from the cells or body fluids of the hosts in which they live. Many live alternately in two hosts. They are placed in the protist phylum *Sporozoa*. In these protozoans, reproduction by spores is accomplished in the following manner: First, the nucleus divides into many small nuclei; then, a small amount of cytoplasm surrounds each nucleus to form a spore. Finally, the protozoan breaks apart and releases these tiny spores. The spores may be enclosed in a resistant wall, or they may be surrounded only by a cell membrane.

Plasmodium (plazz-MODE-ee-um), the parasitic protist that causes malarial fever in man and other warm-blooded animals, is a good example of a spore-forming protozoan. When a female *Anopheles* (uh-NAH-fuh-LEEZ) mosquito bites a person suffering

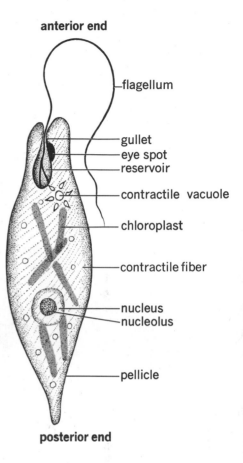

anterior end

flagellum
gullet
eye spot
reservoir
contractile vacuole
chloroplast
contractile fiber
nucleus
nucleolus
pellicle

posterior end

18-6 The structure of a euglena. Contrast this protozoan with a paramecium and an ameba.

Table 18-1 COMPARISON OF THREE PROTOZOANS

	AMEBA	PARAMECIUM	EUGLENA
FORM	variable	slipper-shaped	pear-shaped or oval
LOCOMOTION	pseudopodia (ameboid movement)	cilia	flagellum or euglenoid movement
SPEED	slow	rapid	rapid or slow
FOOD-GETTING	pseudopodia surrounding food	cilia in oral groove, mouth cavity, and gullet	photosynthesis or absorption
DIGESTION	in food vacuole	in food vacuole	in cytoplasm
ABSORPTION	from food vacuole by diffusion and cyclosis	from food vacuole by diffusion and cyclosis	from cytoplasm by diffusion and cyclosis
RESPIRATION	diffusion of O_2 and CO_2 through plasma membrane	diffusion of O_2 and CO_2 through plasma membrane and pellicle	diffusion of O_2 and CO_2 through plasma membrane and pellicle
EXCRETION	through plasma membrane	through plasma membrane and pellicle	through plasma membrane and pellicle
SENSITIVITY	responds to heat, light, contact, chemicals	responds to heat, light, contact, chemicals	eyespot sensitive to light; responds also to heat, contact, chemicals
REPRODUCTION	fission	fission and conjugation	fission

from malaria, some of these protozoans pass into the insect's stomach (Figure 18-7). They grow, reproduce, and work their way into the mosquito's bloodstream, through which they travel to the salivary glands. When an infected mosquito bites another person, the parasite is injected into the bloodstream in a very minute spindle-shaped form. This stage reaches cells of the liver, where the cycle continues 12 to 14 days. The parasites then re-enter the bloodstream where they attack red corpuscles. Within a red corpuscle, the parasite undergoes a series of cell divisions, resulting in ten to twenty asexual spores. The corpuscle is destroyed in the process.

At amazingly regular intervals, depending on the species of *Plasmodium,* the corpuscles burst and the spores are discharged into the bloodstream. Each spore invades another corpuscle and the process continues. Each cycle of spore discharge increases the number of corpuscles involved. It has been estimated that a billion or more corpuscles may be destroyed within two weeks of passage of the parasite from the liver to the bloodstream. The sudden release of spores and waste products from the red corpuscles at regular intervals causes the chills, high fever, and sweating characteristic of malaria.

After several cycles of spore formation, certain of the parasites enter a gamete-forming sexual stage. These structures continue the cycle in the stomach of a mosquito when it draws up blood from a malaria patient.

There are three major species of *Plasmodium* that cause malaria in human beings. One species, *Plasmodium vivax*, causes tertian

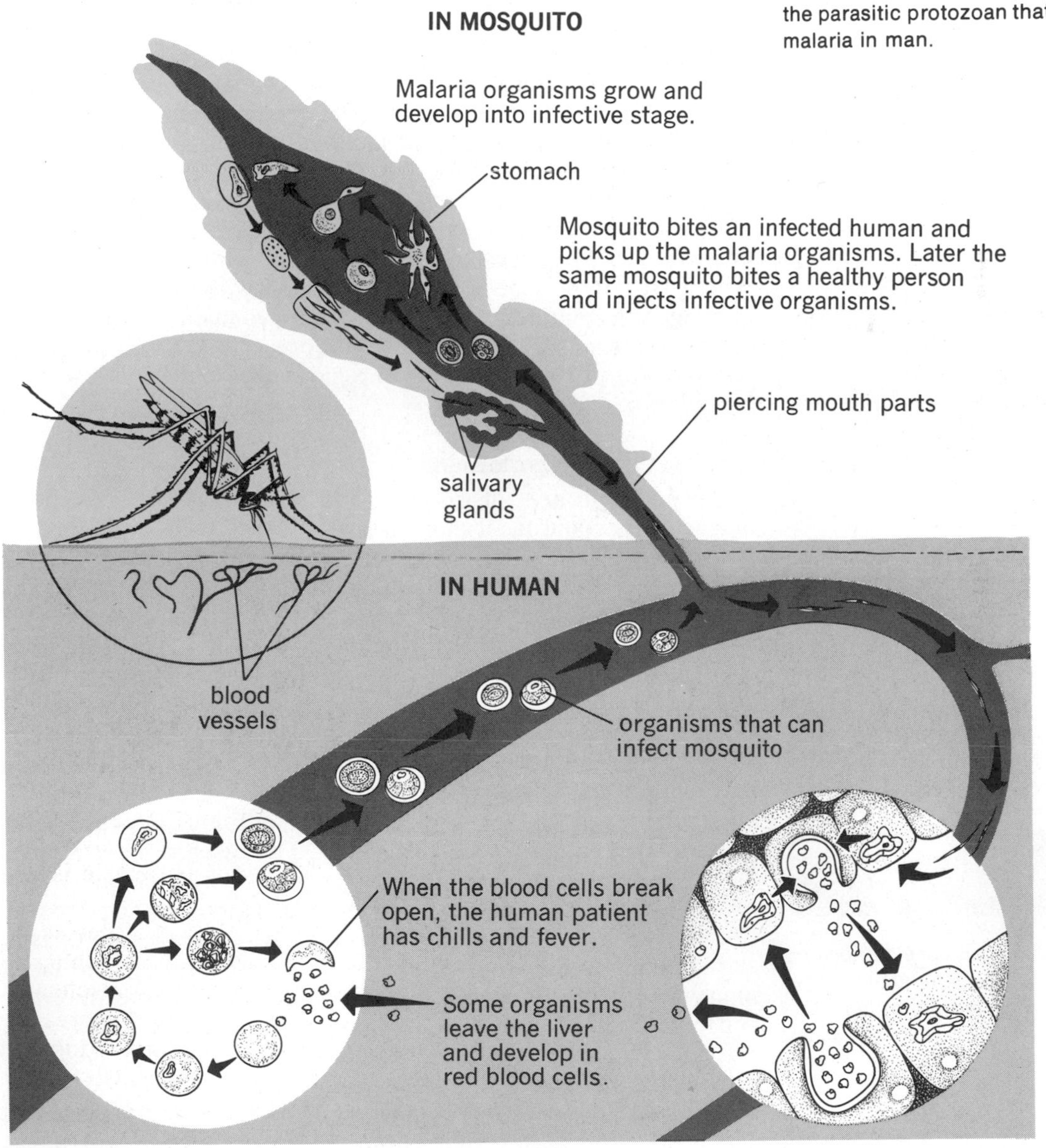

18-7 The life cycle of *Plasmodium*, the parasitic protozoan that causes malaria in man.

IN MOSQUITO

Malaria organisms grow and develop into infective stage.

stomach

Mosquito bites an infected human and picks up the malaria organisms. Later the same mosquito bites a healthy person and injects infective organisms.

piercing mouth parts

salivary glands

IN HUMAN

blood vessels

organisms that can infect mosquito

When the blood cells break open, the human patient has chills and fever.

Some organisms leave the liver and develop in red blood cells.

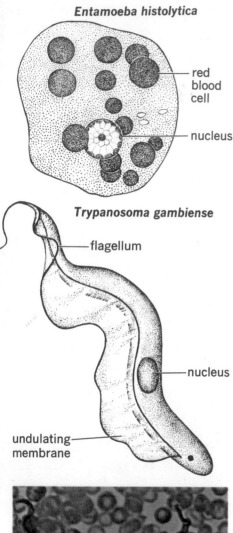

Entamoeba histolytica

red
blood
cell

nucleus

Trypanosoma gambiense

flagellum

nucleus

undulating
membrane

18-8 Two protozoans pathogenic in man—the ameba that causes amebic dysentery (top) and the flagellated protozoan that causes African sleeping sickness. Organisms of the latter type can be seen among the human blood cells in the photomicrograph. (Walter Dawn)

malaria, in which chills and fever occur every forty-eight hours. *Plasmodium falcaparium,* which occurs in tropical areas, causes a more severe and usually fatal form of tertian malaria in which the cycle of chills and fever is irregular or nonexistent. Quartan malaria, in which chills and fever occur in a seventy-two-hour cycle, is caused by organisms of the species *Plasmodium malariae.*

Malaria control involves both prevention and treatment. Spread of the disease can be stopped if malaria patients are isolated from *Anopheles* mosquitoes. This interrupts the life cycle of the parasite and prevents its spread to other people. Eradication of the mosquito is another important control measure. This is difficult to accomplish, especially in malaria infested areas with extensive marshes where the mosquitoes breed.

Several antimalaria drugs are effective in both prevention and treatment. Quinine has been used for many years. Since World War II, other drugs have been produced. One such drug is atabrin (mepacrine). This drug, used extensively in the latter years of World War II, has been largely replaced with newer drugs, including chloroquine, primaquine, and proguanil. Quinine and certain of these newer drugs may be prescribed as a preventive measure for persons traveling to regions where there is a danger of malaria.

Other pathogenic protozoans

Many people are surprised to find that the great majority of human beings and most animals are infected with some type of protozoan. In man and most animals, the usual place for these infections is the intestine, where a flourishing collection of protozoans may be found. If you examine the intestinal contents of a freshly killed animal under the microscope, you will probably see a great many protozoans. Some of them are harmless and may even be helpful, while others live on material in the intestine, just robbing the host of its food. Some may invade the intestinal wall, show up later in the bloodstream, and finally lodge in some other part of the body of the host.

Contaminated food or water can be a source of infection by a parasitic ameba that causes *dysentery*. The disease is most common in the tropics but may also occur in temperate climates. For example, a serious epidemic of amebic dysentery occurred in Chicago in 1933 during the Exposition. More than 1,400 cases and 4 deaths were reported. This infection centers in the large intestine, where the organisms feed on cells of the intestine wall and blood cells, causing bleeding ulcers and abscesses.

Cysts containing inactive ameba cells are excreted in the feces of infected people. For this reason, amebic dysentery is most common in countries where adequate sewage disposal methods are lacking or where human sewage is used as fertilizer. Human carriers who have mild symptoms of infection, if any, are a special problem in controlling amebic dysentery. Several drugs are effective in treating the infection, but proper sanitation is far more important.

Two flagellated protozoans known as trypanosomes are the causative agents of *African sleeping sickness.* These organisms are classified in the phylum *Mastigophora,* along with *Euglena.* Trypanosomes live in the blood of big-game mammals of Africa as well as in many insects. In man they cause a serious and frequently fatal infection.

Trypanosomes are transmitted to human beings by the bite of the *tsetse fly.* The fly, in biting an infected person or animal, withdraws trypanosomes along with blood. After a period of development in the intestine of the fly, the trypanosomes invade the fly's salivary glands and are transmitted when the fly bites and draws blood from the wound of the next victim.

In the human bloodstream, the organisms multiply several weeks or months before the first symptoms of infection appear. Early symptoms include headache, fever, and general fatigue. As the disease progresses, the victim becomes increasingly weak and anemic because of the destruction of blood cells. Later, the trypanosomes invade the cerebrospinal fluid and attack the central nervous system. This results in a prolonged coma and almost certain death.

Several drugs are effective in destroying trypanosomes in the blood if treatment is started early. However, there is little hope for successful treatment after the organisms invade the cerebrospinal fluid. African sleeping sickness is very difficult to control because of the widespread infection of animals in many parts of Africa. The only totally effective measure may be the eradication of the tsetse fly, a measure that usually requires the destruction of infected livestock. However, because the area in which the flies are found is so vast, we are still unable to carry out a total eradication program. Nonetheless, many local and international organizations, including the World Health Organization, are carrying on extensive campaigns to combat African sleeping sickness, as well as malaria, amebic dysentery, and other infectious diseases that affect man so seriously.

The economic importance of the protozoans

Protozoans that live in fresh-water ponds and streams are of great economic importance, for they are the source of food for many small animals. Some salt-water protozoans secrete a hard wall made of calcium carbonate (chalk) or silicon dioxide (silica). These substances may form elaborately beautiful patterns. Members of this group, called the *foraminiferans* and the *radiolarians,* are responsible for the formation of many of the limestone, chalk, and flint deposits throughout the world. As they die, their miniature skeletons fall to the bottom of the sea and there, with billions of others, form a muddy deposit. If, as the earth's surface changes, this deposit dries out, it becomes hard.

Protozoans help to digest food in the intestines of some animals. In cattle they play such a role; and in the intestines of termites, protozoans are chiefly responsible for digesting the woody material the insects eat.

SUMMARY

Protozoans are often referred to as "first animals," "lowly organisms," or "primitive forms of life." But let us not sell this group of protists short. True, they may represent the first animals and, since a single cell constitutes an entire organism, they may be considered primitive forms of life. However, consider what a paramecium cell can do. Without the aid of any other cell, it swims with lashing cilia, washes in food, including other organisms, and digests complex organic substances with enzymes secreted in its cytoplasm. Two contractile vacuoles pump out excess water and maintain its proper water content. It responds to touch and moves around objects in its path. Some unexplained chemical response directs it to areas of abundant food. Its trichocysts are deadly weapons. It maintains cell vigor by periodic conjugation with a paramecium of the opposite mating type.

Could any one of your cells match this performance? Removed from your body and placed in a culture with paramecia, it would soon die. It might even be the next meal for a passing paramecium.

BIOLOGICALLY SPEAKING

pseudopodia	posterior	micronucleus
ameboid movement	mouth pore	fission
ectoplasm	gullet	conjugation
endoplasm	anal pore	euglenoid movement
food vacuole	contractile vacuole	contractile fibers
pellicle	avoiding reaction	reservoir
cilia	trichocysts	eyespot
oral groove	macronucleus	nucleus
anterior		

QUESTIONS FOR REVIEW

1. List four protozoan phyla of the protist kingdom. What characteristics divide the phyla? Give examples of each.
2. Describe ameboid movement.
3. Describe the intake of food by an ameba cell.
4. Explain the vital function of the contractile vacuole.
5. In what ways does the ameba illustrate protoplasmic sensitivity?
6. Describe reproduction in ameba.
7. How might you collect and culture paramecia in the laboratory?
8. Describe locomotion in the paramecium.
9. Describe food-getting in the paramecium.
10. Describe the avoiding reaction in the paramecium.

11. How do trichocysts illustrate cell specialization in the paramecium?
12. Explain how paramecium cells multiply. How are they rejuvenated?
13. Describe two kinds of movement in euglena.
14. Euglena seeks light, while the ameba and paramecium tend to avoid it. What is the adaptive value of this?
15. What are the plantlike and animal-like characteristics of euglena?
16. Describe the life cycle of a *Plasmodium.*
17. Explain the relation of amebic dysentery to sanitary conditions.
18. Why is it so difficult to combat African sleeping sickness?

APPLYING PRINCIPLES AND CONCEPTS

1. Biologists frequently say that understanding the life processes of single-celled protozoans helps them to understand the life processes of complicated organisms like man. Why is this probably true?
2. Compare cell specialization in the ameba, paramecium, and euglena.
3. Once a person has become infected with the malarial parasite, at what stage in the development of the illness would you consider treatment most possible? Why?
4. What useful functions do certain protozoans perform in the animal intestine?
5. In what way are protozoans important in a pond?

RELATED READING

Books

BUCHSBAUM, RALPH, *Animals Without Backbones* (rev. ed.).
The University of Chicago Press, Chicago. 1948. An introduction to the main groups of the invertebrates, written in simple, nontechnical language, with each group used to illustrate some biological principle or some level in the evolution of animals from simple to complex.
BUCHSBAUM, RALPH, and LORUS J. MILNE, *The Lower Animals: Living Invertebrates of the World.*
Doubleday and Co., Inc., Garden City, N.Y. 1960. A natural history of invertebrate animals by two outstanding scientists.
KUDO, RICHARD R., *Protozoology* (5th ed.).
Charles C. Thomas, Publisher, Springfield, Ill. 1966. A very detailed introduction to the common and representative genera of all groups of both free living and parasitic protozoa.
LINDEMANN, EDWARD, *Water Animals for Your Microscope.*
The Macmillan Company, New York. 1967. A guide to identifying and examining the variety of animals found in ponds, lakes, rivers, and at the seashore.
ROWLAND, JOHN, *The Mosquito Man.*
Roy Publishers, Inc., New York. 1963. The story of Sir Ronald Ross and his research that led to success in controlling malaria.

CHAPTER NINETEEN

THE FUNGI

What are fungi?

When we speak of fungi, we refer in a broad sense to a large group of organisms considered to be protists in some classifications and simple plants called *thallophytes* in others. Most species of fungi are multicellular, but fungi never form roots, stems, or leaves. They vary in size from microscopic yeasts to mushrooms and puffballs weighing up to a pound or more.

Fungi vary greatly in form, but they share one important characteristic. All lack chlorophyll and therefore, like animals, cannot photosynthesize their organic food requirements. For this reason, they are nutritionally dependent and, again in contrast with animals, which can move about, must live in close association with an outside food supply. Many are *saprophytes,* living on the remains of dead plants and animals or on cast-off products of organisms. Others are *parasites,* taking their nourishment from the tissues of still-living hosts. Fungi are common in all environments where suitable organic compounds, moisture, and temperature are provided. The great majority are terrestrial, but many are aquatic.

Fungi, as a group, occupy a position of great importance in the living world. In obtaining the food materials and energy they need for life, some cause enormous economic loss to man through spoilage of foodstuffs he has gathered for his own use. Others are agents of the most destructive plant diseases. Certain fungi are parasites of animals, including man. In contrast, certain fungi decompose the organic remains of plants and animals and, in doing so, play a vital role in maintaining soil fertility. Some of those that produce antibiotics are valuable allies in fighting bacterial diseases. In addition, fungi are used extensively in industry—for example, in brewing beer, in bread-making, in cheese-making, and in the production of alcohol. Some are important food items. As you see, whether a fungus is regarded as beneficial or harmful depends on the species and the nature of its activities. Nonetheless, regardless of their roles, fungi as a group are important to us.

Classification of fungi

In the more recent classification, biologists place all of the fungi in two protist phyla. The smaller of these phyla, *Myxomycophyta* (MICK-suh-my-KAH-fuh-tuh), includes the *slime molds.* (These will be discussed in more detail at the end of this chapter.) All other fungi—those that have cell walls—are placed in the phylum *Eumycophyta* (YOO-my-KAH-fuh-tuh), or *true fungi.* The true fungi are, in turn, divided into four classes, based largely on their methods of reproduction. The first three classes, in which sexual reproduction is known to occur, are:

■ *Phycomycetes* (FIKE-oe-my-SEE-teez), the *algalike fungi,* which resemble green algae in certain respects and bear spores in a case known as a ***sporangium*** (Figure 19-1, top).

■ *Ascomycetes* (ASK-oe-my-SEE-teez), the *sac fungi,* which bear spores, usually eight in number, in a saclike structure known as an ***ascus*** (Figure 19-1, center).

■ *Basidiomycetes* (buh-SID-ee-oe-my-SEE-teez), the *club fungi,* which bear spores on a club-shaped structure known as a ***basidium*** (buh-SID-ee-um) (Figure 19-1, bottom).

All those fungi that cannot be placed in the first three classes are grouped very artificially together into the fourth class:

■ *Deuteromycetes* (DOO-tuh-ROE-my-SEE-teez), or *imperfect fungi.*

The true fungi

More than 75,000 different species comprise the four classes of true fungi. Included in this number are such familiar types as molds, mildews, blights, yeasts, morels, rusts, smuts, mushrooms, puffballs, and shelf fungi.

The vegetative bodies of most true fungi are branching filaments called ***hyphae*** (HIFE-ee) (singular, *hypha*). They are usually white or grayish in color. The hyphae of many fungi are divided into cells by cross walls. In other fungi, the hyphae are continuous tubes, lacking cross walls and containing numerous nuclei. The entire mass of hyphae formed by a fungus is known as the ***mycelium*** (my-SEE-lee-um). Although chlorophyll is never present in the hyphae of a mycelium, certain fungi produce green, yellow, orange, red, or blue pigmented substances that give them color.

Fungi secrete various enzymes that penetrate the food supply and digest it externally. The digested food is then absorbed by the hyphae. Fungi have no light requirements, and many thrive best in darkness. Moisture and, for most species, warm temperature are growth requirements. Nearly all true fungi are aerobes, although their oxygen requirements may be low. A few fungi, including yeasts, can tolerate anaerobic conditions, but, as should be expected, they grow faster when oxygen is available.

All true fungi reproduce asexually by forming spores, often in enormous numbers. Spores are dispersed to new environments by air currents, water, or contact with other agents. Most true fungi reproduce sexually as well.

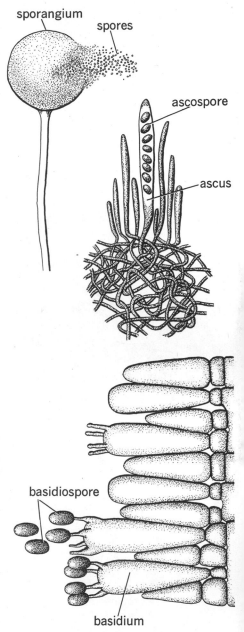

19-1 The classification of fungi is, to a large extent, based on the type of structure associated with spore formation. A sporangium (top) is characteristic of the Phycomycetes; an ascus (center), of the Ascomycetes; and a basidium (bottom), of the Basidiomycetes.

Molds—some of the most familiar fungi

We use the general term *mold* to refer to several kinds of fungi that form a mycelium that often is cottony in appearance. Some molds are Phycomycetes, while others are Ascomycetes. They grow on food, wood, paper, leather, cloth, and many other organic substances. While molds thrive best in warm, moist surroundings, many grow at temperatures near freezing. This makes them a serious problem in cold-storage plants and in home refrigerators.

Bread mold and other Phycomycetes

If you moisten a piece of bread, expose it to the air for several minutes, and place it in a covered culture dish in a warm, dark place, *bread mold (Rhizopus nigricans)* is almost certain to appear within a few days or a week. This familiar mold is a Phycomycete. It starts as a microscopic spore that germinates on the surface of the bread, forming a branching mass of silver, tubular hyphae that lack cross walls. Within a few days, the mycelium usually covers the surface of the bread.

A portion of bread mold viewed with a hand lens or the low power of the microscope reveals several distinct kinds of hyphae comprising the mycelium (Figure 19-2). Those hyphae that spread over the surface of the food supply are called *stolons* (STOLE-onz). At intervals along the stolons, clusters of short, rootlike hyphae called *rhizoids* (RY-zoidz) penetrate the food supply and absorb nutrients and water. Rhizoids secrete digestive enzymes that act on the sugar and starch in the bread. The digested foods are then absorbed into the hyphae of the mold. The flavor, odor, and color spots that mold produces on bread and other foods are caused by chemical changes resulting from the action of these enzymes.

After a few days of growth on the bread surface, black knobs appear among the hyphae of bread mold. Each black knob is a *sporangium,* or spore case, which is produced at the tip of a special aerial reproductive hypha, or *sporangiophore.* Sporangiophores develop in clusters above a mass of rhizoids in bread mold. More than fifty thousand spores develop in a sporangium of bread mold. When the sporangium matures, the thin outer wall disintegrates and releases the spores. The spores are carried away by air currents. If a spore lodges on a food supply in a suitably warm, moist environment, it germinates, forming a hypha that branches and develops into a new mycelium.

Sexual reproduction also occurs in bread mold by a form of *conjugation.* Although *Rhizopus* hyphae look alike, there are two physiologically different mating strains that can be designated as plus and minus. During sexual reproduction, the hyphae form short, specialized side branches (Figure 19-3). If the tip of a branch of a plus strain contacts the tip of a branch of a minus strain, conjugation occurs. Cross walls form a short distance

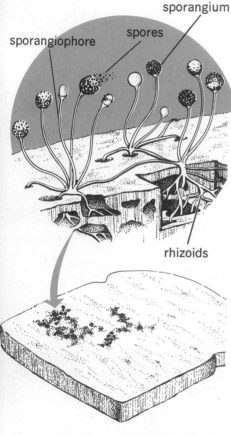

sporangiophore   spores   sporangium

rhizoids

19-2 The structure of bread mold. Note the rootlike rhizoids extending into the bread.

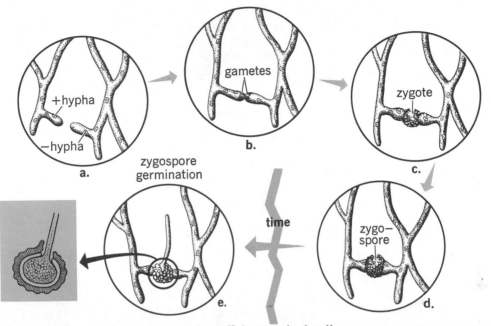

a.
+hypha
−hypha

b.
gametes

c.
zygote

d.
zygo-spore

e.
zygospore
germination

time

behind the tips of the side branches, cutting off the terminal cells that become gametes. Dissolving of the walls at the tips of the end branches allows the two gametes to fuse. Union of the two cells produces a fusion body, or *zygote.* The zygote forms a thick protective wall and matures into a *zygospore,* which enters a period of dormancy. After from one to several months, if conditions for growth are favorable, the zygospore germinates. In some cases, it produces a single sporangiophore that bears a sporangium and spores. In other cases, a short, branching hypha grows from the zygospore. A cluster of rhizoids is produced at the tip of the hypha, above which a sporangiophore, sporangium, and spores develop.

The *water molds,* belonging to the genus *Saprolegnia* (SAP-ruh-LEG-nee-uh), are relatives of the bread mold. They are Phycomycetes and, like bread mold, have tubular hyphae that lack cross walls. Some of these molds are saprophytes and live on the bodies of dead insects and other animals in the water. Hyphae form cottonlike masses on the dead animal (Figure 19-4). Other hyphae penetrate the dead tissues and absorb nourishment. *Saprolegnia* reproduces asexually by forming motile, flagellated spores, known as *zoospores,* in tubular sporangia that develop at the tips of hyphae. In the formation of a sporangium, a terminal portion of a hypha is cut off with a cross wall. Spores are formed from the protoplasmic content of the sporangium. Mature zoospores escape through a pore in the tip of the sporangium and are propelled through the water by two flagella. *Saprolegnia* also reproduces sexually.

19-3 Sexual reproduction in bread mold requires two physiologically different mating strains.

19-4 A cottonlike mass of a water mold, *Saprolegnia,* on a fish. (Allan Roberts)

Other water molds are destructive parasites. They invade the living tissues of fish and other aquatic animals through wounds and irritated areas. Hyphae penetrate the tissues of the host and absorb nourishment. External hyphae form cottony patches. In general, parasitic water molds eventually kill the host animal. They are a constant problem in aquariums, rivers, and lakes.

Blights and downy mildews—destructive plant parasites

One of the most destructive Phycomycetes is the *late blight fungus*. It was this fungus disease that caused the famine in Ireland from 1845 to 1847 by destroying nearly the entire potato crop. Hyphae of the blight fungus grow between cells of the leaves and form short, food-absorbing branches, known as ***haustoria,*** that penetrate cells and absorb nutrients. Other hyphae grow out of the lower side of the leaf through pores and form branching sporangiophores. Oval sporangia are produced at the tips of the sporangiophores (Figure 19-5). In some cases, the entire sporangium germinates and forms a new hypha. In others, the sporangium content forms numerous zoospores, each with two flagella. Zoospores swim to new areas of the leaf in a film of dew or rainwater, spreading the infection rapidly. Finally, the entire top of the plant is destroyed by the fungus. Sporangia that fall to the ground may be washed through the soil to the potato tubers, causing them to rot. The late blight fungus also attacks tomatoes, causing enormous loss.

The *downy mildews* are parasites that attack radishes, mustards, white potatoes, sweet potatoes, cereal grains, sugar cane, tobacco, and lettuce. Like the potato blight fungus, they form hyphae that penetrate the tissues of the host plant and absorb nourishment from the cells. Certain species form downy patches of surface hyphae, usually on the leaves of the host, and send short, branching hyphae deep into the plant tissues.

19-5 Sporangium production and germination in the late blight fungus.

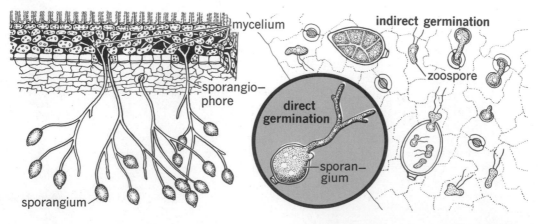

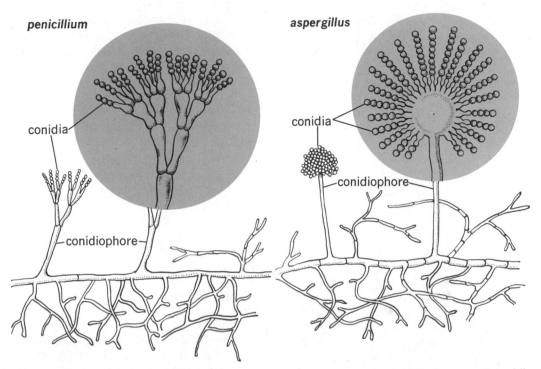

penicillium

aspergillus

conidia

conidiophore

conidia

conidiophore

19-6 Conidiophores and conidia of *Penicillium* and *Aspergillus*.

Blue-green molds and powdery mildews

Molds of the genus *Penicillium* are commonly seen as powdery, bluish-green growths on oranges, lemons, cantaloupes, and other fruits, as well as on bread, meat, leather, and cloth. Molds of the genus *Aspergillus* are common on foods, too. These can be distinguished by the yellow or black rings they form. Both these genera are examples of Ascomycetes and have hyphae with cross walls. Certain of the hyphae penetrate the food source and absorb nourishment. Surface hyphae form branches known as *conidiophores,* on which large numbers of spores are borne. The spores, or *conidia,* are formed in rows at the ends of short branches of the conidiophores (Figure 19-6).

In this group of molds, we find several valuable fungi. Several species of *Penicillium* are used in the processing of cheeses. Cheese manufacturers carefully culture these molds and add them to cheese at a certain point in processing. During the aging period, the hyphae of the mold grow through the cheese, secreting enzymes whose action results in the formation of substances that add distinctive flavors to the cheese. Among the most popular mold cheeses are Roquefort and Camembert. *Penicillium* molds are used in processing both of these cheeses. Another species of *Penicillium* produces penicillin, the first antibiotic discovered, as you learned in Chapter 17.

The *powdery mildews* are also Ascomycetes. They are usually not as destructive as downy mildews. In the early stage of a

19-7 Powdery mildew on a greenhouse rose. (Robert D. Raabe)

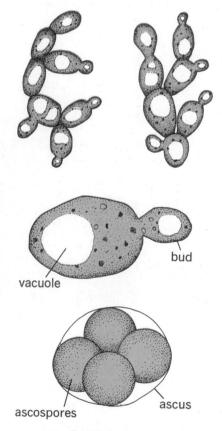

vacuole

bud

ascospores

ascus

19-8 Budding in yeasts. Under ideal growth conditions, a bud develops into a mature yeast cell in about thirty minutes.

powdery mildew infection, a mass of hyphae develop on the surface of the host, usually on a leaf. Short, branching haustoria penetrate the tissues of the leaf and absorb nourishment. Spores are produced at the tips of the surface hyphae, giving the fungus the powdery appearance for which it is named (Figure 19-7). Among the host plants of these mildews are lilacs, grapes, roses, clover, apples, gooseberries, wheat, and barley.

The familiar yeasts

Even though the *yeasts* are one-celled, microscopic organisms, they are classified with the Ascomycetes because they form a recognizable ascus at a certain stage in their life cycles. Each cell is oval or spherical and is bounded by a thin cell wall. Within the cytoplasm of a yeast cell is a prominent vacuole. Special staining will reveal a nucleus close to the vacuole. Under ideal growth conditions, which include a source of nourishment, warmth, and oxygen, yeast cells multiply rapidly by *budding*. A bud starts as a small knob pushing out from the surface of a mother cell (Figure 19-8). As the bud enlarges, the nucleus of the mother cell divides. One nucleus moves into the bud; the other remains in the mother cell. The base of the bud remains attached and, in turn, produces another bud, forming a chain of yeast cells. Additional buds may also develop on the original mother cell.

As conditions become unfavorable for rapid growth and budding, a yeast cell may undergo two nuclear divisions resulting in four nuclei. Each nucleus is surrounded by cytoplasm and a cell wall. We refer to these four cells as *ascospores* and to the yeast cell that formed them as an *ascus*. In the ascospore stage, yeasts are dormant and can endure drying and unfavorable temperatures. It is these live, dormant cells that are packaged and sold commercially as "active yeast." Under favorable conditions, the ascospores germinate and form active yeast cells.

As you may recall from Chapter 7, yeasts carry on anaerobic respiration, or fermentation. In this process, *zymase,* an enzyme secreted by the yeasts, acts to split molecules of simple sugar (glucose or fructose) and thereby to release a relatively small portion of the total energy stored in the sugar molecules for use by the yeast cells in carrying out their life activities. Alcohol and carbon dioxide are metabolic by-products of this process. Certain yeasts secrete additional enzymes that digest complex sugars (maltose and sucrose) to simple sugars that are acted on by zymase. No yeasts, however, produce enzymes that act on starch.

You can observe yeast fermentation by adding commercial yeast to a 10 percent solution of simple sugar in water. This culture should be placed in a cotton-stoppered bottle or flask and kept in a warm place. Within a few hours, the yeast cells multiply by budding to such numbers that they give the solution a cloudy appearance. Tiny bubbles of carbon dioxide rise through the solution, and you can smell the odor of alcohol in the culture.

Various strains of yeast are cultured for industrial use. For example, *baker's yeast* grows rapidly in dough. As it grows, it forms bubbles of carbon dioxide that expand during baking and thereby cause the dough to rise. The small amount of alcohol produced is driven off by the heat. While the growth of baker's yeast is inhibited at an alcohol concentration of from 3 to 5 percent, *wine yeasts* tolerate alcohol up to a concentration of about 14 percent. Higher concentrations of alcohol are obtained by distillation after the fermentation is complete.

Yeasts are important, also, as producers of vitamin B_2, or *riboflavin* (RY-boe-FLAY-vin). This vitamin is essential in normal growth and in the health of the skin, mouth, and eyes. Riboflavin remains in the yeast cells. To obtain it, we must either eat the yeast cells or use a product made of ground yeast.

Wild yeasts are abundant in the air and ferment sugars in natural fruit juices. A few yeasts are pathogenic.

Other Ascomycetes

The *cup fungus* and *morel* are relatives of the yeast. These Ascomycetes are harmless saprophytes. Cup fungi grow on rotting wood and leaves and on organic matter in rich humus soils. Hyphae penetrate the food supply and absorb nourishment from the food material. The white, orange, or red cups are spore-bearing structures composed of tightly massed hyphae. Many of these hyphae end in an elongated sac, or ascus, inside of which there are eight spores.

The morel, or sponge mushroom, is highly prized for its delicious flavor. This is one of the few mushrooms that can be safely eaten (Figure 19-9). It is easily recognized and is not related to the true mushrooms, many of which are inedible or poisonous Basidiomycetes, which are further discussed below.

Some of our most serious plant diseases are caused by parasitic Ascomycetes. Among these diseases are Dutch elm disease, chestnut blight, apple scab, and ergot disease of rye.

The Basidiomycetes

The *Basidiomycetes* are often called *club fungi.* Both of these names refer to a curious club-shaped structure formed at the end of certain hyphae. This *basidium,* as it is called, usually bears four *basidiospores* externally. Four groups of fungi compose the class Basidiomycetes. These include the rust fungi, smut fungi, mushrooms and bracket fungi, and puffballs.

Rust fungi cause many serious plant diseases. There are more than 2,000 known rusts that attack flowering plants and ferns, of which over 250 are grain rusts that are parasitic on wheat, oats, barley, and other cereals. They cause millions of dollars of damage to these crops annually.

Wheat rust is one of the best known of these plant parasites and is a special problem to farmers. It produces four kinds of

19-9 The morel, or sponge mushroom, a highly prized edible fungus. (Alvin E. Steffan, National Audubon Society)

spores in a very complicated life cycle that involves not just one but two host plants.

The rust makes its appearance on wheat during the late spring and early summer months, when the wheat is green and actively growing. The rust forms a mycelium that grows among the cells of the wheat stem and leaves. Tiny blisters appear along the surface of the stems and leaves where clumps of hyphae grow to the surface and discharge their spores (Figure 19-10a). These spores are reddish-orange in color and are all one-celled. Biolo-

19-10 Life cycle of the wheat rust. The two hosts necessary for the completion of the life cycle are the common barberry bush and the wheat plant. Compare this cycle with that of the corn smut. (Inset: courtesy USDA)

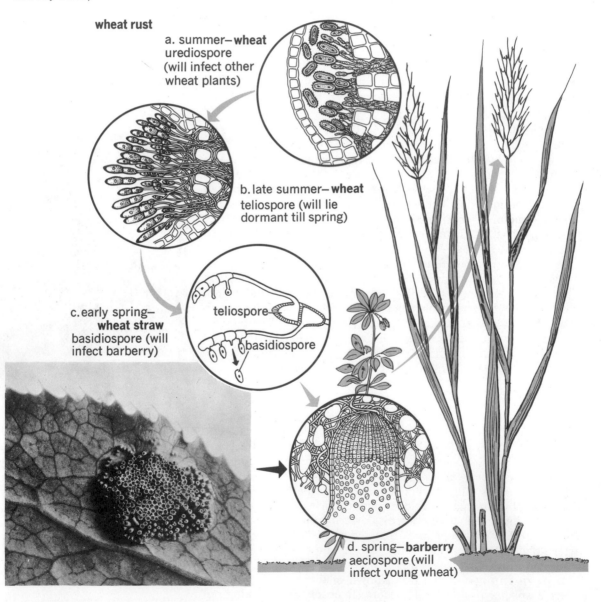

wheat rust

a. summer—**wheat** urediospore (will infect other wheat plants)

b. late summer—**wheat** teliospore (will lie dormant till spring)

c. early spring— **wheat straw** basidiospore (will infect barberry)

teliospore

basidiospore

d. spring—**barberry** aeciospore (will infect young wheat)

gists refer to them as red spores, or *urediospores* (yooh-REE-dee-uh-SPOARZ). These urediospores can reinfect wheat and spread the disease rapidly by lodging on new plants. Later in the summer, when the wheat is ripening and the plants are turning yellow, a second spore stage appears. The same hyphae that produced red spores now produce black spores, or *teliospores* (TEE-lee-uh-SPOARZ). These are two-celled spores with heavy, thick protective walls (Figure 19-10b).

These black spores cannot reinfect wheat. Instead, they remain dormant through the winter on the wheat straw or stubble or on the ground. Early in the spring, both cells of the teliospores germinate, producing four-celled basidia. One *basidiospore* forms on each of the four cells of the basidium (Figure 19-10c). These spores are then carried away by the wind. If a spore lodges on a leaf of the common barberry (not the cultivated Japanese barberry), the life cycle continues. After complicated changes have occurred in the tissues of the barberry leaf, tiny cups appear on the leaf surfaces. These contain rows of *aeciospores* (EE-see-uh-SPOARZ), which drop from the cups and are carried by the wind to young wheat plants as much as five hundred miles away (Figure 19-10d). Infection of the wheat and production of red spores follow, thus completing the life cycle.

Both the wheat and the barberry are necessary for completion of the cycle. Thus, destruction of the common barberry bush is essential in controlling this disease, especially in the northern states, where winters are usually severe. However, in the southern states, where there are mild winters, red spores may not be killed and can reinfect wheat directly.

Another rust, the *cedar-apple rust,* involves the red cedar tree and the apple and its close relatives. The *white pine blister rust* causes serious damage to the white pine tree in one stage of its cycle and lives on the wild currant or gooseberry in another.

Smuts—parasites on cereal grains

The *smuts* attack corn, oats, wheat, rye, barley, and other cereal grasses, causing considerable damage.

Corn smut, one of the most familiar of this group, infects corn plants when they are young. Some weeks later a grayish, slimy swelling appears on the ear, tassel, stem, or leaf (Figure 19-11). These swellings consist of hyphae. As the corn matures, the hyphae produce large numbers of black, sooty spores, which are carried by the wind. These spores may infect nearby corn plants or lie dormant until spring. A germinating spore produces a four-celled, or sometimes three-celled, basidium. Each cell of the basidium produces a single basidiospore, as in the rusts. These basidiospores are carried by the wind to corn plants, where they germinate and start a new smut infection. Corn smut can be controlled by burning infected plants, plowing under the stubble

corn smut

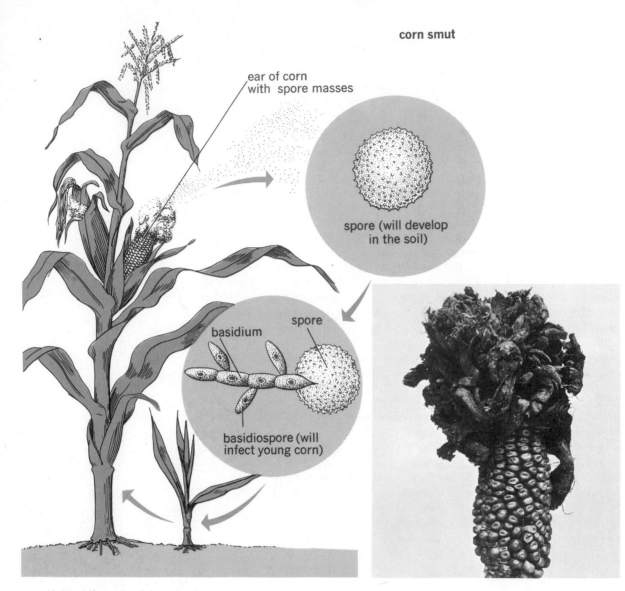

19-11 Life cycle of the corn smut. Trace the life cycle from the point at which masses of hyphae develop in the corn plant until the cycle is completed. (inset: courtesy Illinois Agricultural Experiment Station)

after the corn is cut, and destroying the unused stalks and leaves after the corn is picked.

Mushrooms—the best known of the fungi

We find *mushrooms* in orchards, fields, and woodlands, and we see them popping up suddenly after a warm spring or autumn rain. We seldom see the vegetative mushroom plant. It is a mycelium composed of many silvery hyphae, which thread their way through the soil or the wood of a decaying log or stump. A mushroom may

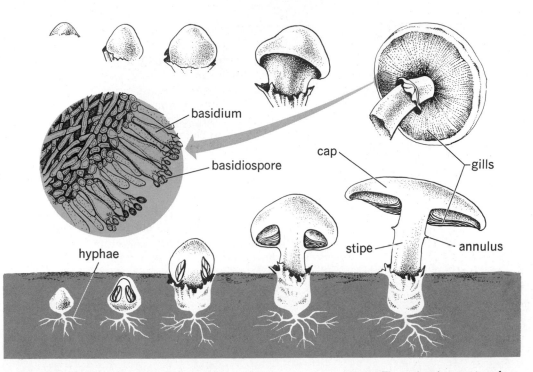

basidium

basidiospore

cap

gills

hyphae

stipe

annulus

19-12 The development of a mushroom. The inset is an enlargement of a gill.

live many years, gradually penetrating more and more area of substrate. Digestive enzymes secreted by the hyphae break down organic substances in the host and change them to forms that can be absorbed and used as nourishment.

At certain seasons, especially in the spring and fall, small knobs develop on the mycelium just below the ground. These consist of masses of tightly packed hyphae. These buttons develop into the familiar spore-bearing structure we recognize as a mushroom (Figure 19-12). The mature mushroom consists of a stalk, or *stipe,* which supports an umbrella-shaped *cap.* While pushing up through the soil, the cap is folded downward around the stipe. After forcing its way through the soil, the cap opens out, leaving a ring, or *annulus,* around the stipe at the point where the cap and stipe were joined.

Most mushrooms contain numerous platelike *gills* on the under-surface of the cap, radiating out from the stipe like the spokes of a wheel. On the outside of each gill and extending all around it are hundreds of small basidia, each of which bears four basidio-spores. It is estimated that a single mushroom may produce as many as ten billion spores. These spores drop from the basidia when mature and are carried by the wind and air currents. Each spore forms a new mushroom mycelium if it happens to land in a new environment favorable for growth. If it does not—and not too many do—the spore dies. Some basidiospores are black, while others are brown, yellow, white, or pink.

19-13   A fairy ring. (Walter Dawn)

19-14   Bracket, or shelf, fungi.
(Herbert Weihrich)

19-15   Puffballs. (Walter Dawn)

Growth characteristics of the fungi

We can learn something about the food relations of fungi from the observation that many fungi grow in circles. In the fairy ring mushrooms, for example, the mycelium of the original plant digests the organic matter available at that spot. As the mycelium expands into the unused organic matter in the soil around it, new mushrooms are produced at the outer edge of this growing ring (Figure 19-13). You can see the same principle in mold growth in a culture dish. The newest part of the mold is on the outside of the circle, and the oldest portion is in the center.

Poisonous and edible mushrooms

The word *toadstool* is frequently used as a popular term for poisonous mushrooms. While many people claim to have methods of distinguishing edible from poisonous varieties, experts tell us there is no certain rule or sign that can be used to distinguish the two types of mushrooms. Frequently, some of the most harmless-looking forms are poisonous and produce severe or even fatal effects if eaten. The only safe advice that can be given is to leave them alone unless you know exactly which forms are edible and which forms are poisonous.

Bracket fungi

The *bracket fungi* are the familiar shelflike growths seen on the stumps or trunks of trees (Figure 19-14). They may be either parasites on living trees or saprophytes, living on dead wood. They are the most destructive of the wood-rotting fungi. The mycelium of the bracket fungi penetrates the woody tissue of the host and causes it to disintegrate internally. The shelflike reproductive body is telltale evidence of the damage that is occurring within the host. The bracket fungi are woody in texture when old and remain attached to the host year after year. New spore-producing hyphae form on the underside of the shelf, forming layers, or rings, of growth. Spores are discharged through tiny pores located on the underside of the shelflike growth.

Puffballs

The *puffballs* resemble mushrooms except that the spores are released into the air only when the reproductive structure dries and splits open. Puffballs are round or pear-shaped growths, usually white in color (Figure 19-15). Members of nearly all species are edible if collected when they are young, before the spores mature. No poisonous puffballs are known to exist.

Imperfect fungi

All those fungi for which only the asexual, or *imperfect,* stage of their life cycle has been observed so far are grouped together as the *Deuteromycetes,* or *imperfect fungi.* This imperfect stage is the stage in which no sexual fusion occurs and only asexual spores are produced. While the majority of imperfect fungi are most like the Ascomycetes, it is only when the sexual, or perfect, stage of a life cycle is discovered that transfer to the appropriate class occurs. Many of the imperfect fungi are of great importance as parasites in man, animals, and plants. Among the plants, some of the diseases of corn, oats, wheat, citrus fruits, tomatoes, cabbage, lettuce, beans, and apples are attributable to this group.

The ringworm fungi produce several skin infections in man, including ringworm, barber's itch (ringworm of the scalp), and the all-too-familiar athlete's foot. Thrush is a serious infection of the mucous membranes of the mouth and throat, caused by an imperfect fungus resembling the yeasts.

Mycorrhiza

For many years, biologists have been investigating interesting natural structural and functional combinations of certain fungi and the roots of certain higher plants known as *mycorrhizas,* or "fungus-roots." While there is a great deal of variation, in general the mycorrhizal fungus is partly in the soil and partly embedded in the mycorrhizal plant root. In turn, in some cases the filaments of the fungus penetrate *between* the cells of the host, and in others, the filaments penetrate *into* the cells of the host. In any case, it is believed that the fungus aids the root in absorbing water and minerals, especially nitrogen compounds. In turn, the fungus takes nourishment from the root tissues.

Slime molds

The classification of this interesting group of organisms has long been a problem to biologists. In many respects, *slime molds* resemble giant masses of amebas and might be considered animal-like. However, at certain times they bear sporangia and spores, much in the manner of a fungus. For these reasons, biologists have referred to them as "slime fungi" and "fungus animals." Much of this confusion is avoided by placing the slime molds in a protist phylum of their own, the *Myxomycophyta.*

The vegetative body of a typical slime mold is a mass of protoplasm containing many nuclei and lacking cell walls. It is called a *plasmodium.* A plasmodium often appears as a slimy, fan-shaped network of living matter (Figure 19-16). This strange mass flows along in an ameboid manner on the soil of a forest floor, over the surfaces of dead leaves, or on a rotting log. As it moves along, it takes in small particles of organic matter and digests them in food vacuoles.

After a period of activity, the plasmodium becomes inactive. Clusters of stalked sporangia, often of delicate design and brilliant color and shaped like tiny feathers, balls, or worms, rise above the plasmodium. Each sporangium bears numerous spores that

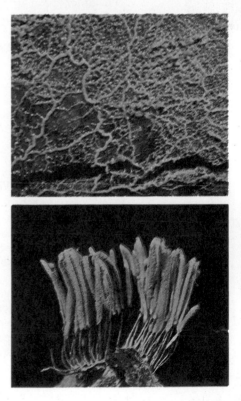

19-16   A plasmodium of a slime mold (top), shown at about two and one-half times natural size. Under certain conditions, the vegetative body becomes inactive, and clusters of stalked sporangia develop. (Hugh Spencer; Thomas Martin, Rapho Guillumette)

are embedded in an intricate network of filaments. As spores are discharged, they are carried by air currents. Spores that lodge in wet surroundings germinate and release one or more tiny cells, each with two flagella. These flagellated cells may act as gametes and fuse in pairs. Each fused pair is called a *fusion body*. The fusion body becomes ameboid and grows rapidly to form a new plasmodium that contains many nuclei. In some cases, many fusion bodies unite and form a plasmodium.

SUMMARY

Since fungi lack chlorophyll and require an organic food source, they are in direct competition with man. Each year, we spend millions of dollars for chemical sprays and dusts to protect our crop plants from fungus diseases. We spend additional millions of dollars in breeding plants that are resistant to fungus attacks. Add to these amounts the enormous loss of foods due to spoilage caused by fungi. It has been estimated that crop losses and food spoilage add more than 600 dollars to the annual grocery bill of the average family of five. It is hard even to estimate the cost of the damage fungi cause in rotting wood and destroying fabrics.

Fungi are not all bad, however. Before you condemn them too severely, think of the bakers, the wine industry, the cheese makers, and the mushroom growers. Remember, too, the antibiotics we extract from certain of the fungi.

Whether good or bad, however, we cannot escape the effects of fungi on our lives.

BIOLOGICALLY SPEAKING

sporangium	sporangiophore	conidia
ascus	conjugation	budding
basidium	zygote	ascospore
hyphae	zygospore	urediospore
mycelium	zoospore	teliospore
stolon	haustoria	aeciospore
rhizoid	conidiophore	plasmodium

QUESTIONS FOR REVIEW

1. What one reproductive characteristic do all fungi have in common?
2. In what way are all fungi nutritionally dependent?
3. List several ways in which fungi may be harmful. In what ways are certain fungi valuable economically?
4. Name the classes of fungi. Which of these classes are included in the true fungi?
5. Describe the mycelium of a true fungus.
6. What different kinds of hyphae make up a bread mold plant?
7. How can bread become inoculated with bread mold even though there may be no molds close by?
8. Describe sexual reproduction in bread mold.
9. Describe asexual reproduction in *Saprolegnia.*
10. Name several plants attacked by downy mildews.

11. Discuss several ways in which *Penicillium* molds are valuable.
12. Discuss budding in yeast.
13. How are the products of yeast fermentation used commercially?
14. Describe the production of four kinds of spores by wheat rust.
15. What stage in the life cycle of corn smut reveals its relationship to the rusts?
16. Describe the structure of a mushroom reproductive body.
17. How are puffballs different structurally from mushrooms?
18. Name several human infections caused by imperfect fungi.
19. Describe the plasmodium of a slime mold.
20. In what specific habitat might you collect a slime mold?

APPLYING PRINCIPLES AND CONCEPTS

1. Mildew and blight diseases of higher plants are more prevalent in some summers than in others. Account for this.
2. Sweet cider will ferment rapidly in a warm place even when it is in a tightly closed container. Explain why.
3. Why are yeast preparations valuable in treating acne and other skin disorders?
4. Why is a severe epidemic of wheat rust likely to follow a mild winter?
5. Explain why certain kinds of mushrooms and molds often grow in a ring.
6. In what respects are slime molds possible links between entirely different protist phyla?

RELATED READING

Books

CHRISTENSEN, CLYDE M., *The Molds and Man* (2nd ed.).
    University of Minnesota Press, Minneapolis. 1965. A general account of the fungi and their impact on man.
FERGUSON, CHARLES, *Illustrated Genera of Wood Decay Fungi.*
    Burgess Publishing Co., Minneapolis. 1963. A technical, ring-bound manual, with 132 pages of helpful information for amateurs.
HUTCHINS, ROSS E., *Plants Without Leaves.*
    Dodd, Mead and Co., New York. 1966. Each plant described, giving method of reproduction, uses, and its odd ways of survival.
KAVALER, LUCY, *The Wonders of Fungi.*
    The John Day Company Inc., New York. 1964. The story of fungi as both destroyer and saver of man, from prehistoric days to the present-day age of wonder drugs.
LARGE, E. C., *The Advance of the Fungi.*
    Dover Publications, Inc., New York. 1962. A history of fungi and their harmful aspects, drawn from scientific writings, past and present.

CHAPTER TWENTY

THE ALGAE

Algae — the "grasses of many waters"

Floating or submerged masses of green, yellowish, or brownish thread-like growth are familiar to anyone who has walked along the shore of a pond or stream or rowed across a lake in the spring, summer, or early fall. You may have called these masses pond scums, water moss, or seaweeds. *Algae* (singular, *alga*) is a better name.

Algae have been called the "grasses of many waters." This is an appropriate name for them. For one thing, they are as *abundant* in aquatic environments as grasses are on land. Algae are the dominant vegetation in rivers and streams, ponds and lakes, and vast stretches of the oceans of the world. They are also as *important* in these environments as grasses are on land. Algae are the principal food producers of aquatic environments. Directly or indirectly, most aquatic organisms depend for food and for oxygen to sustain life on the photosynthesis carried on by algae.

Today, scientists are searching for new sources of food to supply the needs of an ever increasing world population. If we should ever reach the limit of what could be harvested from the land, we could turn to the water, where algae have been feeding the organisms of oceans, lakes, and streams for ages. The "grasses of many waters" await our harvesting as an almost limitless source of nutritious food, if and when they are needed.

What are algae?

If you compare the structures of the algae with those of other types of plants, such as mosses, ferns, and seed plants, you will see that the algae are relatively simple organisms. They lack specialized tissues for conducting water and, for the most part, food. They also lack such usual plant organs as roots, stems, and leaves. However, from the standpoint of reproductive processes and biochemical capabilities, they are quite complex. Further, we must certainly consider them highly successful forms of life, for

20-1 Green algae near the edge of a pond. Algae are the dominant vegetation in the waters of the world. (Grant Heilman)

they have dominated aquatic environments since the Proterozoic era, more than one billion years ago.

Algae are found in a great variety of forms. Many are one-celled and can be seen only with the microscope. These solitary cells may float in the water, settle to the bottom, or swim about with lashing flagella. Other algae, including the giant kelps of the ocean, form plant bodies more than one hundred feet long. Regardless of size, however, the vegetative body of an alga, because it lacks specialized tissues and organs, is called a *thallus* (plural, *thalli*).

In many species of algae, the cells group to form cell colonies. Often, the cells are attached in chainlike, linear groups called *filaments*. Depending on the species, the filaments may be branched or unbranched. Other colonies of algae may be sheetlike, platelike, or spherical. Many of the larger algae, including the seaweeds of the oceans, have broad, ribbonlike thalli composed of thousands of cells.

With few exceptions, algae are autotrophic—they carry on photosynthesis. It is this that distinguishes them, as a group, from the fungi. Glucose, formed during photosynthesis, may be converted to other products in various algae. The cells of many species are surrounded by sheaths of a gelatinous substance secreted by the cells. It is this gelatinous sheath that makes many algae difficult to grasp.

Distribution of algae

Algae are found in nearly all permanent or semipermanent bodies of water. However, the amount of light, water temperature, the oxygen and carbon dioxide supply, and the mineral content of the water are important environmental factors in the distribution of algae.

Some algae grow attached to rocks in the fast-flowing water of rapids and waterfalls. Others thrive in quiet pools and backwaters. Most species float near the surface or thrive in shallow water, where the light intensity is greatest. However, some species live in deeper water. In the oceans, algae are most abundant near the shore, where they can be seen attached to rocks or floating in tidal pools at low tide. However, some marine algae live at depths of five hundred feet or more.

Not all algae are aquatic. Certain species are found on the bark of trees and on the surface of moist soil. Among the most unusual environments in which we find algae are the underside of the blue whale and the hair of the three-toed sloth, a mammal of the South American jungle. Some algae live within other organisms. One unicellular species lives in a *Paramecium.* Other algae live in the bodies of *Hydra,* fresh-water sponges, and aquatic snails. Most often, algae do not damage the organisms in which they live; they are beneficial to them in supplying food they produce by photosynthesis.

Methods of reproduction

Many forms of reproduction are carried on among the algae. Most algae reproduce by more than one method. Frequently, both the kind and the frequency of reproductive processes are determined by environmental conditions and by the period of the growing season.

Reproduction may be either *asexual* or *sexual.* Some algae reproduce only asexually. However, many species reproduce by both methods, asexual and sexual and consequently undergo complicated life cycles.

Many one-celled algae reproduce asexually by cell division, or *fission.* When a unicellular alga divides, two new organisms result. However, cell division in a colonial alga merely increases the size of the colony.

Colonies of algal cells are frequently broken apart by currents of water, passing fish, or animals feeding on them. We refer to this mechanical separation of cells as *fragmentation.* This breaking up of colonies does not increase the number of cells but merely multiplies the number of colonies. Since cells in colonies have little or no dependence on each other, fragments continue to grow if conditions are favorable. Fragmentation occurs frequently and is a major factor in the spread of algae in an aquatic environment.

Many algae reproduce asexually by forming *zoospores*. These motile spores are so named because they have one or more whip-like flagella that enable them to swim freely, as the animal-like protists do. Each zoospore contains at least one nucleus and some cytoplasm of the mother cell from which it was formed; and each is capable of producing a new vegetative cell directly. It is by means of these zoospores that many algae are able to spread over wide areas during a single growing season.

Many algae reproduce sexually by forming *gametes*. As in other organisms, the gametes of algae must fuse and form a zygote in the reproductive process. Many algae produce gametes which are equal in size, alike in form, and usually motile. We call these *isogametes*. Other algae form two types of gametes – male and female. These are called *heterogametes*. The sex of the hetero-gamete can be determined by examining differences in structure – male gametes, or *sperms,* are usually motile and are smaller than the nonmotile female gametes, or *eggs*. Whether gametes are produced in vegetative cells of the thallus or in specialized reproductive cells depends on the kind of alga.

Classification of algae

For many years, biologists classified the algae, together with the fungi, as *thallophytes* and placed them in the plant kingdom. However, because of the lack of tissues and organs in their vegetative bodies, in more recent classifications, algae have been placed in the kingdom Protista.

While all algae contain chlorophyll and are basically green, this color is often masked by other pigments present in their cells. Noticeable variations in the *color* of algae have long been the basis for classifying them into groups. However, color alone is not a reliable basis for classification, since pigments often vary in closely related species. *Cell structure* must also be considered. Still another basis for classification is the nature of *food* accu-mulated in the cells. The forms of *reproduction* also serve to dis-tinguish various groups of algae.

In some recent classifications, the more than thirty thousand species of algae are placed in six major phyla:

- *Cyanophyta* (sy-uh-NAH-fuh-tuh), or *blue-green algae*
- *Chlorophyta* (kloar-AH-fuh-tuh), or *green algae*
- *Chrysophyta* (kruh-SAH-fuh-tuh), or *yellow-green algae,* golden-brown algae, and diatoms
- *Pyrrophyta* (py-RAH-fuh-tuh), mainly dinoflagellates
- *Phaeophyta* (fee-AH-fuh-tuh), or *brown algae*
- *Rhodophyta* (roe-DAH-fuh-tuh), or *red algae*

In other recent classifications, a seventh phylum, *Eugleno-phyta* (yoo-gluh-NAH-fuh-tuh), including *Euglena* and other euglenoid organisms, is placed with the algae.

Blue-green algae

The *blue-green algae* are fundamentally different from all other algae. Together with bacteria, they are considered among the most primitive organisms. The cells of blue-green algae are simple in structure. The nuclear material, which remains near the central region of the cell, is not surrounded by a nuclear membrane. The chlorophyll is not contained in chloroplasts; it is distributed together with other pigments in the outer region of the cytoplasm. None of the blue-green algae reproduce sexually, as do the higher algae. Instead, their usual method of reproduction is simple fission.

In general, blue-green algae are smaller than individuals of the other algal groups. Some blue-green algae are filamentous, while others form slimy masses in which the cells are embedded. You can find blue-green algae in almost every roadside ditch, pond, and stream. Some species live on wet soil and rocks and even thrive in hot springs with a water temperature of as high as 185°F.

Blue-green algae thrive during the hot summer months, especially in water polluted with organic matter. They are a constant problem in drinking water and swimming pools. They give water the odor characteristic of stagnant pools and certain streams. For this reason, biologists regularly check the sources of public water supply for the amount of blue-green algae they contain.

The phylum name *Cyanophyta* refers to a blue pigment, *phycocyanin* (FIKE-oe-SY-uh-nin). This, together with chlorophyll, gives these algae their characteristic blue-green color. Colors of various species range from bright blue-green to almost black. However, a few species contain a red pigment. One of these algae appears periodically in such great abundance in the Red Sea that the name of that sea is believed to stem from the appearance of the water during these periods.

Representative blue-green algae

We have selected four common blue-green algae as representatives of this important group. One of the most curious is *Nostoc* (NOSS-tock). You will find it in mud and sand, usually just at the point where the ripples from a pond or lake strike the shoreline. A *Nostoc* colony looks like a small, gelatinous ball, ranging in size from a pinhead to a marble, and is often likened to a peeled grape (Figure 20-2). The ball is composed of a gelatinous sheath. Embedded in the sheath are many curved and twisted filaments made up of tiny spherical cells, each of which resembles a pearl bead in a necklace. Distributed along the filaments are curious empty cells with thick walls and pores at either side where they join other cells. These are *heterocysts.* Some biologists believe that they may be modified spores that have lost their content and ceased to function. The filaments of *Nostoc* usually fragment between a heterocyst and a vegetative cell. *Nostoc* is, funda-

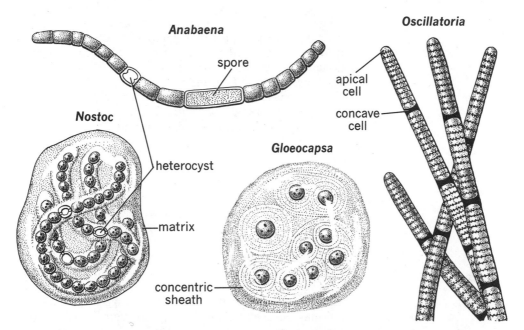

Anabaena

spore

Oscillatoria

apical cell

concave cell

Nostoc

heterocyst

matrix

Gloeocapsa

concentric sheath

20-2 Four rather common blue-green algae. In what ways are they similar? How do they differ?

mentally, a one-celled plant that forms filamentous colonies surrounded by a gelatinous sheath. Each time a cell divides by fission, it forms two small cells of equal size and thus increases the length of the filament. This is the only known method of reproduction in this primitive alga.

Anabaena (AN-uh-BEE-nuh) is a relative of *Nostoc,* as you can see from the similarity in their cells and filaments (Figure 20-2). Filaments of *Anabaena,* however, are solitary and do not occur together as the filaments of *Nostoc* do. Another difference is the production of spores in *Anabaena*. These spores can be recognized easily as enlarged, oval cells in the filaments. Each spore is protected by a thickened wall and contains an abundance of stored food. Each eventually separates from the parent filament and germinates in a new location.

Gloeocapsa (GLEE-oe-KAP-suh) is one of the most primitive blue-green algae (Figure 20-2). It can be found on wet rocks and often grows on moist flowerpots in greenhouses, where it forms a slimy, bluish-green mass. The individual cells of this alga are spherical or oval. The diffused blue and green pigments lie in a zone near the wall. Dark granules can be seen deeper in the cell. Each cell of *Gloeocapsa* secretes a slimy sheath. When a cell divides, each new cell secretes its own sheath within the old one, thus forming characteristic layers of sheaths. A colony of *Gloeocapsa* often contains hundreds of cells, joined by their sheaths in a gelatinous layer.

Oscillatoria (AH-suh-luh-TOAR-ee-uh) is a filamentous blue-green alga composed of many narrow, disk-shaped cells resembling a stack of coins. The tip, or *apical* (AP-ick'l) *cell,* is round on one side in many species. Filaments of *Oscillatoria* sway gently back and forth in the water, a characteristic that gave the alga its name. When one cell in the filament dies, the ones on either side bulge into its place and produce a curious *concave cell* (Figure 20-2). Concave cells are weak places in the filament and cause the plant to break into shorter pieces. *Oscillatoria* becomes abundant in ponds and streams during warm weather.

Green algae

The *green algae* belong to the phylum *Chlorophyta* and vary from one-celled forms to colonial forms composed of a large number of cells. They have a well-defined nucleus enclosed in a nuclear membrane. Both chlorophyll *a* and *b,* which are present in the same ratio as they are in the cells of the seed plants, and other photosynthetic pigments are organized in chloroplasts. These pigments combine to give the algae colors ranging from grass green to yellowish-green. Green algae photosynthesize sugar and convert it to starch for storage in the cells. Reproduction may be asexual (by fission and by asexual motile and non-motile spores) or sexual (by the fusion of isogametes or heterogametes).

Green algae are for the most part fresh-water algae, although certain members of the group live in strange and interesting environments. Many kinds live in the ocean and a few live in bodies of water even higher in salt concentration than the ocean. Some, to the amazement of scientists, thrive in hot springs.

Protococcus, a common green alga

Protococcus is one of the most common of green algae and is an exception among algae in that it does not live in water. Most of you probably do not realize it is an alga when you see it on the trunks of trees. It is also commonly found growing on unpainted wooden buildings and fences and on damp stones. During dry weather you seldom see it, but in wet weather it is very evident. It is more commonly found growing on the north side of tree trunks because bark is more moist on the shaded side.

The cells of *Protococcus* are spherical or somewhat oval (Figure 20-3). Each contains an organized nucleus and a single, large chloroplast. The cells are so small that many thousands cover only a few square inches of bark. They may be carried from tree to tree by birds and insects as well as by the wind during dry weather. Since *Protococcus* is autotrophic, it requires no nourishment from the tree on which it grows.

Reproduction in *Protococcus* is by fission only. Cells may divide in any one of three planes. Following divisions, the cells tend to cling together. This produces the cell groups shown.

20-3 Cells of the green alga *Protococcus* occur singly or in colonies of two or more.

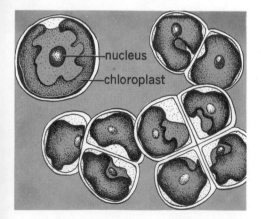

nucleus

chloroplast

Chlorella is a single-celled, spherical alga somewhat resembling *Protococcus*. It has a large, cup-shaped chloroplast. This alga is especially interesting to the biologist because several of its species live in the cells or tissues of protozoans, sponges, and jellyfish. As you learned in Chapter 6, it has also been the subject for much research in the study of photosynthesis as well as in the study of alga cultures as a possible source of food and oxygen.

Spirogyra, a filamentous green alga

Almost any pond or quiet pool will have bright green masses of threadlike *Spirogyra* (SPY-ruh-JY-ruh) during the spring and fall months. The unbranched filaments of this green alga range in length from a few inches to a foot or more. Under a microscope, a thread of *Spirogyra* looks like a series of transparent cells, arranged end to end like tank cars in a train (Figure 20-4, top).

Each cell has one or more spiral chloroplasts that wind from one end of the cell to the other. On the ribbonlike chloroplasts lie small protein bodies called *pyrenoids* (pie-REE-noidz), which are surrounded by a layer of starch grains. The nucleus is embedded in cytoplasm and is suspended near the center of the cell by radiating strands of cytoplasm that appear to be anchored to the pyrenoids. Most of the cytoplasm lies in a layer close to the wall, leaving a large central vacuole that is filled with water and dissolved substances. A thin, gelatinous sheath surrounds each cell and gives the filaments of *Spirogyra* a characteristic slippery feeling.

On a bright day, when photosynthesis takes place rapidly, bubbles of oxygen stream from the cells and collect among the filaments, causing a mass of *Spirogyra* to float to the surface. During the night, the oxygen dissolves in the water and allows the mass to sink. You can see floating masses of *Spirogyra* and other algae on the surfaces of ponds, especially in the afternoon of a bright day.

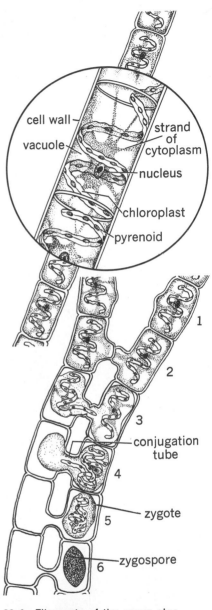

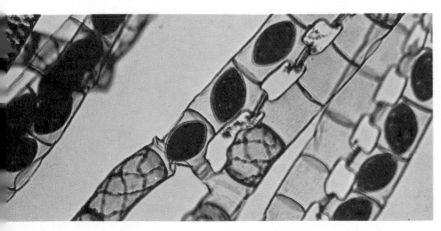

20-4 Filaments of the green alga *Spirogyra*. The structure of a vegetative cell can be seen at the top. Conjugation is shown at the right. Explain what is happening in each of the numbered steps in the sequence. (Bruce Roberts, Rapho Guillumette)

Spirogyra reproduces in two ways. All the cells of a filament undergo fission at certain time intervals. Since the divisions are always crosswise, fission adds to the length of the filaments.

The second method of reproduction is conjugation, which, as you learned in Chapter 18, is sexual. Conjugation occurs when weather conditions are unfavorable for normal growth by fission. You are likely to see it in material collected early in summer, at the end of the spring growing season, and again in fall, with the approach of cold weather.

Conjugation involves two filaments of *Spirogyra* that line up parallel to each other. A small knob grows out from each cell (Figure 20-4, right). Each knob continues to grow until it touches the knob of the cell across from it. Soon the tips of the knobs dissolve and form a passageway between the two cells, which results in a ladderlike arrangement. The content of one cell then flows through the conjugation tube and unites with the content of the other cell to form a spherical or oval *zygote* that soon develops a thick protective wall. It is interesting to note that the content of a cell in one filament moves across to the cell of the other, thus producing one filament of empty cells and another containing rows of zygotes. Since the zygote is a fusion body formed by the union of two nuclei, it has the *diploid* chromosome number.

Soon after conjugation is completed, the thick-walled zygotes, now called *zygospores,* fall from the cells holding them and undergo a rest period. The thick wall protects the zygospore from heat, cold, and dryness. As a zygospore, *Spirogyra* can survive a long, cold winter, a summer drought, or may even be transported from one pond to another by a bird or other animal. When conditions are favorable for growth again, the content of a zygospore resumes activity. The nucleus undergoes two meiotic divisions that result in four nuclei, each containing the *haploid* number of chromosomes. Three of these nuclei disintegrate. The remaining one becomes the nucleus of the new Spirogyra cell that grows from the zygospore and establishes a new filament by successive cell divisions.

The life history of *Ulothrix*

Ulothrix (YOO-luh-THRICKS) is interesting to the biologist because it is typical of a group of green algae with somewhat different reproductive processes. You can find this alga attached to rocks in swift-flowing, shallow water. The short filaments are anchored by a special cell known as a *holdfast,* with finger-like projections. Each cell above the holdfast has a chloroplast shaped like an open ring.

Under ideal growth conditions, a cell of *Ulothrix* undergoes a series of mitotic divisions, followed by a splitting up of the cell content into 2, 4, 8, 16, or 32 oval zoospores (Figure 20-5a).

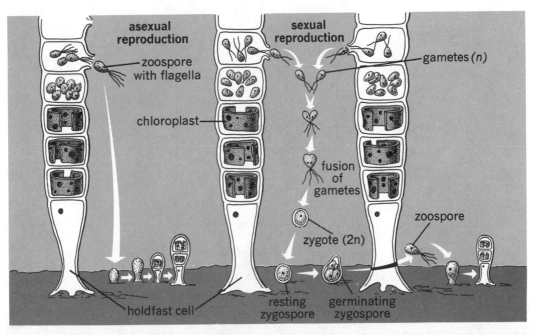

asexual
reproduction

sexual
reproduction

zoospore
with flagella

gametes (n)

chloroplast

fusion
of
gametes

zoospore

zygote (2n)

resting
zygospore

germinating
zygospore

holdfast cell

Each zoospore contains a nucleus and chloroplast and is propelled by four terminal flagella. The zoospores burst out of the encasing mother cell and use their flagella to propel themselves through the water. When a zoospore reaches a suitable place, it grows into a holdfast cell. This holdfast cell divides into two new cells of which the upper one becomes an ordinary *Ulothrix* cell. This cell then forms a filament by further cell divisions.

At other times, *Ulothrix* cells may undergo mitotic divisions and separation of the cell content into 8, 16, 32, or 64 isogametes, which are very similar to zoospores except that they are smaller and have only two flagella. The isogametes leave the parent cell and swim away from the filament (Figure 20-5b). If a gamete meets another gamete that differs from it chemically, the two fuse and produce a zygote that becomes a zygospore. After a rest period, the zygospore undergoes a meiotic division that results in four nuclei, each with the haploid chromosome number, as in *Spyrogyra*. In *Ulothrix,* however, all the nuclei live and, with a portion of the zygospore protoplast, become zoospores. Each zoospore may produce a new holdfast cell and thus give rise to a new filament.

20-5 Asexual and sexual reproduction in the green alga *Ulothrix*. Sexual reproduction involves isogametes. Are the vegetative cells of the filaments diploid (2n) or haploid (n)?

Life history of *Oedogonium*

Oedogonium (EE-duh-GOE-nee-um) is a common green alga found in quiet pools, where it grows attached to rocks, sticks, and other objects. Like *Ulothrix,* the plant consists of an unbranched filament with a basal holdfast cell. Each cell has a

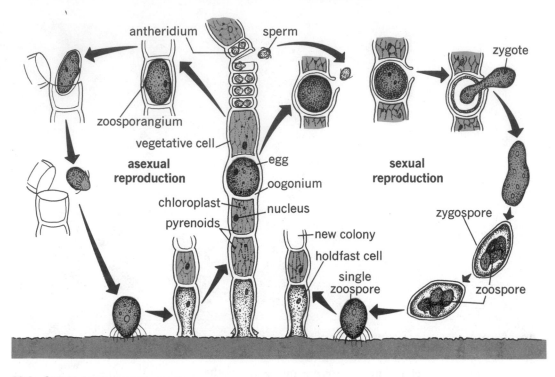

20-6 Sexual and asexual reproduction in the green alga *Oedogonium*. Note that in this alga, sexual reproduction involves heterogametes.

single chloroplast composed of many joined strands. Pyrenoids are numerous (Figure 20-6).

Any cell above the holdfast can convert its content into a large, single zoospore. The zoospore is propelled through the water by its ring of flagella. When a zoospore reaches a suitable location, it settles down and modifies to become a holdfast cell, thus establishing a new filament.

Sexual reproduction involves only certain cells in a filament of *Oedogonium*. In this way, it is different from *Spirogyra* and *Ulothrix*. Special cells along a filament develop into large, single eggs. Each of these special cells is called an *oogonium* (OE-uh-GOE-nee-um) and each develops a tiny opening through its cell wall. In most species of *Oedogonium*, two sperms develop in each of several shortened cells that form in groups in the filament. We call a sperm-producing cell an *antheridium*. A sperm looks like a miniature zoospore. Whether eggs and sperms develop in the same or in different filaments depends on the species of *Oedogonium*. Within a short time (often a few minutes) after a sperm leaves an antheridium, it swims to the pore of an oogonium, enters, and fertilizes the egg. This produces a zygote. When the oogonium wall disintegrates some time later, the zygote escapes, forms a heavy wall and changes into a zygospore, and undergoes a rest period. When conditions are again favorable for growth, the zygospore forms four zoospores by meiosis, which escape and establish new filaments of *Oedogonium. Oedogonium* is considered one of the more advanced algae because of its functionally specialized heterogametes and because of the structural difference between the female gamete, or egg, and the male gamete, or sperm.

A closer look at the reproductive process in algae

In our study of certain fungi and *Spirogyra,* we noted the beginnings of sexual reproduction in the process of conjugation. In *Ulothrix,* we saw a further development in sexual reproduction; that is, gametes are formed that are differentiated sexually although they look alike. In *Oedogonium,* gametes are formed that look distinctly different and are called sperm and egg. Remember that all these forms also undergo asexual reproduction by fission, fragmentation, or spore formation. Thus, they have two ways to reproduce.

Now let us examine the sexual method more closely. In *Ulothrix,* for example, the isogametes, like all gametes, contain the haploid number of chromosomes. The zygote, which results from the fusion of two gametes, is $2n$, or diploid. The diploid stage or generation in the life cycle is called the *sporophyte,* or "spore-producing plant," generation. While in *Ulothrix* this generation is restricted to the diploid zygote, we will find that in the life cycles of the mosses, the ferns, and later the seed plants, this generation is the predominant one.

Soon after fertilization, the zygote undergoes meiotic division, resulting in four nuclei having the n, or haploid, number of chromosomes. These four haploid nuclei become four zoospores. These give rise to new haploid *Ulothrix* filaments. Remember that the filament, besides reproducing asexually, may at times produce gametes. For this reason, it is known as the *gametophyte* (guh-MEET-uh-FITE) generation and is the most conspicuous stage in *Ulothrix.* The gametophyte has the n number of chromosomes, as do the gametes produced from it by mitosis. The occurrence of two distinct stages in the life cycle is known as *alternation of generations.* You will discover, as you study the evolution of plants, that this alternation occurs throughout the plant kingdom, even though the relative predominance of the generations varies.

Desmids

None of the green algae is more beautiful and fascinating to study than the group known as *desmids.* These nonmotile, free-floating algae may be solitary or colonial (Figure 20-7). Some form filaments and are called chain desmids. Desmids and other small floating algae are often spoken of as *plankton* forms. A desmid cell consists of equal halves connected by a narrow isthmus in which the nucleus is situated. Each half contains one

20-7   A living fresh-water desmid. (Walter Dawn)

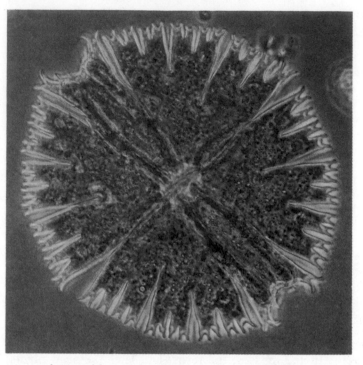

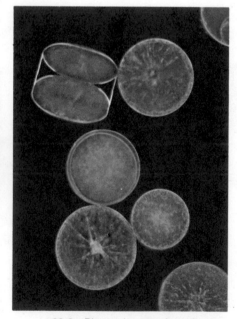

20-8   Photomicrograph of diatoms. As you can see, the details of the intricate design of the shell as well as the shapes and sizes of living diatoms vary with the species. (D. P. Wilson)

or two large chloroplasts. This unique structure makes desmids especially beautiful. They are common in the quiet waters of lakes and ponds.

Diatoms—very common algae in both fresh and salt water

Diatoms are the most abundant members of the phylum *Chrysophyta* and are, probably, the most widely distributed of all algae. More than sixteen thousand species of diatoms have been found in bodies of fresh and salt water. Generally, diatoms are one-celled, nonmotile plankton algae. They may be unicellular or colonial, free-floating or attached to objects or other organisms. The shape and size of diatoms vary with the species. They may be spindle-shaped, oblong, triangular, or round (Figure 20-8). Nearly all diatoms contain chlorophyll, which is localized in chloroplasts. Often, the chlorophyll is masked by a golden-brown pigment. As is characteristic of this phylum, most food reserves are stored as oils rather than carbohydrates. Diatoms are practically the only algae that grow in open seas and next to bacteria, are the most numerous organisms in existence.

The walls of diatoms are composed of an inner layer of pectin and an outer layer of silica (silicon dioxide). The wall is in two sections, or valves, one overlapping the other, like the top and bottom of a pillbox. In a sense, diatoms actually live in glass houses, since silica is the principal ingredient of glass.

The shells of diatoms are attractive, not only because of their many shapes, but also because of the many microscopic lines, ribs, and pores that form intricate designs. The numerous, tiny pores are channels through which the enclosed protoplasm of the diatom contacts its aquatic environment.

The most common method of reproduction among diatoms is mitotic cell division. When cell division occurs, the nucleus, cytoplasm, and chloroplasts divide forming two cell protoplasts. The two valves of the outer shell, each containing a protoplast, separate. A new inner shell is formed over the exposed side of the two protoplasts, resulting in two new cells (Figure 20-9).

When diatoms die, the protoplast and inner pectin wall decompose, leaving the transparent outer silica wall. These walls remain indefinitely and settle to the bottoms of ponds, lakes, or oceans, where they form deposits known as diatomaceous earth. Through the ages, deposits of diatomaceous earth may become amazingly thick. An example is found in California, where a deposit more than 1,400 feet thick settled to the ocean floor in ancient times and is now raised above the water level. Diatomaceous earth is mined and sold for use as an ingredient in various fine abrasive scouring powders and polishes, in filters in the refining of gasoline and sugar, as insulation material, and as material for roads.

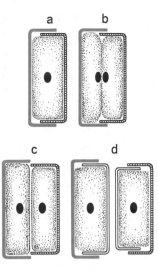

20-9 Diatoms usually reproduce asexually by cell division, as shown.

Dinoflagellates

The phylum *Pyrrophyta,* meaning "fire algae," is the smallest of the algal phyla. It includes a group of unicellular, motile algae known as *dinoflagellates.* Most dinoflagellates live in the oceans, where they are an important source of food for other organisms. A few species are found in fresh water. Certain of the marine dinoflagellates are phosphorescent and emit a strange light that is quite apparent at night if they are present in large numbers.

Dinoflagellates are easily distinguished from other flagellated organisms by the structure of their cell walls. The walls are composed of many-sided cellulose plates and contain two prominent furrows, or grooves. One groove is longitudinal, while the other is transverse and encircles the cell. A flagellum lies in each groove (Figure 20-10). Chloroplasts in the cells of dinoflagellates contain chlorophyll and a yellowish-brown pigment that usually masks the green color.

One of the dinoflagellates, *Gymnodinium,* is responsible for a strange and tragic phenomenon that occurs in warmer oceans at periodic intervals. Cells of *Gymnodinium* contain a red pigment that is toxic to fish and other marine animals. From time to time, conditions in the water seem to favor rapid reproduction of the dinoflagellates and they appear in enormous numbers. The red cast they give the water is called a "red tide." During an outbreak of "red tide" in the Gulf of Mexico, samples of the water were found to contain over two hundred million dino-

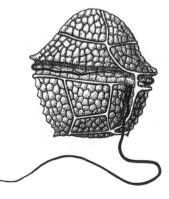

20-10 A dinoflagellate, *(Peridinium).*

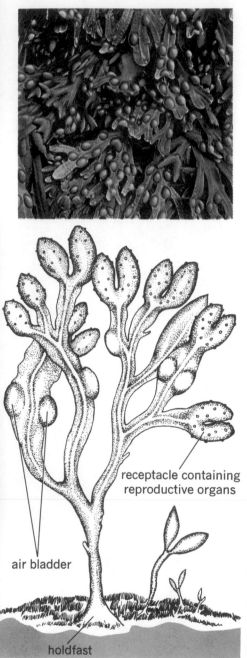

flagellates per gallon. Release of the toxic substance from such large numbers of cells poisons millions of fish and other marine animals, causing enormous economic loss and a severe water-pollution problem. Biologists have not been able to explain why outbreaks of "red tide" occur at periodic intervals. Copper compounds are known to kill the dinoflagellates, but the wide expanse of ocean involved in an outbreak makes treatment difficult.

The brown and red algae

Two phyla of algae, often called *seaweeds,* are represented abundantly in the oceans and seas. The phylum *Phaeophyta* contains more than a thousand species of *brown algae.* These algae are found most frequently in the colder ocean waters, where, attached to rocks, they grow at a water depth of from a few feet to fifty feet or more.

Brown algae vary greatly both in form and size. All are multi-cellular. The vegetative body may be a slender filament or a flat, ribbonlike thallus one hundred feet or more in length. In some respects, the thalli of many brown algae resemble the bodies of higher plants. A large, rootlike holdfast anchors the thallus to a rock or other object in the water. A stemlike stipe supports a flat, leaflike blade. Air-filled bladders on the thalli of many brown algae serve as floats and give them buoyancy in the water. Cells of brown algae contain chlorophylls as well as other pigments, including xanthophyll, carotene, and a golden-brown pigment known as *fucoxanthin,* which usually masks the other pigments. Sugars are the most abundant products of photosynthesis in the brown algae.

Fucus, or rockweed, is a brown alga that is common in the cold waters of the northern Temperate Zone. It can be seen in great quantities attached to rocks in shallow coastal waters when the tide is out (Figure 20-11). The forked thalli, one to three feet in length, float toward the surface at high tide. Enlargements at the tips of many thalli contain reproductive organs. You can see *Fucus* used as packing around oysters, lobsters, crabs, and other seafoods that are shipped from the coast to your markets.

One of the best-known brown algae is *Sargassum,* which floats in large masses in many areas of the Atlantic Ocean. One such area, known as the Sargasso Sea, lies between the Bahamas and the Azores. Here, floating masses of *Sargassum* and other algae spread over an area of more than two million square miles.

Kelps are the largest brown algae. They are common along the coast of western North America. A kelp is composed of several flat blades attached by stalks to a basal holdfast that anchors to a rock. Large kelps attain a length of over one hundred feet. Air bladders on the thalli cause kelps to float.

The more than 2,500 species of *red algae* of the phylum *Rhodophyta* are nearly all marine. Their vegetative bodies may

20-11 *Fucus,* a brown alga that lives along the Atlantic Coast. Inhabiting the intertidal zone, it lies exposed at low tide, but is submerged at high tide. (D. P. Wilson)

be flat and ribbonlike or finely branched and feathery. All red algae contain chlorophyll and are photosynthetic. However, the green pigments are usually masked by a red pigment, *phycoerythrin.* A few species contain the blue pigment *anthocyanin.* Varying amounts of these pigments in different species of red algae produce red, brown, purplish, or green colors. Most red algae are less than one foot in length. While many species live in shallow coastal waters, some are found at depths of more than five hundred feet. They are most abundant in warm, tropical waters. *Chondrus,* or Irish moss, and *Gelidium,* the red alga from which agar is obtained, are better-known red algae as shown in Figure 20-12.

20-12 *Chondrus crispus,* or Irish moss, a marine red alga that grows on rocks along the cold North Atlantic Coast. (Walter Dawn)

The economic importance of the algae

Algae, where they occur, are the chief source of food, and hence of energy, for much of the animal life in those environments. Although many small fish live entirely on algae, one large group of mammals, the whales, also exists primarily on them.

The value of marine algae as soil fertilizers has long been known. If seaweeds are mixed with the soil, they not only add organic matter to it but also replenish the mineral salts that land-growing plants have removed. Algae are particularly rich in iodine, a chemical element essential to plant and animal life.

While algae are useful in certain parts of the world in the preparation of soups, gelatins, and other foods, we are most likely to find them as a part of our ice cream. A sodium compound extracted from algae is added to keep the ice cream smooth. This compound is also used as a stabilizer in chocolate milk and as a thickener in salad dressings. Its use as a cattle-feed supplement results in an increase in milk production. Agar-agar (AH-GAHR-AH-GAHR), produced from red algae from the Indian Ocean, is used in hospitals and laboratories as a gelling agent in culture media for bacteria. Other industrial uses of algal preparations range from cosmetics to leather-finishing.

An interesting new discovery is that algae can be mass-cultured in plastic tubes or tanks. If the culture is supplied with all the best conditions for photosynthesis, such as light, water, and carbon dioxide, the algae multiply rapidly. Periodically, some algae are strained out, dried, and then made ready for use as flour in baking or as a thickening in foods such as soup.

Some algae may become poisonous when they die, and thus pollute the water. This makes the water unfit, not only for human beings, but also for fish and other water life. Great care is taken in fish hatcheries to prevent this. A very weak solution of copper sulfate added to the water will kill blue-green algae. This treatment, however, is not recommended for home aquariums, because there is considerable danger of overdosing, which would kill the fish.

The use of algae in space flights

One of the problems that must be solved before man can make long trips in space is that of disposing of the waste product of man's respiration, carbon dioxide. Recent research indicates that certain forms of algae may be useful in accomplishing this purification of the air in a space ship. Algae, like all organisms containing chlorophyll, use carbon dioxide and give off oxygen as a waste product in the process of photosynthesis. There are certain hot-climate algae that reproduce faster than cold-climate types. They are, in fact, capable of multiplying a thousandfold in twenty-four hours. These types, of course, use up carbon dioxide and produce oxygen faster than ordinary types. It is therefore believed that they might be effective both for purifying the air in a space ship and for supplying some of the oxygen needs of the men operating the craft.

Lichens—curious plant relationships

While *lichens* (LIKE'nz) are often grouped with the fungi, they can also be classified with the algae, because an alga and a fungus make up the plant body of a lichen. While various lichens differ in structure, the alga is usually a green or blue-green form, while the fungus is one of the Ascomycetes.

The plant body of a lichen consists of a mass of fungus hyphae among which the algal cells are scattered (Figure 20-13). Lichens are of three general types. Some form a hard, granular crust *(crustose lichens)*. Others resemble flattened, leathery leaves *(foliose lichens)*. Still others form a network of slender branches *(fruticose lichens)*.

In a lichen, both the alga and the fungus benefit from the association. In fact, neither could survive alone in the environments

20-13 A lichen is actually two kinds of plants—an alga and a fungus—living together symbiotically. (Dennis Brokaw)

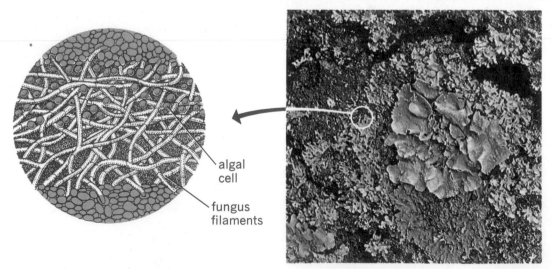

algal cell

fungus filaments

in which lichens live. The fungus depends on the alga for food, produced by photosynthesis. Although the alga is in a sense a "slave," it is protected and kept moist by the fungus hyphae. This condition, in which two organisms benefit from a mutual relationship, is called *symbiosis.*

The biological teamwork illustrated in a lichen allows it to live in places in which few other plants can survive. Many lichens grow on tree trunks. Others cling to rock surfaces far above the timber line in the Alpine zone of high mountains. These lichens are important pioneer plants. Gradually, they cause rock surfaces to crumble and, as they add their own remains season after season, produce organic matter that is the basis for soil. This soil can then support other plants.

Lichens are among the most abundant plants of the windswept areas of the far north. One of these lichens, *Cladonia,* is so valuable as food for reindeer that it is commonly called *reindeer moss.*

SUMMARY

Most people tend to overlook the algae. They pass them off as pond scums or seaweeds and often link them with water pollution. To a biologist, however, they are a fascinating group of plant-like protists occurring in a vast assortment of sizes, shapes, and colors. A jar of algae collected from a pond or stream or an ocean tide pool will provide material for many hours of enjoyable study through the lenses of a microscope.

Most people are not aware, either, of the importance of algae in an aquatic environment. Here, algae are primary food producers and water aerators, supplying oxygen for other forms of life. In a natural aquatic environment, algae thrive in close association with aquatic plants and animals. Either reduction or rapid increase in the algal population upsets this balance, often with serious consequences. As algae die out, fish and other aquatic animal populations decline with them. The environment is no longer productive. On the other hand, an algal population may increase to the point of choking a lake or pond and converting the water into a death trap for fish and other animals when this large accumulation of organic matter decays and ferments.

Man has produced countless problems of these sorts in our water environments by destroying algae or causing conditions that result in their excessive growth. We shall deal with these and other tragic mistakes leading to environmental destruction and pollution in Unit Eight.

BIOLOGICALLY SPEAKING

thallus	apical cell	antheridium
filament	concave cell	sporophyte
fission	pyrenoid	gametophyte
fragmentation	zygospore	alternation of generations
zoospore	holdfast	plankton
heterocyst	oogonium	symbiosis

QUESTIONS FOR REVIEW

1. In what respect are algae simpler in structure than higher plants?
2. How are algae fundamentally different from fungi?
3. Describe various cell groupings in colonial algae.
4. Explain why most algae grow in shallow water or float near the surface in deep water.
5. Describe several unusual habitats of algae.
6. Discuss various methods of reproduction found in algae.
7. Distinguish between isogametes and heterogametes.
8. What characteristics are used in classifying algae into phyla?
9. Name six algal phyla and give some examples of the algae included in each phylum.
10. Why are blue-green algae an especially serious problem in water pollution?
11. Describe the composition of a ball of *Nostoc*.
12. How could you distinguish a single filament of *Anabaena* from a filament of *Nostoc?*
13. List the pigments usually present in cells of green algae.
14. In terms of fission, explain why *Protococcus* forms cell clusters rather than filaments.
15. How is *Spirogyra,* on the basis of its microscopic appearance, easily distinguished from other filamentous green algae?
16. Describe conjugation in *Spirogyra.*
17. How can you distinguish a *Ulothrix* zoospore from a gamete?
18. Describe the formation and union of gametes in *Oedogonium.*
19. What is alternation of generations? Describe its occurrence in *Ulothrix* or *Oedogonium.*
20. What characteristic of desmids distinguishes them from other unicellular algae?
21. Describe several unique characteristics of diatoms.
22. What is diatomaceous earth? List several of its uses.
23. Identify the alga responsible for the "red tide." Account for the destruction of fish during such a period.
24. How do red algae differ in habitat from most brown algae?
25. How does a lichen illustrate symbiosis?

APPLYING PRINCIPLES AND CONCEPTS

1. What evidence can you give that all algae contain chlorophyll even though many are not green?
2. A colony of 50 algal cells is not a 50-celled plant. Explain why.
3. Both spores and gametes are reproductive cells. How are they different?
4. In what respect are the cells of blue-green algae more primitive than those of green algae?
5. Why is conjugation in *Spirogyra* considered a primitive form of sexual reproduction?
6. In what ways is sexual reproduction in *Oedogonium* more efficient and more advanced than the sexual reproduction in *Ulothrix?*

7. Compare *Spirogyra, Ulothrix,* and *Oedogonium* in regard to specialization of cells in a filament.
8. What is believed to be the significance of sexual reproduction in plant evolution?

RELATED READING

Books

Bold, H. C., *Morphology of Plants.*
  Harper and Row, Publishers, New York. 1957. Deals with the structure of plants by phyla, with good material on the algae.
Duddington, Charles L., *Flora of the Sea.*
  Thomas Y. Crowell Company, New York. 1967. The structural evolution of sea-weeds and other ocean algae, with discussions of orders of representative species.
Fink, Bruce, *The Lichen Flora of the United States.*
  The University of Michigan Press, Ann Arbor. 1961. A somewhat technical manual, but still very useful, especially for the identification of individual lichen species.
Prescott, G. W., *How to Know the Fresh-Water Algae.*
  William C. Brown Company, Publishers, Dubuque, Iowa. 1954. An illustrated key for identifying the more common fresh-water algae as to genus, with hundreds of species named and pictured and including numerous aids for their study.

UNIT **4**

MULTICELLULAR PLANTS

Ages ago, modifications probably occurred in certain aquatic plants that allowed them to invade a new environment, the land. Perhaps some ancestral alga growing on moist soil developed an aerial shoot from a filamentous plant body and extended rootlike projections into the soil. Possibly the mosses are the closest living relatives of these first land plants. However, they never became fully adapted to a terrestrial environment. It remained for the seed plants, with vascular tissues for conduction and support and seeds for the protection and nourishment of the embryo plant, to dominate the land environments of the earth.

CHAPTER TWENTY-ONE

MOSSES AND FERNS

The first land plants

Toward the close of the Silurian period and the beginning of the Devonian, land was rising above the shallow, warm seas that had covered the earth for millions of years. This rise occurred during the Paleozoic era, approximately 350 million years ago. The rising land provided a new environment and set the stage for the development of the first land plants (Figure 21-1).

Biologists believe that green algae living in these ancient seas were the ancestors of the first land plants. Evolutionary development in this transition from aquatic to terrestrial life seems to have been along two distinct lines. One line leads to the *Bryophyta,* the phylum that includes the mosses and liverworts. The other line of development resulted in the *Tracheophyta,* or vascular plants, which include the ferns and their relatives as well as the seed plants. If you examine the plants you see around you today, it will be obvious to you that the *tracheophyte* line was far more successful than the other. The *bryophyte* line has been an evolutionary "blind alley." The tracheophyte line, on the other hand, has shown a continuing adaptation through change in structure through the ages. The seed plants, which dominate our vegetation today, are descendants of this line.

The bryophytes

While the bryophytes were among the first land plants, they have never developed more than primitive adaptations to land conditions. For this reason, they cannot compete with seed plants. One of the limitations of bryophytes is their small size. A large land plant must have a root system that can penetrate the soil to reach water supplies and can anchor the aerial plant body. Mosses lack true roots. Their absorbing structures are small and reach only a short distance into the soil. Therefore, they can

Time Scale of Early Land Plants Based on Fossil Records

Era	Period	mosses	liverworts	ferns	club mosses	horsetails
Cenozoic	Quaternary					
	Tertiary					
Mesozoic	Cretaceous					
	Jurassic					
	Triassic					
Paleozoic	Permian					
	Pennsylvanian (Late Carboniferous)					
	Mississippian (Early Carboniferous)					
	Devonian					
	Silurian					
	Ordovician					
	Cambrian					
(Pre-Cambrian)						

anchor only a small plant body. The lack of conducting channels, or vascular tissues, in the stems of mosses limits water conduction to a very short distance, thus further restricting the size to which they can grow.

Another limiting factor in the evolution of bryophytes is their dependence on water for reproduction. In your study of the life cycle of a moss, you will find a stage in which the sperm swims to the egg. Ferns also require water at this stage in the life cycle. However, in the evolution of seed plants from their fern an-

21-1 A time scale of early land plants.

cestors, this dependence on water for reproduction has been eliminated.

In spite of all these structural limitations, the bryophytes, especially the mosses, are widely distributed in the vegetation of the earth. While they occupy land environments from the Arctic to the Antarctic, bryophytes are most abundant in the Temperate and Tropical zones. Although mosses cannot compete with ferns and seed plants, their small plant bodies allow them to live in many environments that larger plants cannot occupy. In the treeless, windswept tundra of the Arctic regions, mosses are dominant plants. They live far above the timber line in the alpine zones of high mountains. In more favorable environments, they grow among ferns and seed plants, which provide the shade and moist environment many species of moss require for survival.

The structure of a moss

The term *moss* is often used loosely and incorrectly. While some mosses are aquatic, most "water mosses" are algae. Reindeer moss is not a moss but a lichen. Spanish moss, which hangs from trees in southern forests, is a seed plant related to the pineapple.

You can recognize true mosses if you examine them closely. You have seen them growing in the cracks of shaded sidewalks, on moist ground under trees, in dense clumps and carpets in deep forests, and on rotting logs and stumps. If you examine a clump or carpet of moss closely, you will find that it is a compact mass of tiny plants. Each plant has a slender stem, usually less than an inch long, with numerous leaves encircling it. The leaves are thin and fragile and are only one cell thick except along the center, or midrib. If you remove a plant from the soil carefully, you will find a cluster of hairlike projections growing from the base of the stem. These are not roots but are *rhizoids* that absorb water and minerals and anchor the plant body. While rhizoids function in a way that is similar to the way roots function, they are cellular filaments, while roots are plant organs composed of specialized tissues. As you examine the rhizoids, you will see why mosses are limited in size.

The familiar moss plant, often referred to as the *leafy stem,* is only one stage in the life cycle of a moss. To find other stages, it is necessary to follow a moss through its life history.

The life cycle of the moss

If you examine the diagram in Figure 21-2, you will see that each moss plant goes through a reproductive cycle in which an asexual, spore-producing stage forms a sexual, gamete-producing stage. This, in turn, forms the spore stage again. You will recognize this as the alternation of generations we discussed in Chapter 20.

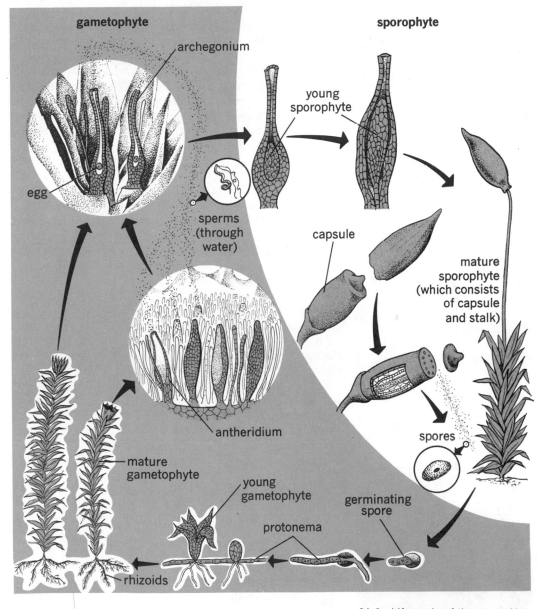

gametophyte

sporophyte

archegonium

young
sporophyte

egg

sperms
(through
water)

capsule

mature
sporophyte
(which consists
of capsule
and stalk)

antheridium

spores

mature
gametophyte

young
gametophyte

germinating
spore

protonema

rhizoids

21-2 Life cycle of the moss. Note that alternation of generations occurs. After fertilization, a sporophyte grows out of, and is parasitic on, the female gametophyte. Spores form at the top of the sporophyte and fall to the ground. These germinate and grow into a new generation of gametophytes which reproduce sexually.

Recall that the gamete-producing plant is the *gametophyte.* Sexual reproductive organs develop at the tips of the leafy stems of gametophyte moss plants, in clusters hidden by leaves. These organs are multicellular in mosses. Depending on the species of moss, male and female organs may be borne on the same plant, on different branches of the same plant, or on separate plants. Sperms are formed in sac-like, short-stalked *antheridia.* A cap at the apex of the antheridium opens at maturity, allowing the sperms to escape. The female organs, or *archegonia,* are borne on short stalks. These organs are often bottle-shaped and consist of a slender *neck* and a swollen base, or *venter,* containing a large, single egg. When the egg is ready for fertilization, a neck canal opens forming a passage-way to the egg. Sperms swim to the archegonium after a rain or even in a film of dew, but they cannot move unless there is some water present. A sperm passes through the neck of the archegonium and fertilizes the egg. This union results in the zygote.

Fertilization starts the *sporophyte* stage, or asexual phase, of the cycle. The zygote always remains in the female organ. Soon, the sporophyte begins to grow and produces a slender *stalk* that grows up and out of the leafy stem of the plant. The top of the stalk swells and becomes a large mass of tissue called a *capsule,* which is covered with a thin hood. As the capsule develops, numerous spore mother cells are formed. Each of these cells undergoes two meiotic divisions resulting in a tetrad of spores. When the spores are ripe, the hood falls off, the capsule opens, and the spores escape. They are carried off by the wind, and when they fall on the ground they begin to grow, provided environmental conditions are right. With this, the asexual, sporophyte stage has ended, and the sexual, gametophyte stage begins. Notice that the spores are dispersed without the aid of water, while the gametes still depend on water for fertilization.

Each spore produces a small, threadlike structure called a *protonema* (PROTE-uh-NEE-muh). All the cells that make up the protonema have chlorophyll and can make their own food. The resemblance of the protonema to an alga is startling and once caused many scientists to class it as a close relative of the algae. Some cells of the protonema produce short buds that grow into new leafy stems. Other threads enter the ground and become the rhizoids. Thus, a new moss plant is formed, and this gametophyte will soon form sex cells.

In the moss life cycle, the sporophyte stalk and capsule which develop from the zygote, are diploid, having the *2n* number of chromosomes. Meiosis occurs in the divisions of spore mother cells in the capsule, resulting in the formation of spores that contain the haploid *(n)* chromosome number. The haploid gametophyte, including the protonema and leafy stems, grows from the spore and produces haploid gametes. Fertilization restores the diploid number of chromosomes in the zygote.

The economic importance of the mosses

The *sphagnum,* or peat-forming, mosses are the most widely used. Sphagnum grows in small lakes and bogs, where it forms floating mats (Figure 21-3). These mats increase in size and thickness each year, as generation after generation occupies the surface of the mat. Plants of previous years decompose slowly, settle to the bottom, and form *peat.* In time, the growth of sphag-

21-3 A clump of *Polytrichum* moss showing stalks and capsules. (Walter Dawn)

num mats in a lake may close the water entirely, causing the lake to enter the bog stage. Eventually, what was once an open lake may become a deep deposit of brownish-black peat. Large sphagnum mats are often invaded by larger plants—rushes, grasses, shrubs, and even trees.

The absorbent quality of sphagnum makes it valuable to the gardener as a *mulch*. It can be worked into the soil or placed on the surface to make the soil loose and to hold water during the usually dry summer months.

Other mosses are important to us as *pioneer plants* in rocky areas. The small amount of soil that collects in cracks on the bare surfaces of cliffs and ledges is sufficient to support moss plants. The rhizoids break down rocks gradually and thus form more soil. As mosses die and decompose season after season, they form enough soil to anchor the roots of larger plants.

The liverworts

Much less familiar than mosses are their relatives, the liverworts. These small plants grow in wet places, often along the banks of streams, around the outlets of springs, on rocky ledges, or in the water. The moist soil and flowerpots in greenhouses are also good places in which to find liverworts. They are thin, leathery leaves laid flat against the ground, anchored by numerous rhizoids on the lower side. One of the common liverworts, *Marchantia,* has a plant body, or ***thallus,*** that resembles a leathery ribbon or tongue (Figure 21-6). The thallus forms Y-shaped branches with deep notches at the tips. The upper surface of the thallus is marked by diamond-shaped plates that cover internal air chambers. A raised pore opens into each chamber. Cells within the chambers contain chloroplasts and give the upper surface a deep green color.

21-4 (Above left) A sphagnum mat on a bog. Note the other plants growing on the mat. (Walter Dawn)

21-5 (Above right) Sphagnum moss. (Walter Dawn)

21-6 *Marchantia,* a common liverwort. (Walter Dawn)

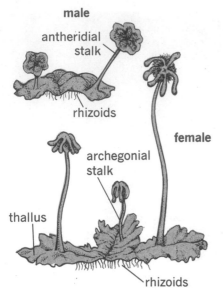

21-7 Female and male liverwort plants.

The thallus of a liverwort is the gametophyte stage of the life cycle and corresponds to the leafy shoot of a moss. The sexes are separate. A female plant bears stalks that rise an inch or two above the thallus. At the tip of each stalk is a head composed, usually, of nine fingerlike projections that bend downward (Figure 21-7). Archegonia similar to those of the moss are borne in rows between the projections. Male thalli bear branches with disklike heads having scalloped edges. Antheridia are embedded in the upper surface. When sperms mature, they swim in rainwater or dew from the antheridia to the archegonia. After fertilization, the zygote forms the sporophyte plant within the archegonium. Spores develop in a capsule that grows out of the archegonium and are discharged much as they are in the mosses.

The tracheophytes—vascular plants

As we indicated earlier in this chapter, all of the plants with *vascular tissues* are placed in the phylum Tracheophyta. Early tracheophytes are believed to have evolved from ancestral algae at about the same time at which the bryophytes were developing along a different line. Living tracheophytes are classified in four subphyla, as shown in the *Investigations* classification table. One of these subphyla contains only two living genera. Two other subphyla are represented by club mosses and by a group known as horsetails. Members of these phyla are often called *fern allies*. The fourth and largest subphylum contains the ferns and seed plants, including the familiar flowering plants. In this chapter, we have limited the discussion of tracheophytes to ferns and their relatives.

21-8 Diorama of a forest typical of the Mississippian and Pennsylvanian periods—the Carboniferous age. (American Museum of Natural History)

Ferns and their relatives were most numerous three hundred million years ago, during the Mississippian and Pennsylvanian periods of the Carboniferous age. There were very few seed plants, and flowering plants, which today are the most abundant seed plants, had not yet appeared. The marshy land and warm climate were ideal for the growth of ferns and other early tracheophytes similar to those shown in Figure 21-8. In that age, ferns were not limited to isolated clumps and patches, as they are today. They were the dominant plants. Tree ferns from thirty to forty feet in height formed dense forests. Among them grew club mosses and horsetails as large as trees. Smaller species, much like those we find today, grew in a dense undercover. Tree ferns — reminders of this once-flourishing age of ferns — can still be seen in Hawaii, Puerto Rico, and other tropical regions (Figure 21-9).

Although no man ever saw those great fern forests, we are reaping the benefits of their existence today. During this age, deep layers of plant remains accumulated in the swampy areas where they grew. Later, the movements of the earth compressed these layers into the coal deposits we mine today. It has been estimated that it took three hundred feet of compressed vegetation to form twenty feet of coal. When we consider what coal has meant to industry and living, we might almost conclude that the advanced civilization of modern America has sprung from the vegetation of millions of years ago.

21-9   Tree ferns in Hawaii. These are small relatives of the tree ferns that grew abundantly some 300 million years ago. (Werner Stoy, Camera Hawaii)

The life cycle of the fern

The familiar fern you see growing in clumps in the woods or garden or as a potted ornamental plant is only one stage in the life cycle of the fern. In the moss, the familiar plant was the gametophyte generation. In the fern, it is the sporophyte.

In all ferns but tree ferns, the stem is underground, creeping horizontally just below the surface (Figure 21-10). This underground stem, or *rhizome,* contains vascular tissues that are similar to those of seed plants. Clusters of roots grow from the rhizome and spread through the soil, anchoring the plant and absorbing water and minerals. The leaves that arise from the rhizome are called *fronds.* The fronds, rhizome, and roots of a fern sporophyte body are comparable to the stalk and capsule of the moss. All cells comprising the sporophyte contain the diploid ($2n$) number.

When fern fronds mature, certain of them develop structures that look like small brown dots or patches on the underside, close to the veins. Each structure is called a *sorus* (plural, *sori*). Some people become alarmed when sori appear on potted fern fronds and attempt to remove them, thinking that the fern has been attacked by scale insects or that it is diseased. Sori are reproductive structures and are a part of the life cycle. The shape and location of sori differ in various families, genera, and species of ferns and are important identifying characteristics.

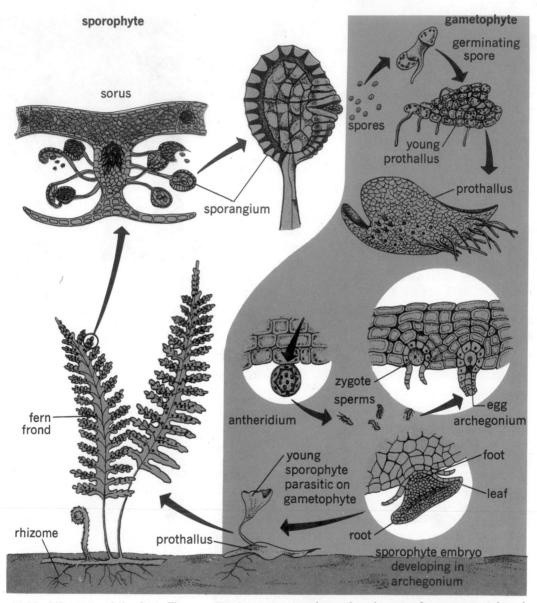

sporophyte

sorus

sporangium

gametophyte

germinating
spore

spores

young
prothallus

prothallus

zygote
sperms

antheridium

egg
archegonium

fern
frond

young
sporophyte
parasitic on
gametophyte

foot

leaf

rhizome

prothallus

root

sporophyte embryo
developing in
archegonium

21-10 Life cycle of the fern. The gametophyte, which is so small you can hardly see it, develops on the ground.

Each sorus consists of a cluster of *sporangia* (singular, *sporangium*), attached by slender stalks. Each sporangium is helmet-shaped and contains sixteen spore mother cells. Each diploid spore mother cell undergoes meiosis, which results in the formation of four spores with the haploid *(n)* chromosome number. Thus, the mature fern sporangium contains sixty-four haploid spores.

When the sporangium matures, a row of special cells along two thirds of its surface dries and begins to shrink. This shrinking stretches the sporangium and ruptures it, discharging the spores into the air. If a spore falls on moist soil, it germinates and forms a vegetative cell—the first cell of the gametophyte generation. A series of cell divisions results in a short filament resembling a filamentous green alga. This stage of the fern is comparable to the moss protonema. One or two colorless rhizoids develop from cells of the filament. Further cell division at the tip, or apex, of the filament broadens it into a sheet of cells. Continued cell division over a period of several weeks produces a heart-shaped *prothallus,* which is the gametophyte fern plant. This stage of the fern, seldom seen, is generally a quarter to half an inch in width. It contains no vascular tissue. All cells of the prothallus contain chloroplasts and are capable of photosynthesis. All contain the haploid *(n)* number of chromosomes. Except for an area near the *apical notch* at the broad end of the heart, the prothallus is one cell in thickness. Near this notch, the prothallus is built up to a thickness of several cells. Rhizoids develop on the underside of the prothallus in its older region, where the cells lie in a single layer. The rhizoids raise the prothallus above the soil surface, anchor it, and absorb water and minerals.

Multicellular sex organs develop on the underside of the prothallus. Archegonia, resembling those of the moss, form in a cluster near the apical notch, where the prothallus is several cells in thickness. The base of an archegonium, containing a single egg in most cases, is embedded in the tissue of the prothallus. Knob-shaped antheridia develop in the basal region of the prothallus, among the rhizoids. In this region, the prothallus is one cell in thickness. Each mature antheridium contains several motile sperms.

When the eggs and sperms are mature, a film of water on the lower side of the prothallus is sufficient to cause the antheridium to rupture and the necks of the archegonia to open. Sperms swim in the water film and enter the archegonia. A single sperm fertilizes each egg, producing a diploid zygote. The zygote is the first cell of the sporophyte generation.

Immediately after fertilization, the zygote undergoes cell division. Further cell divisions result in a mass of tissue that fills the archegonium. Continued enlargement and development produce the *embryo fern,* which consists of a foot anchored in the tissue of the prothallus, a root, a stem, and a leaf. Soon, the young plant becomes established and is independent of the prothallus, which withers and disappears. The stem bears additional fronds, and the familiar fern clump results.

In reviewing the life cycle of a fern, you will find that the sporophyte stage is the more prominent stage, for the sporophyte is larger than the gametophyte and may grow for years. In contrast, the gametophyte prothallus is small and short-lived, surviving only for a few weeks. Indeed, the ferns have evolved far beyond the bryophytes, in which the gametophyte was the prominent stage and the sporophyte was parasitic on it.

21-11 These club mosses (left, *Selaginella;* right, *Lycopodium*) are related to the ferns because of similarity in reproduction. They are often used for Christmas decorations. (Dennis Brokaw; Russ Kinne)

Relatives of the fern

Club mosses *(Selaginella)* and ground pines *(Lycopodium)* are relatives of the fern and are neither mosses nor pines, nor are they related to them. They bear conelike reproductive structures at the tips of certain of their branches (Figure 21-11). These conelike organs contain numerous sporangia from which spores are shed. Leaves of the sporophyte plants are small and scalelike. Certain club mosses grow close to the ground in flat sprays. Others stand erect.

The horsetails *(Equisetum),* shown in Figure 21-12, are also called scouring rushes. They represent a single genus of an ancient plant subphylum. Their ancestors were large plants and grew in abundance over vast areas. They were among the most important coal-forming plants of the Carboniferous age. Today, horsetails can be found, often in dense thickets, in ditches and lowlands, and in the shallow water near the shores of lakes. The stem of a horsetail is an underground rhizome that branches freely and sends up dark green branches that are capable of carrying on photosynthesis. The branches have a gritty, abrasive

21-12 Horsetails or scouring rushes. The horsetail produces bushy, green vegetative shoots and brownish, conelike structures that bear the reproductive organs. Note the scalelike leaves that grow in whorls on the stem. (Walter Dawn; Albert Towle)

texture and were probably used by the pioneers for scouring pots and pans—hence, their common name. The leaves are reduced to scales that encircle the branches at *nodes* that occur at regular intervals. The leaves lack chlorophyll and apparently have no function.

Aerial branches are of two types. Vegetative branches form terminal shoots in tufts. These branches account for the name *horsetail*. Reproductive, or fertile, branches have cones at their tips. The cones contain sporangia in which spores are produced.

Club mosses, ground pines, and horsetails undergo a life cycle similar to that of a fern. The sporophyte is large and prominent, while the gametophyte is inconspicuous.

SUMMARY

Now that you are familiar with the stages of the life cycle of the moss and the fern, perhaps you can understand why biologists consider the bryophytes a side line in evolutionary development while the ferns and other early tracheophytes were a main line leading to the seed plants. You found structural limitations in the moss that prevented it from increasing in size and challenging the much larger tracheophytes in the vegetation of most of the land. The mosses we find today are, probably, like their ancestors of millions of years ago. Yet, even with their limitations, mosses continue to survive and increase their total numbers by occupying rigorous land environments that exclude most of the tracheophytes.

In the fern and its relatives, we find plants with vascular tissues that allow a plant body to grow to a much larger size than a moss. Ferns and their relatives have gradually declined since they dominated the vegetation of the Carboniferous Period, or Coal Age. But as the ferns declined, their relatives, the seed plants, assumed their dominant place in the vegetation of the earth. This is the age of seed plants.

BIOLOGICALLY SPEAKING

tracheophyte	sporophyte	frond
bryophyte	capsule	sorus
rhizoid	protonema	sporangium
leafy stem	thallus	prothallus
gametophyte	vascular tissue	apical notch
antheridia	rhizome	embryo fern
archegonia		node

QUESTIONS FOR REVIEW

1. What two lines seem to have been taken in the development of land plants from algae during the Silurian and Devonian periods?
2. Describe two structural characteristics of bryophytes that limit the size of their plant bodies.
3. Explain why water is necessary for sexual reproduction in bryophytes.
4. Name several organisms that are incorrectly called mosses.
5. Describe the gametophyte of a typical moss.
6. In the life cycle of a moss, which cell is the product of the gametophyte generation and the origin of the sporophyte generation?

7. Describe a sporophyte moss plant.
8. During which stage in the life cycle of a moss does meiosis occur?
9. Which stage in the life cycle of a moss provides evidence of its algal ancestry?
10. Describe the gametophyte of a liverwort such as *Marchantia*.
11. Describe the gamete-producing structures of *Marchantia*.
12. What structural characteristic is common to ferns and seed plants?
13. Describe the sporophyte in the life cycle of a fern.
14. Where are the sporangia located on a fern frond?
15. What stage of a fern develops from a germinating spore? How does this provide a link to its algal ancestry?
16. Where are the sexual organs on a fern prothallus?
17. In what way are ferns, like mosses, dependent on water during sexual reproduction?
18. List several fern relatives, or allies.

APPLYING PRINCIPLES AND CONCEPTS

1. Discuss the lines of development from the algae to the bryophytes and the tracheophytes. Point out those factors that caused one to become an evolutionary "blind alley" and the other to become a biological success.
2. Give possible reasons for the abundance of mosses in regions with extreme climatic conditions; for example, tundra and alpine areas of high mountains.
3. Why must the sporophyte moss plant live as a parasite on the gametophyte?
4. Give possible reasons for the disappearance of fern forests in most parts of the world.
5. Compare the sporophyte plant of a moss and a fern, and explain the ways in which the fern is more advanced.
6. From what you have learned about alternation of generations in the moss and fern, would you expect the plant body of a seed plant, such as an oak tree, to be the gametophyte or sporophyte generation? Explain your answer.

RELATED READING

Books

COULTER, MERLE C., *Story of the Plant Kingdom* (3rd ed., rev.).
  The University of Chicago Press, Chicago. 1964. A general survey of the plant kingdom, arranged by phyla.
HUTCHINS, ROSS E., *Plants Without Leaves.*
  Dodd, Mead and Co., New York. 1966. A description of these plants by a well known biologist, who explains their representative uses and odd ways of survival.
STERLING, DOROTHY, *The Study of Mosses, Ferns and Mushrooms.*
  Doubleday and Company, Inc., Garden City, New York. 1961. The strange story of many species, with detailed descriptions.
WATSON, E. V., *The Structure and Life of the Bryophytes.*
  Hillary House Publishers, Ltd., New York. 1964. The structural features of these plants examined and their evolutionary relationships discussed from the current viewpoint.
WHERRY, EDGAR T., *The Fern Guide.*
  Doubleday and Company, Inc., Garden City, New York. 1961. A convenient reference guide, covering one hundred thirty-five species.

CHAPTER TWENTY-TWO

THE SEED PLANTS

The rise of seed plants

Fossil evidence indicates that the first forests appeared on the earth during the Devonian period of the Paleozoic era (Figure 22-1). Growing among the club mosses, horsetails, and ferns that composed these forests were the first seed plants, known as *seed ferns*. These ancient plants, now long extinct, are believed to have been evolutionary products of the fern and ancestors of the modern seed plants that dominate the vegetation of the land today. Seed ferns were prominent in the swamp forests of the Carboniferous age. During the Permian period of the late Paleozoic era, they are believed to have given rise to other forms of seed plants, including cycads and conifers.

With the dawn of the Mesozoic era some 200 million years ago, the climate of the earth became warmer and drier. Biologists believe that these climatic changes led to the gradual disappearance of the seed ferns during the Triassic period. It seems other seed plants continued to increase in number. The Jurassic period of the Mesozoic era marked the appearance of flowering plants, the most highly evolved seed plants. These plants developed rapidly during the Cretaceous period and have maintained their position of dominance to the present day.

What is a seed plant?

The presence of vascular tissues in seed plants places them in the phylum Tracheophyta, along with club mosses, horsetails, and ferns. Four subphyla compose this large plant phylum. The fourth subphylum, *Pteropsida,* contains three classes, as follows:

■ *Filicineae*—ferns
■ *Gymnospermae*—cone-bearing plants, including seed ferns (extinct), cycads, ginkgoes, and conifers (shown in Figure 22-1)
■ *Angiospermae*—flowering plants

Only **gymnosperms** and **angiosperms** produce seeds and are therefore called **seed plants.** (In former classifications, these plants were separated from ferns in the phylum Spermatophyta.)

Time Scale of Seed Plants Based on Fossil Records

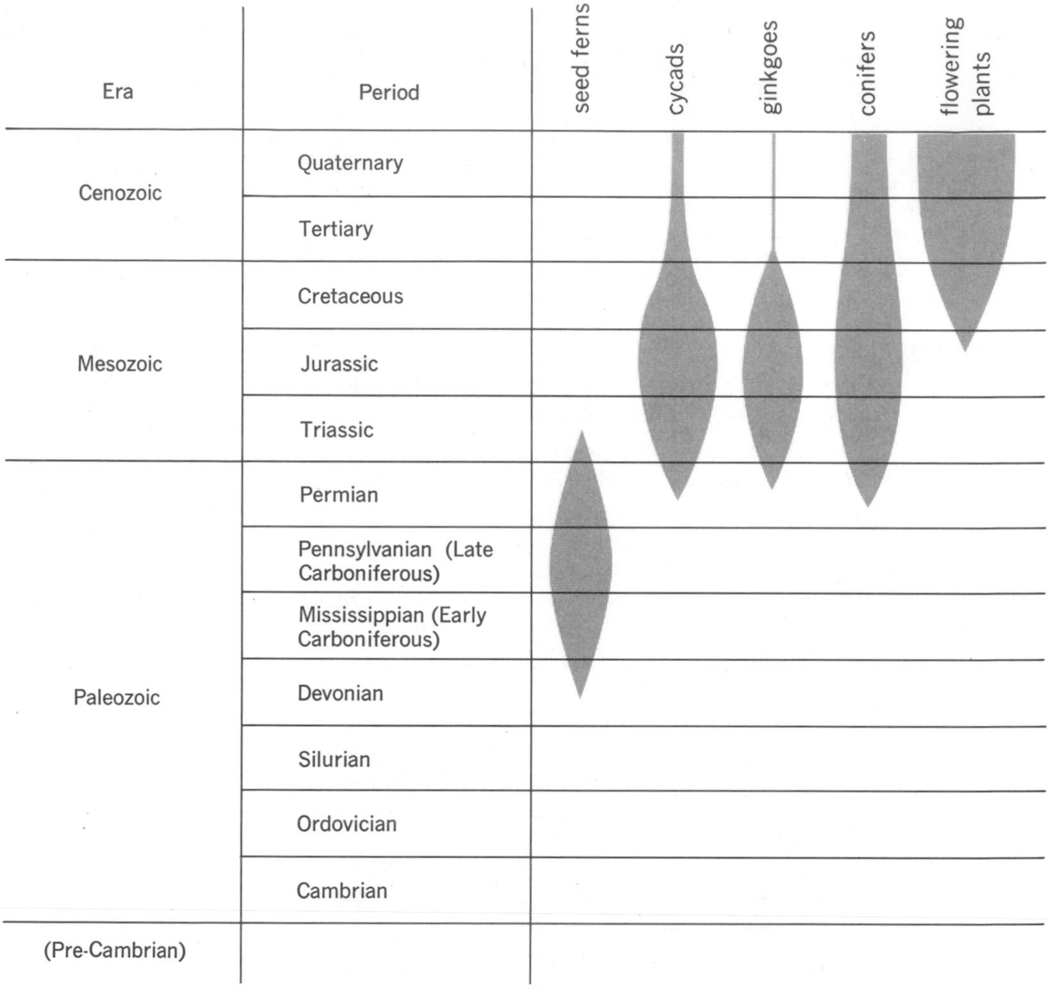

Era	Period	seed ferns	cycads	ginkgoes	conifers	flowering plants
Cenozoic	Quaternary					
	Tertiary					
Mesozoic	Cretaceous					
	Jurassic					
	Triassic					
Paleozoic	Permian					
	Pennsylvanian (Late Carboniferous)					
	Mississippian (Early Carboniferous)					
	Devonian					
	Silurian					
	Ordovician					
	Cambrian					
(Pre-Cambrian)						

22-1 A time scale of seed plants.

In all tracheophytes, the sporophyte generation is prominent. Thus, in your study of seed plants you might compare a pine tree, an apple tree, a rose bush, or a tomato plant with the frond of a fern. However, the most distinctive characteristic of gymnosperms and angiosperms is the production of seeds. This characteristic has been of great importance in the conquest of land environments since the Carboniferous age. What is a seed, and why does it give the plant that produces it a distinct biological advantage in land environments?

A typical *seed* is an embryo plant surrounded by an *endosperm* and covered with one or more protective seed coats. Food stored in the seed nourishes the young plant until the early period of growth has occurred. In a sense, a seed is a packaged plant, ready for delivery. A seed may be blown through the air, float on water, or be carried on the fur of an animal; it may lie dormant for many months. When moisture and temperature conditions are favorable, the seed coat softens, and the young plant begins to grow by sending the root downward into the soil and the shoot upward. Reproduction through the production and dispersal of seeds is highly efficient and is one of the reasons that seed plants have gained dominance in land environments.

The gymnosperms

The gymnosperms as a class are distinguished from angiosperms on the basis of seed development. The name *gymnosperm* means "naked seed" and refers to the fact that the seeds develop in exposed positions on the upper surfaces of scales of cones rather than within the protective wall of the ovary (Figure 22-2). About 750 species of gymnosperms are living today. These species are the survivors of a much larger plant population that flourished in earlier geological periods.

Living gymnosperms are classified in four orders, three of which are represented by only a few remaining species. The order *Cycadales* includes a group of about one hundred species of ancient gymnosperms, often referred to as *cycads*. They are believed to have appeared near the close of the late Carboniferous age and to have flourished during the Triassic and Jurassic periods of the Mesozoic era. Modern cycads resemble palm trees,

22-2 The main difference between the seed of an angiosperm, like the bean, and that of a gymnosperm, like the pine, is that, while they are developing, the bean seed is enclosed in a ripened ovary, or fruit, whereas the pine seed lies exposed on a cone scale.

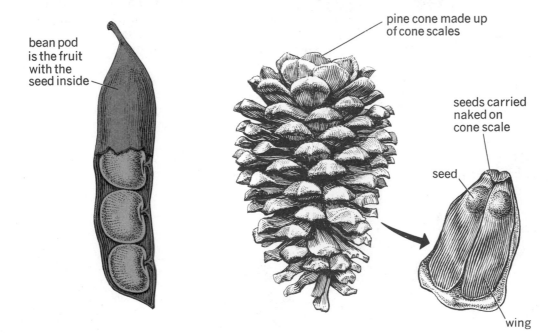

bean pod is the fruit with the seed inside

pine cone made up of cone scales

seeds carried naked on cone scale

seed

wing

22-3 A cycad. While it somewhat resembles a palm tree in appearance, its reproductive structure is like that of gymnosperms. (Albert Towle)

and are often called sago palms (Figure 22-3). They are limited to certain subtropical and tropical regions of Florida, the West Indies and Mexico, and parts of Australia and Africa.

Another order of gymnosperms, the *Ginkgoales,* is represented by a single surviving species, the *ginkgo,* or maidenhair tree *(Ginkgo biloba).* The ginkgo is a large tree, reaching a height of one hundred feet or more, with a trunk diameter of four feet. Its leaves are fan-shaped and two-lobed, two to four inches across, and unlike those of any other tree (Figure 22-4). Their resemblance to the leaflets of the maidenhair fern accounts for the name *maidenhair tree.* Most of the leaves grow in clusters at the tips of spurs spaced along the branches. Some of the branches lack these spurs. Ginkgo trees are either male or female. Male trees produce pollen in short, conelike *catkins.* The seeds are borne on female trees in fleshy coverings that are about an inch in diameter and yellow or orange in color. They are foul-smelling and resemble plums or large cherries. Each contains a single, almond-flavored nut.

During the Triassic period of the Mesozoic era, many species of Ginkgoales were distributed throughout the world. Fossils of these ancient trees have been found in North America, Europe, and Asia. By the glacial age of a million years ago, they had become extinct in all parts of the world except China. Centuries ago, priests of China planted ginkgo trees in temple gardens. Later, they were introduced into Japanese gardens. Today, ginkgo trees are widely planted in yards and parks of the United States, where they grow well in cultivation. If you have a ginkgo tree in your yard, you have one of the rarest of all living plants—a single surviving species of a single genus, the last of a gymnosperm order that flourished at the time of the dinosaurs!

22-4 The Ginkgo, or maidenhair tree. Its leaves resemble the leaflets of the maidenhair fern, hence its second name. (John H. Gerard, National Audubon Society; Walter Dawn)

22-5   A coniferous forest (spruce). (Lincoln Nutting, National Audubon Society)

The conifers

The order *Coniferales* includes such well-known trees as the pines, spruces, firs, Douglas fir, cedars, cypress, larches, and sequoias. The name *conifer* refers to the woody cones in which seeds are borne. Most conifers are trees, although a few, including the yews and certain red cedars, or junipers, are shrubs.

The conifers as a group probably date to the Carboniferous age and have flourished since that time. Today, they are the most widespread and numerous of all gymnosperms. Extensive coniferous forests are found in parts of North America, where they often occupy rocky and sandy soils that will not support forests of broad-leaved trees. Other conifers thrive in swamps and bogs and around the margins of lakes. Coniferous forests are found in Alaska, in Canada and the region of the Great Lakes, along the Appalachian Mountains, in the coastal regions of the South Atlantic and Gulf states, as well as in the Rocky Mountains and the Pacific coastal states. Their contribution to the lumber and forest products industries is enormous. In addition, conifers are planted extensively as ornamental shrubs and trees in yards and parks, as foundation plantings around homes, and as windbreaks around farm and ranch houses. Among all trees, conifers hold the record for height, trunk diameter, and age. A giant "big tree," a species of redwood *(Sequoia gigantea)*, in the Calaveras Grove in California towers three hundred feet above the ground. This tree is over thirty feet in diameter and is estimated to be more than four thousand years old (Figure 22-6). Even this trunk diameter and age are surpassed by a cypress tree, the Big Tree of Tule, growing about two hundred miles south of Mexico City. It is fifty feet in diameter and is estimated to be over five thousand years old.

22-6   General Grant, a magnificent "big tree" in Kings Canyon National Park, California. (R. C. Zink)

Most conifers are evergreen. Their leaves take the form of needles or scales. Exceptions are the larch and bald cypress, which lose their needles each autumn. The cones are of two types—seed and pollen. *Seed cones* are woody and develop over a period of several months. Winged seeds in these cones are borne in pairs on the upper surfaces of shelflike *cone scales* (see Figure 22-2). The sizes and shapes of seed cones vary with the species and are often used as a means of tree identification. Cones range in length from half an inch to two feet or more, in the case of the sugar pine of the Northwest. The red cedar, or juniper, and yew differ from other conifers in producing a single seed surrounded by a fleshy covering that resembles a berry.

Pollen cones are smaller than seed cones and are borne singly or in clusters at the tips of branches. They are yellow or reddish and frequently go unnoticed, since they remain for only a few days after pollen is shed. Most North American conifers bear seed and pollen cones on separate shoots of the same tree.

Angiosperms—the flowering plants

The class Angiospermae includes all *flowering plants*. These most highly evolved plants represent a long line of development that probably branched from the gymnosperm line during the Jurassic period of the Mesozoic era. During the Cretaceous period that followed, angiosperms seemed to have developed rapidly and replaced much of the older vegetation. Today, angiosperms are the dominant plant life; they occur widely in most land areas not occupied by coniferous forests.

There are several reasons for the success of angiosperms. Genetic variations through the ages have produced numerous forms of angiosperms—trees, shrubs, and herbs; upright plants, climbing plants, creeping plants, and floating plants. More than any other group, they can grow and reproduce in a wide variety of environments. Angiosperms thrive in all types of soils and endure most extremes in temperature and rainfall. In their many forms, they inhabit deserts, plains, prairies, marshes, lakes, mountain slopes, and arctic regions. Some require bright light, while others thrive in deep shade.

The flowers, fruits, and seeds of angiosperms have also contributed greatly to their success. Various angiosperms depend on water, wind, insects, and other agents to pollinate their flowers. The fruits that result are distributed to new locations through the actions of a variety of agents. The seeds contain stored food that nourishes the young plant until it is established in a new location. All of these factors have given the angiosperms a great advantage in competing with other kinds of plants, especially in land environments.

The class Angiospermae is divided into two large subclasses, the *Monocotyledonae* and *Dicotyledonae,* which include about

300 families and more than 250,000 species. (Certain of these families are listed in Tables 22-1 and 22-2.) The basis of the division into the two large subclasses is the number of *cotyledons,* or seed leaves, that develop as a part of the embryo plant. A cotyledon is the first leaf of the embryo plant. In many species, it serves as a food reservoir and supplies nourishment to the seedling until it develops green tissue and synthesizes its own food. In other species, a cotyledon serves as the first photosynthetic organ of a seedling. As the names imply, a *monocot* plant bears a single cotyledon while a *dicot* has two. As we study the flowering plants, we shall note other differences between monocots and dicots, including the arrangement of root and stem tissues, the pattern of leaf veins, and the number of flower parts.

The plant body of a flowering plant

Each organ of a flowering plant is highly developed for performing certain activities. The root, stem, and leaf are *vegetative organs.* They perform all the processes necessary for life *except* the formation of seeds.

Table 22-1 SOME FAMILIES OF MONOCOTS

FAMILY	FAMILIAR MEMBERS
cattail (Typhaceae)	common cattail
grass (Gramineae)	cereal grains, bluegrass, sugar cane, bamboo, timothy
sedge (Cyperaceae)	sedges
arum (Araceae)	Indian turnip (Jack-in-the-pulpit), skunk cabbage, calla lily
pineapple (Bromeliaceae)	pineapple, Spanish moss
lily (Liliaceae)	lily, onion, tulip, hyacinth
amaryllis (Amaryllidaceae)	amaryllis
iris (Iridaceae)	flag, iris
orchid (Orchidaceae)	lady's slipper, orchis, orchid
palm (Palmaceae)	coconut palm, date palm, palmetto

Table 22-2   SOME FAMILIES OF DICOTS

FAMILY	FAMILIAR MEMBERS
willow (Salicaceae)	willow, poplar (cottonwood), aspen
walnut (Juglandaceae)	walnut, hickory
birch (Betulaceae)	birch, alder, hazel
beech (Fagaceae)	beech, chestnut, oak
pink (Caryophyllaceae)	pink, carnation, chickweed
water lily (Nymphaeaceae)	water lily, pond lily
crowfoot (Ranunculaceae)	buttercup, hepatica, columbine, delphinium, larkspur
poppy (Papaveraceae)	poppy, bloodroot
mustard (Cruciferae)	mustard, radish, turnip, cress
rose (Rosaceae)	rose, apple, hawthorn, strawberry, pear, peach, plum, cherry
pulse (legume) (Leguminosae)	bean, pea, clover, alfalfa, locust, redbud
flax (Linaceae)	flax
maple (Aceraceae)	maple
mallow (Malvaceae)	marshmallow, hollyhock, hibiscus
parsley (Umbelliferae)	parsley, parsnip, carrot, sweet cicely
heath (Ericaceae)	laurel, rhododendron, azalea, heather, blueberry, cranberry, huckleberry
mint (Labiatae)	catnip, spearmint, peppermint, sage
nightshade (Solanaceae)	tomato, potato, tobacco
figwort (Scrophulariaceae)	mullein, snapdragon, digitalis (foxglove)
composite (Compositae)	dandelion, daisy, sunflower, zinnia, aster, marigold, thistle, dahlia

The *root anchors* the plant in the ground. It spreads through the soil and *absorbs* water and soil minerals. It *conducts* these to the stem for delivery to the leaves. Many roots *store* food substances and return them to the plant as needed.

The *stem* produces the leaves and *displays* them to the light. The stem is a busy thoroughfare, for it *conducts* water and minerals upward and transports downward foods that have been organized in the leaves. Like the root, the stem often serves as a place in which to *store* food. In many plants, green stems are also centers of *photosynthesis*.

The *leaf* is the center of much of the plant's activity. In most plants, it is the chief organ of *photosynthesis*. It also exchanges gases with the atmosphere in the processes of *respiration* and photosynthesis. Much of the water absorbed by the root has a one-way trip through the plant. It escapes from the leaf as water vapor in the process called *transpiration,* which will be discussed in detail in Chapter 23.

After a period of growth, a plant usually reproduces. *Flowers* are organs specialized for sexual *reproduction*. A portion of the flower develops into a *fruit,* which contains the *seeds*. The sequence from flower to fruit, seed, and embryo plant is the most highly evolved and efficient reproductive process in the entire plant kingdom.

Specialized tissues of a seed plant

The various organs of seed plants perform their processes with great efficiency because of the specialized tissues of which they are formed. Generally, each process involves one or more types of modified cells composing the plant body. We shall study these tissues more thoroughly when we discuss each of the plant organs in detail. However, before we deal with any individual organ, you should be familiar with the names of plant tissues and the general functions of each.

Meristematic tissue is composed of small, thin-walled cells that are capable of unlimited reproduction (Figure 22-7). During the growing season, meristematic cells divide almost continuously, forming more meristem. Cells of the meristem mature and form all of the *permanent tissues* composing the plant organs. Plants differ from animals in that growth occurs at the location of meristematic tissues rather than throughout the plant body. We call these areas *growing points*. They are found at the tips of roots and in the buds of stems. The *cambium,* which is responsible for growth in diameter, is a layer of meristematic cells in roots and stems.

Epidermal tissue is a covering layer, usually one cell thick, found on the surfaces of roots, stems, and leaves. Epidermal cells reduce water loss and protect inner tissues from injury. The epidermis of a young root is an absorptive tissue.

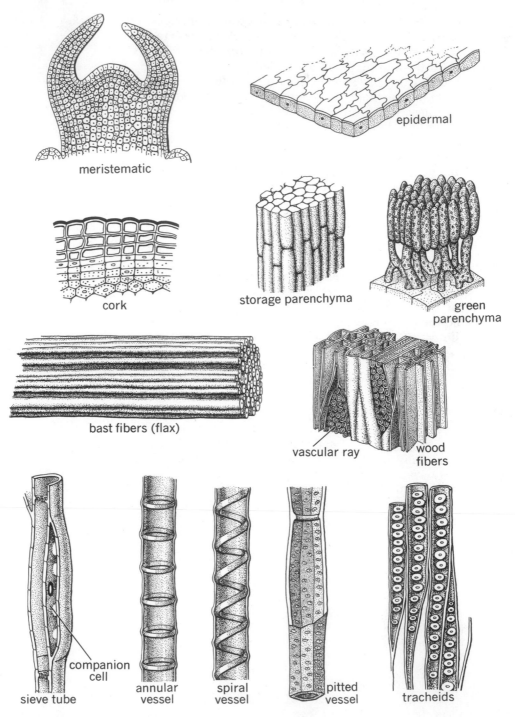

meristematic

epidermal

cork

storage parenchyma

green parenchyma

bast fibers (flax)

vascular ray

wood fibers

companion cell

sieve tube

annular vessel

spiral vessel

pitted vessel

tracheids

22-7 Specialized tissues of a seed plant.

Cork is a covering tissue found on the surfaces of woody roots and stems. Cork cells are usually short-lived but remain as a protective, waterproof covering, usually many cells in thickness.

Parenchyma tissue is composed of thin-walled cells resembling those of meristems. Parenchyma cells are found in flower petals, leaves, and various regions of roots and stems. *Chlorenchyma* is a chlorophyll-containing parenchyma tissue and is the center of photosynthesis in leaves and young stems. It is often called *green parenchyma*. The cells are loosely packed, with numerous intercellular spaces. These spaces are important in the exchange of gases during photosynthesis and respiration. *Storage parenchyma* is colorless. The cells are loosely packed and are larger than those of chlorenchyma. Starches, sugars, and other plant products are stored in this parenchyma tissue.

Strengthening tissues are found in roots and stems and in the stalks and larger veins of leaves. Generally, strengthening tissues are elongated, thick-walled *fibers*. The cells are usually short-lived, but their thick walls remain as supporting structures. *Sclerenchyma* is a strengthening tissue found in various regions of roots, stems, and certain leaves. *Bast* is a form of sclerenchyma tissue composed of elongated fibers with pointed ends. *Wood fibers* resemble bast but have even thicker walls.

Vascular tissues are composed of elongated, tubular cells that serve as channels of conduction. They are of several types. *Sieve tubes* are elongated cells arranged end to end, forming a continuous column. Their end walls are perforated and resemble a sieve or a strainer in a drain. Sieve tubes serve as channels of food conduction, principally downward, in leaves, stems, and roots. *Vessels* are large, tubular cells with thick walls. The end walls of vessels disappear, along with the protoplasm, leaving continuous, tubular channels that may be several feet in length. Vessels are paths of transport for water and minerals that principally move upward in roots and stems and outward through the stalks and veins of leaves. *Tracheids* are similar to vessels but are smaller. The end walls are angular and perforated. They act as channels of water and mineral conduction as well as supporting cells, especially in stems.

Phloem is a complex tissue composed of several simple tissues, including sieve tubes, bast fibers, and parenchyma. In a woody stem, phloem is a bark tissue.

Xylem is also a complex tissue, often called wood. It is composed of vessels, tracheids, fibers, and parenchyma and forms a prominent region in roots and stems.

Herbaceous and woody plants

Biologists often classify plants as *woody* or *herbaceous* on the basis of length of life and tissue makeup of their aerial portions, or stems. A woody plant contains a large amount of strengthening and vascular tissue. Meristematic tissue in growing regions lengthens the stem and roots and increases their diameter each season. Woody plants frequently reach large sizes by adding new tissues season after season. This size increase may continue for centuries. Long after tissues have ceased to function, they remain as supporting structures in the plant body. Woody plants include not only trees and shrubs but many vines as well; these include the wild grape, Virginia creeper, and poison ivy.

Herbaceous, or nonwoody, plants usually have soft stems with the strengthening and vascular tissues reduced to strands or bundles. In temperate climates, the aerial part in some species or, in others, the entire plant lives only one season. Herbaceous plants include most garden flowers and vegetables and many native flowering plants.

Annuals, biennials, and perennials

Plants are also classified on the basis of duration of the plant body. Plants that live for only one season are called *annuals*. These plants, such as the zinnia, marigold, petunia, pansy, bean, pea, and cereal grains, grow from seeds, mature, flower, and bear fruits and seeds in a single growing season. Only the seeds of annuals live from one growing season to the next.

The life cycle of a *biennial* extends through two growing seasons. The roots, stem, and leaves, often in the form of a low ring or rosette, develop during the first year. During the second growing season, the plant produces an aerial stem that bears leaves, flowers, fruit, and seeds (Figure 22-8). The life cycle ends with seed production, after which the plant dies. Beets, carrots, cabbages, parsnips, and turnips are biennial vegetables.

Perennials live more than two seasons, and often for many years. In most cases, they form roots, stems, and leaves the first year but do not flower until the second or third season. Each year, the portion of an herbaceous perennial that is above the

22-8 The pansy is an example of an annual flower. (left) The foxglove is a biennial. The rosette of basal leaves grows the first year. (center) The peony is a perennial. (right) (Jeanne White, National Audubon Society; Walter Dawn; Walter Chandoha)

ground dies. However, the roots and, in many plants, the underground stems remain alive and produce new stems and leaves each season. Delphiniums, lilies, columbines, and irises are herbaceous perennials. Woody perennials include the trees, shrubs, and many vines. Once perennials have started to flower, they usually continue season after season if environmental conditions remain favorable.

SUMMARY

If you ask someone to name a dozen plants, he will probably name a dozen seed plants and most of them, likely, will be flowering plants. These plants are the most conspicuous components of the vegetation of land environments today.

Seed plants are products of a long line of development that extends back to the close of the Carboniferous Period. During the more than 200 million years from that time until today, large groups of seed plants have flourished and died out or dwindled to but a few survivors. The conifers and flowering plants dominate the vegetation of the land today. Conifers form extensive forests, especially in the colder climates and areas of poor soil. In most other regions, flowering plants prevail.

There are several reasons for the success of flowering plants. One, certainly, is their ability to live and reproduce in nearly all kinds of environments. They thrive in deserts, plains, prairies, meadows, bogs, marshes, lowlands, uplands, and mountain slopes. Their numbers include trees, shrubs, vines, and herbs. Many live one season, while others survive for centuries.

In the chapters to follow, you will become acquainted with the organs of a flowering plant and, perhaps, discover why they are dominant in the land vegetation of our age.

BIOLOGICALLY SPEAKING

gymnosperm	dicot	parenchyma tissue
angiosperm	vegetative organ	strengthening tissue
seed plant	root	vascular tissue
seed	stem	phloem
conifer	leaf	xylem
seed cone	flower	woody (plant)
pollen cone	fruit	herbaceous (plant)
flowering plant	meristematic tissue	annual
cotyledon	epidermal tissue	biennial
monocot	cork	perennial

QUESTIONS FOR REVIEW

1. Which group of extinct plants was the probable evolutionary link between ferns and modern seed plants?
2. On what basis are the seed plants divided into the classes Gymnospermae and Angiospermae?
3. Name several parts of the world where cycads can still be found.
4. Explain why we speak of the ginkgo tree as a "living fossil."
5. List several well-known groups or genera of conifers.

6. Locate several coniferous forest areas of North America.
7. What two kinds of cones are produced by conifers in their reproductive cycle?
8. Name the two large subclasses of angiosperms, or flowering plants, and explain the basis for the distinction between these subclasses.
9. List the principal functions of each of the vegetative organs of a flowering plant.
10. Name the reproductive organs of a flowering plant.
11. In what respect is meristematic tissue different from other plant tissues?
12. Name two covering tissues of flowering plants.
13. Identify two kinds of parenchyma tissue and designate the function of each.
14. List several kinds of strengthening tissues found in roots, stems, and leaves.
15. Name three vascular tissues and indicate the function or functions of each.
16. In what respect are xylem and phloem complex tissues?
17. Distinguish between woody and herbaceous stems.
18. On the basis of the duration of the plant body, classify flowering plants into three groups.

APPLYING PRINCIPLES AND CONCEPTS

1. Discuss possible reasons for the disappearance of most gymnosperms and the rise of angiosperms through past ages.
2. Discuss various ways in which angiosperms have had an advantage over other kinds of plants in the struggle for existence.
3. Discuss the advantages a herbaceous plant has over a woody plant in a temperate or arctic environment.
4. Many of our most beautiful garden flowers are annuals. Why are they ideally suited to garden needs?

RELATED READING

Books

DEVLIN, ROBERT M., *Plant Physiology.*
Reinhold Publishing Corp., New York. 1966. Includes recent findings in plant physiology.
Encyclopedia of Life Sciences (eds.), *The World of Plants.*
Doubleday and Company, Inc., Garden City, N.Y. 1965. A beautiful, all-around book on all aspects of botany, written by some thirty leading botanists.
FRISCH, ROSE E., *Plants That Feed the World.*
D. Van Nostrand Co., Princeton, N.J. 1967. An informative account of the flowering plants that, with few exceptions, feed the world.
MILNE, LORUS J., and MARGERY MILNE, *Living Plants of the World.*
Random House, Inc., New York. 1967. A systematic, nontechnical presentation of one hundred fifty seed plant families.
RAY, PETER MARTIN, *The Living Plant.*
Holt, Rinehart and Winston, Inc., New York. 1963. Clarifies the important principles that are known to underlie the activities of land plants.
WILSON, CARL L., and WALTER E. LOOMIS, *Botany.*
Holt, Rinehart and Winston, Inc., New York. 1971. An excellent text in botany which incorporates all the newer knowledge in this field, especially molecular biology.

CHAPTER TWENTY-THREE

THE LEAF AND ITS FUNCTIONS

The leaf—a specialized organ for photosynthesis

The leaf is the center of many activities of a seed plant, the most important of which is photosynthesis. To a great extent, all other processes of the vegetative plant body relate to this essential process. The roots absorb water and minerals, and the stem conducts them to the leaves, where they are used in food manufacture. The stem produces the leaves and displays them to light, the energy source for photosynthesis. The stem and roots receive the products of the chemical activities of the leaves. As you study the structure of a leaf and the tissues that comprise it, you will see what a perfect organ it is for photosynthesis and other related activities.

Usually, a leaf develops and functions for one season only. In evergreens, the leaves remain from one season to the next, but even these plants bear new leaves on young shoots each growing season. Thus, the plant renews its vital food factories frequently, as old leaves fall away.

The structure of a leaf

A leaf is an expanded outgrowth of the stem. In most leaves, there is a flattened, green *blade*. Leaf blades vary in size as well as in form. The edges, or leaf margins, may be smooth (entire), toothed, or indented (lobed). The shapes of the tips and bases of leaf blades also vary in different species. These characteristics are often useful in the identification of plant species.

The blade is attached to the stem by a stalk, or *petiole* (Figure 23-1). Leaflike or scalelike *stipules* are found at the bases of many leaf petioles. These may drop soon after the leaf develops or may remain throughout the season, depending on the species. Some leaves lack petioles and are said to be *sessile*.

The blade is strengthened by numerous *veins* that penetrate the leaf tissues. The larger veins look like ribs and are especially noticeable on the lower side of the blade. In addition to supporting the leaf tissues, the veins are conducting channels through which water, minerals, and foods are moved.

23-1 The Norway maple has a typical dicot leaf. Notice the arrangement of the blade, veins, petiole, and the stipules. (Dr. E. R. Degginger)

23-2 The leaves of the iris, a monocotyledonous plant, have parallel venation. (Walter Dawn)

The arrangement of larger veins tends to fall into one of two patterns that we call forms of *venation.* In monocotyledons, including corn, lilies, irises, and orchids, the larger veins are *parallel* or nearly so (Figure 23-2). In dicotyledons, the veins branch and rebranch, forming a *netted* pattern. The leaves of the sycamore, maple, and many other dicotyledons have several large veins that extend through the blade from the end of the petiole like fingers from the palm of a hand. This pattern is called *palmate* net venation. Other dicotyledons, including the elm, willow, and apple, have a large central vein, or *midrib,* extending through the length of the blade. Smaller veins branch from the midrib and run to the margin like barbs of a feather. This pattern is called *pinnate* net venation.

Simple and compound leaves

When the blade of a leaf is undivided, even though it may be deeply indented, the leaf is classified as *simple.* There are many leaves, however, in which the blade is divided into several parts called *leaflets.* Such leaves are classified as *compound.* The arrangement of leaflets may be palmate or pinnate. Leaves of the buckeye and horse chestnut trees, clover, lupine, poison ivy, and Virginia creeper are *palmately compound.* The leaflets radiate from a common point at the end of the petiole (Figure 23-3).

The leaflets of a *pinnately compound* leaf are attached along a central stalk. The pea, rose, ash, walnut, and hickory have pinnately compound leaves. In certain *bipinnately compound* leaves, including the honey locust and Kentucky coffeetree, the leaflets are further divided into even smaller leaflets.

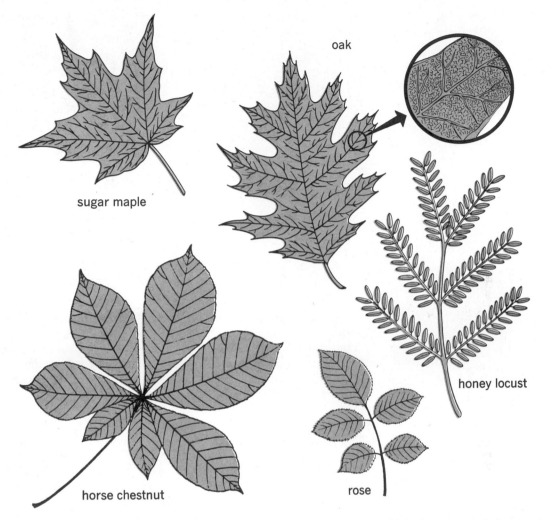

oak

sugar maple

horse chestnut

rose

honey locust

In some cases, it may be difficult to distinguish a simple leaf and a leaflet of a compound leaf. The two can be distinguished readily, however, by observing the location of axillary buds. These buds develop in the axils of leaves—in the angle between the petiole and the stem—but are never found at the bases of stalks of leaflets.

The tissues of a leaf

If you cut across the blade of a leaf and examine the section with a microscope, you will see that three distinct kinds of tissue are clearly visible. The upper and lower surfaces are covered by an *epidermis.* Between the epidermal layers is the photosynthetic tissue of the *mesophyll,* composed of several layers of chlorenchyma cells. Veins of various sizes penetrate the mesophyll

23-3 Leaves vary in form and size of the blade, venation, and compounding and are often used as a basis for identification of plants. Identify the various types of venation and compounding shown in this drawing.

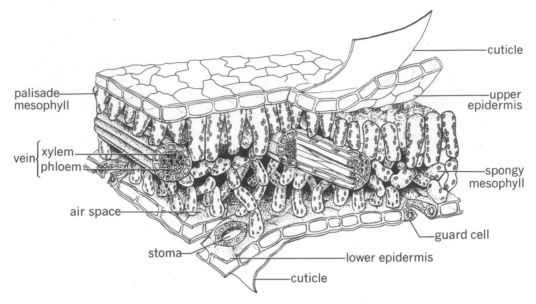

palisade mesophyll

cuticle

upper epidermis

vein { xylem
       phloem

air space

spongy mesophyll

guard cell

stoma

lower epidermis

cuticle

23-4   A cross section of a leaf.

about midway between the upper and lower epidermis (Figure 23-4).

The tissues of a leaf are so modified as to allow penetration of light and exchange of gases between the tissues and the atmosphere. They also provide channels of conduction for a constant flow of materials to and from cells of the mesophyll.

The structure of the epidermis

The upper and lower epidermis are composed of single layers of interlocking cells that usually lack chloroplasts. When viewed in cross section, epidermal cells look like cubes or bricks. However, a surface view of an epidermis shows that the cells are irregular in shape and locked together like pieces of a jigsaw puzzle.

In many leaves, the epidermis is covered by a thin, waxy or varnishlike film called the *cuticle*. The cuticle slows down the passage of water vapor and other gases through the epidermal cells, and thus prevents excessive loss of these substances from the leaf tissues.

Most of the movement of water vapor and other gases into and out of the leaf tissues occurs through *stomata* (singular, *stoma*). These are lens-shaped pores that perforate the epidermis and open into air spaces between cells of the mesophyll. Two bean-shaped *guard cells* surround each stoma opening. Guard cells are modified epidermal cells but, unlike other cells of the epidermis, contain chloroplasts. As can be seen in Figure 23-5, the wall of a guard cell bordering the stoma is thickened. This wall structure is important in changing the shape of the guard cell and opening

23-5   A stoma. Unlike the other cells of the epidermis, the two bean-shaped guard cells that surround each stoma, or opening, contain chloroplasts. (Hugh Spencer)

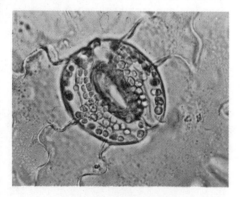

and closing the stoma with changes in water content and resulting turgor. The mechanism of this regulation will be discussed later in the chapter.

A square inch of epidermis may contain as many as several thousand stomata. While they may be present in both the upper and lower epidermis, in most leaves that grow horizontally, they are limited to the lower surface. This is especially true of the leaves of woody plants. Stomata are also found in the epidermis of herbaceous stems and young shoots of woody plants. In vertical leaves, such as those of the iris, stomata are about evenly distributed on both leaf surfaces. Floating leaves have all of the stomata in the upper epidermis.

Many leaves have surfaces that are velvety, fuzzy, or woolly because of the presence of *epidermal hairs*. These hairs are unicellular or multicellular outgrowths of epidermal cells. In some leaves, they are believed to reduce water loss. They may also contain glands that secrete oily or sticky substances on the leaf surface. Those of the nettle contain an irritating poison that causes a burning and stinging sensation when the leaves are touched and points of the hairs break off in the skin.

The mesophyll

Cells of the mesophyll compose all of the leaf blade between the upper and lower epidermis except for the vascular bundles. Mesophyll cells are thin-walled chlorenchyma-type cells that remain alive and active even after they mature. The mesophyll is the principal photosynthetic tissue of the leaf.

The mesophyll of most leaves contains two distinct regions. The cells in the upper region, beneath the upper epidermis, are elongated and lie in vertical rows. This *palisade mesophyll* is so named because the cells resemble rows of stakes in a fence. The activity within the palisade cells reflects an interesting adaptation to light. Numerous chloroplasts stream through palisade cells in a circular path. In the upper region, they receive the greatest exposure to light. As they move downward in the cell, they are partially shaded by other chloroplasts.

Below the palisade cells is a zone of irregular, loosely arranged cells composing the *spongy mesophyll*. Large intercellular spaces in this region provide air passages from the stomata through the tissues of the leaf. Cells of the spongy mesophyll contain chloroplasts but in smaller number than the cells of the palisade region. This explains why the upper surface of most leaves is a darker green than the lower surface is.

Vertical leaves, of course, have no upper or lower surfaces. Often, in these leaves, each epidermis is lined with a zone of palisade mesophyll. When these palisade mesophyll zones occur, they are separated by a zone of spongy mesophyll. However, in other leaves, the mesophyll is not differentiated into zones.

Structure of veins

Veins are composed of varying amounts of conducting and supporting tissues grouped in *fibrovascular bundles.* One or more large bundles enter the blade from the petiole. In a dicotyledonous leaf, large veins branch and rebranch, forming a network that penetrates all regions of the mesophyll. This branching is so extensive that no cell is more than a few cells away from a vein.

A large vein consists of a region of *xylem vessels* on its upper side and a mass of *phloem sieve tubes* on the lower side. As veins subdivide in a dicotyledonous leaf, they become smaller and smaller and the amount of vascular tissue decreases. The smallest veins may contain a single xylem vessel. These veins come to an abrupt dead end in the tissue of the mesophyll.

Small and medium-sized veins are enclosed in a *bundle sheath,* which is a ring of elongated parenchyma cells. Cells of the bundle sheath lie in close contact with the palisade and spongy mesophyll cells. Water and minerals passing from the xylem vessels to the mesophyll cells, as well as foods moving from the mesophyll cells to the phloem sieve tubes, must pass through the bundle sheath. The smallest veins often lack phloem tubes. Cells of the bundle sheaths surrounding these tiny veins transport foods along the vein to the nearest phloem tubes.

Large vascular bundles often contain zones of *sclerenchyma fibers* above and below the vascular tissues and extending to the epidermis. These fibers reinforce the bundles and provide mechanical support in the leaves.

The petiole and leaf traces

A continuous passageway is formed by bundles of conducting tissues that branch from the vascular region of the stem, pass through the petiole, and continue as branches in the veins of the leaf blade. This conducting system originates in the stem, where bundles of xylem and phloem, known as *leaf traces,* grow outward in the region of a node. The number of leaf traces varies from one to many, depending on the species. When a leaf falls, the leaf traces can be seen as bundle scars within the leaf scar.

The leaf and photosynthesis

Photosynthesis as a chemical process was discussed in Chapter 6. That discussion related to cells and chloroplasts, the light-energy source, raw materials, and products. In our study of a leaf, we are here concerned with its adaptations as the chief photosynthetic organ of a seed plant.

Photosynthesis occurs in any green plant tissue, including the cortex cells of herbaceous stems and the young shoots of woody plants. However, the mesophyll cells of the leaf are the principal centers of photosynthesis in most plants. The thin, expanded leaf blade is ideally adapted for exposure of the greatest number of mesophyll cells to light. Water reaches these "food factories" in xylem vessels of the numerous small veins that penetrate the mesophyll. Carbon dioxide, the other raw material for the process, enters the leaf tissue through the stomata from the atmosphere. Oxygen released from the cells during the process escapes to the atmosphere through the same pores.

Glucose formed from PGAL may be used as a fuel in respiration in the leaf cells, converted to starch for temporary storage, or used in the organization of various carbohydrates, fats, proteins, and other compounds.

Translocation of foods from leaves

On a warm, bright day, starch accumulates in a leaf more rapidly than it can be converted to sugar for *translocation,* or movement to other parts of the plant. The starch content usually reaches its peak in about the middle of the afternoon on a bright summer day. Gradually, much of this insoluble starch is digested through enzymatic action and is converted to soluble sugar. It enters the phloem sieve tubes of the veins in water solution and passes through the petiole and into the stem for translocation to places of storage. Translocation continues through the night, clearing most of the mesophyll cells for another day's photosynthetic activity.

Light and leaf arrangement

In the study of photosynthesis in Chapter 6, we discussed various factors that limit the process. Among these influences, both internal and external, are light, temperature, water supply, and carbon dioxide concentration. Of these factors, light is frequently the most critical.

The rate of photosynthesis in plants that normally grow in sunlight is directly related to light intensity—at least, until the intensity reaches one fourth to one half that of full sunlight. Since a plant's survival is dependent on its ability to carry on photosynthesis, it should not surprise you to learn that leaves are usually arranged on a stem so as to expose the blade to the maximum amount of light. Hence, leaves growing singly often are found to form a spiral arrangement. Also, it is usual to find that two leaves growing on opposite sides of a stem in a north-south direction will alternate with leaves arranged in an east-west direction. These arrangements act to minimize the shading of one leaf by another. Exposure of the leaf blade to maximum light is accomplished, further, by bending of the petiole. We refer to the pattern of leaf arrangement for greatest light exposure as *mosaic.*

Respiration in green plants

Photosynthesis occurs only during the sunny hours, but respiration occurs night and day. Respiration in the seed plant is like that of other organisms. Glucose is the principal fuel from which energy is released, and glucose molecules are oxidized by the removal of hydrogen, as we explained in Chapter 6. Oxygen is the ultimate hydrogen acceptor in that it combines with the hydrogen to form water, a respiration by-product. Carbon dioxide is released as the carbon skeleton of the glucose molecule is broken down. During daylight hours, the oxygen released from photosynthesis is more than enough to support respiration. At night, however, atmospheric oxygen enters the leaf and is used in respiration.

We may not be aware of plant respiration, because it does not involve breathing. However, the process is just as important to the plant as to the animal, even though the rate in plants is much lower than it is in animals.

Plants require energy for all of their internal processes. Some energy is used in the streaming of cell protoplasm. In addition, energy is required in the synthesis of carbohydrates other than glucose, and in the synthesis of fats, proteins, and nucleic acids. Growth of the plant requires a constant supply of energy.

Photosynthesis and respiration compared

The relationship between photosynthesis and respiration is interesting. These would almost appear to be opposite chemical reactions. However, this is not the case. The enzymes and chemical reactions involved in the two processes are different. We can say, though, that photosynthesis and respiration are complementary processes. The requirements of one process are products of the other. This is an important aspect of the biochemical balance maintained in a biological society.

The complementary relationship of photosynthesis and respiration becomes more apparent when you compare the matter and energy changes involved in the two processes, as in Table 23-1.

Table 23-1 COMPARISON OF PHOTOSYNTHESIS AND RESPIRATION

PHOTOSYNTHESIS	RESPIRATION
food accumulated	food broken down (oxidized)
energy from sun stored in glucose	energy of glucose released by oxidation
carbon dioxide taken in	carbon dioxide given off
oxygen given off	oxygen taken in
produces glucose from PGAL	produces CO_2 and H_2O
goes on only in light	goes on day and night
occurs only in presence of chlorophyll	occurs in all living cells

As you compare the two processes, notice that matter flows in a cycle. That is, matter can be used over and over. This cycle also involves an endless change of inorganic molecules to organic molecules, followed by a return to the inorganic state. However, while matter cycles, energy does not. Light energy is used in photosynthesis, while heat and useful energy are released in respiration. Energy to support photosynthesis must be received as light. For this reason, life will always depend on the sun for energy.

Perhaps you have wondered if photosynthesis and respiration balance each other in a plant. They may if the plant is in very reduced light, which slows down the rate of photosynthesis greatly. However, in bright sunshine, photosynthesis occurs at a rate ten times or more that of respiration. At night, of course, photosynthesis ceases, while respiration continues. However, the total photosynthesis carried on by a plant far exceeds the total respiration. This places the green plants of the earth in the position of food- and oxygen-producers for the totally dependent heterotrophic plants and animals.

Transpiration in plants

During the growing season, a plant conducts a continuous stream of water up through the roots and stem to the leaves. Some of this water is used in photosynthesis and other processes of the plant. However, biologists estimate that only one percent of the water flowing through a plant is used. The remaining ninety-nine percent escapes from the leaves in the form of water vapor. Most of this water diffuses from the leaf tissues to the atmosphere through the stomata. We call the process in which water is lost from plants *transpiration*.

As water enters a leaf through the veins, it passes from the xylem vessels into cells of the mesophyll. This movement may continue from one cell to another. During transpiration, water evaporates from the wet cell surfaces and diffuses into the inter-cellular spaces of the spongy region as vapor. It then passes to the atmosphere through the stomata. As this water escapes to the atmosphere, more water moves through the veins to replace it. However, the rate of transpiration is regulated by the opening and closing of stomata.

Transpiration can be demonstrated with a potted plant, such as a geranium, and a bell jar, as shown in Figure 23-7. The pot and soil surface are covered with a waterproof material such as waxed paper or metal foil to prevent evaporation. Blue cobalt chloride paper, which turns pink in the presence of water vapor, is used to indicate the loss of water from the leaves or stem. Change in color of the cobalt chloride paper often occurs within fifteen minutes. Blue cobalt chloride paper in the empty bell jar is used as a control.

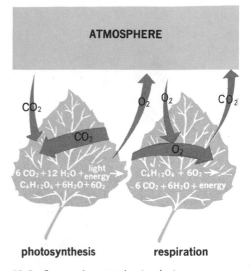

23-6 Gas exchanges due to photosynthesis and respiration. Under normal daylight conditions, five to ten times more carbon dioxide is used in photosynthesis than is produced in respiration. In addition, much more oxygen is released in photosynthesis than is required in respiration.

23-7 Demonstrating transpiration. Water vapor from the geranium causes the indicator paper to turn from blue to pink, whereas the color of the indicator paper in the control jar does not change.

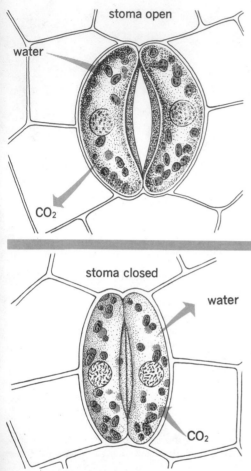

23-8 Changes in the concentration of carbon dioxide and carbonic acid in the guard cells affect enzyme action on sugar and starch resulting in water movement and turgor change that open and close the stoma.

The opening and closing of stomata

It has long been known that variations in turgor pressure cause changes in the shapes of the guard cells. These changes, in turn, cause the stomata to open and close. In most plants, the stomata are open during the day and closed at night. They may also close at certain times during the day, especially in the afternoons of hot days.

As guard cells absorb water from adjacent epidermal cells, their turgor increases. Pressure in the cell bulges the thin outer walls and presses them into adjacent epidermal cells. This pulls the thickened inner walls flanking the stoma into a crescent shape and opens the pore. Outward movement of water from the guard cells reduces turgor pressure, causing the guard cells to shrink. The inner walls straighten and close the stoma.

You might assume that turgor changes in guard cells and the opening and closing of stomata are related directly to the water content of the leaf. If this were true, the stomata would be open at night, when the water content of the leaf tissues is probably greatest, and closed during the day, when a higher rate of evaporation from the leaf reduces the water content. Since this is not the case, we must search further for a possible explanation.

Recent investigation of the mechanism of the opening and closing of the stomata indicates that the concentration of carbon dioxide in the guard cells may be a key to the turgor changes that occur. It has been demonstrated that stomata close when the concentration of carbon dioxide exceeds 0.04 percent and that they open when the concentration decreases below this amount. Other factors in the guard cells seem to be involved. These include: enzyme action in the conversion of sugar to starch and starch to sugar, the formation of carbonic acid from carbon dioxide and water, the effect of an acid medium on enzyme action, and the movement of water to and from the guard cells. With these factors in mind, consider the following series of conditions and changes that may explain the closing and opening of stomata.

During the night, when photosynthesis has ceased due to lack of light, starch-forming enzymes in the guard cells convert sugar to starch. This chemical activity is peculiar to guard cells and differs from the process taking place in the cells of the mesophyll, where starch is changed to sugar and translocated to other parts of the plant during the night. Carbon dioxide released from respiration in the guard cells combines with water to form carbonic acid. Acidity in the guard cells prevents the action of enzymes that convert starch to sugar. As the sugar content is lowered, the guard cells lose water and turgor pressure is reduced, causing the stomata to close.

As light increases in the early morning hours, photosynthesis resumes. Carbon dioxide, released during respiration, is now used in photosynthesis and the acidity of the guard cells is reduced. Decrease in acidity seems to stimulate the action of starch-digesting enzymes. Stored starch is converted to sugar, which dissolves in water in the guard cells. Increase in the sugar content reduces the concentration of water in the solution, resulting in water intake by osmosis. As turgor increases, the guard cells swell and the stoma opens.

Thus, while turgor change in the guard cells is the *direct mechanism* in the opening and closing of stomata, carbon dioxide concentration is the *underlying cause*. Carbon dioxide acts as a chemical regulator in forming carbonic acid and preventing or stimulating starch-forming or starch-digesting enzymes.

Leaf coloration

During the late spring and summer, leaves are green because chlorophyll is present in the chloroplasts. In addition to chlorophyll, the chloroplasts also contain the yellow pigment *xanthophyll* and the orange pigment *carotene*. Chlorophyll, however, masks these two other pigments.

With the coming of fall, the temperature is apt to drop below the level at which chlorophyll formation is possible. Light destroys the remaining chlorophyll, and the previously hidden yellow and orange pigments become apparent.

23-9 A deciduous forest in autumn. Notice the shades of red, orange, and yellow in the different species of trees. (Grant Heilman)

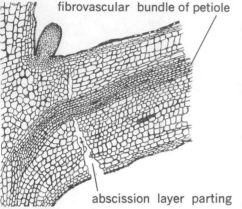

fibrovascular bundle of petiole

abscission layer parting

23-10 During autumn, the cells of the abscission layer separate.

23-11 The aloe leaves (above) are modified for storing water. Some leaflets of the wild pea are modified into slender tendrils. (Dr. William Steere; George Roos, National Audubon Society)

Cool weather also promotes the formation of the red pigment *anthocyanin* in many leaves. This red pigment forms, not in the chloroplasts, but in the cell sap in vacuoles of the leaf cells. It is formed from food materials. This pigment accounts for the red appearance of the leaves of many woody plants during the cool spring and fall seasons.

Brown coloration results from the death of leaf tissues and the production of *tannic acid* inside the leaf.

The falling of leaves

A natural falling of leaves occurs toward the end of the growing season in many woody plants in temperate climates. With the approach of autumn, changes occur in the zone of parenchyma cells, or *abscission layer,* which forms across the base of the petiole (Figure 23-9). Pectin that had previously joined the walls of these parenchyma cells securely is dissolved by enzymes that are formed late in the growing season. The cells separate, leaving the petiole attached to the twig only by its fibrovascular bundles. The slightest jar, gust of wind, or rain may cause the leaf to drop. A thin layer of cork cells seals the scar where the leaf was attached to the twig.

Certain broad-leaved trees, including many oaks, do not drop their leaves in autumn. The leaves remain on the tree in a dried condition and fall gradually throughout the winter and following spring. Evergreens, including most conifers, retain their green leaves through the winter. These leaves drop gradually during the following season, after new shoots and leaves have been produced.

Leaf modifications

While most leaves are either broad and thin or needle-shaped (as in the conifers), some are highly modified for specialized functions. The *succulent leaves* of the century plant *(Agave),* aloes, and sedums are thick and fleshy. Water accumulates in the tissues and supplies the plant during long periods of drought in desert and semidesert conditions. Other desert plants, including the cacti, have succulent stems, and they also have leaves that are reduced to protective *spines* to prevent water loss. *Tendrils* may be leaf petioles, veins, or stipules that coil around supports and anchor climbing plants. They are found in the garden pea, clematis, and greenbrier.

Among the most curious leaf adaptations are those of insectivorous plants, including the sundew, Venus's-flytrap, pitcher plants, and bladderwort. The leaves of these plants are modified in various ways for capturing insects.

The sundews are small plants found in bogs and other wet places. The leaves are borne on long stalks in circular clusters, or rosettes. The rounded leaf blades bear numerous tentacles, or soft spines, on their surfaces and margins. A drop of sticky secretion forms at the tip of each tentacle, causing the leaf to glisten in the sunlight. Insects attracted to the leaves by the sticky secretion are captured and held by the tentacles that bend around them. The secretion contains digestive enzymes that digest insects on the leaf surface.

The Venus's-flytrap, found in wet environments in North and South Carolina, has a rosette of leaves from two to six inches long. Each leaf consists of a flattened petiole and a pocketlike blade composed of two hinged lobes. The margins of the lobes bear a row of stiff, hairlike spines. Numerous hairs on the upper surfaces of the lobes secrete a sweet fluid that attracts insects. This fluid also contains digestive enzymes. When an insect contacts the surface of a lobe, the jaws of the trap close suddenly. The marginal spines interlock and prevent escape. The lobes remain tightly closed for several days, during which the insect is partially digested and absorbed into the leaf tissues. The lobes of the trap then open and are ready to trap other insects.

Pitcher plants grow in wet soils and on mats of sphagnum moss in many regions of North America. Their leaves grow in upright clusters and may be from six inches to several feet in length, depending on the species. The name, pitcher plant, refers to the tubular or vase-shaped leaves with lips that bend outward. Numerous hairs point downward along the inner surfaces of the leaves. Glands near the bases secrete digestive fluids. Frequently, a pool of rainwater collects at the base of each tubular leaf. Insects that crawl into the leaf cannot escape because of the numerous downward pointing hairs. They soon fall into the cavity of the leaf and are digested.

The bladderwort is an aquatic insectivorous plant. Its underwater leaves are modified as bladders that trap insects and other small aquatic animals.

Insectivorous plants contain chlorophyll and synthesize their own carbohydrates. As nutritionally independent organisms, they do not require the substances derived from animal food. However, they usually grow in soils that are deficient in usable nitrogen in the form of nitrates. Hence, the nitrogen-containing products of animal digestion are a valuable nutritional supplement for these insectivorous plants.

23-12 The Venus's flytrap (top) is an insectivorous plant. The plant has special adaptations which allow it to digest insects which become trapped on its sticky surface. The pitcher plant, another common insectivorous plant, has similar adaptations. (Hugh Spencer; Hal H. Harrison from Grant Heilman)

SUMMARY

Those who live in temperate climates eagerly watch for the opening of buds on trees and shrubs and the growth of stems and new leaves in the early spring. This sign of spring marks the end of a long winter dormancy and the start of a new growing season.

To the biologist, the leaf is a plant organ ideally suited to a variety of processes. While they vary in form, size, and function, in most plants, leaves are the primary centers of photosynthesis. The thin, spreading blade exposes large numbers of mesophyll cells to light. Veins penetrating all areas of the mesophyll bring water and minerals to the leaf tissues and receive products of their activities for translocation to other parts of the plant. Numerous spaces among the cells of the mesophyll allow rapid exchange of gases involved in photosynthesis and other cell activities. Numerous stomata in the epidermis open the leaf tissues to the atmosphere for gaseous exchanges.

Processes of the leaf require supporting functions of the root and stem. We shall continue our exploration of the vegetative plant body in the study of roots and stems in the next chapter.

BIOLOGICALLY SPEAKING

blade	epidermis	translocation
petiole	mesophyll	mosaic
stipule	vascular bundle	transpiration
sessile	cuticle	abscission layer
vein	stoma	succulent leaf
venation	guard cell	spine
simple leaf	epidermal hair	tendril
leaflet	fibrovascular bundle	
compound leaf	leaf trace	

QUESTIONS FOR REVIEW

1. How does the thin, broad blade of most leaves adapt them for their function?
2. Name two functions of veins.
3. On the basis of leaf venation, distinguish between monocotyledonous and dicotyledonous plants.
4. Describe two arrangements of leaflets in compound leaves.
5. Of what importance is the cuticle formed on the epidermis of many leaves?
6. Describe the structure of guard cells and their relation to a stoma.
7. Locate and describe two distinct regions of the mesophyll of a leaf that grows horizontally.
8. List the tissues composing a large vein and indicate the function of each.
9. In what form are carbohydrates translocated from cells of the leaf mesophyll?
10. Compare photosynthesis and respiration with regard to glucose, oxygen and carbon dioxide relations, and energy changes that occur.
11. Approximately what percent of the water taken into a plant escapes from the leaves as vapor during transpiration?
12. Describe the changes that occur in leaf stomata during a twenty-four-hour period.
13. What substance involved in photosynthesis is believed to be the underlying cause of the opening and closing of stomata?
14. How is sugar content related to turgor changes in guard cells?
15. Explain why leaves turn yellow, orange, or red with the coming of autumn.
16. Explain the mechanism involved in the dropping of leaves in the fall.
17. How are succulent leaves an adaptation for life in dry environments?
18. Give several examples of insectivorous plants.

APPLYING PRINCIPLES AND CONCEPTS

1. New leaves are produced on woody perennial plants each growing season, although the roots and stem may grow for many seasons. Explain why leaf replacement is necessary in the life of a plant.
2. Cells in the mesophyll are thin-walled and loosely arranged. Why is this important in the activities of the leaf?
3. Cite evidence to indicate that the opening and closing of stomata are related more closely to light and photosynthesis than to the water content of a leaf.
4. Under what conditions might stomata close during the afternoon of a hot, sunny day?
5. Discuss the importance of a leaf mosaic.

RELATED READING

Books

COULTER, MERLE C., and HOWARD J. DITTMER, *The Story of the Plant Kingdom* (3rd ed., rev.).
  The University of Chicago Press, Chicago. 1964. Contains a wealth of information on the functions and activities of the green leaf.
HUTCHINS, ROSS E., *This Is a Leaf.*
  Dodd, Mead and Co., New York. 1962. The fascinating story of leaves—their varied shapes, sizes, and functions.
POOLE, LYNN, and GRAY POOLE, *Insect-eating Plants.*
  Thomas Y. Crowell Company, New York. 1967. Reveals the speciality of these strange plants.
ROBBINS, WILFRED W., and others, *Botany—An Introduction to Plant Science.*
  John Wiley and Sons, Inc., New York. 1964. A useful reference for the study of leaf structure and function.
WENDT, FRITS, and the Editors of *Life* Magazine, *The Plants.*
  Time-Life Books (Time, Inc.), New York. 1963. A picture-essay book that answers most of the questions biology students ask about plants.

CHAPTER TWENTY-FOUR

ROOTS AND STEMS— ORGANS OF ABSORPTION AND CONDUCTION

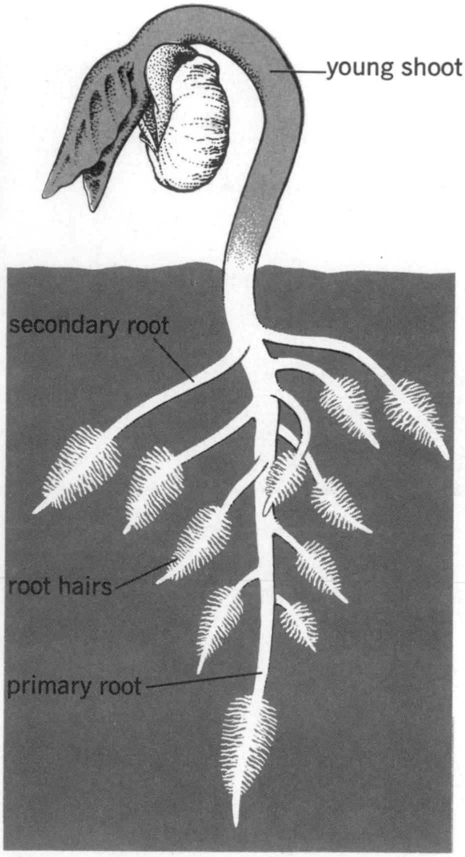

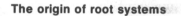

young shoot

secondary root

root hairs

primary root

24-1 Secondary roots branch from the primary root of this bean seedling. Added branching and growth of the secondary roots will occur as the root system develops more fully.

The origin of root systems

When a seed germinates, the first root arises from the lower end, or *radicle,* of the embryo plant and grows downward into the soil. (See Figure 27-14.) This first root is the *primary root*. After a short period of growth, the primary root produces lateral branches, or *secondary roots* (Figure 24-1). Further branching of secondary roots occurs as the plant develops its entire root system.

How does the extent of the root system of a land plant compare with that of its stem and branches? There is no definite rule. Root systems vary in form and extent in different species. Environmental factors such as the physical nature of the soil, soil moisture, and temperature also influence the growth of roots. In general, though, you can estimate that the average land plant has at least as much of its root system below ground as it has stem and branches above ground. However, most roots of a plant's root system penetrate the soil a much shorter distance than the stem extends above the ground. Most roots, even those of large trees, are less than four feet below the surface of the ground. Therefore, the lateral extent of these roots is usually considerably greater than the spread of the branches.

Types of roots and root systems

If a primary root continues to grow and remains the major root of the system, it is known as a *taproot*. Plants with taproots

374

have many advantages. A long taproot is ideal for anchorage. Taproots also reach water supplies deep in the ground. The alfalfa, for example, has a taproot fifteen feet or more in length. This explains why fields of alfalfa remain green during dry periods when shallow-rooted grasses turn yellow and brown. Long taproots also account for oak and hickory forests on dry hillsides and ridges. The taproots of many plants become thick and fleshy and serve as underground storehouses for the food supply of the plant. We grow certain of these plants, including the beet, radish, carrot, turnip, and parsnip, as root crops.

In many plants, including the grasses, the growth of the primary root is rapidly surpassed by the growth of its branches, or secondary roots. Roots of this kind are *fibrous roots.* Fibrous root systems are usually shallow and spread through a large area of soil. They are efficient organs of water and mineral absorption. Fibrous roots are also important as soil binders, anchoring soil particles and preventing erosion by water and wind.

In some plants, secondary roots enlarge and become storage organs similar to taproots. The sweet potato and dahlia are examples of such roots.

The root tip and terminal growth of the root

If a young root is cut off about one-half inch behind the tip, sectioned lengthwise, and prepared for microscopic examination, several important regions can be identified. These are shown in Figure 24-3. The delicate apex is covered by the *root cap.* As the root elongates, the root cap is pushed through the soil. This tears away the outer surface. However, the addition of new cells to the inner surface keeps it in constant repair.

You may wonder how a delicate root tip can force its way through the soil without being completely destroyed. The young root both pushes its way and dissolves its way through the soil. Living root cells give off carbon dioxide as a product of cellular respiration. This carbon dioxide reacts with soil water to form *carbonic acid,* a weak acid that dissolves certain minerals and thereby aids the forward movement of the root tip. When roots grow over limestone rocks, their patterns are often etched into the rocks by this carbonic acid.

A *meristematic region* lies immediately behind the root cap. This is a zone of cell division, or an apical growing point of the root. The length of this region varies from 1.0 to 2.5 millimeters in different roots.

Behind the meristematic region, cell division gradually ceases, and the cells lengthen. This change in the cells marks the *elongation region,* a region usually as long as or longer than the meristematic region. As these cells of the growing root elongate, the root tip is pushed forward. It is also in this region that certain cells mature into vascular tissues of the xylem and phloem.

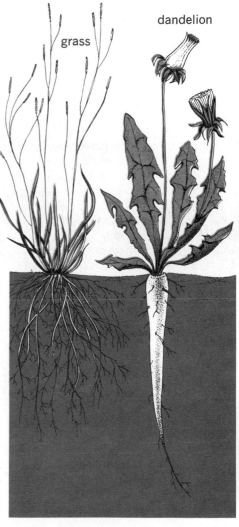

grass   dandelion

24-2 Compare and contrast the fibrous root system of the grass with the taproot system of the dandelion. Which acts as the better soil binder?

24-3 Major regions of a root tip. Note the blunt, thimble-shaped root cap that protects the delicate meristematic region from injury by soil particles.

24-4 In order to locate the elongation region, the young root in *a* was marked with ink at intervals of one millimeter from the tip back. The same root is shown in *b* after 24 hours of growth. Notice the region in which elongation has occurred.

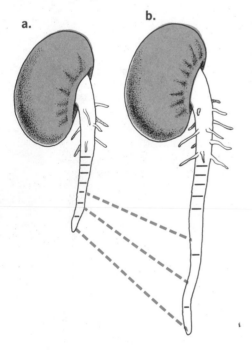

a.   b.

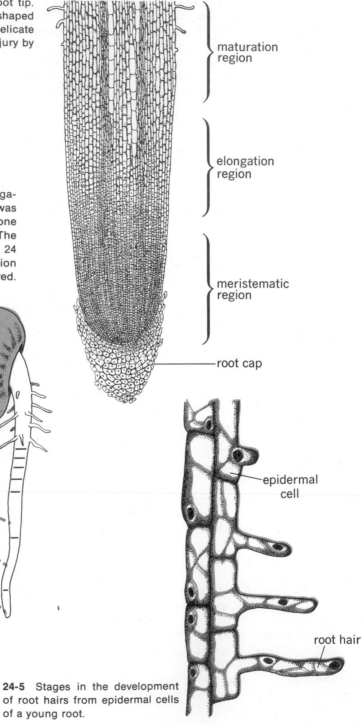

maturation region

elongation region

meristematic region

root cap

epidermal cell

root hair

24-5 Stages in the development of root hairs from epidermal cells of a young root.

Following cell elongation, the cells differentiate into tissues. This occurs in the *maturation region*. One of the most noticeable changes occurs in the outer cell layer that becomes an epidermis. Tiny projections appear on the outer cell surfaces and lengthen to become *root hairs*. While you need a microscope to see their epidermal origin (see Figure 24-5), if you use a hand lens, you can see the root hairs clearly as fuzzy outgrowths extending along a zone approximately 25 to 50 millimeters long. As the root moves forward, new root hairs are continually forming, and older ones farthest from the tip wither away. Thus, with forward progress of the tip, root hairs are constantly extending into new soil areas. As they extend from epidermal cells, they make close contact with the soil and greatly increase the absorptive surface. Water and minerals pass freely through their thin membranes.

Primary tissues of the root

While epidermal cells are forming root hairs, cells deeper in the root are maturing into other specialized tissues that are distinct both in structure and in function. Since the cells originated in the meristematic region and are the first tissues that develop in the young root, we call them *primary tissues*. These tissues occupy three well-defined regions, as shown in the cross section of a young buttercup root in Figure 24-6.

The outermost region, the *epidermis*, is a single layer of thin-walled cells. The absence of root hairs indicates that the section was made behind the root-hair zone. After the root hairs dis-

24-6 The tissues of a young buttercup root. From what tissue does the secondary root arise?

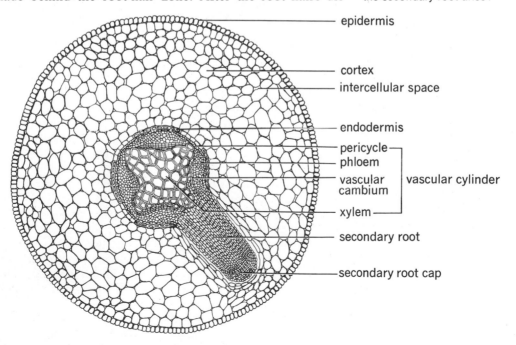

- epidermis
- cortex
- intercellular space
- endodermis
- pericycle
- phloem
- vascular cambium
- xylem
- vascular cylinder
- secondary root
- secondary root cap

appear, the epidermis continues to function as both an absorptive and a protective layer.

Within the epidermis is a thick *cortex* region composed of layers of storage parenchyma cells. The cells of the cortex are rounded and loosely packed, with numerous intercellular spaces. Water and minerals in solution pass through the intercellular spaces as well as through the cells of the cortex of a young root. In many roots, the cortex cells contain starch and other food substances transported from the leaves and stem to the roots for storage. The *endodermis* is a single layer of cells forming the inner boundary of the cortex. The walls of endodermal cells are thickened with waxy materials, especially along the transverse walls of adjacent cells. This prevents the flow of solutions between the cells.

The innermost region, or central core, is the *vascular cylinder*. In many roots, including the buttercup, *xylem* cells are arranged in the form of a star with radiating arms in the center of the vascular cylinder. The thick-walled xylem vessels are channels through which water and minerals are conducted upward through the root to the stem and leaves. *Phloem* cells lie in groups between the arms of xylem. Sieve tubes of the phloem are channels through which food substances are conducted downward through the root.

A very important ring of parenchyma cells, the *pericycle,* lies at the outer edge of the vascular cylinder, adjacent to the endodermis. In most roots, the pericycle is one cell thick. Unlike most other plant tissues, cells of the pericycle retain their capacity for growth and cell division after they have matured. The pericycle is extremely important, because it is from this tissue that secondary roots, as well as other tissues we shall refer to later, arise. Notice the origin of the secondary root pushing its way through the cortex of the buttercup root shown in Figure 24-6.

Secondary growth in roots

Following the maturation of primary tissues, the roots of dicotyledons and conifers undergo further growth resulting from the formation of **secondary tissues.** This growth increases the *diameter* of the root rather than its length.

Certain parenchyma cells lying between the xylem and phloem groups are modified and become meristematic. This important layer, one cell in thickness, is the **vascular cambium.** Cells of the vascular cambium form two kinds of secondary tissues by cell division. When a cambium cell divides, two cells result. One of these cells is still meristematic and remains part of the cambium. Its mate, however, matures. Cells maturing on the inner side of the cambium become *secondary xylem. Secondary phloem* cells mature on the outer side.

The addition of secondary tissues increases the diameter of the vascular cylinder. This expansion crushes the endodermis, cor-

Table 24-1 SUMMARY OF ROOT TISSUES AND THEIR SPECIAL FUNCTIONS

TISSUE	FUNCTION
epidermis	absorption and protection
root hairs	increase in absorptive area of epidermis
cortex	diffusion of water and minerals to vascular cylinder and storage of food and water
endodermis	boundary layer separating cortex from vascular cylinder
vascular cylinder	
pericycle	origin of secondary roots and formation of cork cambium and periderm of fleshy roots
phloem	conduction of food materials downward from leaves and stem
vascular cambium	formation of secondary xylem and phloem, resulting in growth in diameter
xylem	conduction of water and dissolved minerals upward to stem and leaves

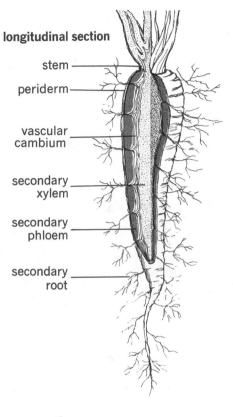

longitudinal section

stem
periderm
vascular cambium
secondary xylem
secondary phloem
secondary root

tex, and epidermis against the soil. Later in secondary growth, the pericycle forms a *cork cambium* that produces layers of cork cells around the vascular cylinder. This cuts off all remaining cells outside of the vascular cylinder. Thus, after secondary thickening, a root ceases to be an organ of absorption but remains an organ of conduction, storage, and anchorage for a plant.

Fleshy roots, such as the carrot shown in Figure 24-7, are composed mostly of secondary xylem and phloem. In this case, however, the secondary tissues formed by the cambium are thin-walled rather than woody and become reservoirs of stored food. This makes the carrot fleshy and edible. The outer covering is a skinlike *periderm* that develops from the pericycle. Secondary roots, formed earlier by the pericycle, extend from the xylem, through the phloem and periderm, and into the soil.

The tissues of a root (listed from outside to inside) and their functions are summarized in Table 24-1.

Modified roots

The root tissues we have described are characteristic, generally, of all roots. However, the structure of roots varies widely with the environment in which they live. Floating plants such as the duckweed and water hyacinth have *aquatic roots* that lack root hairs. The bald cypress tree of southern United States frequently grows in swamps, with its roots submerged in shallow water. Under

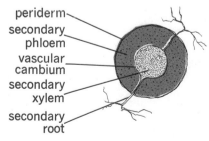

cross section

periderm
secondary phloem
vascular cambium
secondary xylem
secondary root

24-7 Tissues of a fleshy root. As you study the tissues in the longitudinal section of the carrot root, locate the same tissues in the cross section.

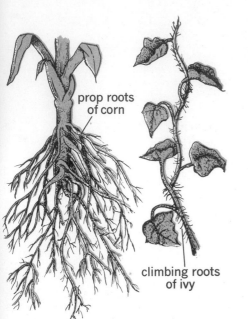

prop roots
of corn

climbing roots
of ivy

24-8 Two types of adventitious roots that develop from stem tissues.

these conditions, the lateral roots produce upright conical growths, known as cypress "knees." These curious structures may aerate the submerged roots, although this has never been established. It is interesting, though, that "knees" do not form when the bald cypress grows in drier places.

Many plants, especially in the tropics, produce *aerial roots*. Tropical orchids live on trees and absorb water from the humid atmosphere, dew, and rainwater. The thick, spongy cortex aids in rapid water absorption. Dust, bark fragments, and other debris that collect around the roots are sufficient to supply the mineral requirements of the plants.

In some plants, *adventitious roots* develop from tissues of the stem or even from the leaves. You may have noticed the circles of roots that grow from the lower joints of a corn stem. These adventitious roots are called *brace roots,* or *prop roots*. They grow into the ground and help the underground roots support the stem. If soil is piled around the stem, additional brace roots develop from the next joint above the soil line. Even the soil roots of the corn are adventitious, since they grow from the stem and replace the short-lived primary root. Brace roots of the banyan tree, a large tree of warm climates, grow downward from the branches and serve as additional trunks in supporting the large, widely spreading limbs.

Poison ivy, English ivy, and other vines produce clusters of adventitious *climbing roots* along their stems. These roots cling to trees or walls and anchor the stems securely. Such plants also have ordinary soil roots that absorb water and minerals.

Plant stems

While the root is growing from one part of the seed embryo, another part of the embryo is developing a stem. The stem and the leaves it will bear compose the *shoot*. In most plants, the stem and leaves are aerial and are the most conspicuous part of the plant.

Stems, like roots, grow in many forms, sizes, and places. The stem of the carrot is nothing more than a short disk at the top of a large taproot. The trunks of large forest trees towering hundreds of feet into the air are also stems. Some stems live but a few weeks; others grow for centuries. Stems may be underground food storehouses or organs that creep horizontally in the upper soil layer.

The stem and roots form a closely integrated whole. Their functions, while distinct and separate, are closely related. Likewise, the tissues composing stems and roots are similar, although both organs have distinguishing structural characteristics.

External structure of a woody stem

The twig of a tree is an ideal subject with which to begin the study of stem anatomy. A dormant twig such as the one shown in Figure 24-9 is especially suitable.

Stems differ from roots in having *nodes.* These are points along the stem where leaves are produced and branches are formed. The intervals between nodes are called *internodes. Leaf scars* at the nodes of dormant twigs mark the places where leaves were attached. They may be circular, oval, shield-shaped, or crescent-shaped, depending on the species. Twigs with an *alternate* leaf arrangement have a single leaf scar at a node. Those with *opposite* leaf arrangement have two leaf scars situated on opposite sides of a node. If there are more than two leaf scars at a node, the leaf arrangement is referred to as *whorled.* Only a few trees, including catalpa, have the whorled leaf arrangement. It is more common in herbaceous plants.

If you examine the surface of a leaf scar with a magnifying lens, you can see tiny dots, or closed pores. These are *bundle scars* and are the ends of strands of conducting tissue through which water and dissolved substances passed to and from the leaf. The shape, number, and arrangement of bundle scars are characteristic of a species and are often used in identifying trees.

Buds are the most noticeable structures of a dormant twig. A large *terminal bud* is usually found at the tip. Along the sides of the twig, situated at the nodes, are smaller *lateral buds.* Most lateral buds are located just above leaf scars. During the growing season, these buds develop in the angle between the leaf stalk and the stem, or in the leaf axil. We call them *axillary buds.* Axillary buds, like leaves, may be alternate, opposite, or whorled in their arrangement at the nodes. Some trees produce an additional *accessory bud* above or beside an axillary bud. These buds usually do not develop unless the axillary bud fails to grow. Overlapping *bud scales* cover the tissues contained in buds and protect them from drying out and from mechanical injury during the dormant period. The number and arrangement of bud scales vary in different species.

When terminal buds swell and their scales drop off at the beginning of a growing season, a series of rings encircling the twig mark the places where the bud scales were fastened. These *bud-scale scars,* at intervals along the branch, show the exact locations of the terminal bud at the start of previous growing seasons. Thus, by starting at the present terminal bud and counting the sets of bud-scale scars along a twig, you can determine the age of the twig.

Along the internodes, especially on young twigs, you can see small, corky areas called *lenticels* on the surface. These are pores that allow the interchange of gases between the internal tissues and the atmosphere.

Some twigs bear characteristic thorns that make them easy to identify. Thorns may be short and broad, long and pointed, or branching. Thorns have different origins in different plants. In the rose, thorns are outgrowths of the epidermis; thorns may also be modified branches, as in the hawthorn or honey locust.

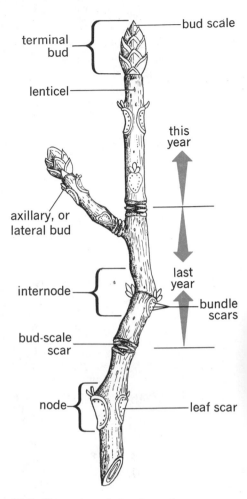

24-9 The external structure of a woody stem.

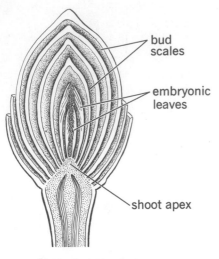

bud
scales

embryonic
leaves

shoot apex

24-10 A longitudinal section through the terminal bud of a woody plant.

Structure of a bud

If you cut a section of a terminal bud lengthwise, as in Figure 24-10, you will see various regions and structures of the embryonic stem. At the tip of the stem, deep inside the bud scales, is a conical mass of meristematic tissue, the **shoot apex.** It is in this region that young cells divide rapidly early in the growing season and form the cells that will mature into the primary tissues of the stem. Cells in the region just below the shoot apex continue dividing, then elongate rapidly. These regions of the embryonic stem are comparable to the corresponding regions of the root tip. Embryonic leaves that have developed from the shoot apex are folded in a cluster inside the bud scales.

Buds vary in their content. Those containing a shoot apex and embryonic leaves are called *leaf buds. Flower buds* contain only embryonic flowers that have developed from the shoot apex. Other buds contain a rudimentary shoot and leaves and flowers, and are called *mixed buds.*

The growth of stems

With the coming of spring in temperate climates, the inner bud scales lengthen rapidly, and the bud opens. Elongation of cells below the shoot apex lengthens the shoot rapidly. Nodes that were compressed in the bud are separated by elongation of cells in the internode. You might compare this rapid growth with the extension of a folded telescope. Within a few weeks, a stem may lengthen from a few inches to several feet. Growth of a shoot from a terminal bud lengthens the stem axis. Growth from the lateral buds forms branches.

Growth from terminal buds is often greater than that from lateral buds. We call this condition **apical dominance.** Biologists have found that chemical substances, known as growth hormones, can *stimulate* the growth of terminal shoots and *retard* or *prevent* growth from lateral buds. We will discuss these hormones and their influence on growth in Chapter 26.

24-11 The spruce (left) illustrates the excurrent growth habit, in which lateral branches arise from the trunk, which acts as a central shaft. In contrast, the oak illustrates the deliquescent growth pattern, in which the trunk is divided into several large main branches. (Herbert Weihrich)

If a young tree has a terminal bud and strong apical dominance, a season's growth of the main stem will far outdistance that of the lateral branches. This produces the *excurrent* growth habit, in which the trunk extends through the tree as a central shaft. Smaller lateral branches are formed at intervals along the trunk. The pine, spruce, hemlock, and redwood are examples of this growth habit. The long, straight trunk makes these types of trees extremely valuable for timber.

The cottonwood, willow, and elm are examples of a different form of branching. These trees either lack apical dominance or do not form terminal buds. The trunk divides into several large branches, producing a spreading, spraylike growth pattern referred to as *deliquescent.*

Primary and secondary growth of woody stems

Two distinct kinds of growth occur in woody stems. *Primary growth* includes the formation of cells in the shoot apex, the growth and lengthening of cells in the young shoot, and the differentiation of cells into primary tissues. Primary tissues form immediately after the cells elongate. They are similar to those of the root and, from outside to inside, include the *epidermis, cortex, primary phloem, primary xylem,* and *pith.* Primary growth lengthens or increases the height of a stem. It occurs rapidly and continues for only a few weeks in most woody stems.

Secondary growth results from the formation of *secondary phloem* and *xylem* by the *vascular cambium.* This growth follows primary growth and increases the diameter of the stem. It continues season after season.

To illustrate both forms of growth, imagine a tree thirty feet in height, twelve inches in diameter, and with its lowest branch six feet from the ground. After ten years, the tree has grown to fifty feet in height and is now sixteen inches in diameter. How far above the ground is the first limb? The answer is *still six feet.* Twigs have lengthened at the tips, but the trunk and branches below the terminal meristematic regions have only increased in diameter.

Regions of a woody stem

If a woody stem is cut crosswise, several regions are clearly visible. The outer region is the *bark.* Inside the bark is the *wood.* During the first year of growth, the *vascular cambium* develops between the bark and wood and adds to the thickness of these regions season after season. The innermost region of the stem is soft *pith.* Pith is a primary tissue; hence, the amount of pith does not increase as the stem grows in diameter. It is often difficult to find the pith in an old woody stem, because it is so small and because of the large amount of wood that has been formed around it.

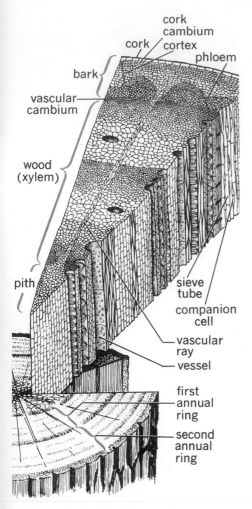

24-12 The tissues of a three-year-old dicotyledonous stem.

Tissues composing bark

The bark of a woody stem includes more than the hard outer covering. It is composed of several tissues, each with a specialized function.

The young twig is covered for a time by a thin epidermis that develops as a primary tissue. During the first growing season, however, cells of the epidermis or the outermost layer of the cortex just beneath it give rise to the *cork cambium.* Repeated divisions of meristematic cells of the cork cambium build up layers of *cork* cells on its outer side. Cork cells are short-lived and soon become air-filled cavities. Cork tissue resists the passage of water and gases from the stem tissues. It also forms a protective barrier against insect attack and invasion of the stem tissues by fungi and other parasitic organisms. In addition, it is an insulator and protects the stem from sudden temperature changes. The thick cork layers of some trees provide protection even from the heat of ground fires. The cork cambium adds new cork on the inside as it is weathered or destroyed on the outside. It also repairs cracks that develop as the stem increases in diameter. We often call cork *outer bark* in order to distinguish it from deeper tissues of the *inner bark.*

Inside the cork and cork cambium is the *cortex,* an inner-bark region composed of thin-walled parenchyma cells. When the stem is young and still covered by an epidermis, these cells contain chloroplasts and carry on photosynthesis. With the formation of cork, the cortex gradually becomes a region of food storage and may disappear as the stem ages.

The inner bark also contains vascular and strengthening tissues. These cells compose the *phloem.* In many stems, including the one shown in Figure 24-12, the phloem cells lie in groups shaped like pyramids. Phloem groups are separated by *phloem rays,* which may be narrow or V-shaped.

Cells of the phloem are of several kinds. Large *sieve tubes* are channels of food conduction, primarily downward through the stem. Smaller *companion cells* lie alongside the sieve tubes. *Phloem parenchyma* cells, which store food temporarily, are scattered among the other phloem cells. Areas of sclerenchyma cells compose *phloem fibers* that strengthen the inner bark.

The vascular cambium

As in the root, the *vascular cambium* is a ring of meristematic cells that forms between the bark and wood during the first growing season. By forming secondary phloem tissues on its outer side and wood, or *xylem,* on its inner side, the vascular cambium increases the diameter of the stem. During each season of cambium activity, many more wood cells than phloem cells are formed. That is why the wood area of a tree is always much greater than its bark thickness.

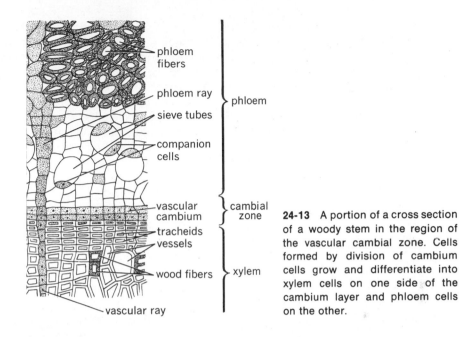

phloem fibers

phloem ray

sieve tubes

companion cells

} phloem

vascular cambium

} cambial zone

tracheids

vessels

wood fibers

} xylem

vascular ray

24-13 A portion of a cross section of a woody stem in the region of the vascular cambial zone. Cells formed by division of cambium cells grow and differentiate into xylem cells on one side of the cambium layer and phloem cells on the other.

Tissues composing wood

The wood includes all of the stem tissues between the vascular cambium and the pith. It is composed of several kinds of xylem cells that function largely in conduction and strengthening of the stem. As in the root, the largest xylem elements are *vessels,* which are long conducting tubes composed of nonliving cells joined end to end. The end walls of cells composing vessels dissolve, forming a continuous passageway through which water and minerals flow upward through the stem. *Tracheids* are smaller than vessels and function both in conduction and support. They are the principal cells composing the wood of conifers. Tracheids are elongated cells with tapering end walls that overlap. Their thick walls contain numerous *pits* through which water flows laterally from one tracheid to another. *Xylem fibers* resemble tracheids but are longer and more tapering. They have little internal cavity, and therefore function in support rather than in conduction. *Vascular rays* extend from the pith through the wood and join phloem rays in the bark. They are sheets or ribbons of parenchyma cells radiating through the wood like the slender spokes of a wheel. Rays vary from one to several cells in width in different kinds of wood. They function as avenues of conduction between the pith, wood, and bark.

As woody stems increase in thickness year after year, the wood formed by the cambium is arranged in layers. Frequently, *spring wood,* produced early in the season, contains numerous large vessels mingled with fibers and tracheids. *Summer wood* often contains fewer large vessels and numerous fibers. In conifers, the tracheids composing spring wood are larger than those of the summer wood. The difference in texture of spring and summer wood results in layers that appear as *annual rings*. By counting these rings, you can determine the age of a tree (Figure 24-14).

As a woody stem increases in diameter year after year, the older layers of wood toward the center cease to function in conduction and receive deposits of resins, gums, tannins, and other substances. These deposits harden the wood and frequently darken its color. This inactive wood constitutes *heartwood*. The active, functioning wood outside of the heartwood is *sapwood* (Figure 24-15).

The pith region of a stem

A central core of *pith* occupies a proportionally large area of a young woody stem but becomes less and less noticeable as the stem increases in diameter. Pith is a parenchyma tissue and functions in the storage of foods, especially in the young stem. Since

24-14 The relatively large tracheids (left) were formed in the spring whereas the smaller tracheids were formed during the summer months of the previous year. Because the layers of spring and summer wood are distinguishable visually, it is possible to see each year's growth as an annual ring. (courtesy International Paper Co.)

24-15 The dark heartwood and the lighter sapwood are clearly distinguishable in this cross section of a tree trunk. How would you determine the age of the tree at the time it was felled? (Allan Roberts)

Table 24-2   SUMMARY OF STRUCTURE AND ACTIVITIES OF A WOODY STEM

REGION	TISSUE OR CELL TYPE	ACTIVITY
bark	epidermis (only on young stems)	protection, reduction of water loss
	cork	protection, prevention of water loss
	cork cambium	production of cork
	cortex (only in young stems)	storage and food manufacture
	sieve tubes	conduction of food, usually downward
	companion cells	uncertain
	phloem fibers	support
vascular cambium	meristematic cells only	formation of phloem, xylem (wood), and rays
wood	xylem vessels	conduction of water and minerals upward
	tracheids	conduction and strengthening of wood
	xylem fibers	support
	vascular rays	conduction laterally
pith	parenchyma	storage

pith is not produced by the cambium, it does not increase in amount as the stem grows in diameter.

Table 24-2 summarizes the tissues of a woody stem and their functions.

Herbaceous dicotyledonous stems

In many respects, the stems of herbaceous dicotyledons such as the clover, alfalfa, buttercup, and bean (shown in Figure 24-16, left) are like woody stems, especially during the first season of growth. The same general regions are present, and the tissues are similar, both in structure and function. The principal difference is in the length of time the cambium functions, and therefore the amount of secondary xylem and phloem produced.

The outer covering is a thin *epidermis*. Cork is usually lacking in herbaceous stems. Many herbaceous stems, including that of the bean, have a *cortex* region just beneath the epidermis. Chlorenchyma cells composing the cortex are centers of photosynthesis and storage.

24-16 Compare and contrast the sections of the herbaceous dicotyledonous (bean) stem and fibrovascular bundle with those of the herbaceous monocotyledonous (corn) stem and bundle.

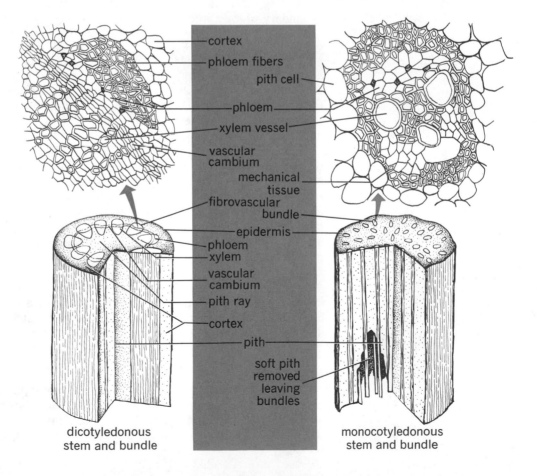

cortex
phloem fibers
pith cell
phloem
xylem vessel
vascular cambium
mechanical tissue
fibrovascular bundle
epidermis
phloem
xylem
vascular cambium
pith ray
cortex
pith
soft pith removed leaving bundles

dicotyledonous stem and bundle

monocotyledonous stem and bundle

The arrangement of xylem and phloem varies in herbaceous stems. In many stems, they occupy groups or strands known as *fibrovascular bundles*. The arrangement of fibrovascular bundles in a ring-shaped zone is characteristic of dicotyledonous stems. The phloem lies on the outer side of a fibrovascular bundle. It is composed of sieve tubes and companion cells. A mass of phloem fibers frequently forms a cap toward the outside of the phloem region. It is this mass of fibers that is removed from the flax stem and used in processing linen thread. Xylem tissue occupies the inner region of a fibrovascular bundle. Thick-walled vessels often lie in rows, separated by parenchyma cells.

A somewhat different arrangement of vascular tissues is found in many herbaceous stems. In these stems, individual fibrovascular bundles cannot be distinguished. Phloem and xylem tissues form continuous cylinders in the fibrovascular zone of the stem.

The development of a cambium is characteristic of most herbaceous dicotyledonous stems. It may develop between the phloem and xylem of each bundle or may form a complete ring separating these tissues. Activity of the cambium adds layers of secondary phloem and xylem.

Monocotyledonous stems

The corn stem shown in Figure 24-16, right, is a typical monocotyledonous stem. It differs from that of an herbaceous dicotyledon in several ways. The fibrovascular bundles are *scattered through the stem*. There is *no cortex*. The stem also *lacks a cambium*.

The corn stem is covered by a thick-walled epidermis. Just beneath the epidermis is a thin layer of sclerenchyma cells that add strength and hardness to the outer region, often called the *rind*. The most abundant tissue is pith, composed of thin-walled, loosely packed parenchyma cells. Numerous fibrovascular bundles are scattered through the pith and are most abundant near the outside of the stem. Each bundle contains an area of phloem sieve tubes and companion cells toward the outside and xylem tissues, including several large vessels, toward the inside. A ring of thick-walled sclerenchyma cells, or *mechanical tissue*, forms a sheath that surrounds the bundle. Since a vascular cambium does not develop in a monocotyledonous stem, growth in diameter is limited and the stem becomes long and slender.

Modified stems

Not all stems are aerial and leaf-bearing. Many plants produce underground stems from which aerial parts arise. They often serve as underground food storehouses and, in many cases, reproduce the plant vegetatively, a function discussed more fully in Chapter 27.

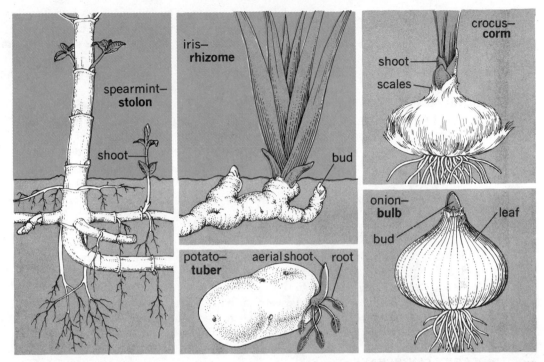

24-17 Five types of modified stems. Why are they stems rather than roots?

The underground stems of the iris, lily of the valley, and water lily are known as *rhizomes*. Most rhizomes are thick and fleshy and grow horizontally near the soil surface. Leaves or leaf-bearing branches are produced at the nodes. Frequently, rhizomes contain starch and other foods.

A *stolon* is similar to a rhizome but is more slender. Stolons form branches at the base of the aerial stem and grow underground for some distance. The tips of stolons may grow upward and form aerial branches. Certain grasses, including crab grass, form stolons and are problems in lawns and flower beds because they spread the plants so rapidly.

The white potato is a familiar example of a *tuber,* a greatly swollen terminal portion of a stolon. The "eyes" of a potato are buds that develop at the nodes. A scar at one end of the potato marks its place of attachment to a stolon.

A *corm* is a short, thick, underground stem enclosed in several thin scale leaves. Its fleshy tissues are a food reservoir. In some plants, a corm produces a new aerial shoot each season. The corm enlarges each year. In others, including the gladiolus, a corm produces only one aerial shoot. During the growing season, lateral buds at nodes of the corm develop into new corms that will grow the following season.

A *bulb* consists of a shortened basal stem from which thick scale leaves grow in circular layers. In a sense, a bulb is a compressed underground shoot. The thick leaves are sites of food storage. The aerial stem grows through the center of the scale leaves from a bud on the shortened stem. The onion is a familiar example of a bulb. Other plants forming bulbs include the lily, hyacinth, daffodil, and tulip.

SUMMARY

Wouldn't it be interesting if a tree had a transparent trunk and you could look in and see the many internal activities on a summer day? Phloem tubes of the inner bark would be conducting a flow of water and dissolved foods downward from the leaves to the roots. Xylem vessels of the sapwood would be conducting a continuous stream of water from the roots upward to the uppermost branches and into the leaves.

Many root and stem tissues are similar. Others are different, as the environments of the two plant organs are different. As you studied the structure of roots and stems, you undoubtedly discovered that each function of these organs involves a specialized tissue. It is the development of these tissues that makes the seed plants the most advanced and successful of all land plants.

What forces are involved in the intake of water by a root and conduction of columns of water 100 feet or more upward to the top of a large tree? What happens to the water when it reaches the leaves? These, and many other questions regarding the water relations of plants will be answered in the next chapter.

BIOLOGICALLY SPEAKING

primary root	epidermis	apical dominance
secondary root	cortex	excurrent
taproot	vascular cylinder	deliquescent
fibrous root	secondary tissues	primary growth
root cap	vascular cambium	secondary growth
meristematic region	shoot	bark
elongation region	node	wood
maturation region	internode	pith
root hair	shoot apex	rhizome
primary tissues		

QUESTIONS FOR REVIEW

1. From the standpoint of origin, distinguish between primary and secondary roots.
2. Describe the general form of a taproot and of a fibrous root system.
3. Locate four regions of a root tip and describe the activities that occur in each.
4. Name and describe the primary tissues of a root as they occur, from outside to inside, and identify each with a specialized function.
5. Locate the vascular cambium in a root.
6. Describe the formation of secondary tissues by the vascular cambium.
7. Give several examples of adventitious roots.
8. Distinguish between terminal, axillary, and accessory buds.
9. From what principal dangers do bud scales protect the shoot apex during the winter season?

10. Classify buds into three types on the basis of the rudimentary structures they contain.
11. Describe two branching habits of trees and explain the relation of each to apical dominance or the lack of it.
12. Name the bark tissues of a woody stem.
13. Describe various cells composing wood.
14. Account for the difference in appearance of heartwood and sapwood.
15. Explain the formation of annual rings in wood.
16. Explain why the pith in a woody stem does not increase in amount as the tree grows in diameter.
17. What chlorophyll-containing tissue composes the cortex of many herbaceous dicotyledonous stems?
18. Describe the arrangement of fibrovascular bundles in the stem of a herbaceous dicotyledon.
19. Describe three structural characteristics of dicotyledonous stems that distinguish them from monocotyledonous stems.
20. Name and describe five forms of underground stems.

APPLYING PRINCIPLES AND CONCEPTS

1. Discuss various advantages of taproot and fibrous root systems.
2. Describe several factors involved in the forward progress of a root tip through the soil.
3. Explain why growth occurs only in certain regions of roots and stems rather than throughout the plant body.
4. Secondary roots originate from cells of the pericycle at the outer edge of the vascular cylinder, adjacent to the xylem. Why is this point of origin of importance?
5. Distinguish between terminal growth and growth in diameter in a woody stem.
6. Using a woody stem as an example, discuss specialized tissues as they relate to functions of the stem.

RELATED READING

Books

GREULACH, VICTORIA A., *Plants, An Introduction to Modern Botany* (3rd ed.).
John Wiley and Sons, Inc., New York. 1966. An interesting, up-to-date account of the more important principles and concepts of botany.
FULLER, HARRY J., and ZANE B. CAROTHERS, *The Plant World* (4th ed.).
Holt, Rinehart and Winston, Inc., New York. 1963. An extremely well-written discussion of the general biological principles of the plant world, including details on the physiology of the flowering plants.
TRANSEAU, E. N., and others, *Textbook of Botany* (rev. ed.).
Harper and Row, Publishers, New York. 1953. An older, but well-written text, useful as supplementary material for class work on roots and stems.

Articles

GROSVENOR, M. B., "World's Tallest Tree Discovered," *National Geographic Magazine*, July, 1964. An interesting article describing the discovery made on the West coast of the United States.

CHAPTER TWENTY-FIVE

WATER RELATIONS IN PLANTS

Plants and water

No factor in the environment of a seed plant is more critical than the water supply. The annual precipitation largely determines whether an area of the world will be a desert or semidesert, a short-grass plain or prairie, or a forest.

In Chapter 3, we referred to water as an inorganic compound essential in many ways to living things. In our discussion of plants and water, we will review some of the reasons why water is essential to the plant.

Water is required for nearly all of the physiological processes of a plant. In the study of photosynthesis, you found that water supplies the hydrogen that combines with carbon dioxide to form sugar. It is equally important in digestion, where it is combined with various foods in the process of *hydrolysis*. It is the medium in which molecules of proteins, fats, and other substances are dispersed in the organization of various colloidal materials that compose cell structures; it is therefore necessary for growth.

Water is a medium of transport in the plant. Soil minerals reach the plant tissues in a water solution. Foods and other organic substances dissolved in water are transported from the leaves and green stem tissues to places of storage. Water pressure in the cells of nonwoody plant tissues makes them firm and rigid. Thus, the plant requires a constant supply of water.

Soil and soil water

Generally, soil is a combination of varying amounts of mineral particles and organic matter. It also contains water and air in the spaces between the particles. In addition to its lifeless components, most soil contains a community of organisms, including bacteria, molds and other fungi, protozoans, and a great variety of worms, insects, and other animals.

Soil texture is determined by the nature and size of its mineral particles. These particles, ranging from the largest to the smallest, may be *gravel, sand, silt* (fine sand), or *clay*. Frequently, these particles are found mixed in a type of soil known as *loam*. Ideal agricultural loam, most valuable for growing most crops, is a mixture of about 40 percent sand, 40 percent silt, and 20 percent clay. These proportions vary, so loams may be sandy loam, silt loam, or clay loam.

Soils also contain varying amounts of organic matter, especially in the upper region, which is often called *topsoil*. This organic matter comes from decaying roots and soil organisms as well as from the remains of plants and animals that decay on the surface. This decomposing organic matter is referred to as *humus*.

Pore spaces that fill with water and air are important in determining the suitability of soil for plant growth. The size and number of pore spaces vary with the nature of the mineral particles and the amount of organic matter present. Sandy soils have large pore spaces, while clay soils are heavy, with the mineral particles packed closely together. Organic matter distributed among the mineral particles loosens the soil and increases the number of pore spaces. The texture and composition of soil are important in determining how rapidly it receives water and how much it can hold and make available for absorption by plant roots. Water passes through sandy soils rapidly because of the large and numerous pore spaces. On the other hand, water moves very slowly through heavy clay soil.

When water enters the soil from the surface, much of it passes through the spaces between the soil particles and reaches the *water table,* or the level at which water is standing in the ground. This *gravitational water* is usually not available to plant roots, for it lies below their reach.

Water that is held in the smaller pore spaces and around soil particles against the force of gravity is called *capillary water.* Following a rain, capillary water moves through the pore spaces from the upper region of the soil to deeper areas. As the soil dries on the surface, capillary water moves upward into the drier region. Even after the movement of capillary water has ceased and the soil surface is dry, a large amount of water is held around the soil particles. It is this capillary water that is available for absorption by roots.

After soil has lost its capillary water and can no longer supply available water to roots, it still contains *hygroscopic water.* This water is held in a molecular film around each soil particle. Hygroscopic water moves from the soil particles to the atmosphere as vapor, but it is never available to plants.

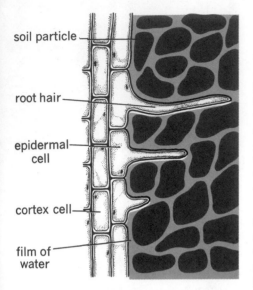

soil particle

root hair

epidermal
cell

cortex cell

film of
water

25-1 Water enters the epidermal cells of the root by active absorption when the plant is transpiring slowly and the moisture content of the soil is high; it enters by passive absorption when water movement through the plant is rapid and the rate of transpiration is high.

Absorption of water by the root

Water normally enters a root through the root hairs of epidermal cells. Figure 25-1 shows root hairs that have pushed through the spaces between soil particles. Normally, capillary water passes from the pore spaces of the soil into the root hairs, and epidermal cells. The mechanisms involved in this water movement are not thoroughly understood. Furthermore, biologists believe that two different mechanisms are involved in water absorption and their relative importance changes as conditions within the plant and in the soil vary. In studying these mechanisms, we must consider water movement through the entire plant to the atmosphere as well as the amount of available capillary water in the soil.

At night, when the stomata are closed, the plant is transpiring at a slower rate, and the moisture content of the soil is high, absorption seems to be primarily by osmosis. The root hairs and epidermal cells contain minerals, food materials, and other substances in water solution. The soil water also contains dissolved minerals, but usually in lower concentration than that of the cell solution. Thus, the concentration of water molecules is greater in the soil than it is in the root hairs and epidermal cells. Water moves into the root by osmosis. Water intake through the mechanism of osmosis is called *active absorption* because it depends on the water and solute concentration of solutions in living epidermal cells.

Water movement continues from cell to cell, from the epidermis through the cortex to the vascular cylinder. During active absorption, the cells of the cortex are turgid because of water pressure. However, pressure in the xylem vessels is lower than that of the cortex, because water is rising toward the stem and leaves. As a result, the cells adjacent to the vessels force water into the moving stream. As water is lost from these cells, additional water moves into them from adjacent cortex cells by osmosis. This process continues across the root to the epidermis. Water movement from the epidermal cells to the cortex results in further absorption from the soil.

A different mechanism seems to be responsible for absorption of water under other conditions inside the plant. During the day, when the stomata are open and conditions in the atmosphere cause rapid transpiration, a large volume of water passes through the plant. Cells of the root cortex and epidermis move water to the xylem vessels rapidly and lose turgor pressure as a result. Under these conditions, the pressure in the root may fall below that of the soil water. This pressure difference forces water into the root cells. This mechanism of water intake is called *passive absorption,* because the forces responsible for it are not forces in the root cells. Rather, they are water loss in the leaves and movement through the stem. Passive absorption accounts for most of the water intake of land plants.

Water loss from a root

How would water movement between the epidermis and root hairs and the soil be affected by a high concentration of soluble salts or other substances in the soil water? In this case, the movement of water caused by osmosis would be from the cell to the soil. This may occur if salt water is poured around roots or if a strong application of soluble mineral fertilizer is applied to a plant. Such a reversal of water movement may continue from cell to cell and kill the root.

Some plants, however, live in areas where the salt concentration of water sometimes varies. Certain plants, called *halophytes,* are adapted to soils with a high salt content—sometimes even higher than that of seawater. Root cells of these plants normally have a high salt concentration. The presence of the salts in the cell solutions lowers the water concentration and water diffusion pressure to below that of the soil. Thus, the root cells absorb water from the salty soil by osmosis and maintain normal turgor. In some plants, the leaves even become coated with salt as transpired water evaporates from them.

The absorption of minerals

Soil water usually contains a low concentration of soil minerals in the form of dissociated ions. Among the ions essential to plants are nitrate, phosphate, sulfate, calcium, potassium, and sodium. These ions reach the plasma membranes of epidermal cells in a water solution. However, there is good evidence that the entry of mineral ions is independent of water intake by absorption.

The concentration of ions inside the root cells is usually greater than it is in the soil water. Thus, *active transport* must be operating in mineral absorption. In other words, the cells must be actively expending energy in absorbing minerals. This process is not to be confused with active absorption of water by a root. The assumption that minerals are taken into root cells by active transport is supported by the fact that roots deprived of the oxygen necessary for respiration stop taking in ions. This would indicate that energy is necessary for the process.

Roots have also been found to be selective in mineral uptake. Some ions are absorbed more readily than others. This selective absorption also applies to different species. For example, various kinds of plants growing in the same soil may absorb different amounts of the available ions. As ions accumulate in the outer root cells, they pass to other cells and eventually reach the xylem, where they are transported in a water solution.

How rapidly do plants absorb soil minerals? A graphic answer was given to this question by a representative of the Oak Ridge Operations, United States Atomic Energy Commission. During an address on the use of radioisotopes in biology, the speaker produced two potted tomato plants. One was watered with distilled water, the other with a solution of radioactive phosphate. In less than forty-five minutes, the water containing the radioactive phosphorous had been absorbed through the root hairs of the tomato plant and had moved through the stem into the leaves. This was proved by removing a leaf and demonstrating the presence of radioactivity with a Geiger counter. The plant had shown no radioactivity at the beginning of the demonstration. The control plant showed none either before or after the demonstration.

Translocation of water and minerals

We can account for the movement of water across the cells of a root by osmosis and its entry into the xylem vessels by citing pressure differences, but what mechanisms are involved in the upward motion of columns of water through the stem? The roots might force water upward through the stem of a small plant, but the rise of water one hundred feet or more through the trunk of a tree is another matter. *Translocation* is the movement of water and other materials considerable distances through a plant. It has been a subject of great interest and extensive study for many years.

At one time, it was thought that living cells bordering on xylem vessels pulsated and pumped water upward. However, this explanation of translocation was discarded when it was demonstrated that even dead stems would conduct water.

Undoubtedly, several mechanisms are involved in water conduction. Some are more important than others. One mechanism that seems to be involved is *root pressure*. The force of root pressure can be demonstrated by cutting off the stem of a well-watered plant and observing the bleeding of water from the severed xylem vessels. If a glass tube is connected to the stem with a piece of rubber tubing, as shown in Figure 25-2, water can be seen to rise in the tube from several inches to a foot or more due to root pressure. This force results from active absorption and turgor pressure in the cortex cells. However, biologists have found that root pressure rarely exceeds a few pounds per square inch. Several times this amount of pressure is required to raise water to the top of a tree one hundred feet or more in height. Furthermore, root pressure is high only when the amount of soil moisture is high and the rate of transpiration is low. Water moves most rapidly through the stem when water loss from the leaves is greatest. It has been found, also, that many plants do not bleed from severed stems. For these reasons, biologists do not consider root pressure a major factor in water conduction.

Figure 25-3 illustrates *capillarity,* another possible mechanism in water conduction. When the end of a small tube is placed in a liquid, the liquid is attracted by surface forces along the sides of the tube and rises to a level above that of water in the container. The smaller the diameter of the tube, the higher the liquid will

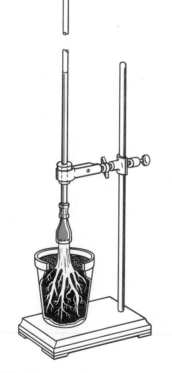

25-2 Demonstrating root pressure. Water rises a considerable distance in a glass tube attached to the stump of an actively growing plant.

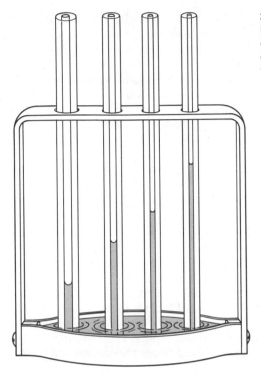

25-3 Demonstrating capillarity. Each of the tubes has a bore of a different size. Explain the difference in heights of the liquids in the tubes.

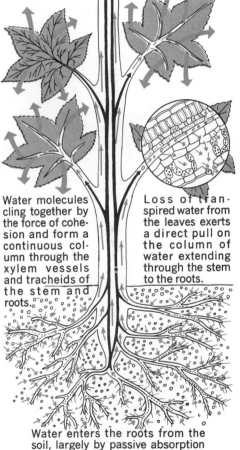

Water molecules cling together by the force of cohesion and form a continuous column through the xylem vessels and tracheids of the stem and roots.

Loss of transpired water from the leaves exerts a direct pull on the column of water extending through the stem to the roots.

Water enters the roots from the soil, largely by passive absorption as water rises through the plant.

25-4 Transpiration-cohesion theory.

rise. In very small capillary tubes, this may be a distance of several feet. If you substitute mentally the many xylem vessels in a stem for capillary tubes, you can see how capillarity might cause water to rise in a stem, at least for a short distance. However, capillarity is now regarded as a minor factor in conduction.

Today, the most acceptable explanation of water translocation is the *transpiration-cohesion theory.* According to this explanation, water is pulled rather than pushed upward through the vessels and tracheids of a stem. Remember that there is a continuous column of water extending from the xylem of the root through the stem, leaf petioles, and veins to cells of the leaf mesophyll. Water passes into the mesophyll cells and evaporates from the cell surfaces as vapor. The water vapor passes through the intercellular spaces and stomata during transpiration.

Loss of water from the leaves exerts a direct pull on the column of water extending through the vessels and tracheids to the roots. The water molecules forming this continuous column have a strong attraction for each other; this attraction is called *cohesion.* Because of this molecular attraction, a cup of water can be filled higher than the brim without spilling. Thus, the transpiration pull in the leaves lifts the entire column upward. The greater the amount of water lost from the leaves, the more rapidly it is pulled through the plant from the roots. When you consider that a giant

Douglas fir or redwood tree lifts columns of water more than three hundred feet above the ground, you have some idea of the enormous force in transpiration pull and cohesion. Biologists have estimated that this force may be as great as 100 atmospheres, or 1,500 pounds per square inch, in some cases.

Mineral ions are carried upward to the leaves with the flow of water through the vessels and tracheids. Certain of these ions may be used in chemical processes in the leaves. Others are translocated from the leaves to other parts of the plant. In this case, the ions move through the phloem.

Translocation of foods

For many years, biologists have known that food materials are translocated from the leaves of the plant to other parts through the phloem sieve tubes. The solute content of sap in phloem tubes has been found to be 10 to 25 percent. Most of this solute content is sucrose, the form in which carbohydrates are translocated from the leaves. Smaller amounts of amino acids, mineral ions, and other substances are also present.

The usual direction of food translocation is downward through the stem and roots. However, the movement may also be upward to flowers and developing fruits. Furthermore, early in the spring, sap containing dissolved foods moves upward through the phloem from places of storage in the roots and lower stem region to the branches and developing buds of woody perennials.

There are several methods of demonstrating that foods are translocated through phloem sieve tubes. *Girdling* is one of the best methods. If a ring of bark is removed down to the cambium, the phloem is destroyed, and food translocation is stopped. The stem and roots below the girdle are deprived of food and gradually die of starvation. The roots may live for from several months to a year or more on reserve food. During this period, they continue to absorb and conduct water and minerals through the xylem. When the food reserves are used up, the roots die of starvation, and the entire tree dies soon after (Figure 25-5). Bark-chewing animals such as mice and rabbits often girdle trees, especially during the winter. Girdling is also done deliberately in order to kill unwanted trees and prevent sprouting from the roots in cultivated areas.

Another, more interesting method of demonstrating translocation of food through phloem sieve tubes was developed by Mittler, a plant physiologist, in 1958. His procedure involved the use of aphids, or plant lice, which feed on the sugar-rich sap from the phloem of young stems. The aphid pierces the bark, or epidermis, with its sharp, tubular mouthpart and inserts it into a phloem sieve tube. Sap oozes through the tube into the aphid's body. Mittler developed a method of anesthetizing aphids while their feeding tubes were inserted. He could then sever the in-

25-5 Photographed at the end of the second season of growth after girdling, you can see that the portion of the trunk above the girdled region has continued to grow, while the portion below the ring has not. The roots of the girdled tree will eventually die of starvation. (from Wilson and Loomis, *Botany,* 4th ed.)

serted tube from the aphid. Drops of sap continued to ooze through the severed tube, and could be collected for analysis in small capillary tubes.

While there is ample evidence that foods are translocated through phloem sieve tubes, the mechanisms involved in the movement are not well understood. Investigators have found that sap moves through the phloem under pressure. They have also measured the rate of flow in isolated sieve tubes and have found it to be about 35 cm per hour. It has also been demonstrated that food translocation is reduced or stopped entirely when oxygen is deficient. This suggests a relationship between food translocation and respiration and energy release and suggests that active transport may be a factor. However, further research is necessary to determine what mechanisms are responsible.

Turgidity in plant cells

As water diffuses into the cytoplasm and vacuoles of plant cells, pressure builds up and presses the protoplast firmly against the wall. We refer to this internal pressure as *turgor pressure*. The firmness of the cell due to turgor pressure is called *turgidity*. You might compare a turgid cell to a plastic bag made rigid by filling it with water. When a plant is fully turgid, the pressure in its cells may be as great as from 15 to 150 pounds per square inch.

Turgor pressure provides mechanical support for nonwoody plant tissues in herbaceous stems, leaves, and flowers. It also permits a delicate seedling to push through soil. Ripening fruits such as tomatoes often split open because of turgor pressure in the cells.

Loss of turgor and wilting

As you learned in Chapter 24, leaf stomata are usually open during the day and closed at night. Thus, they are open when the air is warmest and the relative humidity lowest, and therefore the rate of transpiration is highest. If the soil is well supplied with water, a high rate of absorption will balance a high rate of water loss during transpiration. However, if the rate of transpiration exceeds that of absorption, the plant may be temporarily or permanently injured by water loss.

Temporary wilting occurs when the rate of transpiration exceeds the rate of absorption, even though the soil contains sufficient water to supply the needs of the plant. You have probably seen leaves and soft stems wilt on a hot summer day. In the evening, the air temperature decreases, the relative humidity increases, and the rate of transpiration decreases. The transpiration rate is further reduced by the normal closing of stomata at night. Through the night, absorption continues and gradually builds up the water

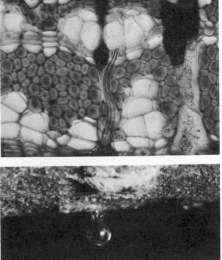

25-6 Demonstrating translocation of food through phloem sieve tubes. An aphid, a drop of sap oozing from its anal opening, feeds on the lower side of a branch. That the aphid is feeding on sap from the phloem can be verified by examination of the photomicrograph, in which the feeding tube can be seen to end in a sieve tube. (courtesy Dr. Martin H. Zimmermann)

25-7 Temporary wilting. During the night, the wilted leaves (left) regain their turgor. (Walter Dawn)

25-8 Guttation along the margin of a leaf. (Walter Dawn)

content of the plant. Leaves and herbaceous stems that were wilted regain their turgor and expand (Figure 25-7). Temporary wilting does not injure a plant unless it occurs often. However, frequent periods of temporary wilting retard normal growth and cause stunting.

Permanent wilting results from exhaustion of available soil water or destruction of absorbing roots. This condition is not corrected by reduction in the rate of transpiration. Permanent wilting occurs during droughts. It can also occur in transplanting unless a sufficient amount of soil is moved to keep the delicate absorbing roots intact.

Guttation

Most of the water lost from a plant escapes by transpiration as vapor. However, some plants force drops of water from their leaves during *guttation*. Guttation usually occurs at night, when the stomata are closed and conditions in the atmosphere have caused reduction of transpiration. Drops of water are forced out of the leaf through special stomata along the margin or at the tip near the ends of veins (Figure 25-8). Guard cells of these stomata are open day and night. Root pressure is believed to be responsible for guttation. Grasses often force out drops of water at the tips of the leaves. Potato, tomato, and strawberry plants show guttation at the leaf margins. Guttation water is not dew. Dew is a condensation of atmospheric moisture on the cool leaf surfaces and on rocks and other objects. Dew forms in a film over the entire leaf surface, whereas guttation water appears only at certain openings.

SUMMARY

What forces are involved in the rise of water through the vessels of a stem, in some instances 100 feet or more? You now have the best answers biologists have been able to find. One force is a push resulting from root pressure. However, this is only a small part of the total force involved. If water isn't forced up from below, then it must be lifted up from above. This pull, involving transpiration and cohesion, seems to be the primary factor involved. As the plant loses water through its leaves, it takes in water through its roots and the entire column is lifted up.

On days when the rate of transpiration is high, water movement through the plant is rapid. At these times, water is literally drawn into the roots and passive absorption occurs. If the rate of transpiration exceeds that of absorption, the leaf cells lose water and temporary wilting occurs. The cells regain their turgor and the leaves stiffen as the rate of transpiration is reduced.

Factors involved in food translocation are not well understood. Active transport involving cell energy may be a primary factor.

BIOLOGICALLY SPEAKING

soil texture	active transport	girdling
pore space	translocation	turgor pressure
gravitational water	root pressure	turgidity
capillary water	capillarity	temporary wilting
hygroscopic water	transpiration-cohesion theory	permanent wilting
active absorption	cohesion	guttation
passive absorption		

QUESTIONS FOR REVIEW

1. List several plant processes that require water.
2. Which component of soil determines its texture?
3. Classify soil types on the basis of texture.
4. Explain why pore spaces in soil are important in absorption of water by a root.
5. Distinguish among gravitational, capillary, and hygroscopic soil water.
6. What forces are involved in active and passive water absorption by root hairs and epidermal cells?
7. Under what conditions may a root lose water to the soil rather than absorb it?
8. What force is believed to be involved in the intake of minerals by a root?
9. List three factors believed to be involved in translocation of water through xylem vessels and tracheids of a stem.
10. Explain how girdling a tree demonstrates that foods are conducted through phloem sieve tubes.
11. In what way is turgor pressure involved in supporting soft plant tissues?
12. Account for temporary wilting of nonwoody plant tissues.
13. Under what conditions can permanent wilting occur?
14. What force in a plant is believed to be responsible for guttation of water?

APPLYING PRINCIPLES AND CONCEPTS

1. Discuss soil texture as it relates to the holding of available water for absorption by roots.

2. Compare the conditions under which roots take in water by active absorption and by passive absorption.
3. Cite evidence to indicate that the intake of mineral ions by a root is independent of water absorption.
4. Compare root pressure, capillarity, and transpiration and cohesion as factors in the translocation of water through the xylem vessels and tracheids of a plant.
5. On what basis might you assume that the factors involved in translocation of foods through phloem sieve tubes are different from those believed to be responsible for the translocation of water through xylem vessels and tracheids?

RELATED READING

Books

HILL, J. BEN, and others, *Botany* (4th ed.).
McGraw-Hill Book Company, New York. 1965. An excellent reference for the study of the relation of water to plants.

JAMES, WILLIAM O., *An Introduction to Plant Physiology* (6th ed.).
Oxford University Press, New York. 1963. A balanced account of the more elementary aspects of plant physiology.

RAY, PETER MARTIN, *The Living Plant.*
Holt, Rinehart and Winston, Inc., New York. 1963. Plants viewed as living organisms—what they do and how they are equipped to do it—with special attention directed to the typical flowering plant.

Articles

GREULACH, V. A., "The Rise of Water in Plants," *Scientific American,* October, 1952. Describes the work of an early investigator in this field, and the work during the following two centuries, which corroborated his guesses.

ZIMMERMANN, MARTIN, "How Sap Moves in Trees," *Scientific American,* March, 1963. A clear presentation on the physiology of translocation.

CHAPTER TWENTY-SIX

PLANT GROWTH AND RESPONSES

Influences on plant growth

To a great extent, the form and size of a plant body are determined by hereditary patterns. Genes exert an influence on the form of the root system, the structure and branching pattern of the stem, the form and arrangement of leaves, the time of reproduction, and the nature of the flowers, fruits, and seeds that will develop.

However, heredity alone does not determine the nature of the plant body. Environmental factors are important external influences on plant growth and development. Among these external influences are light conditions, temperature, atmospheric moisture, soil water, and the mineral content of soil. Any one of these factors may stimulate or limit growth. While all these factors are important, we shall limit our discussion to two of the most important external influences on plant growth: *light* and *temperature*.

Light influences on plant growth

Light has a pronounced effect on plant growth in several ways. As the source of energy for photosynthesis, it has a direct relationship to the food supply. The amount of food available to the tissues, in turn, influences the rate of cell division and the growth of all organs of the plant. As you learned in earlier chapters, light is also necessary for the formation of chlorophyll. The leaves and stems of plants grown in the dark are pale yellow, or *etiolated.* Lack of chlorophyll also reduces or prevents photosynthesis and causes tissue starvation.

Lack of light seems to stimulate elongation of the stem but prevents normal growth and expansion of leaves. Thus, stems of plants grown in the dark are weak and spindly, and leaves are widely spaced and poorly developed (Figure 26-1).

On the other hand, reduced light, as opposed to heavy shade or darkness, may be a growth stimulus. In a moderate degree of shade, the rate of transpiration is reduced more than is the rate

26-1 Compare and contrast the coleus plant on the left, grown under normal light conditions, and that on the right, grown in the dark. (Walter Dawn)

of photosynthesis. This results in an increase in the water content of the growing tissues without a comparable decrease in the rate of food-making. As a result, the stem grows rapidly, and large leaves are produced. As you learned in Chapter 23, light also influences the differentiation of tissues composing the leaf. A leaf exposed to intense sunlight contains more sugar and less water than a shaded leaf of the same plant because of increased rates of photosynthesis and respiration. Elongated palisade cells develop in the mesophyll and may form more than one layer. The amount of conducting tissue increases, and the cuticle thickens. While leaves grown in full sunlight become thicker, the size of the leaf is usually reduced. Leaves grown in shade, on the other hand, contain more water and less food. Reduced light tends to decrease the amount of palisade mesophyll and conducting tissue and increase the amount of spongy mesophyll. Loosely arranged cells of the spongy mesophyll form large air cavities. The surface area of a shade leaf is usually increased.

Photoperiodism

The *duration* of light also has a direct effect on plant growth and on reproduction. The response of plants to varying periods of light and darkness is known as *photoperiodism.*

As you know, the number of hours of daylight and darkness varies with the seasons. On March 21 and September 22, the days and nights are of equal length; that is, each is twelve hours long. From December 21 until June 21, the hours of daylight increase in the Northern Hemisphere. Between June 21 and December 21, the days become shorter.

Extensive studies of plant growth and, especially, reproduction have revealed a striking relationship between flowering and the length of days and nights. Many plants flower in the spring or in the fall, when the days are short and the nights are long. These *short-day* plants include such spring-flowering herbs, shrubs, and trees as the daffodil, tulip, crocus, forsythia, redbud, and dogwood. Correspondingly short days in the late summer and fall are the period in which most chrysanthemums and asters, goldenrods, ragweeds, and poinsettias flower. It is not uncommon for short-day plants such as the forsythia to bloom in the spring and again in the fall, when the hours of daylight, or *photoperiod,* are equal in length.

Another group of flowering plants grows vegetatively when the photoperiod is short and blooms later in the season, when the days are longer. These *long-day* plants flower in the late spring and early summer. Among them are the iris, hollyhock, clover, and such garden vegetables as the beet and radish.

A third group of plants seems to be influenced only slightly, if at all, by the length of days and nights. Flowering starts after a period of vegetative growth, regardless of the length of the photoperiod. Among these *neutral-day* plants are the nasturtium, gaillardia, calendula, marigold, zinnia, snapdragon, carnation, tomato, and garden bean.

Commercial flower and vegetable growers make extensive use of artificial regulation of photoperiods. They have found that the number of hours of darkness may be more important than the hours of light in the flowering of many species. After a period of vegetative growth, flowering can be induced by regulating the photoperiod and hours of darkness.

The influence of temperature on growth

Temperature differences also have a marked influence on plant growth and reproduction. For example, the short days and long nights of early spring and fall are accompanied by lower average temperatures than the days and nights of the late spring and summer months.

Generally, both low and high temperatures tend to affect photosynthesis, translocation, respiration, and transpiration. They also affect growth and reproduction. Between the low and high temperatures at which growth is reduced or ceases entirely, there is an *optimum temperature*. This temperature varies with species, but in most plants it is between 50° and 100°F. Plants may survive at temperatures above or below the optimum, but they will eventually die from prolonged exposure to extremes of heat or cold.

Dormancy in plants

In many regions of the world, there are periods or seasons when the water supply, temperature, or the light conditions are not favorable for active plant growth. During these periods, many plants enter a period of inactivity, or *dormancy*. Woody perennials remain alive above the ground, with their living tissues protected by bark or the scales of winter buds. Herbaceous perennials die to the ground, with only the roots or underground stems surviving through the dormant period. Annual plants do not survive, but the seeds that are produced during a growing season continue the species for the next season.

In many cases, plants grow and flower, then enter a period of dormancy even though conditions are still favorable for growth. For example, daffodils bloom in the spring, then grow for a time. By summer, the leaves have dried, and only the bulb remains alive.

Plant species vary greatly in the lengths of their periods of dormancy. Generally, however, several months of rest are required before growth resumes. In many species, freezing temperatures are necessary to "break" the dormant period and prepare the tissues for growth when external conditions become favorable. In certain plants, drying has the effect of "breaking" dormancy and promoting growth. Gladiolus corms are dug in

26-2 Photoperiodism in a short-day plant. The blooming chrysanthemum was kept alternately, in light for eight hours and in dark for sixteen hours. The other plant only grew vegetatively. (courtesy Department of Floriculture and Ornamental Horticulture, Cornell University, from Wilson and Loomis, *Botany,* 3rd ed.)

the fall and stored in a cool, dry place during the winter. After from six to seven months of dormancy, the corms sprout and are ready for replanting. Similarly, the roots of dahlias are dried and stored from one season to the next.

Plant hormones

To a great extent, plant growth is regulated by *hormones,* which are powerful chemical substances secreted within the plant. Hormones are internal influences, whereas light, temperature, and other environmental factors are external. However, external conditions may influence hormone production in a complex growth-control mechanism.

Hormones differ from other secretions in that they are produced in certain parts of a plant and are translocated to other regions where they influence cell growth and elongation as well as cell division.

The role of auxins

The principal and most widely studied plant hormones are called *auxins.* They act, primarily, in promoting cell elongation and enlargement and in the development of flowers and fruit. Auxins are sometimes called growth hormones, although they act more as *growth regulators,* since their action may be either stimulating or inhibiting. The principal natural auxin produced in plants is *IAA* (indoleacetic acid). This auxin is secreted in largest amounts in growing and developing regions of plant organs. These include the tips of shoots and developing leaves, flowers, and fruits. IAA may also be secreted in cambial cells and in the root tip. From these regions, it is translocated to other parts of the plant.

Different plant tissues react differently to the presence of an auxin. High auxin concentration seems to inhibit stem growth, while lower concentrations act as a stimulus. Furthermore, root tissues are stimulated by a much lower auxin concentration than stem tissues.

Perhaps the best evidence of auxin activity in the tip of a growing shoot is to be seen in experiments with oat seedlings. The primary leaf of the oat and other grasses is enclosed in a protective sheath, or *coleoptile.* As the seedling grows, the leaf normally penetrates the tip of the coleoptile. If the tip of a coleoptile is removed, as shown in Figure 26-3, elongation of the shoot ceases. If the tip is replaced on the decapitated coleoptile, elongation of the cells resumes. This would indicate that auxins secreted in the tip have diffused downward, causing cells to elongate. This assumption can be supported further by placing a coleoptile tip on an agar block into which auxins are absorbed. Placing the agar block on the decapitated coleoptile causes elongation, just as replacing the tip does. If a small agar block containing absorbed

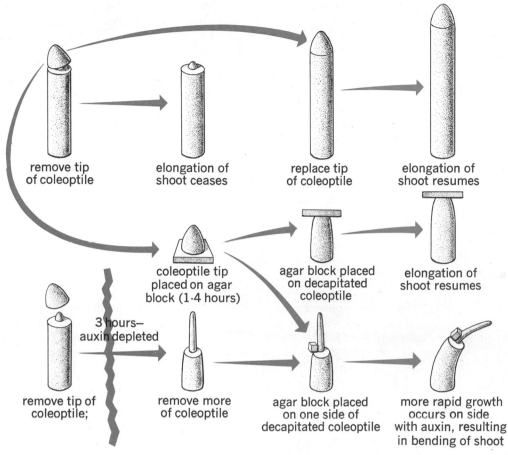

remove tip
of coleoptile

elongation of
shoot ceases

replace tip
of coleoptile

elongation of
shoot resumes

coleoptile tip
placed on agar
block (1-4 hours)

agar block placed
on decapitated
coleoptile

elongation of
shoot resumes

3 hours—
auxin depleted

remove tip of
coleoptile;

remove more
of coleoptile

agar block placed
on one side of
decapitated coleoptile

more rapid growth
occurs on side
with auxin, resulting
in bending of shoot

26-3 Demonstrating the presence and action of auxin in the tip of an oat coleoptile.

auxin is placed on one side of a decapitated coleoptile, cell elongation is stimulated on that side only, resulting in bending of the shoot.

Auxin production is further illustrated in apical dominance. Auxins produced in the region of a terminal bud move downward and inhibit growth of lateral buds. Removal of the terminal bud stimulates growth of the lateral buds. Plants with terminal buds vary in their degree of apical dominance. This may be caused by differences in auxin production or by other factors in combination with auxin production.

Practical uses of auxin preparations

Have you ever used a weed-killer to rid your lawn of dandelions and other broad-leaved weeds? The most common of these is a synthetic auxin, known as 2,4-D. For some reason not entirely understood, 2,4-D in low concentrations (500–1000 ppm) kills broad-leaved plants without injuring grasses.

Other auxin preparations are used to prevent fruit from dropping. This permits the harvesting of a fruit crop at one time. Apparently, auxins prevent the formation of an abscission layer that causes normal dropping when the fruit is ripe.

One of the most interesting uses of auxins by plant growers is the development of fruit without seeds. In this case, auxins seem to stimulate development of the fruit without pollination and fertilization, processes involved in seed formation. By using auxin preparations, growers have been able to produce seedless watermelons, cucumbers, and tomatoes. Fruit development without pollination occurs naturally in the banana and navel orange, both seedless fruits.

Auxin preparations are also used to promote root growth in stem cuttings. The use of these preparations in plant propagation will be discussed in Chapter 27.

Gibberellins and kinins

Another group of growth-regulating substances, the *gibberellins,* are similar to auxins in promoting cell elongation in plants. Gibberellins were first discovered in Japan in 1926, when a connection was established between a fungus *(Gibberella fujikuori)* and the "foolish seedling disease" of rice. Infected plants became greatly elongated due, it was thought, to a substance secreted by the fungus. In 1935, this substance was isolated as a crystalline compound and was named *gibberellin* after the fungus. For many years, biologists believed that gibberellin was produced only by the fungus. Later, they discovered that several gibberellins are natural products in higher plants. The most important of these is gibberellic acid.

Recent studies indicate that gibberellins, in addition to promoting cell elongation, influence growth and development in other ways. They are believed to stimulate flowering as well as fruit development. Evidence also indicates that they promote activity of the cambium in woody plants. In many of their influences, gibberellins may be associated with auxins.

Interesting results are obtained when gibberellin is applied artificially to growing plants. Giant plants result from abnormal cell elongation. Rapid growth may be accompanied by early flowering.

Kinins (also called *cytokinins*) are less familiar regulating substances secreted in plants. They are generally considered to be hormones, although some biologists do not classify them as such, because they are not translocated from the tissues in which they are produced. The best known of the kinins is a substance called *kinetin.*

While auxins and gibberellins promote cell enlargement and development, kinins influence cell division. In addition, they are believed to stimulate leaf growth and the growth of lateral buds.

26-4 Effect of gibberellin on spinach plant growth. A normal and a giant plant. (Dr. M. BuKovac, Michigan State University)

26-5 Phototropism in a coleus plant. The stem is directed toward the light source and the leaves are at nearly right angles to the direction of the light. (Walter Dawn)

Tropisms, growth responses to environmental stimuli

A growth response that is specifically related to an environmental stimulus is called a *tropism* (TROPE-iz'm). Such external stimuli include light, gravity, contact, water, chemicals, and others. A tropic response may be *positive,* toward the stimulus, or *negative,* away from the stimulus. Often, one part of a plant makes a positive response, while the response of another plant part is negative. While tropic responses are purely automatic growth reactions, they usually adapt the plant to an environmental condition and are therefore advantageous.

One of the best-known and most frequently observed tropisms is the growth response to light known as *phototropism* (foe-TOT-ruh-PIZZ'm). You have probably noticed that a plant growing on a window ledge bends toward the light (Figure 26-5). This is a positive response, since the plant bends toward the stimulus. The bending of the stem is a result of uneven elongation of cells. Those on the shaded side, away from the light, elongate more normally than those on the bright side. Biologists have found that auxins control phototropic responses, although the relation of light to the effect of auxins is not clearly understood. One explanation suggests that light reduces the formation of an auxin or reduces its growth-stimulating effect. Another possible explanation is that an auxin is translocated from the bright side to the shaded side. In either case, it has been established that auxins stimulate cell elongation on the shaded side, causing unequal growth and bending toward the light.

Roots, on the other hand, are negative in their phototropic response. If we assume, again, that auxins are more concentrated on the dark side, the auxin must inhibit the elongation of cells.

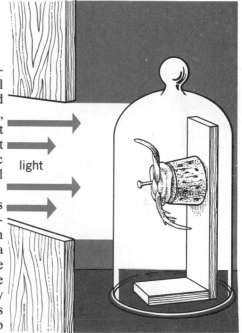

26-6 An explanation of the role of auxins in controlling phototropic responses. Auxin accumulates on the shaded side of the plant, where it causes normal cell elongation in the stem, but prevents normal cell elongation in the root. Because cells on the lighted side of the stem fail to elongate normally, the stem grows toward the light source.

light

More normal cell elongation on the light side would bend the root away from the light stimulus, or toward the dark. We can assume, then, that growth of stem cells is stimulated by auxins, while growth of root cells is inhibited by auxins (Figure 26-6).

The stems and roots of plants also respond in opposite ways to gravity. We refer to this growth behavior as *geotropism* (jee-OT-ruh-PIZZ'm). If you plant a seed upside-down, will the root grow up and the shoot grow down? You probably know that the answer is *no*. The root response is positive, or toward gravity. To account for these geotropic responses, we must again consider auxin.

Assume that a seedling is lying horizontally, as shown in Figure 26-7. As auxin is secreted in the tissues of the tips of the shoot and root, gravity causes it to accumulate on the lower side. The increased auxin concentration on the lower side of the shoot stimulates cell elongation and causes the stem to bend upward. On the lower side of the root, the auxin inhibits cell elongation. Normal cell elongation on the upper side causes the root to bend downward. If a shoot or a root is growing in a vertical position, auxin distribution is equal, and growth continues upward or downward.

Other tropic responses can easily be observed but are not so readily explained. The tendrils of climbing plants wrap around objects they touch (see Figure 23-11). This response to contact is called *thigmotropism* (thig-MOT-ruh-PIZZ'm). Since the tendril grows around an object, the response is positive.

Certain roots are believed to grow toward water. We refer to this growth response as *hydrotropism* (hide-ROT-ruh-PIZZ'm). You may have noticed that willow and cottonwood trees growing along stream banks form masses of roots that extend into the water. These trees and other moisture-loving species often clog drains and sewers by forming masses of roots between the tiles or pipes. It is difficult to explain this root behavior. Contrary to common belief, roots do not "seek" water supplies. Growth may be stimulated when the roots contact water, however. This may explain the masses of roots along the bank of a stream or the clogging of a drain.

The response of roots to chemicals, known as *chemotropism* (keh-MOT-ruh-PIZZ'm), is also difficult to account for. Perhaps, as in the case of water, the concentration of minerals may stimulate root growth rather than serve as a stimulus of attraction.

Nastic movements in plants

Many of the external stimuli that cause tropic responses in plants may also trigger *nastic movements*. In this case, however, the response is independent of the direction of the stimulus. Nastic movements are often seen in the closing of flowers or the folding of leaves at certain times of the day or night or under varying temperature conditions. Nastic movements differ from

26-7 An explanation of the role of auxins in controlling geotropic responses. Auxin accumulates on the lower side of the plant. Because cells on the upper side of the stem fail to elongate normally, the stem grows upward. Normal lengthening of the cells on the upper side of the root, however, causes the root to grow downward.

26-8 Nastic movement in tulips: two flowers during the day (left), and the same flowers during the night that followed. (William Harlow, National Audubon Society)

tropic responses in several ways. Tropic responses are *growth movements,* while nastic responses are *turgor movements.* A nastic movement is not a directed response. There is no directed movement either toward or away from a stimulus. Tropic responses are slow, while nastic movements occur within a few hours or minutes. Furthermore, nastic movements are reversible — that is, a flower that closes may open again. Tropic growth responses are usually more permanent.

Nastic movements can be observed in the opening and closing of flowers under different conditions of light and temperature. The morning-glory provides one of the best examples. During the day, the flowers are open. With the approach of evening and the cooling of the air temperature, the flowers close and remain closed all night. With the coming of daylight, they open again. Similar closing of flowers at night can be seen in the tulip and dandelion (Figure 26-8). The response of flowers of the night-blooming cereus, a cactus, is the opposite. These flowers remain closed throughout the day but always open in the darkness of night.

Leaves also undergo nastic movements. Biologists often call these "sleep" movements. The clover and wood sorrel *(Oxalis)* are good examples. During the day, compound leaves of the clover are open, with the leaflets spread horizontally. During the night, the leaflets fold downward and close the leaf.

Temperature changes may also cause nastic movements. The leaves of the rhododendron, a broad-leafed evergreen shrub, roll inward during cold weather and unroll as the temperature rises.

Flower and leaf movements seem to be caused by rapid changes in cell turgor. These changes may occur in flower petals, leaf blades, or in veins and petioles.

The best example of an almost immediate nastic response to touch is demonstrated by the sensitive plant *Mimosa pudica.* The sensitive plant has compound leaves with numerous small leaflets that are normally in a horizontal position. If a leaf is touched or jarred, the plant will react within three to five seconds. Depending on how widespread the stimulus, some or all the leaflets fold upward, and the petioles droop from the stem (Figure 26-9). The

26-9 Nastic responses to touch in the sensitive plant: normal position of the leaves and leaflets (top), and position of the same leaves and leaflets after they had responded to a touch stimulus. (Walter Dawn)

folding and drooping of leaves and petioles extend in waves from the point of contact. After having folded, they gradually return to their original horizontal positions, and the petioles straighten as turgor is slowly reestablished in the cells. The mechanism of this remarkable folding response is complex. The touch stimulus is received by epidermal hairs and spreads through the phloem sieve tubes of the veins and petiole. At the base of the petiole is a small mass of cells that hold the petiole in an upright position. Sudden loss of turgor in these cells relaxes their support, causing the petiole to droop.

A similar turgor change causes the leaves of the Venus's-fly-trap to close suddenly and trap an insect, as you learned in Chapter 23. The movements that cause flowers such as the sunflower to "follow the sun" are also nastic movements caused by turgor changes.

SUMMARY

If you ever see violets or magnolias blooming in the fall, you shouldn't be too surprised. Compare the length of the days and nights then with those of the spring weeks when these plants normally bloom and you will discover an interesting similarity. Other factors, including the average temperature at the time of the fall blooming may also have an influence, but the hours of daylight and darkness are probably most important. By controlling the hours of light and darkness artificially, plant growers supply poinsettias in full bloom at the Christmas season and blooming lilies at Easter in the northern states and Canada. These are but two examples of the control of growth and flowering by artificial regulation of environmental conditions.

More and more, plant growers are using plant hormone preparations to stimulate root growth, increase the size of plants, advance the time of flowering, and many other useful purposes. A knowledge of plant hormones and their effects is becoming increasingly important in the plant growing industries.

BIOLOGICALLY SPEAKING

etiolated	auxin	phototropism
photoperiodism	growth regulator	geotropism
photoperiod	coleoptile	thigmotropism
optimum temperature	gibberellin	hydrotropism
dormancy	kinin	chemotropism
hormone	tropism	nastic movement

QUESTIONS FOR REVIEW

1. List several environmental factors that influence plant growth.
2. Describe the stem and leaves of a young plant grown in the dark.
3. Give possible reasons for the fact that shade leaves of a plant are larger than sun leaves of the same plant.
4. Give several examples of photoperiodism in flowering plants.
5. What are neutral-day plants?
6. How do commercial growers produce flowers at seasons in which plants would not normally reproduce?
7. What is the optimum temperature range for the growth of the majority of flowering plants?

8. Explain how woody plants, herbaceous perennials, and annuals survive periods of dormancy.
9. Name several external conditions that "break" dormancy in seed plants.
10. How do hormones differ from other secretions in respect to where they are formed and where they act?
11. Why are auxins considered growth regulators rather than growth hormones?
12. Which of the auxins is produced in largest amounts in seed plants?
13. How are gibberellins similar to auxins in their influence on cell growth?
14. Account for the fact that gibberellins were once thought to be formed only by fungi.
15. In what way do kinins influence plant growth?
16. Distinguish between a positive and a negative tropic response.
17. Name five tropisms and identify each with an external stimulus.
18. Distinguish between a nastic movement and a tropic response.
19. Give several examples of nastic flower movements.
20. Why are nastic movements of the leaves of clover and other plants often called "sleep" movements?

APPLYING PRINCIPLES AND CONCEPTS

1. If you cut a twig from a woody stem of peach or willow in the fall, soon after the leaves have fallen, and put it in water in a warm place, the buds probably will not grow as they would have if the cutting had been made in the spring. Give a possible reason for this.
2. Explain why certain plants flower in the spring and again in the fall.
3. Discuss experimental evidence that auxin is secreted near the tip of an oat coleoptile and that it is translocated to the tissues below the tip.
4. Cite experimental evidence in studies of phototropism and geotropism that auxin promotes the elongation of cells in stem tissues but that in the same concentration it inhibits elongation of root cells.
5. Contrast tropic responses as growth movements with nastic responses as turgor movements.

RELATED READING

Books

FOGG, G. E., *The Growth of Plants.*
  Penguin Books, Inc., New York. 1963. Discusses all aspects of plant physiology with special emphasis on plant growth.
GALSTON, ARTHUR W., *The Life of the Green Plant.*
  Prentice-Hall, Inc., Englewood Cliffs, N.J. 1964. A modern picture of the way the green plant functions.
LEOPOLD, A. C., *Auxins and Plant Growth.*
  University of California Press, Berkeley, Calif. 1965. A very useful chapter supplement.
MEYER, BERNARD S., and DONALD B. ANDERSON, *Introduction to Plant Physiology.*
  D. Van Nostrand Co., Princeton, N.J. 1960. A concise introduction to the field of plant physiology.

Articles

STEWARD, F. G., "The Control of Growth in Plant Cells," *Scientific American,* October, 1963. Describes a classic investigation in which isolated cells were stimulated to develop into adult plants.

CHAPTER TWENTY-SEVEN

PLANT REPRODUCTION

Reproductive processes of flowering plants

The life cycle of a flowering plant consists of two stages. One phase of the cycle is the *vegetative stage,* in which roots, stems, and leaves grow and develop. In annuals, vegetative growth occurs in a single season; in perennials, it continues year after year. Many plants reproduce during the vegetative growing season by producing new plants from their roots, stems, or leaves. This type of multiplication, or *propagation,* is *asexual* and is called *vegetative reproduction.*

Growth during the vegetative stage prepares the flowering plant for the *reproductive stage,* in which flowers are produced and sexual reproduction occurs. After the gametes are formed and fertilization has taken place, the fertilized egg gives rise to an embryo that lies within the protective structures of the seed until germination occurs. Flowers, fruits (which contain the seeds), and seeds, then, are products of the activities that occur in the vegetative plant body.

In the study of the reproductive processes of flowering plants, we shall first consider various vegetative methods, both natural and artificial, and then proceed to the flower, fruit, and seed as they are involved in sexual reproduction.

Natural vegetative reproduction

Vegetative reproduction occurs most in woody and herbaceous perennials. One form of natural vegetative propagation can be observed in a strawberry patch. Certain stems of the strawberry, known as *runners,* grow along the ground from the parent plant. Runners form adventitious roots and shoots at the tips, thus giving rise to new plants (Figure 27-1). By this method, a single parent plant can form many new plants in a single season; hence, a patch of strawberries can soon develop from a single plant.

A similar process occurs in the raspberry. The long, aerial stems arch downward and their tips contact the ground. Ad-

27-1 Vegetative reproduction in the strawberry.

runner

adventitious roots

ventitious roots develop, and a shoot arises. A young raspberry plant that soon becomes independent of the parent plant is formed in this process, which is known as *tip layering*.

Vegetative reproduction is also common in plants that form stolons and rhizomes, which we discussed in Chapter 24. These underground stems creep horizontally, forming adventitious roots and shoots at the nodes.

In many cases, aerial shoots arise from horizontal roots growing near the surface. For example, the sumac, a small tree or shrub, forms clusters of young plants from the roots of a parent. Similar groves are formed by the sprouting of roots of the white poplar, osage orange, and lilac.

One of the most unusual types of vegetative propagation is marked by the development of new plants along the margins of leaves of a few kinds of plants. In *Kalanchoe* (commonly called bryophyllum), tiny plants develop between the teeth in the coarsely toothed margin of the thick, fleshy leaves (Figure 27-3). As they increase in size, the tiny plantlets, which arose from meristematic tissue in the leaf margin, drop to the ground and continue to grow. Because of this unusual vegetative reproduction, bryophyllum is a favorite potted plant for the greenhouse or window shelf.

Artificial propagation by cuttings

Plant growers use many methods of propagating plants vegetatively. One of the most common methods is the preparation of

27-2 The fleshy, thickened root of the sweet potato serves naturally as a vegetative reproductive organ. For commercial propagation, the adventitious shoots are removed and planted in fields. (Herbert Weihrich)

27-3 Vegetative reproduction in bryophyllum. Detail of the tiny plants growing along the toothed margin of a fleshy bryophyllum leaf can be seen below. (Jerome Wexler, National Audubon Society)

cuttings. Cuttings are portions of stems, roots, or leaves that are removed and rooted in sand, loose soil, or some other suitable substance. Stem cuttings are most frequently used. A stem cutting may be from several inches to a foot or more in length, including several nodes and lateral buds or leaves. The lower end of the cutting is buried in sand or soil. After several weeks, if the cutting is successful, adventitious roots form at the lower end, and lateral buds, especially near the tip, develop shoots. Many woody plants, including the rose and pussy willow, are easily propagated by this method. Herbaceous plants such as the geranium can also be grown from cuttings, often called *slips*.

In recent years, new methods have been developed that permit more extensive use of cuttings in plant propagation. Various growth-regulating hormones, including indoleacetic acid, initiate cell division and promote root growth. The cuttings are treated by soaking them in a dilute solution of the hormone for from twelve to twenty-four hours or by applying a salve or powder that contains tiny amounts of the hormone. Growth of roots on treated stem cuttings is more rapid if leaves are present to supply sugar to the developing root tissues. By means of hormone treatment, many plants that are difficult to propagate vegetatively are successfully grown from cuttings. Among these are the American holly and many ornamental conifers, including varieties of blue spruce, juniper, and yew.

Propagation by layering

Several forms of *layering* are used in artificial propagation of many plants. In one method, a rose stem is bent to the ground and covered with sand or soil for several inches, leaving only the tip exposed. The buried portion will form roots, and growth of the stem will continue from the tip (Figure 27-4). After the new plant is established, it is cut from the parent rose bush and set out in a new location. This process, known as *simple layering,* is used most widely in propagating climbing roses and other varieties that develop vigorous roots.

Mound layering is used in propagating many woody shrubs, including forsythia, currants, and gooseberries. A shrub is pruned back to promote the growth of many lateral buds near the bases of the stems. The lower part of the shrub is then covered with a mound of loose soil. This promotes the growth of adventitious roots on the young shoots growing from lateral buds (Figure 27-4). After a period of root growth, the stems are removed from the mound and are set out individually.

Certain plants, including the India rubber plant, are propagated by *air layering*. In this process, a branch is deeply cut or girdled and is then surrounded with a ball of wet peat moss in the area of the wound. Adventitious roots grow from the wounded tissue into the wet peat moss. When the roots are well developed, the branch is removed and planted.

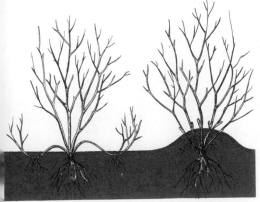

27-4 Simple layering (left) and mound layering.

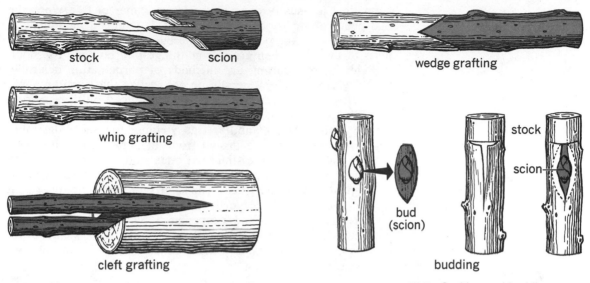

27-5 Grafting and budding.

Grafting and budding

Grafting is one of the most widely used methods of propagating woody plants, especially fruit trees and ornamental trees. You could think of grafting as the splicing together of two stems. The rooted portion is called the *stock*. The cutting that is joined to the stock is the *scion*. The scion is a desirable plant that the grower is propagating on a vigorous stock.

A graft can be successful only if the vascular cambiums of the stock and scion are placed in contact with each other. It is necessary, also, that the graft be made when the scion is dormant so that the cambiums can unite before shoots develop. For this reason, grafts are usually prepared during the winter months. The region of the graft is covered with wax and wrapped with elastic tape to prevent drying out or invasion of the wound by fungi, insects, or other damaging organisms.

Grafts are of several types. The *whip graft* shown in Figure 27-5 is commonly used in grafting apple and other fruit trees. In a *cleft graft,* two or more tapered scions are inserted into a split in the stock in the region of the cambium. A *wedge graft* is made by inserting the tapered end of the scion into a V-shaped cut in the stock.

Budding is similar to grafting except that a bud, rather than a branch, is used as the scion. A vigorous bud is selected and removed together with the piece of bark and active cambium surrounding it. The bud is united with the stock by slipping the piece of bark under the bark of the stock where it has been loosened by a T-shaped cut. This unites the vascular cambiums of scion and stock. The stock is then wrapped with elastic tape to secure the bud and prevent drying out.

Budding is usually done in the late summer or fall. When the bud opens and a shoot develops the following spring, the stock is cut off above the point where the bud was inserted. All aerial parts of the plant then grow from the grafted bud and its shoot.

Budding and grafting are methods of perpetuating desirable hybrid varieties. Many people do not realize that such an apple variety as the Golden Delicious was produced only once from seed. A chance combination of genes produced this variety. It is very unlikely that another tree yielding the same desirable characteristics could be grown from an apple seed. However, the original tree was the source of buds from which grafts could be made. Vegetative propagation does not change the genetic makeup of the young trees. Thus, all Golden Delicious apple trees in the orchards of North America are the same as the original tree, propagated and perpetuated by grafting. The same applies to cultivated varieties of other fruit trees, shade trees, ornamental trees, roses, and many other plants with which you may be familiar.

The production of flowers

Although plants may multiply vegetatively during a growing season, sexual reproduction is essential to the survival of most species. The end result in sexual reproduction is the production of seeds. In the process, genetic traits, usually of different parent plants, are combined and carried into a new generation. Recombination of chromosomes and genes in the union of gametes and formation of a zygote provides the basis for variation. These variations, even though slight, may adapt a species to conditions in a changing environment.

Flowers are the reproductive organs of higher plants. They are branches on which the leaves have been greatly modified and serve as specialized *floral parts*. In flowers, the nodes are very closely spaced, resulting in a compact arrangement of the floral parts. Flowers are specialized for the single function of sexual reproduction. They lack meristematic tissue to continue growth of the shoot. Flowers initiate a sequence of development leading to the formation of the fruit and seed.

There is great variation among flowers. The lily, rose, orchid, and tulip are large, colorful flowers. In contrast, the flowers of the oak, elm, maple, corn, wheat, and lawn grasses are usually unnoticed, although they are equally efficient reproductive organs. The essential parts of flowers are not the showy petals we commonly associate with them.

Floral parts

A typical flower, such as the geranium, apple blossom, snapdragon, sweet pea, or petunia, has four sets of parts (Figure 27-6). These parts grow from a special flower stalk, or *pedicel* (PED-uh-

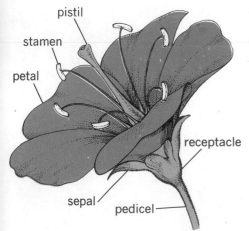

27-6   The floral parts of a complete flower.

SELL), the tip of which is the *receptacle*. The outer ring of floral parts consists of several green, leaflike structures called *sepals* (SEEP'lz). Together, the sepals form the *calyx* (KAY-licks). The sepals cover and protect the rest of the flower in the bud stage. They also help to support the other parts when the bud opens.

Inside the calyx is the *corolla* (kuh-RAH-luh), which usually consists of one or more rows of *petals*. These are often, but not always, brightly colored. The calyx and corolla frequently attract insects, as we shall see later. They may also help to protect the inner parts. In certain flowers, like the tulip, both the calyx and corolla are the same color. It is possible to overlook the fact that both parts are present.

Two kinds of *essential* parts are concerned directly with reproduction: the stamens and the pistil, located in the center of the flower. Each *stamen* (STAME'n) consists of a slender stalk, or *filament*, supporting a knoblike sac called an *anther*. The anther produces various colored grains called *pollen*, which are essential in reproduction.

The *pistil* is usually a flask-shaped organ that consists of an often sticky top called a *stigma*; a slender stalk, or *style*, which supports the stigma; and a swollen base, or *ovary*, which is joined to the receptacle of the flower stalk. Inside the ovary are the *ovules*, which will later become seeds. The ovules are attached to the ovary (either at its base or along its side walls) or to a special axis running through the center of the ovary. Ovules may number from one to several hundred, depending on the kind of flower.

Variation in flower structure

Flowers vary greatly, not only in color, size, and shape, but also in their reproductive structures. Although many plants have both stamens and a pistil on the same flower, some, such as the oaks, squash, and corn, have these structures on separate flowers on the same plant. Other species have flowers bearing pistils on one plant and flowers bearing stamens on another plant. Willows and cottonwoods are well-known and familiar examples of this type.

In studying stems and leaves, you learned that dicots differ from monocots in the arrangement of their vascular tissue. As you might expect, they also differ in the structure of their flowers. Monocots, for example, usually have flower petals and essential parts in threes or multiples of three, while dicots usually have them in fours or fives or their multiples.

Some flowers are not single flowers in the biological sense but, rather, a whole cluster of flowers. The sunflower and daisy, for example, have dense clusters of reproductive flowers in the center, surrounded by so-called petals, which are really another type of flower and which usually attract insects.

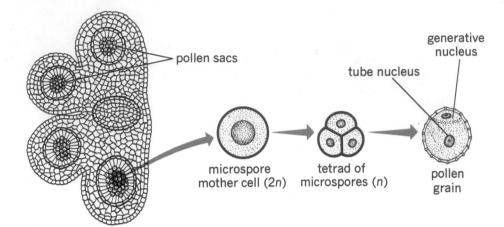

27-7 Development of a pollen grain within the pollen sac of an anther.

The anther and pollen formation

If you examine a cross section of the developing anther of a large stamen, such as that of a lily or a tulip, you can clearly see four chambers, or *pollen sacs* (Figure 27-7). During development of the flower, each pollen sac is filled with cells with large nuclei. These are known as *microspore mother cells*. Microspore mother cells contain pairs of chromosomes, or the diploid number ($2n$). As the anther grows, each microspore mother cell undergoes two meiotic divisions, forming a four-celled tetrad. These cells are called *microspores*. Microspores are haploid and contain the n number of chromosomes.

Each microspore develops into a pollen grain. First, the nucleus divides by mitosis, forming two nuclei. One is designated as a *tube nucleus* and the other as a *generative nucleus*. The wall of the microspore thickens and becomes the protective covering of the pollen grain. At about this time, the anther ripens, and the wall between the paired pollen sacs disintegrates. The pollen sacs burst open, and the mature pollen grains are ready for distribution.

Development of the ovule

While pollen grains are forming in the anthers, changes are occurring in the ovary at the base of the pistil. For the sake of simplicity, we shall describe these changes as they take place in an ovary that contains a single ovule, like the avocado pear. When there are several or many ovules in an ovary, the process is generally the same.

An ovule appears first as a tiny knob on the ovary wall. This swelling contains a single *megaspore mother cell* (Figure 27-8). This cell contains the diploid number ($2n$) of chromosomes. As the ovule grows, it is raised from the ovary wall by a short stalk through which nourishment is received. One or two protective layers form around the ovule and enclose it completely except for

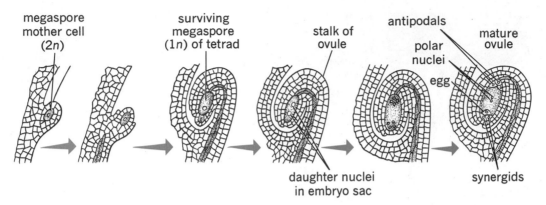

megaspore mother cell (2n)

surviving megaspore (1n) of tetrad

stalk of ovule

antipodals

polar nuclei

mature ovule

egg

daughter nuclei in embryo sac

synergids

27-8 Development of an ovule within the ovary at the base of a pistil.

a tiny pore, or *micropyle.* The micropyle is usually on the lower side of the ovule.

The megaspore mother cell undergoes two meiotic divisions, resulting in a row of four *megaspores.* As in the formation of microspores, this double division reduces the chromosome content of the megaspores to the haploid number (*n*). Of the four megaspores produced in an ovule, one survives, and the other three disintegrate. The surviving megaspore is usually the one farthest from the micropyle. This remaining megaspore enlarges rapidly and forms an oval *embryo sac,* in which further development occurs in the following steps:

■ The megaspore nucleus divides and forms two daughter nuclei. Two additional mitotic divisions result in eight nuclei.

■ Four nuclei migrate to each end of the embryo sac.

■ One nucleus from each group of four, designated as a *polar nucleus,* migrates to the center of the embryo sac.

■ Each nucleus in the groups of three at either end of the embryo sac is enclosed by a thin membrane.

■ One of the cells nearest the micropyle enlarges and becomes the *egg.* The cells on either side of the egg and partially surrounding it are known as *synergids* (si-NURR-jidz).

■ The three cells farthest from the micropyle are known as *antipodals* (an-TIP-uh-duhlz).

The ovule is now ready for fertilization. However, only the egg and the polar nuclei will be involved in this process. Both the synergids and the antipodals are short-lived and have no apparent function. Before fertilization can occur, a pollen grain must be transferred to the stigma of the pistil by one of the various agents in pollination.

Pollination

We can define *pollination* as the transfer of pollen from an anther to the stigma. In some plants, pollen is transferred from anther to stigma in the same flower or to the stigma of another flower on the

27-9 The bee is an ideal pollinator because of its plump, hairy body. (Dr. Edward R. Degginger)

27-10 The pollen-producing flowers of the giant ragweed. The pollen of this wind-pollinated plant causes hay fever in people who are sensitive to it. (Betty Barford, National Audubon Society)

same plant. We refer to this as *self-pollination*. If flowers on two separate plants are involved, the process is called *cross-pollination*. Cross-pollination requires an outside agent. These are primarily insects, wind, and water. Curious adaptations of different kinds of flowers are frequently necessary for the accomplishment of cross-pollination.

Adaptations for pollination

Chief among the insect pollinators are bees. But moths, butterflies, and certain kinds of flies visit flowers regularly and, in so doing, carry on cross-pollination. Insects go to the flower to obtain the sweet nectar secreted deep in the flower by special glands at the base of the petals.

Bees swallow nectar into a special honey stomach, where it is mixed with saliva and converted into honey. When the bees return to the hive, they deposit the honey in the six-sided cells of the comb and later use it as food. The plump, hairy body of the bee makes it an ideal pollinator (Figure 27-9). In reaching the nectar glands at the base of the flower, the bee must rub its hairy body against the anthers. These are usually located near the opening of the flower. When the insect visits the next flower, some of the pollen is sure to rub off on the sticky stigma of the pistil, while a new supply is brushed off from the stamens onto the bee's body.

Brightly colored petals and sweet odors aid insects in locating flowers. Brightly colored stripes located on the petals of some flowers may also serve as guides.

We must include at least one bird in our discussion of agents of pollination. Tiny hummingbirds feed on the nectar of certain flowers. Their long bills and equally long tongues reach down to the nectar glands while the bird hovers over the flower.

The flowers of wind-pollinated plants are much less striking than the flowers of those pollinated by insects. Wind-pollinated flowers are often borne in dense clusters near the ends of branches. As a rule, petals are lacking, and the flowers seldom have any nectar. Frequently, the stamens are long and produce enormous quantities of pollen. The pistils are also long, and the stigmas are large and often sticky or feathery to catch pollen grains that are blown about by the wind. Cottonwood, willow, walnut, corn, oats, and other wind-pollinated plants literally fill the air with pollen when their stamens are ripe.

Growth of the pollen tube and fertilization

Once a pollen grain lodges on the surface of the stigma of the pistil, a chemical from the pistil stimulates the pollen grain to form a *pollen tube* that penetrates the stigma surface (Figure 27-11). As the tube lengthens, it grows through the soft tissue of the style and reaches the micropyle of the ovule. As the generative nucleus moves into the tube, it divides and forms two *sperm nuclei,* or

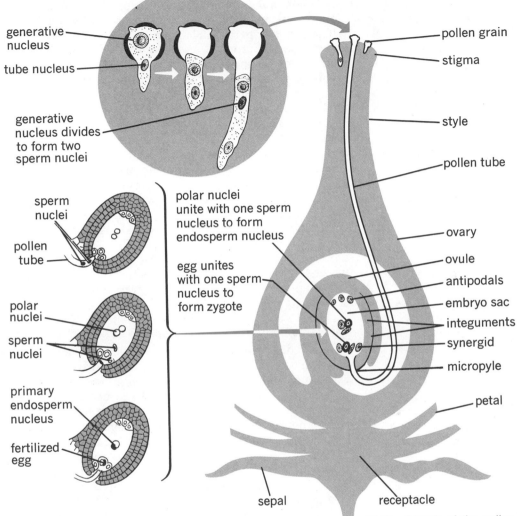

generative nucleus

tube nucleus

generative nucleus divides to form two sperm nuclei

sperm nuclei

pollen tube

polar nuclei

sperm nuclei

primary endosperm nucleus

fertilized egg

polar nuclei unite with one sperm nucleus to form endosperm nucleus

egg unites with one sperm nucleus to form zygote

pollen grain

stigma

style

pollen tube

ovary

ovule

antipodals

embryo sac

integuments

synergid

micropyle

petal

sepal

receptacle

27-11 Growth of the pollen tube and double fertilization.

male gametes. After passing through the micropyle, the pollen tube digests its way through the thin wall of the embryo sac. The tip of the tube ruptures, and the two sperms are discharged into the embryo sac. Meanwhile, the tube nucleus, which is believed by many investigators to play no role in the growth of the pollen tube, degenerates.

One of the two sperm nuclei unites with the egg in fertilization. This produces the fertilized egg, or zygote. Both the egg and the sperm nuclei are haploid. Fertilization restores the diploid chromosome number ($2n$) in the zygote. Since the zygote will form the embryo plant by cell division, the cells of the plant will possess this diploid number until flowers are produced and reduction division occurs in the formation of microspores and megaspores.

The second sperm nucleus unites with the polar nuclei in the embryo sac to produce an *endosperm nucleus*. Since both polar nuclei and the sperm nucleus contain the haploid chromosome number *(n)*, the endosperm nucleus is a triploid *(3n)* structure. We refer to the union of the three nuclei to form the endosperm nucleus as *triple fusion*. These steps are summarized graphically in Figure 27-11.

Immediately after this *double fertilization*—the union of the egg and first sperm nucleus and the fusion of the polar nuclei with the second sperm nucleus—auxins that were produced in the pollen grain and delivered to the ovule through the pollen tube stimulate rapid cell division and tissue growth within the ovule. The zygote undergoes an orderly development into the embryo plant. Meanwhile, the endosperm nucleus gives rise to a mass of tissue that becomes the *endosperm* of the seed. This tissue contains food for the developing embryo plant. In some seeds, the food stored in the endosperm is absorbed by the embryo while the seed is developing. In others, the endosperm remains as a part of the seed at maturity.

The gametophyte and the sporophyte in flowering plants

In our discussion of the algae, the mosses, and the ferns, we noted that some of these organisms have two distinct forms, the gametophyte and the sporophyte, in *an alternation of generations*. In the algae and the mosses, the gametophyte was the most conspicuous form. In the fern, the sporophyte has become the more conspicuous of the two generations. You may never have noticed the small, heart-shaped gametophyte of the fern. In the seed plants, this trend has gone even further. The sporophyte is the plant that we see. The gametophyte has become completely dependent on the sporophyte and is, in fact, microscopic in size. The male gametophyte in seed plants consists only of the pollen tube with its haploid sperm and tube nuclei. The female gametophyte consists of the embryo sac with its haploid egg, polar nuclei, antipodals, and synergids, found in the ovule.

From this description of the gametophytes in seed plants, you can see that the vulnerable egg and sperm nuclei are well protected from drying within the tissues of the sporophyte. This evolutionary trend toward the more conspicuous sporophyte seems to parallel the development of (1) *vascular tissue;* (2) *roots;* (3) *epidermis;* and (4) *stomata,* starting with the ferns and arriving at the most efficient of the land plants, the angiosperms.

From flower to fruit and seed

Fertilization brings a sudden end to the work of the flower. As the sepals, petals, and stamens wither, a group of special hormones causes the plant's full energies to be used in the development of the ovary and the ovules inside. After a few weeks, the ovary and its contents ripen. In many plants, other nearby parts, such as the receptacle or the calyx, enlarge and become part of the fruit; therefore, we can define a fruit as a *ripened ovary, with or without associated parts*. A seed, on the other hand, is a *matured ovule* that is enclosed in the fruit.

Fruits, like the flowers from which they develop, vary greatly in structure. Table 27-1 shows how some of the common fruits are classified according to structure. As you can see, a fruit need not be fleshy, like an apple, a peach, or an orange. A grain (kernel) of corn, a hickory nut, a bean pod with its beans, a sticky burr of burdock, and a cucumber or pumpkin are just as much fruits as the fleshy, juicy type. Thus, the biological meaning of the word *fruit* is quite different from the meaning used in a grocery store.

The relationship between fruits and seeds

Consider for a moment what would happen if seeds all fell to the ground and started to grow close to the parent plant. In a short time, the parent plant would be surrounded by seedlings that would compete both with the parent and among themselves for a limited supply of nutrients, water, and light. As you might imagine, only a few, if any, of the seedlings would survive to maturity.

Many types of seeds, when they mature, are carried to a point some distance away from the parent plant. This *seed dispersal* is accomplished in many ways. In some plants, seed dispersal is a mechanical process, while in others, an outside agent, such as the wind, water, or a bird or some other animal, is involved.

Pods like the bean and pea often twist as they ripen because of changes in the amount of moisture in the air. This causes a strain on the pod, which bursts open suddenly and with enough force to throw its seeds some distance away from the parent plant. The actions of the fruit of the garden balsam, or touch-me-not, constitute another interesting example of *mechanical dispersal* of seeds. When the fruits of this plant are ripe, they open upon the slightest touch and curl upward violently, with the result that the seeds may be thrown several feet. Fruits of the capsule type, such as those of the poppy, do not split open along the sides. Instead, holes form around the top as the ovaries ripen. As the ripened fruit sways back and forth in the breeze on a long and flexible stem, seeds sift out, much as salt does from a saltshaker.

In some cases, such as those we have just discussed, only the seed is dispersed. In many others, the entire fruit, which, as you recall, encloses the seed and protects it while it is developing, plays a less direct role in seed dispersal. The delicious flesh of the apple, grape, or cherry serves as a sort of biological bribe. *Birds and other animals* feed on the fruits. Often, because their

Table 27-1 CLASSIFICATION OF FRUITS

TYPE	STRUCTURE	EXAMPLES
Fleshy Fruits		
pome	outer fleshy layer developed from calyx and receptacle; ovary forms a papery core containing seeds	apple, quince, pear
drupe	ripened ovary becomes two-layered —outer layer fleshy, inner layer hard, forming stone or pit, enclosing one or more seeds	plum, cherry, peach, olive
berry	entire ovary fleshy and often juicy; thin-skinned and containing numerous seeds	tomato, grape, gooseberry
modified berry	like berry, but with tough covering	orange, lemon, cucumber
aggregate fruit	compound fruit composed of many tiny drupes clustered on single receptacle	raspberry, blackberry
accessory fruit	small and hard; scattered over surface of receptacle; edible portion formed from enlarged receptacle	strawberry
multiple fruit	compound fruit formed from several flowers in a cluster	mulberry, pineapple
Dry Fruits (open when ripe)		
pod	ovary wall thin, fruit single-chambered, containing many seeds; splits along one or two lines when ripe	bean, pea, milkweed
capsule	ovary containing several chambers and many seeds; splits open when mature	poppy, iris, cotton, lily
Dry Fruits (remain closed when ripe)		
nut	hard ovary wall enclosing a single seed	hickory nut, acorn, pecan
grain	thin ovary wall fastened firmly to single seed	corn, wheat, oats
achene	similar to grain, but with ovary wall separating from seed	sunflower, dandelion
winged fruit or samara	similar to achene but with prominent wing attached to ovary wall	maple, ash, elm

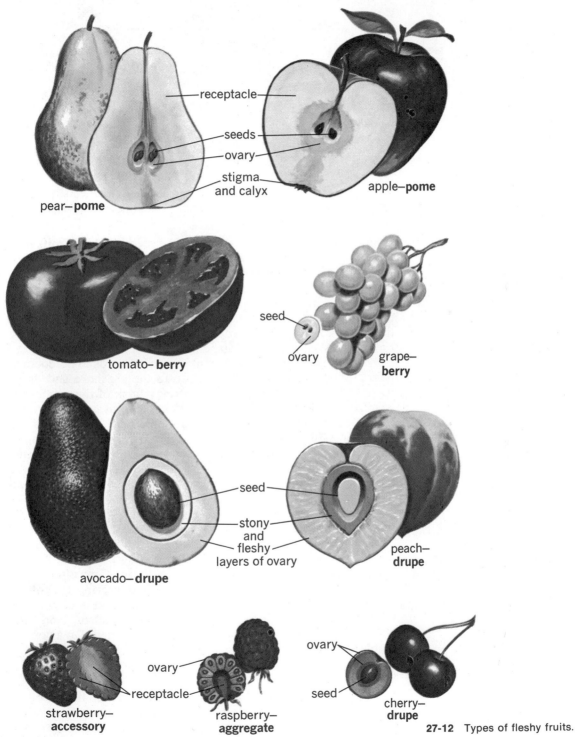

pear–**pome**

receptacle

seeds

ovary

stigma and calyx

apple–**pome**

tomato–**berry**

seed

ovary

grape–**berry**

avocado–**drupe**

seed

stony and fleshy layers of ovary

peach–**drupe**

strawberry–**accessory**

ovary

receptacle

raspberry–**aggregate**

ovary

seed

cherry–**drupe**

27-12 Types of fleshy fruits.

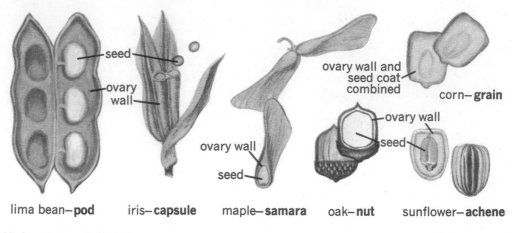

seed

ovary wall

ovary wall

seed

ovary wall and seed coat combined

corn—**grain**

ovary wall

seed

lima bean—**pod**   iris—**capsule**   maple—**samara**   oak—**nut**   sunflower—**achene**

27-13 Various types of dry fruits.

cellulose covers cannot be digested, the seeds of these fruits pass unharmed through the digestive tracts of the animals. It is in this way that the intact seeds are distributed.

Animals aid in fruit and seed dispersal in another way. Many plants produce fruits with stickers or spines that cause the seed to cling to the fur of animals. If you have ever removed beggarticks, stick-tights, and burdocks from your clothes, you have probably assisted in seed dispersal.

Water is the agent of dispersal for many seeds. The coconut palm, for instance, often lives close to the seashore and drops its fruit into the water. The thick, stringy husk of the coconut is waterproof. When the seed germinates, a sprout pushes through one of the three "eyes" at one end of the hard covering of the seed. Grasslike plants known as sedges are among the other plants that may drop their fruits into the water. They, like the coconut, are generally found along the shores of oceans or the banks of rivers and streams where their seeds have found a foothold on the land.

The wind is another agent of dispersal for many fruits and seeds. When the milkweed pod splits open, the wind empties the pod of its seeds, and each, equipped with a tuft of long, loose hairs, is carried to a new location. You have probably blown the fluff off a dandelion or thistle head. The fruits of these plants often travel long distances on their tiny "parachutes." In the spring, the cottonwood tree fills the air as breezes empty its catkins of cottony, tufted seeds. The winged seeds of maple, ash, elm, and pine whirl in the air like tiny propellers, and are scattered to a considerable distance from the place where they developed.

What is a seed?

We defined a seed as a matured ovule and as the final product of plant reproduction. A seed consists of a tiny living plant called the *embryo,* stored food, and the seed coats. The stored food nourishes the young plant from the time it starts to grow until it can produce its own food by photosynthesis. The regions in which food is stored may vary with different seeds. In some seeds, food is stored in thick "seed leaves," the *cotyledons.*

You may have seen thick cotyledons on the stems of such young plants as the green bean or lima bean shortly after they have pushed through the garden soil. The cotyledons are located below the foliage leaves and last for only a few days before they wither and fall off. The number of cotyledons in the seed serves as the basis for the classification of the angiosperms. Monocot plants have only one cotyledon in their seeds, while dicot plants have two.

Not all seeds have the same kind of food stored in the cotyledons. A grain of corn, for example, has its starch and protein stored in the endosperm, while the cotyledon contains oils and proteins. The endosperm, filling much of the corn seed, develops, as you should recall, from the endosperm nucleus. On the other hand, the major part of the bean seed is made up of the two cotyledons, which store a large amount of starch as well as proteins and oil. Some seeds have a large endosperm or none at all. The bean is one of the latter for it has none at all.

Seed coats, which have developed from the wall of the ovule, cover the seed and protect it from drying out and from other dangers before it germinates. Usually there are two seed coats, but some seeds have only one. The outer coat is usually tough and thick. The inner coat is much thinner.

Structure of a bean

The bean seed is usually kidney-shaped (Figure 27-14). The outer seed coat, the *testa,* is smooth and may be white, brown, red, or some other color, depending on the species. The *hilum* (HILE'm), an oval scar on the concave side, marks the place where the bean was attached to the wall of the pod. Near one end of the point of attachment is the tiny pore, the *micropyle.* The pollen tube grew through this tiny opening in the wall of the ovule just before fertilization. The inner seed coat of a bean is a thin, white tissue that is difficult to separate from the testa. Both of these coats have developed from the wall of the ovule.

27-14 The structure of a bean seed.

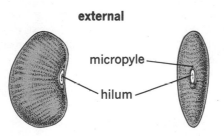

external

micropyle

hilum

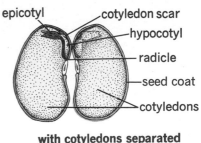

epicotyl

cotyledon scar

hypocotyl

radicle

seed coat

cotyledons

with cotyledons separated

If you soak a dried bean and remove the seed coats, the cotyledons will separate easily. The cotyledons fill the space within the seed coats and are fleshy and not at all leaflike.

Lying between the cotyledons are the other parts of the embryo plant. A fingerlike projection, the *hypocotyl,* fits into a protective pocket of the seed coats. This embryonic stem lies between the cotyledons and the *radicle* (embryonic root).

The *epicotyl* (sometimes called the *plumule*) is that part of the embryo plant that lies above the point of attachment of the cotyledons. It consists of two tiny leaves, folded over each other. Between them lies the minute bud that will later form the plant's terminal bud as the epicotyl develops into the shoot. Both the hypocotyl and the epicotyl will grow rapidly when the seed germinates. The cotyledons supply nourishment to both.

Structure of the corn kernel

Each corn grain is really a complete fruit, and therefore it corresponds to the bean pod and its contents rather than to the individual bean seed. However, there is only one seed in each grain. This one seed completely fills the fruit, the outer coat of the kernel having been formed from the flower's ovary wall. A very thin inner seed coat, only one cell layer thick, is fastened tightly to the outer seed coat. This inner seed coat developed from the wall of the ovule.

The micropyle is covered by the fruit coat, but there is an obvious point of attachment of the corn fruit to the cob. This structure corresponds to the stalk of the bean's flower and is the pathway through which the developing fruit receives its nourishment.

On one side of a grain of corn, there is a light-colored, oval area that marks the location of the embryo. This is plainly visible through the fruit coat. Near the top of the kernel, on the same side as the embryo, is a tiny point, the *silk scar,* where the style was attached.

If you cut a grain of corn lengthwise through the region of the embryo, you can see the internal parts clearly, especially if you put a drop of iodine solution on the cut surface. The endosperm fills much of the seed (Figure 27-15). This part of the seed developed from the endosperm nucleus after fertilization. The endosperm contains starch (which turns blue when treated with iodine) and sugar. The embryo, however, does not contain starch, but it does contain considerable protein. Sweet corn stores sugar in the endosperm, whereas field corn stores starch. For this reason, we eat sweet corn but not field corn.

The embryo, consisting of a very small hypocotyl, an epicotyl, a cotyledon, and a radicle, lies on one side of the corn grain. The radicle points downward toward the point of attachment, and is surrounded by a protective cap. The epicotyl is also protected

external

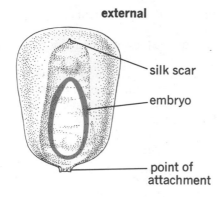

silk scar

embryo

point of
attachment

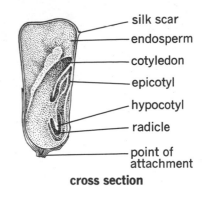

silk scar
endosperm
cotyledon
epicotyl
hypocotyl
radicle
point of
attachment

cross section

27-15 The structure of a kernel
of corn.

by a sheath or cap. The leaves of the epicotyl are rolled, not folded as they are in the bean, into a compact spear.

The corn has only one cotyledon. It is attached to the epicotyl and to one side of the very short hypocotyl, and it lies against the endosperm. During germination, the cotyledon digests and absorbs food from the endosperm. Notice that in the corn grain most of the energy-yielding food is stored outside the embryo rather than in the cotyledon, as we found in the bean (Table 27-2).

Dormancy in seeds

Many seeds go through rest periods before they germinate. Such rest periods, or periods of dormancy, may last for a few weeks, an entire season, or several years. Many plants bear seeds in the fall, and their seeds are normally dormant throughout the winter but germinate during the following spring or summer.

Drought, cold, and heat are all enemies of the embryo, although it is enclosed in protective seed and fruit coats. When conditions

Table 27-2   COMPARISON OF CERTAIN DICOTYLEDONOUS AND MONOCOTYLEDONOUS SEEDS

BEAN	CORN
testa with hilum and micropyle plainly visible	hilum and micropyle covered by a three-layered fruit coat; true seed coat lies inside fruit coat
two cotyledons	one cotyledon
large embryo	small embryo
no endosperm at time of dispersal	large endosperm
epicotyl fairly large	epicotyl rather small
epicotyl leaves folded	epicotyl leaves rolled
fruit a pod with several seeds	fruit a single grain with one seed

are favorable for growth of a particular seed, however, the period of dormancy ends, and *germination,* or sprouting, begins.

While some seeds may lie dormant for several years and still remain alive, there is a limit to the length of this period. Some seeds may live for almost one hundred years in a dormant state and then germinate when conditions become satisfactory. On the other hand, some seeds, like the maple, germinate almost immediately after falling from the tree, with the result that you frequently see a large number of young maples starting to grow under the parent tree in the late spring.

In annuals that grow in colder climates, seeds are the only form in which the plants can survive the winter months. Their period of dormancy normally extends from one growing season to the next. Similarly, the seeds of many perennials lie dormant through the winter months and germinate the following spring or summer.

The ability of seeds to germinate is called *viability*. Seed viability depends on the conditions during dormancy and on the amount of food stored in the cotyledons and endosperm. Cool, dry places are ideal for storing seeds, while warmth and moisture lower viability. Commercial seed growers run viability tests and mark the results on the various lots of seeds they sell. If you check the reported viability test, you can find out what percentage of germination to expect. If a lot of seeds has a viability of 92 percent, you can expect 92 seedlings from each 100 seeds you plant. Remember, however, that viability may vary, since relatively few representative samples are used in each test.

Conditions for germination

For germination, most seeds require moisture, the correct temperature, and oxygen. The amount of each of these required for germination varies greatly in different species.

Seeds of many water plants germinate under water, where there is plenty of moisture, quite an even temperature, and sufficient oxygen dissolved in the water. The seeds of most land plants cannot germinate under water.

Before a seed germinates, it usually absorbs considerable water, causing the seed coats to soften and the seed to swell. But too much warm moisture during the growing season encourages the growth of fungi, which may cause the seeds to decay.

The temperature at which seeds germinate best is also variable. A maple seed can germinate on a cake of ice, but growth will be slow and survival very uncertain under such a condition. Others, like corn, require much higher temperatures, a range of from 60° to 80°F being the most suitable for the majority of seeds.

During germination, the cells of a seedling divide very actively. This increased activity requires a much higher rate of respiration than that of an older plant, and you can see, therefore, why the oxygen supply to a seedling is critical. It is for this reason that the soil in a garden should be loose and the seeds planted near the surface.

As we have already mentioned, much of the food stored in the cotyledons or endosperm of a seed is starch. The starch is changed to sugar through the action of an enzyme known as *amylase,* and the sugar is used by the cells of the growing embryo. This change accounts for the sweetish flavor of sprouting seeds and explains why sugar is extracted from sprouting grain (malt) and why soybean sprouts are sometimes used in cooking.

Growth of the seedling

Because we have seldom stopped to think very much about it, few of us realize what an interesting process germination is. The way in which the seed germinates and the seedling establishes itself varies in different kinds of plants and in the location of the seed during germination. If the seed is lying on the surface, the root must penetrate the soil from above, and the epicotyl will grow freely upward. If, on the other hand, the seed is completely buried, the epicotyl must grow through the soil and unfold its

27-16 Germination of a bean seed.

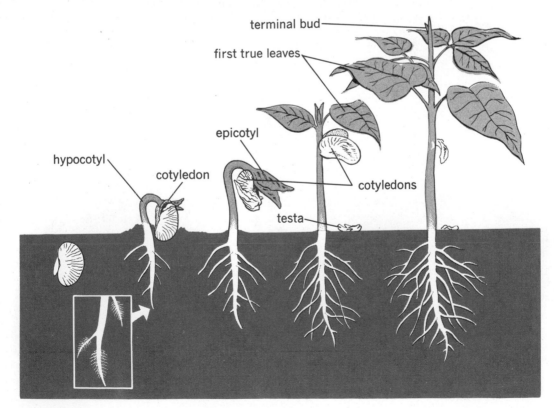

terminal bud

first true leaves

epicotyl

hypocotyl

cotyledon

cotyledons

testa

leaves above the surface while the root grows downward. We shall follow the stages in the germination of a bean seed and a grain of corn and see how this is accomplished.

Figure 27-16 shows the stages in the germination of a bean. After the bean has absorbed water and softened its seed coats, the hypocotyl grows out through the seed coat. The root grows downward and forms the primary root of the seedling, while the hypocotyl is growing upward and forming an arch that pushes its way to the surface. After the hypocotyl arch appears above the ground, it straightens out and lifts the cotyledons upward to form the shoot. Then the minute leaves unfold, forming the first foliage leaves of the plant. These are true leaves and are retained throughout the life of the plant which, in this case, is an annual. The plant dies after its fruit has matured.

The stem lengthens rapidly, developing more leaves, and the small bud that was between the epicotyl leaves of the seed develops as the terminal bud of the plant. The cotyledons remain attached to the stem for a time, below the true leaves. But as the plant becomes better able to supply its own food, the cotyledons wither and finally fall off.

The corn embryo also takes in water after it has been planted, and its root pushes through the softened fruit and seed coats. This

27-17 Germination of a kernel of corn.

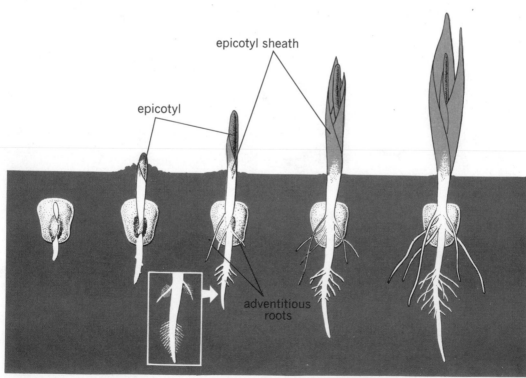

epicotyl sheath

epicotyl

adventitious roots

root forms a temporary primary root. Branch roots that develop from the primary root and later from the bottom of the stem add to this primary root. The leaves of the epicotyl, which are tightly rolled and encased in a sheath, penetrate the surface of the soil. After reaching the surface, the leaves unroll, and the stem continues its growth upward to form the cornstalk. Neither the hypocotyl nor the cotyledon of a corn grain grows above the surface of the soil, unlike the situation in the germination of the bean (Figure 27-17).

SUMMARY

The study of sexual reproduction in this chapter should have given you a new concept of flowers and fruits. A flower, to most people, is attractive and showy. To the biologist, however, it is a stage in the reproductive cycle of a flowering plant in which pollen grains or ovules, or both, are produced. Thus, a flower may be a beautiful rose or lily, a willow catkin, or the young ear and tassel of a corn stalk. Similarly, it may seem strange to call a bean pod, a grain of corn, or a walnut a fruit but all of these meet the biological definition: "a ripened ovary, with or without associated parts."

Did it ever occur to you that you could carry a whole flower garden or a forest in your pocket? Each seed in such an assortment contains an embryo plant, complete with food supply, and "packaged for delivery" in protective seed coats. Given proper conditions, a seed germinates and the embryo plant emerges as a seedling, destined by its genetic make-up to be a garden flower or a giant forest tree.

BIOLOGICALLY SPEAKING

vegetative reproduction	pistil	pollination
cutting	ovule	pollen tube
layering	pollen sac	sperm nucleus
pedicel	microspore mother cell	endosperm nucleus
receptacle	microspore	double fertilization
sepal	tube nucleus	endosperm
grafting	generative nucleus	embryo
budding	megaspore mother cell	testa
floral part	micropyle	hilum
calyx	megaspore	hypocotyl
corolla	embryo sac	radicle
petal	polar nucleus	epicotyl
stamen	synergids	viability
pollen	antipodals	

QUESTIONS FOR REVIEW

1. Give examples of natural vegetative reproduction by runners and tip layering.
2. Describe the use of hormone preparations in promoting the growth of adventitious roots on cuttings.

3. Describe three methods of vegetative propagation by layering.
4. Distinguish between the stock and scion portions of a graft. How must these be placed for the graft to be successful?
5. Name three kinds of grafts used in propagating woody plants.
6. Describe budding as a method of plant propagating.
7. The sepals and petals of a flower are often called accessory parts of a flower. What purpose do they serve?
8. In what respect are the stamens and pistil essential flower parts?
9. List the parts of a stamen and a pistil.
10. How do microspore mother cells differ from microspores in chromosome number?
11. Describe the formation of megaspores in an ovule.
12. Identify the eight nuclei present in the embryo sac at the time of fertilization.
13. Name three common agents of pollination.
14. Describe the growth of the pollen tube after pollination.
15. Describe the fertilization of the egg and the triple fusion that produces the endosperm nucleus.
16. What part of the seed is produced from the zygote? from the endosperm nucleus?
17. What is the biological meaning of the term *fruit*?
18. Describe several ways in which fruits are dispersed.
19. Identify and locate the parts of an embryo plant as illustrated in the bean.
20. Name three conditions necessary for seed germination.
21. List the sequence of stages that occurs in the germination of a bean and early growth of the seedling.
22. How does the young shoot of the corn plant force its way through the soil?

APPLYING PRINCIPLES AND CONCEPTS

1. Discuss the importance of vegetative propagation in the commercial growing of cultivated varieties of fruit trees and ornamental trees and shrubs.
2. Why is it necessary that the cambiums of a stock and scion be adjacent in making a successful graft?
3. Plants produce much larger numbers of pollen grains than ovules. Of what survival value is this?
4. Why is early spring flowering an advantage to wind-pollinated trees?
5. Discuss various characteristics of insect-pollinated flowers that serve as devices for attraction.
6. Discuss the importance of the pollen tube in reproduction of a flowering plant.
7. Compare and contrast the gameotophyte and sporophyte generations of the mosses, ferns, and flowering plants. Discuss the evolutionary significance of the differences.
8. A seed will not germinate unless it has enough water to soften the seed coats. How is this an automatic safeguard against germination during unfavorable conditions?
9. What structural adaptations that favor cross-pollination exist in flowers?

RELATED READING

Books

ANDERSON, A. W., *How We Got Our Flowers.*
Dover Publications, Inc., New York. 1966. A fascinating history of the common flowers of America.

FULLER, HARRY J., and ZANE B. CAROTHERS, *The Plant World* (4th ed.).
Holt, Rinehart and Winston, Inc., New York. 1963. A broad presentation of the main features of structure, physiological activities, and reproduction in the plant world, especially in the flowering plants.

HUTCHINS, ROSS E., *This Is a Flower.*
Dodd, Mead and Co., New York. 1963. An intriguing volume which presents the whole fascinating story of flowers.

U. S. DEPARTMENT OF AGRICULTURE, *Seeds.*
Superintendent of Documents, U.S. Government Printing Office, Washington, D.C. 1961. A volume about seeds—their importance, life processes, production, processing and marketing, and so on.

LIVING SYSTEMS

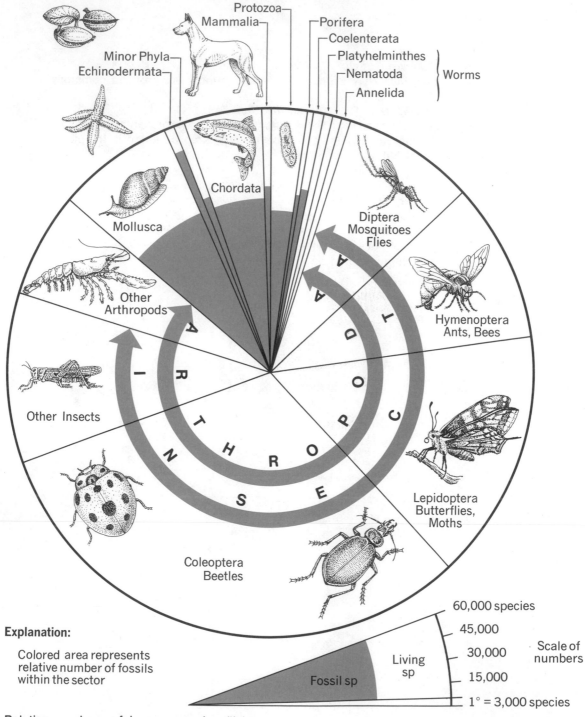

Explanation:

Colored area represents relative number of fossils within the sector

Fossil sp

Living sp

Scale of numbers

60,000 species
45,000
30,000
15,000
1° = 3,000 species

Relative numbers of known species (living and fossil) of various animal phyla.

BIOLOGY OF THE INVERTEBRATES

As you can see from the opposite diagram, more than 95 percent of all members of the animal kingdom are invertebrate organisms; that is, organisms that do not have backbones. As you will find, the invertebrates have a variety of fascinating forms and ways of life. In size, these animals vary from a minute beetle only $1/128$ of an inch (0.2 millimeters) in length to the giant squid which may grow to over 40 feet (12.2 meters). The biologist views the invertebrates with interest because his studies of these organisms have provided many answers to questions about life, and, in turn, have helped man to understand himself.

As you examine these various groups of animals without backbones, you will notice that we start with the simplest organisms and move on to those with increased complexity. We hope this will enable you to understand why some of the common theories about evolution have been proposed.

Are all animals composed of tissues? Do they all have hearts and circulatory systems? What clues can the invertebrates provide us about the origin of the kidney? Why are the insects considered to be a successful group? In your study of the invertebrates you may find answers to these and other important questions about life.

CHAPTER TWENTY-EIGHT

SPONGES AND COELENTERATES

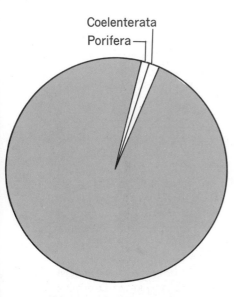

Coelenterata
Porifera

Pie chart showing relative numbers of porifera and coelenterata.

The advantage of association

Do cells living in close association have an advantage over those living independently? Perhaps we can begin to answer this question by reviewing the story of Robinson Crusoe. After Crusoe's ship was wrecked, he found himself isolated on an island. He had to obtain everything necessary for his life and comfort through his own efforts. He had to catch and prepare his food, make his own clothes and shoes, build a shelter, and protect himself from any enemies that might be on the island.

Even though he learned how to do all these things, he could not devote enough time to any one job to excel at it. However, if ten men had been shipwrecked with Crusoe, each could, through experience, have become particularly good at doing a certain job. One could have become a hunter; another, a shelter-builder; and still another, a clothes-maker for the group. The ten could have constituted a small society that functioned more efficiently than Crusoe did as he lived his simple, solitary existence. The many-celled animals that you will begin to study now have just such an advantage over the protists: Their cell specialization permits greater efficiency.

Division of labor and interdependence

As you learned in your study of the cell in Chapter 4, increase in numbers allows for *division of labor*. In multicellular organisms, different cells specialize in performing certain functions for the benefit of all the cells. The modification of a cell to perform a certain activity is called *specialization*. As you study the members of the animal kingdom, you will find that the increase in complexity accompanies an increase in cell specialization. As a result, the animal is better able to adjust to its environment.

Although division of labor allows cells to become specialists, at the same time the cells become more dependent on one another. This is called *interdependence*. In a similar way, a man who devotes his life solely to one task will to some extent lose his ability to perform others. He will then call on specialists for help. The ameba can live independently in a pond. A muscle, nerve, or bone cell, however, cannot live and function normally when it is removed from the body. Cell specialization is carried to a very high degree in the *vertebrates,* the animals with backbones. In this unit, you will study the *invertebrates,* the animals without backbones.

The animal phyla

In man's attempt to study and understand living things, he searches for regularities in the characteristics of living things. For example, the backbone is a characteristic conveniently used to separate the vertebrates from the invertebrates. Since about 95 percent of the animal kingdom is composed of invertebrates, other characteristics must also be used in their classification. As you will learn, these regularities are important to the biologist in many ways. Similarities may indicate evolutionary relationships, but you should also keep in mind that all living things must satisfactorily perform the life functions in order to continue to survive. Therefore, similarities also indicate common solutions to the problems of staying alive. As you read about the various animals, ask yourself: What important structural features place this animal in its group? Where does the animal live? How does it satisfy its organic needs? How does the environment affect it? How does it affect other organisms in the environment?

We shall study ten large animal phyla. As you go along, you may use the Classification Table in BIOLOGY INVESTIGATIONS for reference and review. The characteristics and representative examples of each group are given. You will notice that there are many more than ten animal phyla and that there are certain classes within the ten phyla that we do not discuss. We shall limit our study to the largest and most important phyla, beginning with the sponge phylum, which includes some of the least complex organisms in the animal kingdom.

The sponges

Most sponges are marine, although there are a few fresh-water species. Living sponges vary in color from white, gray, brown, red, orange, and yellow to purple and black (Figure 28-1). They may live singly or in colonies so massed together that they form an encrusting layer over the surface of a rock. Individual sponges vary in size from a fraction of an inch to two yards in diameter.

When you first look at a living sponge, you may conclude, as Aristotle did, that it is an interesting plant. You might possibly change your mind about its being a plant, however, if you put a drop of India ink in the water near the sponge. You would see the ink particles pass into the body of the sponge and reappear

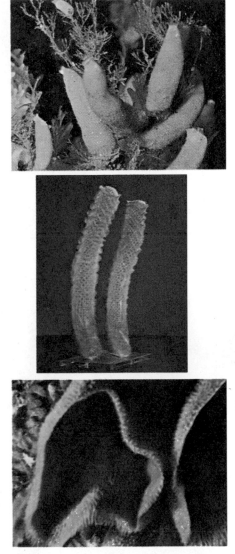

28-1 The appearance of sponges may depend upon the material composing the skeleton. These sponges have a loose needlelike skeleton of calcium carbonate. (top) These beautiful Venus's flower basket skeletons are made of silicon dioxide. (center) Note the beautiful shape of the pink vase sponge. (bottom) (Walter Dawn; American Museum of Natural History; A. Grotell)

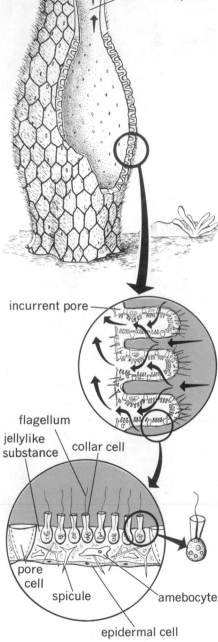

excurrent pore

incurrent pore

flagellum

jellylike substance

collar cell

pore cell

spicule

amebocyte

epidermal cell

28-2 Water is continually drawn into the sponge by the flagella of the collar cells. It passes through small pores into the cavity of the sponge and out through the osculum.

as if they were being forced out from the largest pore. Apparently, something the sponge does sets up currents in the water. If you had time, a good microscope, and several sponges, you could determine that water had passed into the sponge's body through small *incurrent pores,* and out through the larger *excurrent pore* (Figure 28-2). The excurrent pore is also called the *osculum* (OSS-kyooh-lum). Because of their many pores, the sponges are grouped in the phylum *Porifera,* which means "pore-bearing."

We usually think of an animal as being visibly and actively engaged in pursuing, catching, and eating its food. The sponge, however, is *sessile* (SESS-ill), which means that it is permanently attached to a substratum by its base. It must, therefore, draw its food to it. How does it do this? A single sponge 10 cm high and 1 cm in diameter has been observed to pump more than 22 quarts (22.5 liters) of water per day through its body. This is nearly a quart per hour. The sponge regulates the size of the osculum to control the rate of flow or even to stop it completely. Food, in the form of diatoms, small protozoans, bacteria, and organic particles, is carried to the sponge in the water it draws through itself. The sponge acts as a living filter, removing food as well as oxygen from the water. Carbon dioxide and other wastes leave the sponge in the water that passes out through the animal's excurrent pore.

Let us take a closer look at the sponge to see how it is adapted for its submerged, sedentary way of life. A simple sponge consists of a hollow body, whose wall contains many tubes. The wall is formed by two layers of cells separated by a layer that contains a jellylike substance, loose cells, and *spicules* (SPI-kyoolz). The spicules are noncellular skeletal structures that give support to the body of the animal. The classification of the sponges is based on the composition of these spicules. Secreted by living cells, the spicules of some sponges are made of silicon and in others, of calcium carbonate (lime). In a third group of sponges, support is derived from a fibrous network of a tough but flexible substance called *spongin.*

The outside layer, or *epidermis,* is protective. Many cells of the inside layer are unusual in that they contain curious collars with flagella projecting through them. The flagella of these *collar cells* set up the currents that draw water into the sponge. As food particles enter, they are caught by the collars and are digested by enzymes within the collar cells. From here, digested food is absorbed by cells called *amebocytes* because they resemble amebas. The amebocytes wander throughout the jellylike substance between the cell layers, transporting digested food and oxygen to the other cells. These wandering cells also carry wastes and carbon dioxide to the collar cells for disposal. Although sponges are often called loose aggregations of cells, you can see that there is sufficient interdependence to warrant their classification as multicellular organisms.

Reproduction in the sponges

Sponges reproduce sexually, but they also may reproduce asexually in two ways. They may form *buds,* which are groups of cells that enlarge and live attached to the parent for a time, then break off and live independently. Or, during periods of freezing temperatures or drought, groups of cell masses surrounded by a heavy coat of organic matter form and break off from the parent sponge. These are called *gemmules.* They consist of little groups of amebocytes and a few spicules. When favorable conditions return, each gemmule has the capacity to grow into another sponge. Reproduction by gemmules is usually characteristic of fresh-water species.

Sponges reproduce sexually by developing eggs and sperms. The sperms are shed into the water and enter another sponge through the incurrent pores. They are taken into the cytoplasm of the collar cells and are then transferred to the egg by the wandering amebocytes. The fertilized egg develops into a flagellated larva that escapes from the sponge and, after swimming for a while, settles down and grows into a young sponge.

Sponges are able to regenerate missing parts. Because of this ability, sponge-growers are able to increase the number of sponges by cutting them in pieces, which are placed in special growing beds. This is not a method of reproduction normally used by the sponge, but it is utilized by commercial sponge fishermen because it is the fastest way of increasing the sponge population.

Fresh-water sponges are small and of no commercial importance. Some marine sponges with spongin skeletons, especially those in warmer oceans, grow to be very large and are collected by divers and by men using drag hooks. The sponges are then piled on shore or hung on the rigging of the boat until the flesh has decayed. The remaining spongin skeletons are washed, dried, sorted, and sometimes bleached. They are then ready for marketing. Famous sponge fishing grounds include the Mediterranean and Red seas, the waters around the West Indies, and portions of the Gulf Coast of Florida.

The coelenterates

Coelenterates have more specialization than you observed in the sponges. You may have had a pleasant swim in the ocean turned into a painful experience from the stings of a jellyfish. Strange indeed are these pulsating creatures that bob around in the ocean currents and dangle long, stringy tentacles under a floating, inflated sac. The phylum *Coelenterata* (suh-LEN-tuh-RAY-tuh) also includes the hydroids, corals, sea fans, sea anemones, and Portuguese man-of-war. Members of this phylum vary in size from microscopic forms to the largest jellyfish of the North Atlantic, which may reach a diameter of twelve feet. Coelenterates live either as individuals or united to form colonies. All of these animals are aquatic; a few live in fresh water, but most are marine.

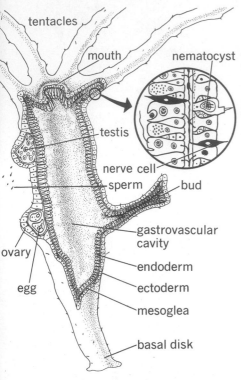

tentacles

mouth

nematocyst

testis

nerve cell

sperm · bud

gastrovascular cavity

ovary

endoderm

egg

ectoderm

mesoglea

basal disk

28-3 The external and internal structures of the hydra can be seen in this diagram. The animal has two layers of cells with a jellylike material between.

28-4 The hydra uses its stinging cells to paralyze its prey. Then the tentacles coordinate their activity to push the food into the body cavity. (Sol Mednick)

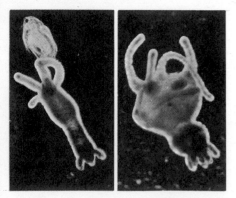

The hydra

The characteristics of this phylum can be seen in *Hydra,* a genus of common fresh-water coelenterates. There are white, green, and brown species, which live in quiet ponds, lakes, and streams. The body of the hydra consists of two cell layers separated by a jellylike material called *mesoglea* (MEZ-uh-GLEE-uh). The outside layer is the *ectoderm;* the inner layer is the *endoderm* (Figure 28-3). The baglike body of the animal has a single opening that is surrounded by *tentacles* bearing stinging cells, each of which contains a structure called a *nematocyst.* Nematocysts are characteristic only of the coelenterates.

Hydras vary in size from less than half an inch, including the tentacles, to about one and a half inches. They attach themselves to rocks or water plants by means of a sticky secretion from the cells of their *basal disks.* Because of their small size, transparency, and habit of contracting into little knobs when disturbed, hydras are often overlooked. Yet they are abundant and are the only really successful organisms among the few members of their phylum that have invaded fresh water. A hydra may leave one place of attachment and float or move to another, or it may secrete a bubble at the base and float to the surface upside down. The hydra sometimes moves by a kind of somewhat odd, type of somersaulting motion.

How the hydra gets food

When a small animal comes in contact with one of the tentacles of a hydra, many nematocysts are explosively discharged, and they pierce the victim's body with tiny, hollow barbs. Since each barb is attached to the tentacle by a thin thread, the combined effect of many threads prevents the escape of the hapless victim (Figure 28-4). At the base of each barb is a small poison sac that discharges its contents through the hollow barb and into the prey, thus paralyzing it.

Once the prey has been paralyzed, the tentacles bend inward and push the prey through the circular mouth and into the body cavity of the hydra. Specialized endodermal cells that line this space function in digestion and absorption. For this reason, the space is called a *gastrovascular cavity.* Some of the cells that line the cavity secrete digestive enzymes that cause the partial breakdown of the prey. The partially digested substances are ingested by the lining cells, in which digestion is completed.

Digestive wastes are expelled through the mouth. The mouth and digestive cavity of the coelenterate are considered to be advances over the feeding structures of the sponge. They allow a greater range in the types and sizes of food that the animal can utilize. Metabolic wastes are discharged by the cells directly into the water. Gaseous exchanges of respiration also occur directly between the water and the cells.

The behavior of the hydra

We have now observed one coelenterate reaction that shows a definite advance over the more primitive sponge. The tentacles of the hydra are coordinated in catching food and pushing it into the mouth. Also, if you touch a tentacle of an extended hydra with a needle, the body and all the tentacles contract suddenly. The stimulus to one tentacle travels to cells of the other tentacles and to the body through a series of nerve cells. This *nerve net* lies in the mesoglea (Figure 28-5). The contraction itself is accomplished by the shortening of slender fibers that lie in the ectoderm. These fibers can be compared to the muscle cells of higher animals. The hydra has no nervous system such as that found in higher animals, and it has no brain. Its nerve cells, in contrast with those of higher animals, conduct impulses in all directions.

28-5 The nerve net of the hydra lies in the jellylike layer between the ectoderm and endoderm. It transmits impulses and allows coordination of the hydra's activities.

Reproduction in the hydra

The hydra accomplishes *asexual reproduction* by forming *buds*. A bud appears first as a knob growing out from the side of the adult as shown in Figure 28-6. Later, this knob develops tentacles, and after a period of growth, it separates from the parent and lives independently. In this method of reproduction, the bud is a small outgrowth of endoderm and ectoderm and is capable of growing into a new organism. Bud formation in the hydra is not at all similar to that of a plant. You may remember that the plant bud is an undeveloped shoot. Its growth results in stem elongation or leaf or flower formation, but it will not become a whole new organism.

Like the sponge, the hydra has remarkable powers of regeneration. If a hydra is cut into pieces, most of the pieces will regenerate the missing parts and become whole animals.

Sexual reproduction usually occurs in autumn. Eggs are produced along the body wall in little swellings called *ovaries;* motile sperm cells are formed in similar structures called *testes.*

The egg is fertilized in the ovary, and the zygote grows into a spherical, many-celled structure with a hard, protective cover. In this stage, it leaves the parent and goes through a rest period before forming a new hydra.

Two ways of life

The body form of a coelenterate is one of two types. The *polyp* form is well illustrated by the hydra, with its tubular body that has a basal disk at one end and tentacles at the other. The bell-shaped, free-swimming form found in the jellyfish is called a *medusa.* A medusa swims by taking water into the cavity of the bell and then forcibly ejecting it. This jet propulsion produces a jerky movement through the water.

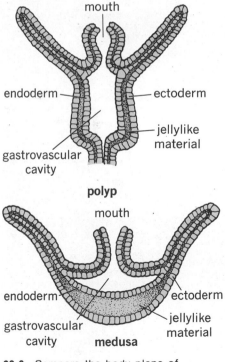

28-6 Compare the body plans of the sessile polyp and the free-swimming medusa.

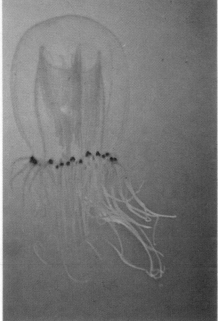

An interesting example of a coelenterate whose life cycle includes both the medusa and the polyp phases is *Aurelia,* a common jellyfish. The sexually reproducing medusa has a scalloped margin from which protective tentacles hang. The male medusas shed sperms into the sea. The sperms may enter the gastrovascular cavity of a female, where they fertilize eggs that have been released by the female. The zygotes are protected for a short time by folds of tissue surrounding the mouth.

When the young are released, they are small, oval-shaped, ciliated swimming larvae called *planulae* (PLAN-yuh-LEE) (singular, *planula*). The planula is the beginning of the asexual phase in the life cycle of *Aurelia* (Figure 28-8). After swimming about for a short period of time, the planula attaches to a rock or seaweed, develops tentacles, and begins feeding. It is then that the coelenterate is considered to have become a polyp. The polyp grows and forms more polyps from buds that develop at the base. In fall and winter, however, the polyp elongates and forms many horizontal divisions until it resembles a pile of saucers. One by one, starting from the top, the "saucers" break loose, swim away, and develop into adult, sexually reproducing medusas. The *Aurelia* is thus one coelenterate that has two different forms—the polyp and the medusa—both of which reproduce.

28-7 Compare these coelenterates with the diagrams in Figure 28-6. (Al Giddings, Bruce Coleman, Inc.; © Woodbridge Williams)

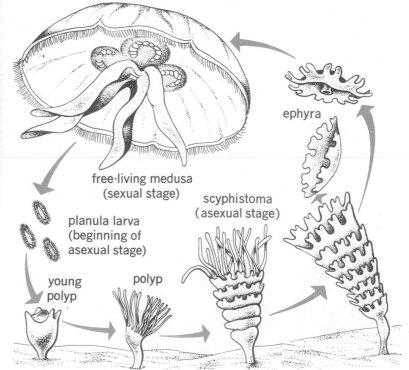

free-living medusa (sexual stage)

planula larva (beginning of asexual stage)

young polyp

polyp

scyphistoma (asexual stage)

ephyra

28-8 Life cycle of *Aurelia*.

Other coelenterates

The coral is the only coelenterate of economic importance. Its body is a small, flowerlike polyp only a fraction of an inch long. Most coral polyps live in colonies and build skeletons of lime, which they extract from the sea. This lime skeleton is firmly cemented to the skeleton of a neighboring polyp. When one animal dies, its skeleton remains and serves for the attachment of another polyp. Lime skeletons of coral thereby increase in size until a single mass may support many thousands of animals living on the surface of the skeletons of their ancestors. In some species, these masses are solid; other corals build delicate, intricate, fan-shaped structures.

Over a period of time, large coral reefs are built up. These are most common in the warm, shallow oceans and are of three types: (1) the *marginal type,* close to the beach; (2) the *barrier type,* in which a ring is formed around an island, with a wide stretch of water between the beach and the reef; and (3) a ring called an *atoll* with an open lagoon in the center.

The Great Barrier Reef off the northern coast of Australia extends about 1,260 miles parallel to the coast. It is about 50 miles wide. The extent to which coral formations exist became apparent during World War II, when vast amounts of coral were used in the construction of airstrips and roads. Coral jewelry can be found in shops all over the world. Some corals are also bleached and dyed various colors so that they can be used in flower arrangements or for decoration in homes.

After storms on the Pacific Coast of the United States, the beaches may be covered with bluish, membranelike animals that measure two to three inches in length. These animals belong to

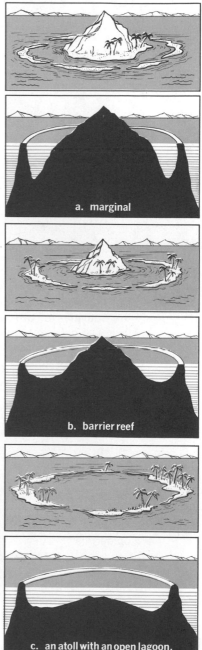

28-9 Three different kinds of coral reefs. a. marginal; b. barrier reef widely separated from the land; c. an atoll with an open lagoon.

a. marginal

b. barrier reef

c. an atoll with an open lagoon.

28-10 Like the Portuguese man-of-war, *Velella* is a passive drifter, traveling with the winds and currents. (D. P. Wilson)

28-11 *Physalia*, the Portuguese man-of-war, is a colonial coelenterate. (George Lower, National Audubon Society)

the genus *Velella;* they are commonly known as purple sails or by-the-wind sailors. Related to the purple sail is *Physalia,* or the Portuguese man-of-war (Figure 28-11). These two organisms are essentially colonies of polyps. Each polyp performs a special function in the colony. A single, large polyp, to which the other polyps are attached and from which they are suspended, acts as a float for the entire colony. This large polyp float keeps the entire coelenterate colony near the surface; wind blowing against the float causes the colony to move through the water. Other specialized polyps digest food caught by the food-getting polyps. Still other polyps specialize in gamete production. Although the Portuguese man-of-war is largely tropical and semitropical, it is found in the Gulf Stream, in which it occasionally drifts as far as the English coast.

SUMMARY

Variation in structure and function of cells in a multicellular organism brings about division of labor, but, at the same time, one cell becomes dependent upon the activities of others. The sponge is the first step in this direction with its two layers of cells and central cavity. The coelenterates are considered to be more advanced than the sponges as they exhibit not only cell differentiation but also coordination of activities made possible by a nerve net. In the coelenterate, a mouth leads to the gastrovascular cavity which is lined with endoderm. Since digestion begins here, larger food can be utilized than in the sponges, and feeding does not have to be continuous.

BIOLOGICALLY SPEAKING

vertebrate	collar cell	tentacle
invertebrate	amebocyte	nematocyst
incurrent pore	bud	basal disk
excurrent pore	gemmule	gastrovascular cavity
osculum	mesoglea	nerve net
spicule	ectoderm	polyp
epidermis	endoderm	medusa

QUESTIONS FOR REVIEW

1. How does the multicellular condition permit efficient division of labor?
2. What are major structural characteristics that distinguish the sponges from the coelenterates?
3. In what kinds of environments would you expect to find specimens of sponges? Where would you expect to find specimens of hydra? Do you

think the size of the animal body has anything to do with the particular habitats of these two groups of organisms?

4. Compare the feeding methods of a sponge and a hydra. How are they alike, and how do they differ?
5. Give examples of regeneration in sponges and coelenterates.
6. In what ways are the *Aurelia* and hydra similar? In what ways are they different?
7. What is the function of the nematocysts?
8. Describe the formation of a coral reef. What are the three types of coral reefs?
9. Why are the purple sail and the Portuguese man-of-war considered to be so much more highly specialized than any of the other coelenterates you have studied thus far?

APPLYING PRINCIPLES AND CONCEPTS

1. Compare the ways in which cell specialization is similar to division of labor in modern civilization.
2. Of what value to a jellyfish is the fact that it is hollow?
3. How is knowledge of the regeneration capabilities of sponges used commercially?
4. How do you account for the fact that some sponges die when exposed briefly to air, even when promptly returned to water?
5. Would you consider a sponge to be a more primitive form of life than a coelenterate? Give reasons for your answer.
6. How do the sponges and coelenterates spread to new habitats? What condition do you believe would be most favorable for their growth?

RELATED READING

Books

BARRINGTON, E. J. W., *Invertebrate Structure and Function.*
Houghton Mifflin Co., Boston. 1970. An excellent reference for concepts in anatomy and physiology as seen in the invertebrates.
BUCHSBAUM, RALPH, *Animals Without Backbones: An Introduction to the Invertebrates* (rev. ed.).
The University of Chicago Press, Chicago. 1948. The main groups of invertebrates, introduced in simple, non-technical language, with excellent drawings and photographs.
GABB, MICHAEL, *The Life of Animals Without Backbones.*
Ginn and Company, Boston. 1966. An overall picture of these animals and their life cycles.
GEORGE, JEAN C., *Spring Comes to the Ocean.*
Thomas Y. Crowell Company, New York. 1966. Describes how some of the myriad forms of undersea life react to spring and propagate their species.
PIMENTEL, RICHARD A., *Invertebrate Identification Manual.*
Van Nostrand-Reinhold Company, New York. 1967. A well illustrated guide to common invertebrates from sponges to insects.
SHEPHERD, ELIZABETH, *Jellyfishes.*
Lothrop, Lee and Shephard Co., New York. 1969. A simple book on the structures and their functions as seen in the jellyfish.

CHAPTER TWENTY-NINE

THE WORMS

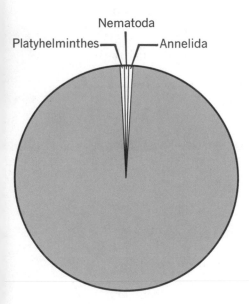

Nematoda

Platyhelminthes———————Annelida

29-1 About 20,000 species of worms have been classified.

Looking for similarities

By now you have observed many different organisms. You have learned the importance of similarity in structure for use in classification. Similarity in the shapes of an organism reveals much more to the biologist. The general form of an organism is referred to as its *symmetry*. The ameba, for example, has no definite shape when it moves about, and is said to be *asymmetrical*. It can orient itself to the environment by moving in any direction with its pseudopods. The radiolarians and *Volvox,* on the other hand, are protists with *spherical symmetry*. They meet the environment on all surfaces of the organism and can be divided into two equal parts by any plane passing through the diameter of the body (Figure 29-2a). A baseball is a good example of a spherically symmetrical object. Organisms with spherical symmetry often lack an efficient method of locomotion and usually float on or near the surface of the water.

For most organisms however, gravitational orientation plays a role in their lives. Consider, for example, the sea anemone. This organism has tentacles at one end that are well suited for reaching around in its immediate environment, and a basal disk at the other end that is well suited for attachment. With this arrangement, the direction of attachment of its body is downward.

Furthermore, the sea anemone has *radial symmetry* (Figure 29-2b). As in most other coelenterates, the sea anemone has a central disk from which the tentacles radiate like spokes of a wheel. Through the mouth and the center of the body passes an imaginary line, the *central axis*. The animal can be divided into two equal parts by any plane that passes through the diameter of the basal disk and the central axis. Some simple sponges, most coelenterates, and most adult echinoderms, which we shall study in Chapter 30, are radially symmetrical.

In the last chapter you read about the functioning of the nerve net in the radially symmetrical hydra. Such a network of nerves allows the animal to sense its environmental changes on all sides. However, the reaction of the hydra to most stimuli is a general overall response of drawing in its tentacles and shortening its body. The same reaction is observed whether the stimulus is a touch on a tentacle or the basal disk. It is observed when a strong chemical is added to the water. It is observed if heat is suddenly applied to one surface. These stimuli are changes in the environment of the hydra, and the reaction of the animal is caused by the spread of impulses over the nerve net. The nerve net is considered by the biologist to be a primitive type of nervous system. The general reaction of the hydra to a strong stimulus is said to be *negative,* since it is a *withdrawal* reaction. As you have observed, the hydra is also capable of a coordinated *positive* reaction when food is present. Radially symmetrical organisms are adapted for a sessile existence.

Actively moving organisms are better adapted for their way of life by having *bilateral symmetry.* This means, literally, "two-sided shape." Only one plane can separate animals with this kind of symmetry into two similar parts. This plane must pass through the longitudinal axis, the center of the back, and the center of the front (Figure 29-2c). This plane would not divide the animal into identical parts. Each piece would be the mirror image of a corresponding piece on the opposite side of the plane. Animals with bilateral symmetry have a definite right and left side. They have an upper, or *dorsal,* surface and a lower, or *ventral,* surface, a definite front, or *anterior,* end and hind, or *posterior,* end. All the vertebrates and many invertebrates have this type of symmetry.

Many organisms with bilateral symmetry have a concentration of nerves and sense organs at the anterior end. Thus, as they move forward, they can sense and more readily react to their environment. You can see why this would be of survival value to the animal.

A group of bilateral organisms

Examine the diagram on page 450, and you will see that the three phyla comprising the worms make up a very small portion of the numbers of known species of organisms. As you will learn,

29-2 Three types of symmetry.

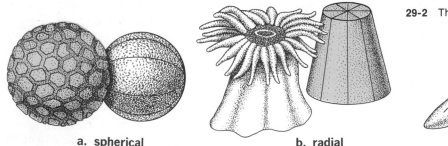

a. spherical   b. radial   c. bilateral

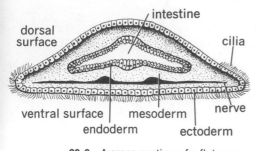

29-3 A cross section of a flatworm showing the relationship of the three layers of cells.

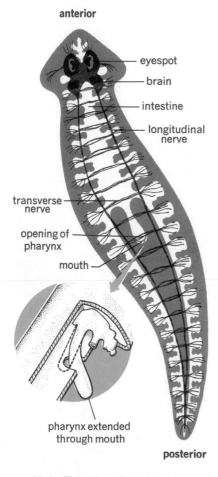

29-4 The digestive and nervous systems of a planarian. The pigmented eyespots can distinguish light and dark.

however, the numbers of individuals play a significant role among living things.

Of the three main worm phyla we will study, the least complex worms are included in the *Platyhelminthes* (PLAT-ee-HEL-MINTH-eez), or flatworm phylum. These flat-bodied animals have three layers of cells: the ectoderm and endoderm, as in the coelenterates, and in addition, a middle layer called the **mesoderm** (Figure 29-3). While the tissues of the two-cell-layered organisms are formed from the ectoderm and the endoderm, organs and organ systems are found in animals containing all three cell layers. This will also be true in all the other animals we shall study, including man. The flatworm phylum is divided into three classes: *Turbellaria, Trematoda,* and *Cestoda.*

A free-living flatworm

The most common examples of the Turbellaria, a group of free-living flatworms, are the *planarians.* (When biologists say that an organism is *free-living,* they mean that it is not parasitic.) Planarians are aquatic, and are found under stones in fresh-water ponds and streams. The next time you are out collecting, tie a string to a piece of liver and leave it on a submerged rock for a few hours. Bring the rock into class and put it into an aquarium filled with pond water. The following day, you will probably find many planarians crawling on the glass of the aquarium.

Planarians range in length from one-fourth to one-half inch and vary in color from black or brown to white. They are bilaterally symmetrical, blunt at the anterior end, and pointed at the posterior end (Figure 29-4). The two **eyespots** on the anterior end are responsible for the planarian's nickname, "the cross-eyed worm." These eyespots are *photosensitive;* that is, light striking them stimulates nerves in this area, and the animal, which usually avoids bright light, responds accordingly.

The **pharynx** is a tube located on the ventral suface of the animal. When extended, this tube sucks up microscopic particles, including tiny organisms. Since planarians clean the water they live in by eating organic matter, they are considered scavengers. When food is drawn into the digestive cavity, it enters any of the three main branches of the intestine and then passes to one of the side branches. Cells lining the intestine ingest the food particles and digest them in food vacuoles. Digested food passes to all body tissues by diffusion. Indigestible waste materials are eliminated through the pharynx and mouth opening. Lacking a separate circulatory system, cellular wastes are collected by tubules that branch throughout the animal. The system has several tiny excretory pores that open on the surface of the worm's body.

Compared to the animals we have studied so far, the nervous system of planarians is well developed. A mass of nerve tissue, the "brain," lies just beneath the eyespots at the anterior end of the organism. Many nerves from the anterior region lead directly to the brain. In a bilaterally symmetrical organism, this arrangement of nerves is very important. As the worm moves, its anterior end is the first to receive stimuli of chemicals, water currents, touch, light, and heat. The planarian is thus able to test the environment it is moving into and, if unfavorable conditions are found to exist, it can move off in another direction. For this reason, the concentration of *receptors* (receivers of stimuli) and nerves in the anterior end of a bilaterally symmetrical organism is of survival value.

The two *longitudinal nerves* run along either side of the body near the ventral surface. These nerve cords are connected by *transverse nerves,* giving the nervous system a ladderlike appearance. Many small nerves go from the surface area to the longitudinal nerves. This type of nervous system gives the planarian coordination of movement and permits it to respond to stimuli on all parts of its body.

When you watch planarians in a glass dish or an aquarium, you will see that they move in two distinct ways. One is a muscular movement whereby the anterior end of the body moves from side to side. The other is a forward gliding motion that is accomplished by almost imperceptible muscular contractions aided by the movement of cilia on the ventral surface.

Reproduction in the planarian is accomplished either asexually by fission or sexually by gametes. Each animal is **hermaphroditic,** which means that both male and female reproductive organs exist within the same body. Cross-fertilization occurs, and the eggs are shed in capsules. The capsules, which usually contain fewer than ten eggs, are often attached to rocks or twigs in the water. In two or three weeks, the eggs hatch, and tiny planarians emerge.

Like many sponges, coelenterates, and plants, many planarians have remarkable powers of regeneration, but there are some interesting differences. Perhaps you have started a geranium plant from a slip. Roots developed on the end of the stem that was pushed into moist sand and leaves developed on the other end. Would the results have been the same no matter which end of the cutting had been in the sand? The answer is no. The stem has organization such that the part nearest the roots of the original plant will develop roots, and the part nearest the top of the plant will only produce leaves. This organization, called a *field of orientation,* cannot be changed even by planting the cutting upside down.

Some planarians will regenerate complete worms from *almost* any piece (Figure 29-5a). The field of orientation indicates that an organism is more than the sum of its parts. If a planarian is cut into anterior and posterior sections, the anterior piece will regenerate a new tail, and the posterior portion will grow a new head. The organization is said to be from anterior to posterior, because there is a small section of the tail that appears not to be capable of regeneration.

A planarian can be made to develop two heads if it is cut longitudinally, as in Figure 29-5b, but it will soon divide completely

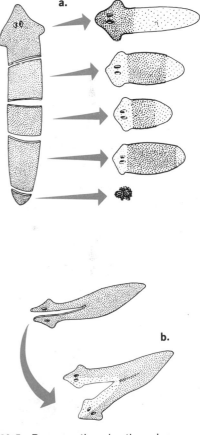

29-5 Regeneration in the planarians shows an organization of orientation. (a) The anterior most part of each section except the last will form a new head. (b) Another animal is shown to form two heads. This animal will complete the separation and become two planarians.

to form two animals. Although biologists are doing research on fields of orientation, we really know very little about the controlling mechanisms. In general, we can say that the more complex the organism, the less is its ability to regenerate lost parts. Crabs can grow new claws, and lizards can regenerate lost tails. Man can produce new skin and some muscles and nerves, but not complete organs.

The parasitic way of life

Parasitic flatworms have no ability to replace lost parts. This is true of most parasitic animals. When we speak of evolution, usually we think of forms increasing in complexity as they adapt to environmental changes. But a parasite living within the body of another animal is faced with problems not at all like those of its free-living relatives. The size of a parasite is limited by the size of its host. If the parasite were to grow to a large size, it would kill its host. Intestinal parasites usually have hooks or suckers by which they can cling to the walls of the host's intestine. This feature prevents them from being swept away with the movement of the intestinal contents. The environment literally bathes the parasite with already digested food from which it merely has to absorb its nutrients. The parasite is protected from being digested itself by a thick, resistant cuticle not found in all of its free-living relatives.

Since certain systems are reduced or lost in the parasitic worms, we say that the worms have *degenerated*. The tapeworm, for example, has no digestive system. This is actually a benefit to the parasite, because more space is available for developing eggs.

Dispersal is a problem to the internal parasite. Larval forms of parasitic worms may be free-living or may live within another organism. Let us look at the life histories of some common parasitic worms.

The flukes

The class Trematoda includes the *flukes,* which are parasites in many animals, including man. Flukes differ from planarians in that adult flukes have no external cilia. Flukes have thick **cuticles** and one or more *suckers,* which they use to cling to the tissues of the host. The anterior sucker surrounds the mouth, which opens to a short pharynx. Although the nervous system of the fluke is similar to that of the planarian, there are, as you might expect, no special sense organs. Most of the flukes have highly developed reproductive systems and are hermaphroditic. The reproductive system occupies a larger part of the fluke's body than it does in the planarian. The fluke also differs from the planarian in having a **uterus.** The uterus is a long, coiled tube in which large numbers of eggs are stored until they are ready to be discharged through the *genital pore.*

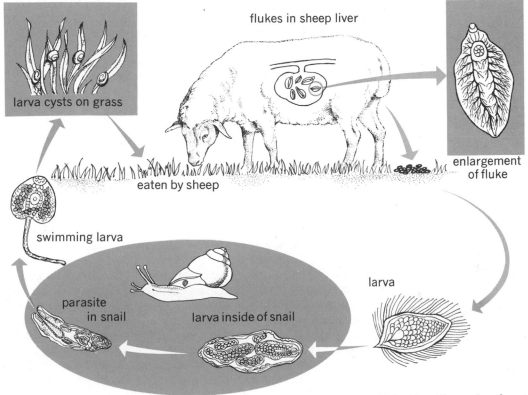

flukes in sheep liver

larva cysts on grass

enlargement
of fluke

eaten by sheep

swimming larva

larva

parasite
in snail

larva inside of snail

29-6 The life cycle of a sheep
liver fluke.

The flukes have complicated life histories. A portion of the
fluke's life cycle is spent in a snail and another portion in one or
more other hosts. The sheep-liver fluke, for instance, lives as an
adult in the liver and gall bladder, usually of a sheep. Eggs pass
from the gall bladder to the intestine. If they fall into water after
they are eliminated by the sheep, they hatch into young worms
called *larvae* (singular, *larva*). These larvae then enter the body
of a particular kind of snail. In the snail, the larvae pass through
several stages, during which they increase in number by asexual
reproduction. They then leave the snail, crawl on blades of grass
growing along the edge of the water, and form *cysts*. If a sheep
eats this cyst, the fluke enters the sheep's liver, and the cycle
starts again (Figure 29-6).

Reactions of the host to the liver fluke may include inflammation, swelling, general sickness, and irritability. In cows, milk
production is reduced. Naturally, fluke infestation renders the
liver unfit for human consumption. Other types of flukes may live
in the blood, intestines, or lungs of animals. Many flukes are found
on the external gills of fish and in cavities of other aquatic vertebrates, living on the epithelial tissue and blood of these hosts.

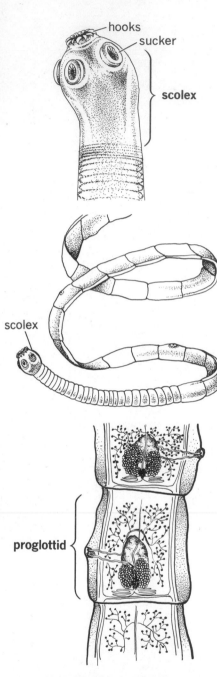

29-7 The structure of a tapeworm. The tapeworm can infect man only when he eats insufficiently cooked meat that is infested with scolex-containing cysts.

Although human fluke infestations are most common in the Orient, they are found in other areas too. Cuba has had several epidemics of human liver fluke. In some areas of the Gulf Coast, economic loss caused by liver-fluke infection of cows, pigs, and sheep has been severe. The best way to control flukes is to eliminate one of the hosts in their life cycle. As a biologist working on this, which host would you try to eliminate?

The tapeworms

The tapeworms, the best known of the parasitic flatworms, are members of the class Cestoda. An adult tapeworm has a flat, ribbonlike body and is grayish-white in color (Figure 29-7). The adult tapeworm has no cilia. The knob-shaped "head," called a *scolex,* is equipped with suckers and in certain species with a ring of hooks. It lacks a mouth and has no digestive structures. The adults are usually found attached to the intestinal wall, bathed in nutrients, and absorb already digested food through their body walls.

Below the slender neck, a number of nearly square sections extend to as great a length as thirty feet. The worm grows by adding new sections. New body sections, or *proglottids* (proe-GLOT-idz), are formed at the anterior end, and the oldest sections are at the posterior end. The proglottids are essentially masses of reproductive organs. Tapeworms are hermaphroditic, and eggs formed in a proglottid are fertilized there. When the eggs mature, the proglottids break off and are eliminated in the host's feces. Proglottids released in this way may be eaten by some animal, such as a pig or a cow. In the body of the new host, the eggs hatch into larvae that burrow into the muscles and form cysts.

Tapeworms enter the human body in the cyst stage if the improperly cooked flesh of an infested animal is eaten. Each cyst contains a fully developed tapeworm scolex. In the human intestine, the scolex is released from the cyst, attaches itself to the intestinal wall, and begins to grow.

Since a tapeworm robs its host of nourishment, the victim may lose weight and vitality. In recent years, human tapeworm has been decreasing because of improved detection and treatment methods and because meat is inspected for tapeworm cysts.

The roundworms

The phylum *Nematoda* (NEM-uh-TOE-duh) has only one class, also called *Nematoda.* Commonly called *roundworms,* nematodes are long, slender, smooth worms, tapered at both ends. They may be as short as $1/125$ of an inch or as long as 4 feet. They occur in soil, fresh water, salt water, and as parasites in plants and animals. The parasitic roundworms include the hookworm, trichina worm, pinworm, whipworm, *Ascaris,* and guinea worm. The nematodes are so abundant that—

If all the matter in the universe except the nematodes were swept away, our world would still be dimly recognizable, and if, as disembodied spirits, we could then investigate it, we should find its mountains, hills, vales, rivers, lakes, and oceans represented by a film of nematodes. The location of towns would be decipherable, since for every massing of human beings there would be a corresponding massing of certain nematodes. Trees would still stand in ghostly rows representing our streets and highways. The location of the various plants and animals would still be decipherable, and, had we sufficient knowledge, in many cases even their species could be determined by an examination of their erstwhile nematode parasites. (Ralph Buchsbaum, *Animals Without Backbones.* The University of Chicago Press, 1948.)

When you consider that more than one third of the human population, mostly in warm regions of the world, are infested with parasitic roundworms, the importance of these worms can hardly be overestimated. Harmless roundworms include the vinegar eel and the numerous useful soil nematodes.

Like the flatworms, the roundworms are bilaterally symmetrical and have three cell layers. They are, however, more complex than the flatworms. Their digestive system is a distinct tube with an opening at each end, housed in a long body that is also a tube (Figure 29-8). This arrangement enables the animal to take food in through an anterior opening, the *mouth,* to digest the food, and to remove from it the usable parts as it passes along the canal. Finally, the undigested material is eliminated through a posterior opening, the *anus.* In the hydra, one opening is used for both the entrance of food and the elimination of undigested substances. The arrangement of a tube within a tube, however, makes possible the orderly progression of material through the digestive tract and greater efficiency in handling it.

Ascaris is a large roundworm that lives in the intestines of pigs, horses, and sometimes man. The females are larger than the males and may reach a length of nearly twelve inches. *Ascaris* eggs enter the human being in contaminated food or water. They do not hatch in the stomach; but they begin to hatch within a few hours when they reach the small intestine. Once hatched, the larvae bore into the intestinal wall to begin a ten-day journey through the body. This journey carries them into the bloodstream and to the lungs. When they reach the lungs, these worms pass into the air passages, up to the throat, and are swallowed, once again passing into the digestive tube. There they grow to maturity in about two and a half months. After fertilization, the eggs are surrounded by a thick, rough shell and passed out through the genital pore of the female worm. A mature female lays about 200,000 eggs each day. These eggs pass out of the body of the host with the feces, and the cycle continues. For the most part, *Ascaris* seems to be relatively harmless in man, although oc-

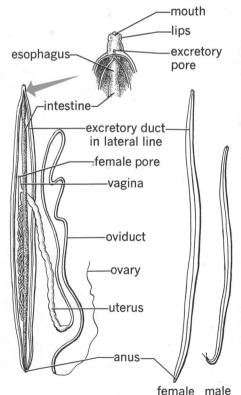

29-8 *Ascaris,* a parasitic roundworm. Side views of a male and female are shown at the right. The dissected worm is a female.

casionally a large number of adult worms twist together, block the intestine, and cause death.

The *hookworm* of the southern states and all semitropical and tropical regions is a far more serious health menace. Larvae develop in the soil and enter the body by boring through the skin of the feet. Then they enter the blood vessels and travel through the heart to the vessels of the lungs. In the lungs, they leave the blood vessels, enter the air passages, and eventually reach the trachea and travel to the throat as *Ascaris* larvae do. They are then swallowed and pass through the stomach to the intestine, where they attach themselves to the wall by means of jaws. In the intestine, the larvae grow to adult worms that suck blood from the vessels in the intestine wall.

Loss of blood lowers the victim's vitality by producing anemia. A typical hookworm victim may be quite sluggish, and his growth may be retarded, although the latter is not always true. The worms reproduce in the intestine, and the fertilized eggs are passed out in the fecal wastes. If these eggs happen to lodge in warm, moist soil, they develop into tiny larvae that can enter the human body through cracks in the feet and eventually travel to the intestine. Thus, three factors are responsible for the spread of this disease: (1) improper disposal of sewage; (2) warm soil; and (3) the practice of going barefoot. Public health agencies have done a remarkable job in reducing the number of cases in the southern part of the United States.

The *trichina worm,* or *Trichinella,* is one of the most dangerous of the parasitic roundworms. This roundworm passes its first stage as a cyst in the muscles of a pig, dog, rat, or cat. If uncooked meat scraps from infested animals are fed to pigs, it is likely that the scraps will contain cysts. In the intestine of a pig that eats uncooked, infested meat scraps, the larvae develop into adult worms, mate, and produce microscopic larvae that pass into the bloodstream and into muscles, where they again form cysts. When a human being eats undercooked, infested pork, the same thing happens: The cysts are released, and the larvae mature in the intestine (Figure 29-9). Each worm discharges into the bloodstream about 1,500 young, which eventually form cysts in the muscles of the human being. This disease is known as *trichinosis.* One method of preventing it is to feed hogs only cooked meat scraps. However, the best way to prevent this disease, as well as other parasitic worm infections, is to cook all meat as thoroughly as possible and thereby avoid any risk.

By this time, the importance and widespread abundance of parasitic worms should be obvious to you. Each of these disease-producing worms is spread and picked up as a result of poor sanitary conditions. The eggs of the worms can be killed by proper sewage disposal. Proper inspection and thorough cook-

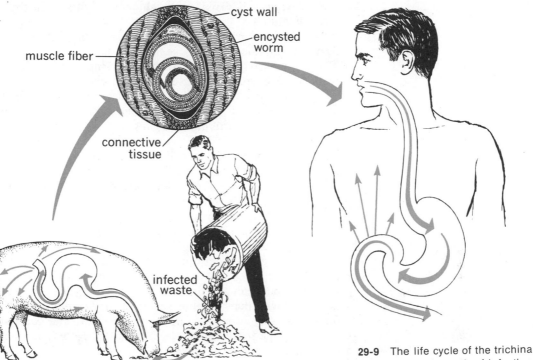

muscle fiber

cyst wall

encysted worm

connective tissue

infected waste

29-9 The life cycle of the trichina worm. Trace the path of infection that leads to man.

ing of meat are other measures that have reduced the spread of these parasites.

The common earthworm

The segmented worms are the most advanced of the worms in body structure. They belong to the phylum *Annelida,* which is commonly divided into four classes (see the Workbook). Most of the segmented worms live in salt water; some live in fresh water; and others, including the common earthworm, live in the soil. The annelids seem to fall midway between the simple protists and the highly complicated arthropods. For this reason, and because they are rather common, many biologists study them closely. They are considered to be typical invertebrates.

If you examine a common earthworm, you will notice immediately that its body consists of many segments. You will also see that the anterior end is darker and more pointed than the posterior end. There is no separate head, nor are there any visible sense organs. The mouth is on the anterior end and is crescent-shaped, lying below a *prostomium* (proe-STOME-ee-um), which is a kind of upper lip. The vertical slit at the posterior end, the *anus,* is the opening of the intestine. The segments, starting with the segment containing the mouth, are often numbered by biologists. On segments 32–37, there is a conspicuous swelling called

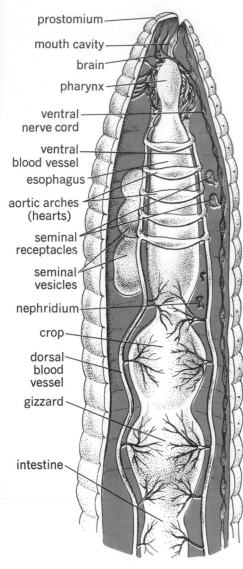

prostomium

mouth cavity

brain

pharynx

ventral nerve cord

ventral blood vessel

esophagus

aortic arches (hearts)

seminal receptacles

seminal vesicles

nephridium

crop

dorsal blood vessel

gizzard

intestine

29-10 The external and internal structure of the earthworm. The anterior portion above is dissected to show the well-developed nervous and circulatory systems. One of the abdominal segments is shown in cross section.

the *clitellum* (kli-TELL-um), which is involved in the animal's reproduction.

Four pairs of bristles, or *setae* (SEE-tee), project from the under-surface and sides of each segment except the first and the last. The setae assist the earthworm in moving and in clinging to the walls of its burrow, as those who hunt night crawlers can testify. The earthworm moves by pushing the setae of its anterior segments into the soil, then shortening its body by using a power-full series of *longitudinal muscles* that stretch from the anterior to the posterior end. The worm then pushes the setae of its posterior segments into the soil, withdraws its anterior setae, and pushes forward through the soil by making itself longer. It does this by constricting the *circular muscles,* which are found around the body at each segment.

As you study the earthworm, you will notice that it consists not only of many cells but also of many kinds of specialized cells. These specialized cells are grouped together. Each group performs the same function; each group, therefore, makes up a *tissue.* The tissues that are grouped together to form larger structures that perform a definite function are *organs.* The earthworm is so arranged that a whole series of organs performs some fundamental body process. These are *systems.* The earthworm has well-developed *digestive, circulatory, excretory, nervous, muscular,* and *reproductive systems.*

The digestive system of the earthworm

Below the prostomium is the *mouth* of the earthworm. There are no jaws or teeth, but the animal uses its muscular *pharynx*

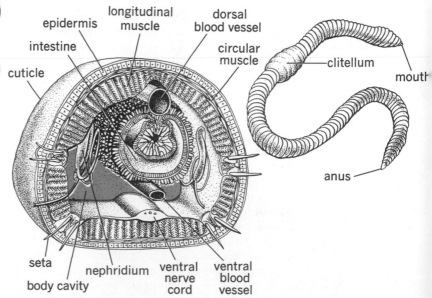

epidermis

intestine

cuticle

longitudinal muscle

dorsal blood vessel

circular muscle

clitellum

mouth

anus

seta

body cavity

nephridium

ventral nerve cord

ventral blood vessel

to suck in soil containing food. The food particles then pass through a long *esophagus* into a round organ called the *crop.* This acts as a temporary storage place for food. From the crop, the food is forced into a very muscular organ called the *gizzard.* The rhythmical contractions of the gizzard cause grains of sand to rub the food particles together, thus grinding the food up. In the *intestine,* which stretches from segment 19 to the end of the worm, complete digestion takes place. Enzymes break down the food chemically, and the blood circulating through the intestine walls absorbs it.

The complex organs of the digestive system of the earthworm take up most of the anterior region of the organism. The earthworm consumes large quantities of soil, which contains organic matter. The useless inorganic matter passes through the system largely unchanged, and is deposited, sometimes on the surface of the ground, in the form of *casts.* Hence, as the earthworm feeds, it loosens the soil, which results in increased air and water penetration, and its wastes add to soil fertility.

The circulatory system of the earthworm

As food is digested, the blood in the circulatory system picks it up for distribution to all the cells of the body. In the simpler animals we have studied so far, the digested food had only to diffuse a short distance in order to reach all the cells of the body. But in higher forms, the distances are greater, and more food material is needed by the many specialized and active cells of the body. In these higher animals, we find a special transportation or distribution tissue – the circulating fluid called *blood.*

The blood of the earthworm moves through a series of closed tubes, or vessels. It flows forward to the anterior end in a *dorsal blood vessel* and moves to the posterior end in a *ventral blood vessel.* Small tubes connect the dorsal and ventral vessels throughout the animal, except in segments 7–11. There the five pairs of connecting tubes are large and muscular. By means of alternate contraction and relaxation, they keep the blood flowing. Not true hearts, they are called *aortic arches.*

Respiration and excretion in the earthworm

The earthworm absorbs oxygen and gives off carbon dioxide through a thin skin. This skin is protected by a thin *cuticle* secreted by the epidermis and kept moist by a slimy mucus also produced by epidermal cells. Recall that a moist surface is necessary for oxygen to be absorbed and carbon dioxide to be given off. If the worm is dried by the sun, it will die because the exchange of gases can no longer take place.

Nitrogen-containing waste materials from cell activities are removed to the outside of the body by little tubes. There are two such structures, called *nephridia,* in each segment except the first three and the last. Each corresponds to a tiny kidney tubule in man.

Earthworm sensitivity

The nervous system coordinates the movements of the animal and sends impulses received from sense organs to certain parts of the body. There is a very small nerve center in segment 3. From it run two nerves that form a connecting collar around the pharynx and join to become a long *ventral nerve cord*. There also are enlarged nerve centers, called *ganglia*, in each segment. Three pairs of nerves, in turn, branch from each ganglion. The earthworm has neither eyes nor ears, but is nevertheless sensitive to light and sound. Certain cells in the skin are sensitive to these stimuli, and the impulses are carried rapidly to the muscles of the earthworm. Think how quickly earthworms can react to a flash of light at night when you hunt them for the next day's fishing!

29-11 Reproduction in the earthworm. Even though the earthworm has the reproductive organs of both sexes, it exchanges its sperms for those of another earthworm. As shown, the sperms travel from the seminal vesicles of one worm to the seminal receptacles of the other.

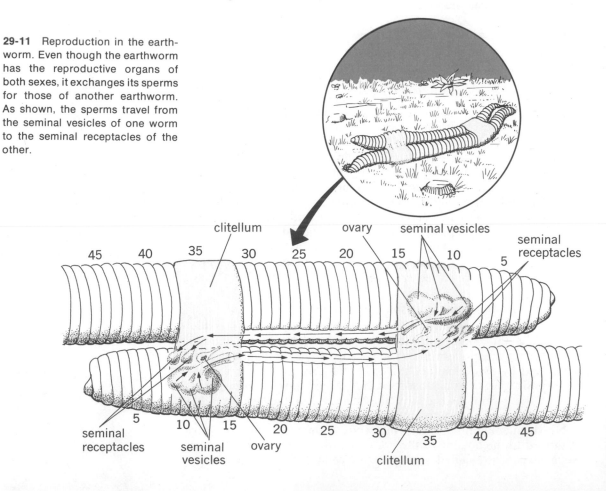

Reproduction in the earthworm

Although earthworms are hermaphroditic, forming both eggs and sperms, the eggs of one worm can be fertilized only by sperms from another worm. *Seminal vesicles* (Figure 29-11) extend from the testis sacs and store sperms produced by two pairs of testes within the sacs. The *seminal receptacles* store sperms from another worm.

Sperms from one worm travel from the seminal vesicles, through openings in segment 15, to the seminal receptacles of another worm, through openings in segments 9 and 10. Here they are stored until eggs are laid. When the eggs mature later, they pass from the ovaries through openings in segment 14, and are deposited in a slime ring secreted by the clitellum. As this ring moves forward, sperms are released from the seminal receptacles, and fertilization occurs. The slime ring slips from the body and becomes the cocoon in which the young worms develop.

The leech—another segmented worm

The *leech* (Figure 29-12), also called the bloodsucker, is an annelid found in streams and ponds. It is an external parasite on fish and other aquatic animals, but it may attach itself to your skin while you are swimming.

29-12 Examples of common worms.

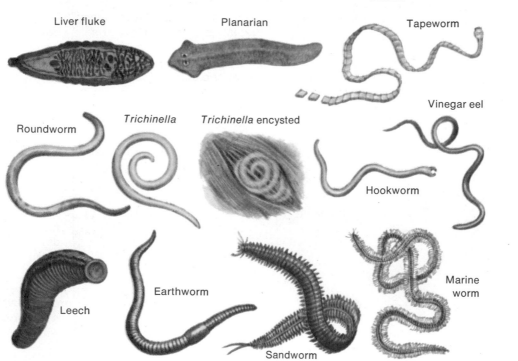

Liver fluke

Planarian

Tapeworm

Roundworm

Trichinella

Trichinella encysted

Vinegar eel

Hookworm

Leech

Earthworm

Sandworm

Marine worm

In sucking blood, a leech attaches itself by the posterior sucker to some vertebrate, applies the anterior sucker to the skin, and makes a wound with the aid of little jaws inside the mouth. The salivary glands of the leech secrete a substance that prevents the clotting of blood while the worm is taking a meal. Leeches were used frequently for medicinal purposes in the Middle Ages and later, when it was thought beneficial to draw blood. Now the salivary substance is extracted and used to slow clotting after surgery.

Table 29-1   SUMMARY OF CHARACTERISTICS OF THE COMMON WORMS

	PLATYHEL-MINTHES	NEMATODA	ANNELIDA
Body Types	flat, unsegmented bodies	round, unsegmented bodies	body divided into segments
Type of Life	many parasitic	some parasitic	majority *not* parasitic
Organization	3 layers, many organ systems	3 layers, many organ systems	3 layers, organs well developed
Digestive System	open at one end only	mouth and anus	mouth and anus
Reproduction	asexually by fission; sexually—hermaphroditic with cross-fertilization	sexual, definite male and female	sexually—hermaphroditic with cross-fertilization
Circulation	none	none	5 pairs of aortic arches, large dorsal vessel, small ventral vessel
Nervous System	2 longitudinal nerve cords	2 nerve cords, one dorsal, one ventral	one large ventral nerve cord with ganglia

SUMMARY

The symmetry of an organism is considered to be of evolutionary significance as it determines an organism's mode of life. Bilateral symmetry of the worms is considered an advance over the sponges and coelenterates, as it is accompanied by a concentration of nerve tissue and sense organs at the anterior end of the organism. Thus, the environment can be tested as the animal moves.

The animals we have discussed in this chapter have three cell layers. The worms inhabit fresh water, salt water, the land, and exist as parasites in a variety of hosts. Examples of degeneration are found in the parasitic forms. The roundworms are much more abundant than you had perhaps realized. They include harmless free-living forms as well as the parasitic hookworm, trichina worm, pinworm, whipworm, *Ascaris,* and guinea worm. The annelids are organized at the system level of development and possess a nervous system which illustrates a significant evolutionary complexity over that of the planarians. The common earthworm and leech are examples of annelids.

BIOLOGICALLY SPEAKING

symmetry	cyst	gizzard
eyespot	scolex	intestine
pharynx	proglottid	blood
cephalization	prostomium	aortic arch
hermaphroditic	clitellum	nephridium
field of orientation	setae	ganglion
cuticle	esophagus	seminal vesicle
uterus	crop	seminal receptacle
larva		

QUESTIONS FOR REVIEW

1. Name and define the types of symmetry. Give an example of each.
2. Discuss regeneration in the planarians.
3. What is the significance of the three layers of cells found in flatworms?
4. How does the planarian test its environment?
5. In what respects are the flatworms more complex than the sponges and coelenterates?
6. In what way does the tapeworm show degeneration?
7. Where are nematodes found?
8. How do the nematodes show an advance over the flatworms?
9. Describe the life cycle of *Ascaris.*
10. How does the trichina worm reach the human body?
11. Describe how the earthworm moves.
12. Trace a particle of food through the digestive system of the earthworm, naming the organs through which it passes.

APPLYING PRINCIPLES AND CONCEPTS

1. Why do we classify the various worms into three separate phyla?
2. What advantage does bilateral symmetry have over spherical symmetry? What advantages does it have over radial symmetry?

3. What is meant by the following statement: An organism is more than the sum of its parts.

4. Trace the path followed by *Ascaris* through the body by naming all the structures through which it passes. At what stages during its development are symptoms of disease most likely to be present? At what stages is treatment most likely to be effective?

5. Compare the advantages of the nervous system of the annelids over that of the planarians.

6. What measures is it important to take in trying to control parasitic worms?

7. Symptoms of tapeworm infestation usually include loss of weight and general fatigue. Account for these conditions.

8. Trichinosis can become a hopeless disease. Why is it almost impossible to treat?

RELATED READING

Books

BUCHSBAUM, RALPH, and LORUS J. MILNE, *The Lower Animals: Living Invertebrates of the World.* Doubleday and Company, Inc., Garden City, New York. 1960. A complete discussion of each animal, with beautiful photographs; a companion to Buchsbaum's *Animals Without Backbones.*

BURT, DAVID R. R., *Platyhelminthes and Parasitism — An Introduction to Parasitology.* American Elsevier Publishing Co., Inc., New York. 1970. A somewhat advanced, but well illustrated introduction to the study of parasitism.

CROLL, W. A., *Ecology of Parasites.* Harvard University Press, Cambridge, Mass., 1966. The life history and economic importance of each parasite, discussed in general terms.

HEGNER, ROBERT W., and E. A. STILES, *College Zoology.* The Macmillan Company, New York. 1959. A comprehensive zoology text that successfully balances structure, function, and principle. Good chapters on invertebrates.

OLSEN, OLIVER W., *Animal Parasites.* Burgess Publishing Company, Minneapolis, Minn. 1962. The biology and life cycles of animal parasites.

CHAPTER THIRTY

MOLLUSKS AND ECHINODERMS

Soft-bodied invertebrates

Undoubtedly, you have heard of clams, oysters, squids, and snails. These animals are all classed in the phylum *Mollusca*. Even though some of these mollusks are called shellfish, they are really not fish at all. In fact, many mollusks do not even have shells. Mollusks live in fresh water, as well as in marine and terrestrial environments. Some are adapted to live buried in sand or mud where the oxygen content may be too low for a more active animal. In abundance of species, the mollusks are surpassed only by the phylum that includes the insects.

Although mollusks and their products have been used for food, money, eating utensils, jewelry, buttons, dyes, tools, and weapons since earliest times, they have been of value to man in still another way. Since the shells of mollusks are hard, the shells or their imprints may remain in mud for thousands of years. Present-day mollusks can be compared with these fossils, thus giving a clue to the changes that have occurred in this form of life. Where layers of shells have accumulated in various strata of the earth, the geologist has been able to utilize them as a tool for dating as well as for reconstructing changes that have taken place in the earth's surface. Aggregations of certain shells or their impressions are also used by engineers to help determine the possibility of finding oil in various regions. What conclusion could you draw if, while you were hiking in the mountains, you found a large deposit of mollusk shells or fossils?

Although many people are not familiar with the internal anatomy of the mollusks, they may collect the shells of these animals. Beautiful shells may be found in homes all over the world. A shell from a mollusk collected on a sandy beach of the Philippines may be found in the living room of a home in Indiana. Some shells of the giant clam of the South Pacific may be found in New York gardens.

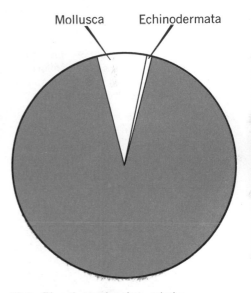

Mollusca   Echinodermata

30-1 Pie chart showing relative numbers of species of mollusks and echinoderms.

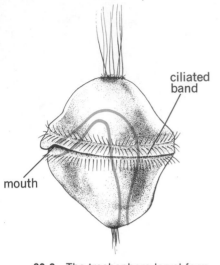

30-2 The trochophore larval form is found in the development of both mollusks and annelids. An outline of the digestive tract is shown in blue.

The binding link of the mollusks

In the first sentence of this Chapter, we stated that the clam, oyster, squid, and snail are placed in the same phylum. Perhaps you wondered why such entirely different organisms are placed in the same group. Biologists not only compare adult forms, but they consider animals with similar larval development to be related.

The animals considered to be mollusks pass through a stage of larval development in which they all look alike. This young mollusk is called a *trochophore* (TROCK-uh-FORE). The trochophore has a tuft of cilia at one end and a ciliated band around its equator (Figure 30-2). In free-swimming forms, the action of these cilia propels the larva through the water and brings food to the mouth. The terrestrial mollusks and many marine forms, however, pass through the trochophore stage while they are confined to the egg capsule. A trochophore stage is also found in the development of annelids. It is because of this that biologists consider the segmented worms and the mollusks to be related.

The general body plan

The body of an adult mollusk consists of a *head, foot,* and *visceral* (VISS-uh-rul) *hump* (Figure 30-3). The *visceral hump* contains the digestive organs, excretory organs, and the heart. It is covered by a *mantle,* a thin membrane that in some species secretes the calcium carbonate, or lime, shell. The *mantle cavity* is formed by the curtainlike mantle that hangs down over the sides and rear of the animal. The *gills,* which are the respiratory organs in the aquatic mollusks, are located in this cavity. Before being carried out of the mollusk, undigested matter passes from the anus into the mantle cavity.

In most mollusks, a current of water passes through the mantle cavity. Since the water carries in oxygen and food as well as carries away carbon dioxide and other wastes, you might conclude that the mollusks with shells have a sanitation problem. However, there is no problem because of the way in which the water circulates through the cavity. Water carrying oxygen and food enters the cavity through an opening called the *incurrent siphon,* a *ventral siphon.* After passing over the gills, the water, now carrying carbon dioxide and wastes, exits through the *excurrent siphon,* a *dorsal siphon.*

The type of shell or shells, if present, is the major characteristic dividing this phylum into classes. Of the five classes into which the mollusk phylum is divided, we will consider three in depth:

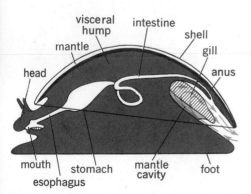

30-3 Although no mollusk living today looks like this drawing, the major characteristics of the phylum are shown. The arrows indicate direction of water flow in the mantle cavity.

■ *Pelecypoda* (PELL-uh-SIP-uh-duh), or hatchet-footed two-shell mollusks

■ *Gastropoda* (gas-TROP-uh-duh), or stomach-footed, single-shell mollusks

■ *Cephalopoda* (SEF-uh-LOP-uh-duh), or head-footed mollusks

30-4 Various mollusks. Giant clam and oyster. Mussels and scallop (below) (*Top,* Douglas Faulkner; *bottom left,* Dennis Brokaw; *bottom right,* Robert C. Hermes, National Audubon Society)

Mollusks with hinged shells

Such mollusks as the clams, oysters, scallops, and mussels are called *bivalves* because their shells are composed of two halves, called *valves.* A hinge connecting the two valves allows for opening and closing during various activities of the animal. Each valve is composed of three distinct layers made of substances secreted by the mantle. The smooth and glistening layer next to the mantle is the *pearly layer.* If a grain of sand or an encysted parasitic worm becomes lodged in the mantle, this layer builds up around the particle and forms a pearl. The middle layer consists of calcium carbonate crystals and is called the *prismatic layer.* The outermost layer, the *horny outer layer,* is very thin. When the shell dries, this layer resembles dried shellac or varnish and can be peeled off. Made of a hornlike organic substance, it protects the middle layer from being dissolved by the small amounts of acid that may be in the water. The hinge that connects the valves is also made of this hornlike secretion.

The scientific name of the group, Pelecypoda, which means "hatchet-footed," refers to the shape the foot assumes during movement. After the foot is extended into the sand, it is spread

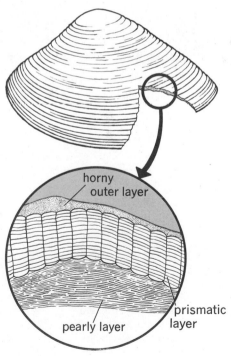

horny outer layer

prismatic layer

pearly layer

30-5 The shells of the bivalve mollusks are composed of three distinct layers.

30-6 Note how its hatchet-shaped foot aids the clam in digging into sand.

out to form a hatchet-shaped anchor (Figure 30-6). Then, when the muscle of the foot contracts, the mollusk is pulled rapidly into the sand or mud. The clam is a rapid digger, as some of you no doubt know!

A clam usually remains partially buried with its valves partly open and its two siphons extended into the water. Water is brought into the mantle cavity through the incurrent siphon (Figure 30-7). The water passes up through the gills and then past the anus, where wastes are excreted. The water then passes out through the excurrent siphon.

As water passes over the gills, two processes occur: (1) oxygen diffuses in, and carbon dioxide diffuses out; and (2) small particles of organic matter stick to a thin layer of mucus on the gills. Cilia on the surfaces of the gills carry the mucus up to the dorsal surface and then forward to the mouth. Animals that feed in this way are said to be *mucus feeders*. Many marine worms also collect their food in a mucus trap. Mucus feeders are scavengers, eating dead and decaying organic matter as well as the numerous microscopic protists that settle to the bottom.

If you have prepared clams for eating, you know that, even after steaming and the clams are agape, two large muscles must be cut before the shells can be pulled apart. These muscles are the anterior and posterior adductor muscles (Figure 30-7) which are capable of holding the shells tightly together when the animal is alive.

Bivalves have a well-developed nervous system with several large ganglia. The edges of the mantle contain sensory cells that are sensitive to contact and light.

Clams, oysters, scallops, and mussels are edible bivalves. However, since they feed on microscopic organisms, some of which

30-7 The structure of a clam. The arrows indicate the direction of water flow through the mantle cavity and the gill (gray area).

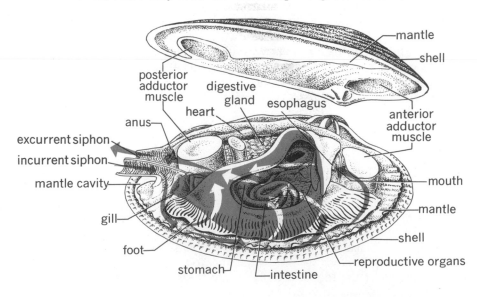

are poisonous to man, and since they are often eaten raw, care must be used in the gathering and selecting of those intended for use as food. Naturally, mollusks are inedible when their digestive tracts contain poisonous protists.

In the region of the East Indies and Australia, an interesting relationship occurs between a type of one-celled alga and the giant clam, which can be a yard and a half long. Immense numbers of these algae live in the mantle of the clam, giving it a green color. The algae manufacture starch and oils, and are often contained in ameboid cells that carry the algae to the digestive area of the clam.

Mollusks with one shell

Land and water snails, slugs, conches, and abalones belong to the group of mollusks called gastropods, which literally means "stomach-footed." Since most gastropods have only one shell, they are also called univalves.

To reconstruct the way in which a land snail might have developed from a hypothetical ancestor, compare Figure 30-3 to Figure 30-8. Try to imagine you are making the following changes in the generalized mollusk: First, rotate the visceral hump halfway around; next, remove the gills; last, put some spirals in the shell—it will now resemble a snail. The land snail is a good mollusk to study, because it is easily obtained and moves around freely while being observed. See if you can find the prominent structures of the land snail, which is diagramed in Figure 30-8.

The familiar land snail has certain adaptations that enable it to live successfully out of the water. Its mantle cavity serves as a modified lung; oxygen diffuses through the thin, membranous lining of the cavity. Since this lining must remain moist for gas exchange to occur, you can understand why snails are active during the evening, night, and early morning hours.

During times when the moisture in the air decreases, and when the temperature falls, the snails remain inactive inside their shells. When the animal retreats in its shell, the opening is closed by a chitinous "door." Water loss is further slowed by the secretion of mucus which hardens to form a seal. The terrestrial snail moves at a pace of about ten feet per hour. Glands in the flat, muscular foot secrete a layer of slime on which the snail travels. Movement is accomplished by a rhythmic, wavelike contraction of the muscles in the foot. The eyes of land snails are on the tips of two tentacles. When touched, the tentacles are drawn in and seem to vanish somewhat as the toe of a stocking vanishes when you turn it inside out.

The *slug* resembles a snail that has lost its shell. It you look closely at a slug, you will see the opening that leads to the mantle cavity, which is used in respiration. By controlling the size of this opening, the slug is able to conserve moisture in the mantle chamber. Like terrestrial snails, slugs are usually active at night,

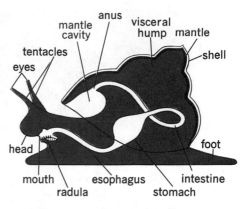

30-8 Compare and contrast the structure of the land snail shown here with that of the generalized mollusk shown in Figure 30-3, and with that of the clam in Figure 30-7.

30-9 Gastropods are terrestial, freshwater, and marine. (*Left,* Louis Quitt, Photo Researchers, Inc.; *center,* Anthony Mercieca, National Audubon Society; *right,* Earl Roberge, Photo Researchers, Inc.; *below,* Walter E. Harvey, National Audubon Society)

30-10 The octopus, a mollusk with well-developed eyes, moves by pulling itself over rocks by its tentacles or by expelling a jet of water from its excurrent siphon. The squid moves by expelling jets of water. (D. P. Wilson; Douglas Faulkner)

leaving trails of slime wherever they go. Many of you already know of the damage done to garden plants by slugs and snails.

Fresh-water snails in an aquarium are excellent subjects in which to observe the feeding mechanism of a gastropod. Watch to see the mouth of the animal open and a tonguelike structure scrape the glass of the aquarium. This structure is the *radula* (RADJ-uh-luh), which literally means "scraper." The radula is like a rasp and actually files algae from the glass sides of the aquarium. Now you can understand how garden snails and slugs, also equipped with radulas, can cause such damage as they feed on plant leaves.

"Head-footed" mollusks

The *cephalopods* (SEF-uh-luh-PODZ) include the octopus, squid, cuttlefish, and chambered nautilus. The octopus has no shell; the nautilus has an external shell; and the squid and the cuttlefish have an internal shell. The giant squid is probably the largest of all invertebrates, sometimes reaching a length of fifty-five feet and a weight of two tons. The octopus has a bad reputation, probably undeserved. As a rule, octopuses are timid creatures. A very large type of octopus, which lives along the Pacific Coast, typically has a body not more than one foot long, but its slender arms sometimes grow to a length of sixteen feet. Today, there are about four hundred species of cephalopods. When one compares this number with the more than ten thousand fossil forms that have been found, one wonders if the cephalopods are to become extinct, as many other organisms have.

30-11 Some examples of common echinoderms are the brittle starfish, common starfish, sand dollar, sea urchin, and sea cucumber. Echinodermata is the only major phylum composed entirely of marine animals. (*Top center,* R. D. Beeman; *others,* Douglas Faulkner)

The spiny-skinned invertebrates

The starfish, brittle star, sea urchin, and sand dollar are common members of the phylum *Echinodermata.* They have hard, radially symmetrical bodies covered with spines. The spines may be long, as in the sea urchin, or very short, as in the sand dollar. All the echinoderms are restricted to the marine environment. They have been collected both from the shallowest tidal pools and from great ocean depths.

The characteristics of the echinoderm phylum can be observed in one of its most familiar members: the starfish. Starfishes typically have five rays, or arms, which radiate from a central disk. In a groove on the lower side of each movable ray are two rows of *tube feet* (Figure 30-11). These are part of a *water-vascular system.* The tube feet are connected to canals that lead through each ray to a circular canal in the central disk. This *ring canal* has an opening to the surface, the *sieve plate,* on the dorsal side. When the starfish presses its tube feet against an object and forces water out of its canals, the feet grip the object firmly by means of suction. Return of water to the canals releases the grip.

The common starfish uses its water-vascular system to open the shells of clams and oysters, its principal foods. The body arches over the prey with the rays bent downward. The valves of the clam or oyster are gripped firmly by the tube feet, and at the same time a steady pull is exerted. The starfish then pushes its stomach out from a small opening in the center of the lower side. The stomach, turned inside out, squeezes through even the tiniest opening between the mollusk's valves. It has been reported that, for some starfishes this opening can be as small as one tenth of a millimeter. The enzymes, secreted by the stomach, then digest the tissues of the mollusk. Digested food is absorbed by the stomach and passes into the digestive glands where further chemical changes occur. After feeding, the starfish retracts its stomach and moves away, leaving behind the empty shells of its prey.

As you may have guessed, oystermen are constantly on the lookout for starfish in clam or oyster beds. An active, adult starfish can destroy from eight to twelve oysters a day. Oystermen, thinking they were destroying the pests, used to tear starfish into pieces and throw the pieces back into the water. Unknowingly, they were multiplying their troubles, for a starfish ray together with a portion of the central disk can regenerate and become an entire new starfish.

Starfish are either male or female. Small ducts lead from the reproductive organs to the outside. During the reproductive season, eggs and sperms are shed to the outside, and fertilization takes place in the water. Female starfish may produce as many as 200 million eggs in a season. The eggs develop into free-swimming, ciliated larvae that have mouths, digestive tracts, and anuses (Figure 30-12). After swimming about for a period of time, the larvae settle to the bottom and gradually change into adults.

Although echinoderms are not an economically important food, some are consumed. Sea urchins are sometimes collected in

30-12 This dorsal view of a dissected starfish shows the location of its internal organs.

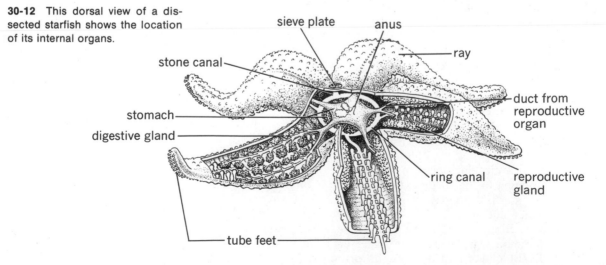

sieve plate

anus

stone canal

ray

stomach

duct from reproductive organ

digestive gland

ring canal

reproductive gland

tube feet

rocky tidepools and their eggs eaten. In the Orient, *sea cucumbers* are collected, dried, and sold as *bêche-de-mer* (BAYSH-duh-MARE) or trepang (tri-PANG) for use in soup.

What is a phylogenetic tree?

Before proceeding with the largest invertebrate phylum, the *Arthropoda,* let us look for some regularities that may be useful in understanding the course of animal evolution. Figure 30-14 is a diagrammatic representation of the way in which a taxonomist might fit shreds of evidence together to explain the evolution of animal diversity. It is called a *phylogenetic tree.* There are many such schemes; they are used to point out relationships based on incomplete fossil records and morphological similarities of existing animals (both larval and adult). The phylogenetic tree represents a hypothesis of origins and relationships of various groups of organisms. It is based on the assumption that the more closely the body plans of two groups resemble each other, the closer their relationship, and the more recent their common ancestor. This is the **principle of homology.**

When we examine the larval forms of the mollusks, for example, we find that they are very similar. We can theorize, then, that they evolved from a primitive ancestor that was like the larval form. But no one can say this for sure, nor has anyone been able to disprove it. Furthermore, if we find that annelids also have a larval form very similar to that of the mollusks, we can postulate that these two apparently different phyla developed from an ancestor of the trochophore type.

It is also interesting to note that even though adult echinoderms are radially symmetrical, many biologists consider the echinoderms much more closely related to the lower chordates than to the radially symmetrical sponges and coelenterates or to the arthropods, which we will consider in Chapter 31. The reason for this stems in part from the fact that the free-swimming larvae of echinoderms are quite different both structurally and embryologically from trochophore larvae.

Notice in Figure 30-14 that there seems to be one main direction of development until structures called **protonephridia** (PROE-toe-nuh-FRID-ee-uh) develop. A protonephron is a very primitive type of "kidney." According to the diagram, then, all the organisms above this point have a protonephron. This is true, but in the development of many animals the protonephron undergoes complicated changes before the adult stage is reached.

One of the characteristics considered in grouping animals is the type of body cavity, or **coelom** (SEEL'm), present in the adult. Among the bilaterally symmetrical organisms, the flatworms do not have a body cavity between the internal organs and the body wall (Figure 30-15). As the body of an animal increases in size, a solid structure like the one found in the flat-

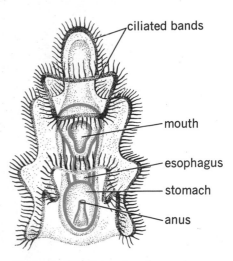

30-13 A free-swimming larva of an echinoderm. What kind of symmetry does it have? Called a *bipinnaria,* it is similar to the larva of some of the lower chordates. Compare it with the trochophore larval type, characteristic of annelids and mollusks, shown in Figure 30-2.

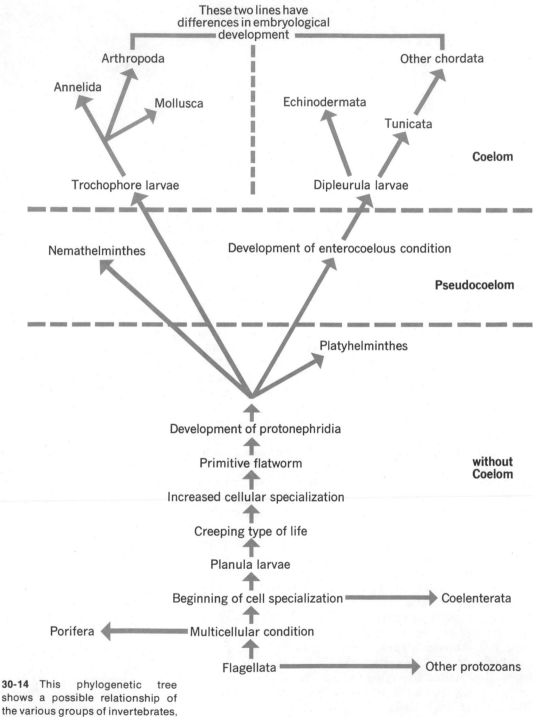

30-14 This phylogenetic tree shows a possible relationship of the various groups of invertebrates, as well as their supposed relationship to the protists and chordates.

30-15 General body plans of organisms of five animal phyla.

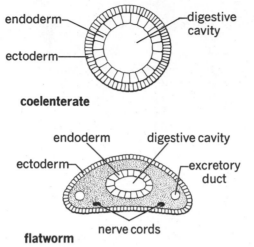

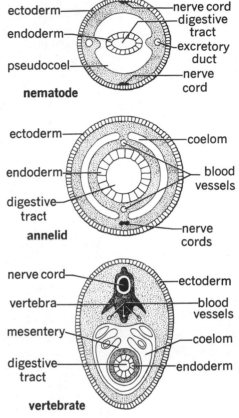

worms would not be physiologically efficient. A fluid-filled cavity, however, allows for looped intestines and aids in circulation of food and oxygen, as well as in waste removal. It should not be surprising, then, to find that the majority of animals have a coelom. The coelom develops within the mesoderm of the embryo and has a lining of specialized covering cells. This lining surrounds the internal organs and covers the inner surface of the body wall. It forms a membranelike structure that suspends the intestine within the coelom.

Between the animals with a coelom and those without, in our phylogenetic tree, are the animals possessing a *pseudocoel* (SOO-doe-SEEL), or false coelom. This cavity is not lined with specialized covering cells and has no suspending membrane. The internal organs are free within the cavity.

SUMMARY

In this chapter you have investigated some of the most prominent inhabitants of the seas and tidepools. The mollusks live in marine, fresh-water, and terrestrial environments, but the echinoderms are exclusively marine.

Fossils have been formed from the shells of mollusks and the skeletons of many echinoderms. These fossils have provided us with keys leading to knowledge of the changes in the earth's surface and its inhabitants.

You have seen some ways in which data are gathered and used to form hypotheses about the relationships and evolution of many diverse forms of life.

BIOLOGICALLY SPEAKING

trochophore	gill	pearly layer
visceral hump	incurrent siphon	prismatic layer
mantle	excurrent siphon	horny outer layer
mantle cavity	valve	radula
water-vascular system	protonephridia	pseudocoel
principle of homology	coelom	

QUESTIONS FOR REVIEW

1. What characteristics of the mollusks are used in dividing the phylum into classes?
2. What are the three characteristics all mollusks have in common?
3. Trace a particle of food from the outside of the clam to the digestive gland, naming the organs through which it passes.
4. What layers are formed in the making of the shell?
5. How does the oyster form a pearl in its shell?
6. In what ways do the cephalopods differ from the gastropods?
7. What characteristics of the echinoderms distinguish them from other invertebrate animals?
8. Describe the movement of the starfish.
9. How does the water-vascular system assist the starfish in eating?

APPLYING PRINCIPLES AND CONCEPTS

1. What structural similarities indicate a relationship between the mollusks and the annelids?
2. What property of mollusks is of interest to the geologist?
3. In what ways have the mollusks been of value to man? In what ways have they been a pest?
4. Shells of mollusks are frequently ground up and used as fertilizer. What are some of the substances that these shells add to the soil?
5. What data are used in making a phylogenetic tree?

RELATED READING

Books

Encyclopedia of Life Sciences (eds.), *The Animal World.*
  Doubleday and Company, Inc., Garden City, New York. 1964. A survey of some amazing facts in the world of biology, from protozoa to mammals.
HALSTEAD, BRUCE W., *Dangerous Marine Animals.*
  Cornell Maritime Press, Inc., Cambridge, Md. 1959. A guide to the marine animals that bite, sting, or shock—none edible, all venomous.
IDYLL, C. P., *Abyss: The Deep Sea and the Creatures that Live in It.*
  Thomas Y. Crowell Company, New York. 1964. An armchair tour of the strange world of the deep sea.
JOHNSTONE, KATHLEEN YERGER, *Collecting Seashells.*
  Grosset and Dunlap Publishers, New York. 1970. A complete guide for beginning and veteran collectors that explores the wonder, beauty and joys of one of the world's most interesting hobbies.
LANE, FRANK W., *Kingdom of the Octopus.*
  Sheridan House, New York. 1960. Describes the life histories of the Cephalopoda.
SCHISGALL, OSCAR, *That Remarkable Creature, The Snail.*
  Julian Messner, New York. 1970. An easy to read, straightforward introduction to freshwater, saltwater, and land snails.

CHAPTER THIRTY-ONE

THE ARTHROPODS

Arthropod characteristics

From the standpoint of numbers, the arthropods are considered a most successful group of animals. The phylum *Arthropoda* includes the familiar insects, spiders, centipedes, millipedes, crayfish, crabs, and lobsters. Arthropods are found everywhere. They serve as food for man, and they compete with man for food. Many live as parasites in or on other organisms, and some transmit disease.

To the casual observer, the graceful butterfly has little in common with the crayfish lurking under a rock in a stream. But careful study of these quite different animals will show that they have much in common. The structural characteristics that indicate that the butterfly is related to the crayfish also indicate its relationship to spiders, scorpions, and centipedes. All members of the phylum Arthropoda, which means "jointed feet," are similar in having the following characteristics:

■ Jointed *appendages,* which include legs and other body outgrowths

■ A hard external skeleton, or *exoskeleton,* composed of a substance called *chitin* (KITE'n), instead of the internal support found in man and the other vertebrates

■ A segmented body; that is, a body with distinct divisions of the exoskeleton

■ A dorsal heart; that is, a heart that is located above the digestive system

■ A ventral nervous system, with the main nerves below the digestive system

The development of the phylum

The fact that the arthropod body is divided into segments suggests a relationship to the annelids. This relationship is often expressed when biologists refer to the "annelid-arthropod" line of development (see the phylogenetic tree shown in Figure 30-14).

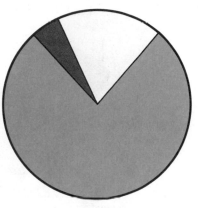

● insect arthropods
● noninsect arthropods
  nonarthropods

31-1 As you can see, more species of animals belong to the phylum Arthropoda than to any other phylum.

479

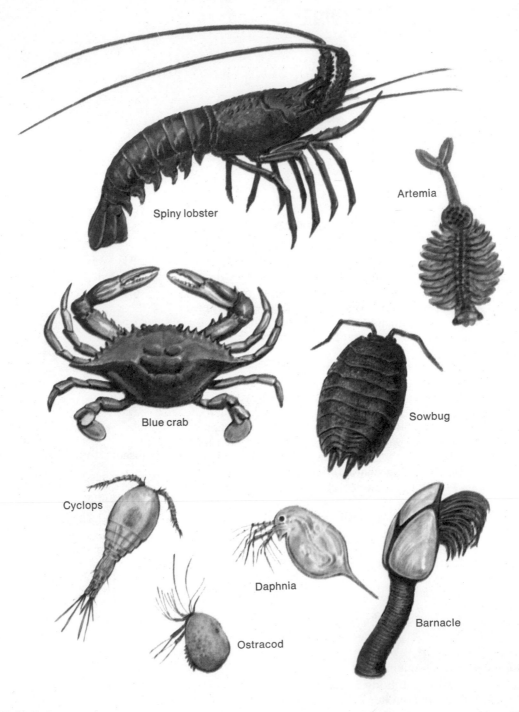

Spiny lobster

Artemia

Blue crab

Sowbug

Cyclops

Daphnia

Ostracod

Barnacle

31-2 Various examples of the classes of arthropods.

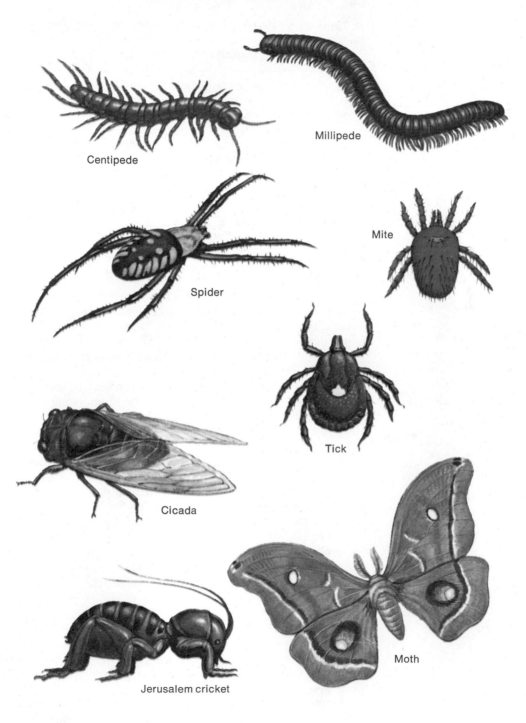

Centipede

Millipede

Spider

Mite

Tick

Cicada

Jerusalem cricket

Moth

Table 31-1   COMPARISON OF ARTHROPODS AND ANNELIDS

ARTHROPOD CHARACTERISTICS	ANNELID CHARACTERISTICS
legs divided by movable joints	no legs
hard, chitinous exoskeleton	flexible cuticle
fewer segments, more highly specialized	many similar segments
dorsal heart	dorsal heart
ventral nervous system with specialized sensory receptors such as eyes and antennae	ventral nervous system but simple sensory receptors
muscles in groups	muscles in sheets

When you compare the characteristics of the arthropods and annelids (Table 31-1), you can understand why the arthropods have been more successful than the annelids. The increased complexity of their sense organs allows for coordinated response to their environments. The grouped muscles and jointed legs permit a greater degree of coordinated movement in the search for food and in escape from enemies. As you see, the characteristics of arthropods permit greater efficiency and adaptability to their environments.

The hard exoskeleton provides a protective advantage over the soft cuticle of the worm. It may seem strange to find the skeleton on the outside of an animal's body. But whether skeletons are internal or external, they serve the same function. They give the body form, protect delicate internal organs, and aid in motion by serving as attachments for muscles. Unlike an internal skeleton, however, an exoskeleton limits the size of an animal. A large exoskeleton with the powerful internal muscles required to move it would crush the animal under its own weight. Flying insects, for example, could never reach the size of birds, because their exoskeletons would be too heavy to be supported by wings. Growth in the size of an organism with an exoskeleton can only take place if *molting,* (whereby the old skeleton is shed and a new one is formed) also takes place. This process will be discussed in detail later in this chapter.

Diversity among the arthropods

In the classification of the arthropods, such widely varied forms as butterflies and crayfish, spiders and centipedes are grouped into classes. This large phylum is commonly divided into five classes, as shown in Table 31-2.

Members of each of these classes have all the fundamental characteristics of arthropods and, in addition, certain characteristics of the class to which they belong. For example, the *Crustacea* (kruss-TAY-shuh) have two pairs of antennae (or "feelers") on the front of the body, two distinct body regions, five pairs of legs, and a chitinous exoskeleton that contains lime. Most have featherlike *gills* for respiration. The *Insecta,* on the other hand, have one pair of antennae, a body composed of three parts, three pairs of legs, often two pairs of wings, and an exoskeleton composed of chitin that lacks lime. They respire by means of air tubes called *tracheae* (TRAY-kee-ee).

The arthropods show a great advance over the other animals we have studied. In our study of the crustaceans as typical arthropods, we shall deal with such animals as the crayfish, lobster, and crab, which are adapted for aquatic life. Division of labor among their various organs is carried to an even greater extent than it is in the earthworm. All of this specialization has resulted in an efficient animal well adapted to our present world. The segmented body is, of course, common to both the arthropods and the annelids. The ventral nervous system first appeared in some of the worms.

Table 31-2 CLASSES OF ARTHROPODS

CLASS	BODY DIVISIONS	APPENDAGES	RESPI- RATION	EXAMPLES
Crustacea	2 — cephalothorax, abdomen	5 pairs of legs	gills	lobster, crab, water flea, sow bug, crayfish
Chilopoda	head and numerous body segments	1 pair of legs on each segment except first one behind head and last 2	tracheae	centipede
Diplopoda	head and numerous body segments	2 pairs of legs on each body segment	tracheae	millipede
Arachnida	2 — cephalothorax, abdomen	4 pairs of legs	tracheae and book lungs	spider, mite, tick, scorpion
Insecta	3 — head, thorax, abdomen	3 pairs of legs; usually 1 or 2 pairs of wings	tracheae	grasshopper, butterfly, bee, dragonfly, moth, beetle

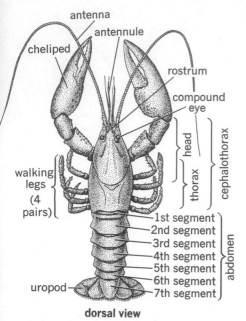

dorsal view

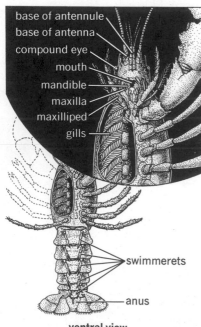

ventral view

31-3 The crayfish, a common crustacean.

The crayfish—a large fresh-water crustacean

Scientists often use the crayfish to observe the features of the class Crustacea because it is large and easily obtained. Crayfish can be found in nearly all rivers, lakes, and streams that contain lime. (Lime is used by the crayfish in hardening its tough, chitinous exoskeleton.)

The body is divided into two regions. The first of these, called the *cephalothorax* (SEF-uh-loe-THORE-acks), includes the *head* and a second region, the *thorax* (Figure 31-3). These are separate in the insects. The *abdomen,* composed of seven movable segments, is posterior to the cephalothorax. Vital parts of the cephalothorax are protected by a shield called the *carapace* (KA-ruh-PACE), which extends forward as a beak, called the *rostrum.*

Specialization in crayfish appendages

The most anterior pair of appendages of the crayfish are the *antennules,* which contain the hearing and equilibrium apparatus. The large *antennae* attach just behind the antennules. They function as organs of touch, taste, and smell. The next appendages are the *mandibles,* or true jaws. These crush and chew food, with the help of two pairs of *maxillae* (mack-SILL-ee), or little jaws. The jaws work from side to side, not up and down. They are actually leglike appendages that are adapted for chewing, and they therefore continue to move horizontally as the legs do.

The first appendages of the thorax are three pairs of *maxillipeds* (mack-SILL-i-PEDZ), or "jaw feet." They hold food during chewing. The next and most obvious structures are the *chelipeds* (KEE-luh-PEDZ), or claw feet, which function in food-getting and protection. The next four pairs of legs are called *walking legs.* Feathery *gills,* the respiratory organs, extend under the carapace, and are attached to all the thoracic appendages.

The abdominal appendages of the crayfish are called *swimmerets.* These appendages are used in swimming. In the female, the last three pairs of swimmerets serve as a place of attachment for eggs. The sixth pair of abdominal appendages is much larger than the first five pairs. It is developed into a flipper, or *uropod* (YOOH-ruh-POD). There is no appendage on the seventh segment, which is reduced to a flat, triangular structure—the *telson.* Strong abdominal muscles can whip the sixth and seventh segments forward, causing the animal to shoot backward very rapidly.

Recall from our discussion in Chapter 13 that structures believed to have developed through time from the same origin but not necessarily having the same function are said to be *homologous.* The appendages of the crayfish are considered to be homologous organs. Each segment except the telson bears a pair of appendages. The antennae and chelipeds of crayfish are homologous to the swimmerets.

Analogous organs are those organs that are similar in function. The wings of the bird and insect are analogous because they both function in flying. They are not homologous, however, because each develops in a different way and from different structures. The gills of the crayfish and the lungs of man are analogous because they both perform the function of respiration. However, they are not homologous, since the gills are developed from the legs, while the lungs are outgrowths of the throat.

Nutrition in the crayfish

Food is held by the maxillae and maxillipeds while it is torn and crushed by the mandibles. It then passes through a short esophagus to the stomach, which is lined with hard, chitinous teeth. These grind the food into smaller particles. When the particles of food are finely ground, they pass through folds of tissue, which act as a strainer, into another portion of the stomach where the ground food is mixed with digestive juices. From here, the digested food passes into the digestive glands, where absorption takes place. Undigested particles pass on through the intestine, instead of entering the digestive glands, and are eliminated through the anus (Figure 31-4). In the crayfish, there are also excretory organs called *green glands;* these lie anterior to the stomach and open near the base of the antennae. These green glands remove wastes from the blood.

Circulatory system

The blood of the crayfish is pumped by the heart into seven large arteries that pour the blood over the major organs of the body. The blood bathes the cells directly by flowing through tissue spaces. Gradually, the blood collects in a large ventral cavity, the *sternal sinus.* Vessels carry the blood from the sternal sinus to the gills where respiration occurs. Next, the blood is carried

31-4 The internal structure of a crayfish. The term *gonad* refers to the ovary in the female and the testis in the male.

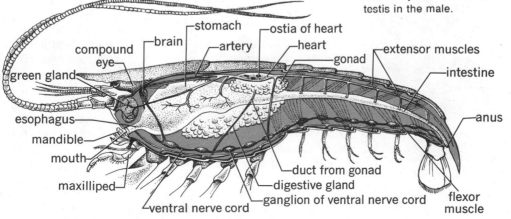

by vessels to another large sinus, the *pericardial sinus,* which surrounds the heart. Blood enters the heart through three pairs of tiny pores, the *ostia.* Upon contraction of the heart, valves close off the ostia to prevent blood from flowing back into the pericardial sinus. This type of circulation is called an **open system** because the blood is not confined to vessels for its entire course through the body. Perhaps you might compare the sternal sinus to the crankcase of an automobile engine which functions as an oil reservoir.

Respiration in the crayfish

Gas exchange in protists is accomplished by diffusion through the plasma membrane; in worms, through the body wall. In crayfish, the thin-walled gills, which are richly supplied with blood vessels, are well adapted for the exchange of oxygen and carbon dioxide between the animal and its aquatic environment.

The gills are protected by the carapace, which extends over them and forms a chamber. Water flows into the chamber under the free edge of the carapace, and is moved forward over the gills through the action of the second maxillae. The gill chamber can hold enough moisture to permit the animal to remain alive for some time after it has been removed from the water.

Nervous system and sense organs

Although the nervous system of the crayfish is similar to that in annelids, it is more specialized. The brain receives nerves from the eyes, antennules, and antennae. From the brain, two large nerves pass ventrally, unite to form a large double ganglion, and then pass posteriorly forming the ventral nerve cord. In each segment, the ventral nerve cord enlarges into a ganglion which supplies nerves to appendages, muscles, and other body organs.

Receptors sensitive to odors and flavors are located in the antennae. The eyes are on either side of the rostrum and are set on short, movable stalks. Since each eye is composed of more than 2000 lenses, it is called a *compound eye.* As you will see, many arthropods have compound eyes. Each unit of the compound eye is like a tiny telescope, directed at a different point in space. Many of these units functioning together would result in a *mosaic image* (see Figure 31-5). This type of vision best functions detecting movement.

Numerous sensory bristles that are sensitive to touch are distributed all over the body, especially on the surfaces of the antennae and other appendages. It is possible that these bristles are also stimulated by sound waves.

The sac at the base of each antennule is also involved with the sense of balance. This sac, called a *statocyst,* is lined with hair-bearing receptor cells. Nerve fibers carry signals from these to the brain. Within each statocyst is a grain of sand that stimulates

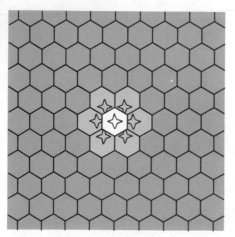

31-5 A representation of how a mosiac image in a compound eye might appear.

the receptor cells. If the crayfish moves so that it is tilted or upside down, the sand grain shifts and stimulates the receptor cells, sending a signal to the brain. The crayfish can react by righting itself.

During molting, the grain of sand is shed along with the old exoskeleton. Usually, the crayfish replaces the lost grain with another sand grain. If, however, for experimental reasons, only iron filings are made available to the crayfish at the time of molting, the crayfish will introduce an iron filing into the statocyst. After the new exoskeleton hardens, the animal is placed into another aquarium. If a magnet is subsequently held above the animal's back, the iron filing will stimulate the hairs on the top of the statocyst. The crayfish will then turn onto its back and stay upside down as long as the magnet is present.

Reproduction and growth in the crayfish

Crayfish usually mate in the fall, at which time the sperms from the male are stored in small receptacles on the lower side of the female's body. The eggs number about one hundred, and are laid in the spring. They are fertilized by the stored sperms as they are laid. Then they are covered with a sticky secretion that causes them to adhere to the swimmerets. Here, these berry-like structures are carried and protected for about six to eight weeks until they hatch. As is true in the development of many invertebrates, the speed of development may depend on external factors of the environment, such as temperature and water conditions. The larvae are quite different from the parent, but after a series of molts they reach the adult body form.

From the time of hatching until the organism reaches adult size, molting occurs at increasingly longer intervals. Most crayfish molt seven times during the first year and about twice a year thereafter. Their average life span is from three to four years.

During the process of molting, the cuticle secretes an enzyme that actually digests the inside of the shell, thus loosening it from the body. The carapace splits along the back, and water is withdrawn from the tissues, causing them to shrink. Next, the animal literally humps itself out of its former skeleton. It also sheds the lining of its stomach, including the teeth. Immediately following this process, water is absorbed, and the animal swells to its new size. As the lime is replaced, the exoskeleton hardens, and the animal cannot increase further in size until it molts again. The crayfish, helpless at the time of molting, usually goes into hiding until the new exoskeleton hardens.

During molting or in battle with enemies, the crayfish often loses or injures appendages. An injured limb is cast off. A double membrane prevents much loss of blood, and a whole new appendage gradually develops to replace the injured member. This process is another example of regeneration of lost parts.

The economic importance of various crustaceans

In one sense, the crayfish is considered to be a beneficial organism, because it acts as a scavenger that readily consumes dead organisms in any condition. In certain parts of the country, however, especially in the Mississippi River basin, crayfish cause extensive damage by making holes in earthen dams and levees and by burrowing in fields, thus destroying cotton and corn crops. Crayfish are used for food in many areas.

Crustacea are adapted for a variety of environments

The edible marine crustaceans are relatively insignificant in terms of numbers. The vast majority of crustaceans consist of minute and even microscopic forms. However, these small crustaceans sometimes occur in such tremendous numbers that the sonar apparatus of ships gives false readings of the bottom depth.

Artemia, the brine shrimp, inhabits tide pools and can survive in an environment of high salt concentration (Figure 31-4). Many tropical fish enthusiasts buy the eggs and raise the shrimp for fish food. *Barnacles* are sessile crustaceans that in larval form settle down on a solid object, produce a shell, and use their feet to kick food into their mouths. Barnacles may accumulate on the hulls of ships in numbers large enough to reduce the speed of the ships by as much as 20 percent. They also clog seawater intake pipes and grow on piers. Here the accumulation of sand and decaying organic matter between the barnacles provides a good medium where bacteria can live and grow. In turn, annelids and other small animals feed on the bacteria. Barnacles, usually associated with mussels, actually change the environment of wharf pilings by providing protection for the living things that grow among them. The force of wave action is lessened, extreme temperature changes are prevented, and moisture is maintained among the mussels preventing drying at low tides. These environmental changes, however, increase the rate of deterioration of the wharves.

The crustaceans are by no means limited to a marine environment. In nearly every sample of fresh water, many tiny, brownish specks will zigzag before your eyes. These organisms may be *Daphnia* (water fleas) or *ostracods*. Ostracods are tiny crustaceans that, in addition to the chitinous exoskeleton, secrete two shells resembling those of clams. Other common minute crustaceans are the *copepods* (KOPE-uh-PODZ), which serve as an important part of the diet of many fish.

Another interesting adaptation of crustaceans can be observed in the sow bugs and pill bugs, which are terrestrial crustaceans. Since they have seven pairs of identical legs, they are called *isopods,* which means "same feet." The isopods lack gills, but they have a series of plates along the ventral surface of the abdomen with tiny tubes through which air can pass. Since these plates must be kept moist for gas exchange to occur, sow bugs and pill bugs are usually found under stones and logs. Hence, even if you do not live near a stream where crayfish can be found, or near the ocean where other crustaceans can be studied, you can probably find isopods near where you live. Terrestrial isopods can be kept alive for weeks in jars with potato or carrot slices available as food.

Centipedes and millipedes

The Chilopoda and the Diplopoda are often grouped together as one class, the *Myriapoda,* which means "many feet." Perhaps you have wondered how a centipede or millipede can operate so many legs without getting them tangled with one another. Such movement is certainly an excellent example of coordination. These curious, wormlike arthropods are often seen racing away with a rippling motion when their hiding places under logs, stones, or pieces of rubbish have suddenly been disturbed.

Centipedes belong to the class *Chilopoda,* and their bodies are composed of many segments. The head bears the antennae and mouthparts; the first body segment bears a pair of poison claws; each succeeding segment except the last two bears one pair of walking legs (Figure 31-6). Centipedes are fast-moving and difficult to capture. Their speed and poison claws aid them in catching prey. In tropical countries, centipedes may measure 12 inches in length, and their bites may be quite poisonous. Some have as many as 173 pairs of legs, but 35 is average.

Millipedes, or "thousand-legs," belong to the class *Diplopoda,* and their bodies are also composed of many segments. As in the centipede, the head bears antennae and mouthparts; there are two pairs of legs on each of the body segments except the last two. The name *Diplopoda,* which means "double feet," reflects this fact. Millipedes, which are largely herbivorous, are frequently slow-moving and are likely to roll into a spiral when they are disturbed.

Spiders—familiar arachnids

Spiders as a group are extremely valuable to man because they destroy harmful insects. They belong to the class *Arachnida* (uh-RACK-nid-uh). Some kinds of spiders, called *orb weavers,* spin tiny silken threads into elaborate webs that are remarkable engineering feats. The web serves as a trap for capturing flying insects. When a victim becomes entangled in the sticky threads of the web, the spider races from its hiding place along the margin. Its bite poisons the prey. When the insect has become partially paralyzed, the spider encases the prey securely in silken threads as it turns it over and over. Other spiders do not spin webs but live a solitary life, stalking their prey as they roam about.

31-6 Compare the centipede (above) with the millipede. In what ways do they differ? (Walter Dawn; Karl H. Maslowski, Photo Researchers, Inc.)

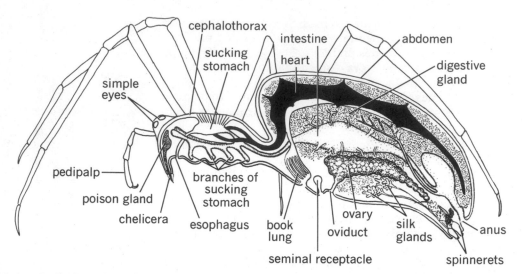

31-7 A longitudinal section of a female spider.

The spider resembles an insect but differs in several respects (Figure 31-7). The spider lacks antennae, and it has only simple eyes. It has eight legs instead of six, and the head and thorax are joined to form a cephalothorax, as in the Crustacea. The first pair of appendages, the *chelicera* (ki-LISS-uh-ruh), serves as poison fangs and to suck the juices from the victim's body. The second pair of head appendages in the spider is called *pedipalps*. These appendages are sensory, and are used especially in reproduction by the male spider.

On the tip of the abdomen in many spiders, there are three pairs of *spinnerets*. Each spinneret consists of hundreds of microscopic tubes through which the fluid silk flows from the silk glands. When the silk passes into the air, it hardens into a thread. This silk is used in making the spider's web, in building cocoons for eggs, in building nests, and as a guide for the spider to find its way back to where it started. Young spiders of many species spin long silken threads that catch in the wind and carry them to distant places. This method of moving is known as *ballooning.*

A pair of air-filled sacs called *book lungs,* situated on the lower side of the abdomen, receives air through slitlike openings. Numerous leaves, or plates, of the book lungs provide a large surface for exposure to air. In addition to book lungs, many spiders have openings in the abdomen that lead to air tubes resembling the tracheae of insects.

The mature male spider can usually be distinguished from the female by size alone. The male is usually the smaller. In species where there is no size difference, the pedipalps of the male are much larger near the tips than those of the female. At maturity, the male transfers sperms to special sacs at the tips of the pedipalps. These sperms are then placed in the *seminal receptacle* of the female, who sometimes devours the smaller male afterward. At the time of egg-laying, the eggs are fertilized as they pass out through the genital pore into a nest or cocoon.

Among the most well-known spiders are the *tarantula* (tuh-RAN-chuh-luh), or banana spider; the *black widow* (Figure 31-8), infamous for its poisonous bite; and the *trapdoor spider* of the western desert regions.

Other arachnids

Spiders are related to many other forms of animal life. *Scorpions,* found in southern and southwestern United States and in all tropical countries, are provided with long, segmented abdomens terminating in venomous stingers. The sting of a scorpion, while painful, is seldom fatal to man. Campers find scorpions annoying because these animals often crawl into empty shoes during the day in order to escape bright light. Scorpions live solitary lives except when mating; after mating, the female often turns on her mate and devours him. The young are brought forth alive and spend the early part of their existence riding on the mother's back.

The *harvestman,* or *daddy-longlegs,* is one of the most useful of the arachnids, because it feeds almost entirely on plant lice. It leads a strictly solitary life, traveling through the fields in search of its prey.

Mites and *ticks* are among the more notorious arachnids, causing considerable damage to man and other animals. They live mostly as parasites on the surfaces of the bodies of chickens, dogs, cattle, man, and other animals. Some forms, like the Rocky Mountain tick, carry disease organisms.

Harvest mites, or *chiggers,* are immature stages of mites that attach themselves to the surface of the skin and insert beaks through which they withdraw blood. They are almost microscopic in size and give no warning of their presence until a swollen area causes great itching and discomfort. After a few days, the sore becomes covered with a scab, and heals.

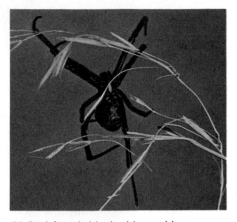

31-8 A female black widow spider. Note the characteristic red hourglass marking on the ventral surface of its round, black abdomen. (Walter Dawn)

SUMMARY

From the standpoint of numbers the arthropods are considered the most successful group of animals. Their diversity has allowed them to live in all environments—fresh-water, marine, and terrestrial. The complexity of the arthropod sense organs allows for coordinated responses to the environment, which is a factor contributing to the success of the group. The hard exoskeleton is a protective advantage over the soft cuticle of the worm, but it also limits the size. Growth can only occur by molting. The segmented body of an arthropod, however, does suggest a relationship to the annelids.

Although you may associate the arthropods with food, they are also important to man as destroyers of crops, stored food, and clothing. Many arthropods are pests and carriers of disease.

BIOLOGICALLY SPEAKING

appendage	rostrum	swimmeret
exoskeleton	compound eye	uropod
molting	antennule	telson
gill	antenna	homologous organs
trachea	mandible	analogous organs
cephalothorax	maxilla	green glands
abdomen	maxilliped	open system
carapace	cheliped	statocyst

QUESTIONS FOR REVIEW

1. What are three external characteristics of an arthropod that distinguish it from other animals?
2. Name the five principal classes of arthropods and give examples of each.
3. In what ways are the arthropods similar to the earthworm?
4. What are some advantages and disadvantages of an exoskeleton?
5. How do the gills in the crayfish carry on the process of respiration?
6. To what stimuli is the crayfish sensitive, and what structures assist in the sensitivity?
7. How does a centipede differ from a millipede?
8. Why are most spiders extremely beneficial to man? What reasons can you give for the fact that many people are genuinely afraid of arachnids?
9. Of what value to the spider is ballooning?
10. What other animals besides the spiders are classified as arachnids?

APPLYING PRINCIPLES AND CONCEPTS

1. Why is it especially important for an armored animal like the crayfish to have long antennae?
2. In which locality do you think the crayfish would be likely to produce weaker exoskeletons, in waters flowing through limestone rock or in waters flowing through granite? Explain.
3. Of what advantage to the young crayfish is its clinging to the adult's swimmerets until after the second molting?
4. Other than reducing a ship's speed, how do barnacles do damage?

RELATED READING

Books

CLOUDSLEY-THOMPSON, J. L., *Spiders, Scorpions, Centipedes, and Mites.*
   Pergamon Press, Inc., New York. 1958. An ecological and natural history study of selected arthropods.
DAVID, EUGENE, *Spiders and How They Live.*
   Prentice-Hall, Inc., Englewood Cliffs, N.J. 1964. A natural history of spiders, particularly those living in the U.S.
SCHMITT, WALDO L., *Crustaceans.*
   The University of Michigan Press, Ann Arbor. 1965. Discusses their structure, diversity, ecology, behavior, and economic value.

SNOW, KEITH R., *The Arachnids.*
Columbia University Press, New York. 1970. An introductory text dealing with the spiders and their relatives—a valuable reference for students interested in this arthropod type.

SUTTON, ANN, and MYRON SUTTON, *The Life of the Desert.*
McGraw-Hill Book Company, New York. 1966. Discusses among other desert animals those arthropods that are able to adapt to and survive in a desert environment.

VONFRISCH, KARL, *Ten Little Housemates.*
Pergamon Press, Inc., New York. 1960. Stories of arthropods sometimes found in the home, including the cockroach and the bedbug.

CHAPTER THIRTY-TWO

INSECTS– FAMILIAR ARTHROPODS

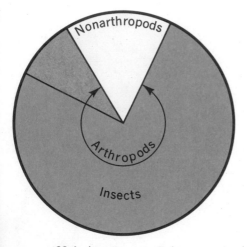

32-1 Insects comprise nearly three quarters of all described species of animals.

Vast numbers of species

Entomology is the study of insects, and biologists who specialize in this field are called *entomologists*. Although more than 675,000 species of insects have been recorded, entomologists regard this number to be not more than half of all insects in existence! As you can see from Figure 32-1, more than three quarters of all classified animals are arthropods, and of these, the class Insecta is by far the largest. The insects are, therefore, the dominant group of animals on the earth today. Many insects produce thousands of offspring. You may have seen pictures of locusts or mayflies, whose swarms can darken the skies. The famous biologist, Thomas Huxley, made the following calculation:

> I will assume that an *Aphis* weighs one-thousandth of a grain, which is certainly vastly under the mark. A quintillion of Aphides will, on this estimate, weigh a quatrillion of grains; consequently, the tenth brood alone, if all its members survive the perils to which they are exposed, contains more substance than 500,000,000 stout men—to say the least, more than the whole population of China!

Of course all offspring do not survive. Fortunately, as you will read in Unit 8 of this book, there are many factors which control this reproductive potential.

What is an insect?

Many people call any small flying or crawling animal a "bug." They are wrong on two counts. First, a true bug is a member of only one order of insects. Second, what some people call a bug may actually be a spider or a centipede. These, of course, are not insects at all.

Insects are arthropods that have three separate body regions: (1) *head,* (2) *thorax,* and (3) *abdomen.* The head bears one pair of antennae and the mouthparts. The thorax of an insect bears three pairs of legs and, if present, the wings. The abdomen has as many as eleven segments and never bears legs. The reproductive structures are usually found on the eighth, ninth, and tenth segments. Respiration in insects occurs by means of branched air tubes called *tracheae*.

Why are insects so successful?

It is estimated that insects have lived on the earth nearly 300 million years. During this time many environmental changes have occurred on the earth. As you have learned, an organism must either adapt to these changes, move, or it will become extinct. Fossils provide us with evidence that many organisms have become extinct. A comparison of these extinct forms to those living today makes us wonder how these changes could have come about. Certainly, diversity is a factor which has allowed the insects to become so successful. Can we explain how the great variety of insects present today could have occurred? You have seen, in this chapter, the great reproductive potential of the insects. Biologists believe that a combination of a high reproductive rate and individual variation occurring over millions of years provides an important key.

These variations do not allow a particular insect to live in all environments, but adjust to changing conditions in a particular environment. Insects as a group, though, are found in a great variety of places—even where extreme environmental conditions exist. Some insects live in the Himalayas at an altitude of 20,000 feet and some in caves deep within the earth's crust. In the Antarctic, insects have been collected where the winter temperatures reach −65 degrees C°; others may live in hot springs where the temperature is +49 degrees C°. Some insects spend part of their lives in crude oil pools around oil wells. Although the marine environment is not well populated with insects, there are some that live on the surface of the ocean. One fly even spends its larval development in brine pools where salt crystals are forming.

The chitinous exoskeleton is a tough, flexible, and lightweight protective armor. Also, the chitin is coated with wax which prevents water loss to the environment. The joints allow movement and are covered by a tough membrane serving to hold the skeletal parts together. The exoskeleton limits the size of a terrestrial organism and most insects are small. However, this also has survival value, as their nutritional requirements are low. Tiny particles of food overlooked by larger animals may provide a feast to an insect. They can find shade beside a small pebble and hide in tiny spaces.

The insects show a great variety in the colors and shapes of the exoskeletons. Thus, they may be protected by colors which blend into the background to make them invisible. Some harmless insects may resemble relatives which are capable of a poisonous

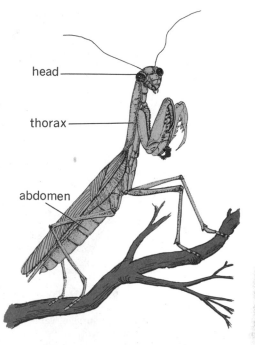

head

thorax

abdomen

32-2 How many insect characteristics can you find on this interesting animal?

32-3 Can you see why camouflage is of adaptive value? (Dr. E. R. Degginger)

32-4 Although this fly looks like a honeybee, it does not sting. Would this be of survival value to the fly? (Walter Dawn)

32-5 This tree hopper resembles a thorn on a bush. (Walter Dawn)

sting (Figure 32-4). Of course, the insects capable of severe stings or bites are protected by the learned behavior of their "would-be" predators. An insect may resemble the plant on which it is living (Figure 32-5).

Also of survival benefit to the insects is the fact that a great variety of foods may be utilized. This is made possible by specialization of the mouthparts. Some insects have chewing mouthparts with strong jaws to grind up leaves. Some have piercing and sucking mouthparts with which they can suck plant juices or, as do the mosquitoes, blood from animals. Some have siphoning tubes for probing flowers to obtain nectar. You have all seen butterflies gathering the nectar from flowers on warm days.

The wings and legs of many insects are developed for swift locomotion. Some insects are adapted for aquatic life and some for burrowing in the ground. Some insects live in colonies; others fight their battles alone. Some, such as the scale insects, have in the process of evolution lost their legs entirely.

Insect diversity

Since there are so many insects, we will not be able to study them all. We can, however, look at the common orders and learn to recognize the characteristics that are used to classify them. These characteristics include types of wings, types of metamorphosis, and types of mouthparts in the adult. Insects are commonly divided into a number of orders varying from twenty-two to twenty-nine. The difference in the number of orders is because some entomologists group two or more orders into one. Table 32-1 and Table 32-2 list sixteen of the most common orders found in the temperate regions.

Insects can easily be found in your home, school, outside in fields, ponds, trees, under stones, and on plants. To make your own insect collection is a particularly valuable project.

A closer look at an insect

By this time, you are aware that a high school course in biology cannot possibly cover the entire subject. When a large group of organisms is studied, we often choose one that shows the characteristics of the group. We cannot say that a grasshopper, for example, is a typical insect, because the diversity of this class is unequaled in the animal kingdom. The grasshopper, however, is large enough to examine without high magnification and is available in large numbers.

The grasshopper is a member of the order *Orthoptera,* which means "straight-winged." The grasshopper's narrowly folded wings are held straight along the body when they are not actually being used in flight. As in all arthropods, the skeleton is external, but it differs from that of the crayfish in that it contains no lime. It consists largely of the light, tough substance called *chitin.*

Since the grasshopper feeds on blades of grass, let us examine the mouthparts to see how they are adapted for this food. The *labrum* is a two-lobed upper lip used in keeping a blade of grass at right angles to the *mandibles*, which are toothed, horizontal jaws (Figure 32-6). Lying posterior to the mandibles are paired *maxillae*. These are accessory jaws that aid in holding and cutting food. The *palpus* is an antennalike sense organ that is part of each maxilla. Posterior to the maxillae is the *labium*, or lower lip, which is also provided with palpi. This organ functions in holding food between the jaws. These mouthparts, like the legs of the crayfish, are believed to be homologous. It is thought that each developed as an appendage of an independent embryonic segment.

Locomotion may be accomplished in several ways

The grasshopper's locomotory appendages are located on the thorax, which is divided into three segments. The head and first pair of walking legs are attached to the *prothorax* (Figure 32-7). The first pair of wings and the second pair of walking legs are attached to the *mesothorax*. The *metathorax* bears the second pair of wings and the jumping legs.

The grasshopper's wings may carry it over dry fields for a short distance or, during longer migrations in swarms, for many miles. The long, narrow, and stiff *anterior* wings protect the delicate underwings when the grasshopper is at rest or walking, but during flight or leaping they act to give lift to the insect. The *posterior* wings are thin and membranous and are supported by many veins. These flying wings are folded like a fan when not in use.

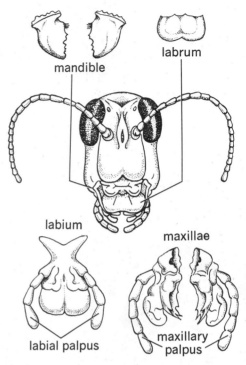

32-6 The mouthparts of a grasshopper.

32-7 The external structure of a grasshopper.

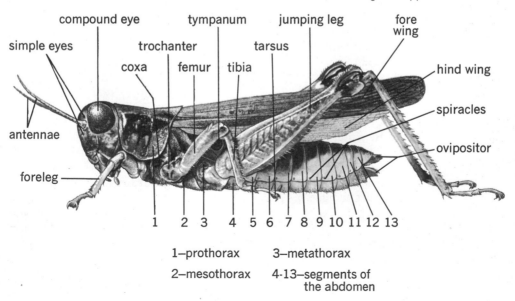

1—prothorax   3—metathorax

2—mesothorax   4-13—segments of the abdomen

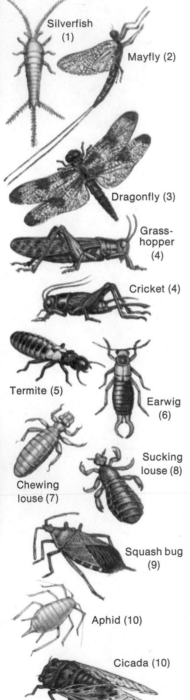

Silverfish (1)

Mayfly (2)

Dragonfly (3)

Grass-hopper (4)

Cricket (4)

Termite (5)

Earwig (6)

Sucking louse (8)

Chewing louse (7)

Squash bug (9)

Aphid (10)

Cicada (10)

Table 32-1   COMMON INSECT ORDERS WITH INCOMPLETE METAMORPHOSIS

ORDER	MOUTHPARTS IN ADULT	ECONOMIC SIGNIFICANCE	EXAMPLES
1. Thysanura ("bristle-tail") 700*	chewing	pests, feeding on starch in bindings & labels of books, clothing, paste in wallpaper	bristletails, silverfish, firebrats
2. Ephemeroptera ("for-a-day—wing") 1,500	adults do not feed	food for many fresh-water fish	mayflies
3. Odonata ("toothed") 4,870	chewing	destroy harmful insects	dragonflies, damselflies
4. Orthoptera ("straight-wing") 22,500	chewing	damage crops, act as pests	grasshoppers, crickets, katydids, locusts, cock-roaches
5. Isoptera ("equal-winged") 1,720	chewing	destroy wood in forests & buildings	termites
6. Dermaptera ("skin-winged") 1,100	chewing	damage crops and garden plants	earwigs
7. Mallophaga ("wool-eater") 2,680	chewing	most are parasitic on birds pests on poultry	chewing lice
8. Anoplura ("unarmed-tail") 250	sucking	parasitic on mammals parasitic on man transmit disease as typhus	sucking lice
9. Hemiptera ("half-winged") 23,000	sucking	damage plants, act as pests, carry disease	squash bugs, all true bugs
10. Homoptera ("like-winged") 32,000	sucking	damage crops and gardens	aphids, mealy bugs, cicada

* Numbers indicate estimated number of species known.

32-8   Common insect orders with incomplete metamorphosis.

Table 32-2 COMMON INSECT ORDERS WITH COMPLETE METAMORPHOSIS

ORDER	MOUTHPARTS IN ADULT	ECONOMIC SIGNIFICANCE	EXAMPLES
11. Neuroptera ("nerve-winged") 4,670	chewing	destroy harmful insects	dobsonfly, lacewing
12. Coleoptera ("sheath-winged") 276,700	sucking or chewing	destroy crops, act as pests, prey on other insects	weevils, ladybugs, ground beetles
13. Lepidoptera ("scale-winged") 112,000	siphoning	pollinate flowers, produce silk damage clothing and crops	butterflies, moths
14. Diptera ("two-winged") 85,000	sucking, piercing or lapping	carry disease, act as pests	flies, mosquitoes, gnats
15. Siphonaptera ("tube-wingless") 1,100	sucking	pests, feed on blood of birds & mammals, transmit disease as bubonic plague	fleas
16. Hymenoptera ("membrane-winged") 103,000	chewing, sucking, or lapping	pollinate flowers, act as pests, parasitize other pests, make honey	bees, wasps, ants

For shorter distances, the large jumping legs are used to escape enemies, to launch into flight, or to search for food. The jumping legs are also used with the walking legs to climb up plants in order to feed on the tender leaves. When the insect is jumping or walking, spines, hooks, and pads of the foot, or *tarsus,* aid in gripping. The long joint next to the tarsus is called the *tibia.* The large muscles for jumping are contained in the *femur,* the heaviest segment of the leg. The *trochanter* (troe-KAN-ter) joins the *coxa* near the body. Together they act like a ball-and-socket joint in providing freedom of motion.

A remarkable respiratory apparatus

Each of the ten segments of the abdomen consists of two curved plates. The upper and lower plates are joined by a tough but flexible membrane that allows the segment to expand and contract in the process of respiration. The flexible membrane

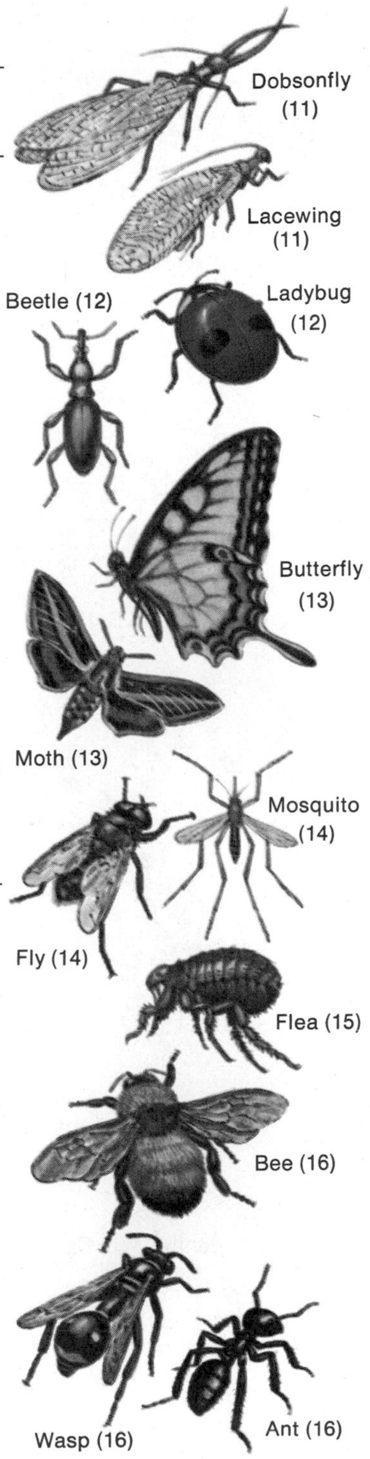

Dobsonfly (11)

Lacewing (11)

Beetle (12)

Ladybug (12)

Butterfly (13)

Moth (13)

Fly (14)

Mosquito (14)

Flea (15)

Bee (16)

Wasp (16)

Ant (16)

32-9 Common insect orders with complete metamorphosis.

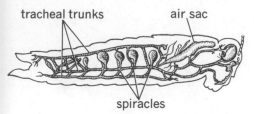

32-10 The respiratory apparatus of a grasshopper.

also joins each segment with the segments anterior and posterior to it, thereby allowing the segments to move.

The first eight of the abdominal segments have pairs of tiny openings, the *spiracles*. Spiracles are also found on the second and third thoracic segments. The openings lead to the tracheae, or air tubes, which form an amazingly complex network inside the animal (Figure 32-10). Air is pumped in and out of the tracheae by action of the wings and movement of the abdomen. Very rapid diffusion of oxygen into the tissues and carbon dioxide into the tracheae accomplishes respiration.

Digestion, excretion, and circulation

The action of the mandibles allows the grasshopper to pinch off bits of grass, which are then sucked into the mouth. An *esophagus* carries the food to a *crop,* where it may be stored for a time (Figure 32-11). As in many animals, *salivary glands* secrete juices that enter the mouth. These juices moisten the food to ease its passage to the crop. You may have seen food regurgitated from the crop when a grasshopper is injured or disturbed, but normally the food passes on to the *gizzard*. Here food is shredded by plates of chitin-bearing teeth.

Partially digested food is screened through thin plates and passes into the large *stomach*. There are several double pouches on the outside of the stomach. These *gastric caeca* produce and pour enzymes into the stomach, where digestion is completed. Digested food is absorbed into the bloodstream through the wall of the stomach. The material remaining in the stomach passes into the *intestine* which is composed of the *colon* and the *rectum,* which terminates at the *anus.*

Cellular wastes picked up by the blood are collected by a series of tubes, called *Malpighian* (mal-PIG-ee-un) *tubules*. These lie in the body cavity among the other organs. The wastes are then concentrated and passed into the last part of the intestine and out through the anus.

32-11 The internal structure of a grasshopper.

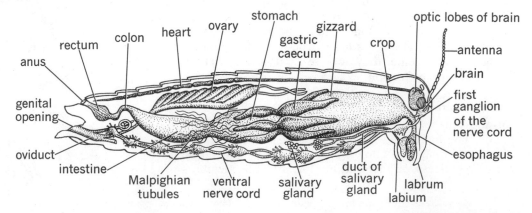

The circulatory system of the grasshopper, like that of the crayfish, is an open system. Blood forced out the anterior end of the dorsal, tubular, muscular *heart* passes through the *aorta* and into the body cavity near the head. As the blood flows toward the posterior region, it gives up nutrients to and takes on waste products from all of the body organs. And finally, it returns to the heart.

The grasshopper's sensory responses

Each side of the first abdominal segment of the grasshopper bears a membrane-covered cavity called the *tympanum.* These sensory organs function in hearing. Touch and smell are perceived by the many-jointed *antennae.* The grasshopper has two kinds of eyes. Figure 32-12 shows the *simple eyes,* located above the base of each antenna and in the groove between them. The large *compound eyes* project from a part of the front and sides of the head and are composed of hundreds of six-sided lenses. The shape, location, and number of lenses seem to adapt the insect for sight in several directions at one time, but the image formed is probably not very sharp. Most insects are considered to be nearsighted, yet they may be able to distinguish colors. We know that night-flying moths seek white flowers, while flies and some other insects are attracted by red and blue.

The stimuli received through the sense organs are then relayed by nerves to certain parts of the body, such as a muscle, which then respond. Nerve centers called *ganglia* act as switches in directing the message to the proper structures for coordinated action. The brain itself is composed of several fused ganglia, with nerves to the eyes, antennae, and other head organs, as well as to the ventral ganglia of the thorax and abdomen. The *optic lobes* are the most prominent part of the brain. The image of your approaching hand is enough to cause the grasshopper to start moving away, and fast! This complicated activity is begun, controlled, and coordinated by the nervous system.

The reproductive organs of grasshoppers

In insects, the sexes are separate. This means that sperm cells are produced in *testes,* found in the male. Egg cells are produced in *ovaries,* found in the female. The male deposits the sperm cells in a special storage pouch, the *seminal receptacle,* of the female during mating. The sperms remain there until the eggs are ready for fertilization.

The extreme posterior segments in the female grasshopper bear two pairs of hard, sharp-pointed organs called *ovipositors* (OE-VUH-POZ-it-erz). With these, the female digs a hole in the ground and deposits her fertilized eggs, which are protected by a gummy substance. One hundred or more eggs are laid in the fall and hatch the next spring.

32-12 The simple and compound eyes of a grasshopper. The compound eyes are made up of hundreds of six-sided lenses (inset). (Walter Dawn; inset: Hugh Spencer)

32-13 Incomplete metamorphosis in the Harlequin cabbage bug. Notice that the newly-hatched young are very similar to the adult. (Dr. C. H. Brett, North Carolina State University)

32-14 Complete metamorphosis. Notice that the newly hatched young of the *Cecropia* moth do not look anything like the adult. Can you think of any survival value in having two such distinct forms? (Herbert Weihrich)

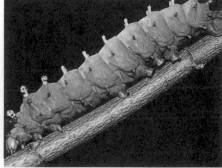

Insect metamorphosis

Most insects undergo several distinct stages during development from egg to adult. Such a series of stages in a life history is called *metamorphosis*. Grasshoppers and their relatives, true bugs, aphids, termites, and many other insects have **incomplete metamorphosis**. This is a three-stage life history consisting of (1) the *egg*, (2) the *nymph*, and (3) the *adult*. A nymph hatches from an egg in a form that resembles that of the adult except for size, absence of wings, and lack of development of the reproductive organs. In most species, nymphs molt five times, each time becoming more like the adult (Figure 32-13).

Butterflies, moths, flies, and beetles are among the insects with **complete metamorphosis**. They pass through four stages in their development: (1) *egg*, (2) *larva*, (3) *pupa*, and (4) *adult*. The larvae that hatch from eggs are segmented and wormlike. Depending on the kind of insect, they are called caterpillars, grubs, or maggots (see Table 32-3). After a period of feeding and rapid growth, the larva enters the pupal stage. This is often thought of as a resting stage because the changes occur within a shell or case, and cannot be seen externally. It is anything but a resting period, however, because in this stage all the tissues of the larva are transformed into those of the adult. The change from caterpillar to butterfly or grub to beetle is truly a marvelous event in nature (Figure 32-14).

In recent years, biologists have been investigating the role of hormones in insect metamorphosis. *Hormones* are chemicals that are produced by a body gland or organ and enter the blood.

Table 32-3   NAMES OF INSECT LARVAE

THE LARVA OF THE		IS CALLED	
	beetle		grub
	fly		maggot
	mosquito		wiggler
	butterfly		caterpillar
	moth		caterpillar

They are then transported to other parts of the body, where they control chemical reactions. In studies of the *Cecropia* moth, it was found that metamorphosis was controlled by hormones. In many insects having complete metamorphosis, the pupal stage occurs during the winter, when the conditions are not favorable for active adult insects. The low temperatures occurring during this pupal stage apparently influence the production of a brain hormone that, in turn, stimulates a gland in the prothorax to produce another hormone. This second hormone brings about the changes from a pupa to an adult moth.

Experiments have been performed in which the brains of *Cecropia* pupae were removed. Such insects did not metamorphose into adult moths. However, when a brain from a chilled pupa was implanted into one that had had its brain removed, metamorphosis did occur. If a pupa is cut in half and the ends are sealed, the anterior half will metamorphose into an adult, while the posterior half will not (Figure 32-15). What does this tell you about the location of the hormone-producing organ?

Some of you may wonder why an unusual cold spell does not cause the young caterpillars to form pupae. The same question occurred to the biologists working with the *Cecropia* moth. They found that if a paired structure, located behind the brain, was removed, the larva would form a pupa. Temperature did not appear to affect this process. It was concluded that the paired structure produces a hormone in the larva that inhibits metamorphosis. This has been called a *juvenile hormone,* since it functions in the larval stage of the insect.

Adaptive value of complete metamorphosis

The Lepidoptera may serve to illustrate a distinct survival value of complete metamorphosis. The sexes are separate, and after mating, eggs are deposited on or near the material that is to be the food of the young. Some eggs pass the winter in this stage, but usually eggs are deposited in the spring and develop into *caterpillars* the following summer (see Figure 32-16).

The caterpillar has three pairs of jointed legs and several pairs of fleshy legs. It eats ravenously, grows large and fat, and molts several times. Because the caterpillar needs tremendous amounts of food to keep up this rapid growth, it is during this period that the insect may do extensive damage.

When fully grown, the caterpillar usually seeks a sheltered spot, hangs with its head down, and becomes very quiet. The body shortens and thickens, and the exoskeleton splits down the back and is shed. The animal now becomes a pupa. The butterfly pupa rests in a hardened case, often brown in color. It is then called a *chrysalis.* The moth larva usually spins a strong case of silk, the *cocoon.*

The Lepidoptera usually spend the winter in the pupal stage. In the spring, the insect emerges, totally changed, as the adult.

32-15 Two of the many experiments performed by Dr. Carroll Williams on metamorphosis in the *Cecropia* moth. When a larva (top) is tied between the thorax and the abdomen, the anterior portion continues to develop but the posterior portion does not. Similar results occur when a pupa is cut in two. Two interdependent hormone centers that control metamorphosis were pinpointed.

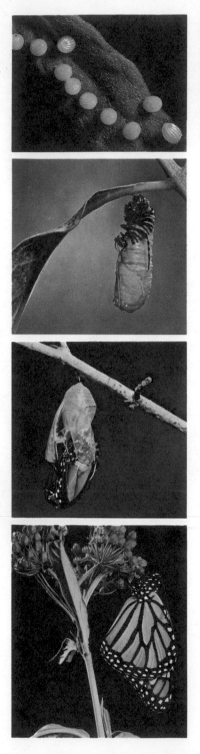

32-16 Life cycle of the Monarch butterfly. Can you name the stages shown? (*Left,* W. Clifford Healy, National Audubon Society; *right,* courtesy American Museum of Natural History)

Notice that in a life history such as that of the Lepidoptera, the larval form has completely different mouthparts, and therefore a completely different diet from that of the adults. The butterfly larva feeds on leaves, while the adult sucks nectar from flowers. A life history such as this allows an insect a varied diet during its life and an abundant supply of usable food during its period of most rapid growth. The adult serves only to continue and disperse the species. There is no competition between the young and adult for food.

Behavior—another survival factor

The success of the insects is also due to behavior more complex than that of the lower invertebrates we have already discussed. This complex behavior has been possible because of the development of (1) *complex sense organs,* (2) *jointed appendages,* and (3) *a brain.* The sense organs receive many kinds of stimuli, both from the immediate environment and from some distance away. These stimuli aid the animal in escaping from enemies, finding food, finding a mate, and locating a place suitable for the development of eggs. The brain is complex enough to allow the insect to integrate impressions from the sense organs and utilize the information in coordinated muscular responses.

With the development of these systems, the insects are able to perform complex, coordinated acts. They are able to modify their environment by building many kinds of homes and nests. The paper wasp builds nests of chewed wood pulp; the mud dauber uses mud; the tarantula hawk, a large wasp, digs a burrow and stocks it with paralyzed insects as food for the carnivorous larvae. Some ants store grain in special passages made for this purpose. The fungus ant builds underground gardens in which it cultivates a certain species of fungus. Before an ant queen takes off on her nuptial flight, she stores some of this fungus in a special pouch in her mouth so she can start the culture in her new home.

As you will see, many insects form societies in which differences in body structure within the species allow for division of labor. Some are adapted for gathering food, some for protection of the home, some for tending the young, and some for reproduction. These insects, which include the bees, ants, wasps, and termites, are called *social insects*. These insects do not reason; they perform their life duties by instinct. That is, they have specialized structures that are used in a behavior pattern that is not taught to them. Animal behavior studies form an interesting field of current research.

The honeybee, a social insect

Many of you may water-ski. Have you ever been towed by a boat that was underpowered? It will drag you through the water, but even with the planing surfaces of your skis at a sharp angle to the water, you cannot get enough lift to ski. If you examine a honeybee, you will see that its wings are very small in comparison with the size of the rest of its body. From this first examination, you might conclude that it is impossible for this insect to fly. However, the bee has very powerful muscles in its large thorax. These muscles move the wings at a sufficiently great speed to make the bee a swift flier with great endurance. The speed of the wings produces the familiar hum.

The queen—the mother of the colony

Perhaps nowhere else in the animal kingdom is individual variation and division of labor more noticeable than in the Hymenoptera. Variation has brought about three distinct forms: the queen, the drone, and the worker (Figure 32-17). The *queen,* which is nearly twice as large as a worker, develops from the special treatment of a fertilized egg. Workers enlarge a wax cell in which the egg is to grow, and when the grublike larva hatches, they feed it with extra portions of *royal jelly,* a high-protein food they secrete. After five days, the larva enters the pupal stage, and is sealed in a large waxen chamber by the workers. The workers have prepared to rear the queen at the correct time. The young queen emerges after the old one has left the hive with a swarm of about half the population. The old queen and her swarm form a new colony, and overcrowding is prevented.

After a few days, the young queen embarks on a flight during which she mates with a *drone* (male bee), receiving several million sperm cells that will serve to fertilize the eggs she produces during the remaining three to five years of her life. On returning to the hive, the queen begins her lifework of laying eggs. As many as one million eggs may be produced in one year. Although we call this bee a queen, she is in no sense the ruler of the hive. Rather, she is its common mother.

The male bee (drone)

Some of the eggs laid by the queen are not fertilized. These haploid eggs develop into the drones (Figure 32-18). These male

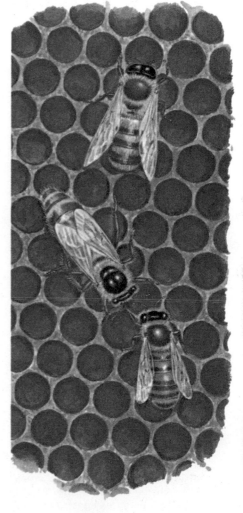

32-17 Three distinct forms of bees: the long slender queen, the large drone, and the smaller worker.

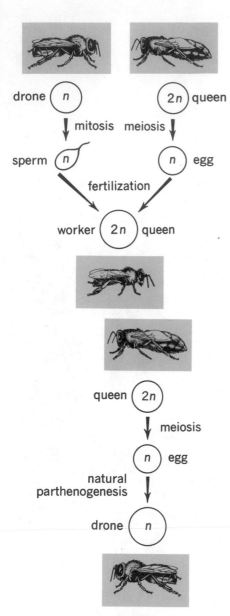

32-18 Sex determination in the honey bee. Note that the drones are haploid because they develop from unfertilized eggs. In contrast, the workers and queens are diploid because they develop from fertilized eggs. If *n* is 16, how many chromosomes are found in the drone? in the worker? in the queen?

bees are larger than the workers but smaller than the queen and have broad, thick bodies, enormous eyes, and very powerful wings. Their tongues are not long enough to obtain nectar, so they have to be fed by the workers. During the summer, a few hundred drones are tolerated in the colony. One of them must function as a mate for the new queen. The rest are of no use in the hive. This easy life, however, has its price. With the coming of the autumn, when honey runs low, the workers will no longer support the drones and will starve them or sting them to death. Their bodies can often be found around hives in the early autumn.

The worker bees

The *worker bees* are by far the most numerous inhabitants of the hive. They are undeveloped females, smaller than drones, and with the ovipositor modified into a sting. This is a complicated organ consisting of two barbed darts operated by strong muscles and enclosed in a sheath. The darts are connected with a gland that secretes the poison that makes a bee sting painful. Except for reproduction, all the varied industries and products of the hive are the business of the worker. They attend and feed the queen and drones. They act as nurses to the hungry larvae, feeding them with partly digested food from their own stomachs. Workers clean the hive of dead bees or foreign matter, and they fan with their wings to ventilate the hive. At all times, thousands of workers are bringing in nectar and pollen as needed for the use of the colony. In summer, the workers literally work themselves to death in three or four weeks, but bees hatched in the fall may live for six months.

The worker has numerous specialized structures that adapt her for her tasks. The labium and maxilla together form an efficient lapping tongue for gathering nectar. On the four last abdominal segments, there are glands that secrete the wax used in comb-making. The legs have various specialized structures, such as an antenna cleaner, pollen packer, pollen basket, and pollen comb.

The structure and products of the hive

The comb is a wonderful structure composed of six-sided cells in two layers. It is arranged with no wasted space and hence affords the greatest storage capacity using the least material. Honey and *beebread,* a food substance made by the worker from pollen and saliva, are stored in the cells of the comb. In a section of the hive called the *brood comb,* the queen places one egg in each cell.

Honey is made from nectar taken into the crop of the bee. Here the sugars are changed to a more easily digestible form and emptied into the comb cells. In these, the honey is left to thicken by evaporation before the cell is sealed. The removal of honey

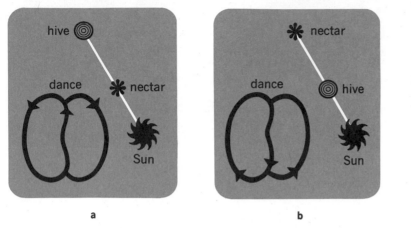

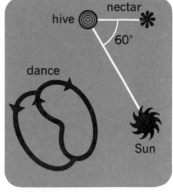

a   b   c

by man does not harm the bees if enough is left for their winter use. About thirty pounds is enough to feed an average colony of forty thousand bees for an ordinary winter.

The language of the bees

Evidence indicates that bees communicate with one another through the carrying out of a complicated set of dances. When a worker returns to the hive, she can inform the other workers of the kind and amount of nectar available, how far away it is, and in what direction. To do this, she dances on the vertical surface of the comb. The dance consists of a series of circles, with the bee making a path through a diameter of the circle each time she passes around one circumference (Figure 32-19). The direction she takes on the diameter indicates the direction of the nectar source in relation to the sun. If, in relation to the hive, the food source is directly toward or away from the sun, the diameter of the circle is in a line perpendicular to the horizon. The dancing worker indicates a direction toward the sun if she traces the diameter while moving up and thus away from gravity. Forming the diameter while moving down indicates that the food source is in a direction away from the sun. If the food is sixty degrees to the left of the sun's direction, the diameter traced is sixty degrees to the left of the vertical, and so on. Notice that this is all the more remarkable because the dancing bee is transposing the angle between the direction of the sun and that of the food source to a vertical angle with respect to gravity.

As the other workers gather around the dancing bee, they obtain the information about the direction of a food source. The type of food that is available is determined by smell. The distance of the food is determined by the number of times the bee waggles her body while tracing the diameter of the circle. The greater the amount of nectar available, the more vigorous is her dance.

32-19 (a) The dancing bee makes the diameter of the "tail-waggling dance" upward on the vertical surface of the comb. This indicates to the other bees that a source of nectar is located in a direction toward the sun. (b) This dance indicates that the nectar is located in a direction away from the sun. (c) Decipher this message.

SUMMARY

In numbers, the insects are the dominant group of animals on the earth today. They have inhabited the earth for millions of years and reproduce at a very rapid rate. Biologists believe these two facts combined with variation have allowed the insects to become so successful. Insects have evolved into a very diverse group adapted to many environmental conditions.

You have seen the survival value of the exoskeleton, small size, coloration, wings, complete metamorphosis, nervous system, and sense organs. The social insects work together to provide their food, raise their young, and even modify their environment.

Because of their numbers and close association with man and his crops and animals, the insects are of great economic importance.

BIOLOGICALLY SPEAKING

entomology	crop	seminal receptacle
labrum	salivary gland	ovipositor
mandible	gizzard	incomplete metamorphosis
maxilla	Malpighian tubule	complete metamorphosis
palpus	aorta	caterpillar
labium	tympanum	chrysalis
prothorax	simple eye	cocoon
mesothorax	ganglion	social insect
metathorax	optic lobe	queen
spiracle	testes	drone
esophagus	ovary	worker bee

QUESTIONS FOR REVIEW

1. How can you support the statement: Insects have evolved into a successful group?
2. Why do the biologists believe the insects were able to evolve with such diversity?
3. What is an insect?
4. What characteristics separate the insect orders?
5. Of what benefit to the insect is the chitinous exoskeleton?
6. What characteristics of the grasshopper make it similar to the crayfish?
7. Most animals have sense organs located on the head. In what respect is the grasshopper an exception to this rule?
8. How do the grasshopper's legs illustrate adaptation?
9. What is the chief difference between the respiratory system of the grass-hopper and that of man?
10. What protection is given the eggs of the grasshopper during winter?
11. Of what survival value is metamorphosis?
12. What advantage do the social insects have over the solitary forms?
13. Name the different types of bees and describe the functions of each in the hive.
14. Do bees communicate with one another in the hive? Explain.

APPLYING PRINCIPLES AND CONCEPTS

1. What reasons can you give for the insects' ability to withstand unusual temperatures, pressures, and other extreme environmental conditions?
2. Is it a fact that insects have inhabited the earth for 300 million years? Explain.
3. What theories can you give to explain why an insect might resemble a thorn on a bush or why a fly might look like a bee?
4. Can a little fly grow to be a larger fly? Explain.
5. Can you suggest any possible mechanisms that account for the change in the amount of juvenile hormone that a *Cecropia* larva may produce? As a biologist studying the hormone balance in metamorphosis, how might you go about testing your hypotheses?
6. What additional experiments might be performed to increase our understanding of communication among the bees?
7. How does sex determination in bees differ from that in humans?

RELATED READING

Books

BARKER, WILLIAM, *Familiar Insects of America.*
Harper and Row, Publishers, New York. 1960. Describes those insects commonly seen in town, country, field, or forest.
CALLAHAN, PHILLIP, *Insect Behavior.*
Four Winds Press, New York. 1970. A highly readable account on many aspects of insect behavior with a special section on possible projects and experiments.
FARB, PETER, and the Editors of *Life* Magazine, *The Insects.*
Time-Life Books (Time, Inc.), New York. 1962. A beautifully colored picture-essay type book that effectively highlights the natural history of a wide range of insects.
HOYT, MURRAY, *The World of Bees.*
Coward-McCann, Inc., New York. 1965. A natural history of the bees, dwelling particularly on them as social insects.
LANHAM, URL, *The Insects.*
Columbia University Press, New York. 1964. A natural history of insects for the amateur.
NEWMAN, L. H., *Man and Insects.*
Doubleday and Company, Inc., Garden City, New York. 1967. A basic book of entomology that points out man's urgent need to learn the insects' role in ecology, and stresses their role in the scheme of things.
ROSS, HERBERT H., *Textbook of Entomology* (3rd ed.).
John Wiley and Sons, Inc., New York. 1965. The fundamental aspects of entomology, organized to give the student a general idea of the entire field.

UNIT **6**

BIOLOGY OF THE VERTEBRATES

This unit introduces the most advanced of all animals, a group that surpasses the others in structural organization and functional efficiency. The group is known as the vertebrates because its members have spinal columns composed of bones called vertebrae. The spinal column alone, however, is not as important as the nerve cord it encases and the highly developed brain to which this nerve cord is joined. A most efficient nervous system and the high-level responses and adjustments this system permits are the key to the biological supremacy of the vertebrates.

CHAPTER THIRTY-THREE

INTRODUCTION TO THE VERTEBRATES

A quick review of invertebrate development

Nature seems to have experimented with several plans of body development in various invertebrate animals. In the protozoans, the specialization of a single cell is carried to the limit. Consider the paramecium with its cilia and trichocysts, its gullet and contractile vacuoles. The "slipper animalcule" is a truly marvelous single-celled organism, but it takes many cells to make a complex animal.

You were introduced to large colonies of cells in the sponges and coelenterates. The beginnings of tissue are found in the ectoderm and endoderm of these animals. In the flatworms and roundworms, there are organs that perform with much greater efficiency such functions as digestion, reproduction, excretion, and response. The segmented worms have a more advanced and efficient tubular digestive tract and an elongated body divided into a series of segments. A well-developed circulatory system transports blood through closed vessels in this group of worms.

The clam and other mollusks have the greatest possible protection for a highly developed, soft-bodied animal. Their shells are both a fort and a prison. Mollusks, however, have not advanced much in many millions of years.

The arthropods combine protection with freedom of movement. But an exoskeleton capable of supporting a very large arthropod would be too heavy to move. Thus, arthropods have remained reasonably small. Through sheer numbers alone they maintain their place of importance.

Chordates, the most complex forms of animal life

One great phylum of the animal kingdom includes the most highly developed animals that have ever inhabited the earth. This phylum, known as *Chordata* (koar-DAY-tuh), comprises four subphyla. By far the largest and most important of these is the subphylum *Vertebrata,* in which we find the fishes, amphibians, reptiles, birds, and mammals, including man.

Before considering vertebrates as a group, however, we shall examine the chordates briefly. Three unique characteristics distinguish chordates from all other animals.

■ The embryos of all chordates have a rod of connective tissue running lengthwise along the dorsal side of the body. We call this supporting rod, or skeletal axis, a *notochord* (the phylum name comes from this term). The more primitive chordates retain the notochord throughout life. Some of the lower vertebrates, including the lamprey, retain the notochord, but it becomes surrounded by cartilage structures of the spinal column. In other vertebrates, the notochord is present in the embryo, but is replaced early in life by the *vertebral column,* or backbone.

■ A second characteristic of all chordates is a tubular *nerve cord* that lies just above the notochord and extends down the dorsal side of the body. The anterior end of the nerve cord enlarges into a *brain.* The remaining portion constitutes the *spinal cord.* Together, the brain and the spinal cord compose the *central nervous system.*

■ At some time in life, all chordates have paired *gill slits* that form openings in the throat. These, like the notochord, disappear early in the development of higher vertebrates, including the reptiles, birds, and mammals. However, they remain in the fish and more primitive vertebrates, including the shark.

While we are familiar with vertebrates as examples of chordates, the three other subphyla represented by marine animals are relatively unknown except to the biologist. These primitive chordates are of great interest to the biologist because they give some idea of what the ancestors of the present-day vertebrates might have been like.

One chordate subphylum (Hemichordata) contains two classes of wormlike animals, the best known of which are the acorn or tongue worms. Another subphylum (Urochordata) is represented by the tunicates, or sea squirts. The fishlike lancelet, or *Amphioxus,* shown in Figure 33-1, probably the best known of the lower chordates, represents the third subphylum (Cephalochordata). This interesting marine chordate, about two inches in length, lives in tropical and temperate coastal waters. It is commonly

33-1 *Amphioxus* retains its dorsal nerve cord, notochord, and gill slits throughout its life. The notochord and gill slits disappear early in the development of higher chordates.

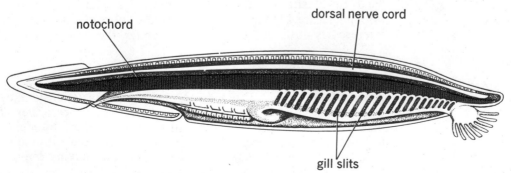

notochord

dorsal nerve cord

gill slits

Time Scale of Vertebrate Classes Based on Fossil Records

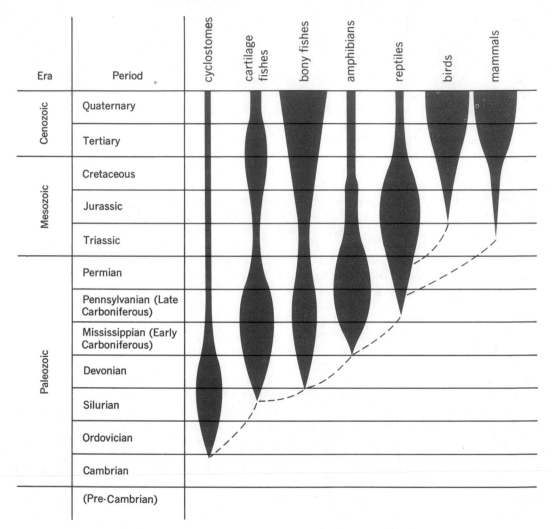

33-2 A time scale of the classes of vertebrate animals.

used to illustrate chordate characteristics that are retained throughout life. However, of the chordate subphyla, our primary interest lies in the *animals with backbones,* the vertebrates.

The rise of vertebrates

When did the earliest vertebrates appear in the animal population of the earth? From what chordate ancestors did they evolve? There is no direct fossil evidence to answer either of these questions. This lack of fossil evidence has led biologists to believe that the ancestors of the vertebrates were soft-bodied chordates whose remains decomposed without leaving impressions in rocks. It is likely that the earliest vertebrates lived during the Ordovician period, some half billion years ago (Figure 33-2). Fossil evidence becomes much more abundant in rocks of the Silurian period and of the Devonian period that followed.

Fossil records indicate that many vertebrate forms appeared and disappeared in successive geological periods. It has been estimated that not more than one percent of the amphibian, reptile, bird, and mammal species living in the Jurassic period have living descendants today. Paleontologists believe that there has been a tremendous turnover in the kinds of animals that have inhabited the earth. Evolution of the vertebrate animals will be discussed more fully when we consider each class individually in the following chapters.

Classes of vertebrates

Modern vertebrates are represented by seven classes. Listed in the order in which they are believed to have evolved and in order of their structural complexity, they are:

■ *Cyclostomata* (sy-kloe-STOME-uh-tuh), the jawless fishes, including the lampreys and hagfishes.
■ *Chondrichthyes* (kon-DRICKTH-ee-EEZ), the sharks, rays, and skates.
■ *Osteichthyes* (oss-tee-ICKTH-ee-EEZ), the bony fishes.
■ *Amphibia* (am-FIB-ee-uh), the frogs, toads, and salamanders.
■ *Reptilia* (rep-TILL-yuh), the snakes, lizards, turtles, and crocodilians.
■ *Aves* (AY-veez), the birds.
■ *Mammalia* (muh-MALE-yuh), the mammals.

As you study these vertebrate classes in the chapters that follow, you will notice a transition from a life in water to a life on land. This appears to have been the direction in the evolution of the vertebrates. As you proceed in your study from one class to another, you will also notice an increase in both the complexity and the efficiency of various body organs. Vertebrates also illustrate an impressive degree of success in both aquatic and terrestrial environments. Lampreys, sharks, and rays still thrive in rivers and lakes, oceans and seas. Fish are abundant in nearly all waters of the earth. Amphibians and reptiles, while greatly reduced in numbers from earlier geological periods, still inhabit both aquatic and land environments. Birds are unchallenged in the air, and mammals are the most highly evolved land animals.

Characteristics of vertebrates

Many structural characteristics distinguish vertebrates from other forms of animal life and give them biological supremacy. Vertebrates possess all of the chordate characteristics as well as the following features:

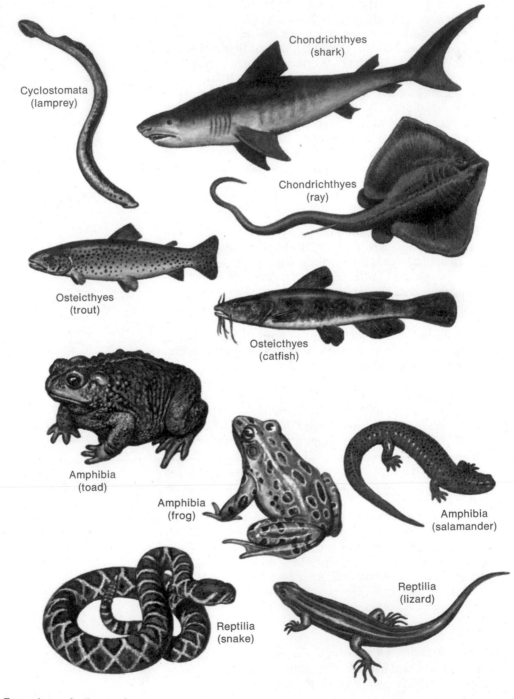

33-3 Examples of the various classes of vertebrates.

Reptilia
(alligator)

Reptilia
(turtle)

Aves
(cardinal)

Aves
(goose)

Mammalia
(horse)

Mammalia
(deer)

Mammalia
(seal)

Mammalia
(chipmunk)

■ An *endoskeleton* (internal framework) composed of *bone* and/or *cartilage* (Figure 33-4). This consists of:

a. A vertebral column, or spine, composed of cartilaginous or bony segments called the *vertebrae*. This backbone constitutes the *axial skeleton.*

b. In most vertebrates, an anterior *pectoral girdle* and a posterior *pelvic girdle* suspended from the spinal column.

c. Limbs in the form of fins, legs, wings, or flippers, never more than two pairs in number, attached to the pectoral and pelvic girdles, forming the *appendicular skeleton.*

■ A body consisting of a head and trunk and, in many vertebrates, a neck and tail region.

■ An efficient closed circulatory system that includes a ventral heart located in the anterior region of the body.

■ Red corpuscles containing hemoglobin.

■ In most vertebrates, twelve pairs of cranial nerves extending from the brain and varying numbers of spinal nerves extending from the spinal cord.

■ Eyes, ears, and nostrils on the head.

■ An alimentary canal, or food tube, elongated and looped or coiled in the body, with a liver and pancreas also present.

Specialized systems of the vertebrate body

The vertebrate systems contain many highly developed organs. These systems include the following:

■ *Integumentary system*—the outer body covering and special outgrowths, such as scales, feathers, or hair, that provide protection.

■ *Muscular system*—muscles attached to bones for body movement; muscles that form the walls of the heart and the digestive organs and blood vessels.

■ *Skeletal system*—the bones and cartilage that make up the body framework.

■ *Digestive system*—the many specialized organs concerned with the preparation of food for use by the body tissues.

■ *Respiratory system*—gills or lungs and related structures used in the exchange of gases between the organism and its external environment.

■ *Circulatory system*—the heart and blood vessels, which function as the transportation system of the body.

■ *Excretory system*—the organs that remove wastes from the body.

■ *Endocrine system*—glands that produce secretions necessary for the normal functioning of the other systems.

■ *Nervous system*—the brain, the spinal cord, nerves, and special sense organs, the most highly developed system of a vertebrate.

■ *Reproductive system*—the male or female organs of reproduction.

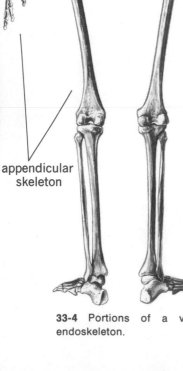

axial skeleton (including vertebrae)

pectoral girdle

pelvic girdle

appendicular skeleton

33-4 Portions of a vertebrate endoskeleton.

Lines of development in the vertebrates

The skeleton shows an interesting development in the vertebrate classes. The lampreys, sharks, and rays have skeletons of cartilage throughout life. Fishes, amphibians, reptiles, birds, and mammals develop bony skeletons. These animals start life with a cartilaginous framework. But early in life, bone cells replace most of the cartilage. Minerals deposited in the bones make them hard and strong.

The classes of vertebrates also show an interesting change from water existence to life on land. The lampreys, sharks, rays, and bony fishes are adapted only for life in water. Their limbs take the form of fins. Their gills absorb dissolved oxygen from the water. Water flows over the gills through gill slits in the throat. After you have studied the frog as a representative amphibian, you will realize that the amphibians represent a transition from water to land. During the tadpole stage, a frog is a fishlike animal with gills and a fin.

The vertebrate heart and brain show great development over those of the invertebrates. The fish heart has two chambers. One chamber receives blood from the body, while the other pumps blood to the gills. The frog has a three-chambered heart and a more complex circulatory system. Birds and mammals have still more complex hearts, consisting of four chambers. One side of the heart receives blood from the body and pumps it to the lungs. The other side receives blood from the lungs and pumps it to the body. This heart is really a double pump. Man's heart likewise consists of four chambers.

Similar advances can be seen in the presence of a definite brain enclosed within a skull, or cranium. One brain region, known as the *cerebrum*, is the center of instinct, emotion, memory, and intelligence. This brain area increases in relative size through the classes of vertebrates. The brain of the mammal has the largest cerebrum in proportion to the size of the body.

The highly developed behavior of vertebrates

Behavior is the way in which an organism responds to stimuli. The type of sense organs, nerve pathways, and organs specialized for nervous control determine the stimuli to which organisms are sensitive and, in many respects, the responses they can make. Protists and plants have no specialized nerve tissue. Their responses are limited to simple tropisms. The nerve net of the hydra enables it to behave as a unit. The sense organs, nerve cords, and ganglia of the higher invertebrates permit even more integrated behavior. In the vertebrates, the highly developed sense organs, brain, and nerves extending to and from all parts of the body provide the basis for complex behavior.

Much of the activity of an animal is inborn. Since such *innate behavior* is inherited, it is reasonable to assume that it is controlled by genes. ***Reflexes*** are simple innate responses. In a reflex, an animal automatically responds in a certain manner to a given stimulus. For example, stimulation of the surface of the eye or eyelid will cause blinking. Reflexes are involuntary; that is, the animal reacts to the stimulus without any control on its part. Generally, reflexes protect the organism from harm. Even man depends on reflex behavior for many of his responses.

The most interesting and the least understood of the innate responses are those that we call ***instincts***. Instincts are complex patterns of unlearned behavior. They are involuntary, since the animal performs them without a deliberate decision.

Self-preservation is a basic instinct in all vertebrates as well as in many invertebrates. In times of danger, an animal will respond to the "flight or fight" instinct of self-preservation. Have you ever cornered an animal that would normally flee? A seemingly harmless animal like a squirrel will bite and claw viciously if it cannot escape from an enemy. Another example of an inborn behavior that is apparently related to self-preservation is the mammal's suckling instinct, the instinct to nurse during the early period of life. The self-preservation instinct causes the tiny bird to pick its way through the shell at the time of hatching.

33-5 Examples of three types of behavior that are highly developed in vertebrate animals; self-preservation, species preservation, and conditioned reaction. The first two are instinctive, the last is learned. (R. D. Estes, Photo Researchers; Walter Dawn; Ringling Brothers, Barnum and Bailey Circus)

A second instinct, that of *species preservation,* directs animal reproduction and care of the young. This is the instinct that, for instance, drives the Pacific salmon up the streams of the Northwest to spawning beds. The adult salmon lose their lives, and a new generation comes downstream to the ocean. This instinct also causes the sunfish to defend its nest against an intruder from which it would normally flee. In responding to an instinct, an animal behaves automatically, without making a decision or choice.

Because of their well-developed nervous systems, vertebrates are capable of learned behavior as well as innate behavior. A *conditioned reaction* is a form of learned behavior common among vertebrates. We say that a reaction becomes conditioned when a particular behavior response continually follows a specific stimulus. A reward, punishment, or threat of punishment reinforces the response. As this chain is continued, the activity becomes more habitual. We develop this level of behavior when we teach a dog to heel at a command or to "shake hands" at a given signal. Even fishes in an aquarium can be trained to go to one special corner of the tank when you approach, if you always feed them in this particular place.

Instinct can be observed in all vertebrates. To a lesser degree, most of the vertebrates are capable of conditioned reactions. *Intelligent behavior,* involving problem-solving, judgment, and decision is a still higher and more complicated level of nervous activity. Birds and most mammals exhibit a limited degree of intelligent behavior. Man, however, is supreme among the vertebrates in development of intelligence. He is also unique among living things in his ability to learn to communicate by symbols, both in speaking and in writing.

SUMMARY

With an understanding of the general characteristics of chordates and vertebrates, the systems of the vertebrate body, and the lines of development in vertebrates, you are ready to investigate the vertebrate classes in greater detail. In this study, you will follow an interesting progression from the lowest to the highest living vertebrates, based on the degree of development of body organs. You will notice this increase in structural and functional efficiency in such organs as the heart, respiratory organs, reproductive organs, and brain.

The progression through the vertebrate classes illustrates another interesting feature. Fossil evidence indicates that vertebrate animals evolved through the geological eras and periods in this order.

BIOLOGICALLY SPEAKING

notochord	cerebrum	self-preservation
nerve cord	innate behavior	species preservation
gill slits	reflex	conditioned reaction
endoskeleton	instinct	intelligent behavior
vertebra		

QUESTIONS FOR REVIEW

1. In what three ways do chordates differ from other animals?
2. Name and give examples of three chordate subphyla other than vertebrates.
3. Account for the lack of fossil evidence of vertebrate ancestry.
4. Name seven classes of vertebrates and give an example of each.
5. Cite evidence to show that vertebrates are highly successful organisms in both aquatic and terrestrial environments.
6. List ten structural characteristics of vertebrates.
7. Name ten organ systems composing the vertebrate body.
8. Which of the various brain regions of a vertebrate is the center of instincts and intelligence?
9. What is the relationship between a stimulus and a response in a nervous reaction?
10. Distinguish between innate behavior and activities on higher levels.
11. In what ways are reflex actions important in the lives of animals?
12. Give an example of (a) a conditioned reaction in a vertebrate and (b) an intelligent act.

APPLYING PRINCIPLES AND CONCEPTS

1. Discuss the development of vertebrates through the various classes, using the skeleton, organs of respiration, heart, and brain as illustrations.
2. Self-preservation and species preservation are instincts. Which is stronger? Give one or more illustrations to support your answer.
3. How can you distinguish instinctive behavior from intelligent behavior in observing the activities of various vertebrate animals?
4. Why are instinct and intelligence more vital to the survival of a vertebrate than to an invertebrate such as a clam, starfish, insect, or crayfish?

RELATED READING

Books

BATES, MARSTON, *Animal Worlds*.
Random House, Inc., New York. 1963. How animals meet varied conditions of life in their respective environments.
DROSCHER, VITUS B., *The Mysterious Senses of Animals*.
E. P. Dutton & Co., Inc., New York. 1965. An account of the mysteries of animal behavior, based on the latest discoveries in this field.
HALSTEAD, L. B., *The Pattern of Vertebrate Evolution*.
W. H. Freeman and Company, San Francisco. 1968. Presents a thought provoking interpretation of the fossil record as it relates to vertebrate development.
HELM, THOMAS, *Shark-Unpredictable Killer of the Sea*.
Dodd, Mead & Co., Inc., New York. 1961. The complete story of sharks, with the origin, history, and description of the most important species.
ROMER, ALFRED S., *The Vertebrate Story*.
The University of Chicago Press, Chicago. 1958. An account of the nature of the vertebrates, from fish to man, giving their structure, function, and ways of life.
VLAHOS, OLIVIA, *Human Beginnings*.
The Viking Press, Inc., New York. 1966. A highly recommended book on evolution of land animals.

CHAPTER THIRTY-FOUR

THE FISHES

Blood-sucking "vampires" of the Great Lakes

About forty years ago, a deadly vertebrate menace made its way from the waters of Lake Ontario through the Welland Canal at Niagara Falls and into Lake Erie. The sea lampreys were invading new waters. At a much earlier time, this species had migrated from coastal waters of the Atlantic Ocean up the St. Lawrence River and had become established in Lake Ontario. Here their movement was stopped by Niagara Falls. But the Welland Canal, built to carry shipping around the falls, gave them passage into Lake Erie. Ten years later, the lamprey hordes had spread through Lake Huron. They traveled through the Straits of Mackinac into Lake Michigan and through the locks at Sault Sainte Marie into Lake Superior.

What sort of creature is this death-dealing sea lamprey? Biologists place it in the class *Cyclostomata,* which means "round-mouthed." (This small class of primitive vertebrates is also sometimes called *Agnatha* (AG-nuh-thuh), which means "jawless.") Ancestors of the cyclostomes are believed to have been the first vertebrates, appearing in the Ordovician period more than 400 million years ago. The cyclostomes, and some other jawless vertebrates long since extinct, increased in number during the Silurian period, then declined during the Devonian period that followed (see Figure 33-2). Today, this once-flourishing group of vertebrates is represented only by the hagfishes and lampreys, the best known of which is the sea lamprey *(Petromyzon).*

The sea lamprey has a slender eel-like body. The mature lamprey reaches a length of about two feet and a weight of about one pound. Its skin is soft and slimy, brownish-green, and blotched or mottled. Paired fins are lacking in the lamprey. Two single fins along the back and a tail fin aid the lamprey in swimming in its usual rippling manner.

The head of a lamprey is curious and quite different from that of a fish. Instead of jaws, the lamprey has a funnel-like mouth lined with sharp, horny teeth. A rasping tongue, also bearing

34-1 The mouth of a sea lamprey (left) is well suited for obtaining food. Note the circular, jawless mouth and rasping teeth. The scar on the trout was caused by a sea lamprey's attack. (Russ Kinne, Photo Researchers; U.S. Fish and Wildlife Service)

teeth, lies in the center of the mouth. Small eyes are situated on either side of the head. Between the eyes, on the top of the head, is a nasal opening that leads to a sac containing nerve endings associated with the sense of smell. Seven oval gill slits, resembling portholes of a ship, lie in a row on each side of the head, behind the eyes. These openings lead to spherical pouches that contain numerous feathery gills. Water moves in and out through the external gill slits of the adult sea lamprey.

During its adult life, the sea lamprey is a very destructive predator. It attaches its sucking mouth to the side of a fish and gouges a hole through the scales with its rasping teeth (Figure 34-1). It feeds on the blood and body fluids of its victim and may even suck out internal organs. When it has killed or weakened a host fish, the lamprey moves on to another. The injury it inflicts, however, is not always fatal, for many fishes are found with scars marking the places where they were attacked by lampreys. It preys mainly on the lake trout, one of the most important commercial fishes of the Great Lakes. When trout are not available, the sea lamprey attacks whitefish, pike, and other species.

The sea lamprey has nearly exterminated the lake trout in Lake Huron and Lake Michigan and has seriously reduced the population in Lake Superior.

Our hope of eliminating the lamprey menace

A knowledge of the spawning habits of the lamprey is helping us to destroy this deadly menace. Sea lampreys reach sexual maturity in the Great Lakes during the months of May and June. At this time, they enter fast-flowing streams that feed the lakes. They lay their eggs in circular depressions in the gravel bottom of cold streams. An average female lays from 25,000 to 100,000 eggs. After about twenty days of development, the eggs hatch into tiny, blind larvae that resemble *Amphioxus* in many respects. The larvae leave the nest and float downstream until they reach quiet water with a mud bottom. Here they burrow into the mud and start a period of inactive life. During this period, the larva lies in a U-shaped burrow and feeds on plant and animal matter drawn into the mouth of the burrow in a current produced by moving cilia. After four or more years in the stream, the larva becomes an adult and starts its journey downstream to the lake. It lives for about one year as an adult, feeding constantly on fish.

Two methods of lamprey control have been used. One frequently used is the lamprey trap, designed to capture the adults as they migrate upstream to spawn. Electrodes charged with electricity are put across the stream in a row. These charge the water and stop the movement of all kinds of aquatic animals. The migrating lampreys and fishes swim along the edge of the charged area into traps, where the lampreys are destroyed. The fishes are caught and put back into the stream above the traps.

A more recent method of lamprey control that has replaced many electric barriers makes use of a selective poison, or *larvicide,* that kills the larvae buried in the streams. It has been used extensively in Lake Superior, where it has reduced the lamprey population by about 80 percent, and is now being applied in Lake Michigan.

Sharks and rays

To the class *Chondrichthyes,* which means "cartilage fishes," belong the sharks, rays, and skates. (This class of fishes is also called *Elasmobranchii,* meaning "plated gills.") The "cartilage fishes" are believed to have appeared early in the Devonian period. Of the fishes that controlled the ancient seas, these have survived in relatively large numbers to the present day.

The shark resembles the true fishes in many ways. Distinguishing characteristics, however, such as *placoid scales,* which have the same origin as the shark's teeth, permit its being placed in a separate class. The body of a shark is torpedo-shaped (Figure 34-2). Its fins resemble those of true fishes. The upper lobe of the tail fin is longer than the lower portion—a characteristic of ancient fishes. This asymmetrical tail fin, called a *heterocercal fin,* forces the head downward as it propels the shark through the water. The shark's mouth is a horizontal, slitlike opening on the ventral side of the head. The jaws of most species are lined with pointed, razor-sharp teeth. Water enters the mouth, passes over the gills on each side of the head, and is forced out through separate pairs of gill slits. Gills, as you probably know, are the special respira-

34-2 A shark. Unlike sea lampreys sharks have jaws, paired nostrils, true teeth, and paired pectoral and pelvic fins and girdles.

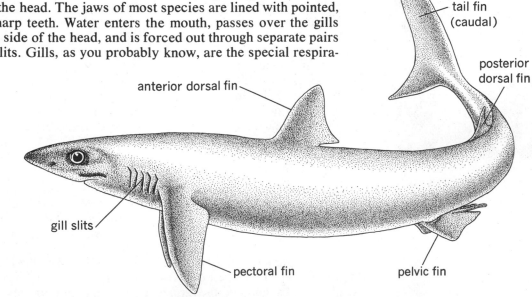

anterior dorsal fin

tail fin (caudal)

posterior dorsal fin

gill slits

pectoral fin

pelvic fin

34-3 The sting ray, grey shark, and banded carpet shark all belong to the same class. (© 1968 Jerry Greenberg; Douglas Faulkner; Allan Power, National Audubon Society)

tory organs of fishes and their relatives. Large, well-developed eyes lie in sockets on either side of the head above the mouth. Paired nostrils on the ventral side of the head anterior to the mouth open into olfactory sacs, which sense odors in the water. The skeletons of sharks and rays are composed of cartilage, as the name of the class implies, rather than bone.

Sharks include the largest living fishes. The whale shark, the giant of sharks, reaches a length of fifty feet or more and a weight of over twenty tons. The great white shark, a man-eater, may exceed forty feet in length.

Reports of attacks by sharks have increased as more people turn to the water for recreation. Scientists are beginning to ask how the shark locates its prey. Does it depend on sight, smell, or both? When opaque plastic eye shields are put over a shark's eyes, the "blinded" shark takes longer to find food. Thus, sight must be important in locating prey. It has been demonstrated that sharks can smell blood for only a few hundred feet but that they can detect sound over some distance. It may be that they are first attracted to possible victims by the sound of splashing.

The rays and skates have broad, flat bodies with whiplike tails. Rays swim gracefully through ocean waters, moving their flat bodies like wings (Figure 34-3). They often lie half buried in the sand of the ocean bottom. The tail has a sharp, barbed spike near the tip that causes a painful wound when it is driven into a victim. Sting rays often come close to shore. The torpedo ray has an excellent means of defense. It consists of certain muscle cells that have been modified into electric organs that can generate a current sufficiently great to shock a man severely. However, this adaptation is probably used mainly for obtaining food.

The skate has a triangular-shaped body with a long, thin tail (Figure 34-4). The triangular pectoral fins, which undulate from front to rear rather than flapping like wings, propel the skate through the water. Two fins attached to the rigid tail act as a steering device.

Skates are well adapted for life on the bottom of the ocean. Water is taken in through two spiracles located on top of the head, just behind the eyes. Then the water passes out through the gill slits underneath. Thus skates avoid taking in debris when they respire.

The true fishes

Biologists put all the true fishes in the class *Osteichthyes,* which means "bony fishes." Their bony skeletons distinguish them from the lampreys, sharks, and rays of modern times. (This class is sometimes called *Teleostomi* (TELL-ee-OST-uh-MEE), which means "complete-mouthed," to distinguish its members from the cyclostomes.) The bony fishes first appeared in the Devonian period, which is often called the Age of Fishes because Devonian rocks contain the greatest numbers and variety of fish fossils. This was a time of rapid evolution for the fishes because their members were so widely distributed. From that time to the present, bony fishes have increased in number steadily (see Figure 33-2). Today, they are the dominant vertebrates in both freshwater and marine environments, and are found in a great variety of forms and sizes.

Bony fishes have *gills* as respiratory organs. Limbs take the form of *fins.* Most fishes have an outer covering of overlapping *scales,* or plates. Fishes are ideally suited to aquatic life. In a wide variety of forms, they live in practically every water environment.

The body of a fish is divided into three regions—head, trunk, and tail. Many people confuse the tail of a fish with the tail fin. The tail is the solid muscular region posterior to the trunk. The tail fin is an outgrowth of this region. In most species, the body is perfectly streamlined—tapered at both ends, or spindle-shaped. The lack of a neck is no disadvantage to a fish. It can turn its body as easily in the water as most other animals can move their heads.

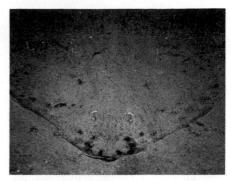

34-4 A skate. The skate is well adapted for living on the bottom of the ocean. What features do you see in the photograph that would support this statement? (Douglas Faulkner)

The body covering of fishes

Scales grow from pockets in the skin and overlap like shingles on a roof. Scales increase in size as a fish grows; a young fish has the same number of scales it will have at maturity. As scales grow, concentric rings are formed. These rings grow closer together in the winter than they do in the summer; thus, it is possible to distinguish seasons of scale growth and determine the age of the fish.

Mucus, secreted by glands in the skin, seeps between the scales and forms a covering that lubricates the body of a fish. This body slime is important in locomotion and in escape from enemies. It is important, too, in protecting the fish from attack by parasitic fungi, bacteria, protozoans, and other organisms. If you handle a live fish with dry hands, you remove some of the mucus and expose the body to parasites. You can avoid harming the fish in this way by wetting your hands before you pick it up.

Many fishes have bright colors, often arranged in lines, bars, or spots. Much of the coloration of fishes is caused by the presence of red, orange, yellow, or black pigment granules in special cells of the skin known as *chromatophores* (krome-AT-uh-foarz). A chromatophore consists of a central body and numerous radiating rays. Dispersal of pigments from the central body into the rays darkens or brightens the color. Concentration of the pigments in the central body causes the color to fade. Coloration may be due, also, to the presence of *guanine crystals,* which reflect light. Guanine crystals are excretory products found in the scales and skin as well as in the eyes and air bladder. Many iridescent colors, including blue and green, are caused by reflection of light from guanine crystals in the scales.

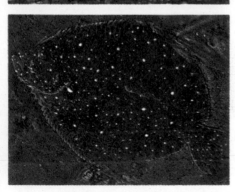

34-5 Flounders are able to blend into almost any environment by involuntarily changing the dispersal or concentration of pigments in their chromatophores. (Russ Kinne, Photo Researchers)

In fish, the ability to change color varies with species, size, and age, as does the time necessary for color change. The flounder probably holds the record for rapidity of color change. Its color pattern can change and match even a checkered or polka-dot background within a few minutes (Figure 34-5). For this reason, the flounder has been a subject of many investigations of pigmentation and color change in fishes. Most other fish change more slowly, often requiring several days.

Color changes are caused by chromatophores rather than guanine crystals. It is believed that the fish changes its color and blends with its surroundings in an involuntary response to visual impressions it has of those surroundings. Nerve impulses to the chromatophores appear to be responsible for the dispersal or concentration of pigments that results in color changes. Aquarists often find that fish placed in new surroundings lose much of their color and regain it in a few days, after they are adjusted to the new environment.

Many fishes illustrate *countershading,* another form of camouflage. Darker pigments on the dorsal side of the body tone down the bright light striking the fish from above. Lighter colors on the ventral side brighten this shaded area. This gives the body a uniform appearance when it is viewed from the side (Figure 34-6). The darker colors on the dorsal side of the fish also blend with the bottom of the body of water or with deep water when the fish is seen from above. The light colors on the ventral side blend with bright light on the surface when the fish is viewed from below.

Head structures of fishes

Although different species of fish vary greatly in body form, many fishes are similar to the yellow perch, shown in Figure 34-7. The head tapers toward the mouth, offering the least possible resistance as the fish moves through the water. The protective covering of the head is in the form of plates instead of scales. The mouth is large and is situated at the extreme anterior end. Carnivorous fishes such as the yellow perch have numerous small, sharp teeth extending from the jawbones and from the roof of the mouth. These teeth slant toward the throat, making it easy for the yellow perch to swallow a prey but hard for the prey to escape. The tongue is fastened to the floor of the mouth and is immovable. It functions as an organ of touch rather than one of taste.

Two nasal cavities lie on the top of the head, anterior to the eyes. Paired nostrils lead to each nasal cavity. The nostrils function in smell only. They do not connect with the throat and are not involved in respiration. The fish has no external openings to the ears, which are embedded in the bones of the skull and probably function as balance organs in addition to receiving vibrations carried by the bones of the skull.

The eyes of most fishes are large and somewhat movable. Eyelids are lacking. The transparent cornea that covers the eye is flattened. The pupils are large in comparison with those of other vertebrates, and admit the greatest possible amount of light. The lens is almost spherical and is moved backward or forward as the eye focuses, such as you focus a camera.

At each side of the head is a crescent-shaped slit that marks the posterior edge of the gill cover, or *operculum* (oe-PURR-kyooh-lum). This hard plate serves as a protective cover for the gills beneath it. By raising the unattached rear edge of the operculum, you can see the four comblike gills lying in a large gill chamber. The edges of the opercula nearly meet on the lower side of the fish, where the head fastens to the trunk at a narrow *isthmus*.

34-6 The yellow perch illustrates the principle of countershading. (Allan Roberts)

Structures of the trunk and tail

Various kinds of fins develop from the trunk and tail. Each fin consists of a double membrane supported by cartilaginous or spiny rays. Fins serve a variety of purposes in the fishes and differ in form in various species.

Two kinds of fins are paired. These are considered homologous with the limbs of other vertebrates. The *pectoral fins* are nearest the head and correspond to the front legs of other vertebrates. Posterior to and below these are the *pelvic fins,* which correspond to hind legs. The paired fins serve as oars when the fish is swimming slowly. They also aid in steering and in maintaining balance when the fish is resting, and are used when the fish moves backward. The *caudal fin* grows from the tail and aids in propelling the fish forward.

34-7 The external structure of a bony fish, the yellow perch.

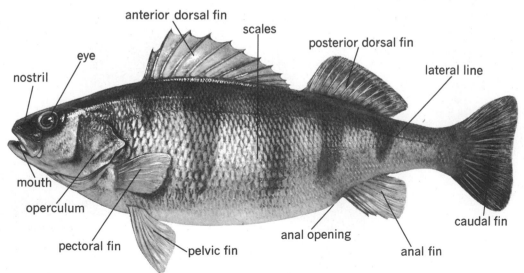

Dorsal fins are situated along the top middle line of the trunk. The anterior, or spiny, dorsal fin of the perch contains sharp projections that aid in defense. The spines of this fin raise upward toward the head, thus making it difficult to swallow the perch tail first. The posterior, or soft, dorsal fin lacks these spines. Both dorsal fins serve as a keel to keep the fish upright while it is swimming. Another single fin, the *anal fin,* grows along the middle line on the lower side. This fin, like the dorsal fin, serves as a keel and helps to maintain balance.

Powerful muscles arranged in zig-zag plates occupy the region of the trunk above the spinal column. A thinner muscle layer lies along the body wall on the sides of the trunk. The tail region is solid muscle with the spine running through most of it.

If you examine the sides of a fish closely, you will notice a row of pitted scales extending from the head to the tail fin. These make up the *lateral line*. Nerve endings and a narrow tube lie under the scales. The line acts as a sense organ, because it is sensitive to low-frequency underwater vibrations as well as to pressure stimuli.

The digestive system of fishes

Many fishes are vegetarians and feed on algae and other water plants. Carnivorous species eat other animals such as frogs, other fish, and a wide variety of invertebrates, including crayfish, worms, and insects. Some fishes, like the bass and pike, swallow fish almost as large as themselves. Especially in carnivorous fishes, the mouth is a large trap for capturing prey.

34-8 The internal structure of a yellow perch.

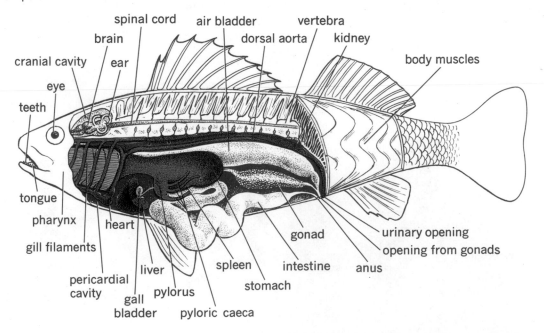

A rudimentary "*tongue*" lies in the floor of the mouth. It does not function in swallowing but serves primarily as an organ of touch. The throat cavity, or *pharynx,* leads to the opening of the short *esophagus* (Figure 34-8). The esophagus, in turn, joins the upper end of the stomach. The *stomach* is in line with the esophagus, thus allowing a large prey to extend from the stomach through the mouth and even protrude for a time as digestion occurs. A rather short *intestine* leads from the lower end of the stomach. Several short finger-like tubes called the *pyloric caeca* (pie-LORE-ick SEEK-uh) extend from the intestine near its junction with the stomach. Digestion continues as food moves through the short loops of intestine. A well-developed *liver* lies close to the stomach. The *pancreas* is difficult to find in a fish dissection. Digested food is absorbed through the intestine wall. Indigestible matter leaves the intestine through the *anus* on the lower side.

The circulatory system of fishes

The blood of a fish is similar to that of other vertebrates. It contains both red and white corpuscles. The **heart** pumps blood through a system of vessels of three types. **Arteries** carry blood from the heart to the gills, then to all other regions of the body. The arteries lead to thin-walled **capillaries,** which penetrate all of the body tissues. The capillaries come together to form larger vessels called **veins,** which return blood to the heart.

The heart lies in the *pericardial cavity* on the lower side of the body just behind the gills. A large vein, the *cardinal vein,* receives blood from various branches coming from the head, trunk and tail, and the liver (Figure 34-9). Just above the heart the cardinal vein enlarges into a thin-walled sac, the *sinus veno-*

34-9 The circulatory system of a fish. Note that the blood flows in a single circuit—from the heart to the gills to the body and to the heart again.

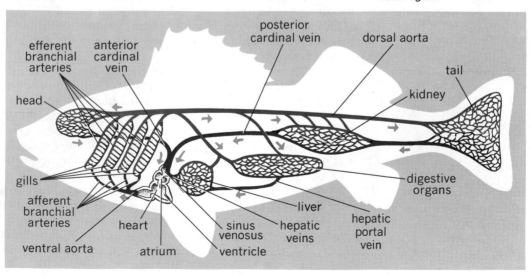

efferent branchial arteries
anterior cardinal vein
posterior cardinal vein
dorsal aorta
tail
kidney
head
gills
afferent branchial arteries
heart
sinus venosus
hepatic veins
liver
hepatic portal vein
digestive organs
ventral aorta
atrium
ventricle

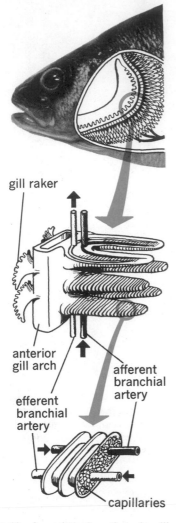

gill raker

anterior
gill arch

efferent
branchial
artery

afferent
branchial
artery

capillaries

34-10 A sectioned portion of a gill filament is shown at the top of the center drawing. A portion of a single filament, much enlarged, is shown at the bottom.

sus. This sac joins the first heart chamber, or *atrium,* often called the *auricle.* From the atrium blood passes into the *ventricle,* the thick-walled, muscular pumping chamber of the heart. Blood is pumped from the ventricle with great force through the *ventral aorta,* leading to the gills. This artery begins with a muscular, bulblike structure, the *bulbus arteriosus,* which is attached to the ventricle. This structure is very noticeable in the fish heart. The ventral aorta branches to the two sets of gills, then rebranches to form *branchial arteries* that lead to the four gills on each side of the head. Another large artery, the *dorsal aorta,* receives blood from the gills and, through its branches, supplies the head, trunk, and tail. Blood returns to the heart through the cardinal vein, thus completing the circulation. Some of the blood returns through veins from the digestive organs and the liver. Various cell wastes are removed as blood circulates through the kidneys.

The blood of a fish passes through the heart once during a complete circulation. The heart receives deoxygenated blood from the body tissues through the cardinal vein and pumps it through the ventral aorta to the gills. In circulating through the gills, the blood discharges carbon dioxide and receives oxygen. This blood, now oxygenated, is received from the gills by the dorsal aorta for circulation to the body tissues.

The gills—organs of respiration

In a bony fish such as the yellow perch, four gills lie in a gill chamber on each side of the head. A *gill* consists of a cartilaginous *arch* to which is attached a double row of thin-walled, thread-like projections called *gill filaments* (Figure 34-10). These filaments are richly provided with capillaries, so that the blood is brought into close contact with the water over a large surface. The gill arches have hard, fingerlike projections called *gill rakers* on the side toward the throat. These prevent food and other particles from reaching the filaments and keep the arches apart to allow free circulation of water.

Blood enters a gill at the base of the arch through the *afferent branchial* (AFF-uhr-int BRANG-kee-uhl) *artery.* Branches of this artery enter each gill filament, where the blood enters a network of capillaries. Here carbon dioxide is discharged from the blood, and oxygen is absorbed through the thin walls of the capillaries and filaments. Oxygenated blood returns to the gill arch and flows out the top of the gill through the *efferent branchial artery* to the dorsal aorta.

The fish requires a continuous flow of water over its gills. Water is drawn into the open mouth as the gill arches expand and enlarge the cavity of the pharynx. The edge of the operculum is pressed against the body as water is drawn in. The mouth is then closed, the gill arches contract, and the rear edge of the operculum is raised, thus forcing the water over the gill filaments and out of the gill chamber around the raised edge of the operculum. The forward motion of the fish when it is swimming aids this process.

The air bladder—pressure organ

A thin-walled sac, the *air bladder,* lies in the upper part of the body cavity of a fish. In fishes that swallow air, such as the lungfish, it connects with the pharynx by a tube. In others, it is inflated with gases (oxygen, nitrogen, and carbon dioxide) that pass into it from the blood. This sac acts as a float and adjusts the weight of the fish so that the weight of the animal equals the weight of the water it displaces. This equilibrium allows the fish to remain at any desired depth in the water with little effort.

Fishes live at various water levels at different seasons of the year. The air bladder adjusts to these variations by losing air to the blood or receiving additional air. When a fish adjusted to deep water is caught and brought to the surface suddenly, the air bladder expands and may push the esophagus and stomach into the mouth. One group of fishes known as *darters* has no air bladder. Darters sink to the bottom after each of their jerky swimming motions.

The nervous system

The nervous system of the fish includes the brain, spinal cord, and the many nerves that lead to all parts of the body. The brain lies in a small bony cavity, the *cranial cavity.* It consists of five distinct parts (Figure 34-11). At the anterior end are the *olfactory lobes,* from which nerves sensitive to odors extend to the nostrils. Behind these lobes are the two lobes of the *cerebrum,* which control the voluntary muscles. Instincts are centered in these lobes. Behind the cerebrum are the *optic lobes,* the largest lobes of the fish's brain. Optic nerves lead from these lobes to the eyes. Behind them lies the *cerebellum,* which coordinates muscular activity, and finally the *medulla oblongata,* which controls the activities of the internal organs. The *spinal cord* passes down the back from the medulla and is encased in the vertebral column. *Cranial nerves* extend from the brain to the sense organs and other head structures. *Spinal nerves* connect the spinal cord with all the various parts of the body.

The fish's brain is not highly developed in comparison with those of higher vertebrates. As you study the brains of other vertebrates, compare them with the fish brain. The same regions are present. There is, however, a gradual increase in the size of the cerebrum in proportion to the other brain regions as vertebrates become more advanced. As the cerebrum increases in size, there is a corresponding increase in nervous activity on higher levels, such as emotional responses, memory, and intelligence.

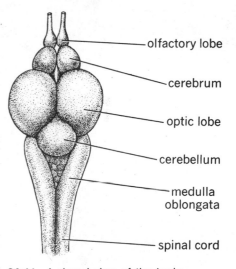

34-11 A dorsal view of the brain of a fish.

Sensations of the fish

The relatively large optic lobes of the brain indicate that fishes have a well-developed sense of sight. However, vision in even moderately deep water is greatly reduced because of insufficient light. Fishes are known to be nearsighted and probably do not see objects clearly at distances greater than one to two feet. Until lately, scientists were not sure whether the fish sees colors or lives in a world of black, white, and shades of gray. However, recent studies indicate that some fish may possibly possess color vision.

Fish lack eardrums and middle ears. However, sound vibrations are transmitted to the sensitive ear structures through the bones of the skull in which the ears are embedded.

Probably the most acute sense in a fish is the olfactory sense or the ability to detect various substances dissolved in the water. This sense is similar to the sense of smell in air-breathing animals. Scientists have conducted extensive experiments to demonstrate the reactions of fishes to odors in the water. It has been found that fishes can distinguish the odors of many water plants, even when these plants are dipped for only a short time in pure water. Similarly, they can detect the odor of hands washed in a stream as well as the odors of many animals, especially mammals. Scientists now believe that odors direct fishes to feeding areas among water plants. It is possible, too, that salmon find the mouths of rivers and streams during the spawning season by the odors of plants living in the fresh water.

Reproduction in the fishes

The reproductive organs, or *gonads,* lie in the posterior region of the body cavity above the digestive organs. The opening from the gonads is just behind the anal opening.

Eggs develop in the *ovaries* of the female over a period of several months. As the eggs enlarge, the ovaries swell and may bulge the sides of the fish. In many fish, including the yellow perch, the paired ovaries fuse into a single ovary during embryonic development. Sperms develop in the paired *testes* of the male. Moments after the female lays her eggs, or **spawns,** the male swims over them and discharges a sperm-containing fluid called **milt.** Sperms swim to the eggs and fertilize them, and the development of the embryos begins. This development may require from a few days to many weeks, depending on the species and the temperature of the water. The developing fish is nourished by a large quantity of nonliving material, the yolk, which is present in the egg. A part of the yolk known as the yolk sac remains attached to the young fish for a short while after it has hatched.

Spawning is not an efficient process in most fishes. Many eggs never receive sperms, and therefore fail to develop. Large numbers of eggs are eaten before hatching by other fish and aquatic animals. After hatching, the young fish are in constant danger of being eaten by cannibalistic fish and other predatory animals. Regardless of the high mortality rate, however, the species survive because of the large numbers of eggs laid. The number varies from about five hundred in the trout to six or seven million in the codfish.

Guppies, mollies, platys, and swordtails are fresh-water tropical fish commonly reared in home aquariums. Unlike most fishes, these fish bear their young alive. The female retains the eggs in her body and receives sperms from the male during mating. The young fish develop internally and are brought forth alive, hence the name *live-bearer*.

Spawning habits of various fishes

The spawning habits and life histories of fishes vary with the species. As a rule, fresh-water fish spawn in the waters where they normally live. Often, though, they swim to a place where the water is shallower to spawn. The shallow water affords not only greater protection but also warmer temperature for development of the eggs.

The yellow perch moves from the deep water of lakes to shallow water near the shore for spawning during the spring months. Eggs numbering several thousands are laid in ribbon-like masses. The sunfish, like many fishes, deposits eggs in a depression made in the bottom of a pool (Figure 34-12). After spawning has taken place, the male guards the nest, fighting off any intruder. The male stickleback makes a curious nest of bits of plants and rubble and drives the female into it for spawning. Then he drives her away and takes charge of the nest and eggs. Channel catfish spawn in holes in a bank of a body of water or in a discarded can or other receptacle they find on the bottom. However, many fish travel long distances to spawning areas— from lakes and rivers into small streams, from fresh-water streams to the ocean, and from the ocean into fresh-water streams.

Among the most interesting fish migrations is that of the eel. This long, slender fish lives in rivers and streams of the Atlantic and Gulf coastal regions and can be found in smaller numbers much farther inland.

Adult eels live in fresh water for from five to ten years or more, depending on the climate and the food supply. When they reach sexual maturity, they begin a final journey downstream to spawning grounds in the Atlantic Ocean, where they began life. The migration starts in the autumn. When they reach the mouths of rivers flowing into the Atlantic Ocean and the Gulf of Mexico, some strange influence directs them to a stretch of the Atlantic north of the West Indies and south of Bermuda. It is believed that they follow warm ocean currents, although both their route

34-12 Nests of the stickleback (top) and sunfish.

and the depth at which they swim remain a mystery. Each female lays several million eggs. Finally, after spawning, both adults die.

The eel larvae that hatch in the spawning grounds in the spring are transparent at first. By the autumn of the first year, the larvae have reached their full size of about two inches and have arrived at the coast of North America. Through the winter months, they change into miniatures of adult eels and are known as *elvers.* The following spring, the young eels enter the mouths of rivers and streams in droves. During the journey upstream, the sexes part company. Males travel only a short distance. The females continue on much farther, in some cases for a thousand miles.

European eels spawn in an area near the spawning grounds of the American eel, but farther east and near the Sargasso Sea. The migration of these eels covers more than three thousand miles of the treacherous waters of the Atlantic Ocean. The return of the larvae and elvers requires several years. It is interesting that American and European eels, even though they spawn in the same general area, always return as larvae to their respective continents.

Of all migrating fish, however, the Pacific salmon have the strongest homing instinct. The migration of the salmon has long been a subject of great interest and scientific investigation. Generation after generation, the various species of Pacific salmon leave the deep waters of the ocean and find their way to the very rivers and branches of rivers in which their parents spawned some three or four years earlier. Salmon migrations start between late spring and early fall, depending on the species and the distance to be traveled. The pink salmon travels only a few miles upstream from the ocean, while the king salmon migrates as far as one thousand miles upstream to cold, fast-flowing streams.

During the migration, the color of the salmon changes from silvery to pink or bright red. A hump develops on the back, and the jaws become hooked. Upon reaching the spawning areas, females lay from three to five thousand eggs over a period of several days and males discharge milt onto the eggs. The adults die soon after spawning. The eggs hatch after thirty to forty days, and the young salmon begin their own slow journey to the sea, where they will live until maturity, then return "home" to spawn and die.

SUMMARY

Since early geological times, fishes have ruled the seas. They remain the dominant vertebrates in most aquatic environments today. The lampreys and hagfishes, sharks and rays, are living examples of the fish populations of ancient seas. Through the ages, bony fishes in a great variety of forms and sizes have gradually replaced most of the more ancient cartilaginous fishes.

In the study of fishes, you found many reasons for their supremacy in the vertebrate populations of aquatic environments. Fins propel the streamlined body through the water with a minimum of resistance. The air bladder, present in most fishes, adjusts buoyancy of the body to various water depths. The scales found on most fishes, together with a coating of mucus, provide a slick body surface and form a barrier against invading parasites, infectious molds, and other organisms. Protective coloration and, in many cases, color changes blend the fish with its environment and increase its chances of survival, both as the hunter and the prey. The mouths of fishes are as varied as their diets and feeding habits.

Blood surging from the heart flows directly to the gills, where oxygen is received and carbon dioxide is discharged into the water through thin capillary walls and membranes of the gill filaments before circulating through the body tissues. The limited vision of most fishes, especially in deep or murky water, is compensated for by a keen olfactory sense. Nerves of the lateral line are sensitive to water pressure and to low-frequency vibrations in the water.

The spawning habits of fishes, including the long migrations of the Pacific salmon and the eel have been subjects of extensive studies by biologists. The instinctive drives directing fish behavior are fascinating to observe but difficult to explain.

BIOLOGICALLY SPEAKING

chromatophore	heart	olfactory lobe
guanine crystal	artery	cerebrum
countershading	capillary	optic lobe
operculum	vein	cerebellum
pectoral fin	pericardial cavity	medulla oblongata
pelvic fin	atrium	spinal cord
caudal fin	ventricle	cranial nerve
dorsal fin	gill	spinal nerve
anal fin	air bladder	spawn
lateral line	cranial cavity	milt

QUESTIONS FOR REVIEW

1. What characteristic of the head of a cyclostome distinguishes it from other fish?
2. Describe the manner in which the sea lamprey attacks its prey.
3. Describe two lamprey-control measures used in the region of the Great Lakes.
4. List several characteristics that distinguish sharks from bony fish.
5. Describe the body regions of a bony fish such as the yellow perch.
6. Of what importance is mucus on the body surface of a fish?
7. What is countershading? How does it camouflage a fish?
8. Locate and describe the sense organs of a fish.
9. Name and locate the fins of a yellow perch and discuss the use of each fin.
10. In the order in which food passes through them, describe the organs of the alimentary canal of a fish.
11. Describe the structure of the fish heart.
12. Trace the path of a drop of blood from the ventral aorta through a gill to the dorsal aorta and describe changes in the blood during its circulation through a gill.

13. Locate and describe the air bladder. What is its function in a fish such as the yellow perch?
14. Name the various regions of the fish brain and give a function associated with each region.
15. Discuss the efficiency of the various sense organs of a fish.
16. Describe external fertilization as it occurs in most fishes.

APPLYING PRINCIPLES AND CONCEPTS

1. Fish are the dominant forms of vertebrate life in aquatic environments. Account for this in terms of structural adaptations.
2. Why does a fish die in the air, even though the air contains more oxygen than the water in which the fish lives?
3. Even though fish lay very large numbers of eggs, they seldom overpopulate the waters in which they live. Give several reasons to account for this.
4. Discuss the homing instinct of fish, using the eel and the salmon as examples.

RELATED READING

Books

BURGESS, ROBERT F., *The Sharks.*
Doubleday & Co., Inc., Garden City, N.Y. 1970. Relates the history of man's fear of sharks and the myths invented about these fish and how these fears are gradually being overcome.
CURTIS, BRIAN, *The Life Story of the Fish.*
Dover Publications Inc., New York. 1961. A simply written book about the fish, from body covering and habits to structure and adaptations.
DEES, LOLA T. (ed.), *The Sea Lamprey.*
U.S. Department of the Interior, Fish and Wildlife Service, Washington, D.C. 1965. The life history and migration of the destructive vertebrate as well as its economic importance.
GROSVENOR, M. B., and editors, *Wondrous World of Fishes.*
National Geographic Society, Washington, D.C. 1965. A colorful account of many types of fishes.
HEARLD, EARL S., *Living Fishes of the World.*
Doubleday & Co., Inc., Garden City, N.Y. 1961. Describes fish not generally covered in fish books; organized along the line of systematic classification.
McCORMICK, HAROLD W., and others, *Shadows in the Sea.*
Chilton Book Company, Philadelphia. 1963. Contains material collected in many parts of the world, mainly concerning sharks, skates, and rays.
OMMANNEY, F. D., *A Draught of Fishes.*
Thomas Y. Crowell Company, New York. 1966. Scientific account of fishes and fishing in all types of water and in all parts of the world.

CHAPTER THIRTY-FIVE

THE AMPHIBIANS

The arrival of the amphibians

Biologists believe that living things were confined to the water for millions of years. Then, at the end of the Devonian period changes occurred that resulted in the evolutionary development of life on land. The modern fishes we discussed in the last chapter all live in the water, of course, and respire by using gills. But do you remember the air bladder they use as an organ of balance? This organ is actually similar to an interesting adaptation that occurred in some of the early bony fishes. These fishes developed lunglike structures, in addition to gills, that enabled them to breathe air. Two types of these primitive *lungfishes* survive even today. They are found in Australia, Africa, and South America—all areas of seasonal drought. Functional lungs are absolutely necessary for their survival. When a drought comes, the lungfish digs a burrow in the bottom of the pool, and it lives there, curled up in a state of inactivity, until the rains come and the water returns.

Jointed, or lobed, fins that somewhat resembled legs were another unique characteristic of the early lungfishes. Until about twenty-five years ago, it was thought that lobe-finned fishes had disappeared by the end of the Mesozoic era. Fossils of lobe-finned fishes have been found in the rocks of the late Paleozoic era and the Mesozoic era, but there is no trace of them in Cenozoic rocks. Until fairly recently, biologists believed that they had become extinct some seventy million years ago. In 1938, however, a native South African commercial fisherman caught a type of fish he had never seen before. It was five feet long and was covered with large, bluish scales. The fisherman gave the strange fish to a local museum, where it was mounted. When it was shown to Dr. J. L. B. Smith of Rhodes University, he recognized it as a *coelacanth* (SEE-luh-KANTH), a lobe-finned fish that was supposed to have been long extinct (Figure 35-1). Since then, many coelacanths have been caught in the waters between Madagascar and Mozambique. They retain lungs as outgrowths of the throat, and their fins resemble crude legs.

35-1 The coelacanth, the only known surviving lobe-finned fish. (courtesy of the American Museum of Natural History)

35-2 Representation of early lobe-finned fishes as they may have looked when they came out onto land (top) and of early amphibians. Their skeletons have been drawn from reconstructed fossil remains. Note the marked similarity between the two types of animals. (courtesy of the American Museum of Natural History)

The development of lungs and leg-like fins in some of the early bony fishes was probably an adaptation for survival. There were great climatic changes in the Devonian period. The ponds and streams often dried up or became stagnant, just as they do in the habitats of the modern lungfishes. Typical fishes could not have survived either condition. A fish that possessed some type of lung, however, could survive both drought and stagnancy. Likewise, a fish that possessed leglike fins could perhaps crawl away from a dried-up pond to a pond that still had water in it. This line of reasoning has led biologists to believe that the primitive lungfishes, with their lobed fins, were transitional forms between true fishes and amphibians (Figure 35-2).

The class name *Amphibia* means, literally, "having two lives." It refers to the fact that although the amphibians have been able to develop some adaptations for life on land, they have never become completely free of water. They must still return to the water for reproduction, because their soft, jellylike eggs would quickly perish on dry land. Furthermore, the young of all amphibians are completely waterbound for a period of time, and few adult amphibians can travel far from shore because they must keep their skins moist.

Amphibians were the dominant vertebrates through the Mississippian and Pennsylvanian periods and into the Permian period. Near the end of the Paleozoic era, they decreased in number. The remaining large forms became extinct in the Triassic period of the Mesozoic era. The amphibians of today, which include only three orders, appeared after this time. They are the few survivors of a once-flourishing vertebrate class.

Characteristics of the amphibians

The water-bound young of amphibians are fishlike. As they become adult, they change to land-dwellers of quite different structure. This series of changes is a metamorphosis, just as is the life history of certain insects. In this transition from water to land forms, many strange combinations of gills and lungs and fins and legs occur. Gills are found on animals with legs, and fins are sometimes found on animals with lungs.

In general, the Amphibia differ from other vertebrate animals in the following ways:

■ Body covered by a thin, smooth, and usually moist skin, without scales, fur, or feathers.
■ Feet, if present, often webbed.
■ Toes soft and lacking claws.
■ Immature or larval forms, vegetarian; adults, usually carnivorous.
■ Respiration by gills, lungs, and skin.
■ Heart, two-chambered in larvae; three-chambered in adults; circulation well developed.

■ Eggs laid in water fertilized externally as soon as laid.
■ Metamorphosis from aquatic larval stage to adult form.

Orders of Amphibia

The order *Apoda* (AP-uh-duh) contains a few surviving legless amphibians of the tropics. These strange, wormlike creatures are often called *caecilians* (si-SILL-ee-anz). A second order, *Caudata* (kaw-DAY-tuh), or *Urodela,* includes amphibians that have tails throughout life. We place the familiar salamanders and newts in this order. The most familiar amphibians are the frogs and toads, members of the order *Salientia* (SAY-lee-EN-chuh), or *Anura.* Frogs and toads are different from other amphibians in that they lack tails in the adult stage. These animals, along with certain of the salamanders, undergo an interesting transition. They change from living an aquatic life as a larva to a semiaquatic or terrestrial life as an adult.

The salamanders

Modern salamanders are very similar to their ancestors. The only evidence of evolution has been the replacement of cartilage by some bone in the skeleton. You are probably familiar with several of the salamanders, although you may have called them lizards because of their general similarity. Both salamanders and lizards have elongated bodies, long tails, and short legs. However, a salamander has soft, moist skin and has no claws on its toes. The lizard has a scale-covered body and claws on its toes; these characteristics of reptiles are almost never found among the amphibians.

Salamanders have very little protection from enemies. A few have skin glands that secrete bad-tasting substances. Others have color pigments that change as they change their surroundings. Salamanders cannot survive in dry environments, which helps us to understand why they are usually found under damp logs and stones or swimming about in water.

Species of salamanders range in length from a few inches to several feet. The giant salamanders are represented in the United States by the *American hellbender,* which reaches a length of two feet or more. This large salamander, with loose grayish or reddish-brown skin, lives in the streams of the eastern United States. One of its relatives, the giant salamander of Japan, grows to five feet in length and is the largest living amphibian.

Another large salamander of the Middle West is the *mud puppy,* or *water dog (Necturus).* Many an unsuspecting fisherman has been startled when he pulled one of these slimy salamanders from a mud-bottom stream in the late evening or night. The mud puppy may reach a length of two feet. It has a flattened, rectangular head, small eyes, a flattened tail, and two pairs of short legs. The most striking feature of its body is three pair of dark red,

35-3 An adult tiger salamander (top) and an axolotl. What larval trait is apparent in the latter? (Allan Roberts)

bushy gills attached at the base of the head just above the front legs. The presence of gills is a larval trait retained throughout life by this particular salamander.

The *tiger salamander,* shown in Figure 35-3, is found in most of the United States. It is one of the larger salamanders, reaching a length of from six to ten inches. The bright yellow bars and blotches on a background color of dark brown give it its name. This salamander lives as an aquatic, gill-breathing larva for about three months, after which it leaves the water and lives on land. Its lungs and thin, moist skin both function in respiration during the land-dwelling stage.

Certain tiger salamanders and others of the same genus remain aquatic throughout life and reproduce while still in the larval stage. Larval salamanders called *axolotls* (ACK-suh-LOT'lz) have long been known in Mexico and the southwestern part of the United States. Their curious characteristic of producing eggs or sperms in the larval stage once led biologists to believe that axolotls should be classified as a separate species of salamander. But experimental feeding of the animals with thyroid-gland tissue or thyroid extracts caused them to metamorphose into adults. This showed that they were larval forms of the tiger salamander and some of its near relatives. Research revealed that the waters in which axolotls live are deficient in iodine, which is essential for the production of a thyroid hormone needed for metamorphosis. Thus, axolotls are reduced to a larval existence because of a thyroid deficiency.

The *spotted salamander* could easily be confused with the tiger salamander, since the two species are similar in size and color. The spotted salamander, however, is shiny black with yellow spots. Its tail is round, while that of the tiger salamander is flattened laterally.

We often speak of small salamanders in the land-dwelling stage as *newts.* The *crimson-spotted newt* is especially interesting because of its "triple life." This small salamander hatches, usually in May, into a gill-breathing aquatic larva. After about two months, it changes to a land-dwelling form with lungs. The coral-red color of the salamander in this stage gives it the name *red eft.* One or two years later, the skin color changes to greenish-olive with crimson spots along the sides. The newt returns to the water and resumes aquatic life, respiring through its skin while under water and using its lungs at the surface.

The salamanders we have discussed are but a few of the many kinds you can find under piles of wet leaves, under rocks in stream beds, in abandoned wells, and in other moist places. Salamanders in the aquatic and land-dwelling stages make ideal specimens for aquariums and moist terrariums. With a little coaxing, these animals will eat meal worms or small insects from your hand.

Toads and frogs

The toads and frogs that lived before the Jurassic period all had elongated bodies and long tails. Biologists believe that the great changes in body form were sudden. The most conspicuous change was the disappearance of the tail in the adult. Other, less obvious changes made these animals better suited to life on land. The hind legs developed an extra joint, and the ankle bones became elongated. These adaptations gave great leaping power to the legs. Although the front legs were short, they were well suited for absorbing the shock of landing from a jump. Modern toads and frogs have wide mouths and front-hinged, sticky tongues that can catch insects with lightning speed. These amphibians have inhabited the earth for more than 200 million years and are found in many habitats in various parts of the world. Biologists consider them among the most highly successful vertebrates.

Although the toads and frogs share many structural similarities, they differ in some respects in their anatomy and behavior. Of all the amphibians, the toad is most able to survive on land. After leaving the water early in life, it never returns except to lay eggs. The toad starts life as a tiny black *tadpole* that soon grows legs, resorbs its tail, and hops onto land as a small, brown, froglike creature with the warty skin characteristic of its kind. Adults of the common toad, *Bufo* (BOO-foe), are usually reddish-brown above and grayish-yellow beneath.

Toads frequently inhabit areas of loose, moist soil where they can dig in and conceal themselves from an enemy or the heat of a summer day. They use their hind legs to dig at a remarkable rate of speed. If the toad is unable to bury itself when it is disturbed, it may crouch close to the ground and remain motionless. The color and texture of its skin blend with the ground and provide excellent camouflage. However, poison glands in the skin probably provide the toad's best defense. The secretion from these glands is apparently irritating and distasteful, especially to mammals, and causes them to leave the toad alone. However, this defense seems not to prevent snakes, the toad's principal enemy, from feeding on them.

The toad has been called the "gardener's friend" because of its diet of insects. Toads are often seen in gardens, partially because of the presence of insects, worms, and other forms of food. However, the loose, moist soil in which toads can easily bury themselves is probably what primarily attracts them to the garden.

The tree frogs of the genus *Hyla* are interesting amphibians. Most of them have amazing protective coloration, and several have the ability to change their color. Members of the genus that live in trees have on each toe a sticky disk that enables them to cling to vertical surfaces. The well-known spring peeper, one

35-4 A toad and a leopard frog. The skin of the toad is somewhat rough and warty, whereas that of the frog is smooth. (Walter Dawn)

member of the genus *Hyla,* lives in swamps and bogs rather than in trees.

Peeper eggs are laid in early spring, and the tiny tadpoles feed on algae and protozoans. The adults eat mosquitoes and gnats, which ought to give these frogs a place in our affection. A curious fact about the tadpole stage of peepers is that the tadpoles often leave the water before the tail is entirely resorbed. Apparently, peepers are able to breathe air earlier in their metamorphosis than most other frogs.

The most common frog in the United States is the *leopard frog,* which inhabits nearly every pond, marsh, and roadside ditch. It frequently travels considerable distances from the water and can be seen hopping through the grass in meadows. Its name comes from the large dark spots, surrounded by yellow or white rings, that cover the grayish-green background color of the skin. The undersurface of the leopard frog is creamy white; therefore, when the frog, resting on the surface of a pond, is viewed from below, it blends with the light sky.

The *bullfrog,* so named because its sound resembles the distant bellowing of a bull, is the most aquatic of all frogs. It seldom leaves the water except to sit on the bank of a lake or pond at night. The color of the bullfrog ranges from green to nearly yellow, although the majority are greenish-brown. The undersurface of the body is grayish-white with numerous dark splotches.

The large, fully webbed hind feet of the bullfrog make it an excellent swimmer. These legs are well developed and ten inches long in large specimens. The bullfrog's diet is quite varied and includes insects, worms, crayfish, small fishes, and even an occasional duckling.

The economic importance of frogs

Insects constitute much of the diet of frogs. If frogs had no other value at all, the service they provide by controlling the insect population would justify their protection. Many states have recognized their value and have passed laws regulating the hunting of frogs and prohibiting their capture during the breeding season.

The large hind legs of the bullfrog are a table delicacy. Frog farms, occupying large marshy areas, supply much of the demand for legs. The smaller species of frogs are widely used by fishermen for bait. As a biological specimen for dissection in the laboratory, the frog has long been a favorite. Since its internal organs are arranged similarly to those of the human body, dissecting a frog is an excellent introduction to human anatomy.

Anatomy of the frog

Facing page 550 you will find a leopard frog as seen by the Trans-Vision process. The first page (Plate I) shows the lower side of the frog. The upper side is shown on the last page (Plate

VIII). As you turn the pages between these plates, you will see the internal organs at various depths of the body. Pages on the right show the ventral sides of the organs. The transparencies on the left show the dorsal sides of the organs.

As we discuss the structure of the frog—its form and body covering, legs, head structure, and internal organs—find the various organs in the plates of the Trans-Vision.

External structure of the frog

The frog's body is short, broad, and angular. It lacks the perfect streamlined form we find in the fishes. For this reason, the frog is not the graceful swimmer the fish is, nor does its hopping compare with the graceful movement of most land animals.

The frog's skin is thin, moist, and loose. It is richly supplied with blood vessels. Glands in the skin secrete mucus, which reaches the surface through tiny tubes. This slimy substance makes the frog difficult to hold. The skin lacks such protective outgrowths as the scales and plates of fishes and reptiles.

Adaptations of the frog's legs

The front legs of the frog are short and weak. Each has four inturned toes with soft, rounded tips. The front feet lack a web and are not used for swimming. The inner toe, or thumb, of a male frog is enlarged, especially during the breeding season. The front legs are used to prop up the body on land and to break the fall after a leap.

The hind legs are enormously developed and adapted in several ways for swimming and leaping. The thigh and calf muscles are very powerful. The ankle region and toes are greatly lengthened, forming a foot that is longer than the lower leg. A broad, flexible *web membrane* connects the five long toes, making the foot an extremely efficient swimming organ. The hind legs fold against the body when the frog is resting on land. From this position the animal can make a sudden leap.

The frog's head and its structure

Probably the most noticeable structures of the frog's head are the eyes. The eyes of frogs and toads are among the most beautiful of the animal kingdom. The colored iris surrounds the elongated, black pupil opening. Muscles attached to the eyeball rotate the eye in its socket. The frog's eyes bulge above the head, but can be pulled into their sockets and pressed against the roof of the mouth. In this position, they help to hold food in the mouth.

When the eyes are pulled down, the upper and lower eyelids fold over them. The frog can float just below the surface with its bulging eyes above water. A third eyelid, the *nictitating* (NICK-tuh-TAY-ting) *membrane,* joins the lower lid. This thin covering keeps the eyeball moist when the frog is on land and serves as a protective covering when it is under water.

35-5 The frog's tongue is especially well adapted for catching insects in that it is both flexible and sticky. Note how it is attached at the front of the mouth.

35-6 The frog's mouth. Its relatively large size is an adaptation for the obtaining of food.

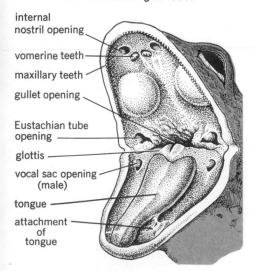

internal nostril opening

vomerine teeth

maxillary teeth

gullet opening

Eustachian tube opening

glottis

vocal sac opening (male)

tongue

attachment of tongue

The nostrils are located far forward on top of the head, allowing the frog to breathe air with all but the top of the head submerged.

The frog has no external ears. The eardrum, or *tympanic* (tim-PAN-ick) *membrane,* lies on the surface of the body just behind the eyes. The cavity of the middle ear lies just below the tympanic membrane. A canal, or *Eustachian* (yoo-STAY-shun) *tube,* connects each middle ear with the mouth cavity. The inner ears are embedded in the skull.

The frog's mouth—an efficient insect trap

The frog's mouth extends literally from ear to ear. If you watch a frog catch a fly, you will discover why the mouth must be so large. It serves as an insect trap. The thick, sticky tongue is attached to the floor of the mouth at the front and has two projections on the free end (Figure 35-5).

When a frog catches an insect, it opens its mouth wide and flips its tongue over and out. The insect is caught on the tongue surface and is thrown against the roof of the mouth. The mouth snaps shut, and the insect is swallowed. This happens so quickly you can hardly see it. Two *vomerine teeth,* projecting from bones of the roof of the mouth, aid in holding the prey. The frog has no teeth on the lower jaw. Small, conical *maxillary teeth,* projecting from the upper jawbone, also aid in holding prey.

Inside the frog's mouth, as shown in Figure 35-6, you can see various openings. The internal nostril openings lie in the roof near the front, on either side of the vomerine teeth. Far back on the sides of the roof of the mouth are the openings of the *Eustachian tubes.* The openings to the *vocal sacs* occupy a corresponding position in the floor of the mouth of a male frog. When a frog croaks, air is forced through these openings into bladderlike sacs that expand between the ears and the shoulders. This action adds resonance and volume to the sound. When the frog croaks under water, air is forced from the lungs, over the vocal cords, into the mouth and back to the lungs. The throat contains two single openings. A large *gullet opening* leads to the stomach. Below the gullet opening is the slitlike *glottis,* the opening to the lungs.

The digestive system of the frog

The diet of the adult leopard frog consists largely of insects and worms. Nevertheless, it can swallow even larger prey because of its large, elastic *gullet.* The short gullet leads to the long, whitish *stomach,* an enlargement of the food tube. The stomach is large at the gullet end and tapers at the lower end. Here it joins the slender, coiled small intestine at a point called the *pylorus.* The stomach content passes into the small intestine through a muscular *pyloric valve.*

The *small intestine* lies in several loops supported by a fanlike membrane, the **mesentery**. The anterior region, which curves upward from the pylorus, is the *duodenum*. The intestine continues as the coiled *ileum*. The small intestine of the frog is proportionally longer than that of the fish. At its lower end, the small intestine leads to a short, broad *colon,* or *large intestine*. The lower end of the large intestine leads to a cavity called the **cloaca** (kloe-AY-kuh). The walls of the cloaca contain openings of the ureters from the kidneys, the urinary bladder, and the oviducts of the female frog. Waste materials and eggs or sperms pass from the cloaca through the *cloacal opening*.

The large, three-lobed *liver* partially covers the stomach. It is a storehouse for digested food and also a digestive gland that secretes bile. The bile collects in the *gall bladder,* which lies between the middle and right lobes of the liver, and passes into the upper small intestine through the *bile duct*. The *pancreas,* a second digestive gland, lies inside the curve of the stomach. Pancreatic fluid and bile pass through the common bile duct into the small intestine. Both of these fluids are necessary for intestinal digestion. *Mucous glands* in the walls of the stomach and intestine secrete mucus, a lubricating fluid. Tiny *gastric glands* in the walls of the stomach secrete gastric fluid, another vital digestive fluid.

We find in the frog a digestive system like that of other vertebrates. A long food tube, or **alimentary canal,** is composed of specialized regions where digestion and absorption of digested food take place. Increased length of the alimentary canal increases the general efficiency of both these processes tremendously.

The respiratory system of the frog

Have you ever wondered how the frog, an air-breather, can stay under water for long periods and lie buried in the mud at the bottom of a pond through a winter hibernation? The skin of the frog and other amphibians is thin and richly supplied with blood vessels. While the frog is in the water, dissolved oxygen passes through the skin to the blood. Carbon dioxide is given off. Respiration through the skin supplies the frog's needs so long as it is quiet. During hibernation, the body processes continue at a very slow rate and the oxygen need is very low. However, body activity such as swimming greatly increases the need for oxygen, and the skin cannot supply enough. The frog then comes to the surface and breathes air.

We inhale and exhale air by increasing and decreasing the size of our chest cavities. This is accomplished through movement of the ribs and diaphragm, a muscular partition at the bottom of the chest cavity. The frog has no diaphragm, and therefore has no chest cavity; nor does the frog have ribs. Air movement is accomplished by changing the volume of and pressure in the mouth

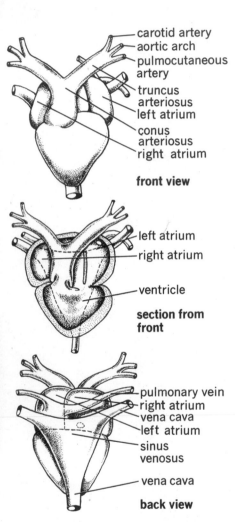

carotid artery
aortic arch
pulmocutaneous
artery
truncus
arteriosus
left atrium
conus
arteriosus
right atrium

front view

left atrium
right atrium

ventricle

**section from
front**

pulmonary vein
right atrium
vena cava
left atrium
sinus
venosus

vena cava

back view

35-7 The frog's heart. Note that three branches of the vena cava lead to the right atrium. The pulmonary veins lead from the lungs to the left atrium. Blood from both these chambers passes through the conus arteriosus. In the back view, you can see where the venae cavae enter the sinus venosus.

cavity. When the frog lowers the floor of its closed mouth, air rushes in through the open nostrils into the partial vacuum. When the floor of the mouth springs up, air passes out through the nostrils.

The lining of the mouth is also well adapted for respiration because it is thin, moist, and richly supplied with blood vessels. At this point, we must distinguish *mouth-breathing* from *lung-breathing*. The frog may pump air in and out of its mouth for some time without using its lungs at all. When the lungs are used, the nostrils are closed by flaps of skin as the floor of the mouth rises. The glottis opens and admits air to the windpipe, or *trachea,* and lungs. Then, with the nostrils still closed, the mouth is thrust down, and air passes from the lungs into the partial vacuum. The upthrust of the mouth immediately following this seems to be greater than usual and forces air back into the lungs. After exchanging air once or twice from mouth to lungs and lungs to mouth, the frog resumes mouth-breathing through the open nostrils.

Thus, the frog depends on its lungs only to supplement mouth-breathing of air. As you might expect, the lungs are small when you compare them with those of higher animals, which depend entirely on lung-breathing. Frog lungs are thin-walled sacs that lack the spongy tissue our lungs have.

The circulatory system

The circulatory system of the frog shows an advance over that of the fish and a step toward the complex system of the higher vertebrates. One of these advances is a three-chambered heart consisting of two *atria (auricles)* and a muscular *ventricle* (Figure 35-7). Deoxygenated blood enters the right atrium from various parts of the body. Blood from the lungs, which is oxygenated when the lungs are in use, enters the left atrium. The atria contract simultaneously and fill the ventricle. Contraction of the ventricle forces blood out through a large vessel, the *conus arteriosus,* which lies against the front side of the heart. This large vessel divides at once into two branches, the right and left *truncus arteriosus,* like the letter Y. Each of these branches divides again into three arches. The anterior pair are the *carotid arches,* which transport blood to the head. The middle pair, or *aortic arches,* bend to the right and left around the heart and join just below the liver to form the *dorsal aorta.* This great artery supplies the muscles, digestive organs, and other body tissues. The posterior *pulmocutaneous arches* form arterial branches that transport blood to the lungs, skin, and mouth.

Blood returning from the body is laden with carbon dioxide and other cell wastes and has been relieved of much of its oxygen. Three large veins, the *venae cavae,* join a triangular, thin-walled sac, the *sinus venosus,* on the back side of the heart.

The sinus venosus, in turn, empties into the right atrium. Part of the blood returning to the heart from the lower parts of the body flows through vessels of the digestive organs and absorbs digested food. This blood flows through the *hepatic portal vein* to the liver on its way to the right atrium. During each complete circulation of blood, some of the blood passes through the kidneys, where water and nitrogen-containing wastes from cell activity are removed. *Pulmonary veins* transport blood from the lungs to the left atrium. When the frog is using its lungs in air-breathing, this blood is oxygenated.

The frog's circulatory system is considered to be more advanced than that of the fish. Blood passes through the two-chambered heart of the fish only once in making a round trip through the body. The three-chambered frog heart receives blood from both the body and the lungs, and it pumps blood to the head and body as well as to the various respiration centers.

The excretory system of the frog

The frog's skin is a vital organ of excretion, since it is here, rather than in the mouth or lungs, that most of the carbon dioxide is discharged from the blood. The liver removes certain wastes and eliminates them with bile or converts them for removal by

35-8 The urogenital organs of a male frog (left) and of a female frog.

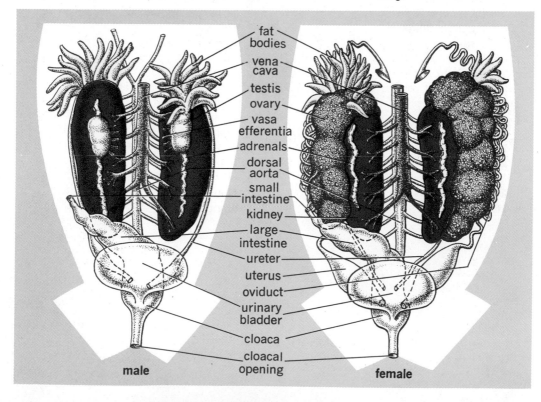

fat
bodies

vena
cava

testis

ovary

vasa
efferentia

adrenals

dorsal
aorta

small
intestine

kidney

large
intestine

ureter

uterus

oviduct

urinary
bladder

cloaca

cloacal
opening

male

female

KEY TO THE STRUCTURES OF THE FROG

1. transverse abdominal muscles
2. vertical abdominal muscles
3. muscles to floor of mouth
4. sockets for attachment of arms
5. shoulder muscles
6. right atrium (auricle) of heart
7. left atrium of heart
8. ventricle of heart
9. great veins to right atrium (auricle)
10. great artery from heart (conus arteriosus)
11. liver
12. stomach
13. pancreas
14. small intestine
15. large intestine (colon)
16. spleen
17. mesentery
18. abdominal vein
19. leg muscles
20. tongue
21. glottis opening
22. trachea
23. lungs
24. sinus venosus
25. pulmonary veins
26. gall bladder
27. bile duct
28. hepatic portal vein
29. sockets for attachment of legs
30. gullet
31. vein from kidneys (posterior vena cava)
32. kidneys
33. dorsal aorta
34. fat bodies
35. ovaries
36. oviducts
37. openings of oviducts
38. egg sac (uterus)
39. urinary bladder
40. cloaca
41. lining of mouth
42. veins from legs to kidneys (renal portal vein)
43. ureters
44. internal nostril openings
45. vomerine teeth
46. teeth of the upper jaw
47. openings of Eustachian tubes
48. eye sockets
49. brain
50. spinal cord
51. spinal nerves

II **LAYER OF SKIN AND MUSCLES REMOVED FROM VENTRAL SIDE OF THE FROG.**
Looking at this layer from the inside, you see the many blood vessels of the skin.
Notice the transverse abdominal muscles (**1**), and the vertical abdominal muscles (**2**).
The large muscles (**3**) which aid in mouth breathing have been cut. The ends which
attach to the floor of the mouth show in the next drawing. In the shoulder are the bones

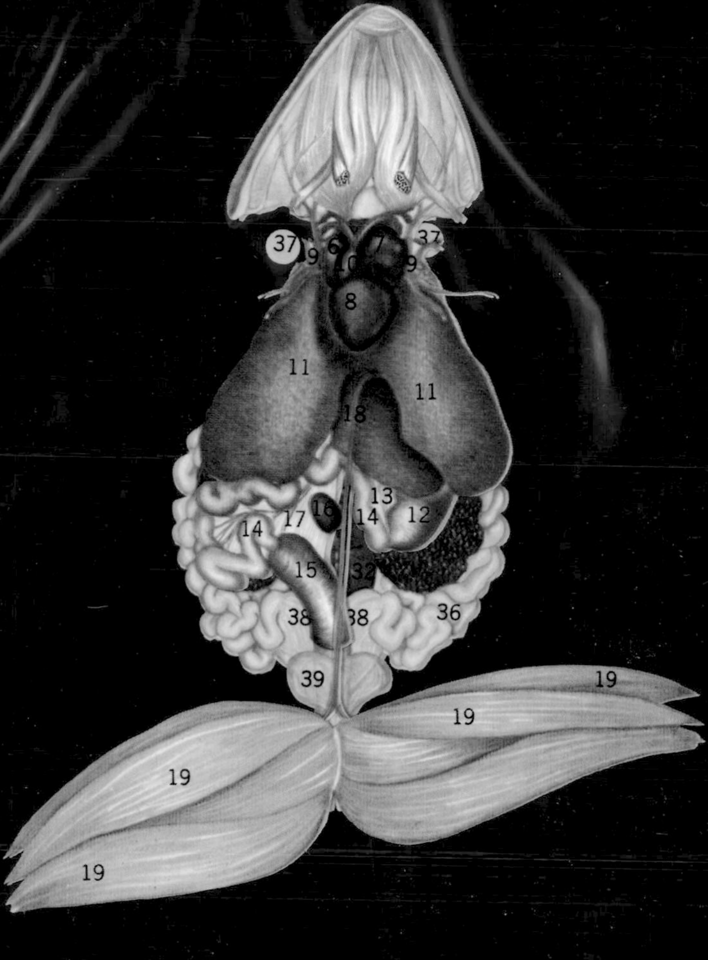

CUTAWAY VIEW SHOWING THE FROG LYING ON ITS BACK WITH FRONT BODY WALL REMOVED. The heart is composed of a right auricle (**6**), a left auricle (**7**), and a ventricle (**8**). Great veins (**9**) carry blood into the heart and a great artery (**10**) carries blood away from the heart. The liver (**11**) covers most of the stomach (**12**) and pancreas (**13**). The small intestine (**14**) leads from the lower end of the stomach to the large intestine (**15**). The spleen (**16**) lies in the thin layers of mesentery (**17**) which fasten the abdominal organs to the body wall. The large abdominal vein (**18**) carries blood from the legs to the liver. Powerful leg muscles (**19**) enable the animal to swim and jump.

VI CUTAWAY VIEW SHOWING DEEPER ORGANS AS SEEN FROM THE BACK. The lining of the mouth (**41**) shows its rich blood supply. You see the gullet (**30**) and stomach (**12**) from the dorsal side. Veins (**42**) carry blood from the legs to the kidneys (**32**). Near these are the ureters (**43**) which carry urine to the cloaca (**40**) which is cut open in the drawing. Urine passes from the cloaca into the urinary bladder (**39**), where it is stored.

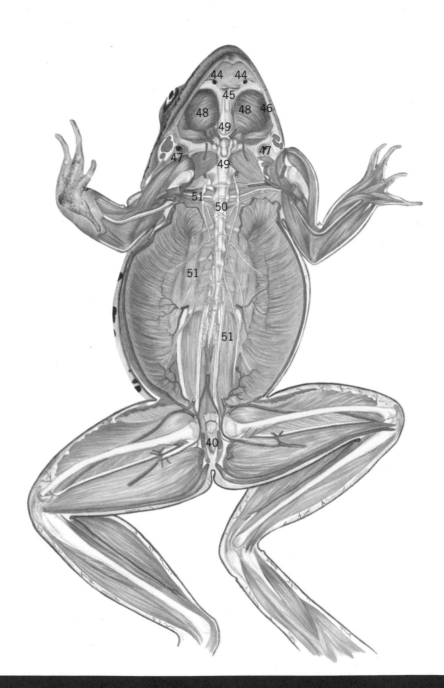

VII CUTAWAY VIEW SHOWING BACK BODY WALL AS SEEN FROM THE FRONT WITH ORGANS AND SOME OF THE LARGE LEG MUSCLES REMOVED. Internal nostril openings (**44**), vomerine teeth (**45**), teeth of the upper jaw (**46**), Eustachian tube opening (**47**) and eye sockets (**48**) can be seen in this view of the head. The cranium and spine are shown as though they were transparent to show the brain (**49**) and spinal cord (**50**). Spinal nerves (**51**) emerge from each side of the spinal cord. Dissection of the lower leg muscles exposes bones and joints, blood vessels, and the dorsal wall of the cloaca (**40**) with ureter and oviduct openings.

the kidneys. The large intestine eliminates undigested food and other wastes. However, the **kidneys** are the principal organs of excretion. They receive wastes from the blood, which flows into them through the *renal arteries* and out through the *renal veins*. The kidneys are large, dark red organs lying on either side of the spine against the back body wall. Urine collects in the kidneys and flows to the cloaca through tiny tubes, the **ureters** (yooh-REET-erz), which you can see in Figure 35-8. The urine may be excreted immediately, or it may be stored after being forced into the urinary bladder through an opening in the cloaca.

The frog's nervous system

The frog's brain shows a considerable advance over that of the fish. *Olfactory lobes* lie at the anterior end of the brain (Figure 35-9). The elongated lobes of the *cerebrum* are proportionally larger than those of the fish. Posterior to these are the prominent *optic lobes*. The *cerebellum* is just behind the optic lobes. In the frog, it is a small band of tissue lying at right angles to the long axis of the brain. The *medulla oblongata* lies posterior to the cerebellum and joins the short, thick *spinal cord*, which extends down the back. Pairs of *spinal nerves* branch from the cord and pass to various parts of the body through openings between the vertebrae. Ten pairs of *cranial nerves* extend from the brain.

The reproductive system

Since the reproductive organs of the frog are internal, it is difficult to distinguish the sexes except during the breeding season, when the thumb of the male is enlarged. The male reproductive organs are two oval, creamy-white or yellowish *testes*. They lie in the back, one on each side of the spine, above the anterior region of the kidneys (Figure 35-8, left). Sperms develop in the testes and pass through tubes, the *vasa efferentia* (VASS-uh EFF-uh-REN-chee-uh), into the kidneys. When the sperms are discharged, they pass through the ureters and on into the cloaca. Some species of frogs have an enlargement, the *seminal vesicle,* at the base of each ureter.

Eggs develop in the female in a pair of large, lobed *ovaries,* which attach along the back above the kidneys (Figure 35-8, right). During the breeding season, the eggs enlarge, burst the thin ovary walls, and are freed into the body cavity. Movement of the abdominal muscles works the eggs toward the anterior end of the body cavity. Here are the funnel-like openings of the long, coiled *oviducts,* which are lined with ciliated cells. The eggs are fanned into the oviduct openings by the cilia. Near their opening into the cloaca, the walls of the oviducts secrete a gelatinous substance that surrounds each egg. At the base of each oviduct is a saclike **uterus** in which the eggs are temporarily stored until they are laid.

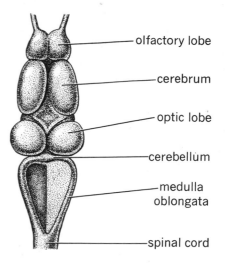

olfactory lobe

cerebrum

optic lobe

cerebellum

medulla oblongata

spinal cord

35-9 A dorsal view of the brain of a frog. Compare it with that of the fish in Chapter 34.

The fertilization and development of the eggs

The female leopard frog usually lays up to about 200 eggs sometime between the first of April and the middle of May. The male clasps the female at the time at which the eggs are laid. The male may also press down on the female, thus helping her to expel the eggs. As the eggs pass from the cloaca of the female into the water, the male spreads sperms over them. As a result of this *direct fertilization,* sperms reach most of the eggs.

The jellylike coat that surrounds each egg swells in the water and binds the eggs together into a rounded, gelatinous mass. In this clump, the eggs look like small beads, each surrounded by a transparent covering. Not only does the jelly protect the eggs from injury, but it also makes them more difficult for a hungry fish to eat. It serves, too, as the first food for the young tadpole.

The frog egg is partly black and partly white. The white portion is the yolk, or stored food material, which will nourish the tadpole during development. The dark portion contains the living protoplasm of the egg and a dark pigment. The yolk is heavier than the rest of the egg. This causes the eggs to float in the water dark side up. The black pigment on the upper side absorbs heat from the sun, while the lighter lower half blends in with the light from the sky and makes the eggs hard to see from below. The gelatinous covering holds much of the heat in the mass. After eight to twenty days, depending on the weather conditions and water temperature, the leopard frog tadpole hatches and wiggles away from the egg mass.

From tadpole to adult – the metamorphosis of the frog

Just after hatching, the tadpole is a tiny, short-bodied creature with a disklike mouth (Figure 35-10). It clings to the egg mass or to a plant. Yolk stored in the body nourishes the young tadpole until it starts to feed. Soon after hatching, the body lengthens, and three pairs of external gills appear at the sides of the head. The tail lengthens and develops a caudal fin. The mouth opens, and the tadpole begins scraping the leaves of water plants with horny lips.

Soon after the tadpole becomes a free swimmer, the horny lip disappears. A long, coiled digestive tract develops, and the tadpole starts living on vegetable scums. Gradually, a flap of skin grows over the gills and leaves a small opening on the left side, through which water passes out of the gill chambers. At this stage, the tadpole is a fishlike animal with a lateral line, fin, two-chambered heart, and a one-circuit circulation. The animal also has a relatively long, spirally coiled intestine.

The change to an adult frog is remarkable. The hind legs appear first. The front legs begin to form at about the same time but do not appear for some time. They remain hidden under the operculum. Soon after the appearance of the front legs, the tadpole starts resorbing (not shedding or eating) its tail. Late in the

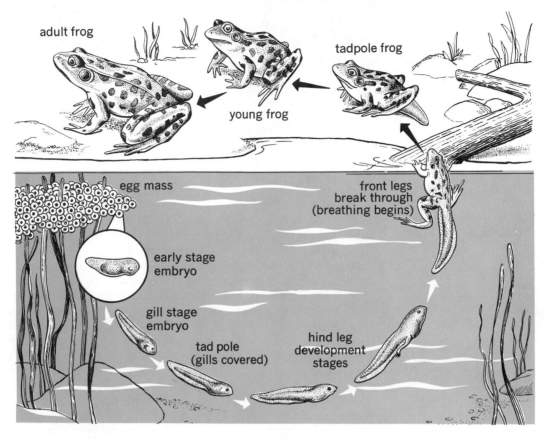

adult frog

young frog

tadpole frog

egg mass

early stage embryo

gill stage embryo

tad pole (gills covered)

hind leg development stages

front legs break through (breathing begins)

metamorphosis, the tadpole's mouth broadens, and teeth develop. While these external changes are taking place, equally important internal changes occur. A saclike chamber, resembling the swim bladder of the fish, forms behind the throat. This divides into two sacs, which become the lungs. The heart develops three chambers, and the gill arteries change to the carotids, aortic arches, and the pulmocutaneous arteries. The gills stop functioning, and the tadpole comes to the surface frequently to gulp air. The thin skin and broad, flat tail still play an important role in respiration during this very critical time in the tadpole's life.

Even before its tail is entirely resorbed, the tadpole leaves the water and comes to land as a young frog. Development from this stage to the full-grown adult frog usually requires about a month. The metamorphosis of the leopard frog varies from sixty to ninety days. Adult leopard frogs usually appear around the first of July. The bullfrog usually spends two winters as a tadpole, and its entire metamorphosis may last as long as three years.

35-10 The history of the frog. The length of time for metamorphosis varies in different species of frogs.

Regeneration in Amphibia

Many amphibians, especially salamanders, have a remarkable ability to *regenerate* lost or injured body parts. A foot, a portion of a limb, or part of the tail may be lost in escaping from an enemy. Such amputated organs can be regenerated rapidly. Frogs and toads in the tadpole stages also have regenerative powers, especially in the early phases. This ability to regenerate disappears as the tadpole matures and is lacking entirely in the adult stage of all genera and species of frogs and toads.

Hibernation and estivation in the frog

The frog is a "cold-blooded" vertebrate, as are the fishes and the reptiles. This does not mean that the blood of these animals is always cold. It means that their body temperatures vary with the temperature of the surroundings. Man maintains a constant average body temperature of about 98.6°F through the regulation of the rate of food oxidation and resulting heat release in the tissues as well as of heat loss from the body surface. The cold-blooded vertebrates carry on much slower oxidation and do not maintain relatively constant body temperatures.

With the coming of fall and the seasonal lowering of temperature, the body temperature of the frog drops to the point where the frog can no longer be very active. It buries itself in the mud at the bottom of a pond or finds shelter in some other protected place in the water. Heart action slows down to a point at which blood hardly circulates in the vessels. The greatly reduced amount of oxygen necessary for life is supplied through the moist surface of the skin. The tissues are kept alive by the slow oxidation of food stored in the liver and in the *fat bodies* attached above the kidneys in most frogs. Nervous activity almost ceases, and the frog lies in a stupor. This is the condition of the frog during *hibernation,* or winter rest. With the coming of spring, the warm days speed up body activity, and the frog gradually resumes the physiological and functional activities of normal life.

The hot summer months bring other problems. Lacking a device for cooling the body, the frog must escape from the extreme heat. It may lie quietly in deep, cool water or bury itself in the mud at the bottom of a pond. This condition of summer inactivity is called *estivation*. Many smaller ponds dry up during midsummer, and the frogs and other cold-blooded animals only survive by burying themselves in the mud and estivating. With the coming of cooler weather and the return of water to the pond, they come out of estivation and continue normal activity until hibernation.

SUMMARY

The vertebrate time scale, shown in Figure 33-2, shows an interesting history of the amphibians. Fossils of the earliest amphibians date back to the close of the Devonian Period, often called the Age of Fishes. This was probably the time when the earliest ancestors of amphibians, the lung fish and lobe-finned fishes, first occupied moist areas of land. This offered a new environment, removed from competition in the seas, and amphibians flourished during the Carboniferous and Permian Periods. However, vertebrates better adapted for life on land developed during the Mesozoic Era, challenged the amphibians, and reduced their numbers greatly.

Amphibians remain today in relatively small numbers, compared to other land animals, especially mammals. They still require water for egg laying and early development. This restricts their distribution on land. Certain of their body structures, both external and internal limit them to moist environments. You can probably list several of these limiting body characteristics. In their "in between" existence, amphibians are surpassed by fish in the water and by other vertebrates you will study on land.

BIOLOGICALLY SPEAKING

newt	mesentery	oviduct
tadpole	cloaca	uterus
nictitating membrane	alimentary canal	fat body
tympanic membrane	trachea	hibernation
Eustachian tube	kidney	estivation
vocal sac	ureter	

QUESTIONS FOR REVIEW

1. Why do biologists believe that the early lobe-finned lungfishes were amphibian ancestors?
2. What characteristics of amphibians distinguish them from other living vertebrates?
3. How many orders of amphibians are represented by living members today?
4. In what ways do salamanders resemble lizards? Name several characteristics that make them different from lizards.
5. Explain why the axolotl does not undergo metamorphosis.
6. Describe the "triple life" of the crimson-spotted newt.
7. Describe the manner in which a frog catches a flying insect.
8. How can a frog croak under water?
9. In the order in which they receive food, name the organs forming the alimentary canal of a frog.
10. Name the chambers of the frog's heart.
11. What three arterial branches carry blood from the great artery leading from the frog's heart?
12. How is urine conducted from the frog's kidneys to the bladder and cloaca?
13. Discuss, in order of occurrence, the changes during the metamorphosis of a frog.

APPLYING PRINCIPLES AND CONCEPTS

1. Explain why biologists believe that "animals developed legs in order to help them find water, not to leave it."
2. Although the amphibians became terrestrial, discuss why they were never successful on land.
3. Discuss the problems in the life of a toad that result from its "in-between" existence.

4. In what respect is the direct fertilization of the frog's eggs more efficient than spawning in fishes?
5. Explain how the frog shows a relationship to the fish in its early development.
6. In what ways are the heart and circulatory system more highly developed in the frog than in fishes?

RELATED READING

Books

CLYMER, ELEANOR, *Search for a Living Fossil.*
   Holt, Rinehart and Winston, Inc., New York. 1963. The actual record of a scientific adventure—the catching of the first and second specimens of the "fossil fish", the coelacanth—supposed to have been extinct for more than 30 million years.

COCHRAN, DORIS, *Living Amphibians of the World.*
   Doubleday & Co., Inc., Garden City, New York. 1963. Three books in one: part one, the caecilians; part two, the salamanders; part three, the other amphibians. A large book with beautiful color photographs.

SMYTH, H. R., *Amphibians and Their Ways.*
   The Macmillan Co., New York. 1962. A nontechnical account of the biology and natural history of amphibians.

CHAPTER THIRTY-SIX

THE REPTILES

The rise of reptiles

It has been difficult for biologists to determine when the first reptiles descended from their amphibian ancestors. Fossil remains are generally limited to hard body structures, such as bones and teeth. Since many of the characteristics that distinguished early reptiles from amphibians were soft body structures, it is often impossible to distinguish the two in fossil form.

Fossils of the earliest reptiles have been found in rocks and other deposits of the early Pennsylvanian period. It is likely that these reptiles had little advantage over the larger and more numerous amphibians of this period. It was probably during the Permian period near the close of the Paleozoic era that more and more land masses rose above the ponds and swamps of the earlier periods and set the stage for the rapid rise of reptiles.

The Triassic period of the Mesozoic era marked the rise of reptiles in many forms and in large numbers. This increase continued through the Jurassic period and well into the Cretaceous period. These periods, spanning more than one hundred million years, are often spoken of as the Age of Reptiles. Giant plant-eating dinosaurs weighing more than thirty tons, flesh-eating dinosaurs, duckbilled dinosaurs, armored dinosaurs, and great flying pterosaurs (pterodactyls) dominated the fauna of the earth. Among these reptiles were the largest land animals that ever lived.

Near the close of the Cretaceous period, however, reptiles declined in number. There were fewer and fewer of the great dinosaurs, and they finally became extinct. Never again were the reptiles dominant over the other vertebrate animals of the earth. What could have caused this decline? We can only guess today. Undoubtedly, there were extreme changes in the topography and climate of the earth. We know that many of the giant plant-eating dinosaurs lived in lakes and swamps. Elevation of the surface of the earth and drainage of these bodies of water may have destroyed many of these dinosaurs. Changes in the

557

Stegosaurus

Brontosaurus

Tyrannosaurus

Pteranodon

36-1 Some reptiles of the Meso-zoic era.

vegetation could have been another reason for their decline. Disappearance of plant-eating dinosaurs would have reduced the food supply of their principal predators, the flesh-eating dinosaurs. Climatic changes may have been another important factor. It would have been difficult for a thirty-ton dinosaur to hibernate during cold weather. However, certain reptiles survived the conditions that brought the age of reptile dominance to a close. Some of those that survived have perpetuated their species with little change since the Cretaceous period. Others became the ancestors of modern reptiles, which have undergone many changes during the seventy million years since the close of the Age of Reptiles. Before studying modern reptiles, however, we shall return to the Triassic, Jurassic, and Cretaceous periods and discuss some of the better-known dinosaurs.

Dating the dinosaurs

Eggs and footprints preserved in rock and fossilized bones are all that remain of the age when dinosaurs roamed the earth. Yet with this evidence, gathered in many parts of the world, and considerable imagination, the paleontologist has been able to piece together a vivid picture of the earth during the time when dinosaurs dominated animal life.

Dinosaur is an appropriate name for these ancient reptiles, for it means "terrible lizard." Many of the dinosaurs were no larger than our larger lizards today. But some of them were giant beasts that would dwarf an elephant. These are the best-known dinosaurs.

The largest of the dinosaurs was the thunder lizard, or *Brontosaurus* (Figure 36-1). This giant measured seventy-five feet in length, about fifteen feet in height, and weighed thirty tons or more. It lived in shallow lakes and marshes and fed on water plants. Its enormously long neck was balanced by an equally long and heavy tail. The plated lizard, while smaller than the thunder lizard, was one of the most heavily armored of the dinosaurs. *Stegosaurus* (STEG-uh-SOAR-us), as this 30-foot monster has been named, had a double row of plates projecting two feet from its back. Pairs of spines near the end of its tail served as deadly weapons with which to lash out at an enemy. *Stegosaurus* had a ridiculously small head with a cranial cavity no larger than that of a small dog. An auxiliary "brain" twenty times larger than the true brain consisted of a mass of nerve tissue formed by the spinal cord in the hip region. This second "brain" is thought to have controlled the seven-foot-long hind legs and ponderous tail of the animal.

The king of dinosaurs was the ferocious tyrant lizard, or *Tyrannosaurus* (ti-RAN-uh-SOAR-us), which was probably the most terrible creature ever to roam the earth. It walked erect on its powerful hind legs and balanced its heavy body with its long tail, much as a kangaroo does. Its front legs were short but powerful, and its long claws could tear most prey to shreds. This giant, flesh-eating reptile was nearly fifty feet in length and towered twenty feet in height. Its powerful jaws were rimmed with double-edged teeth from three to six inches long that could rip the hide even of an armored victim.

Flying reptiles, known as *pterosaurs* (TEH-ruh-SOARZ), lived during the Jurassic period and most of the Cretaceous period. One of the best-known flying reptiles was a pterodactyl known as *Pteranodon,* the giant of them all. It had a body about the size of a turkey and a wing spread of more than twenty-five feet. The wings were membranous and extended from greatly lengthened wrist bones. The skull of *Pteranodon* was over two feet long and extended into a crest on the back. The jaws were elongated into a toothless beak. While *Pteranodon* must have been extremely awkward on land, it could fly and soar gracefully. Biologists believe that pterosaurs fed by swooping down and catching fish swimming near the surfaces of lakes. Like the dinosaurs, the pterosaurs disappeared late in the Cretaceous period. Why did they become extinct? Perhaps at the time of their extinction they could not compete with birds, which were becoming more and more abundant.

36-2 A reconstructed dinosaur skeleton *(Brontosaurus).* (courtesy American Museum of Natural History)

The amniote egg — key to the rise of reptiles

The gradual evolution of reptiles from certain amphibians in the Paleozoic era was an important step forward in vertebrate development. Reptiles emerged as the first vertebrates able to move about on land without having a dependency on water for an aquatic stage or for reproduction. Animals equipped to live entirely on land were probably assured of life with little competition except from one another. What was the key to the success of reptiles on land? The answer was the *amniote egg,* which is believed to have appeared first among the reptiles.

In your study of amphibians, you found that all members of this vertebrate class are bound to water. This was true of amphibians more than a hundred million years ago, and it is true of amphibians today — even the most terrestrial forms, such as frogs and toads. Amphibians must lay their eggs in water, because the eggs lack protective structures that would prevent them from drying out. Furthermore, all amphibians undergo an aquatic larval stage. The amniote egg of the reptile, however, removed both of these biological limitations.

Figure 36-3 shows a diagram of an amniote egg. The egg is proportionally much larger than an amphibian egg. The egg is enclosed in a *shell* that, while porous, prevents rapid water loss and protects the internal structures from injury. Such shell-covered eggs may be concealed in sand or loose soil during development. The egg contains two important *embryonic membranes*. One such membrane, the *amnion,* encloses the embryo. The cavity

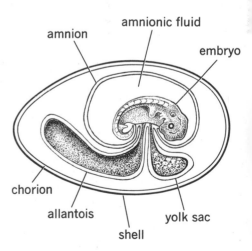

36-3 A cross section of an amniote egg. Locate the specialized embryonic membranes characteristic of this type of egg.

formed by the enveloping amnion fills with a watery amnionic fluid in which the embryo is immersed. This fluid is secreted by the amnion. Thus, the reptile embryo develops in an aqueous environment like that of its amphibian ancestors, but is provided its own pool within the protective structures of the egg. An outer layer of the amnion, known as the *chorion,* lines the shell.

A second embryonic membrane, the *allantois,* pushes outward from the embryo and contacts the chorion. This double membrane thickens and becomes an embryonic organ that functions in absorption, respiration, and excretion. Numerous blood vessels in the allantois join the circulation of the embryo. It is in the numerous small blood vessels of the allantois that oxygen is absorbed and carbon dioxide is given off. The allantois also receives metabolic wastes from the embryo.

The *yolk* of the amniote egg is large and supplies the embryo with nourishment for a longer period of development than would be possible with the food stored in an amphibian's egg. This longer development eliminates a larval stage and provides a direct development from the embryo to the adult reptile.

With the development of the amniote egg, another advance occurred. *Internal fertilization* became an efficient reproductive feature. During the mating process, sperms are introduced into the oviduct of the female and fertilization occurs before the formation of the structures surrounding the egg. The fertilized egg may then be laid or retained within the oviduct until the young reptile hatches internally and is brought forth alive.

Body characteristics of reptiles

While the amniote egg and internal fertilization removed the reptiles from dependence on water for reproduction, several advances in body structure were necessary before they could maintain life in a land environment. We can summarize these characteristics of reptiles as follows:

■ A body covering of scales that protect the skin surface.

■ Dry skin, lacking mucous glands and thickened to prevent water loss.

■ Limbs, if present, with claws on the toes and suited for climbing, digging, or locomotion on land.

■ Well-developed lungs suited to air-breathing and eliminating the need for skin or mouth respiration.

■ Partial or nearly complete division of the heart ventricle, resulting in further separation of oxygenated blood from the lungs and deoxygenated blood from the body tissues. This separation increases the oxygen supply to the tissues and permits a higher degree of body activity.

■ A body temperature varying with that of the environment (a characteristic of cold-blooded animals). In this respect, reptiles are similar to amphibians and lower vertebrates.

Table 37-1 **ORDERS OF LIVING REPTILES**

NAME OF ORDER	REPRESENTATIVES
Rhynchocephalia	*Sphenodon* (tuatara)
Chelonia (Testudinata)	Turtles and tortoises
Squamata	Lizards and snakes
Crocodilia	Alligators, crocodiles, gavials, and caimans

Classification of reptiles

Some 6,000 species represent the class *Reptilia* today. While this may seem a large number, it is only a fraction of the number of species that inhabited the earth at the height of the Age of Reptiles. About 275 reptile species are found in the United States.

Some reptiles are much like those of prehistoric times. Others, considered "modern" reptiles, have become greatly modified in more recent times. Reptiles are most abundant in the tropics, but present geographic distribution indicates that reptiles have migrated to more temperate climates and even into colder parts of the earth. Only the icy regions, high mountain areas, and ocean depths are completely without a reptile population.

Of the sixteen orders of reptiles that once existed, only four have living representatives. One of these orders is represented by a single species that is now nearing extinction. The four remaining reptile orders and representatives of each are listed in Table 37-1.

Sphenodon, a relic of a bygone age

One of the rarest animals on the earth today is the sole surviving species of the order Rhynchocephalia. This ancient reptile, even older than the dinosaurs, is *Sphenodon punctatus* (SFEE-nuh-DON pungk-TAH-tus), or *tuatara* (TOO-ah-TAH-rah). Its relatives disappeared early in the Mesozoic era, probably because they could not compete with more adaptable lizards. This strange survivor of the Age of Reptiles miraculously escaped extinction through the ages in far-off New Zealand and neighboring islands. The tuatara probably survived there because of a total absence of mammals. Once the English settlers had introduced rats, wild pigs, cats, and weasels in New Zealand, the tuatara became extinct on the mainland. Today, the last surviving tuataras are found on a few small islands in the Bay of Plenty and in Cook Strait, off the coast of New Zealand, where they are protected by the government.

Because of a very limited evolution in the Rhynchocephalia, today's tuataras retain many primitive features. Since animals similar to the tuatara lived 170 million years ago, it is clear that the rate of evolution in this order must have been one of the slow-

36-4 *Sphenodon punctatus,* the tuatara of the islands off the coast of New Zealand, is the only surviving species of a once-flourishing order. (Allan Roberts)

est in the vertebrate groups. The tuatara reaches a length of about two feet and resembles a large lizard (Figure 36-4). Its skin is dark olive, marked with numerous light-colored dots. Its eyes resemble those of a cat. The most unusual characteristic of the tuatara is a *parietal* (puh-RY-ut'l) *eye* in the top of its head. Although it is not a functioning sense organ, this strange third eye has the remains of a retina and other eye structures. The tuatara hides in a burrow during the day, coming out at night to feed on insects, worms, and other small animals. The eggs are buried in a shallow depression in the ground, where they remain almost a year before hatching. The female tuatara usually lays from twelve to fourteen eggs. The fact that the tuatara lives well in captivity may make it possible to preserve this rare species in the reptile collections of the world.

Snakes, the most widespread reptiles

Snakes are relative newcomers. As is also true of the early ancestors of the reptiles, there is no adequate fossil record of snakes. But there is little doubt that they evolved from lizard-like reptiles. In fact, the boas and pythons still have vestigial hind limbs, indicating that they descended from animals with legs. Snakes evolved rapidly during the Tertiary period, the same time at which rodents and other small mammals were developing. Snakes are not only the most numerous reptiles today, but they are also the most widely distributed. Snakes are found in water, on the high seas, among rocks, underground, and in trees. They are most abundant in the tropical regions. There are fewer in cooler climates – 126 species in the United States and only 22 species in Canada.

Of the more than two thousand species of snakes in the world, a relatively small number are poisonous. The harm caused by these dangerous snakes is far outweighed by the valuable service all snakes render in destroying large numbers of insects and destructive rodents.

Body structure of a snake

If you examine the elongated body of a snake closely, you can distinguish the head, the trunk, which contains the body cavity, and a tail, which extends beyond the anal opening. As is true of all reptiles, the snake's body is covered with scales. Those on the back and sides of the body are small and oval, thus allowing great flexibility. The heads of many snakes, including our non-poisonous species, are covered with plates. The North American pit vipers, including the rattlesnake, have scale-covered heads. The scales on the lower side of the body form broad plates, known as *scutes*.

Several times each season, during the process of molting, snakes shed the outer layer of scales. As this thin layer loosens, the snake usually hooks a loose portion to a sharp object such as a twig and works its way out. The newly exposed scales are bright and shiny.

Structures of the head

The snake's mouth is large and is provided with a double row of teeth on each side of the upper jaws and a single row in the lower jaw. The numerous conical teeth slant backward toward the throat. The teeth are not used for chewing but are necessary to hold the prey, which is swallowed whole.

The sense of smell is very acute in snakes. Olfactory nerve endings lie in the nasal cavities, which open as paired nostrils near the front of the head. The sense of smell is made more acute with the aid of the curious forked tongue, which is thrust from a sheath in the floor of the mouth close to the front through the small opening that is left when the jaws are closed. The tongue receives dust and other odor-bearing particles from the air and transfers them to tiny pits close to the front of the roof of the mouth. These *Jacobson's organs,* as they are called, contain nerve endings that are highly sensitive to odors.

The snake's eyes have no lids and in this respect are different from those of other reptiles. A transparent scale covers the eye. This becomes cloudy just before molting and causes temporary difficulty with vision. The eyeball can be turned in its orbit. Movement of the lens focuses the eye sharply on objects, especially at close range. Many snakes have round pupils. Biologists have discovered that these snakes are most active in daylight. Others have elliptical pupils, similar to those of cats. These snakes are most active at night.

The ears are embedded in the skull and have no external openings. Thus, the snake cannot hear vibrations transmitted by the air. Instead, the skull bones transmit vibrations resulting from jarring to the highly sensitive ear mechanisms.

Feeding habits of snakes

All snakes feed on living animal prey. No vegetarian snakes are known to exist. We classify snakes into three groups, based on feeding habits.

Many snakes, including most of our nonpoisonous species, merely seize a prey in the mouth and *swallow it alive*. Most of these snakes feed on insects, frogs, toads, lizards, fishes, and other small animals.

The python, boa, king snake, bull snake, and other large-bodied snakes make use of a more specialized method of food-getting. These snakes seize the prey, usually by the head, quickly wrap coils around it, and kill it by *constriction*. The powerful coils squeeze the victim with such force that its chest is compressed and breathing is stopped. In addition, the pressure cuts off the victim's circulation and stops the heart. The shock kills the prey, often without breaking a bone. If after the first constriction, the snake feels a pulse in its victim, it will squeeze again. Swallowing starts immediately after the prey is killed. Biologists have found that warm-blooded animals are killed much more quickly by constriction than cold-blooded prey.

A relatively small number of snakes *poison the prey* before swallowing it. Poisonous snakes secrete *venom* in modified salivary glands at the sides of the head. When a poisonous snake strikes, it thrusts paired fangs into its victim. Venom flows from the poison glands through ducts that enter the fangs (Figure 36-5). The hollow fangs, through which venom flows into the victim, are like hypodermic needles. The amount of venom injected varies with the size and species of snake. Of all snakes, however, the smaller species and the young of larger species have the most concentrated venom. Thus, their bites may be just as dangerous as those of larger poisonous snakes.

Adaptations for swallowing

Snakes eat infrequently. They may go for weeks or, under some conditions, for as long as a year without food. It is amazing, however, to watch a snake swallow a prey four or five times larger than its own body when it does eat. This feat is made possible by several modifications of the jaws. The lower jaws are not joined directly to the skull but are fastened to a separate *quadrate bone,* which acts as a hinge and allows the jaw to drop downward and forward (Figure 36-6). The two halves of the lower jaw are fastened at the front by an *elastic ligament,* which permits each half to operate independently of the other. The numerous slanting teeth hold the prey firmly during the swallowing process.

During swallowing, one side of the lower jaw may pull the prey into the mouth, while the other side is thrust forward for a new grip. By means of this seesaw action, the prey is pulled down the throat, much as you might pull a rope in hand over hand. The snake literally crawls forward around its prey.

The process of swallowing takes so long that special adaptations are necessary so that the snake can breathe while a large prey is in its mouth and throat. The snake's trachea extends along the floor to a glottal opening near the front rim of the lower jaw.

Locomotion in snakes

Several methods of locomotion are found among snakes. The most common is *lateral undulatory movement,* used when the snake is crawling rapidly. The body winds from side to side, forming broad curves. The snake pushes against irregularities, which give it a grip in moving forward. The entire body follows along the same track. This type of movement is also used by water snakes in swimming.

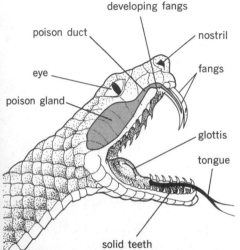

developing fangs

poison duct

nostril

eye

fangs

poison gland

glottis

tongue

solid teeth

36-5 The head of a poisonous snake is well suited for poisoning prey and swallowing it whole. (William M. Partington, National Audubon Society)

In *caterpillar movement,* a snake crawls slowly in a straight line. Scutes are pushed forward in several sections of the body. The posterior edge of each scute grips the ground, while the body is pulled forward by waves of muscular contraction.

A third method, known as *side-winding,* is used by snakes of the sandy desert regions. Here, the body is twisted into S-shaped loops and, except for two or three points of contact, is raised from the ground. The sidewinder "walks" across the sand on these loops.

You may be relieved to learn that most snakes travel at a speed of less than one mile per hour. The fastest ones cannot exceed three miles per hour. You can walk at this speed with ease, and you can run a short distance at a speed of from ten to twenty miles per hour. Thus, you need have no fear of being outrun by a snake!

36-6 What adaptations for swallowing large prey can be seen in this photograph of a snake's skull? (courtesy of the American Museum of Natural History)

Internal organs of a snake

Over three hundred pairs of ribs attach to the vertebrae of a snake's spine. These ribs are set in muscle and are flexible, permitting movement and allowing large prey to pass through the body. In addition, the gullet and stomach are highly elastic, and easily accommodate the whole prey that snakes swallow for food.

The right lung is well developed. The left lung is stunted or entirely missing. Flexible ribs permit expansion of the body wall in breathing. The large heart has a *septum* partially dividing the ventricle into the two chambers.

Since the snake is cold-blooded, the rate of oxidation is much lower than that of a warm-blooded animal; and much less heat is generated in the snake's body. Under resting conditions, a snake is often slightly colder than its surroundings. Also, the heat it does produce is quickly lost since it has no coat of fur or feathers. The reptile is at a disadvantage in being cold-blooded and living on land. In cold regions, it cannot be active during the winter. During hot weather, it must seek shelter during the day, since it has no method of cooling the body below the level of whatever the outdoor temperature happens to be.

Reproduction in snakes

The majority of snakes lay eggs that resemble those of the other reptiles. Each egg, enclosed in a tough white shell, contains stored food to nourish the young snake during its development. The eggs receive no care from the female after being laid and no incubation except the warmth of the sun. Egg-laying snakes are called *oviparous* (oe-VIP-uh-rus); they include the black snake and blue racer.

A smaller group of snakes, including the garter snake and the copperhead, bring forth their young alive, usually in the late

36-7 A female pilot black snake with eggs she has just laid and a mother queen snake with newly born young. Which snake is oviparous and which is ovoviviparous? (Allan Roberts)

summer (Figure 36-7). The eggs are not laid but remain in the uterus, where they develop into young snakes. During development, there is no nourishment provided from the mother's body, as there is in the mammals. Snakes that bring forth their young alive are classed as *ovoviviparous* (OE-voe-vie-VIP-uh-rus) to distinguish them from the higher animals, which are called *viviparous* (vie-VIP-uh-rus) and which nourish their young during development.

Nonpoisonous snakes

The snakes can be divided into two distinct groups on the basis of whether or not they produce toxins. You are probably familiar with the harmless *garter snake, black snake,* and *racers.* We should be interested in protecting these snakes, because they destroy insects, rats, and other rodents. The *king snake* is one of the most valuable snakes, because it eats other snakes and rodents. The *constrictors* of South America, Africa, and Asia are also nonpoisonous snakes.

Poisonous snakes

The poisonous snakes are grouped in four families, all of which have developed specialized teeth or fangs for the injection of poison. The *cobras* are almost entirely limited to the tropics of Africa and Asia. Some of the members of the cobra family are deadly to man. The king cobra of Thailand is the largest of all poisonous snakes. Some species of cobra spit venom at their enemies with surprising accuracy of aim. The fine spray of poison can travel several feet and can cause temporary or permanent blindness if it enters the eye. The *coral snakes* of America are related to the cobras. They are relatively small, with strikingly beautiful coloration. Their venom is very potent, but they cause few fatalities because their short fangs cannot penetrate shoe leather or heavy clothing.

The *sea snakes* are related to the cobra but are usually placed in another family. Although the sea snakes are very poisonous, they seldom harm man, because they are strictly marine creatures. Fishermen, however, fear these snakes, for on rare occasions a specimen is hauled aboard a fishing vessel. Sea snakes inhabit the shallow waters of the East Indies, and one species is frequently seen off the west coast of tropical South America, Central America, and Mexico.

The third group of venomous snakes is composed of the *vipers.* This family includes all those poisonous snakes of Europe, Africa, and Asia that do not belong to the cobra family. The most abundant viper is the common European viper, which is known in England as the adder. The largest number of viper species occurs in Africa.

The fourth family, the *pit vipers,* is distinguished from the true vipers by the presence of highly specialized organs that are sensitive to temperature. These organs are pits that are located on either side of the head, in front of the eyes. Biologists have determined that these pits sense infrared rays. This enables the pit vipers to strike accurately at any prey that produces heat. Since most of the prey of the pit vipers are warm-blooded animals, this adaptation gives them a strong advantage in securing food. Although some pit vipers exist in southern Asia, most species are found in North and South America. The best known are the *copperhead,* the *cottonmouth moccasin* (both of North America), the *fer-de-lance* and *bushmaster* of South America, and the numerous *rattlesnakes* of both continents.

Rattlesnakes are the most widely distributed poisonous snakes in the world. Their range extends from northern Argentina to southern Canada. Of the nineteen or more kinds of rattlesnakes found in the United States, at least twelve species occur in the Southwest. These include the *prairie rattlesnake, western diamondback rattlesnake,* and *horned rattlesnake,* or *sidewinder,* of the desert regions. The range of the *timber rattlesnake* includes most of the eastern United States. The largest of North American rattlesnakes, the *diamondback,* lives in marshy areas of the Southeast. Six- to eight-foot specimens have been captured in southern swamps.

A rattlesnake has a series of dry segments, or rattles, on the end of the tail. When the snake is disturbed, it vibrates these rattles rapidly, causing a whirring sound; this explains why few people are bitten by rattlesnakes even though the snakes are widely distributed. One can usually step away from danger if one hears and recognizes a rattlesnake's warning.

The head of a rattlesnake is large and triangular. The jaws are puffy, because of the presence of poison glands (Figure 36-5). Near the front of the upper jaw is a pair of large, hollow fangs. The fangs, like those of all vipers and pit vipers, are fastened to a bone that is hinged on the upper jaw so that when the snake's mouth is closed, the fangs fold upward against the roof of the mouth. They are pulled down by muscles when the snake opens its mouth to strike. The rattlesnake can strike fiercely to a distance of one third the length of its body or more. The fangs are driven deep into the flesh of the victim, and poison flows from the glands, through the fangs, and into the wound. Both the length of the fangs and the large amount of poison injected make the rattlesnake bite extremely dangerous, especially when the fangs happen to go into a vein.

Rattlesnakes, especially the diamondback, have several economic uses. The skin is used for purses, belts, and other articles. In many regions, the flesh is eaten. The venom is taken from captive specimens and is used for making **antivenin,** a biological product used in treating pit-viper bites.

36-8 Three pit vipers found in the United States: the copperhead (top); the cottonmouth, or water moccasin (center); and the rattlesnake. (top: Allan Roberts; others: Walter Dawn)

Two types of snake venom

The toxic portion of snake venom is made up of various complex protein substances. One type of poison, called *neurotoxin,* affects the parts of the nervous system that control breathing and heart action. The second type, called *hemotoxin,* destroys red blood cells and breaks down the walls of small blood vessels. Both types of toxin are found in the venom of all poisonous snakes, but the proportion varies with the species. The venom of cobras, coral snakes, and sea snakes is usually mostly neurotoxic. Vipers and pit vipers produce venom that tends to be mostly hemotoxic.

The danger from snakebite varies according to the amount of venom injected; the concentration of toxins in the venom; the part of the body receiving the bite; and whether the venom enters the main circulatory system rapidly (through a blood vessel) or slowly (through a muscle or fatty tissue). In the tropics, snakebite is a serious problem. As many as forty thousand people may die annually from snakebite. Cobra bites kill many people in India. In the temperate regions, few people die from snakebite. Still, the bite of any poisonous snake should be regarded with the greatest possible seriousness and treated at once.

The lizards

Of the more than 2,500 species of lizards, only a few are native to the United States. Lizards are chiefly tropical animals, although a few species are found in the colder temperate regions. Many lizards are strange and beautiful. Because of their resemblance to dragons and dinosaurs, they are feared by many people. But most of them are shy, harmless creatures. Of all the fierce-looking animals included in this suborder, only two are poisonous. The lizards have evolved from the Mesozoic era to the present time. Of all the modern reptiles, the lizards have developed the greatest number of adaptations for different environments.

The *iguanas* (i-GWAH-nahz) are considered primitive lizards. They tend to have the typical lizard appearance and proportions; few have evolved any advanced characteristics. Included in this family is the so-called *horned toad,* which has a series of horny spines on its head and back for protection. It lives in the dry plains of the western United States. The spiny skin, which conserves water, is an excellent adaptation for an inhabitant of dry areas. This small lizard survives because its color blends with the sand and spiny cacti of its environment.

While one tropical American iguana grows to a length of seven feet, the best-known lizard of the Western Hemisphere is probably the five-inch-long *anole.* In parts of Florida, this graceful lizard can be found in nearly every tree and shrub. It is widely sold as a pet. Because of its ability to change color, the anole is usually called the *chameleon* (kuh-MEEL-yun), a name that can

36-9 An American "chameleon" *(Anolis)* shedding its skin. Like many other lizards, the *Anolis* eats the shed skin. (Allan Roberts)

properly be applied only to certain Old World lizards. Under the influence of light, temperature, or even its own state of excitement, the anole may be bright green, brown, or gray. The beautiful *collared lizards,* also iguanas, are found from the southern United States through northern South America.

The *geckos* (GECK-oze) are a group of highly specialized lizards. The toes of most geckos are expanded into clinging pads that enable them to run up vertical surfaces and even walk on ceilings. Geckos are the only lizards that can make loud noises. The *skinks* are known for their shiny, cylindrical bodies and generally weak legs. In the forests of Africa and the East Indies, skinks are the most abundant reptiles. In the United States they are exceeded in numbers only by the iguanas. Yet they are not often seen, because they are shy, retiring animals.

The *Gila* (HEE-luh) *monster,* of southwestern United States and Mexico, and its close relative, the *beaded lizard* of Mexico, are the only poisonous lizards. The name *monster* is misleading, for a large Gila monster is less than two feet long. Poison glands are situated in the rear of its lower jaw. Grooved teeth are found in both jaws. The venom does not flow through the teeth (as it does in snakes), but the Gila monster clings tenaciously to its victim and shakes its head from side to side, thus allowing the venom to enter the wound. The toxins of the venom are neurotoxic, affecting the nerves that control the victim's breathing. Even though few human beings die of Gila-monster bites, these lizards should be treated with the same respect given to venomous snakes.

The largest lizards in the world belong to a family called the *monitor lizards.* Monitors are closely related to the ancient lizards thought to be the ancestors of the snakes. Although they do not look like snakes, they have several adaptations found in snakes. Several of them are at home in water. The largest monitor is the *Komodo dragon,* which attains a length of ten feet and may weigh three hundred pounds. It inhabits Komodo Island and

36-10 Examples of lizards. Collared lizard, gila monster, horned lizard, Galápagos iguana. (K. Bobrowsky, Monkmeyer Press; Dr. E. R. Degginger; Stella Grafakos)

36-11 A Komodo dragon, the largest living monitor lizard. (Janet L. Stone, National Audubon Society)

36-12 What difference do you see between the alligators (above) and the Central American crocodile? (E. Ellingsen; Tom McHugh, Photo Researchers)

several neighboring islets in the East Indies. Except for an inability to spout fire, monitor lizards have all of the classical characteristics of the mythical dragons.

The crocodilians

The crocodilians and their extinct relatives reached their supremacy in the late Mesozoic era. They were found in all types of environments, but they thrived only in water. Crocodilians like today's alligators and crocodiles appeared in the Cretaceous period. The modern crocodilians have survived from ancient times because their ancestors developed a means of breathing while in the water—even with their mouths open. Raised nostrils at the end of the snout are connected to the throat by an air passage in the skull. At the back of the mouth is a fleshy valve that prevents water from entering the lungs when the mouth is open. Thus, crocodilians can lie in the water with only their nostrils and eyes above the surface. Unsuspecting animals are often too close to escape by the time they see the predator, or they may be totally unaware of the crocodilian's presence until they are grasped by the animal's powerful jaws.

Compared to the large number of crocodilian species that inhabited the seas in the days of the dinosaurs, only a few survive. About twenty-five living species of *alligators, caimans* (KAY-manz), *crocodiles,* and *gavials* make up the order Crocodilia. All live in the Tropical and Subtropical zones. All are very similar to one another, differing in such small ways as length and width of snout, the arrangement of scales, and the arrangement of teeth. Crocodiles are more aquatic than alligators. They are distinguished from alligators by a triangular head and a pointed snout and, on each side of the lower jaw, a tooth that fits into notches on the outside of the upper jaw.

Alligator hide is of great value for making fine leather, and baby alligators are in great demand as pets. Because of the danger of exterminating the American alligator, the United States has outlawed the collecting of the animals for hide or pets. Most "alligator" hides and baby "alligators" sold in the United States today are in reality the skins and babies of the South American caiman, which is disappearing from large areas of South America. If kept in an aquarium at room temperature and fed regularly on insects, fish, or pieces of raw meat, caimans will live for a long time in captivity.

The turtles

Biologists speak of turtles that live on land as *tortoises*. These hard-shelled, slow-moving turtles have strong feet and claws for walking on land and digging. Most tortoises are vegetarians, living on a variety of plant foods. Several of the hard-shelled, freshwater, edible turtles are known as *terrapins*. These turtles are found in markets in many sections of the country. The large, ocean-dwelling forms with limbs in the form of flippers are *true turtles*. However, for convenience, we shall refer to all of them as turtles.

If you want to know what the earliest reptiles were like, examine a turtle. It has changed but little in the past two hundred million years. This ancient reptile form outdates even the lizards and the dinosaurs and first appeared in the Jurassic period.

The turtle seems to have been nature's experiment with armored vertebrates. The armor, a curious body structure, was highly successful in one sense, since these animals have survived through such a long span of time. However, of the more than seven thousand species of reptiles, only three hundred are turtles.

It is possible that the development of the turtles actually reversed the trend of vertebrates to move from water to land. The ancestors of turtles might have been land animals. Although a few turtles still live on land, most are primarily water-dwellers. But since they are reptiles, all turtles come to the land to lay their eggs.

Several important adaptations have made it possible for turtles to endure for so many years. Most significant was the development of a boxlike pair of shells. The upper and lower shells are connected to each other along the sides, and many turtles can draw their heads and feet back into this armored covering. Although we often think of turtles as slow and stupid, their adaptation for protection at the expense of movement is the main key to their success.

It is not certain just how large the shells of the first turtles were, but some modern turtles are almost entirely enclosed, while others have shells greatly reduced in size. In the soft-shelled turtles, the horny shell has been replaced by a leathery skin, but there is a well-developed bony shell beneath the skin. These turtles have flattened bodies that enable them to lie concealed at the bottom of lakes or ponds. Very long necks enable these active predators to capture unsuspecting prey. The upper shell of the snapping turtle is an excellent cover, but the lower shell is small. Still, the turtle's large head and powerful jaws provide it with

36-13 Various turtles. Galápagos tortoise, Florida box turtle, common snapping turtle, spineless soft-shelled turtle. (E. Ellingsen; John H. Gerard, L. L. Rue III, National Audubon Society)

excellent protection. The typical sea turtles have large upper shells, but the lower shells are somewhat reduced in size. Sea turtles can escape from their enemies because their flippers enable them to swim efficiently. Only when they come ashore to sun themselves or lay their eggs are the sea turtles at the mercy of their enemies. If they are turned on their backs on land, they are totally helpless.

Turtles have valuable adaptations besides their shells. The horny, toothless beak is an efficient shearing mechanism that permits the animal to eat meat or plants. The legs are strong and heavy. Turtles can remain completely submerged in water for long periods of time. Some species have developed very efficient substitutes for gills. Sea turtles are able to absorb oxygen from the water through membraneous areas in both the cloaca and the throat.

Although most turtles live in water, the tortoise has become a land-dweller; some live in dry desert conditions. How tortoises became established on islands far from any mainland is a mystery, but they are found on islands in the Pacific and Indian oceans. The giant tortoises of the Galápagos Islands in the eastern Pacific were extremely plentiful before the arrival of civilized man and his domesticated animals. A number of species of the Galápagos tortoises are found on the islands. When Charles Darwin visited these islands during his trip around the world, he observed the different species. Observations of these animals contributed to the development of Darwin's theory of evolution.

Structure of the turtle

The upper shell, or *carapace,* of a turtle is covered with epidermal plates, or *shields,* arranged in a symmetrical pattern. While these shields vary in number and arrangement in various kinds of turtles, they are the same in all members of a species. The shields vary in color and in markings. Beneath the shields are bony plates that are fused together to form a protective case. The shape and arrangement of these bony plates does not match that of the epidermal shields above them. The lower shell, or *plastron,* has similar epidermal shields covering bony plates. The carapace and plastron join on the sides in a bony *bridge*.

The head of a turtle is generally pointed or triangular. The mouth lacks teeth, but the margins of the jaws form a sharp beak with which the turtle bites off chunks of food. There is a pair of nostrils at the tip of the head, making it possible for a turtle to submerge in the water, leaving only the tip of the head above the surface for breathing air. The eyes are well developed and are protected by three eyelids. In addition to fleshy upper and lower eyelids, the turtle has a transparent nictitating membrane that closes over the eyeball from the front corner of the eye. A smooth tympanic membrane lies just behind the angle of the upper and lower jaws.

The limbs of most turtles are short and, in most species, have five toes provided with claws. The feet vary in the amount of webbing between the toes. The skin covering the limbs is tough and scale-covered. The tails of members of different species of turtles vary greatly in length.

SUMMARY

Fossil evidence indicates that the earliest reptiles evolved from certain amphibians during the Late Carboniferous Period. This was a momentous change in vertebrate life. Amphibians never really adapted to a dry land environment. To this day, their lives are closely bound to water. It remained for the reptiles to conquer the land. This important step in vertebrate development was made possible by such modifications as dry skin and scales, claws, and well-developed lungs. The amniote egg of the reptile eliminated dependence on water for development and for an aquatic larval stage after hatching.

Changes in the land and climate near the close of the Paleozoic Era provided extensive dry land environments suitable for reptiles. During three long Periods of the Mesozoic Era that followed, reptiles abounded in many sizes and forms. Never before, nor since, have vertebrates attained the size of the giant dinosaurs of the Age of Reptiles. For millions of years, these great creatures roamed the earth and dominated the land. Then, at the close of the Mesozoic Era, further changes in the environments of the earth doomed even these giant creatures. Again, vertebrate supremacy shifted and two other vertebrate groups, the birds and mammals, replaced much of the dwindling reptile population.

BIOLOGICALLY SPEAKING

amniote egg	parietal eye	antivenin
shell	Jacobson's organ	neurotoxin
embryonic membrane	venom	hemotoxin
amnion	septum	carapace
chorion	oviparous	shield
allantois	ovoviviparous	plastron
yolk	viviparous	bridge
internal fertilization		

QUESTIONS FOR REVIEW

1. How have biologists been able to date reptiles in various geological periods?
2. Identify the rise of reptiles, the age of reptile supremacy, and the decline of reptiles with geological periods.
3. Name and describe a plant-eating dinosaur, a flesh-eating dinosaur, and a flying reptile.
4. Describe the functions of the shell, chorion, amnion, allantois, and yolk of an amnionic egg.
5. List six body characteristics that enable a reptile to live entirely on land.
6. Name four orders of living reptiles and give an example of each.
7. In what respect is *Sphenodon* a reptile of special interest to the biologist?
8. What use does the snake make of its forked tongue?
9. Describe three methods used by various snakes in capturing prey.

10. Describe several characteristics of the snake's mouth and jaw that represent adaptation for swallowing large prey.
11. Describe three kinds of locomotion in snakes.
12. Distinguish between oviparous and ovoviviparous snakes.
13. Name four families of poisonous snakes of the world.
14. Distinguish the two types of toxin present in snake venom and describe the action of each on a snakebite victim.
15. List several reasons for the success of crocodilians as a reptile group.
16. What adaptations have enabled the turtle to survive through the ages?

APPLYING PRINCIPLES AND CONCEPTS

1. Discuss the amniote egg as the key to the development of land-dwelling vertebrates.
2. Propose a theory to account for the decline of dinosaurs and other large reptiles late in the Cretaceous period.
3. Account for the fact that many unusual and ancient animals are found today only on islands.
4. In what respect is cold-bloodedness a limiting factor in the distribution of reptiles?
5. Discuss several of the body structures of a reptile that you feel represent an advance over those of the amphibians.

RELATED READING

Books

BELLAIRS, ANGUS, *The Life of the Reptiles.*
  Universe Books, New York. 1970. A two volume treatise on all aspects of the reptiles, an excellent reference.
CARR, ARCHIE, and the Editors of *Life* Magazine *The Reptiles.*
  Time-Life Books (Time, Inc.), New York. 1963. Unusual color photographs and much reptile lore in this outstanding Time-Life science book.
COLBERT, EDWIN H., *Dinosaurs: Their Discovery and Their World.*
  E. P. Dutton and Company, Inc., New York. 1967. A comprehensive book on dinosaurs, offering a wealth of scientific information.
MINTON, SHERMAN A., and MADGE R. MINTON. *Venomous Reptiles.*
  Charles Scribner's Sons, New York. 1969. An interesting presentation on all aspects relating to these animals.
MORRIS, RAMONA, and DESMOND MORRIS, *Men and Snakes.*
  McGraw-Hill Book Co., New York. 1965. The full, illustrated story of man's relationship with the snake from the dawn of civilization to the present.
POPE, CLIFFORD H., *The Giant Snakes.*
  Alfred A. Knopf, Inc., New York. 1961. Mainly the natural history of the boa constrictor, the anaconda, and the larger pythons.
PRITCHARD, PETER, C. H., *Living Turtles of the World.*
  T. F. H. Publications, Inc., Jersey City, N.J. 1967. A thorough, well-illustrated guide covering the natural history of turtles.
SEELY, H. G., *Dragons of the Air: An Account of Extinct Flying Reptiles.*
  Dover Publications, Inc., New York. 1967. The anatomy and evolutionary adaptations of the reptiles 60 million years before the first birds.

CHAPTER THIRTY-SEVEN

THE BIRDS

The origin of birds

The birds obviously survived the forces and changes that caused their reptile companions to decline late in the Cretaceous period. Birds have been called "glorified reptiles," and even modern birds show evidence of their reptile ancestry in the form of scales on their feet and claws on their toes. But while even the flying reptiles, or pterosaurs, disappeared entirely more than seventy million years ago, birds have continued to thrive and are today among the most abundant vertebrates on earth, both in the number of species and individuals and in distribution.

The biological success of birds indicates that they were better adapted than the flying reptiles for flight as well as for life on land and on water. Study of a modern bird reveals some of the most remarkable adaptations to be found in any animal. When did this "glorification" of reptiles begin? How rapidly did modifications occur? The first birds are thought to have appeared during the Jurassic period, when the Age of Reptiles was at its height. From this beginning, birds assumed a dominant place through the Cretaceous period and into the Cenozoic era—a position of dominance they have never relinquished.

Archaeopteryx—a fossil of great biological significance

It would seem logical to assume that at some time, millions of years ago, transitional stages between reptiles and birds existed. The modern bird has none of the characteristics of flying reptiles. It seems to have developed along different and more successful lines. Could its ancestors have been lizardlike reptiles? A remarkable discovery of a fossil bird provided an important clue to the answer to this question.

In the late 1800's, an archeologist in Bavaria, Germany, was digging for fossils in a stratum of fine-grained rock dating to the Jurassic period. This sedimentary rock was probably deposited

37-1 *Archaeopteryx.* Observation and interpretation of the fossils of the reptilelike bird, such as that at the top, led to the artist's representation at the bottom. (courtesy of the American Museum of Natural History)

in a shallow lagoon some 150 million years ago. Fortunately for our knowledge and understanding of the Jurassic period, curious, reptilelike birds fell into such lagoons from time to time and sank into the limy deposit at the bottom. Gradually, they became fossilized and were preserved as a record in stone.

Three such fossil birds of the Jurassic period have been discovered, the latest in 1956. *Archaeopteryx* (AHR-kee-OP-tuh-ricks), meaning "ancient bird," is an appropriate name for such birds.

Fortunately, the *Archaeopteryx* fossils included, in addition to a print of the skeleton, excellent impressions of the long flight feathers on the wings and a double row of tail feathers (Figure 37-1). Except for these feather impressions, scientists would doubtless have classified *Archaeopteryx* as a reptile, for many of its characteristics were more those of a lizard than of a bird. However, feathers are unique characteristics of birds and definitely place *Archaeopteryx* in this class.

The body of *Archaeopteryx* was about the size of a crow or small chicken. The jaws and the front of its skull were elongated and narrowed into a beak armed with reptilian teeth. The neck was long and flexible. The back was short and compact, a characteristic important in flight. The lizardlike tail was longer than the body. The modified forelimbs were weak wings supported by slender bones. Typical bird flight feathers grew from the hand and lower arm. Extending beyond the feathered portion of the wing were three clawed fingers. Biologists believe that *Archaeopteryx* used these fingers for grasping and climbing trees, from which it could soar through the air and glide to the ground. The body was supported on strong hind limbs, much like those of a modern bird. The feet were also like those of the modern bird, with three clawed toes extending forward on each foot and one short toe turned backward. The development of the hind limbs indicates that *Archaeopteryx* was more suited to running on land than to flying or gliding through the air.

Archaeopteryx was truly a mixture of reptile and bird characteristics. Its skeleton was largely reptile, although many of its bone modifications were those of a bird. In addition to flight feathers, *Archaeopteryx* had feathers covering much of its body. This has led biologists to believe that it may have been warm-blooded, a definite characteristic of birds.

Apparently, *Archaeopteryx* was a product of several million years of evolution from earlier reptile ancestors. Still more ancient reptilelike birds may be found some day. Until then, however, *Archaeopteryx* will remain the most ancient and primitive of all birds.

During the Cretaceous period, birds advanced far along the evolutionary road that gave rise to modern birds. Their fingers grew together, making the wings stronger. Cretaceous birds still had teeth, but these gradually disappeared, and their mouths developed into horny, toothless beaks or bills. By the beginning of the Cenozoic era, birds had forms much like those of our modern birds. So far as biologists can tell, there has been little or no structural change in birds for more than fifty million years. Most modern birds are structurally similar to one another. That there are so many species of birds today is the result of adaptations to the different environments, diets, and types of life.

Modern birds—highly successful vertebrates

The biological success of a group of organisms can be measured by the number of species and individuals in the group, their adaptation to a wide variety of environmental conditions, and their distribution throughout the world. On the basis of these criteria, we must consider birds, along with bony fishes and mammals, among the most successful vertebrates in modern times.

Birds range in size from the tiny hummingbird to the ostrich. Diets of various birds include nectar, seeds, insects and other small animals, other birds and mammals, and even decaying flesh. Some birds are masters of the air, while others, like the ostrich and penguin, are flightless. Some birds never leave a home environment, while others travel from one pole to the other in migratory flights. An intricately woven bag, a platform of sticks, a hole in a tree, or a depression in the ground can all serve as nests.

Birds capable of flight have an advantage over all other vertebrates. They are able to travel long distances quickly and change environments as conditions require. Many follow the food supply as it varies with changes in season.

In one respect, most birds have surpassed all other vertebrates. They are at home both in the air and on land. Ducks, geese, and other water birds occupy all three environments—a distinction not shared by any other vertebrate.

Characteristics of birds

Even though birds vary greatly in size, color, diet, and other adaptations, they are remarkably similar in body structure and physiological characteristics. We can summarize the characteristics of the vertebrate class *Aves,* which includes all birds, as follows:

■ Body covering of feathers.
■ Bones light, porous, and in certain cases air-filled.
■ Forelimbs developed as wings and, in most birds, used for flight; never for grasping.
■ Body supported by two hind limbs.
■ Mouth in the form of a horny, toothless beak.
■ Heart four-chambered; circulatory system well-developed, with a right aortic arch only.
■ Body temperature constant (warm-blooded).
■ Amniote egg enclosed in a lime-containing shell and usually incubated in a nest.

Gaviiformes
(loon)

Podicipediformes
(grebe)

Pelecaniformes
(pelican)

Ciconiiformes
(heron)

Anseriformes
(duck)

Falconiformes
(hawk)

Galliformes
(quail)

Gruiformes
(rail)

Columbiformes
(mourning dove)

Charadriiformes
(gull)

Strigiformes
(owl)

Cuculiformes
(yellow-billed cuckoo)

Apodiformes
(hummingbird)

Caprimulgiformes
(whipporwill)

Coraciiformes
(kingfisher)

Piciformes
(sapsucker)

Passeriformes
(robin)

37-2   Some examples of common orders of North American birds.

Many structural adaptations are involved in the capability of a bird to fly. Body streamlining is necessary in order to reduce air resistance. Weight is sacrificed without reducing strength. Flight requires enormous muscle power and activity that only birds possess. The fueling of the muscles used in flight requires abundant food and a highly developed digestive system capable of dealing with large quantities of food. The body temperature of birds is higher than that of most animals and averages from 100° to 112°F. Release of this heat energy requires a high metabolic rate. Support of this cellular activity, in turn, requires highly developed respiratory, circulatory, and excretory systems. A brain that is proportionally large in comparison with the body is necessary to control such a high rate of body activity.

We spoke of birds as "glorified reptiles" in introducing this chapter. "Supercharged reptiles" would be more accurate. Birds are truly vertebrate "powerhouses."

Orders of birds

More than 8,500 species of birds have been identified in various environments throughout the world. Classification of this large number of species into orders, families, and genera has been a difficult task for the biologist. The birds of the world are often grouped in 27 orders. The better-known birds of North America represent 17 of these orders, as shown in Figure 37-2.

Several characteristics of birds are used in establishing orders. Among these are the structure of feet and beaks. Modifications in feet and beaks relate closely to the life habits and diets of various birds. Foot structure adapts some birds for diving and others for swimming, wading, perching, running, or carrying prey. Beak modifications determine whether birds are insect eaters, seed eaters, flesh eaters, mud probers, or wood borers. As we discuss foot and beak modifications in this chapter, you may wish to find examples in Figure 37-2.

Structure and functions of feathers

Strange as it may seem, *feathers* are modified scales. They develop from small elevations, or *papillae,* in the skin. The bases of these feather buds sink into pits, the *follicles,* that hold the feathers in the skin. Feather follicles lie in rows or tracts in certain regions of the skin and spread out to cover the featherless areas.

There are several kinds of feathers. Soft *down feathers* form the plumage of newly hatched birds. In older birds, especially in waterfowl, they form an insulation close to the skin. A down feather is composed of a fluffy tuft at the end of a short quill. Down feathers reduce heat loss so efficiently that a bird can fly through cold winter air and still maintain a body temperature of over 100°F. Ducks and geese have a dense covering of down that prevents chilling in the water. The slender, hairlike feathers with tufts on the ends are known as *filoplumes*.

Contour feathers cover the body and round out the angles, giving the bird a smooth outline. They also form an effective shield against injury and provide the coloration so important in the life of a bird. Often the female blends more closely with the surroundings than her brightly colored mate does. *Quill feathers* are large contour feathers that grow in the wing and tail. These large feathers are well designed to provide the lift necessary for flight and for balance and steering in flight.

Figure 37-3 shows the structure of a quill feather. A broad, flat *vane* spreads from a central axis, the *rachis* (RAY-kiss). The rachis ends in a hollow *quill*. If you magnify the vane, you can see the many rays, or *barbs*. Each barb is like a tiny feather with many projections, the *barbules* (little barbs). These are held together with tiny interlocking hooks. This complicated arrangement makes the vane strong, light, and elastic. If a vane is split, the bird shakes its feathers and locks the barbules together again, or it may preen the feather by drawing it through its beak, making it whole again. The rachis is grooved and the quill is hollow. This gives a feather the greatest possible strength with the least possible weight. Nourishment is supplied through an opening at the base of the quill while the feather is growing.

The vane of the wing feather is wider on one side of the rachis than on the other. When the wing strikes the air in a power stroke, the vane turns up and rests against its neighbor. On the return stroke, it is free to turn back. The air passes through the wing as each feather turns slightly on its axis (feathering) and the wing meets less air resistance.

You have probably noticed birds oiling their feathers after a bath or a swim. Many birds take oil from a gland at the base of the tail and spread it over the surface of the feathers. This makes them waterproof. Oil on the feathers is vital to swimming and diving birds such as ducks, geese, swans, loons, and grebes. This oil not only prevents water from penetrating the feathers to the skin, but also makes the birds buoyant and prevents chilling.

Molting in birds

The bird sheds its feathers at least once a year. Feathers, especially those of the wings and tail, may be lost or broken. Since molting usually occurs in the late summer, the bird is provided with new quills before the fall migrations. A second, partial molt often occurs in the spring before the breeding season. This molt provides the bright breeding plumage of many birds. In some species, including the ptarmigan of the far north, two complete seasonal molts occur. The early summer molt provides a plumage that blends with rocks and soil. The fall molt arrays the ptarmigan in snow-white winter plumage.

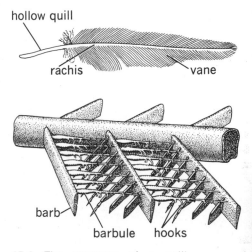

37-3 The structure of a quill feather. The enlargement illustrates a portion of the rachis and vane as seen with a microscope.

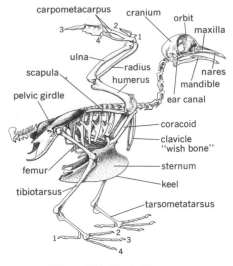

carpometacarpus
cranium
orbit
maxilla
3
2
1
4
ulna
nares
radius
scapula
mandible
humerus
pelvic girdle
ear canal
coracoid
clavicle "wish bone"
sternum
femur
keel
tibiotarsus
tarsometatarsus
1
2
3
4

37-4 A bird's skeleton.

New feathers grow from the same pits from which the old ones were shed. In most species, the wing feathers are shed gradually and in pairs, thus allowing the bird to fly during the molt. Many water birds, including ducks, lose their flight feathers in a group and are unable to fly for a time. This occurs while the birds are nesting and caring for their young. They grow new flight feathers as the young birds are getting their plumage.

Structures of the head

The *skull* of a bird is composed of thin bones that fuse together early in the bird's life. The *cranium* is rounded and forms a large cranial cavity that encases the brain (Figure 37-4). The upper jaw, or *maxilla* and lower jaw, or *mandible,* project forward as the toothless beak. The tongue is small and serves primarily as an organ of touch. The sense of taste is poorly developed in birds, although it is known to be present, because birds reject bad-tasting food. Paired nostrils, or **nares,** open through the beak and admit air to the mouth cavity. The sense of smell is poor.

The *eyes* are large and are deeply set in orbits of the skull. The sense of sight is very keen. It is said that many birds, including hawks and owls, have vision from eight to ten times as keen as that of man. Owls and certain other birds that are active at night have excellent vision in reduced light. The field of vision in a bird is great. The eyes shift focus from close to distant objects rapidly. Birds have a remarkable ability to judge distance both at close range and at great height. A bird can drop out of the sky and land on a rock in a stream, fly through a deep forest without striking a branch, or swoop down on a slender perch. The eyes are protected by an upper and a lower eyelid, as well as by a thin, transparent *nictitating membrane* that can be drawn across the eyeball from the front corner of the eye socket. The nictitating membrane protects the eye and keeps it moist.

Ear canals open on the sides of the head posterior to the eyes. They are covered by tufts of feathers. Eustachian tubes lead from the middle-ear cavities to a single opening in the upper wall of the throat. The sense of hearing is acute in birds, and the ears are especially sensitive to high pitches.

The neck and trunk

The neck of a bird is long and flexible and has vertebrae that slide freely on one another. This freedom of movement is essential in the life of a bird, because the bird cannot use its forelimbs for grasping. The flexibility of its neck permits the bird to turn its head easily to see in all directions. This flexibility also enables the bird to use its beak for feeding, preening its feathers, building nests, and performing other activities.

While the neck is flexible, the bones of the trunk are fused and provide maximum support of the body during flight. Vertebrae of the tail are flexible, thus permitting movement of the tail feathers, which aid in balance and steering during flight.

The breast bone, or *sternum,* of a bird is greatly enlarged and extends over much of the ventral side of the body as shown in Figure 37-4. Ribs attach to the dorsal portion of the sternum. The ventral portion extends downward as a deep ridge, or *keel,* that provides a broad surface for the attachment of breast muscles.

The shoulders and wings

These areas of the bird's body are remarkably adapted for flight. The shoulder, or *pectoral girdle,* consists of three pairs of bones in a tripod arrangement. Narrow shoulder blades, or *scapulae,* lie on either side of the spine above the ribs, and are embedded in muscles of the back. A *coracoid* bone extends from each scapula to the sternum, thus bracing the shoulder from below. Further support is provided by the *clavicles,* which extend from each shoulder and fuse in front of the breastbone, forming a V-shaped "wishbone."

The structure of a bird's wing resembles your own arm in many ways (Figure 13-5, page 204). The upper portion of a bird's wing contains a large, single bone, the *humerus* as in the upper arm of man. The convex head of the humerus fits into a socket at the shoulder. The lower portion, as in the forearm of man, has two bones, the *radius* and the *ulna.* The end section of the wing is greatly modified for flight in a modern bird. Two small wrist, or *carpal,* bones are present. Other bones of the wrist and hand are fused, forming a *carpometacarpus.* The remains of three fingers, or *digits,* are also present. The second digit contains two small bones, or *phalanges.* This digit protrudes from the wing as a short knob and bears a tuft of short feathers. The third digit is longest and contains three phalanges. Only a single bone of the fourth digit remains. The first and fifth digits, found in many other vertebrates, are lacking in the bird.

The longest flight feathers, known as *primaries,* grow from the "hand and fingers." *Secondary* feathers grow from the "forearm," and *tertiaries* grow from the "upper arm." The quills of these feathers are covered with smaller feathers known as *coverts.* The outline of the wing as a whole is concave on the lower side, thick on the forward edge, and thin and flexible on the rear edge—a perfect design for flight.

Motion of the wings in flight

We might compare the motion of a bird's wings in flight to a horizontal figure eight—down and back, up and forward. The down stroke is the power stroke. The upward movement returns the wing to position for the next power stroke. These two actions of the wing require two sets of muscles, arranged in layers on the breast. You may have noticed that these layers separate on

37-5 The ostrich (above) and the penguins are flightless birds. (Rue III, National Audubon Society; Roger T. Peterson, Photo Researchers)

the breast of a chicken. The tougher muscles of the outer layer pull the wing down in a power stroke. Those of the more tender inner layer raise the wing for the next stroke.

Flightless birds

In many parts of the world, there are birds that have lost the ability to fly. The best known are the ostrich of Africa, the rhea of South America, the kiwi of New Zealand, the cassowary of Australia, and the penguin of the Southern Hemisphere.

Most flightless birds have succeeded in life only because they live in areas that are free of predators. Several, like the giant elephant bird of Madagascar, the dodo of the Mascarene Islands, and the moas of New Zealand, became extinct only after the arrival of man and his domestic animals. Others, like the ostrich, the rheas, and the cassowaries, survive among predators only because they have exceptionally strong legs and an extremely keen sense of vision.

Penguins are interesting birds. Although they are flightless, their wings are well developed. They use them for swimming through the water. Their webbed feet serve them well as rudders. Penguins have the keeled breast that most flightless birds lack. This adaptation may be related to the development of muscles used in swimming.

The pelvic girdle and legs

Three pairs of bones are firmly fused and united with the lower spine to form the *pelvic girdle* of the bird. The legs attach at the hip joint, high on the pelvic girdle, and provide excellent balance. This balance is important in bipedal locomotion in the form of walking, running, hopping, wading, swimming, or diving.

The upper leg, or thigh, contains a large, single bone, the *femur.* The upper end of the femur joins the pelvic girdle in a ball-and-socket joint. The lower leg contains two bones. The large bone is the *tibia,* or "drumstick." A second bone, the *fibula,* is reduced to a slender spine that lies alongside the tibia. Bones of the ankle and foot are fused in an elongated *tarsometatarsus,* which is often confused with the leg. The bird walks, literally, on its toes. In most birds, the first toe is short and is turned backward. The second (toward the inside of the foot), third, and fourth toes are directed forward. All of the toes end in claws. All of the toes contain *phalanges.* It is interesting to note that the number of phalanges in each toe is one more than the number of the toe.

Bird's feet, like their beaks, vary greatly in structure and are important factors in determining the activities of various species (Figure 37-2). In addition to providing locomotion, the feet are used in food-getting and in other activities such as nest-building. The feet of many ground birds are adapted for scratching in search of food. Swimming and diving birds have webbed or lobed feet. The legs are set far back on the body; this, although it makes them awkward on land, provides excellent propulsion in the water. Birds with very long legs and long toes are well adapted for wading. Hawks and other birds of prey have strong feet with long, sharp claws or talons for grasping their prey. The feet of woodpeckers have two toes turned backward and two pointing forward; thus the woodpecker is ideally suited to clinging to the side of a tree.

The digestive system of the bird

The maintenance of a constant, high body temperature and the tremendous muscle exertion during flight require a high rate of body metabolism. To meet these energy needs, birds consume a large volume of food—the fuel for the cellular metabolic "furnaces." Hence, to say that someone eats like a bird is actually to say that he is a glutton. Furthermore, flight and the other activities of birds require that birds eat when food is available and store it for use between feedings. Food is swallowed whole and, in the case of many birds, especially seed-eaters, is difficult to digest. Thus, birds require a well-developed and efficient digestive system.

Food passes from the *mouth cavity* down a long *esophagus* that enlarges into a *crop* at the base of the neck (Figure 37-6). Here the food is stored and moistened. From the crop, food passes into the first division of the stomach, known as the *proventriculus*. This organ has thick, glandular walls that secrete gastric fluid, which mixes with food as it passes through. Food then enters the second stomach region, the *gizzard*. The thick, muscular walls of the gizzard, aided by small stones in many birds, churn and grind the food. The ground food mass and digestive fluids pass from the gizzard through the *pyloric valve* into the slender, coiled *intestine*. At its lower end, the intestine joins the short *rectum*. Two blind pouches, or *caeca*, branch off at this junction. The rectum leads to the *cloaca*, which opens to the outside at the vent, or *cloacal opening*. Ducts from the kidneys and genital organs also open into the cloaca.

The two-lobed *liver* is large and may or may not have a *gall bladder* on the lower side, depending on the species of bird. Bile is poured into the small intestine through two ducts. The *pancreas* lies along the U-shaped portion of the intestine and pours its secretion into the intestine through three ducts.

Respiration, circulation, and excretion in the bird

The *lungs* of a bird lie in the back against the ribs, in the anterior region of the body cavity. The capacity of the lungs is greatly increased by a system of *air sacs* that extend from the

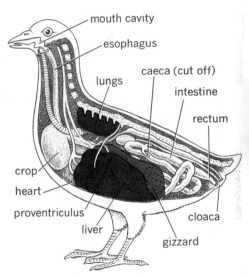

37-6 The internal organs of a bird.

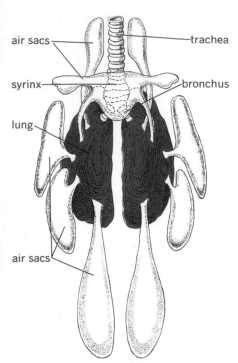

37-7 The respiratory system of birds is adapted to accomplish the rapid oxidation of food and release of energy needed by birds for their high degree of activity.

37-8 A bird's brain. Contrast the size of the cerebrum to that of the fish and frog. What does the size of the optic lobes indicate about the sense of sight?

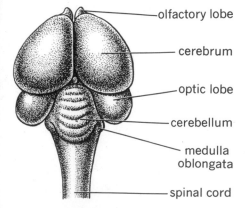

lungs into the chest area and the abdomen and connect with cavities in the larger bones (Figure 37-7).

Air is drawn through the nostrils in the beak and down the *trachea* and its lower divisions, or *bronchi,* to the lungs and air sacs by relaxation of the thoracic and abdominal muscles. Contraction of these muscles forces air out. Though the lungs are small, a rapid rate of respiration fills them often and supplies the blood with the great amount of oxygen necessary to carry on a high rate of oxidation in the body tissues.

The bird's respiratory system is also its principal temperature-regulating system. It has no sweat glands and cannot eliminate heat through its skin. Most excess heat is discharged from the body through the lungs. The air sacs are believed to assist in heat elimination. You may have noticed that birds often pant on hot days. At such times, the insulation provided by feathers is more a liability than an asset.

The lungs also supply air for singing. The bird's song is not produced in the throat but, rather, at the base of the trachea, where it divides into the bronchi. Here, the song box, or *syrinx,* a delicate and highly adjustable structure, is located.

The *heart* of a bird is large, powerful, and four-chambered. The bird heart is far superior to that of any vertebrate you have studied. It consists of two thin-walled *atria* and two muscular *ventricles.* The right side of this four-chambered heart receives blood from the body and pumps it to the lungs. Blood returns from the lungs to the left side and is pumped to the body. It is interesting to note that although amphibians and reptiles had both right and left aortic arches, the bird has only a *right aortic arch.* In birds, this continues as the aorta.

The hearts of birds beat at amazing rates. The rate, ranging from a resting beat to a beat under exertion, is 135 to 570 heartbeats per minute for the mourning dove, 350 to 900 for the English sparrow, and 480 to 1,000 for the chickadee. The rate of activity of the heart may partially account for the relatively short lives of most birds.

The *kidneys* are dark brown, three-lobed organs lying along the back. They excrete *uric acid,* a waste product of cell activity. This metabolic waste is discharged with very little water through the ureters into the cloaca, where it is eliminated with intestinal waste. The absence of a urinary bladder and the lack of temporary storage of urine are considered to be adaptations for flight, because they result in a significant decrease in weight.

The nervous system

The brain of the bird is large and broad and completely fills the cranial cavity. The *olfactory lobes* are proportionally small, indicating a poorly developed sense of smell (Figure 37-8). The hemispheres of the *cerebrum* are the largest of any animal we have discussed so far. The highly developed instincts of birds and the control of their voluntary muscles center in this brain region. The *optic lobes* are also large, thus accounting for the excellent vision found in birds. The large *cerebellum* explains the high degree of muscle coordination. The *medulla oblongata* joins the spinal cord, which extends down the back, encased in vertebrae.

The reproductive system

The oval *testes* of the male bird lie in the back in about the same position in which we found them in the male frog. Small tubes, the *vasa deferentia,* lead from the testes to openings in the cloaca. In many birds, the vasa deferentia are dilated at the lower ends, forming saclike *seminal vesicles.* Sperms travel from the testes through the vasa deferentia, and are stored in the seminal vesicles until mating occurs. During mating, sperms are transferred from the cloaca of the male to the cloaca of the female.

In most birds, the female reproductive system consists of a single left ovary in which eggs develop and a long, coiled oviduct that leads to the cloaca (Figure 37-9). The disappearance of the right ovary and oviduct early in life is another modification that results in decreased weight, again an adaptation well suited for flight.

If you examine a hen you are preparing for dinner, you may find a mass of *yolks* of various sizes in the ovary, which lies in the dorsal region of the body cavity. Each yolk is part of an *egg cell,* which also contains a nucleus and cytoplasm. As the egg cell matures, the yolk, which contains stored foods, increases in size so much that it pushes the nucleus and cytoplasm to the surface. When a yolk has reached full size, the egg cell breaks out of the ovary, and is drawn into the upper end of the oviduct by lashing cilia. Following mating, sperms travel up the oviduct from the cloaca. Fertilization occurs in the upper region of the oviduct before additional egg structures form around the egg cell. Fertilization occurs about forty hours before the egg is laid.

As the egg cell passes down the oviduct, it is surrounded by layers of *albumen,* or egg white (Figure 37-10). The egg cell is suspended by strands of *chalaza,* which extend into the albumen toward the ends of the egg. A double *shell membrane* forms around the albumen. Finally, the membranes are surrounded by the shell, which is secreted by *lime-producing glands* in the lower part of the oviduct. Shortly after this, the egg is laid.

The egg cell is the only living part of an egg. The protein of the albumen and the proteins and lipids (primarily oils) of the yolk are stored nourishment for the developing bird. The shell and shell membranes protect the egg and reduce water loss. However, they must be porous enough to admit air for respiration during incubation of the embryo. The pores are also large enough to admit bacteria, which accounts for the spoilage of

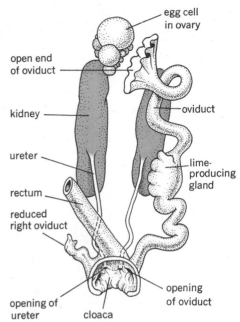

37-9 The reproductive system of a hen. Note the immature eggs in the ovary.

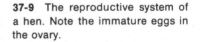

37-10 A chicken egg. The egg cell, including the yolk, is held in place by the chalaza.

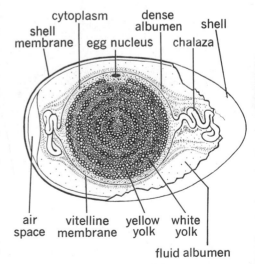

eggs, especially in warm weather. Fertilization of the egg cell is not necessary for the formation and laying of eggs. In fact, care is taken to be sure that the eggs you buy are infertile.

Incubation and development of a bird

While development of the embryo begins soon after fertilization, it continues after the egg is laid only if the egg is kept continuously warm, or *incubated.* The female usually sits on the eggs, although in some species her mate shares in this responsibility. The egg-incubation temperature of most birds is slightly above 100°F. In addition to providing body warmth, the nesting bird turns the eggs with her beak at intervals. The time of incubation varies from thirteen to fifteen days in smaller birds to as many as from forty to fifty days in larger birds. The chicken's egg is incubated for twenty-one days. Duck eggs require twenty-eight days.

In a fertile egg, cell division occurs in the area of the nucleus. As the developing embryo enlarges with continued cell division, it first becomes a plate of cells on the surface of the yolk (Figure 37-11). As growth progresses, a membrane grows from the digestive tract of the developing embryo and surrounds the yolk. This is the *yolk sac.* It contains a network of blood vessels and produces digestive enzymes that change the yolk substance to a soluble form. These nourishing materials are transported through the circulation of the yolk sac to the embryo. As the embryo increases in size, the yolk decreases in size, and is entirely absorbed by the time of hatching.

As was true in the reptiles, another membrane develops from the body wall and completely encloses the embryo. This is the *amnion.* The cavity in which the embryo lies is filled with *amnionic fluid* secreted by the membrane.

A third membrane, the *allantois,* grows from the digestive tract of the embryo and forms a vascular, saclike organ that functions in respiration and receives waste materials from the embryo. The allantois comes in contact with the *chorion,* an embryonic membrane that lies against the double shell membrane. Thus, the development of the bird is like that of the reptile. In fact, it has been said that "the reptiles invented the amnionic egg, and the birds inherited it."

Just before hatching, the baby bird absorbs the last of the yolk. Hatching begins as the bird picks a hole through the shell. The hole is enlarged and finally cuts the shell nearly in two, after which the young bird pushes the halves of the shell apart and works its way out.

Egg number and parental care in birds

The way in which the parent bird (or birds) cares for the young and the degree of development of the young at the time of hatching determine the number of eggs laid by various bird species.

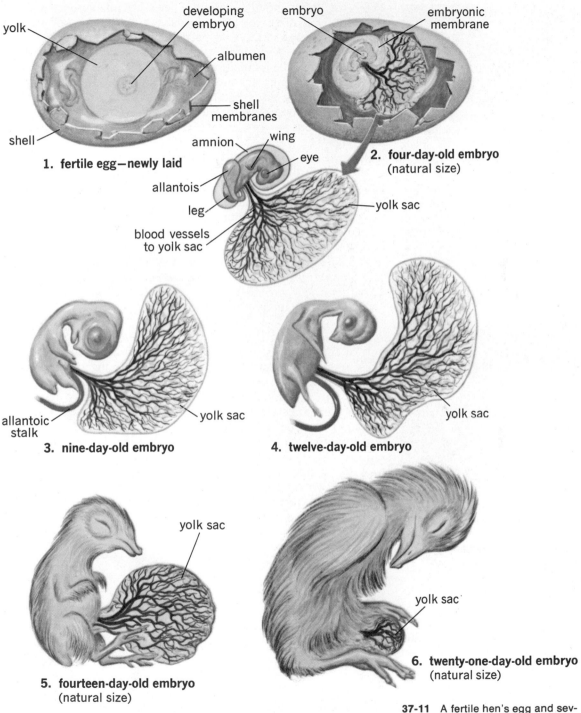

yolk

developing
embryo

albumen

shell
membranes

shell

1. fertile egg—newly laid

amnion

wing

eye

allantois

leg

blood vessels
to yolk sac

embryo

embryonic
membrane

2. four-day-old embryo
(natural size)

yolk sac

allantoic
stalk

yolk sac

3. nine-day-old embryo

yolk sac

4. twelve-day-old embryo

yolk sac

5. fourteen-day-old embryo
(natural size)

yolk sac

6. twenty-one-day-old embryo
(natural size)

37-11 A fertile hen's egg and several of the stages in the development of a chicken embryo.

37-12 The chicken, a precocial bird, has young that are able to follow it and feed themselves almost immediately after hatching. (R. Van Nostrand, National Audubon Society)

Precocial birds, including ducks, geese, quail, grouse, turkeys, chickens, and other fowl-like birds, lay many eggs, often from twelve to twenty. Precocial birds, except for the wood duck, nest on or near the ground. Incubation does not begin until the last egg is laid. The incubation period is longer than it is in smaller birds, and the young are more fully developed when they hatch. All of the young birds hatch within a few hours of each other, and they are soon able to leave the nest. These birds are able to follow their parents and feed themselves immediately after hatching. The young birds find warmth and protection under the breast and wings of the parent bird.

Altricial birds like the robin, thrush, sparrow, and warbler usually lay fewer than six eggs. Incubation begins when the last egg is laid. The incubation period in these birds is less than two weeks. All of the young birds hatch at about the same time and are weak, helpless, and nearly featherless. They must be kept warm and fed in the nest until they are feathered and ready to fly. The supplying of food for two to six nestling birds is a constant activity for one or both parents. However, the amount of parental care provided increases the chances of survival and insures continuation of the species with a smaller number of young.

Hawks and owls usually lay from one to four eggs at intervals of several days. Each egg is incubated as soon as it is laid. This results in a "stairstep" family. The young of these birds are also fed in the nest until they are feathered and able to fly.

Most birds instinctively defend their eggs and young when they are endangered. However, the methods used are almost as varied as the types of birds. Some species that nest on the ground will remain on the nest until they are nearly stepped on. The killdeer, a ground-nesting bird, will feign a broken wing and lure an intruder away from the nest. Normally shy birds will swoop down on an intruder and draw attention from the nest. At the other extreme is the cowbird, which lays its eggs in other birds' nests and depends on the "foster parents" to incubate the eggs and rear the young.

37-13 The wood thrush is an altricial bird that feeds the young in the nest until they are feathered and able to fly. (Allan Roberts)

37-14 A broad-winged hawk and its "stairstep" family. The oldest of the offspring is in the foreground. How can you distinguish between it and the youngest offspring to its right? (Ron Austing, Photo Researchers)

SUMMARY

For centuries, man has watched the birds soar gracefully through the air and wished that he could fly. A few daring inventors even constructed artificial wings to attach to arms. You can imagine their disappointment when they flapped these wings and failed to rise off the ground. Efforts to glide from a high perch resulted only in crash landings.

Compare the anatomy of a bird with the structure of your own body and you will discover why man cannot fly. Our bones are too heavy and our muscles are too weak for flight. The human heart could not endure even the resting pulse of many bird hearts. The energy supply for flight would keep us at the dinner table all of our waking hours.

There are advantages in not being able to fly, however. None of us would trade our 70 or more years of life expectancy for the three to five years of most birds. The bird is a "living power house," but it soon "burns out."

Human ingenuity has realized the dream of man to fly. We can enjoy a "bird's-eye" view of our earth from the window of an airliner and leave the flying to powerful engines.

BIOLOGICALLY SPEAKING

feather	keel	albumen
papilla	crop	chalaza
follicle	air sacs	shell membrane
nare	bronchus	lime-producing glands
ear canal	syrinx	incubate
sternum	uric acid	yolk sac

QUESTIONS FOR REVIEW

1. What structural characteristics of modern birds are evidence of reptile ancestory?
2. List the reptile and the bird characteristics of *Archaeopteryx*.
3. On the basis of what criteria do biologists consider birds highly successful vertebrates?
4. List eight characteristics that distinguish birds from other vertebrates.
5. Name and describe the functions of the different kinds of feathers composing the plumage of birds.
6. Describe the structure of a feather.
7. Describe various adaptations of the beaks of birds.
8. Rate the following senses of a bird as good or poor: smell, taste, sight, and hearing.
9. Why is a long, flexible neck important in the life of a bird?
10. Of what importance is the greatly enlarged sternum of a bird?
11. Describe the tripod arrangement of bones forming the pectoral girdle.
12. Describe the bone structure of the wing.
13. Distinguish between primary, secondary, tertiary, and covert feathers.
14. Describe the bone structure of the leg and foot.
15. List the organs of the alimentary canal in the order in which food passes through them.
16. Of what use are the air sacs extending from the lungs of a bird?

17. In what respect is the bird heart different from that of any other vertebrate you have studied?
18. Why is the excretion of uric acid rather than urine considered to be an adaptation for flight?
19. Describe the formation of an egg as it passes from the ovary to the cloaca.
20. Why is internal fertilization necessary?
21. Review the embryonic membranes formed during the development of a bird.
22. Distinguish between precocial and altricial birds.
23. Account for the "stairstep" family of a hawk or owl.

APPLYING PRINCIPLES AND CONCEPTS

1. From what you have learned about birds, give possible reasons why they have survived in large numbers while flying reptiles disappeared millions of years ago.
2. What structural advances are associated with warm-bloodedness in birds?
3. In terms of what you have learned about mutations, variations, and survival, relate the various forms of beaks and feet to the diet of birds.
4. What physiological problem results from feathers as insulators in birds? How is it solved?
5. Account for the limited distribution of flightless birds.
6. In what ways does the digestive system compensate for the toothless beak of a bird?
7. Discuss the relationship between parental care and the number of eggs laid by various birds.

RELATED READING

Books

ALLEN, GLOVER, M., *Birds and Their Attributes.*
Dover Publications, Inc., New York. 1962. A general, introductory survey of birds—their structure, habits, and relation to man.
FLANAGAN, GERALDINE LUX, *Window Into an Egg.*
William R. Scott, Inc., New York. 1969. Using excellent photographs, probes the day by day development of a chicken—the 21 day wonder.
GREENWAY, JAMES C., *Extinct and Vanishing Birds of the World.*
Dover Publications, New York. 1967. A timely discussion of birds that have been and those that are about to become extinct.
GRIFFIN, DON R., *Bird Migration.*
Doubleday & Co., Inc., Garden City, N.Y. 1964. A review of the work done in solving the riddle of migration, and what the scientists believe to be the likely solution.
LANYON, WESELY E., *Biology of the Birds.*
Natural History Press (distr., Doubleday & Co.), Garden City, New York. 1963. An inexpensive nontechnical book on the principles of bird biology.
REED, C. A., *North American Birds' Eggs.*
Dover Publications, Inc., New York. 1965. A fine book back in print after a long absence, with nomenclature revised; details and photographs on range, habit, and life history of 566 species' eggs.
WETMORE, ALEX, *Song and Garden Birds of North America.*
National Geographic Society, Washington, D.C. 1964. Portrayal in color of 327 species, all fully described.

CHAPTER THIRTY-EIGHT

THE MAMMALS

The rise of mammals

Compared to other animals, mammals are a recent form of life. Fossils indicate that the earliest mammals appeared during Jurassic times and probably lived during the period of *Archaeopteryx* and other early birds. The ancestors of mammals are believed to have been reptiles. However, fossil evidence of mammal-like reptiles is very scanty and is limited, largely, to jaws and teeth found in Jurassic rock.

Biologists believe that Mesozoic mammals of the Jurassic and Cretaceous periods were small and unimportant in the animal population of the earth. Reptiles were still the dominant vertebrates in both land and water environments. But even these small, early mammals were important in one respect. In possessing certain body characteristics, they had a biological destiny that would raise them to a position of supremacy among animals in a future age.

About seventy million years ago, the Mesozoic era came to a close and with it ended the Age of Reptiles. It was as though nature had closed one chapter of the book of life and opened another—one entitled "The Cenozoic Era, the Age of Mammals."

With the dawn of the Cenozoic era, continents were uplifted, and mountain ranges and high plateaus arose. Shallow seas and swamps disappeared, and the water collected in great oceans. With changes in the land came changes in climate. No longer was the earth uniformly warm and humid. Climatic zones evolved in which parts of the earth remained hot and humid, while other parts became temperate and frigid. Seasonal temperature changes developed in many parts of the earth. Most reptiles could not adjust to such changes and disappeared. However, the many varied land environments provided ideal surroundings for the sudden increase of mammals in a wide variety of forms (see Figure 33-2, page 514). Through the seven epochs of the Cenozoic era, mammals established their numbers and became the dominant land-dwelling vertebrates of the earth.

38-1 Cenozoic scene, showing
early mammals.

Extinct mammals

Through the early epochs of the Cenozoic era, mammals increased in size. The evolution of the horse is one of the best examples and is well known to science. An early ancestor of the modern horse lived in the woodlands of North America during the Eocene epoch. This horse *(Eohippus)* was four-toed and about the size of a fox terrier (Figure 38-2). During the Oligocene epoch that followed, the horse had become a three-toed mammal about the size of a sheep. The horses of the next two epochs, the Miocene and Pliocene, approached the size of a modern horse but were still three-toed. They were grazing animals that could run rapidly over the grasslands on which they lived. The modern one-toed horse *(Equus)* descended from the three-toed horse early in the Pleistocene epoch, or Ice Age. It is interesting to note that while the evolution of the horse occurred in North America the horse became extinct as a North American native mammal several thousands of years ago. However, early in the Pleistocene epoch, the horse migrated from North America to Europe, Asia, and Africa. It was returned to the New World by early Spanish explorers.

Other extinct mammals of earlier epochs include the wooly rhinoceros, an ancient camel lacking the hump of the modern camel, a straight-horned bison, and a large wild pig. Flesh-eating mammals, including great bear dogs, giant short-faced bears, and ferocious saber-toothed cats, destroyed large numbers of hoofed mammals. During the million years of the Ice Age, mastodons and mammoths migrated to Africa from North America, Europe, and Asia. These early elephants have long been extinct, but their descendants are still to be found in Asia and Africa.

Diversity among the mammals

Mammals, unlike birds, cannot change environments rapidly. Their movements on land are much more restricted. Changes in an environment over a long period of time are accompanied by changes in the mammals occupying an area. These changes are brought about through the mechanisms of natural variation and selection in terms of survival of the fittest. This basis for change in members of a population, known as *adaptive radiation,* was discussed in Chapter 13. The same mechanism has brought about changes in mammalian structure in another way. Mammals migrate from one area to another, where environmental conditions may be somewhat different. As mutations occur, they may provide a survival advantage that will be continued in offspring. This is a basis for the many structural differences that are present in mammals today. Thus, we find the twelve thousand or more species of mammals distributed over the earth in a wide variety of forms and sizes, from tiny mice and shrews to elephants and whales; creatures of the forests, grasslands, and oceans; plant-eaters, flesh-eaters, and insect-eaters.

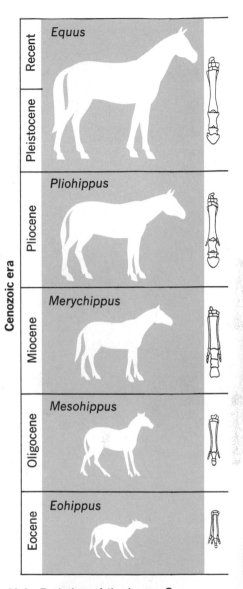

38-2 Evolution of the horse. Our knowledge of the history of horses is based on a rather complete fossil record. *Eohippus,* or "dawn horse," is believed to have developed from an ancestor with five toes on each foot.

wooly rhinoceros

wooly mammoth

Characteristics of mammals

While mammals differ widely in body structure, *with a few exceptions* they have certain characteristics in common. The following are the characteristics of the class *Mammalia:*

■ Body covered with hair.
■ Viviparous; young nourished during development in the uterus of the mother.
■ Young nourished after birth with milk secreted by the ***mammary glands*** of the mother, the characteristic for which the class is named.
■ Lung-breathing throughout life.
■ A ***diaphragm*** (breathing muscle) separating the thoracic (chest) cavity and abdominal cavity.
■ Heart four-chambered, with left aortic arch only.
■ Warm-blooded.
■ Seven cervical (neck) vertebrae in most species.
■ Two pairs of limbs for locomotion in most species.
■ Cerebrum and cerebellum of brain highly developed.

saber-toothed cat

Orders of mammals

The mammals of the world are classified in as many as eighteen orders. Fourteen of these orders, which include most of the mammals, are shown in Figure 38-9, page 602. As you look at examples of these orders, you may wonder what an elephant, a horse, a beaver, and a bat have in common. As mammals, they all have the characteristics we listed for the class. The many ways in which they are different establish the characteristics of the orders in which biologists place them. The structural differences that separate the orders of mammals show the extent to which these animals have become adapted to greatly different environments, diets, and ways of life over a period of more than 100 million years. Before starting a survey of mammalian orders, we shall consider some of the characteristics the biologists use in grouping them.

You will find three lines of mammalian development relating to *reproduction.* We can separate two orders of mammals from the remaining orders on this basis.

giant short-faced bear

straight-horned bison

38-3 Some mammals of the Plio-cene epoch that have become extinct as conditions changed.

38-4 The four types of mamma-lian teeth. Notice how these types are adapted for different functions in the rat, cat, and horse.

The *teeth* of mammals provide a basis for establishing several orders of mammals and, also, indicate feeding habits. The presence of specialized teeth distinguishes mammals, as a group, from reptiles. Reptiles have numerous pointed, conical teeth that are basically alike. Mammalian teeth are of a definite number, depending on the species, and are of several types, as shown in Figure 38-4. *Incisor* teeth serve for cutting or gnawing and are well-developed in certain mammalian orders. *Canine* (cuspid) teeth are enlarged in many mammals and are used for tearing flesh. *Premolar* (bicuspid) and *molar* teeth are used for grinding, especially plant foods.

Foot structure also varies among mammals, as shown in Figure 38-5. In one type of foot posture, the entire palm or sole rests on the ground. Man has such a foot, as does the bear. Many mammals, including the cat and dog, walk on the toes with the heel raised. The horse walks on one toe with the entire foot raised.

There are also variations in the horny outgrowths of the fingers and toes of mammals. In some cases, they are *nails,* as in man. Many mammals, including cats and dogs, have *claws.* The horse, cow, deer, and many other mammals have *hoofs.* These are, actually, modified nails. Two orders of mammals are distinguished on the basis of the number of toes and the structure of their hoofs.

38-5 Mammalian foot modification. Note the relationship of the calcaneous in each foot.

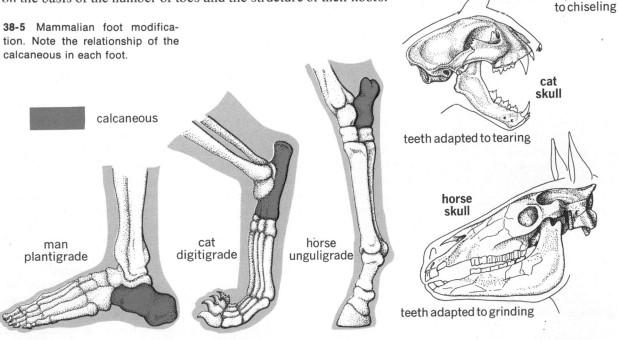

incisor (dog)

canine (dog)

molar (horse)

enamel

dentine

pulp cavity

premolar (human)

rat skull

teeth adapted to chiseling

cat skull

teeth adapted to tearing

horse skull

teeth adapted to grinding

calcaneous

man plantigrade

cat digitigrade

horse unguligrade

Three lines of mammalian development

Three different types of mammals have descended from a common ancestral form that lived early in the Jurassic period. In one direct line of development from this ancestor are the *egg-laying mammals,* or **monotremes,** which are represented today by only a few living species. Another branch took a different direction. This branch seems to have divided at the close of the Jurassic period into two major branches. One led to the *pouched mammals,* or **marsupials.** These mammals, while more numerous than monotremes, are of secondary importance in the mammal population today. The most important line of development led to the *placental mammals.* These mammals are the most highly evolved organisms on earth. The Cenozoic era was the Age of Mammals. The Recent epoch, in which we live, is truly the Age of Placental Mammals.

The survey of mammalian orders that follows will illustrate the relative importance of egg-laying, pouched, and placental mammals in the population of the earth today.

Egg-laying mammals—the monotremes

The most primitive of mammals compose the order *Monotremata* and are represented by the *duckbilled platypus* and the *spiny anteaters.* These curious animals are found in remote parts of Australia and New Guinea. Here, removed from larger predators that would have destroyed them ages ago, they have existed for many millions of years.

Like their reptile ancestors, the monotremes lay eggs. Other similarities between monotremes and reptiles include the presence of a cloaca in which both the intestine and urinogenital systems terminate and a lack of external ears.

The platypus is about twelve to eighteen inches long. It has waterproof fur, a horny, ducklike bill, and feet modified as paddles (Figure 38-6). Its home is a burrow along a stream bank, extending several feet into the bank and ending in a grass-filled nest. The platypus uses its beak for probing in the mud of stream bottoms in search of worms and grubs. The platypus usually lays two or three eggs. The eggs are retained in the body of the female for some time before they are laid in the nest. After the female lays them, she clutches them to her body and rolls herself into a ball to incubate them. After hatching, the young are nourished on milk secreted by sweat glands analogous to the mammary glands of other mammals. The platypus has no nipples; the young lick the milk from the body hair.

Spiny anteaters are found in the deep forests of Australia. They are well protected by a body covering of sharp spines. The elongated, toothless jaws form a long, tubular snout, well adapted for probing into ant hills. Spiny anteaters lay two eggs at a time. These eggs are incubated in a brood pouch on the lower side of

38-6 The duckbilled platypus, a monotreme. Perhaps one of the most striking similarities between the reptiles and the platypus is that the male platypus has a poison spur like a pit viper's fang on eacn of its hind legs. (Jerry Cooke)

the female. The eggs remain in the pouch several weeks before hatching.

Pouched mammals—the marsupials

Another order of primitive mammals, *Marsupialia,* is represented by a relatively small number of species, mostly in remote parts of the world. Marsupials differ from other mammals in the unique way in which they care for their young. The marsupial egg does not contain enough yolk to nourish the unborn young to full term, and the mother's body cannot supply nourishment directly during development. Hence, marsupials are born prematurely and in a helpless condition. Immediately after birth, the young, which in the case of the *opossum* are smaller than honeybees, find their way into the mother's pouch, in which the mammary glands are located. There, each attaches by mouth to a nipple through which milk is pumped from the mammary glands.

The opossum is the only marsupial found in North America. While it is a common mammal, you seldom see it, because it sleeps during the day. At night, it roams the countryside in search of food—small birds and mammals, eggs, and insects. Two or three times a year, the opossum bears a litter of from six to as many as twenty young. The hairless, helpless young are born after a period of development of only thirteen days. They remain in the mother's pouch for about two months. The young opossums adjust to life outside the pouch gradually, venturing out at times and returning at others.

During the early epochs of the Cenozoic era, marsupials were found in all parts of the world. Biologists believe that, at one time, nearly every kind of placental mammal of modern times had a marsupial counterpart. Some were bearlike, while others resembled dogs. One marsupial of the Pliocene epoch was a large flesh-eater that bore a striking resemblance to the saber-toothed cat. Other marsupials resembled modern moles, mice, and groundhogs. We referred to certain of these marsupials as examples of convergent evolution in Chapter 13.

During that same span of time, marsupials were direct competitors of placental mammals, and apparently held their own for many millions of years. However, they were no match for placental mammals and gradually lost in the struggle for survival. Today, they are found in abundance only in parts of the world that were never accessible to placental mammals. Australia, New Guinea and other nearby islands in the Pacific Ocean are the haven for marsupials today. In Australia, the *kangaroo* and a small species known as the *wallaby* live in the grasslands. They resemble the deer and antelope in head structure and grazing habits. The *koala bear* lives in the forests of Australia, where it feeds only on the leaves of the eucalyptus tree. Other modern marsupials of Australia include the rodentlike *wombat*

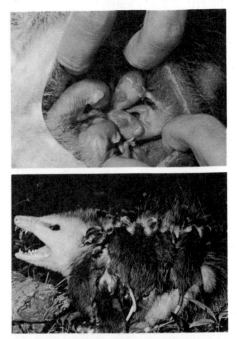

38-7 Opossums. After birth the helpless young find their way into the mother's pouch where they remain until they are mature enough to venture out. When they become too large to fit in the pouch, but are not yet large enough to leave her, they are carried about on her back. (Allan Roberts)

and the squirrel-like *phalanger*. On the island of Tasmania, south of Australia, two curious flesh-eating marsupials are found. The *Tasmanian devil* resembles a raccoon, while the *Tasmanian wolf* resembles a wild dog.

Placental mammals

Ninety-five percent of the mammals living today are *placental mammals*. These mammals differ from the monotremes and marsupials in how they are nourished prior to birth. Early in its development, the embryo of a placental mammal becomes embedded in the wall of the uterus. The *chorion,* one of the embryonic membranes, grows from the embryo and spreads over the surface of the uterine wall. The chorion and adjacent tissue of the uterus compose the **placenta**. Oxygen, nutrients, and other substances required by the embryo and carbon dioxide and other waste products from the embryo are exchanged through the placenta. The embryonic development of a placental mammal will be discussed more fully in Chapter 47.

The placental mammal has a distinct advantage over the marsupials. Because of a considerably longer period of development, the young are born in a more advanced condition. Even the most helpless newborn placental mammals—for example, mice, rabbits, kittens, and puppies—are far more advanced than newborn marsupials. Within a week or two, these mammals are ready to move about with their parents. Hoofed mammals, including the cow, horse, deer, sheep, and pig, are born in a still more advanced condition. These mammals are able to walk within a few hours after birth. The whale is even more advanced; a newborn whale is able to swim immediately after birth.

The period of uterine development is called the **gestation period**. The length of gestation, or *pregnancy,* varies in placental mammals. Generally, the longer the gestation period, the more advanced and larger the newborn mammal is at birth. The gestation periods of some of the more familiar mammals are shown in Table 38-1.

38-8 The lion is an example of a placental mammal. (Dr. E. R. Degginger)

Table 38-1 GESTATION PERIODS

MAMMAL	PERIOD	MAMMAL	PERIOD
mouse	21 days	man	40 weeks
rabbit	30 days	cow	41 weeks
cat	63 days	horse	48 weeks
dog	63 days	whale	20 months
pig	120 days	elephant	20 to 22 months

The insect-eating mammals

Members of the order *Insectivora,* represented by the shrews and moles, were probably the first placental mammals. All of today's higher forms of mammals probably evolved from this order. The insectivores never became very large. Their brains are small and their teeth very primitive. The *shrew* is the smallest of all mammals and resembles both a mole and a mouse. It is noted for its high metabolic rate, which gives it a ravenous appetite for insects, mice, and even other shrews. Shrews are seldom seen, because they run along tunnels in the grass and hide easily under leaves. Such secretive habits have probably been responsible for the survival of shrews.

The *mole* is well adapted for life under the ground. Its powerful limbs have long claws for digging, and its tiny eyes are covered with skin. Its long nose is adapted for rooting out grubs and worms in the soil. While the mole is valuable in destroying many harmful beetle grubs, it is a pest in lawns and golf courses because it digs up the turf.

The flying mammals

The *bats* are classified in a different order from the insectivores, even though most of them eat insects. The order *Chiroptera* includes more than six hundred species of bats. These are the only mammals that have developed structures for flight. They have greatly lengthened finger bones covered by membranes. These structures probably evolved during the Eocene age, and there has been little change in the bats to modern times. Bats fly mostly at night, when insects are abundant in the air. They spend the days hanging in caves or hollow trees. Most are nearly helpless on the ground, because their hind limbs are so poorly developed and their forelimbs are proportionally so long. Bats are most numerous in the tropics, but they occur in all temperate zones as well.

Vampire bats live in tropical America. They live on the blood of other mammals, including horses and cattle. The "blood letting" front teeth of vampire bats cut the skin and the bat laps up the oozing blood.

The gnawing mammals

Rats and *mice* are among the most common members of the order *Rodentia.* Others include the *squirrel, woodchuck, prairie dog, chipmunk,* and *gopher.* The *beaver* is the largest North American rodent. The *muskrat* is another fur-bearing rodent.

Rodents are doubtless the most successful group of mammals. They outnumber all other mammals combined, and are found in nearly every area of the world and in all climates. Most rodents are terrestrial, tree-living, or burrowing forms, but the *beaver* and *muskrat* live a semi-aquatic existence, and the *flying*

38-9 Examples of some of the better known orders of mammals.

Sirenia
(sea cow)

Proboscidea
(elephant)

Carnivora
(mountain lion)

Perissodactyla
(horse)

Artiodactyla
(bison)

Primates
(monkey)

squirrel, with its ability to glide easily from tree to tree, is well suited for its arboreal life.

Why have rodents been successful? Most of them are small, which allows them to live in environments not suitable to larger animals. They have a rapid rate of reproduction, which enables them to occupy new areas and adapt quickly to changing conditions. There is very little specialization in the body build. All have strong, chisel-shaped teeth. These teeth have sharp edges that become even sharper with use, because the front edge is harder than the back edge, causing the biting surface to wear at an angle. The forelimbs of rodents are adapted for running, climbing, and food-getting.

The rodentlike mammals

For many years, biologists considered the rabbits, hares, and pikas to be rodents because they have enlarged incisor teeth used for gnawing. These mammals, however, now placed in the order *Lagomorpha,* have four incisor teeth in each jaw rather than two, as in rodents. They grind plant foods with a sideways motion of the lower jaw.

Pikas (PEEK-uhz) are small, with short legs and short ears. The *hares* and *rabbits* have long hind legs that enable them to leap great distances. The forelimbs are shorter and absorb the shock of landing. The long ears give them an acute sense of hearing.

The cottontail rabbit is the most widely hunted mammal of the United States and supplies more flesh than any other wild mammal. But even though preyed upon by man, predatory birds, and other animals, these rabbits have been able to hold their own in most localities. The jack rabbit is common in the broad expanses of the western prairies and plains. It reaches a length of nearly thirty inches and has characteristic long ears and large, powerful hind legs. Both its sight and hearing are especially keen, enabling it to escape from its enemies.

The toothless mammals

Armadillos, sloths, and great anteaters belong to an order of mammals that are toothless, or nearly so. These members of the order *Edentata* have relatively small brains and large claws, and some have nine neck vertebrae instead of the seven usually found in mammals. The *armadillos,* because of their protective armor, have survived since the Tertiary period. Their diet consists of insects, carrion, bird eggs, grubs, worms, birds, and other small animals. Various species of armadillo live in our southwestern states, Mexico, and Central and South America. Our North American species is known as the nine-banded armadillo. It hides in a burrow during the day and spends its nights digging in the ground for insects. The young of the armadillo, having started life as a single fertilized egg, are identical quadruplets.

The *tree sloths* are odd creatures of the jungles of Central and South America. They spend most of their time hanging upside down from the branches of trees by their greatly elongated limbs and two or three hooked claws. They feed on the leaves of the trees in which they live. The hair of some species is colored green by the algae that live in it.

Anteaters probably evolved from the same ancestors as the tree sloths. The skull of the anteater is greatly elongated, ending in a long snout. Inside is a sticky tongue that can be extended to lap up the termites on which the animal lives. The greatly enlarged, curved claws on its front feet are well adapted for digging in a termite's nest. Locomotion, however, is difficult. The animal walks and runs primarily on the knuckles of its front feet.

The aquatic mammals

Members of the order *Cetacea*—the whales, including the dolphins and porpoises—apparently evolved from some land mammal during the Tertiary period. They are in many ways well adapted to life in the oceans. The torpedo-shaped body and fish-like tail and forelimbs provide for locomotion. These animals continue to use lungs for breathing, but they can take in great quantities of air and hold their breath for a long time. Thus, they can remain submerged for some time. The young are born with the ability to swim, and their mothers push them to the surface immediately after birth for their first breath of air. Unlike man, they are able to withstand sudden changes in water pressure and can dive to great depths.

The *blue whale* is the largest living animal and probably the largest that ever lived. Specimens may reach a length of 100 feet or more and weigh as much as 150 tons. The head of the *sperm whale* contains an enormous reservoir of oil, which is used commercially as a lubricant and as a base for cosmetic creams. The most valuable product of this whale is *ambergris,* a secretion of the intestine. Its principal use is in the manufacture of perfumes.

Dolphins are any of several small whales, usually under fourteen feet in length, found in the sea throughout the world. You probably have recently read accounts of the work researchers are doing in investigating the intelligence of the bottle-nosed dolphin and how these aquatic mammals communicate with one another. *Porpoises* are closely related to the dolphins, but are of a different family. Generally smaller than the dolphins, porpoises have a blunt, rounded head. Like the dolphins, porpoises are found in all the oceans of the world. The harbor porpoise may be especially familiar, because it is common along the coasts of the United States. Traveling in herds, these mammals often swim close to moving ships and delight the passengers with their graceful leaps. Their food consists primarily of fishes, which they catch with their tooth-lined jaws.

Another order, *Sirenia,* includes a mammal that lives in fresh water: the *sea cow,* or manatee. These sirenians probably became adapted for aquatic living during the Eocene epoch. Sea cows reach a weight of one ton. The large head resembles that of a walrus. The body is streamlined, and the hind limbs are lacking. The forelimbs of the sea cow are modified to form flippers. The tail is rounded and not notched like that of a whale. Sea cows have tough skins with sparse coverings of hair. Sea cows are found primarily in the rivers and warm, brackish waters where the rivers empty into the sea along the Atlantic coast of South America. Sirenians spend most of their time feeding on aquatic plants.

The trunk-nosed mammals

Only two species of elephants remain as representatives of the order of trunk-nosed mammals, or *Proboscidea.* During glacial and preglacial times, at least thirty species of elephant-like mammals lived in Asia, Europe, Africa, and North America. Biologists have not been able to explain why most of these elephants became extinct.

Elephants are the largest of all living land-dwellers. They may reach a weight of seven tons or more. The *Asiatic elephant* is the familiar performing elephant of the circus. In many parts of the world, it is used as a beast of burden. The *African elephant* is a taller, more slender animal with sloping forehead and enormous ears. African elephants travel in herds in the deepest parts of Africa, and are not as easily domesticated as their Indian cousins.

The flesh-eating mammals

The flesh-eating mammals, which are in the order *Carnivora,* evolved in Tertiary times, probably from the insectivores. All of the **carnivores** have strong jaws with enlarged canine teeth for piercing tough skin and other teeth for crushing bone. They have a high level of intelligence, which may enable them to outwit their prey. They have a well-developed sense of smell, and the sense of sight is usually keen. Powerful bodies and limbs and the presence of claws contribute to their ability to overcome and devour other animals.

The dog and cat families of this order are not as well represented in America as in other lands. The *mountain lion* was found over most of North America at one time, but civilization has driven it to the remote regions of the Southwest. The chief harm it does lies in its destruction of livestock, especially young horses. Both the *bay lynx* (links), or bobcat, and the *Canadian lynx* live in deep forests and are seldom seen. The *jaguars* of

South America, the *lions* of Africa, and the *tigers* of Asia are all well-known members of the cat family.

The *gray wolf,* or timber wolf, is most frequently found in the northern forests and may be dangerous during the winter, when it runs in packs. The *coyote* (KY-OTE), a prairie wolf, has been more successful than its larger cousin in surviving the effects of civilization. It is still abundant on the western plains. However, in some regions, too many have been destroyed, and their natural prey, including rodents and jack rabbits, have become pests. It has therefore been necessary to import coyotes into these regions.

The *red fox* ranges over much of the United States. It is sly and a very fast runner for short distances. The *gray fox,* somewhat larger than the red fox, lives in the warmer regions of the South.

The *raccoon,* with its black mask and long, ringed tail, is a favorite of many people because if it is captured young, it makes a nice pet. The raccoon prefers fish and clams as its food but will eat other things as well if these are unavailable. It has a habit of "washing" its food before eating, probably to moisten the food rather than to clean it.

The weasel family includes some of the most bloodthirsty carnivores and some of the most valuable fur-bearing mammals. The *mink* especially is prized for its fur. These long-bodied, short-legged animals live along streams. The *ermine* is an Arctic weasel that grows a coat of white fur (except for a black-tipped tail) in winter but is brown in summer. The largest and most destructive member of the family is the *wolverine* of the northern forests. Its body, including its bushy tail, is about 36 inches long and weighs between 30 and 35 pounds.

The *bears* were the last carnivores to evolve, but they have changed little since Pleistocene times. Bears have teeth that are specialized for eating plant substances as well as meat. The molar teeth are elongated, and the enamel of the crown is wrinkled. Bears are widely distributed, but are not found in Africa or Australia.

Sea lions, walruses, and *seals* are water-living carnivores. Their bodies are streamlined for swimming, but they have never developed the finlike appendages of whales. They have webbed feet, the front ones serving for balance and stability and the rear ones turned in such a way as to provide an efficient means of propulsion. Walruses probably evolved from sea-lion ancestors. Their canine teeth have become long tusks, and they have developed broad molars to facilitate crushing and grinding the oysters and other mollusks on which they feed.

The hoofed mammals

Man has lived in close association with the hoofed mammals since prehistoric times. The *goat* was probably the first **ungulate** to be domesticated. For ages, man has depended on the horse, camel, ox, llama, and other hoofed mammals as beasts of burden. The *cow, pig,* and *sheep* are our principal food animals. *Deer, elk, caribou, moose,* and *antelope* are our most important big-game ungulates.

The teeth of these herbivores, or plant-eaters, are well adapted to cropping and grinding grasses, leaves, and stems. Some of the hoofed mammals have elongated limbs and feet that enable them to cover hard ground rapidly. They walk on the tips of their toes, with the wrist and ankle off the ground. Hoofs, which are modified toenails, help to absorb the shock of running.

About sixteen families of ungulates evolved, probably from the insectivores, in the Cenozoic era. Using the number of toes as a basis, biologists classify the hoofed mammals into two orders. Ungulates with an odd number of toes are placed in the order *Perissodactyla,* while those with an even number of toes are placed in the order *Artiodactyla.* Of the ungulates with an odd number of toes, only the *horses* continue to flourish; the *tapir* and *rhinoceros,* also in this group, are not nearly as plentiful. In contrast, the ungulates with an even number of toes have become very numerous and are probably at their peak today. These include the *pig, hippopotamus,* and the cud-chewers, or **ruminants** (ROO-muh-nants). The *cow, bison, sheep, goat, antelope, camel, llama, giraffe, deer, elk, caribou,* and *moose* are all ruminants. These animals have four-chambered stomachs. While grazing, they eat large quantities of food, which passes into a large paunch, or **rumen,** the first of the stomach divisions. Here food is stored for later chewing. Later, the food is forced back into the mouth for leisurely chewing as a cud. After thorough chewing, the cud is swallowed into the second stomach, where digestion begins. From there, it passes through the other stomach regions to the intestine.

The erect mammals

Superior brain development places the order *Primates,* or erect mammals, at the top of all groups of living organisms. Primates have well-developed arms and hands. Their fingers are used for grasping, and one or more fingers or toes are equipped with nails. Most primates can walk erect if necessary. Primates live in South America, Africa, and the warm regions of Asia. They tend to live in trees; only man lives exclusively on the ground. The teeth of the primates are less specialized than those in any other group of mammals. Primates feed on plants and flesh. Eyesight is well developed in the primates, but the sense of smell is poorly developed.

The apes are most like man in body structure. They have no tails. The arms are longer than the legs. When these animals walk, their feet tend to turn in. These primates, especially the chimpanzee, respond to a very high degree of training. Among the primates, we find the following animals and groupings of animals:

■ The *gorilla,* the largest of the apes, lives in Africa. It walks on two feet and is one of the most powerful animals.

■ The *chimpanzee* also lives in Africa. It is smaller than the gorilla and quite intelligent.

■ The *orangutan* lives in the East Indies. It is a droll animal with red hair.

■ The *gibbon* is a long-armed type found in Asia.

■ The *Old World monkeys* have long tails but do not use them in climbing. They sit erect. Food is stored in cheek pouches. The baboon, an Old World monkey, has a long, doglike nose.

■ *New World monkeys* have flat, long tails used for grasping. The septum between the nostrils is wide, and these monkeys lack cheek pouches.

■ *Marmosets* are small primates ranging from Central America to South America. They resemble squirrels in appearance and activity.

■ *Lemurs* are primates found in Madagascar.

Structural advantages of the mammals

In your survey of mammalian orders, you have undoubtedly noticed the diverse lines along which mammals have developed. Many adaptations have occurred in the limbs, especially the forelimbs. We found limbs for digging, hanging, flying, running, defense, and capturing prey. These limb modifications have equipped mammals for life in nearly all the environments of the earth. Their numbers compete successfully with the fish in aquatic environments, with amphibians and reptiles on land, and with the birds in the treetops.

Variations in tooth structure are nearly as pronounced as those in limbs. Well-developed incisors serve as chisels in the mouths of rodents. Greatly enlarged canine teeth form the flesh-tearing fangs of cats and dogs and other carnivores. Large molars provide the grinding surfaces for grazing animals like the deer, the cow, and the horse. These modifications and others, including antlers and horns, and protective coloration provided by pigmented hair that blends with the environment, are of great importance in the distribution of mammals. However, to find the reason for the supremacy of mammals, we need to take a closer look at the mammal's internal development.

Regulations of the internal environment

In the body of the mammal, we find tissue specialization and organ development at the most complex level in all of life. With this specialization, however, comes increasing cell interdependence. Mammals could never have reached their high level of development without a controlled internal environment.

We refer to mammals as well as birds as warm-blooded verte-brates. Except during periods of hibernation or inactivity, the body temperature remains at a nearly constant level. The main-tenance of this uniform internal temperature requires the com-bined activity of many highly developed organs and organ systems. We might begin with the release of heat during respira-tion in the tissues. Here we find the rate of cell metabolism elevated far above that of the cold-blooded vertebrates, whose body temperature fluctuates with that of the environment. In-crease in the metabolic rate requires an increase in the supply of nutrients and oxygen to the cells. Highly efficient digestive organs and well-developed lungs are necessary to supply these needs. The increased rate of metabolism results in more meta-bolic wastes, which must be removed from the cells constantly. Supply and waste removal for the billions of cells composing the body of a mammal require an extensive transport system and a powerful heart to force blood through its many miles of vessels. Kidneys that surpass those of all other animals in func-tional efficiency remove the waste products of cell metabolism in a complicated system of blood filters.

We find the basis for mammalian structure and function in the lower vertebrates you have studied. In fact, we can trace the development of each organ system through stages repre-sented in these lower forms.

Development of heart and lungs of a mammal

In many ways the efficiency of the transport system is the key to the development of warm-blooded animals and regula-tion of the internal environment. However, in considering the heart and circulatory system of the mammal we must also include the organs of respiration, for these two systems are closely related. Let us think back to the fish. Here a relatively small heart composed of two chambers is sufficient to force blood through a circulatory system consisting of a single circuit. One ventricle provides sufficient pressure to force blood through the gills and to the vessels of the body without returning to the heart between the two circulations. Gills are relatively simple organs of respiration that function efficiently in a water environment. Since the metabolic rate of a fish is much below that of the higher vertebrates and varies with the water temperature, a two-cham-bered heart is adequate. However, neither gills nor two-cham-bered hearts are suitable for life on land.

In the amphibians, especially in the adult stages of frogs, toads, and salamanders, the heart has undergone a structural change that permits life in a land environment. Here we find a separa-tion of the upper heart chambers into two atria. The presence of these two chambers provides separation of deoxygenated blood from the body from oxygenated blood from the lungs. However,

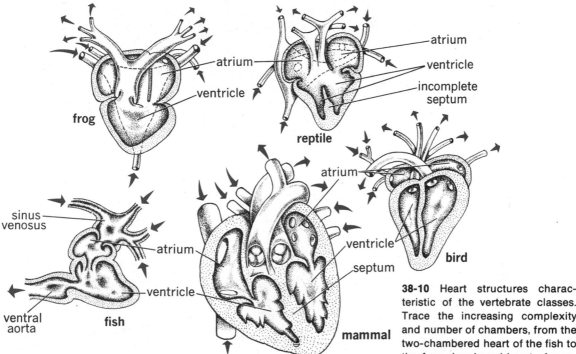

38-10 Heart structures characteristic of the vertebrate classes. Trace the increasing complexity and number of chambers, from the two-chambered heart of the fish to the four-chambered heart of man.

mixing occurs in the single ventricle, which forces blood both to the body tissues and to the respiratory organs. Such a heart could not supply the requirements of a bird or a mammal. For one thing, the pumping of mixed blood to the body tissue could not supply the oxygen requirement of the mammal. Furthermore, the lungs of amphibians are not adequate to supply the needs of a bird or a mammal. Remember that amphibians still require skin respiration. This function of the skin has been lost in the bird and mammal through increase in thickness and development of the body covering.

The reptile heart is still more efficient than that of the amphibian, because the ventricle is partially divided by a partition, or septum. The mammalian heart, like that of a bird, eliminates all mixing of blood because of a complete septum forming two separate ventricles. Thus, it is really a double pump. The right side receives the deoxygenated blood from the body and pumps it to the lungs, and the left side receives oxygenated blood from the lungs and forces it throughout the entire body.

Development of the nervous system

It is primarily the high degree of development and specialization of the nervous system that has placed the mammals above all other forms of life. The brain of the mammal is larger in proportion to the body weight than that of any other animal. This high development of the mammalian nervous system could have occurred only in a closely regulated internal environment. Of all the body tissues, nerve tissue is probably the most dependent and the most rapidly affected by internal changes.

We can learn a great deal about the behavior and mental abilities of the vertebrates by comparing their brain structures. The brains of the most primitive vertebrates consist of three main regions: a *forebrain* composed of the olfactory lobes, or bulbs, and the two hemispheres of the cerebrum; the *midbrain* containing the optic lobes; and the *hindbrain* composed of the cerebellum and medulla. All vertebrates have these brain regions, but their relative sizes vary (Figure 38-11). When we compare the brain structures of the various vertebrates, we see an evolutionary pattern in which the relative size of the cerebrum increases

38-11 Compare the size of the cerebrum in the mammalian brain with that of the brains characteristic of the other vertebrate classes.

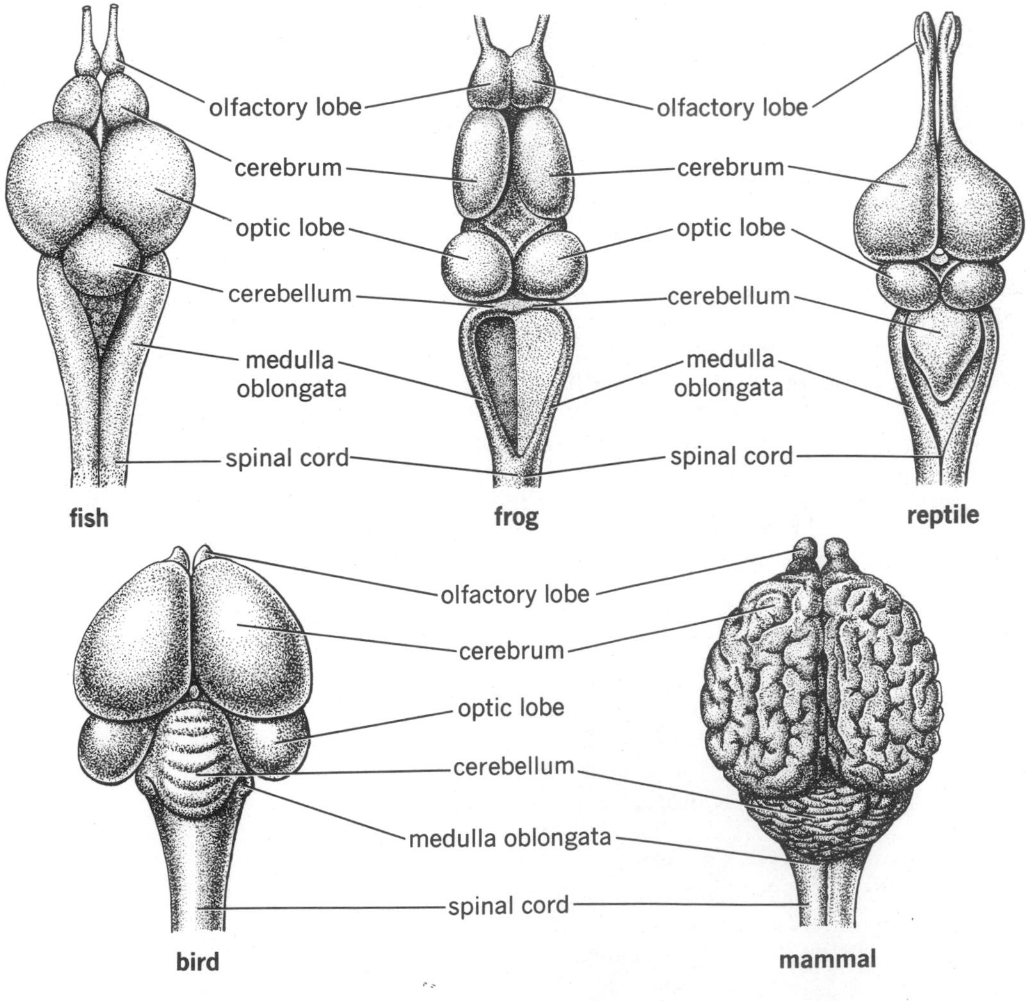

fish frog reptile

bird mammal

and reaches its highest development in the mammal. You will recall from your studies of the brain of the fish that the cerebral hemispheres are relatively small. Large olfactory lobes extend from the anterior region of the cerebrum, indicating that the sense of smell is well developed in the fish. The largest brain regions in the fish are the optic lobes. In the amphibian, the cerebral hemispheres are proportionally larger. This increase in the relative size of the cerebrum continues in the reptiles and also in the birds.

It is interesting that various vertebrates differ also in the relative sizes of the cerebellum. This brain area functions in muscle coordination. The cerebellum is relatively large in most fish but is greatly reduced in amphibians. In the birds and mammals, it is large and well developed.

In the mammalian brain, the cerebral hemisphere fills most of the cranial cavity and is spread over the other brain regions. The olfactory area consists of two lobes extending from the anterior regions on the lower side of the cerebrum. The optic region lies in the posterior region of the cerebral hemisphere. The cerebellum lies below and posterior to the cerebrum. The high development of this brain region accounts for the excellent muscle coordination of the mammal. The short medulla oblongata, which controls vital body processes, lies below the cerebrum and extends to the spinal cord. Cranial nerves extend from the brain to the sense organs and other head structures. The spinal nerves extend from the spinal cord to and from all body regions.

Vertebrate reproduction and parental care

Another line of development that has occurred in the vertebrates has been an increase in the efficiency of reproduction and a corresponding decrease in the number of young. Let us return to a consideration of the fish, where species are preserved only because of the enormous number of eggs produced. Depending on the species, a fish may lay from a few hundred to several million eggs. You will recall that fertilization is external. The male fish swims over the eggs and discharges milt containing sperms. Many eggs are never fertilized. Many are eaten before they hatch. In most fishes, the young are given little or no parental care after they are hatched. As a result, most of the young fall victim to predators before reaching maturity.

Amphibians lay smaller numbers of eggs, which are fertilized directly as they are laid. This insures the development of a much greater proportion of the eggs laid. As in the fish, however, amphibians must deposit eggs in water, since they have no protection against drying out.

An abrupt change occurs in the reptiles, the first vertebrates to lay eggs on land. This advance requires internal fertilization and the enclosure of the egg with protective shells. Similar eggs are produced by birds. With the greater protection supplied by the eggs of reptiles and birds, the number of eggs produced is greatly reduced. Parental care, especially among the birds, reduces the mortality rate of offspring greatly.

In all mammals but the monotremes, development is internal. As you learned in your study of marsupials, their young are born prematurely. However, the pouch of the female provides a protective "incubator" where development can continue until the young reach a more advanced condition.

Vertebrate reproduction reaches its highest level in the placental mammal, where uterine attachment permits a longer period of development. The birth of more fully developed young greatly increases the chances of survival.

The dependence of a young mammal on its mother's milk requires a period of *parental care*. In most mammals, this parental care is provided by the female. Parental care requires a reduction in the number of offspring. This number is usually proportionate to the length of the period of parental care. Mice, rabbits, and other small mammals mature rapidly. Several litters, numbering ten or more young, are born each season. Black bears give birth to one or two cubs in mid-winter. When the cubs are born, they are nearly hairless, have closed eyes, and weigh less than half a pound. By the time spring rouses the mother, the

38-12 These bear cubs are cared for by their mother until they are old enough to care for themselves. (J. R. Simmon, Photo Researchers)

cubs have grown sufficiently to roam with the mother and learn to find food and to climb trees when the mother warns them of danger. The horse, cow, deer, elk, moose, and other hoofed mammals usually bear one or two young and care for them for six months or more.

Of all the mammals, the human being is most helpless at birth and requires the greatest parental care for the longest period of time. The period between birth and reproductive maturity, averaging about twelve years for females and fourteen years for males, is also the longest of any mammal. Much of the period of parental care is devoted to training, in which the child learns to walk, develop skill in using its hands, talk, think, and reason.

SUMMARY

Would you consider the mammals the most successful forms of animal life? Wouldn't that depend on what is meant by "successful"? If success is based on the number of species and individuals, we would have to place the Arthropods first. We speak of our age as the "Age of Mammals," yet mammals have never conquered the air and offer little competition for fish in the oceans. We would be correct in saying that mammals are the most successful land-dwelling vertebrates. We can say, also, that the organs and systems of a mammal are the most highly developed of any form of life.

What are some of the reasons for the success of mammals as land animals? One is the maintaining of a constant body temperature, or warm-bloodedness. This characteristic, together with protective fur or hair, has enabled mammals to occupy environments from the tropics to the polar regions. Placental mammals bear smaller numbers of young in a more advanced stage of development than other animals. They provide nourishment as milk and, in most cases, care for their young for some time after birth. This reduces mortality of the offspring greatly. Perhaps the greatest reason for success of the mammals is the high development of the brain and sense organs. This has given the mammal a great advantage in behavioral adjustments to environmental conditions.

BIOLOGICALLY SPEAKING

adaptive radiation	placental mammal	ruminant
mammary gland	placenta	rumen
diaphragm	gestation period	forebrain
monotreme	carnivore	midbrain
marsupial	ungulate	hindbrain

QUESTIONS FOR REVIEW

1. List several extinct mammals.
2. Describe several stages in the evolution of the horse during the Cenozoic era.
3. List ten characteristics of mammals that distinguish them as a group from other vertebrates.
4. List several characteristics that are used as a basis for classifying mammals into orders.

5. What evidence of reptile ancestry is shown in the monotremes?
6. Explain the importance of the pouch in a marsupial.
7. Which marsupial is believed to be the most ancient North American mammal?
8. In what way is the length of the gestation period generally related to the size and degree of development of a mammal at birth?
9. Why is it appropriate to list the order Insectivora first among the orders of placental mammals?
10. In what way are members of the order Chiroptera unique among mammals?
11. Rodents vary greatly in size. What characteristic do they have in common?
12. How do the rodentlike mammals of the order Lagomorpha differ from their relatives the rodents?
13. List several members of the order Edentata and describe their widely varying habitats and diets.
14. Describe adaptations of cetaceans and sirenians for aquatic life.
15. Name two surviving species of elephants.
16. List several North American carnivores belonging to different families.
17. On what basis are North American ungulates separated into two orders?
18. Distinguish between a ruminant and a nonruminant.
19. On what basis might we consider primates the most highly developed mammals?
20. List several structural adaptations of mammals that make them well suited for life in diverse environments.
21. Which class of vertebrates is most like the mammal in heart structure?
22. Which brain region is most highly developed in a mammal?

APPLYING PRINCIPLES AND CONCEPTS

1. Discuss several environmental changes during the Cenozoic era that probably favored the survival of mammals and caused the decline of reptiles.
2. Explain why mammals have been more subject to adaptive radiation than birds and many other vertebrates.
3. Discuss three lines of mammalian development and account for the supremacy of placental mammals today.
4. Give possible evidence to support the theory that cetaceans and sirenians "went to sea" ages ago.
5. Discuss the importance of regulation of the internal environment to the biological success of mammals.
6. Compare the reproductive processes, relative number of offspring, and parental care in the mammal with those in other vertebrates.

RELATED READING

Books

BOORER, MICHEAL, *Mammals of the World*.
  Grossett and Dunlap Publications, New York. 1967. Presents a concise survey of evolution, overall structure and classification of mammals. Mammals are then dealt with in a systematic sequence.
BOURLIERRE, FRANCOIS, *The Natural History of Mammals* (3rd ed.).
  Alfred A. Knopf, Inc., New York. 1964. A comprehensive picture of what science has discovered about the life and the ancestors of the mammals.

CARRINGTON, RICHARD, and the Editors of *Life* Magazine, *The Mammals.*
Time-Life Books (Time, Inc.), New York. 1963. Tells in words and color pictures how mammals solve the complex problems of life.

COLE, SONIA, and M. HOWARD, *Animal Ancestors.*
E. P. Dutton and Co., Inc., New York. 1964. The family trees of familiar animals, traced back to earliest ancestors.

EIMERL, SAREL, and others, *The Primates.*
Time-Life Books (Time, Inc.), New York. 1965. Emphasizes the social behavior of the primates, as an important aspect of this select group of vertebrates.

RIEDMAN, SARAH R., and ELTON T. GUSTAFSON, *Home Is the Sea for Whales.*
Rand McNally and Co., Chicago. 1966. An excellent scientifically accurate study of Cetaceans; traces the evolution of the whales.

SANDERSON, IVAN T., *The Monkey Kingdom: An Introduction to the Primates.*
Chilton Book Co., Philadelphia. 1963. Contains an up-to-date, detailed list of the known living primates, with their scientific names and a comprehensive account of the various aspects of the species.

VAN GELDER, RICHARD G., *Biology of Mammals.*
Charles Scribner's Sons, New York. 1969. A clear description of how mammals are born, develop, how they find shelter, food, water, and how they defend and protect themselves.

Article

Life Education Reprint, No. 26, "The Private Life of the Primates" (originally published in *Life,* March 8, 1963). Examines the natural behavior of the suborders of primates, which may give evolutionary clues to the origin of man's own society.

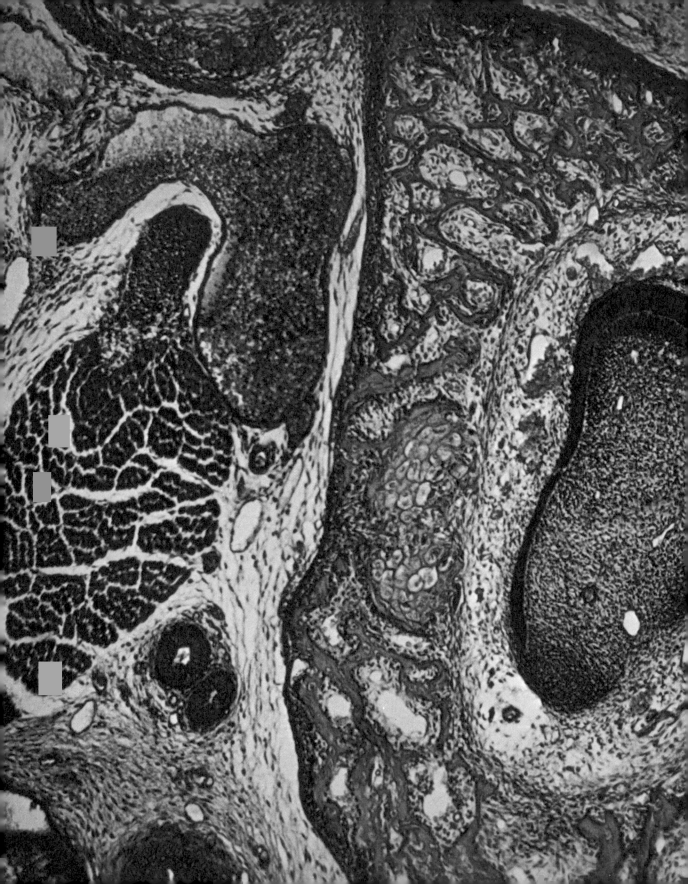

UNIT 7

THE BIOLOGY OF MAN

Unlike many other vertebrate animals, man is not structurally adapted for skill in climbing, running, flying, and swimming. Although he does possess these skills, they are not as well developed as in certain other animals. However, with the application of man's intelligence to direct him, combined with his upright posture and his ability to manipulate tools, he has achieved wonders. Man has devised spoken and written languages, found a means of controlling fire, built submarine vessels to carry him to great ocean depths, developed aircraft capable of carrying hundreds of passengers at great speeds. He has designed spacecraft to carry him to the moon in order to satisfy his curiosity and test his hypotheses about this marvelous universe.

Man's highly developed brain is the organ that has made him more advanced than any other living organism. The human brain is larger in proportion to body size than the brain of any other mammal and is more highly specialized. This brain has enabled man to think out problems and reach satisfactory solutions, to acquire skills by constant practice, and to form habits more easily. Man has even extended the range of his brain by building computers with memory and speed facilities exceeding his own powers. It is no surprise, then, that we should wonder and hypothesize about our own history on earth.

CHAPTER THIRTY-NINE

THE HISTORY OF MAN

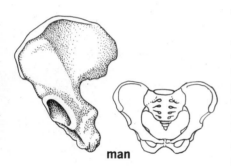

man

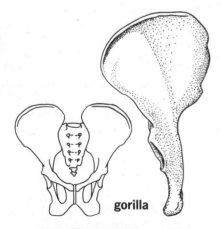

gorilla

39-1 The broad blade of the human pelvis forms support for the weight of the internal organs. Its posterior projection serves as attachment for the large muscles used in walking. Compare the human pelvis to that of the gorilla.

A search for clues

Quite naturally man's intellect and curiosity cause him to wonder about his own history. We are aware of some changes that have occurred in the earth and its inhabitants. We wonder what life was like for early man. *Anthropology* is the study of man and the societies in which he groups himself, and one who specializes in this field is an *anthropologist*. As anthropologists search the past, they are faced with the problem of what distinguishes a true man from a man-like ape. Characteristics used for this distinction are: upright posture, use of tools, possession of a brain distinctly larger than that of an ape and teeth like those of man as we know him today.

The ability to walk erect is indicated by bone structure of the pelvis. As shown in Figure 39-1, the blade of the pelvis of man is broad to support the internal organs, and it has a large flange which projects to the rear. This flange strengthens the pelvis and serves as an attachment for the large muscles used in walking.

A study of the environment of primitive man also helps to piece together man's development. What did early man eat? Wherever the anthropologist finds evidences of primitive man, he searches for fossil remains of animals that might have been used for food. Charred bones indicate that man had learned to use fire to cook meat. The type of food indicates the climate and nature of the surroundings. The anthropologist is also interested in evidence of tools, since their use is unique to man. If similar tools are found over a wide area, some type of communication must have occurred. The material used and the complexity of the tools are indications of how highly developed the men were.

The use of tools is enhanced by another physical characteristic unique to man. The action of man's thumb opposes that of the fingers, so his hand is ideally suited for grasping. Naturally, his upright posture frees his hands, so they can be used

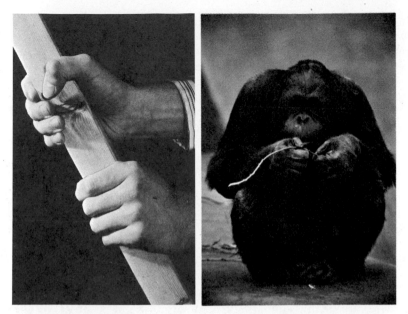

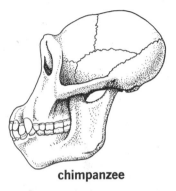

39-2 The man's hands are well adapted for a firm grip, but are more suitable for manipulation of tools than those of the ape. (John King; Kenneth Fink, National Audubon Society)

chimpanzee

man

39-3 Comparison of the skulls of the chimpanzee and man.

entirely for manipulation. Because of their many sensory receptors, the hands are highly sensitive to touch and are grooved so that they can grip smooth objects.

The development and use of tools by man is also aided by the structure and location of his eyes. Both man's eyes see the same vision from slightly different angles, resulting in a sense of depth.

The brain size of an organism can be determined from surprisingly small fragments of a skull. Reconstruction of fossils, as well as comparison of skulls of modern man to that of the ape, reveal important differences. The front part of the brain is highly developed in man. As you will see in a later chapter, this is associated with the many thought processes which make man unique. The ape, unlike man, has a much smaller brain

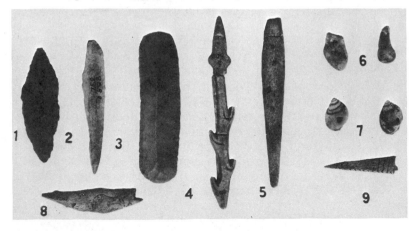

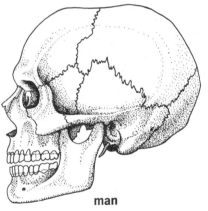

39-4 Some examples of the tools used by early man. Can you determine how each tool was used? (American Museum of Natural History)

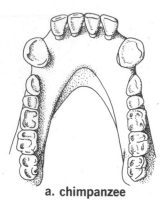

a. chimpanzee

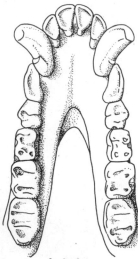

b. baboon

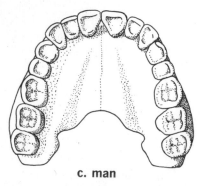

c. man

39-5 The difference in the teeth of the chimpanzee, baboon, and man is the size of the canine teeth.

capacity, hardly any frontal bulge, and a heavier bone structure. The proportion of the face to the brain is also much larger in the ape than in man. These observations serve as clues to the anthropologist when he is attempting to characterize a skull.

The characteristic that has probably made the most difference in man's development is his capacity to use symbols in the form of the written and spoken word. This ability to communicate with others of the same species by symbols and to pass learning on from generation to generation has allowed man to undergo a rapid cultural evolution not possible in other species.

The types of teeth and their arrangement in the bone are useful in characterizing many fossils. A careful comparison of the living forms aids the anthropologist in classifying the fossils. Notice that, although the numbers of teeth are the same (Figure 39-5), the shapes of the jaws, the teeth, and the spacing are very different.

Man and mammals

Although man is distinctly different from all other mammals, there are many similarities. The organs and systems of the human body closely resemble those of other mammals. The structural similarity of man and the other primates is especially striking. Almost every anatomical detail is similar. You can observe this similarity in the skeleton and muscles, tooth structure, position and structure of the eyes, form of the hands and feet, and even in the facial expressions. Similarities in the internal structures of man and the other primates are equally striking. They can be seen in the structure of the heart and blood vessels, lungs, digestive organs, excretory organs, glands, and nearly all other internal organs.

Even the chemical secretions of man and other mammals are similar, and the digestive enzymes are the same. The insulin used to save the lives of diabetics is extracted from beef or hog pancreas. The Rh factor, so important in matching human blood types, was discovered originally in the blood of the rhesus monkey.

Thoughts about man's development

On December 27, 1831, the ship *HMS Beagle* left Plymouth, England on a voyage to study the plants, animals, geography, geology, and cultures of the world. Charles Darwin, a young man of twenty-two years was on board as a naturalist. The five year voyage carried Darwin to Brazil, Argentina, Tierra del Fuego, Chili, Peru, the Galápagos Islands, Tahiti, New Zealand, and Australia. During this time, Darwin collected and classified many interesting organisms and sent specimens back to England for further study. In 1871, Charles Darwin published his famous book entitled *The Descent of Man*. Darwin proposed that the same forces operating to bring about changes in plants and animals could also affect man and his development. Biologists classify man as a primate because of his many structural similarities to the monkey, gorilla, chimpanzee, orangutan, and gibbon. In his book, Darwin pointed out these similarities, and some people interpreted his comparisons as suggesting that man evolved from monkeys. Actually, this was an unfortunate misinterpretation that was far from Darwin's intent. As you know, biologists believe that the less specialized primitive forms can most easily move into new environments and evolve into new species. The orangutans and chimpanzees of today are highly specialized forms that may have evolved from more primitive ancestors. Likewise, it is hypothesized that modern man has probably evolved from primitive, more generalized ancestors.

If we were able to trace the history of the primates back perhaps ten million years, we might find a generalized primate that was the common ancestor of both modern man and the modern primates. It is believed, however, that the two lines of descent separated at a very early date and might have gradually evolved into the forms of today.

At the time Darwin lived, no remains of early man had been found. Since then, hundreds of bones have been unearthed. They have been dated and studied by anthropologists, who specialize in the history of man. This evidence still does not provide us with a complete picture of man's development, but the fossils serve as clues from which hypotheses can be formed. A fossil skull indicates the size of the brain, shape of the head, and age of the man at the time of death. The jaw structure indicates whether its owner had the capacity for speech. The presence of the capacity, however, does not necessarily mean that a language was used.

39-6 This watercolor by George Richmond shows Charles Darwin in 1840 after his return to England from the cruise of the *Beagle*. (Derrick Witty, George Rainbird Ltd., courtesy of the Royal College of Surgeons)

Dating fossils

In Chapter 13, we discussed the dating of fossils by the determination of the relative ages of the layers of the earth. The *radiocarbon method* is also used to date fossils. This method was made possible by the discovery in 1930 of a radioactive isotope of carbon with an atomic weight of 14 instead of 12. In living organisms, there is about one carbon-14 atom to one trillion carbon-12 atoms.

Radioactive isotopes spontaneously give off radiations at a constant rate, changing eventually to the stable form of the element. The time required for half of a sample of a radioactive isotope to decay to the stable form is called the *half-life*. The half-life of carbon-14 is 5,568 years.

So long as an organism is living, it incorporates carbon compounds into its body, but at death this process ceases. Thus, if a sample of bone is examined and found to contain half the amount of carbon-14 that occurs in organisms today, its age would be estimated at 5,568 years. The radiocarbon method is considered to be accurate when used in dating materials that are as much as 50,000 years old.

The potassium-argon clock

Recently, another method of dating has been developed. Similar to the carbon-14 technique, the *potassium-argon method* also depends on a slow but constant change in certain atoms. Potassium-40 breaks down into calcium-40 and argon-40 at a very slow but constant rate. Although this method of dating is quite accurate and will enable us to date materials that are millions of years old, it is not without shortcomings. The method involves the measurement of argon-40 present in a rock sample. Therefore, when the potassium-argon method is to be employed in dating fossils, it is necessary to find a rock that was formed at about the same time at which the organism was alive. The amount of argon-40 present in the rock will, then, determine its age. You will soon see how this dating procedure has provided us with valuable information about primitive man.

Putting together the pieces

In 1924, a perceptive worker in a South African quarry noticed a small fossilized skull that had been dislodged by blasting. The fossil was given to a South African medical school, where Professor Raymond Dart, an anatomist, recognized its resemblance to human skulls. Dr. Dart named the primate *Australopithecus* (aw-STRAY-lope-ith-EE-kuss) *africanus,* which means "southern ape from Africa." Since then, other bones of this primate have been found.

Dr. and Mrs. L. S. B. Leakey have spent nearly 35 years excavating and studying fossils. Much of their effort has centered in an area of the Olduvai Gorge in Tanganyika. They first found many pebble tools thought to have been used for cutting instruments. Then, in 1959, the Leakeys unearthed some teeth; this discovery led them to excavate an area believed to be a campsite of ancient man. Since then, skull fragments have been found here and pieced together. Bones from a foot, fingers, and a lower jaw with well-preserved teeth have provided us with clues about the structure of the primate that was named *Zinjanthropus* (zin-JAN-thrup-uss). By using the potassium-argon method, scientists at the University of California have placed its age at 1,750,000 years.

The bones of *Zinjanthropus* and *Australopithecus* are similar and look like those of modern apes. The pointed, fanglike teeth of modern apes are not found in these forms, however. Furthermore, the shape of the pelvis and opening for the spinal

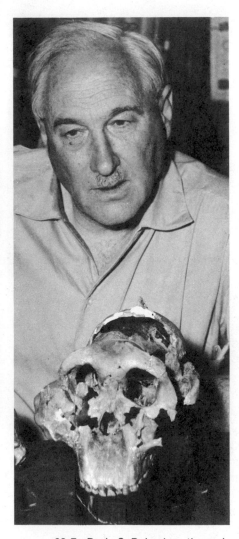

39-7 Dr. L. S. B. Leakey, the eminent British anthropologist who unearthed the remains of the earth's earliest known manlike creature in East Africa. The skull of *Zinjanthropus* is believed to be 1,750,000 years old. (Camera Pix, Rapho Guillumette)

cord in the skull indicate an upright posture. Anthropologists are not agreed as to whether these forms should be placed in the ape family or in the family that includes man.

In mid-1964, Dr. Leakey published an account of the discovery of the remains of still another primitive manlike form in the Olduvai Gorge. He has named this form *Homo habilis,* and dating evidence indicates that it lived at the same time as *Zinjanthropus. Homo habilis,* however, is more manlike. He seems to have walked and run erect, to have had a well-opposed thumb, and to have eaten both meat and plant foods, like modern man. Dr. Leakey now believes that *Homo habilis* was the ancestor of modern man, while *Zinjanthropus* and *Australopithecus* were evolutionary dead ends. Dr. Leakey's theory indicates that the genus Homo was present on earth about 1,200,000 years earlier than anthropologists had previously thought.

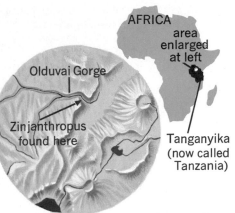

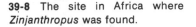

39-8 The site in Africa where *Zinjanthropus* was found.

Early forms of man

In 1891, a part of a skull, a piece of a jaw, and an upper leg bone were discovered in an excavation on the island of Java. These remains lay in a deposit of sand and gravel transported by an early glacier, so the period could be roughly dated. Similar remains were found in the same general region in 1937. Java man, as he was first called, is believed to have walked erect, so he is now called *Pithecanthropus erectus.* He is thought to have lived 500,000 years ago. He had a slanting forehead and heavy brow ridges. The skull of Java man indicates that his brain, while only about half the size of modern man's, was more than one third larger than that of the present-day gorilla. Anthropologists believe that Java man learned to make use of crude stone weapons and of fire. *Pithecanthropus* remains have also been discovered in excavations of ancient caves near Peking. Many of the skulls of Peking man that have been found are broken near the base, suggesting that he was a headhunter.

Neanderthal man

More is known about *Neanderthal* (nee-AN-der-thawl) *man,* who lived in Europe, Asia Minor, Siberia, and Northern Africa. Scientists believe that Neanderthal man disappeared about 25,000 years ago, near the close of the glacial age. Information about him has been acquired from careful study of almost one hundred skeletons, many of them nearly complete. He was about five feet tall, and his bone structure indicates that he was powerfully built. His facial features were coarse. Like *Pithecanthropus,* his forehead sloped backward from heavy brow ridges. His mouth was large, and he had little chin.

Neanderthal man lived in caves from which he journeyed on hunting expeditions in search of the hairy mammoth, saber-toothed cat, and wooly rhinoceros. His brain was as large as

39-9 Prehistoric manlike primates and modern man. In what ways do their skulls differ? In what ways are they similar?

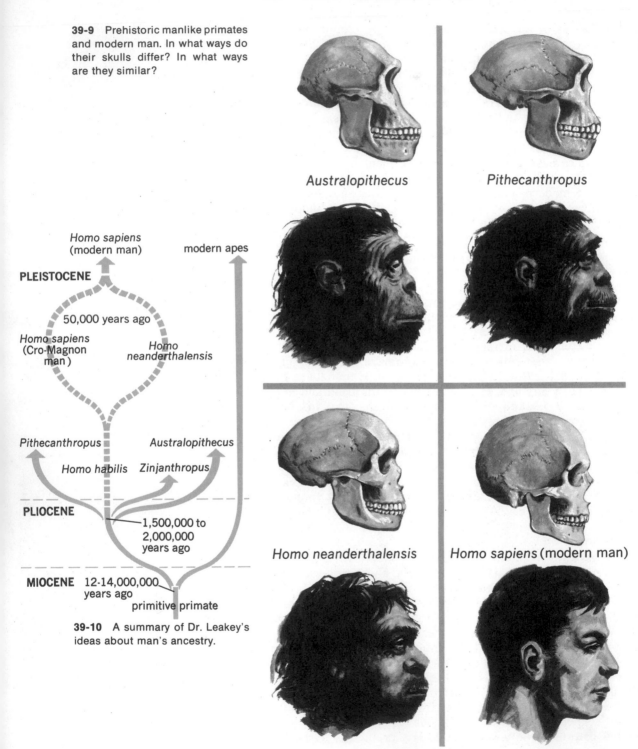

Australopithecus

Pithecanthropus

Homo neanderthalensis

Homo sapiens (modern man)

Homo sapiens (modern man)

modern apes

PLEISTOCENE

50,000 years ago

Homo sapiens (Cro-Magnon man)

Homo neanderthalensis

Pithecanthropus

Australopithecus

Homo habilis

Zinjanthropus

PLIOCENE

1,500,000 to 2,000,000 years ago

MIOCENE   12-14,000,000 years ago

primitive primate

39-10 A summary of Dr. Leakey's ideas about man's ancestry.

or larger than that of modern man. Neanderthal man used stone tools and weapons, made use of fire, buried his dead, and lived in a family group.

Cro-Magnon man

Anthropologists place *Cro-Magnon man* in the same species as modern man, *Homo sapiens* (HOE-moe SAPE-ee-enz). Cro-Magnon man lived in Europe, especially in France and Spain, about fifty thousand years ago. He had a high forehead and a well-developed chin, and lacked the heavy brow ridges of most of the more primitive men. Several caves along the southern coast of France have yielded Cro-Magnon weapons of stone and bone as well as skeletons. The walls of these caves bear beautiful drawings of animals of the region. These drawings have been valuable in dating the period.

Anthropologists believe that Cro-Magnon man, living at the same time as Neanderthal man and being superior to him in intelligence, may have exterminated him. There is also evidence that Neanderthal man mixed with Cro-Magnon man and in time lost his identity.

Modern man

All people living today belong to the species *Homo sapiens.* Anthropologists sometimes divide this single species into several racial groups, according to features that are common within each group. They recognize, however, that there is great variation within a racial group and that all men are more alike than they are different.

Modern man dominates the biological world. A comparison of modern man and ancient man shows the triumph of "brain over brawn." Few people today could match the strength of primitive man, nor could they endure the hardships of life in primitive times. Yet modern man, while he may have lost some of his physical development through the ages, lives three or four times as long as primitive man. Modern man has used his great capacity for learning along with his hands to construct the civilization in which we live.

Yet, even in today's "modern world" cultures exist which we consider primitive and which give us an insight into the habits of early man. Civilization is gradually forcing these people to change their life styles as contacts become more frequent and the areas for a primitive existence become smaller. Anthropologists studying the primitive cultures today hope to use their data to give greater understanding to the unearthed fossils and encampments of early man.

39-11 Women from a Bushman tribe in Africa at a temporary camp. (Arthur Tress)

SUMMARY

Although anthropologists cannot state when man first inhabited the earth, they believe man-like creatures existed nearly two million years ago. Radioactive carbon and potassium are used to date fossils and the rocks from which the fossils have been collected. Careful study of fossils and sites of habitation indicate whether the creature had upright posture and was able to use tools. Other important characteristics used are the size of the brain and structure of the teeth.

The chemical secretions of man and other mammals are similar, and the structural similarity of man and the other primates is especially striking. The significant differences of man from other mammals, however, have allowed him to become biologically superior. His intelligence has provided him with the ability to communicate and transmit information from one generation to another. His inquisitiveness and imagination have enabled him to theorize about his origin and development. Charles Darwin, based on his observations and thinking, has suggested that the same forces bringing about changes in populations and individuals have also acted in the development of man.

All people living today are of the species *Homo sapiens,* but a study of primitive cultures may provide us with clues about the life of early man.

BIOLOGICALLY SPEAKING

anthropology	half-life	Neanderthal man
radiocarbon method	potassium-argon method	Cro-Magnon man

QUESTIONS FOR REVIEW

1. What characteristics separate man from other animals?
2. What characteristics are used to distinguish true man from an ape-like primate?
3. What did Charles Darwin believe about man's development, and what was the misunderstanding about his theory?
4. Why is the finding of an old skull important?
5. What can be learned from a study of the lower jaw and teeth?
6. How can we determine the type of environment of a given area thousands of years ago?
7. How can we determine the diet of primitive man?
8. Describe two methods of dating fossils.
9. Describe *Australopithecus.*
10. Describe *Pithecanthropus erectus.*
11. Why do anthropologists study primitive man in a modern world?

APPLYING PRINCIPLES AND CONCEPTS

1. Do you think that tropical forests would yield good fossils? Explain your answer.
2. In what ways can the intelligence of primitive man be judged from fossil evidence?
3. Explain the theory that reduction of the canine teeth is related to tool-making.

4. Compare the skull of modern man with the skulls of Java, Neanderthal, and Cro-Magnon men.
5. Did Cro-Magnon man have any methods of communication? Explain.
6. How could agriculture have developed from a settled life?

RELATED READING

Books

BALDWIN, GORDON C., *Calendars to the Past.*
W. W. Norton and Company, New York. 1967. Tells how archeologists determine the age of their discoveries.

CLARK, ROBIN, *The Diversity of Man.*
Roy Publishers, Inc., New York. 1964. The latest findings on and inquiries into the question of race similarities and differences.

CLYMER, ELEANOR, *The Case of the Missing Link.*
Basic Books, Inc., Publishers, New York. 1962. Story of the dramatic search for clues to man's antiquity.

GREGORY, W. K., *Our Face From Fish to Man.*
G. P. Putnam's Sons, New York. 1965. Account of the development of the human face.

HOWELL, F. CLARK, and the Editors of *Life* Magazine, *Early Man.*
Time-Life Books (Time, Inc.), New York. 1965. A picture-essay book covering the day-to-day problems and general conditions of life in the ancient times.

LEHRMAN, ROBERT L., *Race, Evolution and Mankind.*
Basic Books, Inc., Publishers, New York. 1966. An intelligible review of the current scientific knowledge of race and its social implications in the light of recent findings in anthropology, psychology, and population genetics.

POOLE, LYNN, and GRAY POOLE, *Carbon-14 and Other Scientific Methods That Date the Past.*
McGraw-Hill Book Company, New York. 1961. Explanation of today's methods for dating ancient objects and natural substances.

SILVERBERG, ROBERT, *The Morning of Mankind.*
New York Graphic Society, Ltd., Greenwich, Conn. 1967. Traces man from the earliest Stone Age toolmakers to the complex civilization of the Bronze Age.

Article

PAYNE, MELVIN, "The Leakeys of Africa: In Search of Prehistoric Man," *National Geographic Magazine,* February, 1965. Tells how the remains of earliest man were finally found in the Olduvai Gorge in Tanganyika, Africa by a man and his wife who have devoted their lives to this search.

CHAPTER FORTY

THE BODY FRAMEWORK

Groups of cells

Like that of every other multicellular organism, the human body is composed of cells and their products. Groups of cells having similar structures and performing similar functions make up the various tissues. For convenience we can divide the body tissues into four groups:

■ *Connective tissues* lie between groups of nerve and muscle cells. They fill up spaces in the body that are not occupied by specialized cells, and form protective layers. Connective tissue also binds together many softer tissues and gives them strength and firmness. Fibers in the walls of organs, tendons of muscles and ligaments binding bones, and the bones themselves are all types of connective tissue. Table 40-1 lists the many types of connective tissues composing the human body.

■ *Muscle tissues* function in movement and will be discussed later in this chapter.

■ *Nervous tissues* coordinate movement and inform an organism about the environment. They will be considered further in Chapter 44.

■ *Epithelial tissues* cover the body surfaces, both inside and outside. For example, certain flat epithelial cells cover the blood vessels and heart. Another type of epithelium lines the stomach. Some cells of this lining are modified to secrete mucus and other stomach secretions. Still another epithelial tissue forms the skin, and another the ciliated lining of the trachea.

Tissues are organized into organs and systems

Familiar examples of organs in the human body include the arms, legs, ears, eyes, heart, liver, and lungs. Each of these organs is specialized to perform a definite function or a group of related functions involving several different tissues. The arms, for example, are composed of tissues of all four groups. All these work together to perform such acts as grasping, writing, and sewing.

you

are a community of systems which continuously interacts from deep sleep to violent effort; even the simplest function involves more than one system

more complex than a computer, the intricate coordination of your body's processes and anatomy are so efficient and adaptable that the human race continues to survive despite the problems of our planet

use this Trans-Vision® unit as an overview as you study in detail the many concepts found in your text and reference books; the anatomical plates are coordinated with the following systems:

INTEGUMENTARY
MUSCULAR
RESPIRATORY
DIGESTIVE
CIRCULATORY
NERVOUS
ENDOCRINE
EXCRETORY
SKELETAL

COVER BY R. T. HANDVILLE

INTEGUMENTARY SYSTEM
(skin, hair, nails)

skin layers	description	function
epidermis	outer layer, epithelial cells—flattened, dead, scale-like. Transparent over eye. Pleated over knuckles, elbows, knees	protects live tissues beneath by being rubbed off, produces callus, contains hair and nails; protects against bacterial invasion, injury and drying out of tissues; produces melanin (skin color)
dermis	middle layer, tough, fibrous connective tissue, rich supply of blood and lymph vessels, nerves, sweat and oil glands, capillaries, muscle and hair follicles	responds to temperature change, touch, pressure, pain, emotion and infection; excretes perspiration and oil
fatty layer (subcutaneous	bottom layer full of fat cells, blood vessels and nerve endings	provides blood to the upper layers, last layer between skin and deeper structures

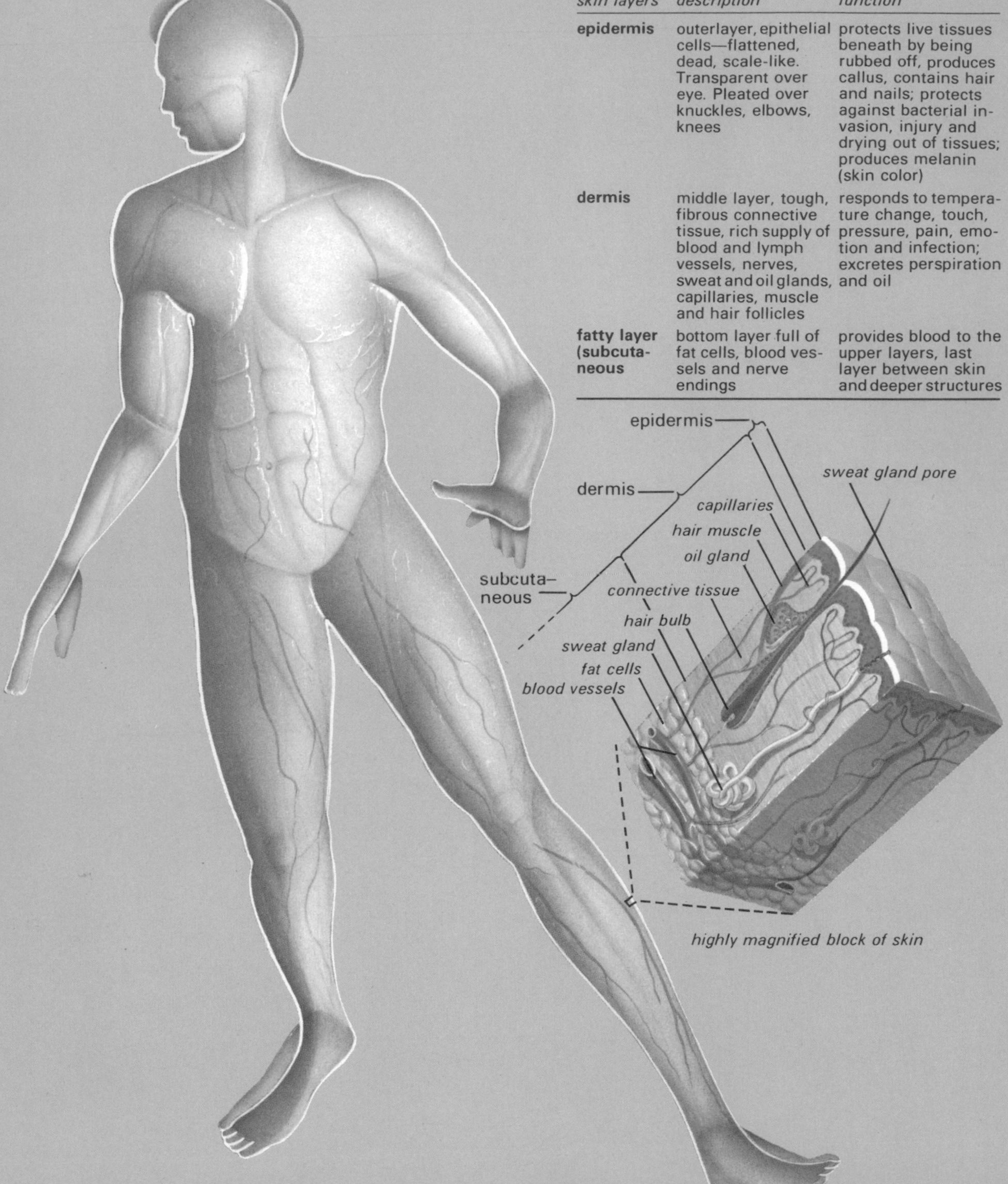

highly magnified block of skin

MUSCULAR SYSTEM

muscle types	description	function
smooth muscle	each cell is an elongated spindle containing one nucleus; found in stomach and intestinal walls	contracts in waves to churn food or pass it along through the digestive tract
	found in artery walls	during stress, muscles constrict walls which raises blood pressure
	involuntary	not controlled at will
skeletal muscle	each fiber is a long cylinder with tapering ends, contains many nuclei; fibers are shorter than most muscles, bound in small bundles by connective tissue sheaths; heavier sheath encloses entire muscle which is attached by inelastic tendons to bones; arranged in pairs—	voluntary movement such as walking, running, grasping, lifting, bending
	flexors *extensors*	bend joints straighten them
	voluntary	controlled at will
cardiac muscle	found only in the heart, cells form branching, interlacing network; plasma membranes separate cardiac muscle cells	muscles contract, squeezing heart chambers and forcing blood out through vessels
	involuntary	not controlled at will

sterno-cleido-mastoid
turns head sideways

major pectoral
moves arm across chest

deltoid
raises arm

triceps
extends forearm

biceps
flexes forearm

extensors of hand

throat muscles
swallowing

minor pectoral
assists major pectoral

intercostals
breathing

serratus
moves shoulder forward

flexors of hand

flexors of thigh
raise leg

adductors
move legs together

quadriceps femoris
extends leg

abdominal rectus
flexes trunk, raises pelvis, compresses abdomen

gastrocnemius
extends or depresses foot in walking or standing on tiptoe

soleus
acts with gastrocnemius

extensors of toes

anterior tibial
flexes foot at ankle

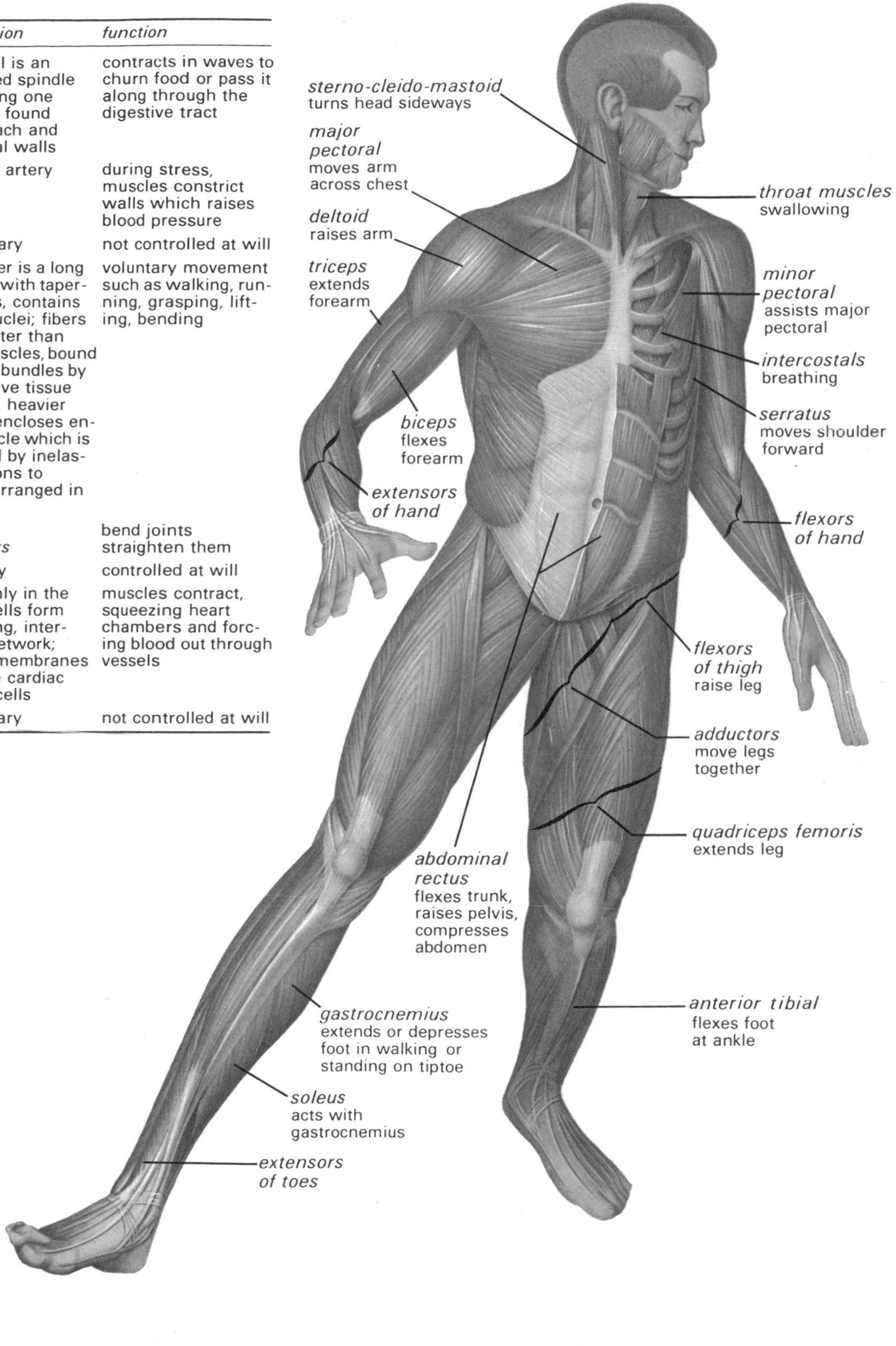

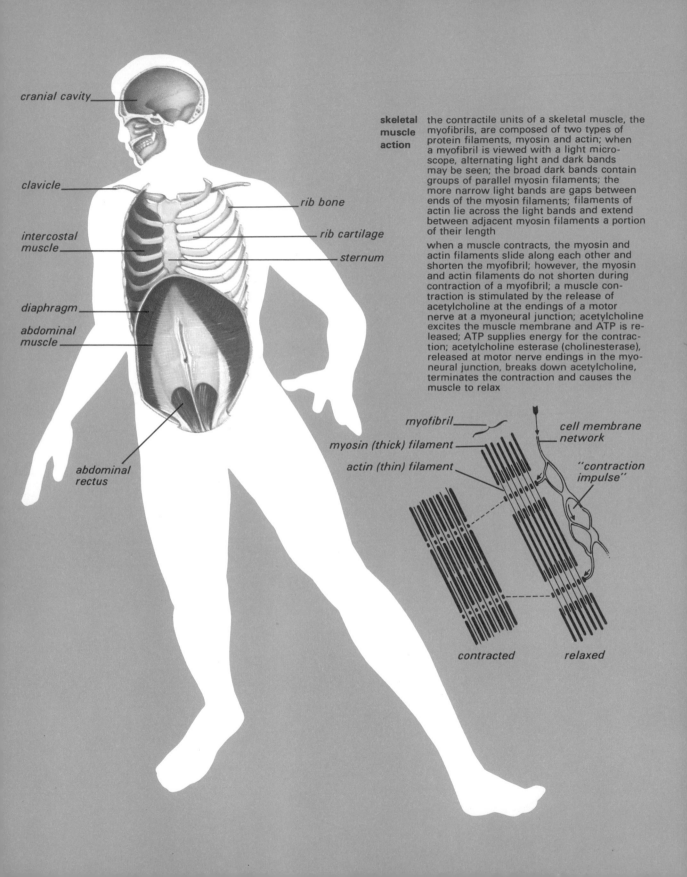

cranial cavity

clavicle

intercostal muscle

diaphragm

abdominal muscle

abdominal rectus

rib bone

rib cartilage

sternum

skeletal muscle action

the contractile units of a skeletal muscle, the myofibrils, are composed of two types of protein filaments, myosin and actin; when a myofibril is viewed with a light microscope, alternating light and dark bands may be seen; the broad dark bands contain groups of parallel myosin filaments; the more narrow light bands are gaps between ends of the myosin filaments; filaments of actin lie across the light bands and extend between adjacent myosin filaments a portion of their length

when a muscle contracts, the myosin and actin filaments slide along each other and shorten the myofibril; however, the myosin and actin filaments do not shorten during contraction of a myofibril; a muscle contraction is stimulated by the release of acetylcholine at the endings of a motor nerve at a myoneural junction; acetylcholine excites the muscle membrane and ATP is released; ATP supplies energy for the contraction; acetylcholine esterase (cholinesterase), released at motor nerve endings in the myoneural junction, breaks down acetylcholine, terminates the contraction and causes the muscle to relax

myofibril

myosin (thick) filament

actin (thin) filament

cell membrane network

"contraction impulse"

contracted

relaxed

RESPIRATORY SYSTEM
(intake of oxygen, elimination of carbon dioxide)

component	description	function
nostrils (1)	two nose openings divided by a wall called the septum	contain hair for filtering dirt; air intake begins here and through the mouth
nasal passages (2)	from nostrils to pharynx	warms air, adds moisture on the way to pharynx
pharynx (3)	muscular throat cavity, extends from nasal cavity to soft palate (uvula)	common passageway for breathing and digestion; air passes here to the trachea
trachea (4)	windpipe, a tube supported by horse-shoe shaped rings of cartilage, closed by epiglottis (cartilaginous flap) during swallowing	air passage to lungs lined with cilia in constant motion, carries dust upward toward mouth as air descends to bronchi
bronchi (5)	trachea branches into two passages; each bronchus extends to one lung, subdivides into countless bronchial tubes	air moves through bronchi into bronchioles inside the lungs
bronchioles	countless subdivisions of bronchial tubes, ending in air sacs made up of protrusions called *alveoli*; each sac is surrounded by capillaries (tiny blood vessels)	air sac and capillary walls are thin and moist, permit *external respiration*—the exchange of oxygen from air *to blood* with carbon dioxide and water *from blood* to air
right lung (3 lobes) (6) left lung (2 lobes) (7)	two organs composed of spongy tissue surrounded by a double membrane, consisting mainly of bronchioles with tiny air sacs, blood vessels and capillaries; total surface area about 2,000 square feet.	supply oxygen by way of the blood to millions of body cells having no direct access to oxygen; return waste gases and water vapor to the atmosphere
pulmonary artery	artery from the heart to the lungs	brings blood with reduced oxygen and increased carbon dioxide and water to the lungs, diffusing through the capillary network
pulmonary veins	veins from the lungs to the heart	returns blood rich in oxygen to the heart for the tissues; exchange of gases between blood and body tissues is called *internal respiration*

diaphragm (8)   *pericardial cavity* (9)

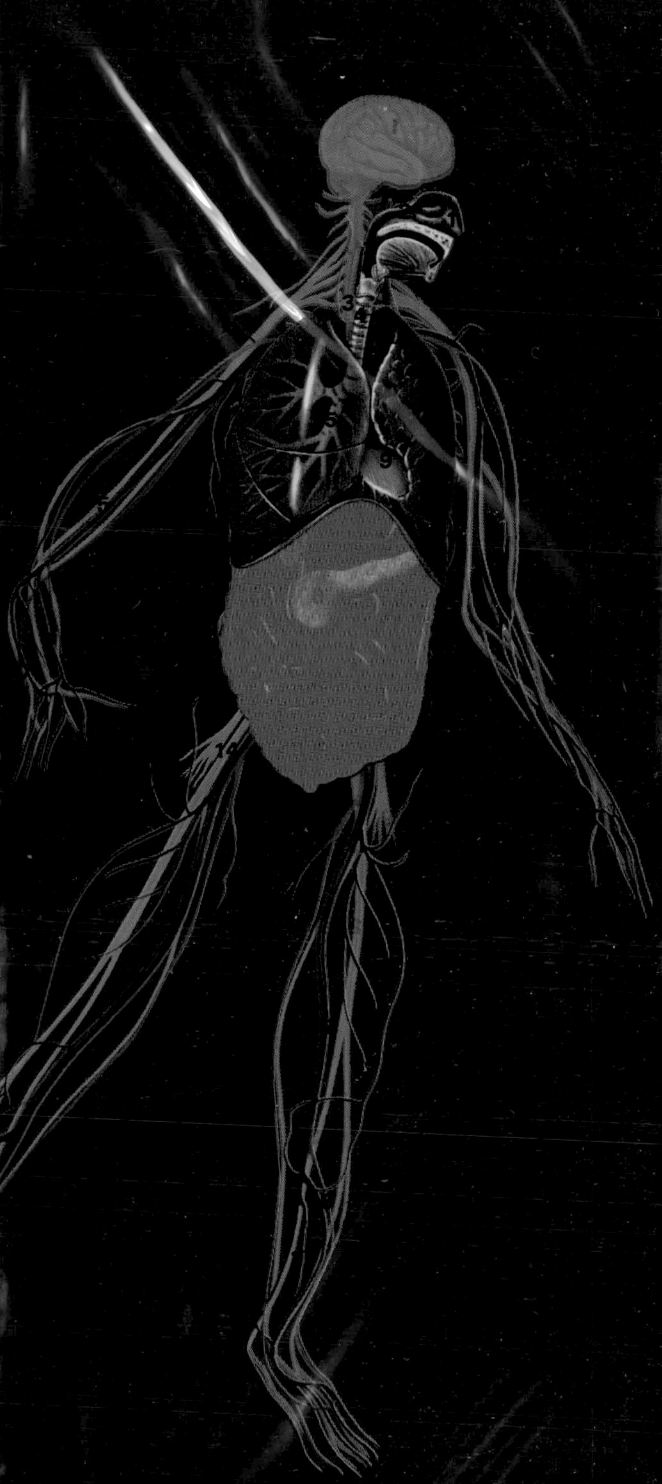

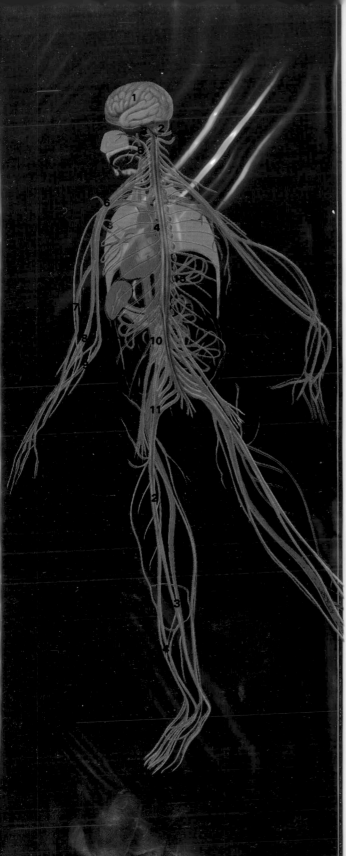

NERVOUS SYSTEM
(central, peripheral, autonomic)

component	description	function
central nervous system (1)-(4)	*brain*—about three pounds of soft nervous tissue composed of:	control center for body's activities
	cerebrum—largest region consisting of two hemispheres; outer surface is the *cerebral cortex*	coordinates voluntary muscle movement such as seeing, hearing, touching, tasting, smelling, intelligence, emotional centers
	cerebellum—below back of cerebrum	assists cerebrum in coordinating voluntary movements
	brain stem—consists of several regions including the *medulla*	controls breathing, heart muscles, gland secretion
	spinal cord—extension of brain from medulla through protective bony arches and almost full length of spine	relays nerve impulses to and from the brain
peripheral nervous system (5)-(14)	*cranial nerves*—twelve pairs	communicates with special sensory organs, head structures and viscera
	spinal nerves—thirty-one pairs branch off spinal cord between bones of spine	carry impulses to and from spinal cord
autonomic nervous system (15)	completely involuntary, composed of two nerve systems:	
	sympathetic—two nerve cords left and right of spinal column with sympathetic ganglia at *solar plexus*, near heart, in abdomen and neck	regulates heart action, ductless gland secretion, arterial blood supply, smooth muscle action of stomach and intestines, other internal organs
	parasympathetic—nerve fibers extending from the brain stem and sacral region of the spinal cord	check and balance system with the sympathetic system

cerebrum (1)

cerebellum (2)

medulla oblongata (3)

spinal cord (4)

cervical spinal nerves (5)

brachial plexus (6)

radial nerve (7)

median nerve (8)

ulnar nerve (9)

lumbar plexus (10)

lumbo-sacral plexus (11)

sciatic nerve (12)

tibial nerve (13)

peroneal nerves (14)

sympathetic ganglion (15)

SKELETAL SYSTEM

(front view with front ribs removed)

functions

(1) provides body form and surface to which muscles attach

(2) provides greatest amount of support with least amount of weight

(3) provides protection for delicate organs such as brain, trachea, lungs, heart and esophagus

(4) provides a storehouse for minerals

(5) manufactures red blood cells and certain white blood cells

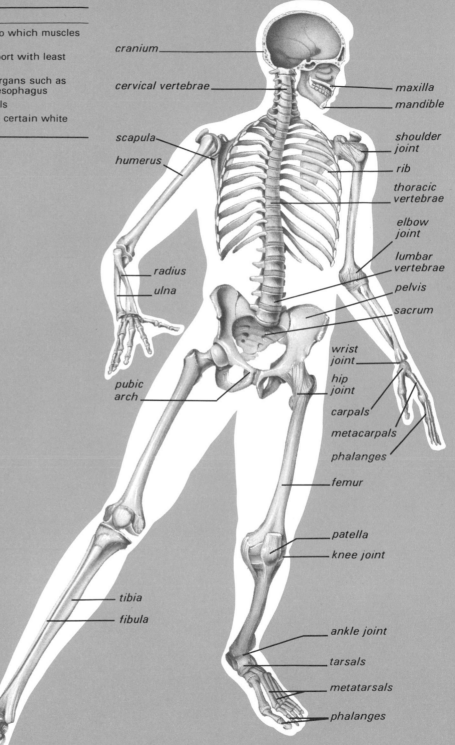

cranium

cervical vertebrae

maxilla

mandible

shoulder joint

scapula

humerus

rib

thoracic vertebrae

elbow joint

lumbar vertebrae

pelvis

radius

sacrum

ulna

wrist joint

hip joint

pubic arch

carpals

metacarpals

phalanges

femur

patella

knee joint

tibia

fibula

ankle joint

tarsals

metatarsals

phalanges

ANATOMICAL ART BY LEONARD D. DANK

MODERN BIOLOGY (1973) OTTO & TOWLE, HOLT, RINEHART & WINSTON, INC., NYC

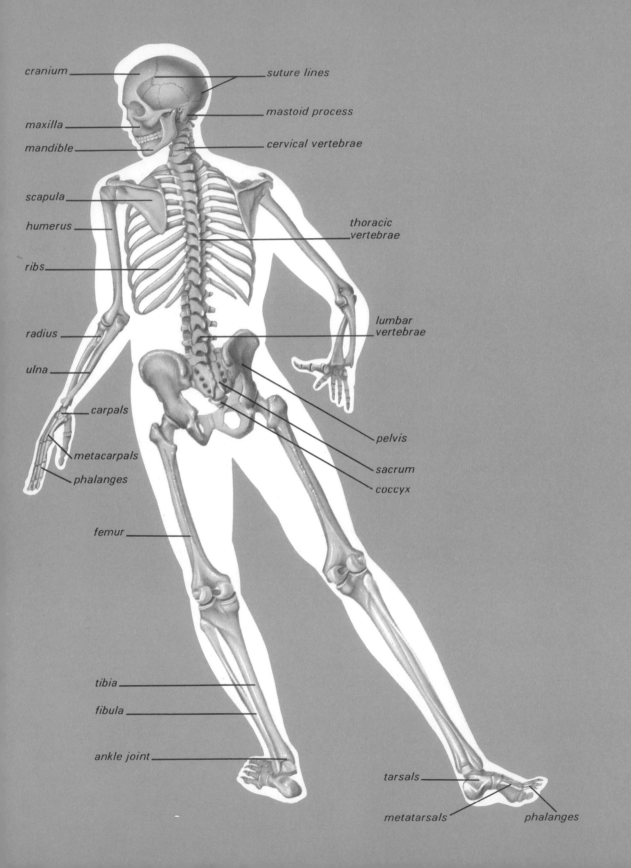

cranium

suture lines

mastoid process

maxilla

cervical vertebrae

mandible

scapula

thoracic
vertebrae

humerus

ribs

radius

lumbar
vertebrae

ulna

carpals

metacarpals

pelvis

phalanges

sacrum

coccyx

femur

tibia

fibula

ankle joint

tarsals

metatarsals

phalanges

ENDOCRINE SYSTEM
(ductless glands—body regulators)

gland	hormone	function of hormone
thyroid	thyroid hormone	accelerates the rate of metabolism
para-thyroids	parathormone	controls the use of calcium in the tissues
pituitary anterior lobe	somatotropic hormone	regulates growth of the skeleton
	gonadotropic hormone	influences development of sex organs and hormone secretion of the ovaries and testes
	ACTH	stimulates secretion of hormones by the cortex of the adrenals
	lactogenic hormone (female)	stimulates secretion of milk by mammary glands
	thyrotropic hormone	stimulates activity of thyroid
pituitary posterior lobe	oxytocin	regulates blood pressure and stimulates smooth muscles
	ADH (antidiuretic) hormone	controls water resorption in the kidneys
adrenal cortex	cortin (a hormone complex)	influences metabolism, salt, and water balance; controls production of certain white corpuscles and structure of connective tissue
adrenal medulla	epinephrine or adrenalin	causes blood vessel changes, increase in heart action, stimulates liver and nervous system
pancreas with the islets of Langerhans	insulin	enables liver to store sugar and regulates sugar oxidation in tissues
ovaries follicular cells (female)	estrogen	produces female secondary sex characteristics; influences adult female body functions
	progesterone	maintains growth of the mucous lining of the uterus
testes interstitial cells (male)	testosterone	produces male secondary sex characteristics

EXCRETORY SYSTEM
(cell waste removal)

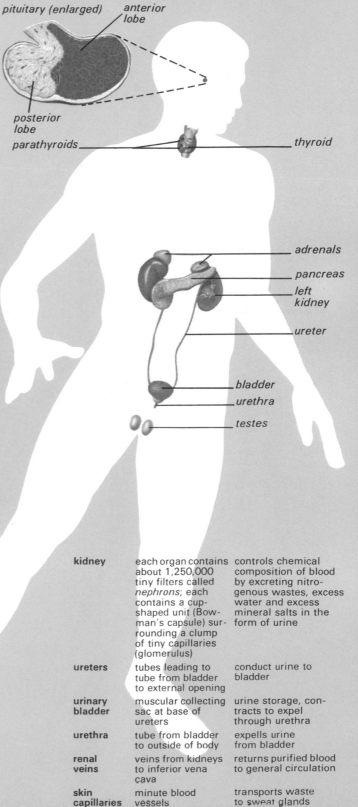

pituitary (enlarged)   anterior lobe

posterior lobe

parathyroids

thyroid

adrenals

pancreas

left kidney

ureter

bladder

urethra

testes

components	description	function
renal arteries	branch directly from the aorta into each kidney, re-branch into maze of tiny arterioles penetrating all areas	supply blood to the kidneys for cleansing
kidney	each organ contains about 1,250,000 tiny filters called nephrons; each contains a cup-shaped unit (Bowman's capsule) surrounding a clump of tiny capillaries (glomerulus)	controls chemical composition of blood by excreting nitrogenous wastes, excess water and excess mineral salts in the form of urine
ureters	tubes leading to tube from bladder to external opening	conduct urine to bladder
urinary bladder	muscular collecting sac at base of ureters	urine storage, contracts to expel through urethra
urethra	tube from bladder to outside of body	expels urine from bladder
renal veins	veins from kidneys to inferior vena cava	returns purified blood to general circulation
skin capillaries	minute blood vessels	transports waste to sweat glands

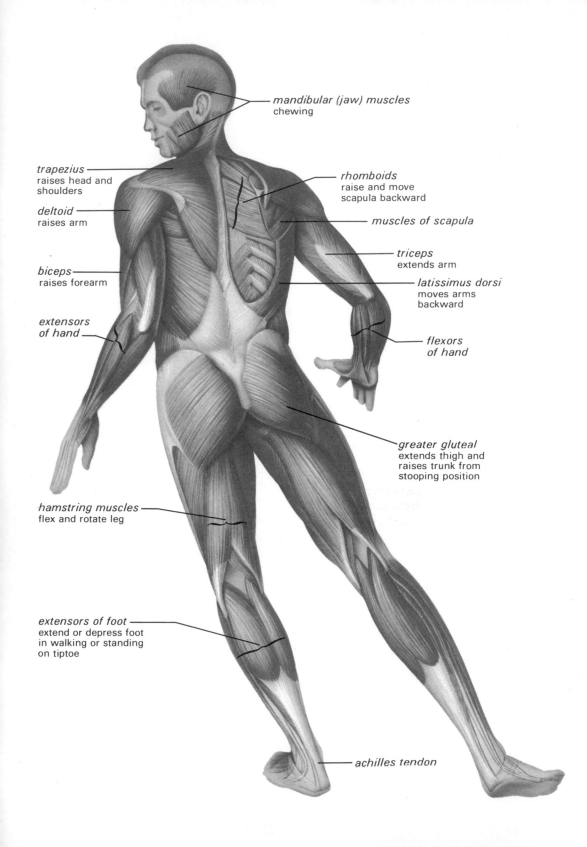

mandibular (jaw) muscles
chewing

trapezius
raises head and
shoulders

rhomboids
raise and move
scapula backward

deltoid
raises arm

muscles of scapula

triceps
extends arm

biceps
raises forearm

latissimus dorsi
moves arms
backward

*extensors
of hand*

*flexors
of hand*

greater gluteal
extends thigh and
raises trunk from
stooping position

hamstring muscles
flex and rotate leg

extensors of foot
extend or depress foot
in walking or standing
on tiptoe

achilles tendon

Table 40-1 TISSUES IN THE HUMAN BODY

TISSUE	OCCURRENCE	FUNCTION
I. CONNECTIVE TISSUES A. bone	skeleton	composes framework for movement, support, and protection; serves as a storehouse for minerals; manufactures blood cells
B. cartilage	outer ears, ends of long bones, larynx, tip of nose, between vertebrae, juncture of ribs and breastbone, trachea	acts as cushion, lends rigidity to structures that lack bones, provides slippery surface to some joints
C. dense fibrous connective tissue		
1. regularly arranged	tendons, ligaments	joins muscles to bones or bone to bone to aid in movement
2. irregularly arranged	membrane around bone (periosteum), one of the membranes around spinal cord and brain (dura mater), inner layer of skin	provides protection and carries blood supply
D. loose fibrous connective tissue		
1. fibroelastic (elastic—strong, closely woven)	capsules of organs	holds organ together
2. fibroareolar (areolar—loosely woven)	facial area beneath skin	acts as filler tissue
3. reticular	surrounding individual cells and muscle fibers	acts as filler tissue
4. adipose	around organs, beneath skin	cushions and insulates, stores fat
E. liquid tissue		
1. blood	in heart and vessels (arteries and veins)	has essential part in respiration, nutrition, excretion, regulation of body temperature, protection from disease
2. lymph	fluid in tissue spaces between cells, cerebrospinal fluid	bathes the cells, has part in nutrition and protection from disease
II. MUSCLE TISSUES A. smooth	in internal organs	produces either voluntary or involuntary movement
B. skeletal	attached to bones, tendons, and other muscles	
C. cardiac	in heart	
III. NERVOUS TISSUES	brain, spinal cord, nerves	carries impulses that cause muscles to contract, carries messages to brain to inform individual about the environment
IV. EPITHELIAL TISSUES	covering surface of body (skin), lining nose, throat, and windpipe, lining all of digestive tract, many glands	provides protection, produces secretions

Organs grouped together to perform a specific function make up a *system*. Even the simplest function requires the use of more than one system. Movement, for example, involves bones, nerves, muscles, and blood vessels. For convenience, however, we shall divide the body into the following ten systems:

- *skeletal*—bone and cartilage;
- *muscular*—muscles;
- *digestive*—teeth, mouth, esophagus, stomach, intestines, liver, pancreas;
- *circulatory*—heart, arteries, veins, capillaries;
- *excretory*—kidneys and bladder;
- *integumentary*—skin and hair;
- *respiratory*—nasal passages, trachea, bronchi, lungs;
- *nervous*—brain, spinal cord, nerves, eyes, ears;
- *endocrine*—ductless glands;
- *reproductive*—testes, ovaries, uterus, oviducts.

The body regions in man

The general form of the human body is similar to that of the other vertebrate animals. It includes the limbs (in the form of arms and legs), the head, neck, and trunk. The head includes the **cranial cavity,** which is formed by the bones of the skull and safely encloses the brain. The head also contains the organs of several senses. These are located close to the brain, to which they transmit impulses.

The **thoracic cavity** is formed by the ribs, breastbone, and spine. It encloses the lungs, the trachea, the heart, and the esophagus. A dome-shaped partition, the **diaphragm,** separates the thoracic cavity from the **abdominal cavity,** which is included in the lower part of the trunk. Inside the abdominal cavity are the stomach, liver, pancreas, intestines, spleen, kidneys, and, in the female, the ovaries. While the abdominal organs lack the bony protection of the cranial and thoracic cavities, they are protected by the vertebral column along the back and by layers of skin and muscle on the front.

The body framework

In a model airplane, the framework of the body is usually built first. Then come the covering and painting, and finally the motor, wheels, and accessories are added. The strength of the entire structure depends on the framework to which all the other parts are fastened. Man and the other vertebrates, like the model airplane, have a very efficient system of support in the form of an internal skeleton, or **endoskeleton.** You will recall that the arthropods have exoskeletons. Man's bony framework gives him the greatest support with the least amount of weight. This framework also permits more efficient movement than any other type of framework. The animal with an internal skeleton is, however, at one great disadvantage. It lacks much of the protection against injury from the outside that is provided by an external skeleton. Many soft parts of the body are exposed. Consequently, the organism must rely on its nervous system and sense organs to make up for the protection the skeleton does not provide.

The functions of the skeleton

The bones of the body serve the following functions: (1) to provide support and form; (2) to provide a surface to which the muscles attach; (3) to provide protection for delicate organs; (4) to provide a storehouse for minerals; and (5) to manufacture red blood cells and certain white blood cells. Many of the 206 bones composing the human skeleton have more than one function. For example, the vertebral column, the shoulder girdle, the hip girdle, the bones of the legs, and those of the arms both support the body and give it definite form. Some of these bones also have muscles attached to them, permitting the many types of movement. Certain delicate organs lie under special protective bones. Examples are the brain, which is encased by the cranial bones; the heart, which lies under the sternum; and the lungs, which are protected by the ribs.

The development of bone

We use the expression "dry as a bone" and assume that living bone is like a dried-out bone. Actually, living bone is far from dry. It is moist and active and requires nourishment just as any living organ does. True, part of what we call bone is nonliving, for bone tissue is a peculiar combination of living cells and their products and mineral deposits.

Among some of the lower vertebrates the skeleton is composed entirely of *cartilage,* which lasts throughout their lives and results in a tough, flexible skeleton. In the early stages of the development of the human embryo, the skeleton is also composed almost entirely of cartilage.

After about the second month of development, however, certain of the cartilage cells disappear and are replaced by bone cells. Such cells remove calcium phosphate and calcium carbonate from the blood and deposit these minerals to form the bone structure. This process is called *ossification* (oss-i-fi-KAY-shun) and occurs throughout childhood. Not all cartilage undergoes ossification. Permanent cartilage is found in the end of the nose, the external ear, and the walls of the voice box and trachea.

Not all bones, however, are preceded by cartilage. The flat bones, such as those of the skull and sternum, are formed from membrane layers that later undergo ossification. The bones of the skull in the newborn infant are not fused to form a solid cranium. Later in his development, the margins of the individual bones unite to form irregular seams called *sutures* (Figure 40-1).

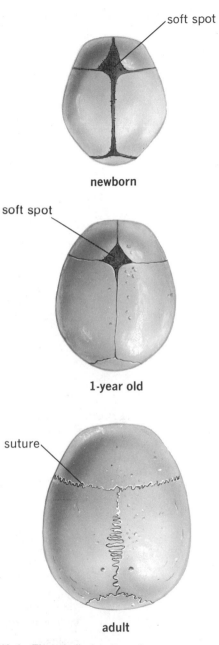

soft spot

newborn

soft spot

1-year old

suture

adult

40-1 The skull develops from a membrane that forms separate plates of bone. By the time adulthood is reached, the margins of these bony plates have united.

Since ossification involves the deposit of calcium compounds between the bone cells, it results in an increase in the strength of the bone. This deposition cannot occur unless the proper minerals are present. Calcium compounds enter the body with food and are carried to the bone tissues by the blood. Diet, especially in childhood, is therefore an important factor in governing mineral deposition in bone. Milk, the natural food of all young mammals, is the ideal source of calcium compounds. Developing bone tissue must assimilate the minerals after they have been supplied by a proper diet. Certain vitamins, especially vitamin D, are necessary for the normal growth of bone. We shall study these in Chapter 41 under vitamins.

Bones grow along lines of stress. This means that they become heaviest and strongest where the strain is greatest. It is important to keep this fact in mind in dealing with bone fractures. If a broken bone is protected by a cast and is unused during the period of repair, the fact that it is under no stress delays healing. For example, if a leg bone is broken, the patient is provided with a walking cast, which puts a broken bone under limited stress during the healing period and speeds up the repair process. If, on the other hand, a limb is paralyzed or made useless, the minerals are reabsorbed by the blood and deposited elsewhere.

The structure of a bone

If a long bone, such as a bone from the leg or thigh, is cut lengthwise, several distinct regions can be seen (Figure 40-2). The outer covering is a tough membrane called the *periosteum*. This membrane aids in nourishing the bone (because of its rich

40-2 The structure of the human femur shown at increasingly greater magnification.

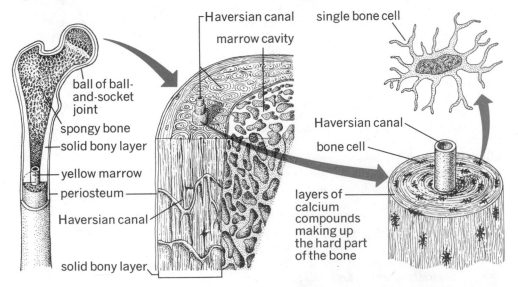

Haversian canal

marrow cavity

single bone cell

ball of ball-and-socket joint

spongy bone

solid bony layer

yellow marrow

periosteum

Haversian canal

solid bony layer

Haversian canal

bone cell

layers of calcium compounds making up the hard part of the bone

blood supply) and in repairing injuries; it also provides a sur-face to which muscles are attached. Beneath the periosteum is a *bony layer* containing the deposits of mineral matter. This layer varies in hardness from an extremely hard material in the mid-dle region to a porous and spongy material at the ends. The bony layer is penetrated by numerous channels, the *Haversian canals,* which form a network extending throughout the region. These canals carry nourishment to the living cells of the bony layer through blood vessels that connect with those of the outer membrane.

Many bones have hollow interiors and contain a soft tissue called marrow. The marrow is richly supplied with nerves and blood vessels. There are two distinct types of marrow. The *red marrow* is found in flat bones such as the ribs and sternum, as well as in the ends of long bones and vertebrae. It is active in forming the red corpuscles and most of the white corpuscles of the blood. The *yellow marrow* fills the central cavity of long bones and extends into the Haversian canals of the bony layer. It is normally inactive and primarily composed of fat cells but may produce corpuscles in time of great blood loss and in cer-tain blood diseases.

The smaller bones are solid rather than hollow and vary considerably in the amount of spongy bone tissue present. Although they are solid, they are completely penetrated with blood vessels.

The joints of the body

The point at which two separate bones meet is called a *joint.* The various bones of the human body are connected by several different kinds of joints (Figure 40-3). The elbow is an example of a *hinge joint.* Such a joint moves as a hinge in one plane only but can give great power because there is little danger of twist-ing. When the biceps muscle of the upper arm contracts, the lower arm is pulled upward only. The knee is another example of a hinge joint. The hip and shoulder joints are examples of *ball-and-socket joints.* Here the bone of the upper arm ends in a ball that fits into a socket of the shoulder girdle. Such a joint has the advantage of movement in any direction within the limits imposed by the muscles. The hip joint is similar to that of the shoulder, with a ball on the end of the *femur,* or thighbone, fitting into a socket of the hipbone, the *pelvis.* Ball-and-socket and hinge joints are held in place by tough strands of connec-tive tissue called *ligaments.* Ligaments can be stretched with exercise, thus loosening joints and permitting freer movement.

The ribs are attached to the *vertebrae* by joints that are only *partially movable.* Long strands of cartilage attach the ribs to the breastbone in front to allow for chest expansion during breathing. The junction between the spine and pelvis, the sacro-

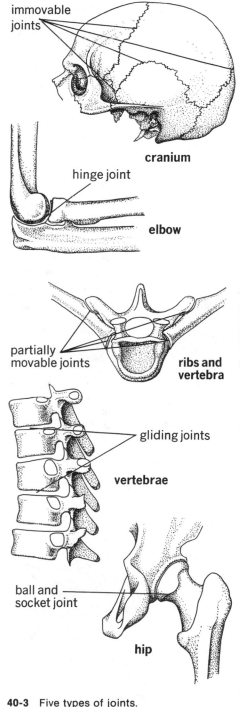

40-3 Five types of joints.

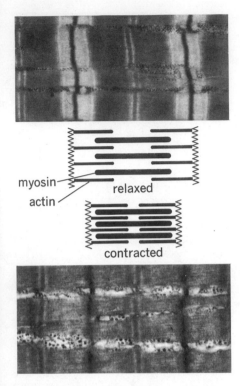

myosin
actin
relaxed

contracted

40-4 A representation of the current concept of muscle contraction. Compare and contrast the relaxed muscle tissue (top) with the contracted muscle tissue (bottom). (Lewis Koster)

40-5 A motor unit. (courtesy of General Biological Supply House, Inc.)

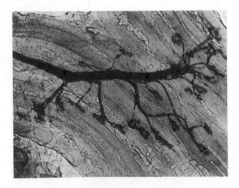

iliac joint, provides a well-known example of partial movability. It is frequently injured in sudden falls. Some joints, such as those in adult skull bones, are *immovable*. Other joints include the *angular joints* of the wrists and ankles, the *gliding joints* of the vertebrae, and the *pivot joint* of the head on the spine.

The inner surfaces of the joints are covered with layers of permanent cartilage. A secretion called *synovial* (suh-NOVE-ee-uhl) *fluid* serves to lubricate the joints. In some joints, such as the knee and shoulder, a sac called the *bursa* serves as a cushion between the bones.

How muscles produce movement

Bones, even in a living body, have no power to move by themselves. Muscle cells, however, are specialists in motion because of their ability to contract. Grouped in bundles, these cells accomplish such mechanical activities as walking, grasping, breathing, heartbeat, and movement of the digestive organs. There are about four hundred different muscles in the human body, making up about one half of the body weight.

Muscle tissues of man and other higher animals are composed of bundles of long, slender cells often called muscle *fibers*. Each of these consists of numerous fine threads called *myofibrils,* which lie parallel and run lengthwise in the fiber. The myofibrils, in turn, are bundles of two distinct kinds of still smaller protein filaments—thick *myosin* filaments and thin *actin* filaments. These are arranged in the myofibril in a definite pattern (Figure 40-4).

When muscle cells, or fibers, are supplied with energy from ATP and activated by a nerve impulse, they contract. The actual mechanism by which muscle cells contract is not known, but a widely accepted current theory states that the thick and the thin protein filaments slide past each other in a way that shortens the myofibrils.

Each nerve cell that carries impulses to a muscle branches to supply a small number of muscle fibers. The nerve cell and the individual fibers it stimulates comprise what is called a *motor unit* (Figure 40-5). Each fiber, when stimulated to contract, will do so to its fullest. The factor that determines whether a movement will be a very delicate, precise one or a very forceful one is, therefore, the number of motor units that are called into action. Muscle fibers may also be stimulated to contract by heat, light, chemicals, pressure, and electricity.

Types of muscle cells

The three types of vertebrate muscle cells are shown in Figure 40-6. *Smooth muscle* forms the walls of many internal organs. Each cell in smooth muscle is an elongated spindle con-

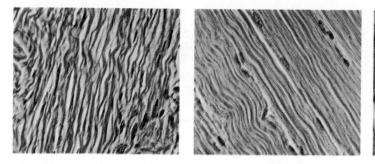
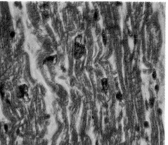

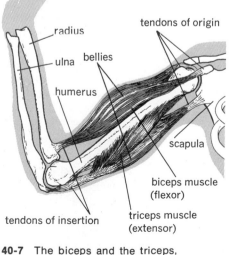

taining one nucleus usually situated near the center of the cell. The stomach and intestinal walls contain layers of smooth muscle cells that contract in waves to churn food or pass it along through the digestive tract. Artery walls also contain layers of smooth muscle. Impulses from the nervous system cause the artery walls to constrict and raise the blood pressure during danger or emotional upset. All action of smooth muscle is controlled by parts of the nervous system over which we have no conscious control; smooth muscle is, therefore, a type of *involuntary muscle.*

The **skeletal muscles** can be controlled at will and are, therefore, *voluntary muscles.* Each fiber is a long cylinder with tapering ends, and each contains many nuclei situated near the periphery of the cell throughout its length. Skeletal muscle fibers do not run the entire length of most muscles but are bound together in small bundles by connective tissue sheaths. These small bundles are then held together by a heavier sheath that encloses the entire muscle. This structure gives most voluntary muscles a spindle shape.

Some skeletal muscles attach directly to bones, some attach to other muscles, and some attach to bones by inelastic **tendons** extending from the tapered end. These tendons are dense bands of fibrous connective tissue. Since muscles are organs of movement, they must attach at two points. The **origin** is stationary while the **insertion** is the attachment of the muscle on the movable part. Be sure to note the origin and the insertion of the *biceps* in Figure 40-7.

The skeletal muscles that move joints of the trunk and limbs are always arranged in pairs. Muscles that bend joints are called *flexors,* while those that straighten them are called *extensors.* For instance, when you bend your elbow joint, the tendon of the contracted biceps muscle raises the radius bone of the forearm. The other end of this muscle is securely anchored at the shoulder. During this contraction, you can feel the biceps muscle swell on the front side of your upper arm. The extensor muscle involved in this movement is called the *triceps.* It is on the back side of the upper arm. When you lower the arm, the

40-6 Three types of vertebrate muscle cells: smooth (left), skeletal (center), and cardiac (right). (Walter Dawn)

radius

ulna

humerus

bellies

tendons of origin

scapula

biceps muscle (flexor)

triceps muscle (extensor)

tendons of insertion

40-7 The biceps and the triceps, two opposing muscles.

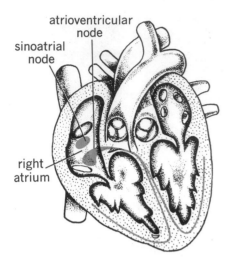

sinoatrial
node

atrioventricular
node

right
atrium

40-8 The conducting pathways in the heart. The impulse for the heartbeat originates in the sinoatrial node. It is carried through the muscle to the atrioventricular node, from which it is relayed through the muscles of the lower chambers.

triceps contracts, and the biceps relaxes. If you straighten your arm completely, you can feel this muscle contract.

Even when a joint is not being moved, flexor and extensor muscles oppose each other in a state of slight contraction called *tone*. Increased use of muscles results in enlargement and increased tone. When they are totally unused, muscles become weak and flabby, decrease in size, and lose tone.

Cardiac muscle, the involuntary muscle found in the heart, is made up of the body's third kind of contractile cell. In the past, it was thought that cardiac muscle cells, like the skeletal muscles, contained many nuclei. Now, however, the electron microscope clearly shows us that this is not true. Plasma membranes separate the cardiac muscle cells. The cells form a branching, interlacing network. Thus, when cardiac-muscle fibers contract, the chambers of the heart are squeezed, and blood is forced out through the vessels.

The action of heart muscle is unlike that of any other muscle. The heartbeat *originates* in the wall of the right atrium of the heart in a small mass of tissue called the *sinoatrial* (SIE-noe-AY-tree-uhl) *node* (Figure 40-8). From this pacemaker, the beat is carried through the muscle of the upper chambers to another node, the *atrioventricular* (AY-tree-oe-ven-TRICK-yoo-ler) *node,* where it is relayed through the muscles of the lower chambers. Conduction of the beat from cell to cell through the heart-muscle mass results in the characteristic rhythmic wave of contraction.

SUMMARY

The general form of the human body is similar to that of other vertebrate animals. Cells and their products form the tissues of which various organs are composed. Although some lower vertebrates have cartilaginous skeletons, most of the cartilage in man is replaced by bone. The internal skeleton gives the body form, acts as levers for muscles, and protects the more delicate organs.

Muscles produce movement. Those that are attached to bones by means of tendons are striated and voluntary. Smooth muscles form layers in the walls of such internal organs as the stomach, intestines, and the arteries. Their control is involuntary. The muscle cells controlling contraction of the heart are also involuntary.

BIOLOGICALLY SPEAKING

connective tissues	cartilage	motor unit
epithelial tissues	ossification	smooth muscle
muscle tissues	periosteum	skeletal muscle
nervous tissues	bony layer	tendon
cranial cavity	Haversian canal	origin
thoracic cavity	red marrow	insertion
diaphragm	yellow marrow	tone
abdominal cavity	joint	sinoatrial node
endoskeleton	ligament	atrioventricular node

QUESTIONS FOR REVIEW

1. How do the tissues, organs, and systems of the human body illustrate division of labor?
2. What are the three body cavities? By what structures are they enclosed? Name some of the organs found in each.
3. What are the principal functions of bones? Give an example of a bone serving each purpose.
4. What are some important functions of the Haversian canals?
5. Describe some of the tissues surrounding a joint. What are their functions?
6. Describe the various kinds of joints found in the body.
7. How do we classify muscle cells as to appearance, control, and location?
8. Why are muscles often found in opposing pairs in the body?
9. Describe the current concept of muscle contraction.

APPLYING PRINCIPLES AND CONCEPTS

1. Explain the importance of a highly developed nervous system in an organism with an internal skeleton.
2. Proper diet alone does not insure good teeth and healthy bones. What other factors are involved?
3. How does a walking cast speed up the repair of a bone fracture?
4. Why is it possible that hearts removed from some lower animals and placed in nutrient solutions keep beating?

RELATED READING

Books

HILL, A. V., *First and Last Experiments in Muscle Mechanics.*
  Cambridge University Press, Cambridge, England. 1970. A personal account of a life-time's work in muscle physiology—appealing to a student interested in physiology.
HORROBIN, DAVID F., *The Human Organism—An Introduction to Physiology.*
  Basic Books, Inc., Publishers, New York. 1966. A research scientist's introduction to basic chemistry and the physical processes involved in body functions.
McCULLOCH, GORDON, *Man and His Body.*
  The Natural History Press, Garden City, New York. 1967. An excellent collection of drawings, pictures, diagrams, and discussion on all aspects of human physiology.
NOURSE, ALAN, and the Editors of *Life* Magazine, *The Body.*
  Time-Life Books (Time, Inc.), New York. 1964. The construction and function of the human body made clear in a picture-essay book.
SAMACHSON, JOSEPH, *The Armor Within Us: The Story of Bone.*
  Rand McNally and Co., Chicago. 1966. A nontechnical, lively account of body bone—its structure, function, and use in the body.
SCHNEIDER, LEO, *You and Your Cells.*
  Harcourt, Brace, and World, Inc., New York. 1964. Delves into the structure and function of various organisms, including man.
ZIM, HERBERT S., *Bones.*
  William Morrow and Co., New York. 1969. A simple, but basic, introduction to a complex subject, quite useful to a beginning biologist.

CHAPTER FORTY-ONE

NUTRITION

41-1 A nourishing lunch. (John King)

What is food?

A short while after you eat a sandwich and drink a glass of milk, these substances are in your tissues. The process of preparing foods to enter the body occurs in a thirty-foot tube, the alimentary canal. Several enzymes break down the bread, beef, milk, and butter into smaller molecules of glucose, amino acids, and fatty acids which can then pass into the bloodstream. *Food* consists of nutritive substances taken into an organism for growth, work, or repair, and for maintaining vital processes. This is another way of saying that you eat to take in energy to be active, to grow, and to maintain your body. According to this definition, water, minerals, and vitamins would be considered as food, as well as carbohydrates, fats, and proteins. Table 41-1 lists the food substances, their functions, and their sources.

Perhaps this would be a good time for you to review the chemical structure and importance to the living organism of carbohydrates, fats, and proteins, as outlined in Chapter 3.

The many uses of water

Water is inorganic and does not yield energy to the tissues. However, it is so vital in the maintenance of life that a person deprived of it dies sooner than he would if deprived of other types of food.

If you weigh 100 pounds, your body contains between 60 and 70 pounds of water. Much of this water is organized into your body protoplasm and in the spaces between the cells. The fluid part of the blood, called *plasma,* is 91 to 92 percent water. Water is essential in the plasma as a solvent for the food and waste products that are transported to and from the body tissues. Water serves further as a solvent in the movement of dissolved foods from the digestive tract to the blood. Tissue wastes from the skin and kidneys are dissolved in water. The kidneys alone pass two to five pints of water daily.

Table 41-1 FOOD SUBSTANCES

SUBSTANCE	ESSENTIAL FOR	SOURCE
A. Inorganic compound		
Water	composition of protoplasm, tissue fluid, and blood; dissolving substances	all foods (released during oxidation)
B. Mineral salts		
sodium compounds	blood and other body tissues	table salt, vegetables
calcium compounds	deposition in bones and teeth, heart and nerve action, clotting of blood	milk, whole-grain cereals, vegetables, meats
phosphorus compounds	deposition in bones and teeth; formation of ATP, nucleic acids	milk, whole-grain cereals, vegetables, meats
magnesium	muscle and nerve action	vegetables
potassium compounds	blood and cell activities growth	vegetables
iron compounds	formation of red blood corpuscles	leafy vegetables, liver, meats, raisins, prunes
iodine	secretion by thyroid gland	seafoods, water, iodized salt
C. Complex organic substances		
Vitamins	regulation of body processes, prevention of deficiency diseases	various foods, especially milk, butter, lean meats, fruits, leafy vegetables; also made synthetically
D. Organic nutrients		
carbohydrates	energy (stored as fat or glycogen) bulk in diet	cereals, bread, pastries, tapioca, fruits, vegetables
fats	energy (stored as fat or glycogen)	butter, cream, lard, oils, cheese, oleomargarine, nuts, meats
proteins	growth, maintenance, and repair of protoplasm	lean meats, eggs, milk, wheat, beans, peas, cheese

The loss of water by sweat also aids in the regulation of heat loss from the body. *Evaporation,* or the change of a substance from a liquid to a gas, requires heat. Therefore, when perspiration evaporates from the body surface, heat resulting from internal oxidation is lost.

The necessary water loss from the body must be balanced by an adequate intake of water. Water requirements are met in three ways: (1) some water is present in the food you eat; (2) some is formed as a by-product of oxidation reactions and dehydration synthesis in the cells; and (3) some is consumed as drinking water.

If adequate water intake does not occur, water is lost, first from the intercellular spaces, and then from the cells themselves. The protoplasm then becomes more solid, the cell cannot function, and dies. This water loss is part of the process called *dehydration.*

The importance of mineral salts in the body

Table salt, or *sodium chloride,* is consumed directly and in considerable quantities in the diet. Other mineral salts are also present in food. Since salts are lost in perspiration, persons exposed to excessive heat over long intervals must either increase the salt in the diet or supplement the normal diet with salt tablets.

Animals require *calcium* and *phosphorus* in greater abundance than other mineral elements. Calcium is necessary for proper functioning of plasma membranes, while phosphorus is a component of ATP, DNA, and RNA. Calcium phosphate is important in the formation of bones and teeth. These two elements form about 5 percent of animal tissue when they are combined with other elements in proteins. Milk is an ideal source of these two elements; other sources include whole-grain cereals, meat, and fish. Calcium is also needed to insure the proper clotting of blood and together with *magnesium* is essential to nerve and muscle action. *Potassium compounds* are necessary for growth.

Iron compounds are essential for the formation of red blood corpuscles. Meats, green vegetables, and certain fruits, such as plums or prunes and raisins, are important sources of iron in the diet. *Iodine salts* are essential in the formation of the thyroid gland's secretion. Iodine can be obtained from drinking water or by eating seafoods.

Minerals are vital to the body in many ways. Each of them, however, must be in a compound form before it can be used by the body. Eating chemically pure elements such as sodium or chlorine would be fatal. When these elements are in a compound form, such as sodium chloride, however, they are harmless and in fact essential to the body.

Vitamins—organic compounds essential to proper body functioning

In 1911, Dr. Casimir Funk found that certain substances that are not ordinary nutrients are present in very small amounts in foods. These substances seemed to be necessary for normal growth and body activity and in the prevention of certain diseases called *deficiency diseases*. Dr. Funk called these substances *vitamins*. Vitamins were first designated by letters – A, B, C, and so on. Later, it was discovered that certain vitamins thought to be simple were made up of many different components (for example, the vitamin-B complex). Then such names as B_1, B_2, and so forth were adopted. Today, most of the vitamins have names that indicate their chemical composition, although letters are. still used as a means of easy and simple reference. *Vitamins are organic substances that are indispensable for life but are not required as a source of energy.* They act as catalysts, and in this respect they are similar to the digestive enzymes. The best sources, functions, and deficiency symptoms of the better-known vitamins are summarized in Table 41-2.

In early times, when sailors went on long journeys, they lived for months on preserved food. It often happened that many became ill with scurvy and died from hemorrhages of the gums and internal organs. The British sailors found that simply including citrus fruits in their diets prevented scurvy. Since barrels of limes were carried aboard their ships thereafter, the sailors earned the name "limeys." It was the vitamin C in the citrus fruit that prevented the disease. Apparently, man cannot manufacture this substance, so it is necessary to include it in his diet. Some mammals, however, such as the rat and the hamster, can synthesize sufficient amounts of ascorbic acid (vitamin C).

Some vitamins can be stored in the body, while others must be supplied constantly because the excess in the diet is excreted in the urine. Vitamin D can be produced in the skin. Other vitamins or their precursors must be supplied by the diet or taken in the form of extracts, if the normal diet lacks them. But the best source of vitamins is a balanced diet.

Synthetic vitamins

Most of the vitamins listed in Table 41-2 can be purchased in highly concentrated synthetic form. These preparations are important in supplementing the natural vitamins of the diet when deficiency occurs. However, even with all the publicity given commercial vitamin preparations, remember that a normal, balanced diet is much more desirable for good health than supplementary doses of vitamins. Your doctor can diagnose vitamin deficiency and prescribe concentrated vitamins if he thinks you need them. If a proper diet is followed, additional vitamins are probably a waste of money and are unnecessary for the average person.

What are organic nutrients?

We call carbohydrates, fats, and proteins *organic nutrients* because they are originally formed by living cells and contain

Table 41-2 FUNCTIONS AND IMPORTANT SOURCES OF VITAMINS

VITAMINS	BEST SOURCES	ESSENTIAL FOR	DEFICIENCY SYMPTOMS
vitamin A (oil soluble)	fish-liver oils liver and kidney green and yellow vegetables yellow fruit tomatoes butter egg yolk	growth health of the eyes structure and functions of the cells of the skin and mucous membranes	retarded growth night blindness susceptibility to infections changes in skin and membranes defective tooth formation
thiamin (B_1) (water soluble)	seafood meat soybeans milk whole grain green vegetables fowl	growth carbohydrate metabolism functioning of the heart, nerves, and muscles	retarded growth loss of appetite and weight nerve disorders less resistance to fatigue faulty digestion (beriberi)
riboflavin (B_2 or G) (water soluble)	meat soybeans milk green vegetables eggs fowl yeast	growth health of the skin and mouth carbohydrate metabolism functioning of the eyes	retarded growth dimness of vision inflammation of the tongue premature aging intolerance to light
niacin (water soluble)	meat fowl fish peanut butter potatoes whole grain tomatoes leafy vegetables	growth carbohydrate metabolism functioning of the stomach and intestines functioning of the nervous system	smoothness of the tongue skin eruptions digestive disturbances mental disorders (pellagra)
vitamin B_{12} (water soluble)	green vegetables liver	preventing pernicious anemia	a reduction in number of red blood cells
ascorbic acid (C) (water soluble)	citrus fruit other fruit tomatoes leafy vegetables	growth maintaining strength of the blood vessels development of teeth gum health	sore gums hemorrhages around the bones tendency to bruise easily (scurvy)
vitamin D (oil soluble)	fish-liver oil liver fortified milk eggs irradiated foods	growth regulating calcium and phosphorus metabolism building and maintaining bones, teeth	soft bones poor development of teeth dental decay (rickets)
tocopherol (E) (oil soluble)	wheat-germ oil leafy vegetables milk butter	normal reproduction	(undetermined)
vitamin K (oil soluble)	green vegetables soybean oil tomatoes	normal clotting of the blood normal liver functions	hemorrhages

the element carbon. Carbohydrates and fats supply energy. The tissue-building value of foods can only be measured by observing growth in animals when they are fed. But the energy value can be measured in heat units called *food Calories,* or *nutritional Calories.* One food Calorie is the amount of heat required to raise the temperature of 1,000 grams, or 1 kilogram (the mass of about one quart), of water 1 centigrade degree. This amount of heat is 1,000 times greater than the "small" calorie used by the physical scientist. Thus, one food Calorie is equivalent to 1,000 "small" calories, which is the same as 1 kilocalorie (1 kcal).

Many people fail to realize that when they are talking about how many "calories" there are in a certain amount of food, they are actually referring to the number of kilocalories, not the number of "small" calories. For example, a piece of apple pie that is said to contain 310 "calories" really contains 310 food Calories, or kilocalories. This is a large amount of energy. It is approximately equivalent to the amount of energy in one pound of coal or one-sixth cup of gasoline.

The number of food Calories required in an average day's activity varies with the kind of activity and with the age and body build of the person concerned. A daily requirement of 2,500 to 3,500 food Calories is probably above average.

More than half of your total diet is carbohydrate food. Regardless of this high percentage, the accumulated carbohydrate reserve in your body is less than one percent of your total weight. This is evidence that carbohydrates are primarily fuel foods and that they are oxidized rapidly to supply the energy required for body activity.

The significance of carbohydrates in the diet

Many different kinds and forms of *carbohydrate* foods are included in the average diet. Some are easily digested and are transported to the tissues with little chemical change. Others require more chemical simplification for tissue use. One group, while indigestible, provides necessary bulk, or roughage, in the diet. All carbohydrates that are digestible reach the body tissues as *glucose,* or *dextrose,* but carbohydrates need not be reduced to glucose to be absorbed in the intestines.

Many simple sugars are present in the foods you eat. These include glucose, fructose, and galactose. As you learned in Chapter 3, these sugars are classed as monosaccharides because they consist of single hexose molecules with the chemical formula $C_6H_{12}O_6$. These sugars are quick-energy sources because they require little or no chemical change before they are absorbed by the blood from the digestive organs.

Sucrose (cane sugar), *lactose* (milk sugar), and *maltose* (malt sugar) are *disaccharides,* composed of two hexose units. These double sugars undergo *hydrolysis* (see Chapter 3) and are thereby broken into simple sugar molecules for absorption.

Starches, or polysaccharides, compose a large part of the carbohydrate portion of the diet. They are abundant in cereal grains such as wheat, corn, rye, barley, oats, and rice, in addition to potatoes and tapioca.

As you learned in Chapter 3, starches are composed of large chains of glucose units, each represented as $C_6H_{10}O_5$. Starch digestion involves the addition of a water molecule to a glucose unit. Maltose is formed during starch digestion. Later, this double sugar is reduced to glucose and absorbed by the blood, which carries it to the body tissues.

Much of the glucose received by the blood and transported to the liver is converted temporarily to animal starch, or *glycogen.* As glucose is oxidized in the body tissues, glycogen is changed back to glucose (dextrose) and released into the bloodstream. In this way, the level of blood sugar is maintained. Were it not for this action of the liver, we would constantly have to eat small quantities of carbohydrate foods.

Celluloses are complex carbohydrates present in the cell walls of vegetable foods. These materials cannot be digested in the human system. As roughage in the digestive system, however, they expand the intestines and stimulate muscle contractions of the walls, resulting in movement of the food content. This muscular activity is necessary for normal digestion.

41-2 A burning Brazil nut—a graphic demonstration of the large amount of chemical energy stored in a small amount of food. (John King)

Fats are highly concentrated energy foods

Fats and *oils* yield more than twice as much energy as carbohydrates. Common sources of these foods in the diet include butter, cream, cheese, oleomargarine, lard and other shortenings, vegetable oils, and meats.

Fats are slowly hydrolyzed through enzyme action during digestion. This happens in a series of chemical reactions in which three water molecules combine with each fat molecule to yield one molecule of *glycerin* (glycerol) and three molecules of *fatty acids.*

Body fats are formed by the conversion of excess carbohydrates. The chief storehouses of body fat are the tissue spaces beneath the skin, the region of the kidneys, and the liver. Excess body fat is detrimental to health. For this reason, both the carbohydrate and fat content of the diet should be carefully regulated.

Proteins and their uses

You learned in Chapter 3 that *proteins* are complex organic molecules that occur in an almost unlimited variety of chemical structures. They are composed of large numbers of units known as *amino acids.* The proteins you consume as food are foreign to your body and cannot be used in your tissues. However, reduced to amino acids during digestion, they supply the units required by your cells in synthesizing your own specific protein molecules. Both growth and repair of body substances depend on protein intake in your diet.

Certain of the amino acid molecules absorbed by the blood are not used in cell protein synthesis. These are broken down by chemical activity of the liver, called *deamination,* into two parts. One part contains carbon and is sent to the tissues as glucose. The other part, the nitrogen-containing part, is synthesized as urea, a waste product received by the blood and transported to the kidneys for excretion in the urine. This urea is added to that formed during the breakdown of tissue protein.

The most valuable protein sources in the diet include lean meat, eggs (albumen), milk (casein), cheese, whole wheat (gluten), beans, and corn. These are body-building foods, essential in growth during childhood and young adult life and in the maintenance and repair of protoplasm in the mature years.

The phases of digestion

There are two reasons why tissues cannot use most foods in the forms in which you eat them. First, many substances are insoluble in water and could not enter the plasma membranes of the cells even if they reached them. Second, these foods are too complex chemically for tissues to use, either in oxidation or for growth and repair by protein synthesis. Digestion brings about changes in both of these conditions, with the result that cells can absorb and use the products. Thus, in *digestion* complex foods are broken down into smaller molecules of water-soluble substances that can be used by the body cells.

The first part of the change that occurs during digestion is *mechanical.* This phase involves the chewing of food in the mouth and the constant churning and mixing action brought about by the muscular movement of the walls of the digestive organs. The breakdown of food into small particles and the thorough mixing with various juices aid the second phase of digestion, which is *chemical.* This phase is accomplished by digestive enzymes that are present in various secretions produced by the digestive glands. In studying the entire process of digestion, you will find it helpful to refer to the Trans-Vision of the human torso between pages 630 and 631.

The digestive system includes the organs that form the *alimentary canal,* or food tube (Figure 41-3). It also includes those organs that do not actually receive undigested food but that act on foods in the alimentary canal by means of secretions delivered to the canal by various ducts. *Ducts* are tubes extending from certain glands into the digestive organs.

The mouth—the first digestive structure

The mouth is an organ of sensation and of speech, but its chief function is to prepare food for digestion. The *hard palate*

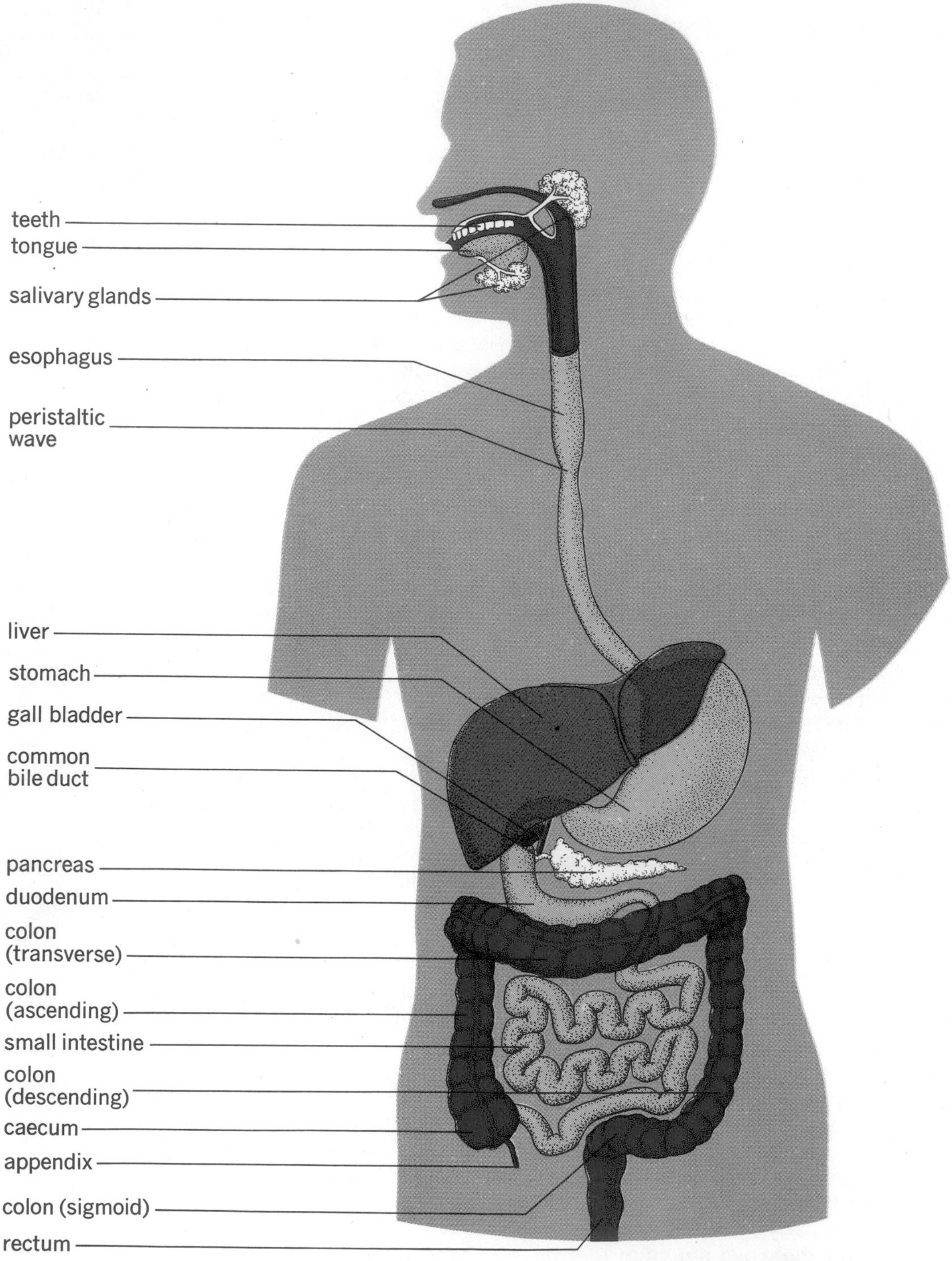

teeth

tongue

salivary glands

esophagus

peristaltic
wave

liver

stomach

gall bladder

common
bile duct

pancreas

duodenum

colon
(transverse)

colon
(ascending)

small intestine

colon
(descending)

caecum

appendix

colon (sigmoid)

rectum

41-3 The organs of digestion in
the human body.

forms the roof of the mouth in the chewing area. It consists of a bony structure covered with several membranes. The *soft palate* lies just behind the hard palate. It is formed by folded membranes that extend from the rear portion of the hard palate and fasten along the sides of the tongue. You can see a knob-like extension of the soft palate called the *uvula* (YOO-vyuh-luh) in a mouth that is opened wide (Figure 41-4).

The back of the mouth opens into a muscular cavity called the *pharynx* (FA-ringks). This cavity extends upward, above the soft palate, to the nasal cavity. The soft palate partly separates the nasal cavity from the mouth cavity and extends into the pharynx, somewhat like a curtain, as you can see in Figure 41-5. The inside of the cheeks forms the side walls of the mouth cavity. The cheek linings are mucous membranes, containing numerous mucous glands. *Mucus,* a lubricating secretion, mixes with food in the mouth and aids in chewing and swallowing. The lining of the mouth turns outward to form the lips.

The salivary glands

The *parotid glands* are the largest of the *salivary glands*. One lies on each side of the face below and in front of the ears. Ducts from these glands empty *saliva* into the mouth, opposite the second upper molars. The disease called mumps is an infection of the parotid glands that causes swelling and irritation. The *submaxillary glands* lie within the angles of the lower jaws. The *sublingual glands* are embedded in the mucous membranes in the floor of the mouth, under the tongue (Figure 41-6). Ducts from both of these glands open into the floor of the mouth under the tongue. The smell of food, the sight of it, the presence of it in the mouth, and the taste of it stimulate the secretion of saliva. In other words, your mouth "waters."

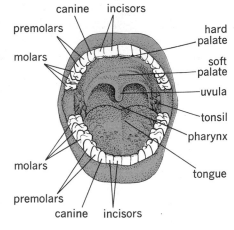

41-4 Digestion begins in the mouth, the first organ of the alimentary canal.

41-5 A section of the head showing structures of the mouth and throat (below left).

41-6 A section of the head showing location of the three pairs of salivary glands (below right).

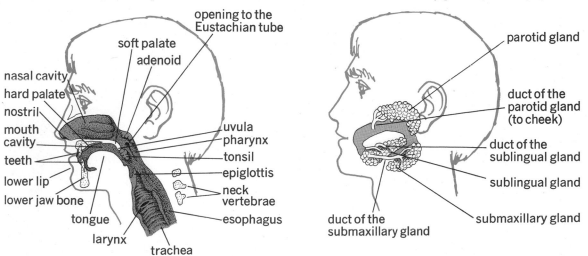

The tongue and its functions

The tongue lies in the floor of the mouth and extends into the throat. This muscular organ performs several different functions:

■ It acts as an organ of taste. Scattered over the surface of the tongue are numerous tiny projections. These projections contain *taste buds,* which have nerve endings at their bases. When food is mixed with mucus and saliva, it makes contact with the taste buds, thus stimulating the nerve endings.

■ The tongue aids in chewing by keeping the food between the teeth.

■ During swallowing, food is worked toward the back of the tongue. When the tongue is jerked downward, food lodges in the pharynx and passes into the esophagus opening. The opening of the trachea is closed by the pressure of the tongue, and breathing ceases for a moment during the process of swallowing.

■ The tongue keeps the inner surface of the teeth clean, because you roll it around in your mouth.

■ The tongue is essential in speech. In the formation of word sounds, the tongue acts together with the lips, teeth, and hard palate. Without such interaction, these sounds could not be formed into words.

The types of teeth

Starting between the two front teeth and counting back, your permanent teeth are arranged in the following order: The first two are the flat *incisors,* with sharp edges for cutting food (Figure 41-4). Next, near the corner of your lips, there is a large conical *canine,* or cuspid, tooth. These teeth are often called eye teeth, although they have no connection with the eyes. Behind the canine teeth are two *premolars,* or bicuspids. Next come three *molars* (two, if you have not cut your wisdom teeth). The premolars and molars have flat surfaces and are adapted for grinding and crushing. Many jaws are too small to provide space for the third molars, or wisdom teeth. In such jaws, they may grow in crooked, lodge against the second molars, or remain impacted.

The structure of the teeth

A tooth is composed of three general regions. The exposed portion above the gum line is called the *crown*. A narrow portion at the gum line is called the *neck,* while the *root* is encased in a socket in the jawbone and holds the tooth securely in place. Roots vary in form in the different kinds of teeth. They may be long and single, or they may consist of two, three, or four projections. The crown is covered with a hard white substance, the *enamel.* The covering of the root is called *cementum;* this holds the tooth firmly together. The root is anchored firmly in the jaw socket by the fibrous *periodontal membrane.*

If you cut a tooth lengthwise, you can see the *dentine* beneath the protective layers of enamel and cementum (Figure 41-7). Dentine is a softer substance than enamel and forms the bulk of the tooth. The *pulp cavity* lies inside the dentine area.

The structure of the esophagus and stomach

After a food mass is ground between the teeth, rolled on the tongue, and mixed with saliva, it passes through the pharynx to the *esophagus.* This is a tube about a foot long that connects the mouth to the stomach. Food travels to the stomach with the aid of layers of smooth muscle in the wall of the esophagus. One layer is circular and squeezes inward. The other layer is longitudinal and contracts in a wave that travels downward, pushing the food ahead of it.

The *stomach* lies in the upper left region of the abdominal cavity just below the diaphragm. The stomach walls contain three layers of smooth muscle, each arranged differently. One layer is longitudinal, one is circular, and one is angled, or oblique. Contraction of the smooth muscle fibers of the various layers in different directions causes a twisting, squeezing, and churning movement of the stomach.

The lining of the stomach is a thick, wrinkled membrane in which numerous *gastric glands* are embedded. Each of these glands is a tiny tube with an opening that leads into the stomach. The walls of each gland are lined with secretory cells. There are three kinds of glands. One kind secretes an enzyme; a second secretes hydrochloric acid; and a third, mucus. Together, these secretions form the *gastric fluid.*

Food usually remains in the stomach for from two to three hours. During this period, rhythmic contractions of the stomach muscles churn the food back and forth in a circular path. This action separates the food particles and mixes them thoroughly with the stomach secretions. At the completion of stomach digestion, the valve at the intestinal end, the *pyloric valve,* opens and closes several times. With each opening of the valve, food moves into the small intestine. Finally, the stomach is relieved of its contents and begins a period of rest. After several hours without food, the stomach starts contracting again. These contractions cause the sensation of hunger.

The small intestine

When food leaves the stomach through the pyloric valve, it enters the *small intestine,* a tube about one inch in diameter and twenty-three feet long. This part of the alimentary canal is the most vital of all digestive organs. The upper ten inches of the small intestine are called the *duodenum* (DOO-uh-DEE-num). The duodenum curves upward, then backward and to the right, beneath the liver. Beyond the duodenum is a second and much longer region, the *jejunum* (ji-JOO-num). This portion, about

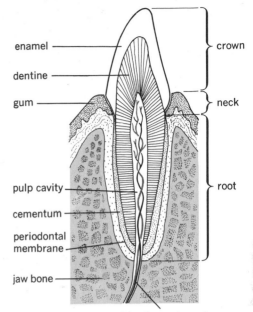

enamel
dentine
gum
crown
neck
pulp cavity
cementum
periodontal membrane
jaw bone
root
blood vessels and nerve fibers to pulp cavity

41-7 The structure of a canine tooth.

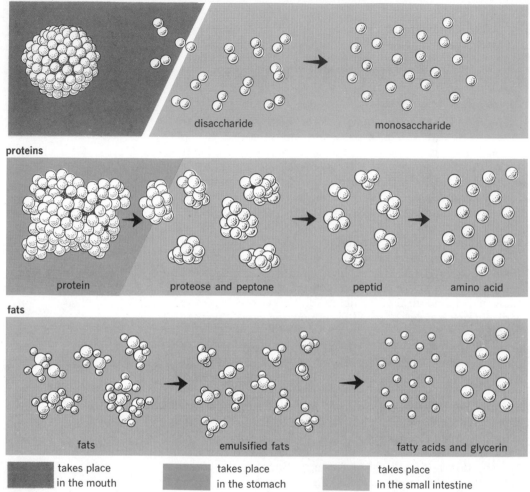

carbohydrates

disaccharide   monosaccharide

proteins

protein   proteose and peptone   peptid   amino acid

fats

fats   emulsified fats   fatty acids and glycerin

takes place in the mouth

takes place in the stomach

takes place in the small intestine

41-8 Phases in the digestion of carbohydrates, proteins, and fats. Why can they not be used by the tissues of the body in the form in which they are eaten?

seven feet long, is less coiled than the other regions. The lower portion of the small intestine, called the *ileum,* is about fifteen feet long and coils through the abdominal cavity before joining the large intestine.

Embedded in the mucous lining of the small intestine are many tiny *intestinal glands*. These glands secrete *intestinal fluid,* containing four enzymes, which passes into the small intestine.

The liver—the largest gland in the body

The *liver* weighs about three and one-half pounds and is a dark chocolate color. It lies in the upper right region of the abdominal cavity and secretes *bile*. Bile is a brownish-green fluid that passes from the liver through a series of *bile ducts* that form a Y. As bile is secreted in the liver, it passes down one branch of the Y, then travels up the other branch to the *gall bladder.* Here the bile is stored and concentrated as part of the water is removed. The base of the Y is the common bile duct, which carries bile from the gall bladder to the duodenum. If the common bile duct becomes clogged by a gallstone or a plug of mucus, bile enters the bloodstream and causes *jaundice,* a yellowing of the eyes and skin.

The pancreas and pancreatic fluid

The *pancreas* is a many-lobed, long, whitish gland, quite similar in general appearance to a salivary gland. It lies behind the stomach and the upper end of the small intestine, against the back wall of the abdominal cavity, and it performs two entirely different functions. The production of insulin by the pancreas will be discussed in Chapter 46. *Pancreatic fluid,* a digestive secretion, passes into the small intestine through the *pancreatic duct,* which leads to a common opening with the bile duct in the wall of the duodenum.

The large intestine, or colon

The small intestine ends at a junction with the large intestine, or *colon,* in the lower right region of the abdominal cavity. Below the point of junction is the *caecum,* a blind end of the large intestine. The *vermiform appendix* is a fingerlike outgrowth of the caecum. *Appendicitis* is an inflamed condition of the appendix resulting from infection.

The colon is usually five to six feet long and about three inches in diameter. It forms an inverted U in the abdominal cavity. The *ascending colon* runs upward along the right side, where it curves abruptly to the left to form the *transverse colon.* This portion extends across the upper region of the abdominal cavity. Another curve leads to the *descending colon* on the left side. At its lower end the descending colon becomes the *sigmoid colon,* so called because of its S shape. The *rectum* is a muscular cavity at the end of the large intestine. The lower end of the rectum forms the **anal opening.** A valvelike muscle in the lower end of the rectum controls the elimination of intestinal waste.

The chemical phases of digestion

As foods move through the organs of the alimentary canal, a series of chemical changes occur in the step-by-step process of simplification. Each of these changes requires a specific enzyme in a digestive secretion. The chemical changes generally involve hydrolysis, since water molecules interact with molecules of the various food materials. Digestive enzymes are therefore hydrolytic enzymes, each associated with the splitting of specific molecules. Digestion in the alimentary canal is necessarily extracellular. That is, enzymatic action takes place outside the cells rather than inside, as it does in some other forms.

Digestion in the mouth

The chemical action on food begins in the mouth, where an enzyme in saliva begins the hydrolysis of starch. Saliva is a thin, alkaline secretion of the salivary glands. It is more than 95 percent water and contains mineral salts, lubricating mucus, and the enzyme *ptyalin* (TIE-uh-lin), sometimes called *salivary amylase* (AM-i-LACE). This enzyme converts cooked starch to maltose, a disaccharide. It is necessary to cook starchy foods such as potatoes in order to burst the cellulose cell walls. This allows the ptyalin to contact the starch grains. Because of the short time food is in the mouth, starch digestion is seldom completed when food is swallowed. However, ptyalin continues to act in the stomach.

The action of gastric fluid

The principal enzyme in gastric fluid is *pepsin,* sometimes called *gastric protease.* This enzyme acts on protein, splitting the complex molecules into simpler groups of amino acids known as *peptones* and *proteoses.* This is the first in a series of chemical changes involved in protein digestion.

Hydrochloric acid, in addition to providing the proper medium for the action of pepsin, dissolves insoluble minerals and kills many bacteria that enter the stomach with food. It also regulates the action of the pyloric valve, which opens at the completion of stomach digestion and allows food to pass to the small intestine.

The food passing from the stomach to the small intestine contains the following: (1) fats, unchanged; (2) sugars, unchanged; (3) the starches that were not acted on by ptyalin; (4) maltose formed by the action of ptyalin; (5) coagulated milk casein; (6) those proteins that were unchanged by the pepsin of the gastric fluid; and (7) peptones and proteoses formed from pepsin acting on protein.

Functions of the liver and bile

The liver performs several vital functions. In receiving glucose from the blood and changing it to glycogen, it serves as a chemical factory. It serves also as a storehouse in holding reserve carbohydrates as glycogen. In acting on amino acids and forming urea, it is an organ of excretion.

As a digestive gland, the liver secretes bile, which acts on food in the small intestine. In the formation of bile, the liver plays a part in using what might otherwise be discarded as waste. Part of the bile is formed from worn-out hemoglobin that the blood system can no longer use. Bile has the following important characteristics:

■ It is partially a waste substance containing material from dead red blood corpuscles filtered from the bloodstream by the liver.

■ It increases the digestive action of lipase, an enzyme produced in the pancreas, by acting to break globules of fat into smaller, more easily acted on droplets.
■ It is a specific activator of lipase.

Actually, bile is not a digestive secretion. In the splitting of large fat particles into smaller ones, a milky colloid called an *emulsion* is produced. In this form, pancreatic fluid can act on fats more readily.

The role of the pancreas in digestion

Pancreatic fluid acts on all three classes of organic nutrients. Pancreatic fluid contains the following three enzymes: (1) *trypsin;* (2) *amylase;* and (3) *lipase* (LIP-ace). Trypsin continues the breakdown of proteins that began in the stomach by changing peptones and proteoses into still simpler amino acid groups called *peptides*. In addition, trypsin may act on proteins that were not simplified during digestion in the stomach. Peptides are not the final product of protein digestion, because one additional step is necessary to form the amino acids used in protein synthesis by the cells. Amylase duplicates the action of the ptyalin in saliva by changing starch into maltose. This is how the potatoes you do not chew enough are changed into sugar. Lipase splits fat into *fatty acids* and *glycerin,* both of which can be absorbed by the body cells.

Digestion in the small intestine

The intestinal fluid secreted by the intestinal glands is highly alkaline and contains four principal enzymes: (1) *erepsin;* (2) *maltase;* (3) *lactase;* and (4) *sucrase*. Erepsin completes protein digestion by changing peptides, formed by the pancreatic fluid, into amino acids. Maltase splits the disaccharide maltose into the monosaccharide glucose, the final product of carbohydrate digestion. Lactase has a similar action on lactose, or milk sugar, in changing it into glucose and galactose. Sucrase acts on sucrose and changes it into the simple sugars glucose and fructose.

Thus, with the combined action of bile, pancreatic fluid, and intestinal fluid in the small intestine, all three classes of foods are completely digested. As soluble substances in the form of simple sugars, fatty acids and glycerin, and amino acids, they leave the digestive system and enter the blood and lymph.

Absorption in the small intestine

A magnified portion of the small intestine shows that its irregular lining gives rise to great numbers of fingerlike projections called *villi*. These projections are so numerous that they give a velvety appearance to the intestinal lining. Within the villi are *blood vessels* and branching lymph vessels called *lacteals* (Figure 41-9). The villi bring blood and lymph close

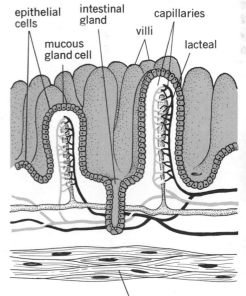

epithelial cells intestinal gland capillaries villi lacteal mucous gland cell

muscle of intestinal wall

41-9 The absorption surface of the small intestine is greatly increased by the villi.

to the digested food and increase the absorption surface of the intestine enormously. Absorption is increased further by a constant swaying motion of the villi through the intestinal content.

Glycerin and fatty acids enter the villi and are carried away by the lymph. They eventually reach the general circulation and travel to the tissues. Monosaccharides and amino acids, however, enter the blood vessels of the villi. From the villi, they are carried directly to the liver through the portal vein.

Water absorption in the large intestine

The large intestine receives a watery mass of undigestible food bulk from the small intestine. As this mass progresses

Table 41-3 SUMMARY OF DIGESTION

PLACE OF DIGESTION	GLANDS	SECRETION	ENZYMES	DIGESTIVE ACTIVITY
mouth	salivary	saliva	ptyalin	changes starch to maltose, lubricates
	mucous	mucus		lubricates
esophagus	mucous	mucus		lubricates
stomach	gastric	gastric fluid	pepsin	changes proteins to peptones and proteoses
		hydrochloric acid		activates pepsin; dissolves minerals; kills bacteria
	mucous	mucus		lubricates
small intestine	liver	bile		emulsifies fats; activates lipase
	pancreas	pancreatic fluid	trypsin	changes proteins, peptones, and proteoses to peptides
			amylase	changes starch to maltose
			lipase	changes fats to fatty acids and glycerin
	intestinal glands	intestinal fluid	erepsin	changes peptides to amino acids
			maltase	changes maltose to glucose
			lactase	changes lactose to glucose and galactose
			sucrase	changes sucrose to glucose and fructose
	mucous	mucus		lubricates
large intestine (colon)	mucous	mucus		lubricates

through the colon, much of the water is absorbed and taken into the tissues. The remaining intestinal content, or *feces* (FEE-seez), becomes more solid as the water is absorbed. The feces pass into the rectum, from which they are eventually eliminated through the anal opening.

Table 41-3 summarizes the digestive processes that occur in the various organs.

SUMMARY

The human body requires a variety of complex organic nutrients as well as water, minerals, and vitamins. The organic nutrients are divided into three groups: carbohydrates, fats, and proteins.

The digestive system is a tube divided into various regions. Each particular region of the tube is a specialized organ adapted for performing certain phases of the digestive process. Many glands pour their enzymatic secretions into the digestive tract. These enzymes cause the chemical changes in foods, while muscular contractions bring about the mechanical changes.

BIOLOGICALLY SPEAKING

food	digestion	pyloric valve
plasma	alimentary canal	small intestine
deficiency disease	pharynx	liver
vitamin	mucus	gall bladder
organic nutrients	salivary glands	pancreas
food Calorie	stomach	villi

QUESTIONS FOR REVIEW

1. What are the functions of foods?
2. What are some important functions of water in the body? How is it obtained?
3. In what two general ways must foods be changed during digestion?
4. List, in order, the divisions of the alimentary canal. What digestive processes occur in each?
5. Discuss five or more ways in which the tongue is used.
6. Name the regions that can be distinguished in a tooth cut lengthwise.
7. Why is it especially important that you chew bread and potatoes thoroughly?
8. Suppose that a person had a glass of milk and a sandwich consisting of bread, butter, and ham. Tell what would happen to each of these foods as it was digested.
9. Name two important functions of the large intestine.

APPLYING PRINCIPLES AND CONCEPTS

1. Explain how a vitamin deficiency is possible even if an adequate amount of all the vitamins is taken daily.
2. Why is food acid in the stomach and alkaline in the small intestine?
3. Explain how interference with the rhythmic waves of the walls of the large intestine may cause either constipation or diarrhea.
4. Why is it easier to digest sour milk than fresh milk?

RELATED READING

Books

AMES, GERALD, and ROSE WYLER, *Food and Life.*
Creative Education Press, New York. 1966. A fine book on nutrition.

FABER, DORIS, *The Miracle of Vitamins.*
G. P. Putnam's Sons, New York. 1964. How man's curiosity about nutrition led to the establishment of the value of minerals and vitamins.

MICKELSON, OLAF, *Nutrition Science and You.*
McGraw-Hill Book Company, New York. 1964. Story of nutrition science in action.

PIKE, RUTH L., and MYRTLE L. BROWN, *Nutrition: An Integrated Approach.*
John Wiley and Son Inc., New York. 1967. An advanced book which integrates Biochemistry, Physiology, and Nutrition at the cellular level with application to the nutrient needs of the complex organism.

SIMEONS, A. J. W., MD., *Food: Facts, Foibles, and Fables.*
Funk and Wagnalls, New York. 1968. An unusual account of the development of man's feeding from prehistoric times to present day.

VONHALLER, ALBERT, *The Vitamin Hunters.*
Chilton Book Company, Philadelphia. 1962. Engrossing story of the scientists who discovered the causes and cures of nutritional disease.

CHAPTER FORTY-TWO

TRANSPORT AND EXCRETION

The transport system

The flow of nutritive fluids, waste materials, and water in living organisms is called *circulation.* The system can be called the *circulatory system,* or *transport system.*

The sponges accomplish circulation by literally pumping the ocean into their bodies! The seawater supplies each cell with its individual oxygen needs and washes its wastes away. Actually, the cells in man's body are bathed in a fluid with a salt content very much like seawater. We call this solution *tissue fluid.*

However, the circulatory system in man is more complex than that of the invertebrates. Man produces his own "seawater" and adds other vital substances to it. Then this fluid is piped through his body and circulated with a pump—the heart. If the pump stops working, man's cells are in the same predicament in which a sponge would be if it were thrown up on the beach.

Blood—a fluid tissue

Blood, the transporting medium for all body substances, is a peculiar type of connective tissue in that the cells are scattered among the nonliving substances composing the fluid portion. The average person has about 12 pints of blood, which composes about 9 percent of the body weight. The blood consists of a fluid portion, or *plasma,* and the blood cells, or *solid components.* The plasma is a straw-colored, sticky liquid of which nine tenths is water. The proteins in the plasma give it its sticky quality. One of them, *fibrinogen* (fy-BRIN-uh-jin), is essential in the clotting of blood. Another is *serum albumin,* which is necessary to normal blood tissue relationships during absorption. The third is *serum globulin,* which gives rise to antibodies that provide immunity to various diseases. *Prothrombin,* an

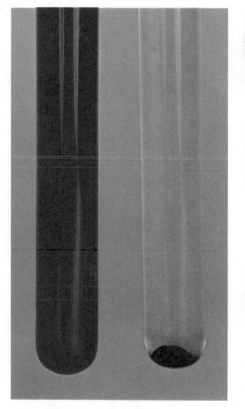

42-1 Fresh blood in a test tube and blood in a test tube after settling and centrifugation. (Percy W. Brooks)

enzyme found in plasma, is produced in the liver in the presence of vitamin K. It is inactive normally but functions during clotting. The proteins in the serum also function to give the blood a thickness, or *viscosity,* that aids in maintaining pressure within the vessels.

The following materials are also present in plasma:

■ *Inorganic minerals,* dissolved in water, give plasma a salt content of approximately 1 percent, while that of seawater is approximately 3 percent. These compounds include carbonates, chlorides, and phosphates of the elements calcium, sodium, magnesium, and potassium. They are absolutely essential to the blood and to the normal functioning of body tissues. Without calcium compounds, blood would not clot in a wound.

■ *Digested foods* are present in plasma in the form of glucose, fatty acids and glycerin, and amino acids. These are transported to the liver and other places of storage and to the body tissues.

■ *Nitrogenous* (ny-TROJ-uh-nus) *wastes,* resulting from protein metabolism in tissues, and *urea,* produced largely in the liver during the breakdown of amino acids, travel in the plasma to the organs of excretion.

The solid components of blood

The red corpuscles (red blood cells, or erythrocytes), the white corpuscles (white blood cells, or leucocytes), and the platelets (thrombocytes) are the three solid components of blood (Figure 42-2). The *red corpuscles* are shaped like disks, both sides of which are concave. Sometimes they travel in the blood in rows that resemble stacks of coins, although they may separate and float individually. The red cells are so small that ten million of them can be spread out in one square inch. They are so numerous that placed side by side they would cover an area of 3,500 square yards. It is estimated that the blood of a normal person contains twenty-five trillion red blood cells, or enough to go around the earth four times at the equator, if they were laid side by side. The pigment in the red blood cell is *hemoglobin.* This protein substance gives blood its red color and is essential to life.

The erythrocytes are produced by the *red marrow* of such bones as the ribs, vertebrae, and skull. In children, the ends of the long bones also function in this manner. During their development, the red blood cells are large, colorless, and have large nuclei. Normally, by the time they are to be released into the bloodstream, they have lost the nuclei and have accumulated hemoglobin. The average life span of a red corpuscle is 20 to 120 days, after which time it is removed by the liver or the spleen. At the same time certain valuable compounds are released into the bloodstream and used in the manufacture of new red blood cells.

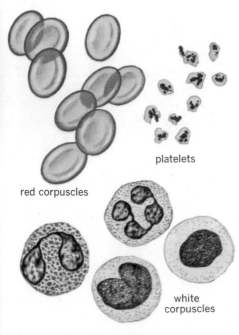

platelets

red corpuscles

white corpuscles

42-2 The solid components of blood include red corpuscles, white corpuscles, and platelets.

What do the erythrocytes do?

The pigment hemoglobin is a complex iron-containing protein that is within the cell membrane of a red blood cell. It is the chemical element iron that gives hemoglobin the ability to carry oxygen. Perhaps you have seen an iron nail turn red with rust. It has oxidized, which means that it has combined with oxygen from the air. The iron of hemoglobin combines with oxygen in the lungs, but there is an important difference. The iron of the rusty nail does not easily give up its oxygen. The iron in hemoglobin, however, gives up its oxygen at the proper time and place in the body. The erythrocytes are bright red when their pigment is combined with oxygen. In the tissues, the oxygen is given up, and part of the carbon dioxide that has formed in the tissues then combines with hemoglobin. In this way, much of the carbon dioxide is carried to the lungs, where it is released, and the cycle is repeated.

The white blood cells

Most *white corpuscles* are larger than red blood cells and differ from them in three ways:
- White corpuscles have nuclei.
- White corpuscles do not contain hemoglobin and are therefore nearly colorless.
- Some white corpuscles are capable of ameboid movement.

The white blood cells are less numerous than the red cells, the ratio being about one white cell to every six hundred red cells. White corpuscles are formed in the red bone marrow and in the lymph glands. Normally, there are about eight thousand white corpuscles in one cubic millimeter of blood. The red cells normally number four and a half to five million in one cubic millimeter of blood.

The white blood cells that can move about are able to ooze through the capillary walls into the tissue spaces. Here they engulf solid materials, including bacteria, and thus are an important defense of the body against infection. Whenever an infection develops in the tissues, the white-cell count may go from 8,000 to more than 25,000 per cubic millimeter. White corpuscles collect in the area of an infection and destroy bacteria. The remains of dead bacteria, white corpuscles, and tissue fluid is what is known as *pus*.

The platelets

Platelets are irregularly shaped, colorless bodies, much smaller than the red corpuscles. They are probably formed in the red bone marrow. Platelets are not capable of moving on their own but float along in the bloodstream. They have an important function in the formation of a blood clot.

Table 42-1 SUMMARY OF COMPOSITION OF BLOOD

PLASMA	SOLID COMPONENTS
water	red corpuscles
proteins	white corpuscles
fibrinogen	platelets
serum albumin,	
globulin	
digested foods	
mineral salts	
organic nutrients	
cell wastes	

A summary of the components found in the blood is given in Table 42-1 and an overview of its functions as a transporting medium is given in Table 42-2.

How blood clots

When you cut small blood vessels in a minor wound, blood oozes out. Such an injury is not alarming, because a clot will soon form and the blood flow will stop. You probably take this for granted, without considering what would happen if the flow did not stop.

Clotting results from the chemical and physical changes in the blood. When blood leaves a vessel, the platelets disintegrate and release *thromboplastin*. This substance reacts with *prothrombin* and with *calcium* to form *thrombin*. The thrombin changes *fibrinogen,* a blood protein, to *fibrin*. The tiny threads

Table 42-2 BLOOD AS A TRANSPORTING MEDIUM

TRANSPORTA-TION OF	FROM	TO	FOR THE PURPOSE OF
digested food	digestive organs and liver	tissues	growth and repair of cells, supplying energy, and regulating life processes
cell wastes	active tissues	lungs, kidneys, and skin	excretion
water	digestive organs	kidneys, skin, and lungs	excretion and equalization of body fluids
oxygen	lungs	tissues	oxidation
heat	tissues	skin	equalization of the body temperature
secretions	ductless glands	various organs, glands	regulation of body activities

of fibrin form a network that traps blood cells, thus forming a clot, and preventing further escape of blood (Figure 42-3). These trapped corpuscles dry out and form a scab. Healing takes place as the edges of the wound grow toward the center. If any of the substances mentioned above is not present, clotting will not occur. Clotting can be summarized as:

(1) thromboplastin + prothrombin + calcium $\rightarrow$ thrombin
(2) thrombin + fibrinogen $\rightarrow$ fibrin

If blood vessels are broken under the skin, a discolored area, known as a *bruise,* may appear as clotting occurs. Gradually, the clotted blood is absorbed, and the color of the bruise changes and finally disappears.

Blood transfusions

Conditions like hemorrhage, wound shock, severe burns, and various illnesses may require blood transfusions. If whole blood is used, the patient receives both the necessary plasma and blood cells. However, the blood of the donor must be typed and matched with that of the patient. The matching is done by adding a drop of a test serum to a drop of the donor's blood. If the red cells *agglutinate,* or clump together, the bloods are incompatible (Figure 42-4). If, however, the red cells remain in suspension, the samples are compatible, and it is possible to perform the transfusion. Blood types are designated A, B, AB, and O. You can see why using the wrong type might result in serious blood reactions, clotting, and death of the patient. This would be a good time to review the inheritance of blood types.

Often, a patient needs an immediate increase in the volume of liquid in the bloodstream and does not require additional

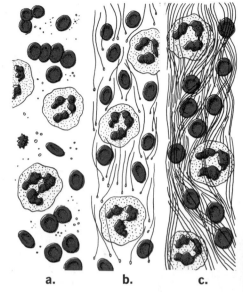

a. **b.** **c.**

42-3 The microscopic changes that occur during the clotting of blood: (a) before clotting begins; (b) formation of threads of fibrin; (c) shortening of the fibrin threads and trapping of blood cells.

42-4 Blood can be typed by using test serum from type A and type B bloods. (Fundamental Photos, Community Blood Council)

serum from	natural antibodies in serum	antigen in red blood cells of blood being tested			
		A	AB	B	O
type A	anti-B				
type B	anti-A				

blood cells. This condition is called *shock*. The red corpuscles form rapidly if the blood volume is maintained. At such times, plasma may be transfused in preference to whole blood. Typing is not necessary when plasma is used, because of the absence of cells.

The Rh factor in blood

A factor in blood independent of the A, B, AB, and O blood groups is called the ***Rh factor,*** after the *rhesus* (REE-sus) *monkey* in which it was discovered. About 85 percent of the people in the United States have this factor in their blood and are designated as *Rh positive.* The other 15 percent are *Rh negative.* The Rh factor, like blood types, is inherited and is actually any one of six protein substances called ***antigens.***

If a patient is Rh negative and receives Rh-positive blood in a transfusion, he produces an antibody against the factor. This particular antibody causes the corpuscles of the Rh-positive blood to agglutinate and to dissolve. There is little danger during the first transfusion, because the antibody is not present when the Rh-positive blood is added. But the patient now builds antibodies against this Rh-positive blood, so that a second transfusion may result in serious or even fatal complications.

How the Rh factor may affect childbirth

Complications from the Rh factor occur with childbearing in about one in three or four hundred mothers. When the mother is Rh negative and the father is Rh positive, the child may inherit the Rh-positive factor from the father. During development, blood from the child, containing the factor, may seep into the mother's circulation through tiny ruptures in the membranes that normally separate the two circulations. Blood from the mother seeps into the child through the same channels.

Since such seepage is uncommon, many Rh-negative mothers bear normal Rh-positive children. However, if seepage occurs again with a second Rh-positive child, the antibody in the mother's blood, produced in the first pregnancy, enters the child's circulation and causes serious damage. Occasionally the child dies before birth. But if the damage to the child is not too extensive, an immediate transfusion after birth may save its life. Sometimes the child's blood is almost entirely removed and replaced by transfused blood. Blood used in such a complete transfusion is Rh negative but does not contain the antibody. In other words, the donor has never received positive blood and his blood is not sensitized against the factor.

The structure of the heart

The heart is a cone-shaped, muscular organ situated under the breastbone and between the lungs. It is enclosed in a sac

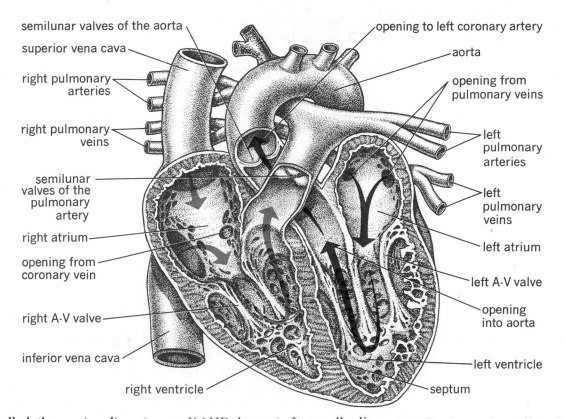

semilunar valves of the aorta

superior vena cava

right pulmonary arteries

right pulmonary veins

semilunar valves of the pulmonary artery

right atrium

opening from coronary vein

right A-V valve

inferior vena cava

right ventricle

opening to left coronary artery

aorta

opening from pulmonary veins

left pulmonary arteries

left pulmonary veins

left atrium

left A-V valve

opening into aorta

left ventricle

septum

called the *pericardium* (PEH-RI-KAHR-dee-um). It usually lies a little to the left of the midline of the chest cavity, with its point extending downward and to the left between the fifth and sixth ribs. Since the beat is strongest near the tip, many people have the mistaken idea that the entire heart is on the left side.

The heart is composed of two sides, right and left. The two halves are entirely separated by a wall called the *septum.* Each half is composed of two chambers, a relatively thin-walled *atrium* and a thick, muscular *ventricle.* The two atria act as reservoirs for the blood entering the heart. Both contract at the same time, filling the two ventricles rapidly. Next, the thick, muscular walls of the ventricles contract, forcing the blood out through the great arteries.

Flow of blood from the ventricles under pressure and maintenance of pressure in the arteries between beats require two sets of one-way heart valves. The valves between the atria and ventricles are called the *atrioventricular valves,* or *a-v valves.* They are flaplike structures that are anchored to the floor of the ventricles by tendonlike strands (Figure 42-5). Blood passes freely through these valves into the ventricles. The valves cannot be opened from the lower side, however, because of the tendons anchoring them. Thus, blood is unable to flow back-

42-5 The human heart. Note the location of the valves to the heart chambers and blood vessels. The arrows indicate the direction of blood flow.

ward into the atria during contraction of the ventricles. Other valves, called the *semilunar valves,* or *s-l valves,* are located at the openings of the arteries. These cuplike valves are opened by the force of blood passing from the ventricles into the arteries, and they prevent blood from returning to the ventricles.

Circulation of blood through the heart

Blood first enters the right atrium of the heart by way of the *superior vena cava* (VAY-nuh KAH-vuh) and the *inferior vena cava.* The superior vena cava carries blood from the head and upper parts of the body. The inferior vena cava returns blood from the lower body regions. From the right atrium, blood then passes through the right a-v valve into the right ventricle. When the right ventricle contracts, blood is forced through a set of s-l valves into the *pulmonary artery,* which carries the blood to the lungs. After the blood has passed through the lungs, it is returned to the heart through the right and left *pulmonary veins.* These vessels open into the left atrium, from which the blood passes through the left a-v valve into the left ventricle. From here, blood passes out the *aorta* (ay-OAR-tuh) and is distributed to all parts of the body.

Although the heart is an organ filled with blood, its muscle layers are too thick to be nourished by this blood. The cells receive special nourishment through arteries called *coronary arteries.* There is an enlargement of the aorta at the point where it leaves the heart. This is called the *aortic sinus.* From here, the right and left coronary arteries branch off. These arteries curve downward around each side of the heart, sending off smaller vessels that penetrate the heart muscle.

The heart, a highly efficient pump

A complete cycle of heart activity, or beat, consists of two phases. During one part of the cycle, called *systole,* the ventricles contract and force blood into the arteries. During the other part, called *diastole,* the ventricles relax and receive blood from the atria.

The sounds you hear in a stethoscope when you listen to a normal heart sound like the syllables "lub" and "dup" repeated over and over in perfect rhythm. The "lub" is the sound of the contraction of the muscles of the ventricles and the closing of the a-v valves during systole. The "dup" is the closing of the semilunar valves at the bases of the arteries during diastole.

At rest the heart of an average adult beats about 70 times per minute. During strenuous work or exercise, the heart rate may be as high as 180 beats per minute.

With the body at rest, the heart pumps about 10½ pints of blood per minute. Your body contains 12 pints of blood, so all the blood makes a complete circulation through the body in slightly over a minute. However, mild exercise such as walk-ing speeds the heart output to about 20 pints per minute, and strenuous exercise may increase it to as much as 42 pints per minute. If you were to use a hand pump to move blood, you could not possibly keep up with a heart under exertion. This will give you some idea of the efficiency of this organ, which weighs only one fourth of a pound and works every minute of every day and night of your life.

The blood vessels

Blood moves in a system of tubes of varying sizes (Figure 42-6); these tubes have been classified as follows: *arteries* and *arterioles,* vessels carrying blood *away* from the heart; *capillaries,* very small, thin-walled vessels; and *veins* and *venules,* vessels carrying blood back *toward* the heart.

The aorta branches into several large arteries. These arteries branch and become arterioles. As the arterioles branch, they soon become capillaries, which are the smallest vessels in the body. After passing through a tissue or organ, the capillaries come together to form venules. As the venules join, they be-come veins, which take the blood toward the inferior or superior vena cava and into the right atrium of the heart.

Arteries have elastic, muscular walls and smooth linings. Because of their elasticity, arteries can expand and absorb part of the great pressure resulting from contraction of the ventricles at systole. The pressure in the aorta leading from the left ventricle is greater than that in the pulmonary artery pumped by the smaller right ventricle. If the aorta were cut, blood would spurt out in a stream of six feet or more. When the ventricles contract, arterial pressure is greatest and is called *systolic pressure.* The elasticity of the artery walls maintains part of this pressure while the ventricles are at rest. This is the time of lowest pres-sure in the arteries, or *diastolic pressure.* The bulge in an artery wall caused by systolic pressure can be felt in the wrist or any part of the body where an artery is near the surface. This *pulse* has the same rhythm that the heartbeats have.

What are capillaries?

As arterioles penetrate the tissues, they branch into capillaries (Figure 42-6). Capillaries differ from arterioles in that their walls are only one-cell-layer thick. Capillaries are only slightly greater in diameter than the red blood cells. Red corpuscles must pass through them in single file and may even be pressed out of shape by the capillary walls.

Dissolved foods, waste products, and gases pass freely through the thin walls of capillaries and in and out of the tissue spaces. Tiny openings in the walls are penetrated by white corpuscles as they leave the bloodstream and enter the tissue spaces. In

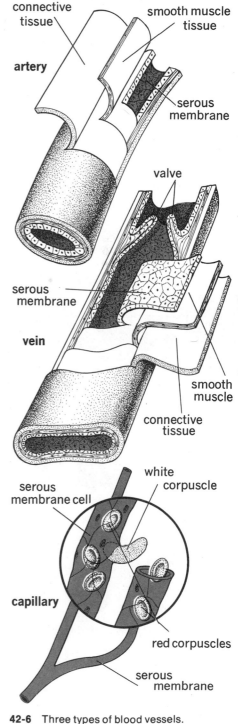

42-6 Three types of blood vessels.

the capillaries, too, part of the plasma diffuses from the blood and becomes tissue fluid. Thus, all the vital relationships between the blood and the tissues occur in the capillaries and not in arteries and veins.

Vein structure and function

On leaving an organ, capillaries unite to form veins. Veins carry dark red blood—that is, blood containing less oxygen. In the skin, the veins have a bluish color because the skin contains a yellow pigment that changes the appearance of the dark red blood. The walls of veins are thinner and less muscular than those of arteries, and their internal diameter is proportionally larger. Many of the larger ones are provided with cuplike valves that prevent the backward flow of blood.

Veins have no pulse wave, and the blood pressure within them is much lower than that in arteries. Blood pressure resulting from heart action is almost completely lost as blood passes through the capillaries. Blood from the head may return to the heart with the aid of gravity, but in the body regions below the level of the heart, other factors are required. Venous flow from these regions is aided by the working muscles and by respiration movements.

Circulations in the body

A four-chambered heart, such as that in the human body, is really a double pump in which the two sides work in unison. Each side pumps blood through a major division of the circulatory system. The right side of the heart receives dark, deoxygenated blood from the body and pumps it through the arteries of the *pulmonary circulation* (Figure 42-7). The great pulmonary artery, extending from the right ventricle, sends a branch to each lung (see the Trans-Vision between pages 630 and 631). These arteries in turn branch within the lungs, forming a vast number of arterioles. Here the blood discharges carbon dioxide and water and receives oxygen. Oxygenated blood, now bright scarlet in color, leaves the lungs and returns to the left atrium of the heart through the pulmonary veins.

Oxygenated blood passes through the left chambers of the heart and out the aorta under great pressure. The blood is now in the *systemic circulation,* which supplies the body tissues. This extensive circulation includes all of the arteries that branch from the aorta, the capillaries that penetrate the body tissues, and the vast number of veins that lead to the venae cavae. The systemic circulation also includes several shorter circulations that supply or drain special organs of the body.

The *coronary circulation,* referred to in the discussion of the heart muscle, supplies the heart itself. This short but vital circulation begins at the aorta and ends where the coronary veins

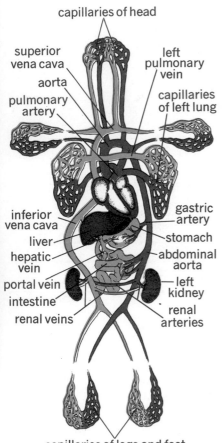

capillaries of head

superior vena cava

aorta

pulmonary artery

left pulmonary vein

capillaries of left lung

inferior vena cava

liver

hepatic vein

portal vein

intestine

renal veins

gastric artery

stomach

abdominal aorta

left kidney

renal arteries

capillaries of legs and feet

42-7 A diagrammatic representation of the various circulations in the human body. Trace the pulmonary, systemic, coronary, renal, and portal circulations.

empty into the right atrium. Every beat of the heart depends on the free flow of blood through the coronary vessels.

The *renal circulation* starts where a *renal artery* branches from the aorta to each kidney. It includes the capillaries that penetrate the kidney tissue and the *renal veins,* which return blood from the kidneys to the inferior vena cava. Blood on this route nourishes the kidneys and discharges water, salts, and nitrogenous cell wastes. Thus, even though it is low in oxygen content, blood in the renal veins is the purest blood in the body.

The *portal circulation* includes an extensive system of veins that lead from the spleen, stomach, pancreas, small intestine, and colon. The large veins of the portal circulation unite to form the portal vein, which enters the liver. Blood flowing from the digestive organs transports digested food and water. Blood laden with food for the body tissues flows from the liver in the hepatic veins, which in turn empty into the inferior vena cava, thus ending this vital branch of the systemic circulation.

Return of tissue fluid to the circulation

The tissue fluid that bathes the cells is collected in tubes and is then called *lymph* (limf). These tiny lymph vessels join one another and become larger lymph vessels in the same way in which capillaries join to form venules. Lymph nodes, which are enlargements in the lymph vessels, are located along the vessels much like beads on a string. In these lymph nodes, the lymph tubes break up into many fine vessels once again. Here certain white corpuscles collect and destroy bacteria that may be in the lymph. The lymph glands, then, act to strain or to purify the lymph before returning it to the blood. The greatest concentrations of these lymph nodes are in the neck, the armpit, the bend of the arm, and the groin. Often when there is an infection in the hand or arm, the lymph nodes of the armpit swell and become painful. Both the *tonsils* and *adenoids* in the throat are merely masses of lymphatic tissue that often become inflamed during childhood and have to be removed surgically.

The lymph of the right side of the head, neck, and right arm enters into a larger vessel named the *right lymphatic duct*. This vessel returns the lymph to the blood by opening into the *right subclavian vein*. The lymphatics from the rest of the body drain into the *thoracic duct,* which in turn empties into the *left subclavian vein* (Figure 42-8).

In the walls of the larger lymph vessels there are valves to control the flow of lymph. These valves are similar in structure to those found in the veins. In inactive tissues, lymph flows very slowly or is completely stagnant. When activity increases, the fluid flows faster. The return of lymph to the bloodstream is aided by the contracting movements of many of the body muscles.

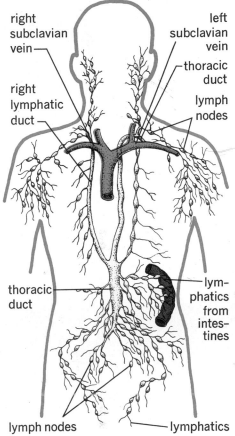

42-8 The lymphatic system returns tissue fluid to the bloodstream.

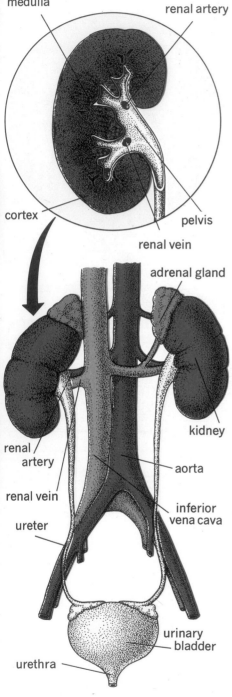

medulla

renal artery

cortex

pelvis

renal vein

adrenal gland

kidney

renal artery

aorta

renal vein

inferior vena cava

ureter

urinary bladder

urethra

42-9 The kidney is an efficient filtering organ.

Removal of wastes from the circulating fluids

The oxidation of foods involved in metabolism produces waste products that must be given off by the body in the process called *excretion.* In protein metabolism, waste products result from the separation of the carbon and nitrogen parts of amino acids before oxidation of the carbon part. Other waste products result from the synthesis of proteins from amino acids during growth processes. These nonprotein *nitrogenous wastes* include *urea* and *uric acid.*

Any great accumulation of wastes in the tissues, especially nonprotein nitrogens, causes rapid tissue poisoning, starvation, and eventually suffocation. Tissues filled with waste products can absorb neither food nor oxygen. Fever, convulsions, coma, and death are inevitable if nonprotein wastes do not leave the tissues.

Further complications arise if mineral acids and salts accumulate in the body because of excretory failure. This accumulation disturbs certain delicate acid-base balances in the body and also upsets the osmotic relationships between blood and lymph and the tissues. When excess salts are held in the tissues, water accumulates and causes swelling.

One-celled organisms and animals like the sponge and jellyfish discharge their cell wastes directly into a water environment. However, when many millions of cells form an organism, as in higher animals, the removal of cell waste products becomes a complicated process involving many organs. Each cell discharges its waste materials into the tissue fluid, which in turn reaches the bloodstream. The blood transports the cell wastes to excretory organs such as the kidneys and skin for elimination.

The kidneys—the principal excretory organs

The *kidneys* are bean-shaped organs, about the size of your clenched fist. They lie on either side of the spine, in the small of the back. Deep layers of fat around them form a protective covering (see the Trans-Vision between pages 630 and 631). If you cut a kidney lengthwise (Figure 42-9), you can see several different regions. The firm outer region that composes about one third of the kidney tissue is called the *cortex.* The inner two thirds, or *medulla,* contains conical projections called *pyramids.* The points of the pyramids extend into a saclike cavity, the *pelvis* of the kidney. The pelvis, in turn, leads into a long, narrow tube called the *ureter.* The two ureters (one for each kidney) empty into the *urinary bladder.*

Each kidney contains about 1,250,000 tiny filters called *nephrons.* The function of these nephrons is to control the chemical composition of blood. Each nephron consists of a small, cup-shaped structure called a *Bowman's capsule* (Figure 42-10). A tiny, winding *tubule* comes from each capsule. This tubule becomes very narrow as it straightens out and goes toward the renal pelvis. The tubule widens out again into a loop called *Henle's loop* and goes back into the cortex. Once this tubule has passed back into the cortex, it becomes very crooked again and then enters a larger, straight tube called the *collecting tubule.* The collecting tubule is a straight tube that receives the tubules of many nephrons. It carries fluid to the renal pelvis. If all these tubules were straightened out and put end to end, they would extend more than 200 miles.

How does the nephron function?

Blood enters each kidney through a large *renal artery,* which branches directly from the aorta. It is the largest artery in the body in proportion to the size of the organ it supplies. In the kidney, the renal artery branches and rebranches to form a maze of tiny arterioles, which penetrate all areas of the cortex.

Each arteriole ends in a coiled, knoblike mass of capillaries, the *glomerulus* (glom-ERR-yuh-lus). Each glomerulus fills the cuplike depression of the Bowman's capsule. In the first stage of removal of waste from the blood, far too much of the blood content leaves the bloodstream and enters the Bowman's capsule. However, this is corrected in a second stage, in which valuable substances return to the blood. The first stage takes place in the coiled capillaries of the glomeruli. Here, water, nitrogenous wastes, glucose, and mineral salts pass through the walls of the capillaries and into the surrounding capsule by a process called *filtration*. This solution resembles blood plasma without blood proteins. Complete loss of this much water, glucose, and minerals would be fatal. However, after this fluid leaves the capsule through the tubules, it passes a network of capillaries. Here, many of the substances are reabsorbed into the blood by active transport. Only nitrogenous wastes, excess water, and excess mineral salts pass through the tubules to the pelvis of the kidney as *urine*.

Some recent studies of kidney function indicate that for every 100 milliliters of fluid that pass from the blood in the glomeruli into the capsules, 99 milliliters are reabsorbed. The urine passes from the pelvis of each kidney through the ureters to the urinary bladder. Blood leaves the kidneys through the *renal veins* and returns to the general circulation by way of the *inferior vena cava.* As we said earlier, the blood in these veins, while it is deoxygenated, is the purest blood in the body.

After urine leaves the kidneys through the ureters, it collects in the muscular urinary bladder. Contraction of the urinary bladder at intervals expels the urine through the *urethra* (yooh-REETH-ruh). The two kidneys have tremendous reserve power. When one is removed, its mate becomes enlarged and assumes the normal function of two kidneys.

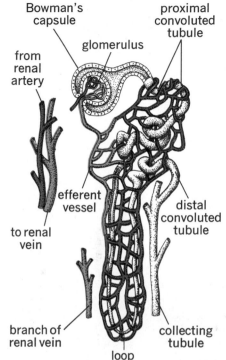

42-10 The structure of a nephron. Note the close relationship of the tubules and blood vessels by which materials are reabsorbed into the blood.

Another role of the kidneys

Although you may think of the kidneys as organs functioning mainly in the elimination of urea and other wastes of protein metabolism, they have another important role. Substances in the blood which reach a higher than normal concentration are excreted. Excess sugars, acids or bases, and water will be eliminated from the blood by the kidneys. The concentration of salts in the body fluid is also maintained by the regulating activities of the kidneys. You can realize the importance of this by recalling the effect of changes in osmotic concentration in the environment of a cell. Proper functioning of the kidneys, then, maintains a constant environment of the body's cells. This, you may remember, is called *homeostasis*.

The skin—a supplementary excretory organ

The skin helps the kidneys in the excretion of water, salts, and some urea, in the form of *perspiration*. This fluid is, however, much more important in regulating body temperature than it is as an excretory substance.

The skin consists of an outer portion, or *epidermis*, composed of many layers of epithelial cells (Figure 42-11). The outer cells, or *horny layer*, are flattened, dead, and scalelike. The inner

42-11 Can you identify the structures in this diagram of a highly magnified, thin section of skin?

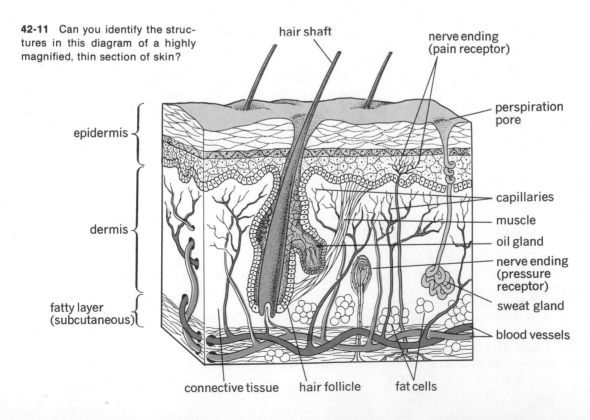

ones, or *germinative layer,* are more active and larger. The epidermis serves largely for protection of the active tissues beneath it. It is rubbed off constantly, but active cells in the lower layers replace cells as fast as they are lost. Friction and pressure on the epidermis stimulate cell division and may produce a *callus* more than a hundred cells thick. Hair and nails are special outgrowths of the epidermis. The *dermis* lies under the epidermis. It is a thick, active layer, composed of tough, fibrous connective tissue, richly supplied with blood and lymph vessels, nerves, sweat glands, and oil glands.

The functions of the skin

The following are among the varied functions of the skin:
■ Protection of the body from mechanical injury and bacterial invasion.
■ Protection of the inner tissues against drying out. The skin, aided by oil glands, is nearly waterproof. Little water passes through it, except outward through the pores.
■ Location of the *receptors* that respond to touch, pressure, pain, and temperature.
■ Excretion of wastes present in sweat.
■ Control of the loss of body heat through the evaporation of sweat.

This last statement needs further explanation. In an earlier discussion about water and its uses, we mentioned that heat is used during the change of liquid water to water vapor. Thus, as sweat evaporates from the body surface, heat is withdrawn from the outer tissues. The skin is literally an automatic radiator. It is richly supplied with blood containing body heat withdrawn from the tissues. As the body temperature rises, the skin becomes more flushed with blood, and heat is conducted to the surface. At the same time, secretion of sweat increases and bathes the skin. This increases the rate of evaporation and the amount of heat loss.

Other organs of excretion

During expiration, the *lungs* excrete carbon dioxide and considerable water vapor. The excretory function of the *liver* in forming urea has been discussed earlier. The bile stored in the *gall bladder* is also a waste-containing substance.

The *large intestine* removes undigested food. This, however, is not cell excretion in the strict sense, since the food refuse collected there has never actually been absorbed into the tissues.

SUMMARY

The circulatory system is the transportation system of the body and its vehicle is blood. Blood is a fluid tissue, composed of plasma and solid components. Plasma contains water, blood proteins, prothrombin, inorganic substances, digested foods, and cell wastes. The solid parts of the blood are of three types: red corpuscles, white corpuscles, and platelets. Red corpuscles are essential for carrying oxygen to the body cells and carrying carbon dioxide away as a waste product. White corpuscles aid in fighting disease bacteria. Platelets are an essential factor in the process of blood clotting.

The heart is a pump that forces blood through the arteries to all parts of the body. It consists of two atria, which receive blood from the veins, and two ventricles, which force blood through the arteries by contractions. Arteries carry blood from the heart to the tissues, and veins return it. The arterial and venous systems are connected by countless microscopic networks of capillaries.

Once part of the plasma has seeped into the tissue spaces, it is collected as lymph and filtered by the lymph nodes. Then it is returned to the bloodstream.

Various wastes that result from metabolism are removed from the body through actions of the kidneys, skin, lungs, liver, and large intestine. The kidneys, the most vital organs of excretion, serve as blood filters. They are responsible for the removal of practically all the nitrogenous wastes resulting from protein metabolism, excess water, and mineral acids and salts. The skin also has a complex role. It excretes large quantities of water as sweat, and functions as the radiator of the body in eliminating heat during evaporation of sweat.

BIOLOGICALLY SPEAKING

circulation	systole	portal circulation
blood	diastole	lymph
red corpuscle	artery	excretion
white corpuscle	capillary	kidney
platelet	vein	ureter
Rh factor	pulse	nephron
antigen	pulmonary circulation	urine
atrium	systemic circulation	urethra
ventricle	coronary circulation	epidermis
aorta	renal circulation	dermis

QUESTIONS FOR REVIEW

1. What is blood?
2. What are the origins of the various blood cells?
3. What condition in the body does a high white-blood count usually indicate?
4. What are the various steps in the clotting of blood?
5. Why is plasma more quickly and easily used in a transfusion than whole blood? Which method do you think is better and why?
6. Trace the path of a drop of blood from the right atrium to the aorta.
7. Why can you feel the pulse in an artery and not a vein?
8. What is tissue fluid? How does it get back to the bloodstream?
9. How does lymph differ from blood?
10. In what way do the kidneys regulate blood content?
11. What are the differences in composition between the glomerular fluid and the urine that finally leaves the kidneys?

APPLYING PRINCIPLES AND CONCEPTS

1. What is the basis for the saying that "a man is as young as his arteries"?
2. Alcohol dilates the arteries in the skin. What would be its effect, then, on the temperature control of the body?
3. In an Rh-negative patient, why might a second transfusion with Rh-positive blood be fatal, even though the first transfusion with Rh-positive blood caused no complications?
4. How does the manufacture of red blood cells demonstrate conservation of resources by the body?
5. Why is increased salt intake recommended in hot weather?
6. How do the kidneys aid in maintaining water balance of the body?

RELATED READING

Books

HARE, P. J., *The Skin.*
St. Martin's Press, Inc., New York. 1966. A thorough book about skin, that of the animal as well as man's.

HARRISON, WILLIAM C., *Dr. William Harvey and the Discovery of Circulation.*
The Macmillan Company, New York. 1967. An easy-to-read book about the life and research of the doctor who persevered in his ideas.

HERBERT, DON, and FULVIO BARDOSSI, *Secret in the White Cell.*
Harper and Row, Publishers, New York. 1969. A case history of a biological search into the manner in which white blood cells intercept and destroy harmful organisms.

RIEDMAN, SARAH R., *Your Blood and You.*
Abelard-Schuman Limited, New York. 1963. All about blood and its relationship to the working of the body and the supply line that serves it.

SEEMAN, BERNARD, *The River of Life.*
W. W. Norton and Co., Inc., New York. 1961. The development of man's knowledge of blood, traced from earliest times to the present.

SNIVELY, WILLIAM D., *Sea Within Us.*
J. B. Lippincott Co., Philadelphia. 1960. Our dependence on the precarious balance of fluid and electrolytes in our bodies.

SNIVELY, WILLIAM D., and JAN THUERBACH, *Sea of Life.*
David McKay Company, Inc., New York. 1969. The amazing story of the sea within our bodies and how it keeps us alive.

WOODBURN, JOHN H., *Know Your Skin.*
G. P. Putnam's Sons, New York. 1967. Describes the wonders of the human skin and its intricate functions.

CHAPTER FORTY-THREE

RESPIRATION AND ENERGY EXCHANGE

Respiration—a life process common to all living things

A constant supply of energy is required by all cells. Although some cells are able to live in an anaerobic environment, most would soon die without oxygen. Cellular respiration was discussed in Chapter 6. We can define *respiration* as the intake of oxygen and the elimination of carbon dioxide associated with energy release in living cells. In simple organisms such as the protists, sponges, and jellyfish, the cells are in direct contact with the environment. The exchange of gases between the cells and their surroundings occurs directly. However, as organisms become more complicated, the cells are no longer in contact with the external environment of the organism.

The external environment of the cells in your body is much more stable than that of a protozoan in a pond. The blood and tissue fluid which bathes the cells provide a constant supply of oxygen and nutrients. In turn, these tissue fluids remove the cellular wastes. Furthermore, circulation in warm-blooded animals distributes heat to maintain a fairly constant temperature in which the cells can carry out the life processes.

Two phases of respiration in man

External respiration is the exchange of gases between the atmosphere and the blood. This process occurs in the lungs. *Internal respiration* is the exchange of gases between the blood or tissue fluid and the cells. *Breathing* is the mechanical process involved in getting air (containing oxygen) into the lungs, and air (containing waste gases) out of the lungs.

We can divide the organs concerned with breathing and external respiration into two groups. The first group includes the passages through which air travels in reaching the bloodstream: the nostrils, nasal passages, pharynx, trachea, bronchi, bronchial tubes, and lungs (Figure 43-1). The second group is concerned with the mechanics of breathing, that is, with changing the size

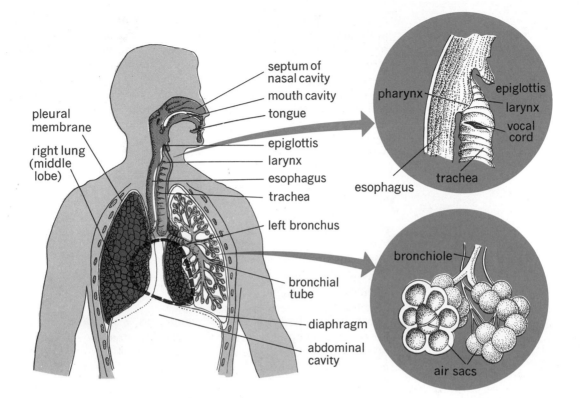

pleural membrane

right lung (middle lobe)

septum of nasal cavity

mouth cavity

tongue

epiglottis

larynx

esophagus

trachea

left bronchus

bronchial tube

diaphragm

abdominal cavity

pharynx

epiglottis

larynx

vocal cord

trachea

esophagus

bronchiole

air sacs

43-1 The organs concerned with breathing and external respiration in man. In the lungs, the air passages increase in number, decrease in size, and end in tiny air sacs.

of the chest cavity. This group includes the ribs and rib muscles, the diaphragm, and the abdominal muscles.

The nose and nasal passages

Air enters the nose in two streams, because the nostrils are separated by the *septum*. From the nostrils, air enters the nasal passages, which lie above the mouth cavity. The nostrils contain hairs that aid in filtering dirt out of the air. Other foreign particles may lodge on the moist mucous membranes in the nasal passages. The length of the nasal passages warms the air and adds moisture to it before it enters the trachea. All these advantages of nasal-breathing are lost in mouth-breathing.

The trachea

From the nasal cavity, the air passes through the *pharynx* and enters the windpipe, or *trachea*. The upper end of the trachea is protected by a cartilaginous flap, the *epiglottis*. During swallowing, the end of the trachea is closed by the epiglottis. At other times, the trachea remains open to permit breathing. The *larynx* (LA-ringks), or Adam's apple, is the enlarged upper end of the trachea. Inside it are the *vocal cords*. The walls of the trachea are supported by horseshoe-shaped rings of cartilage that hold it open for the free passage of air. The trachea and its branches are lined with cilia. These are in constant motion and carry dust or dirt taken in with air upward toward the mouth. This dust, mixed with mucus, is removed when you cough, sneeze, or clear your throat.

In smog-laden air, the cilia cannot carry away all the particles, and many become lodged in the lungs. It is because of this that the lungs of a person from the country look different from the lungs of one who lives in an industrial area. White blood cells are able to ingest a certain number of the foreign particles breathed into the lungs, and some may become embedded in the walls of the air sacs. Others may become lodged in the bronchioles, bronchi, and trachea and then coughed up. Air pollution in the form of smoke is one of the present-day problems of our society.

The bronchi and air sacs

At its lower end, the trachea divides into two branches called *bronchi*. One bronchus extends to each lung and subdivides into countless small *bronchial tubes*. These, in turn, divide into many small tubes called *bronchioles,* which end in *air sacs*. These air sacs are made up of protrusions called *alveoli* and compose most of the lung tissue. The walls of the air sacs are very thin and elastic. Through these thin walls, gases are exchanged between the capillaries and the air sacs. Thus, the lungs provide enough surface to supply air by way of the blood for the needs of millions

43-2 This posterior view of the lungs and heart shows the branches of the pulmonary arteries and the pulmonary veins. Which vessels carry the oxygenated blood?

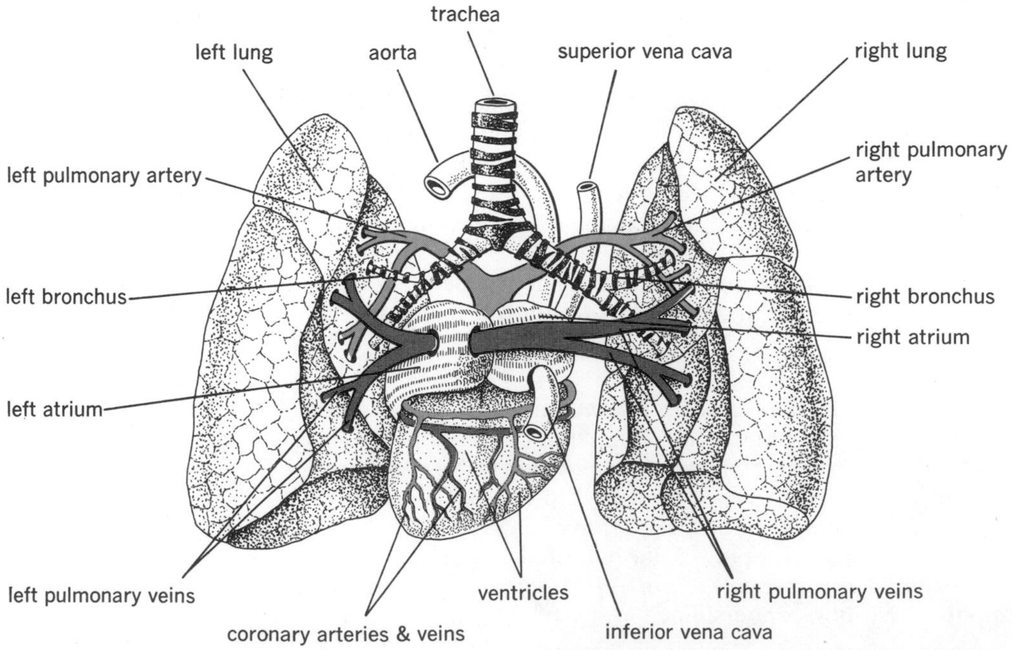

trachea
left lung
aorta
superior vena cava
right lung
left pulmonary artery
right pulmonary artery
left bronchus
right bronchus
right atrium
left atrium
left pulmonary veins
ventricles
right pulmonary veins
coronary arteries & veins
inferior vena cava

of body cells having no direct access to air. The total area of the air sacs in the lungs is about two thousand square feet, or more than one hundred times the body's surface area.

The *lungs* fill the body cavity from the shoulders to the diaphragm, except for the space occupied by the heart, trachea, esophagus, and blood vessels. The lungs are spongy and consist mainly of the bronchioles and air sacs and an extensive network of blood vessels and capillaries, held together by connective tissue. The lungs are covered by a double *pleural membrane*. One part adheres tightly to the lungs, and the other covers the inside of the thoracic cavity. These membranes secrete mucus that acts as a lubricant, permitting the lungs to move freely in the chest during breathing.

The mechanics of breathing

Many people suppose that the lungs draw in air, expand, and bulge the chest. Actually, this is the opposite of what happens. The lungs contain no muscle tissue and cannot expand or contract of their own accord. They are spongy, air-filled sacs, anchored in the chest cavity. Breathing is accomplished through changes in size and air pressure of the chest cavity (Figure 43-3).

Breathing movements

Inspiration, or intake of air, occurs when the chest cavity is increased in size and therefore decreased in pressure. Enlargement of the chest cavity involves the following movements:

1. The rib muscles contract and pull the ribs upward and outward. To inhale with force, you carry this action even further with the aid of the shoulder muscles.

2. The muscles of the resting, dome-shaped diaphragm contract. This action straightens and lowers the diaphragm, and increases the size of the chest cavity from below.

3. The abdominal muscles relax and allow compression of the abdominal organs by the diaphragm.

The enlargement of the chest cavity results in decrease of the air pressure within. In an equalizing of outer and inner pressures, air passes down the trachea and inflates the lungs (Figure 43-4).

Expiration, or the expelling of air from the lungs, results when the chest cavity is reduced in size. The action involves the following four movements:

1. The rib muscles relax and allow the ribs to spring back.

2. The diaphragm relaxes and rises to resume its original position.

3. The compressed abdominal organs push up against the diaphragm. The vigor of this action is increased during forced exhalation by contraction of the abdominal muscles.

4. The elastic lung tissues, stretched while the lungs are full, shrink and force air out.

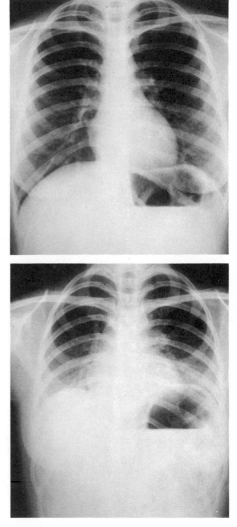

43-3 These X-rays show the chest during exaggerated breathing: inhalation (top); exhalation (bottom).

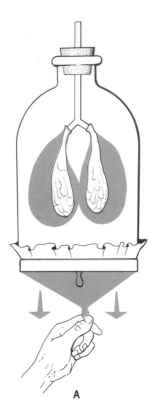

A

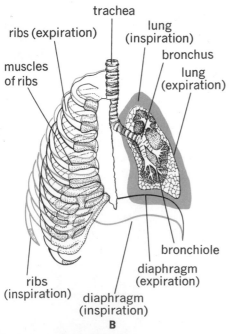

ribs (expiration)

trachea

lung (inspiration)

muscles of ribs

bronchus

lung (expiration)

bronchiole

ribs (inspiration)

diaphragm (expiration)

diaphragm (inspiration)

B

The control of breathing

The factors that control breathing and breathing rate are both nervous and chemical. Nerves lead from the lungs, diaphragm, and rib muscles to a respiratory control center at the base of the brain. When the lungs expand in inspiration, impulses pass from nerve endings in the above tissues along the nerves leading to the control center. The control center, in turn, sends impulses to the rib muscles and diaphragm, causing them to return to their resting position in the act of expiration. When expiration is complete, the process is reversed. The control center sends impulses that cause the rib muscles and diaphragm to contract, and hence the lungs to expand, and the cycle starts again. Inspiration and expiration occur from sixteen to twenty-four times a minute, depending on the activity, position, and age of the body.

The air capacity of the lungs

When the lungs are completely filled they hold about three hundred cubic inches of air. But only about thirty cubic inches are involved each time we inhale and exhale. The air involved in normal, relaxed breathing is called *tidal air*. Forced breathing increases the amount of air movement.

To illustrate the effects of forced breathing, inhale normally without forcing. Your lungs now contain about two hundred cubic inches of air. Now exhale normally. You have moved about thirty cubic inches of tidal air from the lungs. Now, without inhaling again, force out all the air you can. You have now exhaled an additional one hundred cubic inches of *supplemental air*. The lungs now contain about seventy cubic inches of *residual air,* which you cannot force out.

When you inhale normally again, you replace the supplemental and the tidal air, or about one hundred thirty cubic inches. If you inhale with force, you can add one hundred cubic inches of *complemental air,* raising the total capacity of the lungs to about three hundred cubic inches. We can say that the *total capacity* of the lungs consists of the sum of these volumes (see Table 43-1).

TABLE 43-1 LUNG CAPACITY

tidal air	30 cu. in.
supplemental air	100 cu. in.
complemental air	100 cu. in.
residual air	70 cu. in.
total capacity	300 cu. in.

43-4 Breathing—a mechanical process. When the rubber sheet (above) is pulled downward, the pressure is decreased. Air enters through the Y-tube and inflates the balloons. Note the similarity between the model and the representation of the thorax.

43-5 The mouth-to-mouth method of artificial breathing. *First,* tilt the head back so the chin is pointing upward (1). Pull the jaw into a jutting-out position (2 and 3). *Second,* open your mouth wide and place it tightly over the victim's mouth. At the same time, pinch the victim's nostrils shut (4 and 5). *Third,* remove your mouth, turn your head to the side, and listen for the return rush of air that indicates air exchange. *Fourth,* if you are not getting air exchange, recheck the head and jaw position. If you still do not get air exchange, turn the victim on his side and administer several sharp blows between the shoulder blades to dislodge any foreign matter. (redrawn from American Red Cross diagram).

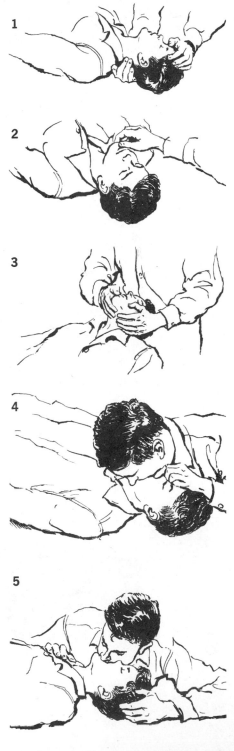

Artificial respiration

A short time after a person stops breathing, a serious condition develops—the level of oxygen drops below that required for normal cell activity. *Artificial respiration* is simply a method of artificially forcing the lungs to inspire and expire air rhythmically. It can be described more accurately as *artificial breathing.* A method of artificial breathing now strongly recommended by the Red Cross is the *mouth-to-mouth method,* illustrated in Figure 43-5.

The air—our source of oxygen

Although air may vary slightly in its composition, its major gases are nitrogen, oxygen, and carbon dioxide (see Table 43-2). These will be discussed in a later chapter, but we should examine some properties of these gases which are vital to life. The weight of air causes a pressure on all living things. At sea-level this force may be expressed as 14.7 pounds per square inch. We are not aware of this pressure because it is nearly equal from all directions. At 5,000 feet, the atmospheric pressure is about 12.3 pounds per square inch, and at 18,000 feet, 7.3 pounds per square inch (or about one-half of that at sea level). Thus, at 18,000 feet the molecules of the gases have spread apart, but the percent composition of nitrogen, oxygen, and carbon dioxide may be the same as that at sea-level.

Molecules of air can also be compressed as when you pump up a bicycle or an automobile tire. Scuba divers use tanks into which air has been pumped to create pressures of 2250 pounds per square inch. The gases in the tank have the same propor-

TABLE 43-2 COMPOSITION OF NORMAL ATMOSPHERE

	Percentage of volume
Nitrogen	78.03
Oxygen	20.99
Argon	0.94
Carbon dioxide	0.03
Hydrogen, neon, helium	0.01

tions as the air we breathe, so you see, the composition of air does not change by changing the pressure.

Another property of a gas can be illustrated by a bottle of soda. Soda water is made by forcing carbon dioxide into solution under a high pressure and then capping the bottle tightly. When the cap is removed, the carbon dioxide begins to bubble out of the water and go into the air where it spreads out (diffuses) among the other molecules. After a period of time, the air in the bottle will have a concentration of carbon dioxide equal to that in air (0.03%). This diffusion process can be compared to that of other chemical molecules going into solution.

Gases are also capable of diffusing through membranes. If two samples of oxygen, for example, of different concentrations are placed, one on each side of a membrane, diffusion will occur until the concentrations on the two sides of the membrane are equal. This property of gases is important to unicellular organisms, and, as you will see, to multicellular organisms as well.

Another property of gases which is vital to life is that they are not equally soluble in all solutions. The solubility of carbon dioxide is about 30 times greater than that of oxygen when the two gases are under identical conditions. Therefore, when protozoa, fish, and other aquatic organisms give off carbon dioxide, it dissolves quickly and diffuses rapidly through the water.

Furthermore, solubility of oxygen, nitrogen, and carbon dioxide in a particular solution will vary with temperature. Warm water will hold less dissolved gases than will cold water. Perhaps you have observed tiny bubbles in a glass of water left on the sink. At a certain temperature (when the glass of water was warming) the water contained all the dissolved gases it could hold. Further warming caused the bubbles to form, thus maintaining an equilibrium with the atmospheric pressure.

Gas exchange in the lungs

The *pulmonary artery* brings dark red (deoxygenated) blood to the lungs. There it divides into an extensive network of capillaries, completely surrounding each air sac (Figure 43-6). Since the air and the blood in the lungs contain gases in different concentrations, diffusion occurs through the thin moist walls of both air sacs and capillaries. Oxygen diffuses from the air into the blood and carbon dioxide and water diffuse from the blood to the air.

Transport of oxygen

Oxygen is not very soluble in the plasma, and at man's body temperature (98.6°F) it is even less soluble than it would be at lower temperatures. If oxygen were carried only in the plasma,

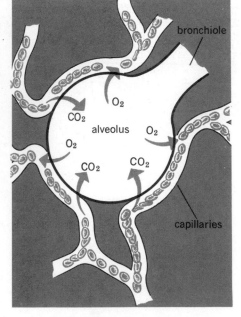

43-6 The relationship between alveoli and capillaries. A cluster of alveoli make up each air sac at the end of a bronchiole.

man would require more than 424 pints of blood to survive! As you read in the last chapter, the average volume of blood in a human is 12 pints. This is possible because of the respiratory pigment, hemoglobin, contained in the erythrocytes.

Hemoglobin has such an affinity for oxygen that when inspiration fills the air sacs, nearly all the oxygen is taken into the blood. At sea-level, the blood leaving the lungs is about 97 percent saturated. Each 100 ml of blood leaving the lungs contains 19.4 ml of oxygen.

When the blood reaches the tissues where the concentration of oxygen is low, the hemoglobin releases its oxygen. The oxygen then diffuses into the tissue fluid where it reaches the cells. The affinity of hemoglobin for oxygen decreases with increasing acidity. This is an important characteristic, because during violent exercise, lactic acid is produced by the active muscle cells. This increases the acidity of the blood and causes the hemoglobin to release more of its oxygen than it would normally. Depending on muscular activity, blood returning to the lungs may contain from 8 to 14 ml of oxygen in each 100 ml of blood.

Transport of carbon dioxide

As we have discussed earlier, the energy requirements of cells utilize oxygen and release carbon dioxide. The carbon dioxide, as mentioned, is much more soluble than oxygen, so it readily diffuses through the tissue fluid and into the capillaries. In the blood, only about 20 percent of the carbon dioxide is carried by hemoglobin, and about 10 percent as dissolved carbon dioxide. Most of the carbon dioxide passes through the plasma and into the red blood cell. A specific enzyme in the erythrocyte unites the carbon dioxide with water to form carbonic acid:

$$CO_2 + H_2O \xrightarrow{\text{enzyme}} H_2CO_3$$

This weak acid then rapidly ionizes to form hydrogen ions and bicarbonate ions:

$$H_2CO_3 \longrightarrow H^+ + HCO_3^-$$

When the blood reaches the lung, the reverse process occurs. Carbon dioxide and water in the lung capillaries have a higher concentration than in the connecting air sacs. Therefore, they diffuse outward into the area of lower concentration in the air sacs and are exhaled. The hemoglobin containing erythrocytes are now ready to pick up oxygen and return to the tissues.

Oxygen debt

During times of muscular exertion, the need for oxygen in the tissues is greater than the body can possibly supply. The

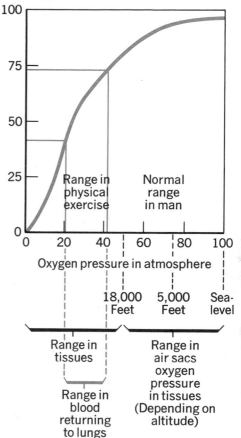

43-7 When hemoglobin reaches equilibrium with the atmospheric oxygen in the air sacs, it is about 97 percent saturated. When blood reaches the tissues where the amount of oxygen is low, diffusion occurs and the hemoglobin gives up its oxygen.

lungs cannot take in oxygen, nor can the blood deliver it rapidly enough. All of us, at some time or another, have had the experience of running or swimming in a race, playing tennis, climbing a mountain, or running to board a train or a bus. Perhaps you can remember feeling, for a second, that you just couldn't finish that race or catch that bus. But with a final surge of reserve energy, you succeeded. You felt limp, lightheaded, and completely exhausted. You may recall that your heart pounded and your breathing was deep and rapid. After twenty or thirty minutes, you were probably ready for another race, another tennis match, or the long hike back.

What happens in one's body to cause these changes, and how is this rapid recovery brought about? During mild exercise, the supply of oxygen meets the demands of the cells. During strenuous exercise, however, there is not enough oxygen. Then, respiration becomes anaerobic, and pyruvic acid becomes the hydrogen acceptor. This produces *lactic acid,* which accumulates in the tissues and causes fatigue. Like carbon dioxide, lactic acid signals the respiratory center in the brain. Breathing becomes rapid, and the heart speeds up in order to supply the tissues with enough oxygen. As you learned above, hemoglobin releases more oxygen because of the acidity of the blood. But even with these efforts, the lactic acid accumulates in the body, and the body is in a state of *oxygen debt.* During a half-hour rest, some of the accumulated lactic acid is oxidized, and some is converted to glycogen. Carbon dioxide and excess water are excreted, and the debt is paid. The body is then ready for more exercise.

Body metabolism and its measurement

In Unit 1, you learned that the sum of all the processes occuring in a cell or an organism is called metabolism. The constructive phase of metabolism includes carbohydrate and protein synthesis, while the destructive phase includes oxidation and energy release. The rate of metabolism increases in proportion to the increase in the activity of the body. This activity may be muscular, as in walking, running, or some other form of exertion; or it may be mental. Other factors governing the metabolic rate include exposure to cold and activity of the digestive organs during digestion of food. One way to measure the metabolic rate of the body is to measure the rate of oxidation by determining the amount of heat given off from the body surface. This can be measured by a device called a *calorimeter.*

The person to be tested enters a closed compartment that is equipped to measure accurately all the heat given off by his body. He may lie quietly in bed during the process, or he may sit in a chair or exercise vigorously, depending on the nature of the activity to be tested. The amount of heat energy given off during each type of activity is a direct indication of the rate of oxidation in the body tissues. Calorimeter tests are important in determining the energy needs of various individuals in order to adjust a diet to their specific requirements.

Even when the body seems completely inactive, as it does in sleep, respiration, oxidation, and energy release are continuing. With the cessation of muscular and, to a great extent, nervous activity, the rate of oxidation is greatly reduced. The activities required to maintain the body and to supply energy necessary to support the basic life processes are included in the term *basal metabolism*. The rate at which these activities occur is called the *basal metabolic rate,* or *BMR.*

BMR can be determined by means of the calorimeter test. Another method, widely used in hospitals, measures the amount of oxygen consumed in a definite period. The patient rests for at least an hour before the test. The test is usually run in the morning, and the patient is instructed to eat no food until after the test is completed. After a rest period during which the body is completely relaxed, the nose is plugged to prevent breathing from the atmosphere. A mouthpiece connected to a tank of oxygen is fitted into the mouth. Thus, all oxygen inhaled during the test period comes from the measured tank. The amount of oxygen used from the tank is recorded on a graph. From these data, the rate of oxidation is determined. The BMR is calculated from the rate of oxidation in the tissues during complete rest.

External influences on breathing and respiration

External factors such as temperature, moisture in the air, and oxygen and carbon dioxide content of air are very important influences on the rate of breathing and respiration. Certain of these factors are involved in *ventilation*. Stuffiness in a room is caused mainly by increase in the temperature and moisture content of the air, rather than by accumulation of carbon dioxide and decrease of the oxygen content. Movement of air in a ventilating system increases the flow of air over the body surfaces and speeds up the evaporation of perspiration. Modern air-conditioning systems circulate air and remove moisture.

The air in most homes, especially those equipped with central heating systems, becomes too dry during the cold months. This condition dries out mucous membranes and lowers their resistance to infection. For this reason, the moisture content of the air should be kept as high as possible by means of humidifiers or other devices.

Many people carry to extremes the ventilation of bedrooms at night. Your body requires less oxygen while you are asleep than at any other time. If the windows are open too much during cold weather, your body may chill during the night. There is little logic in piling covers on a bed to keep part of the body warm, while at the same time chilling the exposed parts with cold air from an open window.

Carbon-monoxide poisoning

Far too often we read of people who have died in a closed garage where an automobile engine was running or in a house filled with gas from an open stove burner or a defective furnace. The cause of death is given as *carbon-monoxide poisoning*. Actually, the death is not caused by poisoning but by *tissue suffocation*. Carbon monoxide will not support life. Yet it combines with the hemoglobin of the blood 250 times more readily than oxygen does. As a result, the blood becomes loaded with carbon monoxide, and its oxygen-combining power decreases. As tissues suffer from oxygen starvation, the victim becomes light-headed and ceases to care about his condition. Soon paralysis sets in, and he cannot move even if he wants to. Death follows from tissue suffocation.

Respiration problems at high altitudes

In a sense, we are living at the bottom of a large sea of air. With increased altitude, the atmospheric pressure is reduced. We have discussed the important factor of the pressure exerted by the weight of air. Although we are not normally aware of this pressure, most of us have experienced our ears "popping" during an altitude change in either an automobile or an airplane. This is due to an adjustment in the middle ear to equalize the pressure.

You have read about the importance of pressure in determining the amount of oxygen in a given volume and how the oxygen combines with the hemoglobin in the blood. This is why mountain climbers and airplane pilots experience increasing difficulty in breathing and progressive weakness as they increase their altitude. At elevations near 12,000 feet, many people fatigue easily.

When an airplane approaches an altitude of 20,000 feet, the pressure becomes so reduced that the pilot experiences difficulty in seeing and hearing. This condition, called *hypoxia,* is the result of oxygen starvation of the tissues. It causes death if not corrected within a short time. Hypoxia can be avoided by equipping the pilot with an oxygen tank and a mask.

Passengers in modern airliners can fly at high altitudes in the safety and comfort of pressurized cabins. In these cabins, an internal pressure and oxygen content equivalent to an altitude of approximately 5,000 feet is maintained.

Respiration—a vital problem in space travel

At sixteen miles above the earth, the density of the air is only about 4 percent of its density at sea level; beyond an altitude of seventy miles, there is practically no atmosphere and therefore practically no oxygen. Hence, astronauts encounter the problems of high altitudes to a much greater degree than pilots do.

43-8 The pressure suit planned for use in all manned Apollo missions. The suit is designed to provide comfortable levels of pressure and oxygen during the prelaunch, launch, and reentry phases of the missions. (courtesy of NASA)

As part of a broad investigation directed toward solving these problems, scientists studied a tribe of Indians who live at high altitudes in the mountains of Peru. These Indians were of particular interest because they are able to carry on normal physical activities that would quickly exhaust a healthy person accustomed only to living at sea level. The research revealed that these tribesmen have greater lung capacity and a higher red blood cell count than do people who live at lower altitudes.

Additional research revealed that the bodies of men trained in an atmosphere with a reduced oxygen level, such as astronauts, become similarly adapted to the reduced level of oxygen. These men can not only exist in an atmosphere in which the oxygen content is about half that of the atmosphere at sea level, but can also be fairly efficient in carrying out normal tasks in such an environment.

SUMMARY

Respiration involves the exchange of gases between living matter and its surroundings. External respiration and internal respiration are concerned respectively with the actual exchange of gases between the lungs and blood and between the blood or tissue fluid and the cells. The movement of air in and out of lungs is accomplished by the mechanical process known as breathing, which consists of inspiration and expiration.

During the chemical process of oxidation, foods are broken down and energy is released. In the lower forms of life individual cells are in direct contact with their surroundings. In higher animals blood is the conducting medium between the body tissues and respiratory organs in contact with the outer environment. Atmospheric pressure and the concentration of gases in the air play important roles in the diffusion of gases through membranes. Another property of gases which is vital to the life processes is their solubility in water and body fluids.

Metabolism includes respiration, oxidation, and the growth processes. The rate at which these processes occur during rest is expressed as the basal metabolic rate.

BIOLOGICALLY SPEAKING

respiration	epiglottis	alveoli
external respiration	larynx	pleural membrane
internal respiration	bronchi	inspiration
breathing	bronchial tubes	expiration
pharynx	bronchioles	artificial respiration
trachea	air sacs	basal metabolism

QUESTIONS FOR REVIEW

1. What are the differences between respiration and breathing?
2. What properties of gases aid the body in respiration?
3. Describe gas exchange in the lungs, naming the structures involved and explaining why the exchange occurs.
4. How do pressure changes within the chest cavity cause inspiration and expiration?

5. What characteristic of hemoglobin is vital to the respiratory processes of the cells in our bodies?
6. Why is the fact that carbon dioxide is so soluble in water of vital importance to unicellular organisms and multicellular organisms alike?
7. How is carbon dioxide carried in the blood from the tissues to the lungs?
8. What is the purpose of artificial respiration?
9. How do you build up an oxygen debt? How is it repaid?
10. Define BMR and give two ways in which it can be measured.
11. Explain the physiology of carbon-monoxide poisoning.
12. Compare respiration problems encountered on a high mountain to those in space travel.

APPLYING PRINCIPLES AND CONCEPTS

1. If plants produce oxygen in photosynthesis, how do you explain that they also respire?
2. What changes do you think would occur in the blood if you were to hold your breath for a period of time? if you were to breathe rapidly and deeply for a period of time?
3. People who live in dry climates, such as the southwestern parts of our country, report that high temperatures there are easier to tolerate than the same temperatures in more humid areas. Why?
4. Explain the decompression procedure used when divers come up from great depths.
5. What differences would you find in the blood of a person living at high altitudes of the Andes compared to a person living at sea-level?

RELATED READING

Books

ASIMOV, ISAAC, *The Human Body: Its Structure and Operation.*
Houghton Mifflin Company, Boston. 1963. The structural parts of the body are listed and described, with each compared and contrasted with similar parts of other creatures.

MILLER, A. T. MD., *Energy Metabolism.*
F. A. Davis Co., Philadelphia. 1968. Presents the basic concepts of energy metabolism as first applied to the cell and then to the utilization of energy by organisms.

WEART, EDITH L., *The Story of Your Respiratory System.*
Coward-McCann Inc., New York. 1964. Diagrammatic explanation of the respiratory system and the functions of its different parts.

CHAPTER FORTY-FOUR

BODY CONTROLS

Who did the planning?

Although the photographs below both illustrate coordination and complex behavior patterns, there are rather obvious differences. The chimp has been conditioned to sit quietly, and when the circus director starts the motorcycle, he goes around in a circle until the director stops the motor. The young man riding his motor through the country roads must make complex decisions about the road, its turns, hills, and avoid obstacles. He starts his ride, enjoys the passing scenery, and, perhaps, plans to arrive home at dinner time when he can communicate his day's activities to friends or family. You can imagine the results if a chimp were placed on a motorcycle and sent down a country road!

44-1 What skills do these two possess in common? (Ringling Brothers, Barnum and Bailey Circus; Harley Davidson)

As you have learned, the ability to perform certain functions may be determined by structure. The bear may walk upright for a short time, but its bone and muscular structure is not adapted for this posture. It has vocal cords, but their structure does not make it possible for the formation of spoken words. Differences in the structure or the size of the brain may also determine the way in which an organism is able to react to its environment.

The nervous system of man

The brain and spinal cord comprise the *central nervous system*. They communicate with all parts of the body by means of the nerves of the *peripheral* (puh-RIFF-uh-rul) *system*. Another division is the *autonomic* (Aw-tuh-NOM-ick) *system*, which regulates certain vital functions of the body almost independently of the central nervous system.

Nerve cells are called *neurons*. Each has a rounded, star-shaped, or irregularly-shaped cell body that contains a nucleus and cytoplasm. Each neuron is able to act as a link in the communication system of the body because of its structure. Thread-like processes, often called *nerve fibers,* extend from the cell body (Figure 44-2). Messages, called *impulses,* travel along the nerve fibers. In some of the lower animals, impulses can travel in either direction, but in man, fibers named *dendrites* carry impulses to the cell body, and *axons* carry impulses away from the cell body. The number of dendrites entering a cell body may range from one to 200, but a single axon leaves the nerve cell body.

44-2 The structure of a typical motor neuron. Note, however, that the axon of a real motor neuron is considerably longer than shown in proportion to the rest of the neuron.

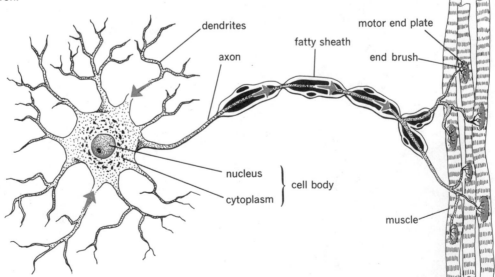

dendrites

axon

fatty sheath

motor end plate

end brush

nucleus

cytoplasm

cell body

muscle

Since cell bodies are located within the central nervous system (or ganglia within the autonomic nervous system, as you will see), the lengths of the dendrites and axons will vary. We perceive our environment through our senses, and it is the dendrites that convey impulses to the central nervous system. Therefore, the peripheral dendrite fibers are referred to as *sensory neurons*. When you stub your toe, the impulse conveying the feeling of pain travels through a dendrite to the nerve cell body located within the spinal cord. Before you wiggle your toe to be certain it is all right, impulses have traveled from the nerve cell body in the spinal cord through the axon to the muscles in your toe. The nerve cell involved in the action of the toe's movement is a *motor* neuron. Can you visualize the length of a single cell which supplies the foot of a giraffe?

Similar to smaller wires bound to form cables, the nerve cell fibers are grouped together as *nerves*. In the peripheral nervous system any nerve composed of only the fibers of motor neurons is a *motor nerve*. Any nerve composed only of the fibers of sensory neurons is a *sensory nerve*. Many nerves contain both motor and sensory fibers and are *mixed nerves*. Most axons are surrounded by a fatty sheath made by specialized cells which wrap around the axon. The fatty sheath seems to act as an insulator.

Connections between neurons are made in the central nervous system, or in ganglia. The processes of one neuron never touch those of another neuron. The spaces between endings of neuron processes are called *synapses* (SIN-aps-iz). Impulses must pass over these synapses as they travel from one neuron to another. Furthermore, an impulse never travels from one motor neuron to another. Nor do impulses travel from one sensory neuron to another.

Nerve impulses

A *nerve impulse* is known to be an electrochemical impulse, which brings about a change in the nerve fiber. It is not a flow of electricity, for nerve impulses travel much more slowly than electricity does. A nerve impulse travels at a rate of about 300 feet per second, while electricity travels at a rate of 186,000 miles per second. Also, when a nerve impulse passes along a nerve, carbon dioxide is liberated, which indicates that a chemical reaction is involved.

A nerve which is not conducting an impulse has a positive charge outside the cell membrane and a negative charge inside (see Figure 44-3). This resting neuron is said to be *polarized*. At the time an impulse passes along a particular point on the neuron, the polarity reverses and the outer surface of the membrane becomes negatively charged, and the inner surface positively charged. The resting state is next obtained when the original polarization occurs. The impulse, then, travels along the

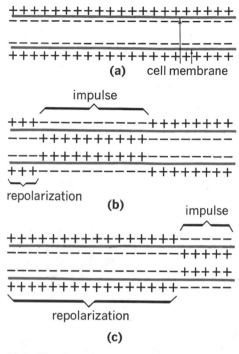

(a)   cell membrane

impulse

repolarization
(b)

impulse

repolarization
(c)

44-3 The impulse travels along the neuron and can be measured as a change in electrical charge: (A) resting neuron; (B) impulse conducted as reverse of polarity; (C) original polarity restored.

nerve as a series of these changes in the charge on either side of the membrane.

For a long time, scientists were in doubt as to how a nerve impulse causes a muscle to contract. Now we know that the stimulation is indirect. An impulse traveling along the axon of a motor neuron ends at the *motor end plate* at the tips of the brushlike structures. Here the impulse causes the release of a minute amount of a chemical called *acetylcholine* (uh-SEET'l-KOE-leen). This substance transmits the impulse to muscle fibers, which begin the process of contraction we discussed in Chapter 40. Following a brief period of contraction, the nerve releases another substance, *cholinesterase* (KOLE-in-ESS-tuh-RACE), which neutralizes acetylcholine and causes the muscle fibers to relax. This process takes 0.1 second or less.

The production of a chemical at the ends of axons also transmits an impulse across a synapse. Axon endings may produce either acetylcholine or an adrenalin compound which stimulates the dendrite and begins an impulse. Neutralization of the acetylcholine or adrenalin compound prepares the synapse for transmitting another impulse.

The brain and its membranes

The brain (Figure 44-4) is probably the most highly specialized organ of the human body. It weighs about three pounds and fills the cranial cavity. It is composed of soft nervous tissue covered by three membranes, together known as the **meninges** (muh-NIN-jeez). The inner membrane, or *pia mater* (PIE-uh MAY-tur), is richly supplied with blood vessels that carry food and oxygen to the brain cells. It is a delicate membrane that closely adheres to the surface of the brain. The middle membrane, or *arachnoid* (uh-RACK-noid) *mater,* consists of fibrous and elastic tissue. This membrane does not dip down into the grooves of the brain but bridges them. The space between these two membranes is filled with a clear liquid, the *cerebrospinal fluid,* which, as the name implies, is also found around the spinal cord. The outermost layer of protective membranes is a thick, strong, fibrous lining, the *dura mater.* Besides being a membrane of the brain, this layer serves as a lining for the inside of the cranium. The meninges protect the brain from jarring by acting as a cushion. A *concussion* is a brain bruise resulting from a violent jar that causes damage in spite of the protective meninges. These three membranes extend down the spinal column to cover and protect the spinal cord.

Cavities of the brain

There are four spaces called ventricles within the brain. Two *lateral ventricles* open into the *third ventricle,* which leads to the *fourth ventricle.* From the fourth ventricle, the cavity is

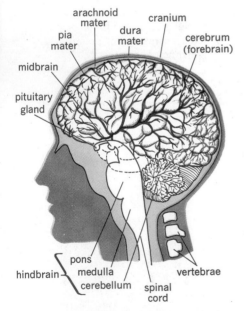

44-4 A longitudinal section of the brain, showing the regions and the meninges.

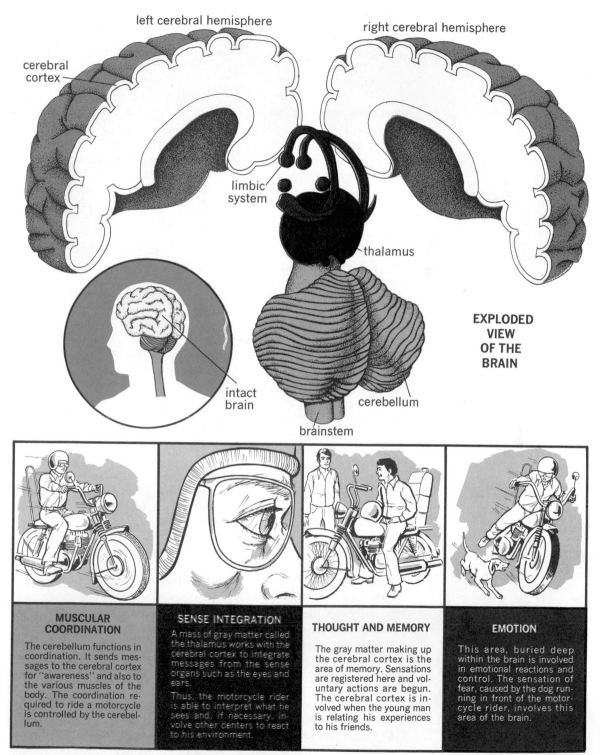

left cerebral hemisphere

right cerebral hemisphere

cerebral
cortex

limbic
system

thalamus

intact
brain

cerebellum

brainstem

**EXPLODED
VIEW
OF THE
BRAIN**

**MUSCULAR
COORDINATION**

The cerebellum functions in
coordination. It sends messages to the cerebral cortex
for "awareness" and also to
the various muscles of the
body. The coordination required to ride a motorcycle
is controlled by the cerebellum.

SENSE INTEGRATION

A mass of gray matter called
the thalamus works with the
cerebral cortex to integrate
messages from the sense
organs such as the eyes and
ears.

Thus, the motorcycle rider
is able to interpret what he
sees and, if necessary, involve other centers to react
to his environment.

THOUGHT AND MEMORY

The gray matter making up
the cerebral cortex is the
area of memory. Sensations
are registered here and voluntary actions are begun.
The cerebral cortex is involved when the young man
is relating his experiences
to his friends.

EMOTION

This area, buried deep
within the brain is involved
in emotional reactions and
control. The sensation of
fear, caused by the dog running in front of the motorcycle rider, involves this
area of the brain.

44-5 A color-coded diagram of the brain and some of its functions.

continuous with the *subarachnoid* (sᴜʙ-uh-RACK-noid) *space* and the central canal of the spinal cord. The cavities of the brain and the central canal are lined with ciliated epithelium, which keeps the cerebrospinal fluid with which the cavities are filled in motion.

The cerebrum — the largest of the brain regions

The region of the brain called the *cerebrum* is proportionally larger in man than in any other animal. It consists of two halves, or hemispheres, securely joined by tough fibers and nerve tracts. The outer surface, or *cortex,* is deeply folded in irregular wrinkles and furrows, the *convolutions,* which greatly increase the surface area of the cerebrum. Deeper grooves divide the cerebral cortex into lobes (Figure 44-6).

The cerebral cortex is composed of countless numbers of neurons. We frequently call this area gray matter because of the color of these cells. The cerebrum below the cortex is composed of white matter, formed by masses of fibers covered by sheaths and extending from the neurons of the cortex to other parts of the body.

The functions of the cerebrum

Different activities are controlled by specific regions of the cerebrum. Some areas of the cerebral cortex are *motor areas,* which means that they are centers that control voluntary movement. Starting at the top of the lobes and working downward, the motor area of the cerebrum controls the muscles of the legs, trunk, arms, shoulders, neck, face, and tongue. Some of the areas of the cerebral cortex are *sensory areas,* which means that the various senses, such as seeing, hearing, touching, tasting, and smelling, are interpreted here. For example, we interpret what our eyes see in the vision center of the *occipital* (ock-SIP-it'l) *lobes.* If these lobes were destroyed, we would not be able to see anything, although our eyes might be perfect. We know also that the *frontal lobes* are centers of emotion, judgment, will power, and self-control. These functions, however, are shared by other areas of the cerebral cortex.

The things we see, hear, and feel are registered as impressions in the sensory areas of the cerebral cortex. The things we do are controlled by the motor areas. These areas are, in turn, connected by a vast number of *association areas.* Thoughts are the result of associations of impressions. Your intellectual capacity is determined by the ability of your cerebral cortex to register impressions, the activity of your association areas, and the sum of your past experiences.

The functions of the cerebellum

The *cerebellum* is a structure lying below the back of the cerebrum. Like the cerebrum, it is composed of hemispheres, but its

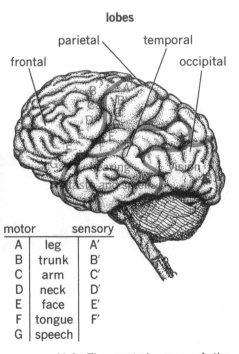

lobes

parietal   temporal

frontal   occipital

motor		sensory
A	leg	A′
B	trunk	B′
C	arm	C′
D	neck	D′
E	face	E′
F	tongue	F′
G	speech	

44-6 The control areas of the cerebrum. Note that some are involved with the origin of action and others with the termination, or interpretation, of senses.

convolutions are shallower and more regular than those of the cerebrum. The surface of the cerebellum is composed of gray matter. Its inner structure is largely white matter, although it contains some areas of gray matter. Bundles of nerve fibers connect the cerebellum with the rest of the nervous system.

In a sense, the cerebellum acts as an assistant to the cerebrum in controlling muscular activity. Nervous impulses do not originate in it, nor can one control its activities. The chief function of the cerebellum is to coordinate the muscular activities of the body. Thus, without the help of the cerebellum, the impulses from the cerebrum would produce uncoordinated motions.

The cerebellum functions further in strengthening impulses to the muscles. This action is a little like picking up a weak radio or television signal and amplifying it before broadcasting it.

Another function of the cerebellum is to maintain tone in muscles. The cerebellum cannot originate a muscular contraction, but it can cause the muscles to remain in a state of partial contraction. You are not aware of this, because the cerebellum operates below the level of consciousness.

The cerebellum functions also in the maintenance of balance. In this activity, it is assisted by impulses from the eyes and from the organs of equilibrium of the inner ears. Impulses from both these organs inform the cerebellum of your position in relation to your surroundings. The cerebellum, in turn, maintains muscular contractions necessary to balance your body.

The brain stem

Nerve fibers from the cerebrum and cerebellum enter the brain stem, an enlargement at the base of the brain. The lowest portion of the stem, the *medulla oblongata* (muh-DULL-uh OB-long-GAH-tuh), is located at the base of the skull and protrudes from the skull slightly where it joins the spinal cord. Another part of the brain stem is the *pons* (ponz), which receives stimuli from the facial area.

There are twelve pairs of *cranial nerves* connected to the brain. These are part of the peripheral nervous system and act as direct connections with certain important organs of the body. One pair, for example, connects the eyes with the brain. Another cranial nerve connects the brain with the lungs, heart, and abdominal organs.

The medulla oblongata controls the activity of the internal organs. The respiratory control center we discussed in the last chapter is located here. Heart action, muscular action of the walls of the digestive organs, secretion in the glands, and other automatic activities are also controlled by the medulla oblongata.

The spinal cord and spinal nerves

The *spinal cord* extends from the medulla oblongata through the protective bony arch of each vertebra, almost the length of

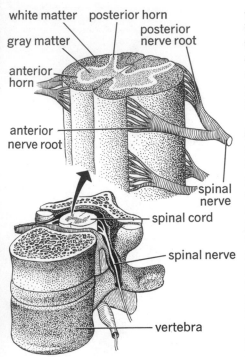

white matter   posterior horn

posterior nerve root

gray matter

anterior horn

anterior nerve root

spinal nerve

spinal cord

spinal nerve

vertebra

44-7 The structures of the spinal cord and their relationship to the vertebral column. The white matter is located in the outer region of the spinal cord.

44-8 Trace the path of the impulse from the receptor to the effector, the muscle. Why are reflex actions said to be automatic?

the spine. Its outer region is white matter, made up of great numbers of nerve fibers covered by sheaths. Neurons composing the gray matter lie inside the white matter in a shape like that of a butterfly with spread wings (Figure 44-7). The pointed tips of the wings of gray matter are called *horns*. The *posterior* pair points toward the back of the cord, while the *anterior* pair points toward the front of the cord.

Thirty-one pairs of **spinal nerves** branch off the cord between the bones of the spine. Along with the cranial nerves, these nerves and their branches form the peripheral nervous system. One member of a pair goes to the right side of the body. Its mate goes to the left side. Spinal nerve branches begin in the neck and continue all the way along the cord. These large cables are mixed nerves. Of their many fibers, some are sensory fibers that carry impulses into the spinal cord, while others are motor fibers that lead impulses away from it. Each spinal nerve divides just outside the cord. The sensory fibers that carry impulses from the body into the spinal cord branch to the posterior horns of the gray matter. This branch of each spinal nerve has a **ganglion** (near its point of entry to the cord), in which the sensory cell bodies are found. The other branch at the junction leads from the anterior horns of the spinal cord, in which the motor cell bodies are located. The motor fibers of this branch carry impulses from the spinal cord to the body.

If the spinal cord were cut, all parts of the body controlled by nerves below the point of severance would be totally paralyzed. Such an injury might be compared to cutting the main cable to a telephone exchange.

Nervous reactions

Nervous reactions vary greatly in form and complexity. The simplest of these is the **reflex action** (Figure 44-8). It is an auto-

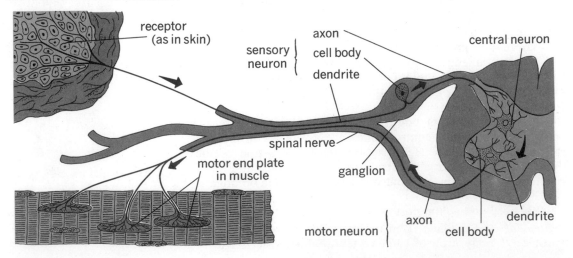

receptor (as in skin)

axon

central neuron

sensory neuron

cell body

dendrite

spinal nerve

ganglion

motor end plate in muscle

axon

dendrite

cell body

motor neuron

matic reaction involving the spinal cord or the brain. The knee jerk is an excellent example of a simple reflex action. If you allow your leg to swing freely and strike the area just below the kneecap with a narrow object, the foot jerks upward. This reaction is entirely automatic. Striking the knee results in the stimulation of a sensory neuron in the lower leg. An impulse travels along the dendrite to the spinal cord. Here the impulse travels to a central neuron. In turn, an impulse from this neuron stimulates a motor neuron extending to the leg muscles, causing a jerk. The entire reflex takes only a split second.

When you touch a hot object, your hand jerks away almost instantly. This reflex action is similar to the knee jerk. After the reflex is completed, the impulse reaches the brain and registers pain. If the muscle response had been delayed until the pain impulse had reached the brain and a motor impulse traveled down the spinal cord from the cerebral motor area, the burn injury would have been much greater. This rapid reaction indicates one value of the simple reflex. Other reflex actions include sneezing, coughing, blinking the eyes when the cornea is touched, laughing when tickled, and jumping when frightened.

The autonomic nervous system

The *autonomic nervous system* is entirely involuntary and automatic. It is composed of two parts, one of which is called the *sympathetic system.* This system includes two rows of nerve tissue, or cords, which lie on either side of the spinal column. Each cord has ganglia, which contain the bodies of neurons. The largest of the sympathetic ganglia is the *solar plexus,* located just below the diaphragm. Another is near the heart; a third is in the lower part of the abdomen; and a fourth is in the neck. Fibers from the sympathetic nerve cords enter the spinal cord and connect with it and with the brain, as well as with one another. The sympathetic nervous system helps to regulate heart action, the secretion of ductless glands, blood supply in the arteries, the action of smooth muscles of the stomach and intestine, and the activity of other internal organs (Figure 44-10).

The *parasympathetic system* opposes the sympathetic system and thus maintains a system of checks and balances. The principal nerve of the parasympathetic system is the *vagus nerve,* a cranial nerve that extends from the medulla oblongata, through the neck, to the chest and abdomen. The check-and-balance system is illustrated by the fact that the sympathetic system acts to speed up heart action, while the vagus nerve acts to slow it down.

The sensations of the skin

The terminal branches of dendrites of sensory neurons in the skin end in special sensory end organs called *receptors.* Some receptors are many-celled; some consist of only one spe-

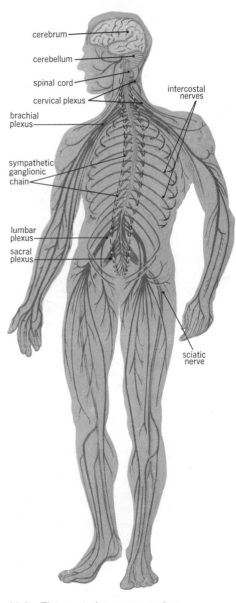

44-9 The central nervous system of the major nerve trunks of the peripheral nervous system. For clarity, only a few of the branches from the major nerve trunks and nerves are included.

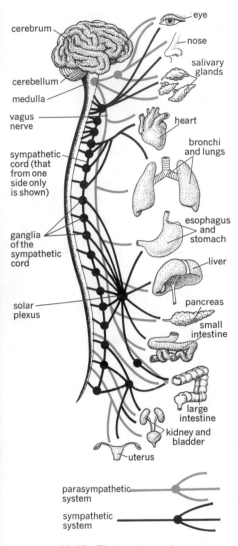

cerebrum

eye

nose

salivary glands

cerebellum

medulla

vagus nerve

heart

bronchi and lungs

sympathetic cord (that from one side only is shown)

esophagus and stomach

ganglia of the sympathetic cord

liver

pancreas

solar plexus

small intestine

large intestine

kidney and bladder

uterus

parasympathetic system

sympathetic system

44-10 The autonomic nervous system regulates the internal organs of the body. What are the functions of its two divisions?

44-11 The five types of receptors found in the skin.

cialized cell; and some are the bare nerve endings themselves. Each receptor is suited to receive only one type of stimulus and to start impulses to the central nervous system. The skin has five different types of receptors. Certain of these respond to touch, while others receive stimuli of pressure, pain, heat, or cold (Figure 44-11).

Normally, no one receptor reacts to more than one stimulus, and thus the five sensations of the skin are distinct and different. The *pain* receptor, for example, is a bare dendrite. If the stimulus is strong enough, a pain receptor will react to mechanical, thermal, electrical, or chemical stimuli. The sensation of pain is a protective device that signals a threat of injury to the body. Pain receptors are distributed throughout the skin.

The sensory nerves of the skin are distributed unevenly over the skin area and lie at different depths in the skin. For instance, if you move the point of your pencil over your skin very lightly, you stimulate only the nerves of *touch*. The receptors for touch are close to the surface of the skin in the region of the hair sockets. The fingertips, the forehead, and the tip of the tongue contain an abundance of receptors that respond to touch.

Receptors that respond to *pressure* lie deeper in the skin. If you press the pencil point against the skin, you feel both pressure and touch. Since the nerves are deeper, a pressure stimulus must be stronger than a touch stimulus. You may think that there is no difference between touch and pressure. But the fact that you can distinguish the mere touching of an object from a firm grip on it indicates that separate nerves are involved.

Heat and *cold* stimulate different receptors. This is an interesting protective adaptation of the body. Actually, cold is not an active condition. Cold results from a reduction in heat energy. If both great heat and intense cold stimulated a single receptor, we would be unable to differentiate between the two. In turn, we would be unable to react appropriately to either. However, since some receptors are stimulated by heat and others by the absence of it, we are constantly aware of and can react to both conditions.

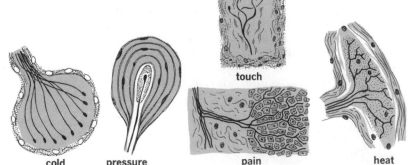

touch

cold

pressure

pain

heat

The sense of taste

Taste results from the chemical stimulation of certain nerve endings. Since nearly all animals prefer some food substances to others, we must assume that they can distinguish different chemical substances. The sense of taste in man is centered in the *taste buds* of the tongue. These flask-shaped structures, containing groups of nerve endings, lie in the front area of the tongue, along its sides, and near the back. Foods mixed with saliva and mucus enter the pores of the taste buds and stimulate the hairlike nerve endings (Figure 44-12).

Our sense of taste is poorly developed. We recognize only four common flavors: *sour, sweet, salty,* and *bitter.* Taste buds are distributed unevenly over the surface of the tongue. Those sensitive to sweet flavors are at the tip of the tongue. Doesn't candy taste sweeter when you lick it than when you chew it far back in the mouth? The tip of the tongue is also sensitive to salty flavors. You taste sour substances along the sides of the tongue. Bitter flavors are detected at the back of the tongue, which explains why a bitter substance does not taste bitter at first. If a substance is both bitter and sweet, you sense the sweetness first, then the bitterness. Substances such as pepper and some other spices have no distinct flavor, but they irritate the entire tongue and produce a burning sensation.

Much of the sensation we call taste is really smell. When you chew foods, vapors enter the inner openings of the nose and reach nerve endings of smell. If the external nasal openings are plugged up, many foods lack the flavor we associate with them. Under such conditions, onions and apples have an almost identical sweet flavor. You probably have noticed the loss of what you thought was taste sensation when you had a head cold and temporarily lost your sense of smell.

The sense of smell

Like taste, smell results from the chemical stimulation of nerves, except that odors are in the form of gases. The nasal passages are arranged in three tiers, or layers, of cavities, separated by bony layers called *turbinates.* The upper turbinate contains branched endings of the **olfactory nerve,** which is a cranial nerve (Figure 44-13). Stimulation of these endings by odors results in the sensation of smell. Receptors that are exposed to a particular odor over a long period of time become deadened to it, although they are receptive to other odors. Nurses are not usually aware of the odor of iodoform in a hospital, but a visitor is.

The structure of the human ear

Our *ears* and those of other mammals are wonderfully complex organs. The external ear opens into an *auditory canal* em-

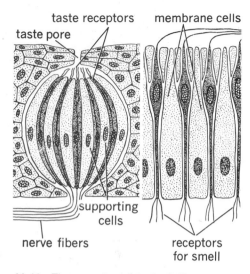

44-12 The receptor of taste (left) and smell (right), as they appear under a microscope.

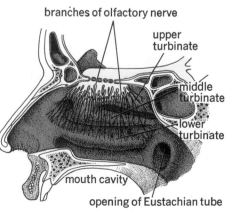

44-13 The surface of the inner wall of the nose. What function does the Eustachian tube perform?

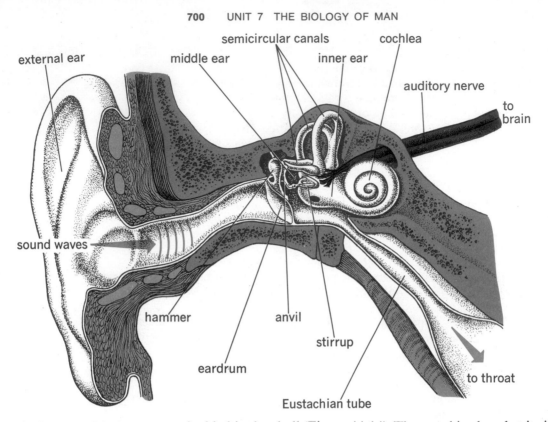

semicircular canals

cochlea

external ear

middle ear

inner ear

auditory nerve

to brain

sound waves

hammer

anvil

stirrup

eardrum

to throat

Eustachian tube

44-14 Structure of the human ear.

bedded in the skull (Figure 44-14). The canal is closed at its inner end by the *eardrum,* or *tympanic membrane,* which separates it from the middle ear. The middle ear connects with the throat through the *Eustachian* (yoo-STAY-shun) *tube.* This connection equalizes the pressure in the middle ear and the atmosphere.

When the connection between the middle ear and throat is blocked by a cold that involves the Eustachian tube, the pressures outside and inside do not equalize. For this reason, divers and fliers do not work when they have colds. The outside pressure increases during a dive. But with a blocked Eustachian tube, the middle-ear pressure would not be equalized, and the difference might burst the eardrum. The flier's situation would, of course, be the reverse; that is, the pressure would be less outside than in the middle ear.

Three tiny bones, the *hammer, anvil,* and *stirrup,* form a chain across the middle ear. They extend from the inner face of the eardrum to a similar membrane that covers the *oval window,* the opening to the inner ear.

The inner ear is composed of two general parts. The *cochlea* (KOCK-lee-uh) is a spiral passage resembling a snail shell. It is filled with a liquid and is lined with nerve endings that receive the sound impressions. The **auditory nerve** leads from the cochlea to the brain. The *semicircular canals* consist of three loop-shaped tubes, each at right angles to the other.

How we hear

An object vibrating in air produces regions in which the air molecules are squeezed together (compressions) and regions in which they are farther apart (rarefactions). This regular pattern produced by any vibrating object in air or other matter is called a sound wave. When sound waves reach the ear, they pass through the external ear into the auditory canal and to the eardrum. Here they cause the eardrum to vibrate in the same pattern as the compressions and rarefactions of the sound wave. The vibration of the eardrum, in turn, causes the hammer, anvil, and stirrup bones of the middle ear to vibrate. The vibration is transmitted to the membrane of the inner ear and sets the fluid in the cochlea in motion. Vibration of the fluid, in turn, stimulates the nerve endings in the cochlea. Impulses travel through the auditory nerve to the cerebrum, where the sensation of sound is perceived. If the auditory region of the cerebrum ceases to function, a person cannot hear, even though his ear mechanisms receive vibrations normally.

The sense of balance

Our sense of equilibrium, or balance, is centered in the semicircular canals (see Figure 44-14) of the inner ear. These canals lie at right angles to one another in three different planes. Their position has been compared to the parts of a chair. One canal lies in the plane of the seat, another in the plane of the back, and a third in the plane of the arms.

The semicircular canals contain a great number of receptors and a fluid similar to that of the cochlea. When the head changes position, the fluid rocks in the canals and stimulates these receptors. Impulses travel from them through a branch of the cerebellum. Thus, the brain is made aware of the position of the head. Since the canals lie in three planes, any change in position of the head moves the fluid in one or more of them. If you spin around rapidly, the fluid is forced to one end, and impulses travel to the brain. When you stop spinning, the fluid rushes back the other way, giving you the sensation of twirling in the opposite direction, so you feel dizzy. Regular, rhythmic motions produce unpleasant sensations that involve the whole body. These sensations are what we call *motion sickness*. Disease of the semicircular canals results in temporary or permanent dizziness and loss of equilibrium.

The structure of the human eye

The normal *eye* is spherical and slightly flattened from front to back (Figure 44-15). The wall of the eyeball is composed of three distinct layers. The outer, *sclerotic* (skluh-ROT-ick) *layer* is tough and white. This layer shows as the white of the

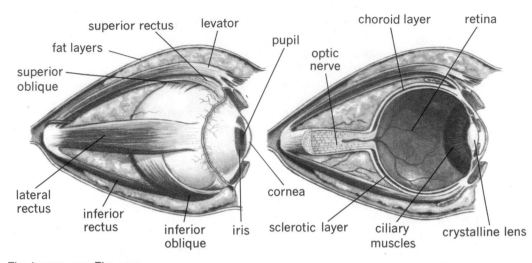

superior rectus   levator   pupil   choroid layer   retina
fat layers   optic nerve
superior oblique
lateral rectus   inferior rectus   inferior oblique   iris   cornea   sclerotic layer   ciliary muscles   crystalline lens

44-15 The human eye. The muscle and socket are shown at the left; the various internal structures can be seen in the cutaway diagram at the right.

eye. It bulges and becomes transparent in front, and is called the *cornea*.

The middle, *choroid layer* is richly supplied with blood vessels. It completely encloses the eye except in front, where there is a hole. This small opening, lying behind the center of the cornea, is the *pupil*. Around the pupil, the choroid contains pigmented cells. This area is the *iris* and may be colored blue, brown, hazel, or green. Change in size of the circular pupil is accomplished by muscles in the iris. This adjustment in size of the pupil opening to the intensity of light is an automatic reflex. When the light is reduced, the pupil becomes large, or dilates. In bright light it becomes small, or constricts. The eye doctor uses drops to block the automatic iris reflex so that he can use a bright pinpoint of light to see inside the eye. This is the only place in the body where the blood vessels can actually be seen. The black pigment of the choroid layer can also be seen inside the eye. This black pigmentation prevents reflection of light rays within the eye.

A convex, *crystalline lens* lies behind the pupil opening of the iris. The lens is supported by the *ciliary muscles* fastened to the choroid layer. Contraction of these muscles changes the shape of the lens. In this way, light rays are focused on the retina.

The space between the lens and the cornea is filled with a thin, watery substance, the *aqueous humor*. A thicker, jellylike, transparent substance, the *vitreous humor*, fills the interior of the eyeball. This fluid helps to keep the eyeball firm.

The structure of the retina

The inner layer of the eye is the most complicated and delicate of the eye layers. This layer, the *retina*, is less than $1/80$ inch thick. Yet it is composed of seven layers of cells, receptors, ganglia, and nerve fibers. The function of all the structures of the eye is to focus light on the retina. The specialized receptors that are stimulated by light are called ***photoreceptors*** and are of two types, *cones* and *rods* (Figure 44-16). The photoreceptors lie deep in the retina, pointing toward the back surface of the eyeball. Impulses from the cones and rods travel through a series of short nerves with brushlike endings to ganglia near the front part of the retina. More than half a million nerve fibers lead from the ganglia over the surface of the retina to a large cranial nerve, the ***optic nerve***. There are no rods or cones at the point where the end of the optic nerve joins the retina. Since there is no vision at this point, it is called the *blind spot*.

An optic nerve extends from the back of each eyeball to the vision center in the occipital lobe of the cerebrum. Some of the fibers cross as they lead to the cerebrum. This means that some of the impulses from your right eye, for example, go to the left occipital lobe and that some go to the right occipital lobe. Thus, what you see with each eye is interpreted in both lobes.

How we see

The cones of the retina are sensitive to bright light and are responsible for color vision. When we see in daylight, light rays pass through the cornea, aqueous humor, pupil, lens, and the vitreous humor to the retina. However, the lens focuses the rays particularly on a small, sensitive portion of the retina called the *fovea,* where the cones are especially abundant. This

44-16 The structure of the back of the eye. The shapes and arrangement of the rods and cones are shown in greater detail in the enlargement.

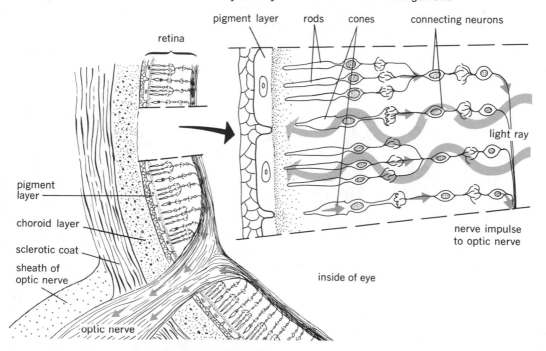

pigment layer   rods   cones   connecting neurons

retina

light ray

pigment layer

choroid layer

sclerotic coat

sheath of optic nerve

nerve impulse to optic nerve

inside of eye

optic nerve

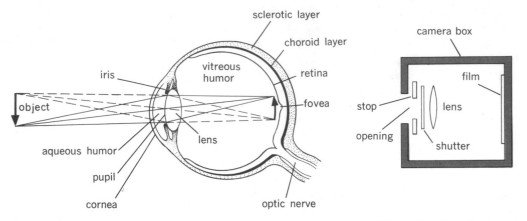

44-17 The human eye and a camera. After studying the two, indicate exactly how the parts of the camera are similar to those of the eye, and how the parts differ. Why is the image reversed and inverted on the retina?

is the point at which we see an object clearly. Cones outside the fovea register vision only indistinctly. Thus, if you focus your eyes on an object, you see it clearly, while all other objects in your field of vision lack detail.

During the late evening or at night, the light is too reduced to stimulate the cones, but is sufficient to stimulate the rods. However, while the rods are able to act under conditions of reduced light, they are unable to distinguish among colors.

Rods produce a substance called *visual purple,* which is necessary for their proper functioning. Bright light fades visual purple, which causes the rods to be insensitive to further stimulation by light. This explains why when you leave a bright room at night, you are temporarily night-blind. As visual purple is restored, the rods begin to function, and you can see objects in dim light. Interestingly, we can now explain why, although you can see an object "out of the corner of your eye" at night, when you focus on it, it disappears: The fovea, the portion of the retina where we see most clearly, contains many cones but no rods.

The human eye contains fewer rods than many animal eyes, so our night vision is relatively poor. The cat, deer, and owl see well at night because they have many rods. The owl, however, lacks cones in its eyes and therefore is day-blind.

Eye movement and protection

The eye rests in its socket against layers of fat that serve as cushions. Movements of the eyeball are accomplished by pairs of muscles that attach to its sides and extend back into the eye socket (Figure 44-15). The sclerotic layer is supplied with nerve endings that register pain when a foreign object touches it. The eye is further protected by its location deep in the recesses of the eye socket, by bony ridges, by the eyelids, and by the tear glands that keep its surface moist. Tears wash over the eye and drain into the tear duct in the lower corner of the eye socket, which leads to the nasal cavity. Because tears contain an antibacterial enzyme, they are mildly antiseptic.

SUMMARY

Comparisons of a chimp and a man riding a motorcycle might, at first, appear to indicate similar abilities. Actually, the thought processes involved in each instance are quite different. They both balance on the moving vehicle, but the chimp has been trained to sit on the seat until stopped by the circus director. The man has full control of his speed, direction, and duration. He can perceive the environment in advance and make allowances for obstacles. If the motor failed, the man may be able to make repairs and continue on his way.

The brain and spinal cord compose the central nervous system. They communicate with all parts of the body by nerves. The cerebrum controls conscious activities. It is the center of intelligence and contains both sensory and motor areas. Impulses from the cerebral motor area pass through the cerebellum, where coordination of impulses takes place. The medulla oblongata controls the activity of internal organs and is the control center of respiration.

Sensory nerves carry impulses from their receptors in sense organs to the central nervous system. The numerous minute sense organs of the skin respond to touch, pressure, pain, heat, and cold. The endings involved in the sense of smell contact odors that have reached the upper turbinate region of the nasal passages. The ears are highly developed sense organs which receive air-borne vibrations and carry them to the receptors in the cochlea in the inner ear. They also contain the semicircular canals, which control our sense of equilibrium.

The eye is the most highly specialized of the sense organs. It receives light rays through the pupil and directs them to the retina by means of the lens. In the retina, the photoreceptors send impulses through the optic nerve to the visual center in the occipital lobes of the brain.

BIOLOGICALLY SPEAKING

central nervous system
neuron
synapse
nerve
meninges
cerebrum
cerebellum
medulla oblongata
pons

cranial nerves
spinal cord
spinal nerve
ganglion
reflex action
autonomic nervous
    system
sympathetic nervous
    system

parasympathetic nervous
    system
receptor
olfactory nerve
ear
auditory nerve
eye
photoreceptors
optic nerve

QUESTIONS FOR REVIEW

1. Compare the reactions of a motorcycle-riding chimp in a circus with those of a young man riding on a country road.
2. Name the three main divisions of the nervous system and state the functions of each.
3. Why are peripheral nerves that contain only axons considered to be motor nerves?
4. What is a nerve impulse?
5. What occurs at the endings of a motor neuron that causes a muscle fiber to contract? What causes it to relax?
6. Name the parts of the brain and state the functions of each.
7. In what way is the autonomic nervous system really two systems?

8. Name the five sensations of the skin. In what ways are the receptors different?
9. Account for the fact that we think we distinguish more than the five tastes the tongue can perceive.
10. Describe how a sound wave in the air stimulates the receptors in the cochlea.
11. How can an infection in the middle ear produce temporary deafness?
12. Describe the movements of the head that would be necessary to stimulate each semicircular canal separately.
13. Why is our night vision relatively poor compared to night vision of an owl?
14. Why is it that you can see an object out of the corner of your eye at night, but when you focus on it, it disappears?

APPLYING PRINCIPLES AND CONCEPTS

1. What is intelligence?
2. Explain the activity that occurs after a chicken's head has been cut off.
3. Explain the fact that the sympathetic nervous system is sometimes called "the system for fight or flight."
4. How would you go about designing an experiment to prove or disprove that the eye receives an image upside down and that the brain interprets it oppositely?

RELATED READING

Books

ADLER, IRVING, and RUTH ADLER, *Taste, Touch and Smell.*
The John Day Company, Inc., New York. 1967. A survey of the three main senses of man.

ASIMOV, ISAAC, *The Human Brain: Its Capacities and Functions.*
Houghton Mifflin Company, Boston. 1963. Considers the intricate organization of each of the smoothly integrated parts of the brain as it functions in the body.

BEGBIE, G. HUGH, *Seeing and the Eye.*
Natural History Press, Garden City, New York. 1969. Takes the reader through the entire complex relating to vision including the physics of light, anatomy of the eye and the function of the brain.

FROMAN, ROBERT, *The Many Human Senses.*
Little, Brown and Company, Boston. 1966. The story of how researchers discovered that man has more than five obvious senses.

GILMOUR, ANN, and JAMES GILMOUR, *Understanding Your Senses.*
Frederick Warne and Co., Inc., New York. 1963. Easy-to-perform experiments show how messages from the outer world enter the gateways of our mind through our senses.

HORROBIN, DAVID F., *The Communication Systems of the Body.*
Basic Books, Inc., Publishers, New York. 1964. How man and animals utilize their vitally important communication systems to coordinate information from the outside world with their own actions and the activities of their various organs, muscles, etc.

MOORE, JOSEPHINE C., *Neuroanatomy Simplified.*
Kendall-Hunt Publishing Co., Dubuque, Iowa. 1969. A somewhat advanced discussion of the anatomy of the nervous system including excellent diagrams.

CHAPTER FORTY-FIVE

TOBACCO, ALCOHOL, AND DRUGS

Three social and health problems

Problems resulting from the use of alcohol and drugs are as old as civilization. They are more acute today because automobiles, firearms, and other mechanical devices become implements of destruction in the hands of an intoxicated or drug-influenced person.

We shall consider tobacco, alcohol, and nonmedical uses of drugs together, because they are all harmful substances when used habitually. Young people who see a large number of adults smoking without any *apparent* effect may arrive at the wrong conclusions, because the long-term effects of tobacco are not always visible. While there is no real justification for the use of tobacco, its influence on the human mechanism is not to be compared to the social, emotional, and mental damage resulting from the habitual and excessive use of alcohol or narcotics.

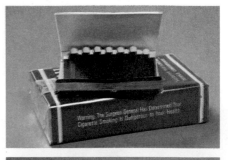

45-1 All these may contribute to the complicaton of one's problems. (John King; Bureau of Narcotics and Dangerous Drugs)

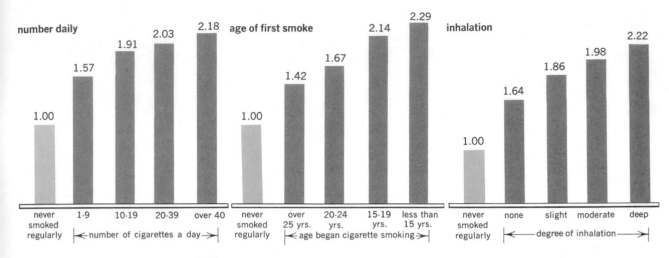

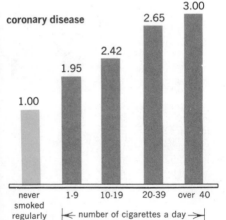

45-2 These graphs show ratios of death rates in the age group 40–69. The death rate for non-smokers is given in each case as 1.00; the other numbers are ratios to this base number. (adapted from a report by E. C. Hammond)

In this chapter, we will discuss some of the effects of these substances on the body. You, as an individual, will then have to form an opinion and make a decision about your own personal habits.

Tobacco—the nation's leading habit

More than seventy million people in the United States use tobacco in some form. The great majority of these are cigarette smokers. The smoker actually becomes a slave to two habits: the smoking habit and the tobacco habit. The first involves reaching for a cigarette at regular intervals, lighting it, and going through the various movements associated with smoking. Heavy smokers often light a second cigarette even before finishing the first one. They also acquire a physiological craving for the nicotine in tobacco—the tobacco habit.

Most young people who start smoking feel that it makes them seem more mature. Yet, if they asked the advice of an older person who has smoked for several years, his advice would undoubtedly be not to start. Certainly several things should be considered before deliberately starting a practice as habit-forming and as dangerous to the health as smoking.

The effects of smoking

In 1964, the Public Health Service published a report on the effects of smoking based on experiments with animals, clinical and autopsy studies in man, and studies of the occurrence of disease in the population. Included in the report were these findings:

■ *Tissue damage.* Secretions of lung tissue from thousands of smokers have been examined after the smokers' deaths. Even in individuals who did not die of cancer, abnormal cells were found in the lungs. Enlarged and ruptured alveoli and thickened arterioles were observed. In the trachea and bronchi, cilia and the protective cells of the mucosa had been destroyed. Remember that these structures normally cleanse and lubricate the respiratory tract and help to prevent infection.

■ *Elevated death rate.* For the purpose of this study, the number of deaths among a large sample of nonsmokers was compared with the number of deaths among a similar sample of smokers. There were 70 percent more deaths from all causes among smokers than among nonsmokers. The greater number of deaths from certain diseases among smokers was particularly marked. There were 1000 percent more deaths from lung cancer and 500 percent more from chronic bronchitis and the degenerative lung disease called emphysema. The death rate was also considerably higher for cancer of the tongue, larynx, and esophagus, for peptic ulcer, and for circulatory diseases.

■ *Elevation of the death rate with an increase in the amount smoked.* In general, the greater the number of cigarettes smoked, the higher the death rate. For men who smoke less than ten cigarettes a day, the death rate is about 40 percent higher than it is for nonsmokers. For those who smoke forty or more, it is 120 percent higher. The same kind of relationship exists between the number of years of smoking and the death rate.

It should be obvious from these findings that smoking is a health hazard. To put the figures on lung cancer in another way, 95 percent of the victims of lung cancer are heavy smokers. One half of one percent are nonsmokers. Lip cancer has definitely been traced to irritation from pipes or cigars. Lung and lip cancer, however, do not account for a large percentage of deaths in this country. Heart and circulatory diseases are the number-one cause of death in the United States. The death rate for these diseases is 200 percent higher among smokers than among nonsmokers.

Aside from its long-term effects, smoking has many short-term disadvantages. It is an expensive habit and gets more so every year. It is messy and often annoying to other people. Smoking stains the teeth and fingers and makes the breath unpleasant. It causes constant throat irritation, stomach discomfort, nervousness, and sometimes headaches.

Alcohol in the body

The alcohol in beverages is *ethyl alcohol,* or ethanol. As you recall, it is produced by the action of yeast on sugars. Two general sources of the sugars used are fruits and grains. The term *grain alcohol* stems from the use of a mash of ground corn, rye, barley, wheat, or other cereal grains.

Alcohol does not have to be changed in form or composition in the stomach before absorption occurs. It starts to enter the blood within two minutes after it is swallowed, and is rapidly delivered to the tissues, where it is absorbed by the cells. This absorption is even more rapid when the stomach is empty than when the stomach contains food.

In the cells, oxidation begins immediately, and large amounts of heat are released. The body tissues oxidize alcohol at the rate of approximately one ounce in three hours. Because this rate is constant and does not depend on the energy needs of the body, alcohol has no value as a food. The excess heat produced from the oxidation of alcohol raises the temperature of the blood. This, in turn, stimulates the heat-control center in the brain, which responds by causing increased circulation to the skin. There the heat radiates from the body. The increased circulation to the skin causes the characteristic rosy skin tone, or alcoholic flush. Also, since the receptors of heat are in the skin, the rush of blood to the skin gives a false impression of warmth. Actually, the internal organs are being deprived of an adequate blood supply.

Some effects of alcohol on the body organs

Not all the alcohol in the body is oxidized. Part is released into the lungs as vapor, causing an alcoholic breath odor. Some reaches the skin, and is added to perspiration in the sweat glands. Part passes into the kidneys and is carried out of the body in the urine.

Since alcohol is absorbed by all the body organs, they are all affected by its presence, but some seem to be affected more than others. The oxidation of alcohol produces water, which is excreted in large quantities by the skin during heat elimination. Not only do the tissues become dehydrated, but also the loss of water causes a greater concentration of nitrogenous wastes in the kidneys and interferes with normal elimination.

Vitamin-deficiency diseases are common among alcoholics, who often starve themselves during long periods of excessive drinking. In addition, during these fasts the liver is deprived of its stored food, and it swells as the carbohydrates are replaced by fats. This condition, known as *fatty liver,* is found in 75 percent of alcoholics. Continuation of these fasts may lead to a more serious degeneration of the liver, called *cirrhosis* (si-ROE-siss). In this disease, the fatty liver shrinks and hardens as the fats are used. Although excessive use of alcohol is not the only cause of cirrhosis, this condition occurs eight times more frequently in alcoholics than in other people.

Another organ frequently affected by excessive use of alcohol is the stomach. Alcohol causes an increase in stomach secretions, which often leads to *gastritis,* a painful inflammation of the stomach lining.

The effects of alcohol on the nervous system

Alcohol is a *depressant;* that is, it has an anesthetic effect on the nervous system. You may have heard it called a stimulant because its anesthetic effect releases inhibitions. Its overall effect, however, is exactly the opposite of a stimulant.

The first effects of alcohol occur in the cortex of the brain. Loss of judgment, will power, and self-control occur. Cares seem to vanish, and the person becomes gay and light-headed. Influence on the frontal lobe alters emotional control and may lead to a feeling of great joy, shown by foolish laughter, or to sadness and weeping. As the effects of alcohol progress through the brain tissue, the vision and speech areas of the cerebrum become involved. There is blurred or double vision, inability to judge distance, and slurred speech.

As the cerebellum becomes involved, coordination of the muscles is affected. The victim becomes dizzy when standing, and if he is able to walk at all, he does so with a clumsy, staggering gait.

In the final stages of drunkenness, a person becomes completely helpless. The brain cortex ceases activity, resulting in complete unconsciousness. The skin becomes pale, cold, and clammy. Heart action, digestive action, and respiration slow down, and the victim lies unconscious and near death.

Alcoholism is a disease

The disease called *alcoholism* is characterized by an abnormal, chronic dependence on alcohol. It may begin with occasional social drinking. As distressing situations and problems arise and life seems temporarily unpleasant, the individual uses alcohol as an escape from reality. The problems, however, remain unsolved. With loss of judgment and will power, the chances of solving these problems are further reduced. The alcoholic then resorts to solitary drinking for the effects of alcohol alone. The destruction of his personality has begun.

In about one in ten, alcoholism progresses to the seriousness of *alcohol psychosis,* an acute form of mental illness in which the victim must be admitted to a mental hospital. The cause of alcohol psychosis is not fully understood. Part of the condition may be caused by the toxic effect of alcohol on nerve tissue. This would account for general deterioration of parts of the brain. Another cause is deficiency in the B-complex vitamins essential for normal nervous activity.

The victim of alcohol psychosis becomes confused to the point that he may not be able to recognize members of his family or even know who he is. This confusion is accompanied by terrifying hallucinations, mostly involving visual horror, and uncontrollable trembling; hence the name of this state, *delirium tremens,* or D.T.'s. Usually in these hallucinations, moving animals are about to attack the patient. In some cases of alcohol psychosis, the patient suffers a loss of memory of recent events. The patient may turn to inventing stories to fill the gap in his thinking resulting from his memory loss. The treatment of

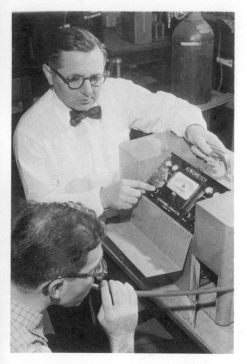

45-3 An alcometer is used to determine the alcoholic content of the breath. (Wide World Photos)

alcohol psychosis involves not only psychotherapy but also rigid regulation of the diet and vitamin supplement.

Alcohol is the tool and not the underlying cause of alcoholism. Thus, the alcoholic seeking a cure must first find the problem that underlies his drinking. Then he must attempt to solve the problem instead of resorting to alcohol as an escape from it. Sympathetic understanding and the cooperation of his family and friends help greatly in overcoming the problem.

Most states have special agencies to deal with alcoholism. They carry on studies and research and furnish educational materials to those interested in learning more about alcoholism. This constructive attitude toward the alcohol problem is reflected further in the changing attitude of many courts. An increasing number of judges, magistrates, and court officers view the alcoholic as a person in need of psychiatric counsel and medical treatment rather than as one to be punished for his difficulty.

Many private organizations and religious groups have worked for years to deal with this problem. Among the most widely effective is Alcoholics Anonymous, a voluntary organization started in 1935. Since the establishment of AA, about 350,000 alcoholics have, with the help of other AA members, been able to stop their drinking.

Alcohol and the length of life

Life insurance companies ask applicants about their use of alcohol and drugs; this is evidence that heavy drinkers are poorer risks than those who abstain. It is difficult to say that limited or moderate use of alcohol shortens life. But no one can deny that even moderate drinking of alcoholic beverages increases the possibility of accidental death. It also lowers body resistance and increases the possibility of death from infectious disease, especially tuberculosis. There is no question that heavy drinking shortens life considerably.

Alcohol and society

The effects of alcohol involve much more than damage to the habitual drinker himself. His family and society pay a price for his shortsightedness. Often an alcoholic will neglect his family to satisfy his desire for alcohol. Child-neglect, loss of job, divorce, and other acute domestic problems frequently result.

Alcohol results in behavior without intelligent control. Many people become hostile and aggressive after several drinks. Fights are not uncommon, and other violent acts are often associated with drinking. Some criminals fortify their courage by drinking and become brutal in crimes of assault, robbery, burglary, sex crimes, and murder. Alcohol, therefore, plays a significant role in a large proportion of crimes.

Alcohol and driving

Important experiments have recently been carried on in Pennsylvania to test thoroughly, under actual road conditions, the relationship between drinking and driving a car. Motorists who were given measured amounts of alcohol but who were not drunk (all but one passed the standard police sobriety tests) were found to make all sorts of errors that could lead to accidents. Not only did most of these drivers have a slower braking reaction time, but they were also inaccurate in performance. Yet every one of these drivers thought he was doing well. The fundamental trouble, graphically proved by psychological tests, was found to be the *impairment of judgment after only one or two drinks*.

Therefore, it is not surprising to find that alcohol is a factor in a large proportion of all fatal traffic accidents. In accidents involving only one vehicle 70% of the drivers killed had been drinking. In multi-vehicle accidents, alcohol had been consumed by 50% of the drivers killed. Also, alcohol was found to be a contributing factor in more than 50% of pedestrians killed in traffic accidents. The effects of alcohol on a driver of an automobile or a motorcycle are:

■ To increase his reaction time. The driver takes longer to put on the brakes or swerve to avoid a collision.

■ To impair his vision and distance judgment.

■ To make him inattentive. His mind is on everything but driving.

■ To make it more difficult for him to associate danger signals with danger. Stop lights, stop signs, and railroad flashers do not have their ordinary impact on him.

■ To give him a false feeling of security. The drunken driver thinks he is the best driver on the road.

■ To make him discourteous. The drunken driver seldom gives right of way. He is often bullheaded and hostile.

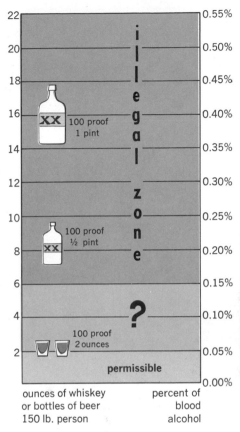

ounces of whiskey
or bottles of beer
150 lb. person

percent of
blood
alcohol

45-4 Drinking limits for U.S. drivers. It is clearly shown by this diagram that alcohol and driving do not mix. Most states have set a blood alcohol content of 0.15 percent as the legal limit for drivers. A higher percentage is proof of drunkenness.

45-5 "If you drive, don't drink; if you drink, don't drive." Traffic officials estimate that one third of all automobile accidents are either the direct or indirect result of drinking. (G. E. Arnold, Black Star)

Drugs

As you know, substances may enter the body through the intestinal walls or through the thin membranes of the air sacs in the lungs. The various substances provide the environment for the cells to carry out vital processes necessary to the well-being of the organism. When anything interfers with the normal functioning of the body's cells, the organism is ill. Your body has its own remarkable mechanisms for defense against pathogenic organisms and for the healing of wounds. However, there are times when the body's recovery may be aided with the administration of *drugs*. A **drug** is any substance used as a medicine or used in making medicines, for internal or external use. A physician may prescribe a drug which kills a pathogenic organism, or speeds up a healing process, or to treat symptoms to make the patient feel better. When a medical doctor prescribes a drug, however, he is familiar with its effects on various tissues of the body. The physician knows that many medications have adverse effects when taken for a long period of time or in large quantities.

Many medications can be purchased without a prescription and are *over-the-counter* drugs. They are labeled with their proper use, dosage, and caution against improper use. While many people think of common aspirin as a harmless substance, it is capable of causing severe disturbances and even death. More than 100 children die each year from an overdose of aspirin. Aspirin may cause headaches, ringing in the ears, dizziness, difficulty in hearing, dimness in vision, mental confusion, sweating, thirst, nausea, vomiting, and diarrhea. Furthermore, the amount of aspirin causing ill effects will vary with different persons.

Sources of drugs

The witch doctor of primitive man administered to the ills of the tribe. He often had great power and mixed superstition with his treatment. However, he probably learned through trial and error that preparations made of specific plants relieved itching of skin rashes or speeded up the healing of wounds. Often the plants used by the witch doctor were known to members of the tribe but thought to be poisonous—as well they might have been in improper dosages or with improper extraction.

The cinchona tree in South America, for example, was regarded by the natives as being very toxic. Its first recorded use, however, was in 1630. The bark of the tree was used to treat fever of a Spanish official in Ecuador. His fever disappeared and recovery from malaria was complete. Nine years later it was used again with such miraculous results that bark from these trees was brought to Spain. We now know that the drug, *quinine,* was the effective substance. Drugs of plant origin, such as quinine, are called *natural drugs*. Other drugs, such as sulfanilamide, some vitamins, as well as many others, are made in laboratories and are called *synthetic drugs*.

It may surprise you to learn that naturally occurring drugs are found in coffee, tea, and cocoa, and the *Kola* nut from which cola drinks are made. These popular beverages contain substances which prevent sleep and cause emotional excitement. These substances and others which cause similar responses are called **stimulating drugs**. Coffee and tea are widely consumed and it is easy to form the habit of the "coffee break."

Habit and addiction

The habit of consuming a drug over a period of time may result in a need for the substance. The individual's dependence on the drug is called **addiction**. *Psychological addiction* is an emotional or mental dependence on a drug. If the habit of smoking or drinking coffee is suddenly stopped, for example, symptoms of craving and nervousness appear. *Physical addiction* is a physiological dependence on a drug: that is, the body requires a continuing supply. A person may be both psychologically and physically addicted to a drug. If the supply of a drug is cut off, a person with a physiological dependence will show *withdrawal symptoms*. These symptoms may be severe and include sleeplessness, difficulty in breathing, irregular heart action, and acute suffering. Mental symptoms include depression and derangement. Withdrawal sickness may be severe enough to cause death. The longer an addict uses such an addicting drug, the greater amount he must take to ward off the withdrawal symptoms. Thus, his body develops a *tolerance* for the drug.

Narcotics

A **narcotic** is a drug which produces sleep or stupor and relieves pain. Substances are designated as such by the Federal Government. They include opium, cocaine, their derivatives, and the synthetic compounds that produce similar physiological results. The sale of narcotics, except on a doctor's prescription, is illegal.

Opium is the source of a family of narcotic drugs. It is extracted from the juice of the white poppy. *Morphine* and *codeine* are derived from opium, while *heroin* is a synthetic compound prepared from morphine. Morphine is used to reduce pain. Codeine has similar uses and is an ingredient in special kinds of medicines, including some cough syrups.

Cocaine is a narcotic drug extracted from the leaves of the South American coco plant (not connected with the beverage cocoa). Cocaine deadens skin and mucous membranes. A doctor may use it to deaden the area around a wound before he cleanses it or takes stitches. When it is taken internally, cocaine causes a temporary stimulation of the nervous system and a feeling of pleasure. Later, however, the victim is seized by a feeling of great fear and may even become violent.

Chemical and drug misuse—a road to self-destruction

Chemicals or drugs are misused when taken into the body for purposes other than medical, or when their use influences behavior so that lives are endangered. Drug misuse often stems from a desire to escape from problems. A person may take a shot of heroin and find that he feels better about his problems. Suddenly his troubles don't seem to be important, and he seems to be in a dream-world of his own. Nothing terrible seems to have happened, but soon the individual feels the effects wearing off and wants another shot. After a short period of time, the momentary pleasures seem to be gone. The sense of doing something illegal, dangerous, and exciting has disappeared, but a craving has built up and he needs more. As tolerance to the narcotic develops, four to ten times the original dose is needed, and a physical dependence on the drug is established. When the effects of the drug begin to wear off, he feels irritable, his eyes water, and the other withdrawal symptoms occur. His temperature and blood pressure increase, he sweats, and may experience vomiting and diarrhea. As his body cells now need the drug, the addict's activities are directed to a desperate end of getting rid of these terrible feelings and pains by taking another shot.

This self-destructive process now requires increasing doses and increasing frequencies. Four to five shots daily are not uncommon. Heroin is so dangerous and addictive that the possession or use in the United States, even for medical purposes, is illegal. Therefore, its price is high. He may require amounts up to one hundred dollars a day. Money may be obtained from stealing and pawning items taken from family and friends. He may even become a "pusher" to obtain the drug for himself. The addict does not care about his appearance or health habits, and his resistance to disease is lowered.

Volatile chemicals

Kerosene, paint thinner, model airplane glue, household cement, gasoline, lighter fluid, and other solvents are not drugs. However, some young people have sought quick thrills by breathing in such chemicals. Inhalation of the volatile chemical causes it to diffuse into the bloodstream. Since the amount of oxygen is diminished, a dizziness is perceived. Loss of coordination, slurred speech, blurred vision, loss of color vision, ringing in the ears, and nausea also develop. Drowsiness, unconsciousness, and death may follow. However, in cases where an individual has breathed these toxic fumes many times, permanent damage has occurred to the brain, nerves, liver, kidney, and bones.

Marijuana

Cannabis sativa is the name of the Indian hemp. Its flowers and leaves and seeds contain *marijuana* (MA-ri-WAH-nuh), and it has no uses in modern medicine. This plant grows wild in regions of temperate climates in many parts of the world. It grows as a weed in many of our mid-western states. Marijuana often receives considerable attention in the newspapers, as its use is illegal. Although marijuana may be ingested, it is usually smoked. The marijuana user develops an emotional addiction to the effects of the drug but does not experience physical addiction. The effects of using marijuana are unpredictable, since they vary with different individuals. Some users become violent and bring danger to themselves and others. Some are unable to judge distances and are hazards on the highways. Many marijuana users turn to opiates and become physically dependent on narcotics.

The barbiturate problem

The *barbiturates* are synthetic drugs that act as tranquilizers or sedatives. A *sedative* is a drug used by doctors to induce sleep, to reduce emotional anxiety, and to relieve pain. They are commonly called *sleeping pills*. Hundreds of thousands of people in this country use barbiturates habitually. Barbiturates are not addictive in the sense in which narcotic drugs are, but victims develop a psychological dependence on them and may even have withdrawal symptoms when they are deprived of them.

Aside from dependence, there are two severe hazards of barbiturates: They are frequently used for suicide, and they can be lethal when combined with alcohol in the body. When barbiturates and alcohol are taken together, each one doubles the effect of the other. You have undoubtedly read of tragic deaths resulting from this combination.

Stimulants

The *amphetamines* (am-FET-uh-MEENZ) are a class of stimulant drugs used in *pep pills*. Their effects are similar to those of cocaine. While pep pills are sometimes taken by truck drivers and night workers to keep them awake, they impair judgment and vision and may even produce hallucinations. Such people would be a hazard around machinery or on the highway. These pills are also sometimes taken by students to help them to stay awake for long hours while they are cramming for tests. However, the result is nervousness and confusion, followed by a severe letdown.

Dexedrine is an amphetamine frequently prescribed by physicians to aid overweight patients. However, when taken in excess, the user becomes agitated, restless, and talkative. Sometimes he becomes argumentative, and his skin flushes as he perspires. His pupils dilate and the blood pressure increases. He may even develop delusions and believe he has enemies. The amphetamines do not produce physical dependence or withdrawal symptoms, but excessive use causes a mental and emotional dependence. The individual may feel as though he "needs" it to get through each day. Then, at night, he cannot sleep and may attempt to relax with barbiturates. Therefore, he may become an addict.

Another drug problem is created by the use of *bromides*. These compounds are salts that have a sedative effect on the nervous system. They are components of many "nerve tonics" and headache remedies, and people who take them regularly develop an emotional dependence on their quieting effects. Repeated use of bromides, however, causes slow poisoning and damage to the nervous system.

Psychedelic, or "mind-expanding," drugs

Psychedelic drugs are a class of compounds that affect the mind by changing the perception of all the senses, the interpretation of space and time, and the rate and content of thought. Hallucinations or visions are often described by persons under the influence of these drugs. The most common psychedelic drugs are *LSD, mescaline,* and *psilocybin.* Medical investigation into the effects these substances have on the body may prove to be valuable. Mescaline has been used in the treatment of certain types of mental illness. It is extracted from the tops of the mescal, or peyote, a spineless cactus that grows in the American Southwest. Peyote buttons, which are dried tops of the mescal, are chewed for mescaline's narcotic effect, especially during religious ceremonies, by the Indians of that area. Psilocybin is extracted from certain mushrooms.

LSD is d-lysergic acid diethylamide, and it is so powerful that a single ounce provides 300,000 doses. Some individuals have had noticeable effects with as little as 20 to 30 micrograms (millionths of a gram) although usual doses are around 200 micrograms. It is usually taken orally, but may be sniffed in powdered form or injected in solution. Since it is such a strong substance, it is often diluted and put on sugar cubes, candies, biscuits, and even cloth or blotter surfaces. Once it is absorbed into the bloodstream it easily diffuses into the brain. If taken during pregnancy, this powerful substance diffuses through the placenta and into the fetus. Effects of this drug usually begin within an hour after it is taken and last from 8 to 12 hours.

Responses to this drug vary so there may be an intense negative experience of fear or terror to the point of panic. Emotional instability, hallucinations, and feelings of extreme joy may suddenly be replaced by profound depression. Strong emotional feelings of reliving past experiences may also occur.

Another experience created by this drug is a feeling that problems are suddenly viewed from a different perspective. The drug influenced person may feel that he is standing off in space and looking at himself and his problems from afar. Sensory impressions with vision are most often affected so that sounds may create mental images. Objects such as tables, flowers, or stones may appear to glow, pulsate, and become alive. Accompanying these hallucinations may be a feeling that he is melting into the universe. He has the impression that he is suddenly becoming the chair in which he may be sitting.

Physical changes accompanying these various emotional changes brought about by the drug may also vary with the emotional state of the individual. The heart rate and blood pressure may increase, the pupils dilate, alternating periods of chill and fever occur with a flushing or pallor. Trembling and nausea also occur frequently. The physical changes accompanying fear may be extreme and self-inflicted injury is not uncommon.

Although some of these feelings may sound intriguing, there are obvious dangers involved. Since this drug is obtained illegally, one can never be certain of its purity. Therefore, because of the extremely small amount bringing about these changes, overdoses are common. The manner in which LSD alters the brain cells is not known. However, "flashbacks" of the experience may occur days, weeks, or months after the drug has been taken. Furthermore, each individual reacts differently to this chemical and permanent mental derangement is sometimes produced after only one LSD experience. Complete mental breakdown and suicides have been attributed to this drug. As you can now understand, children born to LSD users have been born with defects due to the drug.

When psychedelic drugs are obtained and used illegally for "kicks," they are highly dangerous. We can learn to appreciate the beauty of music, of a sunset, or of a forest park through our natural senses. Attempts to enhance emotions or appreciation through the use of drugs is a foolish risk. It often ends in a devastating experience for society, families, and one's own mental stability.

Why do people become addicts?

People may become narcotic addicts in several ways. Some people may become addicted after an illness in which a narcotic drug was used medically to relieve pain. Highly nervous or distressed people may use certain narcotics as depressants by prescription and continue purchase of the drugs illegally after their medical treatment has ended. Still others deal with dope peddlers who are associated with organized and unlawful sale of narcotics, especially in large cities. Many young people are curious and seek thrills. They may "join the crowd" in their experimentation without realizing the danger involved. Some people begin taking drugs to express dissatisfaction with their life, parents, school, or employers. We have already discussed the use of drugs as an attempt to escape problems.

Treatment of addicts

While we have no way of knowing the exact number of drug addicted people in the United States, the number of heroin addicts alone has been estimated at 250,000. For most of these people, crime is a necessary way of life. They are forced to steal to pay the enormous prices of the illegal narcotics traffic. Of the addicts who are treated in the two Federal Hospitals, few can avoid drugs after their release. Many of these addicts are teen-agers, some as young as thirteen and fourteen.

Narcotic addicts may now be treated in treatment centers, hospitals, and clinics. The *Methadone maintenance* treatment is a new approach to the problem. Once the patient agrees to stop taking heroin, Methadone mixed with fruit-juice is given to him in prescribed doses at regular intervals. Methadone prevents withdrawal symptoms and reduces the craving for heroin. Many oppose the use of Methadone as it is also addicting. However, an individual being treated with Methadone may return to function in society and gain his self-respect. Scientists are searching to find a nonaddicting drug that would substitute for heroin.

Once an individual has developed a physiological dependence on a drug several things can happen to him. He can ask for treatment, which is often unsuccessful. He can be arrested for illegal possession of narcotics. He can die from an overdose, an infected needle, or suicide. The only way to avoid these tragedies is never to start taking a narcotic.

SUMMARY

While tobacco is not addicting in the sense that narcotic drugs are, the various ways in which cigarette smoking is dangerous to health are becoming more obvious with each year of further research.

Alcohol is a depressant that has no relation to the energy needs of the body. Its habitual use may lead to alcoholism, various organic diseases, and even death.

The medical use of drugs has done much to relieve the suffering and speed up the healing process of the ill. However, self-medication can often lead to a drug habit in which an individual may become mentally dependent on its effects. Narcotic drugs cause a physical dependency so a person keeps taking it to avoid the symptoms that occur when it is stopped. Since the body develops a tolerance to the narcotic, increasing doses are necessary until the entire life of the addict is occupied with finding the drug. In effect, this activity keeps the individual from being a productive member of society. In fact, he becomes a burden to his fellow-man. Thus, whatever the motive for drug experimentation, the price is high and the problems to one's self, his friends, family, and society are many.

BIOLOGICALLY SPEAKING

fatty liver	alcohol psychosis	sedative
cirrhosis	delirium tremens	stimulating drugs
gastritis	drug	barbiturates
depressant	addiction	amphetamines
alcoholism	narcotic	psychedelic drugs

QUESTIONS FOR REVIEW

1. What are some findings concerning the death rate among smokers?
2. What are some short-term disadvantages of smoking?
3. Why is alcohol not considered a food?
4. What happens when alcohol enters the body?
5. What organs of the body are especially affected by alcohol?
6. Why is alcohol not considered a stimulant?
7. Explain the progressive effects of alcohol on the nervous system.
8. Why is it dangerous for a person who has had an alcoholic drink to drive an automobile?
9. Name several drugs that fit the federal government's definition of a narcotic drug.
10. How can the physical addiction of a narcotic develop without the individual realizing what is happening?
11. Why is it dangerous for a person who has consumed alcoholic beverages to take sleeping pills?
12. How could a person develop a mental dependence on both amphetamines and barbiturates? Can you suggest any alternatives to such a dependency?
13. In what ways might the psychedelic drugs be harmful?
14. What factors may induce a person to begin misusing drugs or chemicals which are harmful when taken into the body? Can you suggest any alternatives?

APPLYING PRINCIPLES AND CONCEPTS

1. Why is inhaling smoke from a cigarette more injurious than not inhaling it?
2. Why is drinking alcohol on an empty stomach more injurious than drinking it with meals or after eating?
3. Why does the presence of alcohol in the body give a person a feeling of warmth?
4. Why do life insurance companies always ask applicants the extent of their drinking and smoking habits?
5. Explain the possible relationship between drug addiction and juvenile delinquency. How do drug addicts and alcoholics show weakness?
6. If marijuana is not a narcotic, why do you think it has been declared illegal?

RELATED READING

Books

BACON, M. K., and M. B. JONES, *Teen-Age Drinking.*
    Thomas Y. Crowell Company, New York. 1968. Has up-to-date facts and figures. Discusses tribal and social drinking customs, and taboos.

CAIN, DR. ARTHUR H., *Young People and Smoking.*
The John Day Company, Inc., New York. 1964. After ten years of research, Dr. Cain presents the facts about teen-age smoking and smokers.

HOWAT, A., and G. HOWAT, *The Story of Health.*
Pergamon Press, Inc., New York. 1967. Discusses health and medicine from early times to the present.

LIEBERMAN, MARK, *The Dopebook—All About Drugs.*
Praeger Publishers, New York. 1971. Using ordinary language, explains the chemistry of drugs, what they do inside the body and the results of their activity.

LUTHER, TERRY L., and DANIEL HORN. *To Smoke or Not to Smoke.*
Lothrop, Lee and Shepard Co., New York. 1969. Discusses the beginnings and the continuance of the habit and how dependency may lead to disability and death.

OCHSNER, ALTON, *Smoking: Your Choice Between Life and Death.*
Simon and Schuster, New York. 1970. A comprehensive presentation of all the ramifications of smoking. Includes latest information on smoking and health.

SOLOMON, DAVID (ed.), *LSD: The Consciousness Expanding Drug.*
G. P. Putnam's Sons, New York. 1964. A revealing book on the controversial drug and its uses, both good and bad.

TAYLOR, NORMAN, *Narcotics: Nature's Dangerous Gifts.*
Dell Publishing Co., Inc., New York. 1963. The history of man's use of certain plants in order to take flight from his everyday problems.

VERMES, HAL, and JEAN VERMES, *Helping Youth Avoid Four Great Dangers: Smoking, Drinking, VD, and Narcotics.*
Association Press, New York. 1965. By two specialists in translating professional and technical answers to young people's problems into easy reading. A very desirable, easy-to-read book for use in biology and health classes.

Article

HAMMOND, E. CUYLER, "The Effects of Smoking," *Scientific American,* July, 1962. New biological studies help explain how tobacco and smoke damage the body tissues.

CHAPTER FORTY-SIX

BODY REGULATORS

What are the ductless glands?

You are already familiar with some glands, such as the salivary glands of the mouth and the gastric glands of the stomach. These pour secretions into the digestive tract through ducts. The ductless glands, which we shall study in this chapter, are entirely different from the digestive glands. The term *ductless* indicates that no ducts lead from these glands; their secretions enter the bloodstream directly. With blood as a transporting medium, these secretions reach every part of the body and influence all the organs. Ductless glands are also called *endocrine* (EN-doe-krin) *glands*.

We call the secretions of ductless glands *hormones*. These chemicals are formed from substances taken from the blood. They regulate the activities of all the body processes. Thus, the circulatory system is vital to the endocrine system, both in supplying the raw materials and in delivering the finished product. For the most part, the endocrine glands are small, but their size does not reflect the vital influence they exert on the body.

As you study the various glands, you may want to refer to Figure 46-1 and Table 46-1.

The thyroid gland and its hormone

You are probably more familiar with the *thyroid* than with any of the other endocrine glands. The gland is relatively large and lies close to the body surface, in the neck near the junction of the lower part of the larynx and the trachea. The thyroid consists of two lobes connected by an isthmus (Figure 46-2). The lobes lie on either side of the trachea and extend upward along the sides of the larynx. The isthmus extends across the front surface of the trachea. As you can see, the complete thyroid gland somewhat resembles a butterfly with spread wings.

The thyroid hormone contains a substance called *thyroxine,* which has the highest relative concentration of iodine of any

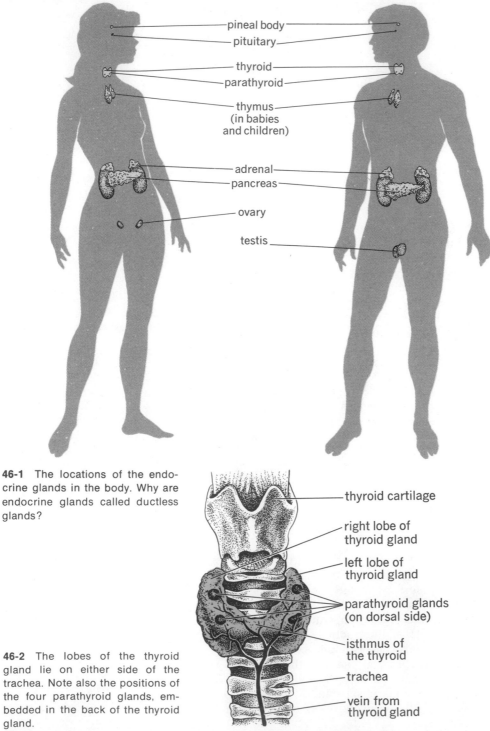

pineal body
pituitary
thyroid
parathyroid
thymus
(in babies
and children)
adrenal
pancreas
ovary
testis

thyroid cartilage

right lobe of
thyroid gland

left lobe of
thyroid gland

parathyroid glands
(on dorsal side)

isthmus of
the thyroid

trachea

vein from
thyroid gland

46-1 The locations of the endocrine glands in the body. Why are endocrine glands called ductless glands?

46-2 The lobes of the thyroid gland lie on either side of the trachea. Note also the positions of the four parathyroid glands, embedded in the back of the thyroid gland.

Table 46-1   DUCTLESS GLANDS AND THEIR SECRETIONS

GLAND	LOCATION	HORMONE	FUNCTION OF HORMONE
thyroid	neck, below larynx	thyroid hormone	accelerates the rate of metabolism
parathyroids	back surface of thyroid lobes	parathormone	controls the use of calcium in the tissues
pituitary anterior lobe	base of brain	somatotropic hormone	regulates growth of the skeleton
		gonadotropic hormone	influences development of sex organs and hormone secretion of the ovaries and testes
		ACTH	stimulates secretion of hormones by the cortex of the adrenals
		lactogenic hormone	stimulates secretion of milk by mammary glands
		thyrotrophic hormone	stimulates activity of the thyroid
posterior lobe		oxytocin	regulates blood pressure and stimulates smooth muscles
		vasopressin	controls water resorption in the kidneys
adrenal cortex	above kidneys	cortin (a hormone complex)	regulates metabolism, salt, and water balance controls production of certain white corpuscles and structure of connective tissue
medulla		epinephrine, or adrenalin	causes constriction of blood vessels, increase in heart action and output; stimulates liver and nervous system
pancreas	below and behind stomach		
islets of Langerhans		insulin	enables liver to store sugar and regulates sugar oxidation in tissues
ovaries follicular cells	pelvis	estrogen	produces female secondary sex characteristics; influences adult female body functions
		progesterone	maintains growth of the mucous lining of the uterus
testes interstitial cells	below pelvis	testosterone	produces male secondary sex characteristics

substance in the body. Commercially, the thyroid hormone is extracted from the thyroid glands of sheep. After it is purified, it is called *thyroid extract* and is used in treating thyroid disorders. Thyroid extract is the least expensive of all commercial endocrine preparations.

The thyroid and metabolism

Because the thyroid hormone regulates certain metabolic processes—especially those related to growth and oxidation—the maintenance of a normal level of thyroid hormone is essential for normal body functions.

Overactivity of the thyroid gland causes a condition known as *hyperthyroidism.* The condition is marked by an increased rate of oxidation and body temperature. The rate of heart action is increased, and blood pressure is elevated. Sweating when the body should be cool is a common symptom of this condition. Extreme nervousness and irritability are also symptoms. Some victims develop characteristic bulging eyes and a staring expression.

Hyperthyroidism used to be treated solely by surgery, but recently treatment with a drug called *thiouracil* (THIE-oe-YOOHR-uh-SILL) has been found to be effective. Another treatment for hyperthyroidism consists of dosage with radioactive iodine. This is picked up by the gland as ordinary iodine would be. It then bombards the gland with radioactivity and destroys some of the gland tissue, as surgery formerly did.

Underactivity of the thyroid gland results in a condition called *hypothyroidism.* This condition is characterized by symptoms opposite to those that mark hyperthyroidism. The rate of oxidation is decreased, and activity of the nervous system is reduced. This produces characteristic physical and mental retardation. Heart action decreases and in many cases the heart enlarges. Both overactivity and underactivity of the thyroid gland can be determined by measuring the rate of basal metabolism. Hypothyroidism can be treated with thyroid extract.

If the thyroid is defective during infancy, *cretinism* results. This condition is characterized by stunted physical and mental development. The face usually becomes bloated, the lips greatly enlarged, and the tongue thick and protruding from the mouth. If the cretin passes from infancy to childhood without thyroid extract treatment, the dwarfism and mental deficiency can never be corrected.

If the thyroid slows down during adult life, *myxedema* (MICK-suh-DEE-muh) results. This causes coarsening of the features and swollen eyelids. Often mental ability suffers. Like cretinism, myxedema can be corrected with thyroid extract if treatment is started early.

Iodine deficiency is the major cause of enlargement of the thyroid gland, known as *simple goiter.* This condition is rare

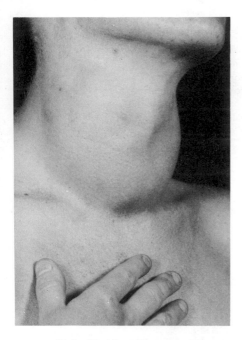

46-3 Simple goiter can be prevented by the addition of iodine to the diet. Have you noticed the label on packages of salt? (Percy W. Brooks)

along the seacoast, where people eat an abundance of seafoods containing iodine. It is more common in mountainous regions and in the Great Lakes basin, where the iodine content of the soil is low. The addition of iodine compounds to table salt and to the water supply in certain regions is an adequate preventive measure.

In specific instances a doctor may want to measure the activity of the thyroid gland. Since metabolism and oxygen consumption indicate thyroid activity, the basal metabolic rate may be measured. More accurate tests, however, determine the rate at which iodine is taken into the gland. The amount of iodine in the blood may be measured and used to indicate this, but many medical centers also use radioactive iodine. A small amount of radioactive iodine is given to the patient by mouth. Later, at a specific time interval, a counter is placed over the thyroid gland (see Figure 46-4). The rate of concentration of the radioactive iodine in the thyroid tissue determines the activity of the gland.

The parathyroid glands

The *parathyroids* are four small glands embedded in the back of the thyroid, two in each lobe (Figure 46-2). Their secretion, *parathormone,* controls the use of calcium in the body. Bone growth, muscle tone, and normal nervous activity are absolutely dependent on a constant, stable calcium balance.

The pituitary gland

The small *pituitary,* a gland about the size of an acorn, lies at the base of the brain. It was once called the "master gland," because its secretions influence the activities of all other glands. It is now known that other glands, especially the thyroid and adrenals, in turn influence the pituitary.

The pituitary gland consists of two lobes, anterior and posterior. The *anterior lobe* secretes several different hormones. One of these, the *somatotropic* (soe-muh-toe-TROP-ick) *hormone,* or *growth hormone,* regulates the growth of the skeleton. Other secretions of the anterior lobe of the pituitary gland, the *gonadotropic* (goe-NAD-uh-TROP-ick) *hormones,* influence the development of the reproductive organs. These also influence the hormone secretion of the ovaries and testes. The gonadotropic hormones together with the sex hormones cause the sweeping changes that occur during adolescence, when the child becomes an adult.

Other secretions of the anterior lobe of the pituitary gland include hormones that stimulate the secretion of milk in the mammary glands *(lactogenic hormone)* and the activity of the thyroid gland *(thyrotropic hormone). ACTH (adrenocorticotropic hormone)* is another secretion of the anterior lobe of the pituitary gland; it stimulates the outer part, or cortex, of the

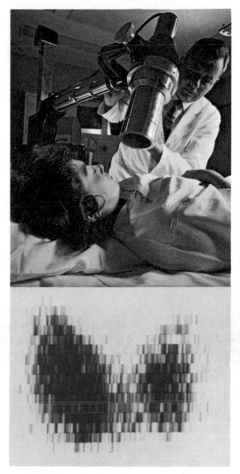

46-4 Measurement of uptake of radioactive iodine in the thyroid gland of a patient. The counter in the upper picture records the areas of concentration of iodine in the gland. A photograph of the record is shown in the lower picture. (Brookhaven National Lab.; Percy W. Brooks)

adrenal glands. ACTH has been used in the treatment of leukemia and, more successfully, in the treatment of arthritis. Good results in the treatment of asthma and other allergies with ACTH have also been reported. Even though ACTH may not permanently cure these diseases, its use may lead the way to the discovery of their actual causes.

The *posterior lobe* of the pituitary gland produces two hormones: *oxytocin,* which helps regulate the blood pressure and stimulates smooth muscle; and *vasopressin,* which controls water resorption in the kidneys. Oxytocin is administered during childbirth to cause contraction of the uterus.

Vasopressin deficiency causes a condition called *diabetes insipidus* (in-SIP-uh-dus), in which large quantities of water are eliminated through the kidneys. This disease should not be confused with true diabetes, which we shall discuss in connection with the pancreas.

Disorders of the pituitary gland

The most frequent disorder of the pituitary gland involves the somatotropic hormone. If an oversecretion of this hormone occurs during the growing years, a *giant* may result. There are circus giants over 8 feet tall who weigh over 300 pounds and wear size 30 shoes. If the oversecretion occurs during adult life, the bones of the face and hands thicken, because they cannot grow in length. The organs and the soft tissues enlarge tremendously. This condition is known as *acromegaly* (ACK-roe-MEG-uh-lee). Victims of this disorder have greatly enlarged jaw bones, noses, and hands and fingers.

Somatotropic hormone deficiency results in a pituitary dwarf, or *midget.* Such individuals are perfectly proportioned people in miniature. They are quite different from the thyroid dwarf in that they have normal intelligence.

The glands of emergency

The **adrenal glands,** also called *suprarenals,* are located on top of each kidney (Figure 46-6). They are composed of an outer region, the *cortex,* and an inner part, the *medulla.* The adrenal cortex, unlike the adrenal medulla, is absolutely essential for life. It secretes a hormone complex called *cortin.* These hormones are responsible for the control of certain phases of carbohydrate, fat, and protein metabolism as well as of the salt and water balance in the body. The adrenal cortex also yields hormones that control the production of some types of white corpuscles and the structure of connective tissue.

Addison's disease results from damage or destruction of the adrenal cortex, which often occurs as a result of tuberculosis. Symptoms include fatigue, nausea, loss of weight, general circulation failure, and changed skin color. Treatment with the compound in cortin called *cortisone* is effective.

46-5 In terms of the amount of somatotropic hormone produced, what is the nature of the malfunction of the anterior lobe of the pituitary gland that results in *midgetism? In giantism?* (UPI)

The medulla secretes a hormone called *epinephrine* (EP-uh-NEFF-reen), or *adrenalin*. The adrenal glands have been called the glands of emergency because of the action of this hormone. Many people have performed superhuman feats of strength during periods of anger or fright, with the help of epinephrine. This strength of desperation results from a series of rapid changes in body activity:

■ The person becomes pale, because of constriction of the blood vessels in the skin. The rapid movement of blood from the body surfaces reduces loss of blood if there is a surface wound. It also increases the blood supply to the muscles, brain, heart, and other vital organs.

■ The blood pressure rises, because of constriction of surface blood vessels.

■ The heart action and output are increased.

■ The liver releases some of its stored sugar and provides material for increased body activity and oxidation.

The pancreas

The production of pancreatic fluid in connection with digestion is only part of the function of the *pancreas*. Special groups of cells, called *islets of Langerhans*, secrete the hormone *insulin*. This hormone enables the liver to store sugar as glycogen and regulates the oxidation of sugar.

A person who lacks insulin cannot store or oxidize sugar efficiently. Thus, the tissues are deprived of food, and sugar collects in the blood. As the blood sugar rises, some of it is excreted in the urine. Doctors call this condition *diabetes mellitus* (MELL-uh-TUSS). Diabetes mellitus, however, is probably

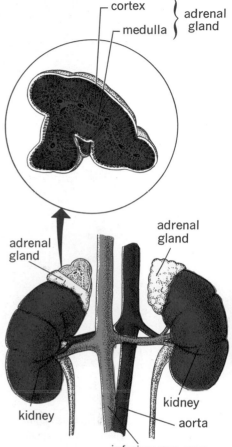

46-6 Note the position of the adrenal glands on the kidneys. What two important hormones do these glands secrete?

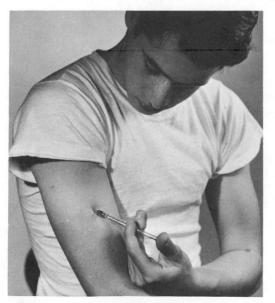

46-7 A diabetic learns to inject himself with insulin and also to control his diet so that he can lead a normal life. (New York Diabetes Association, Inc.)

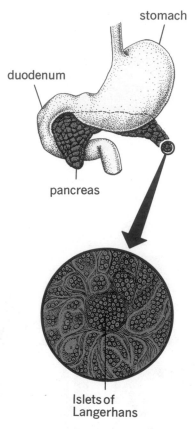

46-8 The pancreas is both an endocrine gland and a digestive gland located in the curvature of the duodenum.

not merely simple failure of the islet cells of the pancreas to produce insulin. The pituitary, thyroid, and adrenal glands, as well as the liver, are known to play an important part in the disease. Body weight also influences the appearance of this condition. Diabetes mellitus is definitely hereditary. If you have it in your family, regular periodic checkups for the level of blood sugar should be made by your family doctor. There is no cause for alarm if the disease appears, for, once discovered, it can usually be controlled. In fact, if treatment is begun in the disease's early stages, the patient can lead a normal life.

Production of excess insulin results in a condition called *hypoglycemia,* which means, literally, "low blood sugar." Excess insulin in the blood causes sugar that should be delivered to the cells to be stored in the liver. The fatigue caused by hypoglycemia is treated by a diet controlled in its carbohydrate content.

The ovaries and testes

The *ovaries* of the female and the *testes* of the male have dual functions. We think of them primarily as organs for the production of eggs and sperms. However, certain cells of the ovaries and the testes serve as ductless glands. These ovary cells secrete the female hormones *estrogen* (EST-ruh-jin) and *progesterone* (proe-JESS-tuh-RONE). Special cells of the testes produce the male hormone *testosterone* (tes-TOSS-tuh-RONE). This hormone can now be produced artificially and is used in treating sex hormone disturbances in both males and females. Furthermore, the production of this hormone is not limited to the testes. It is secreted by the cortex of the adrenal glands in both males and females. In the female, the estrogen secreted in the ovaries normally neutralizes the effects of the testosterone from the adrenal glands. However, if the estrogen secretion in the ovaries is reduced, the female may become mannish. Similarly, reduced production of testosterone in the testes of the male can result in feminine tendencies. Thus, different individuals may represent various degrees of maleness and femaleness.

Sex hormones control the development of the secondary sex characteristics that appear in the change from childhood to adulthood. These changes first appear with the maturation of the ovaries and testes during the period called *puberty.* In the animal world, these characteristics may appear as the large comb of the rooster, the bright plumage of most male birds, and the antlers of the deer. Many secondary characteristics are appearing and have appeared in your own body. As a boy approaches puberty, his voice cracks and then deepens. His beard appears, together with a general increase in body hair. The chest broadens and deepens. Rapid growth of the long bones adds to his height. As a girl matures, her breasts develop and hips broaden because of the formation of fat deposits under the skin, and menstruation begins. These physical changes in both boys and girls are accompanied by sweeping mental and emotional changes. Compare your present personality with that of a child ten to twelve years old.

Testosterone is also responsible for the normal sex drive in the male. *Castration,* or the removal of the testes, has been used for thousands of years to modify the behavior of domestic animals. If the male is castrated when it is young, not only is it sterilized, but it also fails to develop into a typical male. Its behavior is more docile, and it stores more fat. For these reasons, the livestock breeder uses castration to increase the weight of his livestock.

The pineal body and thymus

The *pineal body* is a mass of tissue about the size of a pea, located at the base of the brain. It lies directly behind the junction of the spinal cord and brain tissue. This body may or may not be a ductless gland. No hormone secretion from the pineal body has been discovered, and we do not know what its function is.

The *thymus,* too, may be a ductless gland, but no endocrine secretion has been identified with it. It lies just above the heart, under the breastbone. The thymus of a newborn infant weighs less than one half ounce. It increases in size and reaches its maximum size between the ages of twelve and fourteen. When it is at its maximum size, the thymus gland usually weighs about one ounce, or twice its original weight. During adulthood, it gradually grows smaller and finally shrinks to its original size. The thymus is a center of the production of cells called *lymphocytes.* Recent research indicates that these cells may be the parent cells of those that produce antibodies in the lymph nodes and spleen. Thus, the thymus may be important in the body's defenses against disease.

Dynamic balance in the endocrine glands

We have seen that too much or too little of a hormone can upset the balance that the endocrine glands normally maintain in the body. We have also seen that one gland can influence the activity of others. Besides the influence of glands on one another, there are two other factors operating to produce the delicate checks-and-balances system in body chemistry.

In the first of these factors, called *feedback,* the accumulation of a substance in the blood automatically cuts down the production of the substance by the endocrine gland. For example, you will remember that parathormone regulates the level of calcium in the body. The concentration of calcium, in turn, regulates the production of parathormone. When the calcium level in the body drops, the secretion of parathormone increases to restore the calcium level. When the proper level is reached, the calcium influences the parathyroids to decrease their secretion. This same kind of feedback occurs in the other glands and between various glands. In this way, a balanced state is automatically maintained in a body that is functioning normally.

The endocrine glands are also affected by the activity of the nervous system, which acts as a monitor of both internal and external conditions. The adrenal medulla, for example, may be stimulated to produce epinephrine as the need is signaled by the sympathetic nervous system. Nervous control and feedback are further examples of homeostatic mechanisms operating in the body to maintain a steady state in the face of constantly changing conditions.

SUMMARY

Ductless glands secrete hormones directly into the bloodstream. These hormones influence body metabolism, growth, mental capacity, chemical balance in the body fluids, and many other functions. Glands are controlled by feedback and nervous control, as well as by one another. Homeostasis, therefore, is maintained by a delicate balance. Illness occurs if this balance is upset either by a malfunction of one of the endocrine glands, or by taking an excess of any hormone into the body.

BIOLOGICALLY SPEAKING

endocrine gland	adrenal gland	progesterone
hormone	epinephrine	testosterone
thyroid	pancreas	puberty
hyperthyroidism	insulin	pineal body
hypothyroidism	diabetes mellitus	thymus gland
parathyroid gland	estrogen	feedback
pituitary gland		

QUESTIONS FOR REVIEW

1. What role does the blood play in the function of endocrine glands?
2. How does the thyroid gland regulate the rate of metabolism?
3. Explain how hyperthyroidism and hypothyroidism can affect personality.
4. In what ways do the pituitary and thyroid glands influence growth?
5. How does the pituitary gland affect the sex glands?
6. Compare the body characteristics of a thyroid dwarf and a pituitary dwarf. How do they differ, and in what ways are they similar?
7. What is ACTH? What is its function?
8. In what ways are puberty and adolescence a result of glandular activity?
9. Why does sugar appear in the urine of a diabetic?

APPLYING PRINCIPLES AND CONCEPTS

1. Which gland has a hormone that may influence intelligence?
2. How do you account for the fact that the heartbeat of a basketball player increases a great deal before the game as well as during the game?

3. Why is a study of the endocrine glands often carried on at the same time as a study of the nervous system?
4. What hormone injected into the bloodstream of a male rat will often result in a mothering instinct? Why?
5. Discuss dynamic balance in the endocrine system that results from feedback.

RELATED READING

Books

DUBOS, RENE, and the Editors of *Life* Magazine, *Health and Disease.*
Time-Life Books (Time, Inc.), New York. 1965. Discusses health and disease in a way that shows the quantity as well as quality of life.
GREENE, RAYMOND, *Human Hormones.*
McGraw-Hill Book Company, New York. 1970. A description of the hormones themselves and a description of the effects they have on various parts of the body.
HAMBURGER, JOEL T., *Your Thyroid Gland—Fact and Fiction.*
Charles C. Thomas, Publisher, Springfield, Illinois. 1970. Presents in layman's language the basic principles of anatomy and physiology of the thyroid gland and its associated problems.
INGLIS, BRIAN, *A History of Medicine.*
The World Publishing Co., New York. 1965. A history of the medical practices of every period of civilization, including the treatment of disease and deformity.
MAISEL, ALBERT Q., *The Hormone Quest.*
Random House, Inc., New York. 1965. Describes one of this century's most exciting advances in human welfare.
ROWLAND, JOHN, *The Insulin Man.*
Roy Publishers, Inc., New York. 1965. The story of Sir Frederick Banting and his search for a weapon against diabetes.
WEART, EDITH L., *The Story of Your Glands.*
Coward-McCann, Inc., New York. 1963. Fascinating, scientific answers given to many questions about glands.

CHAPTER FORTY-SEVEN

REPRODUCTION AND DEVELOPMENT

The significance of sexual reproduction

At this time, it might be a good idea to review the significance of the process of sexual reproduction, which you first studied in Unit 1. Sexual reproduction is the union of two gametes to form a zygote that is capable of growing into an organism resembling the parents. Gamete production occurs in the gonads. During the meiotic division in gamete formation, the chromosome pairs separate, resulting in haploid cells. The mechanism of the separation of chromosome pairs assures variation in the genetic composition of the gametes. When fertilization forms the zygote, the diploid number of chromosomes is restored, with each parent contributing one homologous chromosome to each pair. The genes then begin their influence on the development of a new organism. All the zygotes of a species contain genes that control the development of an individual and thereby insure that each individual has the broad characteristics of the species. At the

47-1 Offspring generally resemble their parents. (Hugh Rogers, Monkmeyer Press)

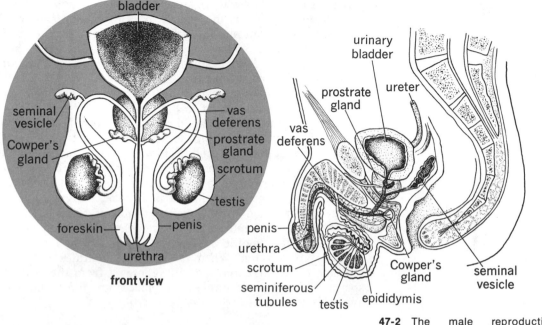

front view

47-2 The male reproductive system.

same time, the combination of genes for unlike characteristics produces offspring that vary from both parents.

Organisms that reproduce asexually, in a one-parent system, do not have this possibility of variation that results from the combination of unlike genes from two parents. Remember that variations in offspring sometimes result in favorable adaptations that improve the species. Although favorable mutations may occur occasionally in asexual organisms, the ability of such organisms to adapt as a group is greatly lessened by the one-parent system.

The male reproductive system

The male gonads are the *testes;* they are located outside the body in a pouch of skin called the *scrotum* (Figure 47-2). Although these paired organs produce a male hormone that controls the development of secondary sex characteristics (see Chapter 46), the testes have the other important function of producing sperms. The highly coiled tubes within the testes are the *seminiferous tubules* (SEM-i-NIFF-a-russ TOO-byoolz). Through the action of ciliated cells of the seminiferous tubules, the haploid sperms, formed by meiotic divisions of the cells in the testes, are carried to the *epididymis* (EP-uh-DID-uh-miss) for storage. The *vas deferens* (VASS DEF-uh-renz) is a duct that carries the sperms from the epididymis past the *seminal vesicle.* A short tube connects the seminal vesicle to the urethra by passing through the *prostate gland* and another small organ called

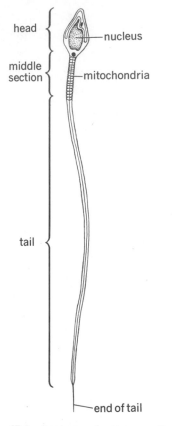

head { nucleus

middle section { mitochondria

tail {

end of tail

47-3 Anatomy of a sperm cell.

47-4 A human ovum and sperms at fertilization. (American Museum of Natural History)

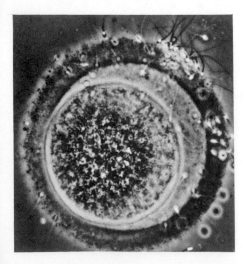

Cowper's gland. These three structures add their activating secretions to the sperms as the sperms pass by. The fertilizing fluid consisting of sperms and fluids from the seminal vesicle, prostate gland, and Cowper's gland is called *semen.* In the male, the urethra serves as a duct for the passage of semen; it also carries urine from the urinary bladder for excretion.

Figure 47-3 shows the parts of a human sperm. The male gamete is very small as compared to the ovum (see Figure 47-4). The head is a flattened, oval-shaped part containing the nucleus with its haploid number of chromosomes. The tail acts as a flagellum to provide motility. The middle piece contains many mitochondria. As you may remember from Chapter 6, cellular respiration occurs in the mitochondria as energy is provided for the cell's activity. The male gamete is a small, active cell, and its energy is furnished by the semen which contains a high concentration of a simple sugar (fructose). It has been estimated that 130 million motile sperms are necessary to insure fertilization of one ovum. When the sperm penetrates an ovum at the time of fertilization, the tail separates from the rest of the sperm. The head and connecting piece enter the ovum and the zygote is formed.

The female reproductive system

The paired *ovaries* of the female are located in the abdominal cavity on either side of the midline. They are about 1¼ inches long and ⅝ inch wide. The two ovaries are not connected directly to the oviducts, or *Fallopian tubes.* When an ovum is released, the motion of ciliated cells lining the Fallopian tubes causes it to be drawn into one of the tubes. Then the ovum passes into the *uterus,* or womb (Figure 47-5).

The uterus is a hollow, thick-walled, muscular organ. The mucous membrane lining the uterus contains small glands and many capillaries. If the ovum is not fertilized, it passes through the narrow neck of the uterus, called the *cervix,* and into the *vagina,* from which it is discharged.

The ovarian and uterine cycle

The development of the ovum and of the uterus are coordinated by hormones. The human ovaries usually produce only one egg during each twenty-eight-day cycle of activity. The mass of ovarian cells producing an ovum forms a *follicle* (Figure 47-6). The cycle is controlled by a hormone called the *follicle-stimulating hormone,* or *FSH,* which is produced in the anterior lobe of the pituitary gland. As the egg reaches maturity, the follicle becomes filled with a fluid containing the hormone *estrogen.*

After the ovum has been discharged, the follicle becomes yellowish in color; it is now called the *corpus luteum.* The development of the corpus luteum is controlled by another hormone of the pituitary gland—the *luteinizing hormone,* or *LH.* The

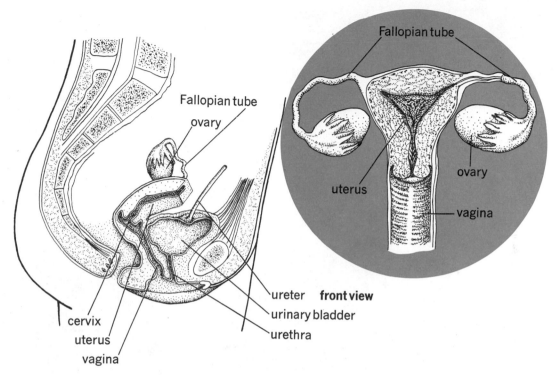

front view

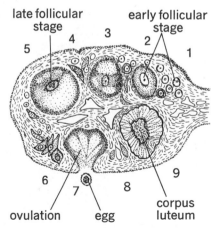

47-5 The female reproductive system.

corpus luteum, in turn, produces another hormone, ***progesterone***. The function of progesterone is to maintain the growth of the mucous lining of the uterus. If the ovum is not fertilized, however, the corpus luteum degenerates, progesterone production ceases, and the inside membrane of the uterus sloughs off. The breakdown and discharge of the soft uterine tissues and the unfertilized egg is called ***menstruation***.

The events of ovum production correlated with changes in the mucous membranes of the uterus are shown in Figure 47-7.

The uterine cycle consists of four recognizable stages: (1) *menstruation,* averaging about five days; (2) the *follicle stage,* occurring from the end of menstruation to the release of the ovum—about ten to fourteen days; (3) *ovulation,* the release of a mature ovum from the ovary; and (4) the *corpus luteum stage,* lasting from ovulation to menstruation—about ten to fourteen days.

Fertilization of the ovum

As soon as a sperm enters the ovum, a membrane forms around the ***zygote***. This is called the *fertilization membrane,* and its formation prevents other sperms from entering the ovum. Fertilization usually occurs in one of the Fallopian tubes, and it brings about several important changes. The corpus luteum of the ovary continues to develop and produce progesterone, which acts on the uterus. The membrane of the uterus continues to

47-6 A representation of a section through the human ovary. In stages 1–5, the ovum is shown maturing. In stage 6, ovulation is shown taking place. In stages 7–9, the corpus luteum is shown forming.

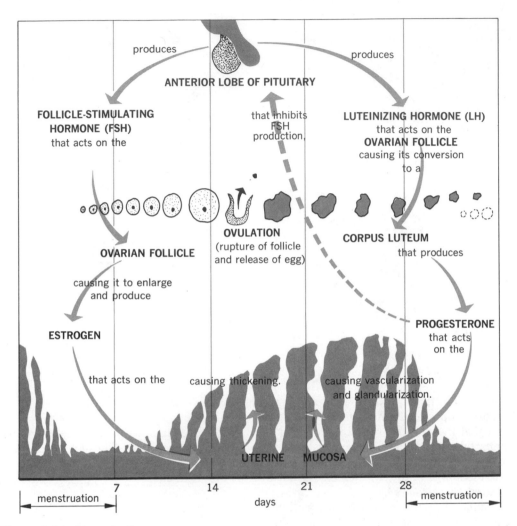

produces

ANTERIOR LOBE OF PITUITARY

produces

**FOLLICLE-STIMULATING
HORMONE (FSH)**
that acts on the

that inhibits
FSH
production,

LUTEINIZING HORMONE (LH)
that acts on the
OVARIAN FOLLICLE
causing its conversion
to a

OVULATION
(rupture of follicle
and release of egg)

CORPUS LUTEUM
that produces

OVARIAN FOLLICLE

causing it to enlarge
and produce

PROGESTERONE
that acts
on the

ESTROGEN

that acts on the

causing thickening.

causing vascularization
and glandularization.

UTERINE   MUCOSA

7   14   21   28

menstruation   menstruation

days

47-7 The relationship of the
pituitary gland to the uterine cycle.

thicken, and many small glands and capillaries form throughout
the tissue in preparation for the zygote.

The zygote reaches the uterus in from three to five days, and
during this time it continues its growth. Following the reestablishment of the diploid number of chromosomes, the zygote
begins a series of mitotic divisions, and in a few days it consists
of a mass of cells. During the early stages in the development
of the zygote, the energy is furnished by nutrients stored in the
large ovum.

Development of the zygote

The term *embryo* refers to a developing organism whose body
form has not yet acquired the characteristics that make it recognizable as a member of an individual species of animal. For the

human being, this stage lasts from six to eight weeks after fertilization. At the end of six weeks, the embryo is about two thirds
of an inch long, but its growth rate increases, and human characteristics become obvious. From this time until birth, it is called
a *fetus.*

Repeated division of the fertilized ovum results in a sphere of
cells containing a large, fluid-filled cavity (Figure 47-8). An
inner cell mass develops at one pole, and some of these cells
will become the embryo. At this time, the entire sphere of cells,
called a *blastocyst,* is still not attached to the uterine wall. Cells
from the inner cell mass will form the ectoderm, endoderm, and
mesoderm. These cell layers are collectively referred to as
primary germ layers, because they form the various tissues and
organs of the body. Structures of the body formed from the
specific germ layers are summarized in Table 47-1.

Attachment of the embryo

At the same time at which the embryo is passing down the
Fallopian tube toward the uterus, a membrane forms around the
mass of dividing cells. This is the first of several membranes

Table 47-1   STRUCTURES FORMED FROM SPECIFIC PRIMITIVE GERM LAYERS

ECTODERM
skin and skin glands
hair
most cartilage
nervous system
pituitary gland
lining of mouth to the pharynx
part of the lining of rectum
adrenal medulla

MESODERM
connective tissue
bone
most muscles
kidneys and ducts
gonads and ducts
blood, blood vessels, heart, and lymphatics

ENDODERM
lining of alimentary canal from pharynx to rectum
thyroid and parathyroids
trachea and lungs
bladder

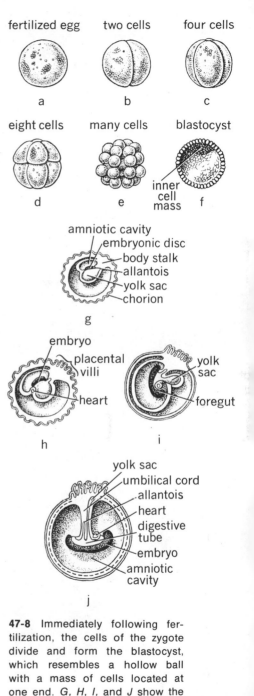

47-8 Immediately following fertilization, the cells of the zygote
divide and form the blastocyst,
which resembles a hollow ball
with a mass of cells located at
one end. *G, H, I,* and *J* show the
development of the extraembryonic membranes that accompany
the development of the human
embryo.

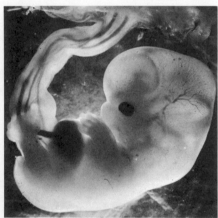

that will form, and since it will not become part of the embryo itself, it is called an **extraembryonic membrane.** The first membrane, the *chorion* (KORE-ee-on), forms many small, fingerlike projections, the *chorionic villi.* Enzymes produced by the chorionic villi enable the villi to sink into the uterine membrane to make close contact with the capillaries and provide nourishment for the embryo.

Soon another extraembryonic membrane, the *amnion,* develops. The fluid in the cavity formed by the amnion protects the developing embryo from mechanical injury and keeps it moist. The *yolk sac* is the third extraembryonic membrane. In animals that hatch from eggs, this provides the food for the embryo. The human yolk sac is small and not significant. The *allantois* (uh-LAN-toe-iss) is the fourth extraembryonic membrane, and it is present in man for only a short time. In birds and reptiles, this membrane serves as an embryonic lung.

Once the chorionic villi become embedded in the uterine wall, capillaries break down and form blood sinuses around the villi. There is no direct connection between the blood of the mother and the embryo, but food and waste exchange occurs by diffusion through the thin membranes. In the uterus, the area of the chorionic villi and maternal blood supply forms the *placenta.*

With growth, the area that attached the embryo to the yolk sac and allantois lengthens to form the **umbilical cord.** Two umbilical arteries, one umbilical vein, and the allantoic duct connect the developing embryo to the placenta.

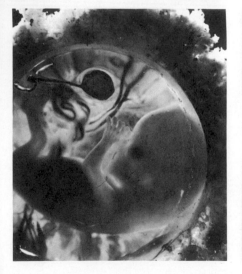

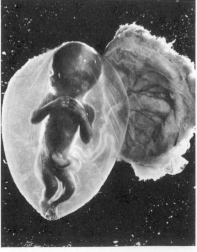

47-9 Development of the human embryo and fetus. Starting above, development has progressed for five weeks, six and a half weeks, eleven weeks, and sixteen weeks. (© Lennart Nilsson)

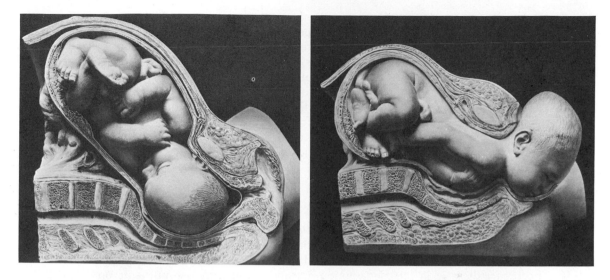

47-10 Some forty weeks after fertilization of an ovum, uterine contractions begin, and a baby is born. (Maternity Center Assoc. of New York)

Birth of the child

The period of fetal development ends with the birth of the child approximately forty weeks after fertilization of the ovum. The smooth muscles of the uterus begin to contract, and when the membrane of the amnion breaks, the fluid is discharged through the vagina. Muscles of the cervix of the uterus and the vagina relax to increase the size of the opening. Further uterine contraction forces the child from the uterus. At this time, the umbilical cord still attaches the baby to the placenta. To prevent the child from losing blood through the umbilical vessels, the cord is tied and cut on the side of the placenta. The *navel* is a scar on the abdomen that marks the location of the attachment of the umbilical cord to the fetus. Shortly after the child is born, the placenta and remains of the amnion, now called the afterbirth, are expelled.

Until the time of birth, the baby receives its nourishment and oxygen through the placenta. During development, movement of the thoracic muscles of the fetus draws fluid into the lungs, which aids in their expansion. The first cries of the infant remove the fluid and fill the lungs with air.

Another highly important change occurs at the time of birth. During fetal life, the blood does not circulate through the lungs. Instead, as blood leaves the right ventricle, it goes through a vessel called the *ductus arteriosus,* which takes the blood to the aorta. In this way, the lungs are short-circuited during the life of the fetus. At birth, however, the ductus arteriosus closes off, and the blood flows through the pulmonary arteries to the lungs. This aids the lungs in their expansion as the baby begins to breathe for himself.

Occasionally the ductus arteriosus fails to close off completely. When this happens, not all the blood gets to the lungs. Thus, the oxygen content of the blood is below normal. In severe cases, babies with this condition appear to have a bluish cast to their skin. These babies are called *blue babies*. Often the vessel will, in time, close off naturally, but sometimes surgery has to be performed and the vessel tied.

Also during fetal circulation there is an opening between the atria. It is called the *foramen ovale*. A membrane grows over this opening, and it is normally completely closed shortly after birth. Failure of this opening to close could also result in mixing of the blood from the atria and cause a "blue baby."

SUMMARY

In placental animals, such as the human being, internal fertilization occurs. Sexual reproduction provides for genetic combination causing offspring to vary from their parents. The testes produce the small gametes, named sperms, which become motile when released. During activity, the sperms use nutrients provided by the semen. Compared to a sperm, the ovum is a very large cell containing nutrients for the first period of development of the zygote. A balance of hormones controlling the development of the uterine wall prepares it to receive the embryo. By the time the zygote reaches the uterus, it consists of many cells and an extraembryonic membrane that aids in attachment to the uterus. The total dependence of the fetus on the mother terminates when the offspring is born.

BIOLOGICALLY SPEAKING

testis	vagina	embryo
semen	LH	fetus
ovary	estrogen	primary germ layers
Fallopian tube	corpus luteum	zygote
FSH	extraembryonic membranes	placenta
uterus	progesterone	umbilical cord
cervix	menstruation	

QUESTIONS FOR REVIEW

1. Of what advantage to a species is sexual reproduction?
2. Describe the passage of sperms and the production of semen in the male reproductive system.
3. How does the quantity of sperm in the human being compare to the quantity of ova? Explain the significance of this variation.
4. Describe the passage of an ovum in the female reproductive system.
5. Describe the effect of the sex hormones on the reproductive cycle in the female.
6. Name and define the four stages in the uterine cycle.
7. Describe the changes occurring in the formation of the embryo from the zygote.
8. Name and identify the functions of the extraembryonic membranes.

9. Name several structures that are produced from the primary germ layers.
10. Name the components of the umbilical cord.
11. What important change occurs in the circulatory system after birth?

APPLYING PRINCIPLES AND CONCEPTS

1. Discuss the importance of size and other structural differences in a human ovum and sperm.
2. Why must fertilization occur in a Fallopian tube before the ovum moves into the uterus?
3. Explain why hormone imbalance that alters the normal function of the pituitary gland might result in interruption of ovarian function.
4. If an egg is fertilized, menstruation does not occur, and the lining of the uterus is prepared for implanting of the embryo. Account for this in terms of hormone secretion.
5. Discuss the importance of the ductus arteriosis as a bypass from the pulmonary circulation to the systemic circulation in a fetal heart.

RELATED READING

Books

ALLAN, FRANK D., *Essentials of Human Embryology, 2nd ed.*
  Oxford University Press, New York. 1969. A difficult, but excellent reference for the student interested in the more detailed aspects of human embryology.
BALINSKY, B. I., *Embryology.*
  W. B. Saunders Co., Philadelphia. 1965. A textbook suitable for high school biology at the reading level of Modern Biology.
DUBLIN, LOUIS I., *Factbook on Man from Birth to Death* (2nd ed.).
  The Macmillan Co., New York. 1965. A useful assembly of human statistics embracing a broad range of topics.
JULIAN, CLOYD J., and ELIZABETH NOLAND JACKSON, *Modern Sex Education* (paperback).
  Holt, Rinehart and Winston, Inc., New York. 1967. A presentation of the biological facts in relation to behaviorial and psychological problems of young people.
PEMBERTON, LOIS, *The Stork Didn't Bring You.*
  Thomas Nelson and Sons, New York. 1965. Presents the facts of life for teenagers in simple, straightforward, language.
TANNER, JAMES M., *Growth.*
  Time-Life Books (Time, Inc.), New York. 1965. The text delineates and the pictures trace the timetable of human development.

UNIT **8**

ECOLOGICAL RELATIONSHIPS

So far, we have analyzed a great variety of organisms as units of life. We have studied their nutritional requirements, their cellular composition, their organs and their systems. As you are aware, no organism is self-sufficient, but is dependent on other organisms in the environment. This unit will discuss the whole living world—its forests, grasslands, deserts, lakes, and oceans. In all these communities, organisms with like requirements interact with conditions of the environment and with each other in a dynamic system. The exploration of these communities is one of the most fascinating areas of biology.

As a vital component of the living world, man's influence on his surroundings is of great significance. Also, as we are becoming increasingly aware, the influence is not always favorable. However, man's capacity to think, reason, observe, compare, communicate, and learn from experience will enable him to find solutions to many problems facing him today. Some of you will become leaders in man's struggle to survive in a beautiful world. Certainly, each one of you will have an opportunity to contribute to the preservation of a healthy environment.

CHAPTER FORTY-EIGHT

INTRODUCTION TO ECOLOGY

The science of the environment

No plant or animal is independent of its environment. Organisms are products of their surroundings. If the organisms are to survive, these surroundings must provide conditions suitable for maintaining life and carrying on all the life activities. The study of the relationships of living things to their environment and to each other is the branch of biology known as *ecology*. The word means, literally, "the study of houses." The field or pond or forest in which a plant or animal lives out its days is in a very real sense its "house."

Ecologists work in every region of the earth—from equatorial forest to polar outpost and from ocean depths to mountain-top—studying the relationships of plants and animals to their environments. Any forest, field, pond, lake, or ocean is an outdoor laboratory to the ecologist.

The biosphere—the layer of life

The area in which life on our planet is possible is called the *biosphere*. Life exists only a few feet below the surface of the earth, on its surface, in the oceans and other bodies of water, and in the lower atmosphere. This comparatively thin layer of life is affected by many factors. Radiations from the sun constitute the most important single factor affecting the biosphere.

As you already know, solar energy enables green plants to manufacture food. Some of the sun's energy is converted to heat when it reaches the earth, and in this form it is also vital to the existence of many organisms. Heat creates the earth's winds, and it causes the evaporation of water that later returns to the surface as rain or snow. Although the entire biosphere may be considered one gigantic biological system, it will be more convenient for us to consider it as a complex series of smaller units.

Ecosystems as units of the biosphere

Any stable environment in which living and nonliving things interact, and in which materials are used over and over again, is called an *ecosystem*. An ecosystem may be a jar with a fish, snail, water plants, water, and sand, or it may be a man in his space ship. Forests, rivers, coral reefs, and ponds are examples of larger ecosystems.

The lake shown in Figure 48-1 is an ecosystem. Although the diagram does not include all the organisms actually found in a lake, it does show typical ones. By becoming familiar with the various organisms in and around such a body of water, you will see how many factors affect the lake and its surrounding area.

By examining Figure 48-1, you will notice the labels on the different organisms living there. We refer to the life in an ecosystem as the *biotic community*. You will immediately be able to observe some of the food relationships, but there are also many factors affecting this ecosystem that are not as obvious. The number of individuals of each kind is not indicated. A group of individuals of any one kind of organism in a given ecosystem is called a *population*. Hence, in discussing a population, we must identify the kinds of individuals that make it up and define their limits in the ecosystem. In a lake, for instance, we may refer to the plant-eating fish population in, say, the summer of 1972. An even greater number of different species would be involved if we referred to the insect population of the lake in the same summer.

In discussing the various parts of the ecosystem, we have not mentioned the non-living component. This is called the *physical environment*. As you will see, it has an important in-

48-1 A fresh-water lake is an example of an ecosystem.

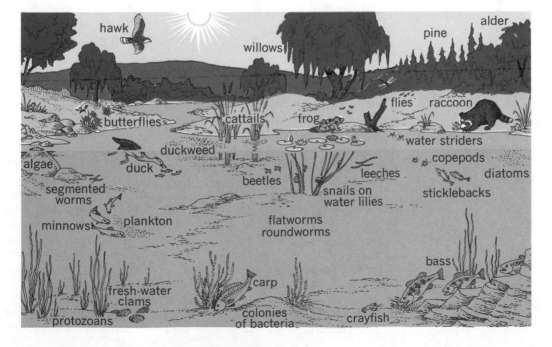

fluence on living things. The ecosystem, therefore, can be more completely represented in the following way:

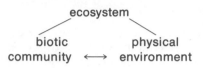

Three kinds of relationships occur in an ecosystem: (1) the interactions in the biotic community; (2) the interaction between the biotic community and the physical environment; and (3) the interactions among the physical factors of the environment.

Interaction in the biotic community

By examining the ecosystem of the lake, you will be able to find many ways in which organisms are dependent on one another. How many organisms can you identify that compete for food, light, oxygen, or merely a place in which to grow? Competition within any one population is usually greater than that between different populations, because organisms of the same species have almost identical requirements. Male stickleback fish, for example, not only compete for food, but they also vie for a suitable nesting site and for a female. Cattails compete with one another for space, soil, and a suitable depth for growth. They would not compete as strongly with duckweed, for example. This small plant, floating on the surface of water, does not have the same requirements.

Interaction between the biotic community and the physical environment

Sunlight, soil, and temperature are some of those physical factors of the environment that constantly interact with the biotic community. As you know, all green plants require sunlight for photosynthesis. In a lake, the plants are limited by the depths to which sunlight can penetrate. This depth depends on the clearness of the water. When this clearness is altered by a heavy growth of algae or by mud stirred up from the bottom, the water may become turbid. In the clearest of lakes, light may penetrate to an average depth of seventy-five feet for the year.

The remains of dead plants that grow near the shore can alter the nature of the bottom. Over a period of years, the area of a lake can be substantially reduced by the accumulation of dead shoreline plants. Organic acids and pigments from decomposing shoreline plants often affect the composition and color of the water in a lake or stream.

Water temperature may not be a significant factor to organisms living in deep lakes. At great depths, temperatures vary only slightly. But variations in daily and seasonal temperatures have a very great effect on the organisms living in shallow lakes or along the shore. These effects will be discussed further in the next chapter.

Another important factor influencing the distribution of plants and animals in a lake is man. Although man does not live in a lake, his influence over its biotic and physical environment should not be overlooked. Careless pollution of the water by industrial wastes or sewage can seriously alter both the balance and distribution of living things and the chemical nature of the water. Man, in satisfying his own interests has changed the courses of rivers and built dams to create vast lakes. He has also changed rivers into open sewers, and has even altered coastlines. Many of the far-reaching effects of man's activities were not even considered until large aquatic wastelands had been produced.

Interactions within the physical environment

The interactions within the physical environment may be temporary, as they are, for example, when a cloud cover reduces the light intensity. More permanent changes occur when a flood carries debris into a lake. Even the shape of the lake may be slightly altered by deposited materials. Some interactions in the physical environment may be permanent, as they are when an earthquake alters the course of the stream so that it bypasses the lake altogether. New springs may carry chemicals into the lake to alter the chemistry of the water. In many areas, soda and sulfur springs may affect the chemical composition of the water.

The water cycle, a physical cycle

Closely associated with the interrelationships of the biotic and physical environment, but exclusive to none, are the chemical and physical cycles. You have already learned much about the roles of water, oxygen, carbon, and nitrogen in the maintenance of the living condition. Now let us see how they return to the atmosphere and become available again to organisms.

The *water cycle* is a continuous movement of water from the atmosphere to the earth and from the earth back to the atmosphere. The movement from the atmosphere to the earth is called *precipitation*. Eventually this water returns to the atmosphere through *evaporation* (Figure 48-3). When it rains, some of the water evaporates while falling, and some evaporates quickly from the surface of the ground. Much of the rainwater runs along the surface of the ground and travels from rivulet to stream to river. This *runoff water* eventually reaches a pond, lake, or the ocean. Water evaporates constantly from the surfaces of this collection system. Large amounts of precipitation normally enter the soil to become *ground water*. This water may reach a pond, lake, or ocean through springs or underground streams. Or, it may move upward through the soil during dry periods and pass again into the atmosphere as water

48-2 This lake was created as one result of an earthquake in Yellowstone National Park in 1962. This represents a permanent change in the physical environment. (Jen & Des Bartlett, Bruce Coleman, Inc.)

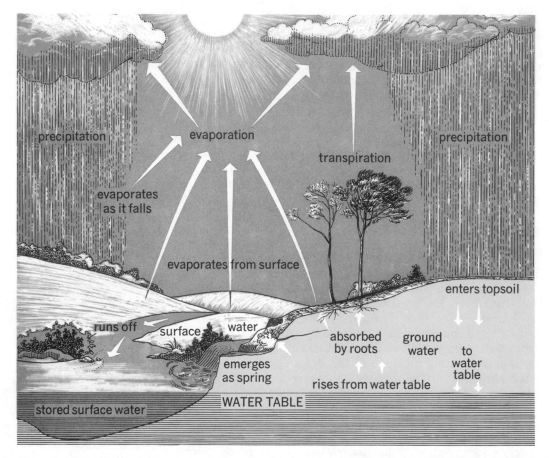

precipitation

evaporation

precipitation

transpiration

evaporates
as it falls

evaporates from surface

enters topsoil

runs off   surface   water

absorbed
by roots

ground
water

to
water
table

emerges
as spring

rises from water table

stored surface water

WATER TABLE

48-3 Trace the steps in the water cycle as represented above.

vapor. As warm air containing water vapor rises through the atmosphere, it cools. The vapor then condenses into droplets of water, forming clouds. The droplets collect to form drops, which fall from the clouds as rain. Snow is formed when the vapor condenses at a temperature below the freezing point of water.

Ground water and its movement

Topsoil acts as a sponge, receiving and holding water from precipitation. Some of this water moves downward into the sub-soil and fills the spaces around the rock particles. The upper level of soil that is saturated with water is called the *water table* —the depth at which water is standing in the ground. If the rainfall is heavy, most of the water may run off the surface instead of penetrating into the saturated topsoil.

The depth of the water table depends on the amount of precipitation, the condition of the soil surface for receiving water, the nature of the rock layers under the soil, and the proximity of large bodies of water. Where depressions occur, as in basins of lakes and ponds, the water table may be above the surface.

Plants do not wilt between rains, because water moves from the water table up through the soil by capillary action. Much of this water is absorbed by roots and passed to the atmosphere during transpiration. Some of it reaches the surface of the soil, where it evaporates into the atmosphere. This movement of water upward from the water table is an important part of the water cycle. However, the largest proportion of the water enters the atmosphere by evaporation from the oceans and smaller bodies of water.

The role of living things in the water cycle

To a certain extent, both plants and animals are involved in the water cycle. Plants take in water through their roots by absorption and give off water vapor from their leaves in transpiration. Animals are involved to the extent that they drink water and give off a certain amount of water vapor in exhalation. The amount of water that cycles through living things is small, however, compared to the amount that cycles through bodies of water, especially the oceans.

The carbon-oxygen cycle

Respiration and photosynthesis are the two basic life processes involved in the *carbon-oxygen cycle* (Figure 48-4). Land-

48-4 The carbon-oxygen cycle. In photosynthesis, green plants use carbon dioxide and release oxygen. Oxygen is used by animals and plants in respiration, and also in the burning of fuels. Carbon dioxide is released in both processes, as well as in the process of decay.

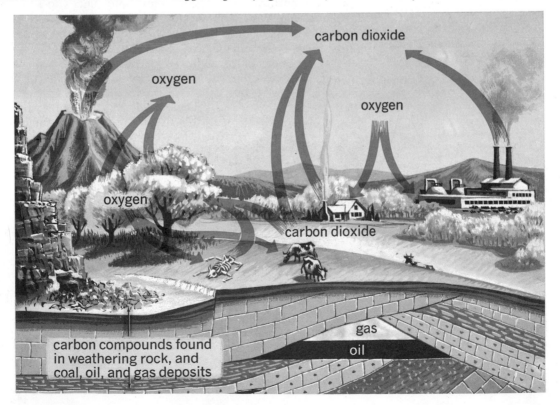

carbon dioxide

oxygen

oxygen

oxygen

carbon dioxide

carbon compounds found in weathering rock, and coal, oil, and gas deposits

gas

oil

dwelling organisms take oxygen directly from the atmosphere, while aquatic and marine organisms use oxygen that is dissolved in the water. The oxygen that is chemically united with hydrogen in water molecules is not available for respiration because organisms lack the ability to decompose water for this purpose.

During respiration, compounds containing carbon are oxidized to form carbon dioxide, and this gas is released to the environment.

During photosynthesis, green plants take water and carbon dioxide from the environment. In this process, as you recall, water molecules are decomposed so that their hydrogen atoms can be combined with carbon dioxide to form carbohydrates. Oxygen is released to the environment as a byproduct of photosynthesis. The atmosphere normally contains about 21 percent oxygen and 0.04 percent carbon dioxide. These percentages are fairly constant, indicating the efficiency of living things in maintaining the cycle.

Another portion of the carbon-oxygen cycle relates to the organic compounds synthesized by plants and animals from the carbohydrates produced in photosynthesis. Plants produce proteins and other protoplasm-forming substances. Animals that eat plants may synthesize other organic substances, and carnivorous animals, in turn, resynthesize these substances to suit their own needs. The carbon in these compounds is retained in the bodies of organisms until they die. It is then released as carbon dioxide when their remains decompose after death.

Smaller amounts of oxygen and carbon are involved in the formation and decomposition of rocks and of mineral fuels such as coal and petroleum. These two elements are also involved in chemical changes when fuels are burned and when volcanoes erupt. However, the major contributing factors to the cycle are respiration and photosynthesis.

The nitrogen cycle

The *nitrogen cycle* involves green plants and several kinds of bacteria. It may or may not involve animals. As you read about the various steps in the nitrogen cycle in the next two paragraphs, follow them in the diagram shown in Figure 48-5.

Let us begin with the green plant and the formation of protein. The plant's roots absorb *nitrates,* a group of soil minerals. These compounds contain nitrogen in chemical combination with oxygen and usually with sodium or potassium. In building proteins, the plant adds nitrogen from nitrates to the carbon, hydrogen, and oxygen that have been organized in an organic compound during photosynthesis. Sulfur and phosphorus may be added from other soil minerals.

Proteins are used in forming plant or animal protoplasm. This happens when one animal eats a plant or another animal.

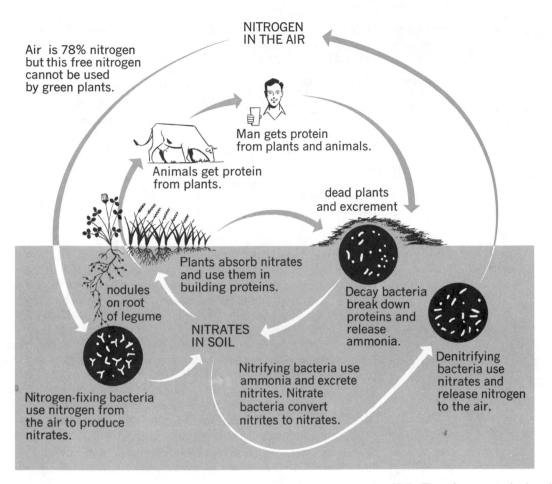

NITROGEN
IN THE AIR

Air is 78% nitrogen
but this free nitrogen
cannot be used
by green plants.

Man gets protein
from plants and animals.

Animals get protein
from plants.

dead plants
and excrement

nodules
on root
of legume

Plants absorb nitrates
and use them in
building proteins.

Decay bacteria
break down
proteins and
release
ammonia.

NITRATES
IN SOIL

Denitrifying
bacteria use
nitrates and
release nitrogen
to the air,

Nitrogen-fixing bacteria
use nitrogen from
the air to produce
nitrates.

Nitrifying bacteria use
ammonia and excrete
nitrites. Nitrate
bacteria convert
nitrites to nitrates.

48-5 The nitrogen cycle. In addition to the steps shown, some free nitrogen in the air is changed to nitrates by lightning and carried to the soil by rain or snow.

Each time an animal consumes protein in food, digestion separates its amino acids, which are then combined to form a protein characteristic of the organism in which it is formed.

However, not all of the protein consumed by an animal is converted into protoplasm. Some of it may be oxidized to produce energy. In this event, nitrogen compounds may be excreted by the animal. As you recall, this is one of the common functions of kidneys in higher animals. These excreted compounds may be further decomposed by bacterial action in the soil or water.

When an organism dies, decay by bacterial action begins. Nitrogen is released from the decaying protein in combination with hydrogen as *ammonia*. We refer to this part of the nitrogen cycle as **ammonification.** Other kinds of bacteria, living in the soil, oxidize ammonia and form *nitrites,* which cannot be absorbed by plant roots. Further oxidation by still other bacteria results in the formation of *nitrates,* the mineral compounds from which green plants receive their necessary nitro-

root nodules

nitrogen-fixing bacteria from root nodule cells

cells of root nodule with nitrogen-fixing bacteria in cytoplasm

48-6 Nitrogen-fixing bacteria are found in the root nodules of legumes as shown in the photograph and illustrated in the diagram. Of what benefits are the plant and the bacteria to each other? (Hugh Spencer)

gen. We refer to the chemical process of nitrate formation by bacteria as *nitrification.* Thus, in starting with nitrates and ending with nitrates, we complete the nitrogen cycle.

The role of the atmosphere in the nitrogen cycle

As you probably know from your previous science courses, the atmosphere is composed of about 78 percent nitrogen. Is this pure nitrogen involved in any way in the nitrogen cycle? It is, but in a rather roundabout way. Atmospheric nitrogen cannot be used by green plants in their chemical activities. However, two groups of bacteria can oxidize free nitrogen and form nitrites and nitrates in the soil. One of these groups of bacteria lives in the soil. The other lives on the roots of clover, alfalfa, and other members of the legume family in a close relationship with the plant. The legume is the host, since the bacteria live within its tissues. These remarkable bacteria receive sugar from the host and use it in oxidizing free nitrogen to nitrates. The nitrates, formed within the cells of the roots of the host, can be absorbed and used in protein formation. Thus, the bacteria are of great benefit to the host plant. We call this important process *nitrogen fixation.*

Fortunately, legumes accumulate more than enough nitrates to meet their own requirements. The excess builds up the nitrogen content of the soil. When a farmer plants clover or alfalfa in a field as part of a crop-rotation schedule, he knows he is building up his soil from almost unlimited supplies of atmospheric nitrogen. Whether or not he realizes that he is doing this by raising nitrogen-fixing bacteria is not important. It is important, however, that he knows he will receive the greatest benefit if he plows the clover back into the soil at the end of the growing season.

One phase of the nitrogen cycle, however, is unfavorable to agriculture. Certain bacteria liberate nitrogen by breaking down ammonia, nitrites, and nitrates in a process called *denitrification.* In this way, some nitrogen may be lost from the soil. Fortunately, denitrification does not occur in well-drained, cultivated soil, since denitrifying bacteria are anaerobic. That is, they live in an environment that has little or no oxygen present. They thrive in soils that are waterlogged, or packed so tightly that air cannot easily penetrate.

Biological balance in nature

Organisms are associated with one another in a living society. Any biotic community is a complex of societies composed of many kinds of populations living together in very close association. The number of individuals in a population may fluctuate from season to season, or from year to year, but within limits. The relationship of population densities of various species in a biotic community is often referred to as the *balance of nature.*

As you will see, there are many factors interacting to maintain this "balance."

Populations change

As you have already seen, physical and biotic interreactions affect not only individuals but groups of organisms as well. Populations vary in numbers of individuals from season to season or from year to year. An ecologist measures the changing populations by counting and recording individuals from a large area at various times. This measure of *population density* is expressed as numbers of individuals in a definite area at a specific time.

Density studies provide valuable information. If we know the reasons for periodic fluctuations in the numbers and kinds of grasses on which cattle graze, we can use natural pastures without destroying them. All effective game laws are based on population studies. Game and fish populations are carefully harvested to insure future generations the pleasure of hunting and fishing.

How are density studies made? Obviously, it would not be practical to count all the clover plants in a field or on a hillside or to count all the bass in a lake. Instead, the ecologist selects several areas that are large enough to provide an accurate representative sample. Then he actually counts the individuals in the selected areas. Under natural conditions, organisms are not distributed evenly. Such studies help us answer such questions as the following: What factors are responsible for the uneven distribution? Why do populations vary from year to year? Why might the use of DDT in a California farmer's field affect the crab population along the Pacific Coast? If the ecologist can find reasons for changes in population densities, it may be possible to restore stability where it has been upset.

Factors causing fluctuations in population density

If organisms are placed in a new habitat that is favorable for their development, the population will increase, slowly at first. In the next stage, the reproductive rate will be increased, and the numbers of individuals will multiply rapidly. Next, the rate of population increase will level off. If the ecosystem is balanced, the number of individuals will then fluctuate within a range called the average size of the population. Figure 48-7 shows the S-shaped curve representing this type of population growth.

The leveling off of a population at the top of the growth curve is of interest to the biologist. The factors maintaining this range of numbers originate in two places: one within the population itself and one without. If you were to begin a paramecium culture and make density counts at regular intervals, you would observe the S-shaped curve. After a time, however, the population density would drop rapidly. At the beginning, ample food

48-7 The S-shaped curve (A) represents the growth of a population when a new habitat is opened to an organism. Once the organism is established in the ecosystem, the population density remains relatively stable, marked by only minor fluctuations (B). If a population is kept in a culture in which the medium is not changed, after a time the population rapidly becomes extinct (C).

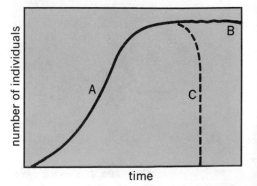

would be available. As the size of the population would increase, the rate of reproduction would increase, because more
individuals would be dividing at any given time. Soon the density
would reach a peak as the available food and oxygen were
used. Then, the density of paramecia would exert an inhibitory
effect on growth and reproduction within the population. An
accumulation of metabolic wastes would begin to alter the environment, and the individuals would begin to die more rapidly
than they were produced. The end result in the culture would
be a reduction in the population to zero.

In a larger system, these same factors exert an influence on
the numbers of individuals but serve to limit the population
rather than to bring about its extinction. For example, if the
reproductive rate of trout in a certain lake were higher than
the death rate, the trout population would increase. As competition for food becomes more severe, many trout might die of
starvation or become stunted because of lack of food. In such
a weakened state, the trout would be more susceptible to parasites and disease, which would bring about a further reduction
in numbers.

As you can see, factors such as reproductive rate, death rate,
parasitism, disease, and predation depend, to an extent, on
the population density at any given time. The ecologist calls
these *density-dependent factors.*

Other factors, however, may not be related to the population density of a species. These are called *density-independent
factors* and include such changes in the physical environment
as weather, temperature, humidity, daily and seasonal changes,
and available energy. The density-independent factors exert
a constant influence on a species regardless of the population
density. An example of this can be observed in areas where
the first frosts of the fall or winter season reduce the insect
populations no matter how dense or sparse they may be. The
categorizing of the factors influencing populations is only a
convenient way for the ecologist to analyze the changes he
observes.

Overcrowding and undercrowding

You can see some disadvantages to an individual organism
in overcrowded conditions. At the same time, there may be
some advantages to consider. A large band of monkeys, for
example, would provide many sentinels to warn of impending
danger. A large flock of birds may present a confusing picture
to a predatory hawk. A large herd of deer would offer more
protection from wolves to its members than a small group. In
some cases, a large number of organisms may be able to modify
the environment favorably. Honeybees sometimes gather forces
to fan the hive on a warm day in order to prevent the melting
of the comb.

48-8 The owl is a natural enemy
of rodents, upon which it preys.
(Thornhill, National Audubon Society)

Undercrowding, however, has drawbacks. A low population level reduces the reproductive potential of a species by making it more difficult for its members to find mates. The defensive strength of the population may be decreased. In highly social groups such as the termites, ants, and bees, efficiency may be lost when the population is decreased.

Let us consider the events in an open-field community. A meadow mouse is running along a pathway through the grasses in search of seeds. One might expect that by eating seeds, the mouse would reduce the next generation of grasses and other open-field plants. But since plants produce a far greater number of seeds than is needed for maintaining the plant population, plant numbers are not necessarily reduced. An owl swoops down and catches the mouse. The owl is one of the *natural enemies* of mice and other small mammals. We speak of it as a *predator* (PRED-i-ter) because it preys on other animals. Suppose there were no owls or other natural enemies of mice. Soon the mice would overrun the field. They might, then, actually reduce the numbers of grass plants and starve by virtue of their own numbers. This has happened many times in areas where animals have become too numerous.

In a forest, a field, a lowland marsh, a rocky meadow on a mountaintop—wherever life is found—there exists a close relationship between plant and animal and between prey and predator. Natural enemies play a vital part in maintaining the density of populations.

Keep in mind, however, that the relationships between organisms are not static. They may be affected by seasonal and annual fluctuations or by disease or man. If the owl population should increase rapidly in a given area, for example, the rodent population would decrease. With the lowering of the number of rodents, the grasses might increase in number and variety within the community. Many owls, however, would starve, and the owl population would decrease. After a year or two, the rodent population would build up to the level where it could support more owls, and the total animal-plant population would return to proportions similar to those existing before.

Human population growth

We have been discussing the factors controlling the density of populations; the same factors affect the human population. The rate of population increase is determined by the number of daily births and deaths. The number of babies born in the world every day has been estimated at 270,000; the number of deaths, at 142,000. Thus, the population of the world is believed to be increasing at the rate of 128,000 every day. However, as you observed in Figure 48-9, the rate of growth does not remain constant. The estimated world population (Figure 48-9) shows a similar trend. Projection of this graph indi-

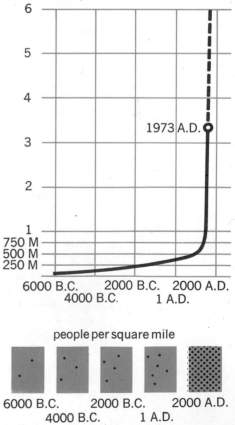

48-9 The graph shows that the rate of human population growth does not remain constant.

THE POPULATION EXPLOSION

Billions of People

1973 A.D.

6000 B.C.   2000 B.C.   2000 A.D.
       4000 B.C.   1 A.D.

people per square mile

6000 B.C.   2000 B.C.   2000 A.D.
    4000 B.C.   1 A.D.

cates that in forty-five years the world population will exceed 50 billion people. At our present rate of food production this number of human species could not be supported. In fact, there are many societies in today's world where the food demand is insufficient. More than 10,000 people in the world are dying from starvation and malnutrition every day. The same factors limiting populations of other organisms are in operation for man. We may find ways to increase our food production, but as you will read in Chapter 52, we must also find better methods of waste disposal.

SUMMARY

To understand biology, we must know about the interrelationships of plants and animals to one another and to their environment. The study of these relationships is called ecology. Water, carbon-oxygen, and nitrogen cycles are of vital importance to living things. Although the relative numbers of individuals in a specific population may vary from time to time, the fluctuation is within a range. The many factors affecting population densities are also applicable to man. Overcrowding and undernourishment occur in some societies of the human species today, but with the increasing rate of population density the problems are likely to become more severe throughout the world in the future.

BIOLOGICALLY SPEAKING

biosphere	runoff water	nitrification
ecosystem	ground water	nitrogen fixation
biotic community	water table	denitrification
population	carbon-oxygen cycle	population density
water cycle	nitrogen cycle	natural enemies
precipitation	ammonification	predator
evaporation		

QUESTIONS FOR REVIEW

1. What is the biosphere?
2. How would you go about making a density study?
3. What factors determine the size of the population of an organism at any time?
4. What does an ecologist mean when he speaks of interactions in the ecosystem?
5. What interactions occur between the biotic community and the physical environment?
6. What interactions occur among the physical environmental factors?
7. Describe the water cycle and discuss its importance.
8. Summarize the carbon-oxygen cycle and its value to living things.
9. Why is the nitrogen cycle so important in biology?
10. Generally, how do density-dependent factors differ from density-independent factors?
11. In what ways can a knowledge of ecology be useful to us?

APPLYING PRINCIPLES AND CONCEPTS

1. Describe an ecosystem near your home.
2. Biologists sometimes speak of a "closed ecosystem." Explain why this might not be an accurate term.
3. Discuss possible reasons for annual variations of any specific population.
4. Discuss the methods an ecologist uses in determining the density of populations of organisms in an ecosystem.
5. Discuss the differences between the environment of the shore region of a lake and that of the deep-water bottom region.
6. Are the factors controlling human populations the same throughout the world?

RELATED READING

Books

AMOS, WILLIAM H., *The Infinite River*.
Random House, New York. 1970. An account of the hydrologic cycle along with a description of the great variety and multitude of forms that exist in water and the implications of its deterioration as a suitable environment.

BILLINGTON, ELIZABETH T., *Understanding Ecology*.
Frederick Warne and Co., Inc., New York. 1968. A simple presentation on the fundamentals of ecology.

BOYD, MILDRED, *The Silent Cities: Civilization Lost and Found*.
Criterion Books Inc., New York. 1967. Broad in scope and clearly written. Ecological facts included to explain the decline of past civilizations and their principle cities.

DARLING, LOIS and LOUIS DARLING, *A Place in the Sun*.
William Morrow and Company, New York. 1968. A remarkably thorough and well written description of ecology. Generously illustrated with drawings and diagrams.

FARB, PETER, *Ecology*.
Time-Life Books (Time-Inc.,), New York. 1969. Explains the delicate balance and the important part that ecology plays in this balance.

HIRSCH, S. CARL, *The Living Community*.
The Viking Press, Inc., New York. 1966. The basic concept of ecology traced back to the ideas of Thoreau and Darwin; also shows how biological needs and environment can determine what species will succeed in a particular region.

HYLANDER, CLARENCE J., *Wildlife Community: From the Tundra to the Tropics in North America*.
Houghton Mifflin Company, Boston. 1966. A basic, introductory book to the study of ecology, by a well-known author.

NICKELSBURG, JANET, *Ecology*.
J. B. Lippincott Company, Philadelphia, 1969. A clear account of the principles of ecology designed to give the student a good understanding of the complexities of the subject.

NIERING, DR. WILLIAM A., *The Life of the Marsh*.
McGraw-Hill Book Company, New York. 1967. Fifth volume in Our Living World of Nature series. Ecological treatment of the wetlands of North America, placing emphasis on the vital interrelationships between plants and animals and their physical environment.

CHAPTER FORTY-NINE

THE HABITAT

49-1 What is the habitat of this animal? What is its niche? (Dr. E. R. Degginger)

The address and occupation of an organism

You are aware of the many general conditions under which plants and animals live. Conditions in a forest are quite different from those in an open field bordering a forest. The environment of a ravine or valley is unlike that of a hillside or mountaintop. Some of these differences are obvious; others are not. Still, each factor of an environment has a critical influence on the plants and animals that live there.

The *habitat* of an organism is the place in which it lives. There are many habitats within the ecosystem of a lake. The habitat of the bullfrog is quite different from that of the bass, and yet both contribute to the complex structure of the ecosystem. In fact, a bass may occasionally eat a bullfrog if their habitats should happen to overlap. However, you would not go out to the middle of a lake to find a bullfrog.

The "address" of an organism is its habitat, but what might be called its "occupation" is its *niche*. Within a biotic community, there are many ways in which organisms can "earn a living." A lake, for example, may contain tiny suspended organisms of many species, called *plankton*. It is the niche of many small fishes to feed on these organisms. Larger fishes may occupy the same habitat as the small fishes but are in a different niche because they feed on the small fishes. Different organisms that occupy the same niche are in competition with one another. For example, where they occur together, wolves and mountain lions occupy the same niche and compete for the animals on which they prey.

Limiting factors in the habitat

If an organism is to live in a certain habitat, it must be able to obtain the materials it needs for growth and reproduction. Anything that is essential to an organism and for which there is competition is called a *limiting factor*. Cattails growing along the shore of a lake, for example, require a marshy condition where the water is not too deep. In a lake, therefore, the area where the bottom is soft and the water is shallow is a limiting factor for cattails. They compete within that area and cannot live beyond it.

Since the space available to cattails is limited, it is a density-independent factor. The shallow-water shoreline does not limit the cattail population, however, until the density fills the space. If a suitable lake were found in which no cattails were growing, the planting of a few specimens could be observed. In time, the S-shaped curve you saw in the last chapter (Figure 48-7) could be plotted.

As you are probably realizing, the presence and continuing success of an organism may depend on a very complex and specific set of conditions. Just as a deficiency of any kind may limit the survival of an organism, so will an excess. An organism's ability to withstand a variety of environmental conditions is called its *tolerance*. A knowledge of the extremes of tolerance of organisms to various conditions will help us to understand why they live where they do. Many organisms, for example, live in estuaries where rivers carry fresh water into oceans or bays of salt water. The periodic rise and fall of tides and the variation in amounts of fresh water in the river during storms cause great fluctuation in the salt concentration of the water. The many kinds of worms, clams, oysters, fishes, and barnacles that live in such an area have a wide tolerance to water of varying salt concentration. Some of the deeper marine species of corals, sponges, sea urchins, and fishes would perish under such conditions, because they have a much narrower tolerance to changes in salt concentration.

As you have already seen, the habitats of organisms are governed by two sets of factors—physical and biotic. If an organism is to survive in a particular habitat, both sets of factors must be such that they allow the organism to carry on its life processes. Thus, geographic distribution of a species is governed by its limits of tolerance. We would expect to find a species concentrated in areas where the conditions are best for it. In their efforts to learn more about living things, biologists may deliberately subject organisms to a variety of conditions designed to reveal the organisms' limits of tolerance. Let us examine some of the important physical factors that limit the habitats of plants and animals.

Soil and its origin

Think of the earth as a gigantic ball of rock. Soil lies in a thin film on the surface of this great ball. Season after season, running water, freezing and thawing, wind, and other forces of nature crumble the rocks and form gravel, sand, or clay. These materials become mineral soil, or *subsoil*. In most regions, the subsoil forms a layer several feet thick, a layer that represents thousands of years of the slow disintegration of rock.

The organic part of the soil comes from the slow decay of roots, stems, leaves, and other vegetable materials, and the remains of animals. We refer to the organic remains of land plants as *humus,* while sphagnum moss and other aquatic plants form *peat* in lakes and bogs.

Organic matter and mineral matter from the subsoil combine to form *topsoil,* or *loam*. Topsoil is the most vital part of soil, the nutritional zone of plants both large and small. It forms very slowly, at a rate of about one inch in five hundred years.

Topsoil supports great numbers of bacteria, molds, and other fungi, which we call the *soil flora*. Activities of the many soil organisms are essential to fertility of the soil. Decay, ammonia production, nitrate formation, and many other chemical processes condition topsoil for the growth of higher plants.

If you examine a soil profile along a bank or the side of a ditch, you can see the dark topsoil and the lighter subsoil beneath. Under natural conditions, a small quantity of topsoil washes away or is blown away each season. This is replaced by additional topsoil formed by decaying vegetation added to the upper surface. Thus, topsoil formation is a continuous process. Remember, however, that it is an extremely slow one.

Soil—a physical factor of the environment

Soil is more than just dirt that covers the earth. It is one of the most important factors of an environment. Careful examination shows that soil varies greatly in different localities. The plant and animal life it supports varies accordingly. Some soils are compact because they are composed mostly of *clay,* while others are loose because they contain mostly *sand*. The particles of *silt* are intermediate in size between those of clay and sand. *Loam* is a mixture of clay, sand, and organic matter. Sandy soils may support pine forests in Michigan, New Jersey, Georgia, or eastern Texas. Heavy loam supports beech and maple forests in Ohio and Indiana. Waterlogged soils of bogs and swamps provide ideal conditions for larch, white cedar, and cypress forests. The rocky, shallow soils of certain mountain slopes produce luxuriant forests of redwood, yellow pine, and spruce in our western states.

A *sour soil* is one that is acid. A *sweet soil* is alkaline. The degree of acidity or alkalinity in soil is an important factor in plant growth. For many plants, a neutral point midway between the acid and alkaline is best. Under cultivation, soil tends to become more acid. To correct this condition, lime is often worked into the soil. Such plants as beets, spinach, lettuce, cauliflower, onions, peas, alfalfa, and clover do not grow well

49-2 In this redwood forest, fallen trees are reduced to a form that can be reused. What natural processes return minerals to the soil? (Albert Towle)

in acid soils. Plants like rhododendrons, azaleas, and blueberries, however, grow better in sour soil, and lime is harmful to them.

Salty and alkaline conditions of soil lower the productivity and value of much agricultural land in the United States. An estimated one fourth of our twenty-nine million acres of irrigated land contains an excess of soluble salts of sodium, calcium, and magnesium. High concentrations of these elements reduce the rate at which plants absorb water, and as a result, their growth is considerably retarded.

The character of a soil is always changing. In some places, rocks are breaking down to form more soil. This breaking down is caused by the action of weather, by chemical disintegrations, and by plants growing on the rocks. In other places, the mineral content of the soil is being depleted because of the quantities of salts removed by plants through their roots. Certain soils may be enriched through the decay of layers of vegetation, while other soils are becoming exhausted because of heavy crop production and failure to replace the lost minerals. Much useful farm land is ruined by bad soil care. As soils change, plants and animals must find other suitable habitats.

Temperature—an important controlling factor of the environment

In temperate regions of the earth, including most of North America, temperatures vary considerably. They range from narrow fluctuations between day and night to the much more extreme differences of summer and winter. Many animals, including fish, amphibians, and reptiles, do not maintain a constant body temperature. These are the cold-blooded, or *poikilothermic* (poy-KILL-uh-THURM-ick), animals. Their body temperature fluctuates with that of their environment. The birds and mammals maintain a fairly constant body temperature regardless of their surroundings. These are the warm-blooded, or *homoiothermic* (huh-MOY-uh-THURM-ick), animals. Thus, as you might expect, the warm-blooded animals can extend their habitats over a wide range of temperatures.

On a cold morning, a snake may crawl slowly out on a flat rock and lie in the sun. When its body temperature has increased, the snake becomes more active. Similarly, when the temperature is low, a butterfly may fan its wings for several minutes to warm up before flying. A meadow mouse living in the same area as the snake and the butterfly, however, can wake up on a cold morning and dart about actively.

Any organism must be able to adjust to the slight variations between day and night. But seasonal variations between winter and summer present a much greater problem. Most trees and shrubs in temperate regions flourish through the fairly warm weather of spring, summer, and fall. Then they enter a period of inactivity, or ***dormant period,*** through the colder months. Leaves may fall, and sap may move to parts of the plant that are not injured by freezing. The leaves of the pine, spruce, and other evergreen trees remain throughout the winter, even though most activity in the plant has stopped. The aerial portion of nonwoody plants may die and then grow again in the spring from dormant roots, stems, or seeds.

Many birds of the Far North migrate into the northern regions of the United States during the winter months. Meanwhile, summer residents of these same areas have migrated into southern areas, even as far south as the tropics of South America.

Other animals may burrow under the ground or look for a protected cave, where they become inactive or sleep during the cold of the winter. Some desert animals remain in their burrows during the heat of the day. A more detailed discussion of the adjustments of animals to seasonal and daily temperature variations will be found in Chapter 50.

49-3 Xerophytes, such as this cactus, are well suited for living in a desert habitat. (E. Ellingsen)

Water is essential to life

Probably no environmental factor is more important to living things than water. The habitats of plants and animals vary from a complete water environment to a sun-parched desert. The ways in which various plants and animals meet the universal need for water is always of interest to the biologist.

Oceans, lakes, rivers, streams, and ponds contain plants and animals that need a water environment. Water-dwelling organisms are said to be ***aquatic.*** Those living only in salt water are called ***marine.*** The bodies of such organisms are adapted to perform all their functions in water. Removed to land, even in the wettest surroundings, aquatic and marine organisms soon die. Land plants require less water than aquatic or marine forms. Rainfall is a major factor in controlling the growth and reproduction of these ***terrestrial,*** or land-living, plants.

Ecologists classify plants that grow entirely or partially submerged in water as *hydrophytes.* Included among these plants are pond lilies, cattails, bulrushes, eelgrass, and cranberries. Plants that occupy neither extremely wet nor extremely dry surroundings are classified as *mesophytes* (MEZ-uh-FITES). The trees of the hardwood forests of the central and eastern states are mesophytes, as are most of the flowers and vegetables we cultivate in our gardens. In general, mesophytes have well-developed roots and extensive leaf areas.

The driest environments—semideserts and true deserts—are occupied by *xerophytes* (ZEAR-uh-FITES). These plants have extensive root systems for absorbing water and a greatly reduced leaf area to cut evaporation to the minimum. Cacti are examples of xerophytes whose leaves are reduced to spines and whose thick stems are adapted for water storage (Figure 49-3).

Light, an additional critical factor in the environment

As you already know, light is essential to all green plants in food-making. But we find certain plants and animals living normal lives in environments of total darkness. Blind fishes with undeveloped eyes live in underground streams and rivers in the Mammoth Cave of Kentucky and other dark places. Likewise, deep-sea fishes live at depths to which light cannot penetrate. Many bacteria live without light, and are killed by long exposure to direct sunlight. Careful study of these organisms living in darkness, however, shows that all but a few species of bacteria depend indirectly on light for existence. They all require food and its stored energy in order to carry on their activities. This food can be traced to the green plant and its food-making processes, which are, of course, dependent on light.

Light conditions vary from place to place. Deep valleys, the floor of a forest, or the north side of a hill are places where plants and animals with low light requirements can thrive. Here we find snails, toads, and salamanders, as well as ferns and mosses. Open fields, southern slopes, deserts, and other exposed places offer ideal situations for plants that need full sunlight. Along with these plants, we find rabbits, groundhogs, coyotes, badgers, prairie dogs, ground squirrels, horned toads, and many other animals.

The atmosphere, a chemical storehouse

The air around us has an important direct effect on living things. With the exception of the anaerobic bacteria and a few other organisms, all living things must have free oxygen for life. This oxygen may be taken directly from the atmosphere or from water as a dissolved gas.

Deep-sea life has a greater oxygen problem than other forms of life. Since water receives its supply of oxygen from air, the oxygen content of water decreases with depth. The ocean is over 35,000 feet deep in the Mindanao Deep off the Philippine Islands. Near Guam, in the Marianas Trench, a depth of 35,800 feet has been measured. Although most deep-sea fishes live in the ocean only down to a mile below the surface, many organisms thrive at the bottom of these deeps.

Plants and animals that live in the soil are most abundant near the surface. The depth to which life can penetrate the soil is partly limited by food supply and certainly by oxygen supply.

Air movement, caused by fluctuations in atmospheric pressure, also has a direct influence on living things. Storms may destroy plants and drive animals to shelter. Air currents and winds have a much greater effect on life than most of us realize. Winds greatly increase the rate of evaporation of water. Plants and animals whose habitats are windy plains, prairies, and mountainous regions must not only withstand the wind but must also survive the accompanying loss of water by evaporation. On mountains,

49-4 Note that the branches on this cypress tree extend in only one direction. What factor is responsible for this? (Albert Towle)

high winds force trees to grow close to the ground and to form their branches only on the protected side (Figure 49-4). Winds also cause a reduction in size of leaves and an increase in root systems.

The varied surroundings provided by land formations

The physical features of the earth, called *topography,* have a great influence on living things. Changes of the earth's topography are caused by such events as erosion, volcanic eruptions, earthquakes, and flooding. As the earth changes, environments change. As a result, plants and animals must migrate to new and more favorable areas. They are then replaced in their old location by other organisms more suited to the conditions they left behind.

Nutritional relationships—important biotic factors

The way in which living things affect one another is equally as important as the effect of physical factors. Many of the biotic relationships involve food. The autotrophs require only inorganic nutrients from the environment to synthesize organic compounds, so we call them *food-producers*. In a lake, these food-producers are of three kinds: the emergent, rooted plants like cattails and water lilies found near the shore; the submerged plants, like eelgrass and hornwort; and the suspended algae that form the *phytoplankton*. Microscopic examination of a few drops of pond or lake water usually reveals hundreds of one-celled algae.

At certain times of the year, the phytoplankton may increase so much that the lake water turns to a dark green color. At the time of such algal "blooms," the small crustaceans such as ostracods and copepods feed well. Since these organisms feed on plants, they are called *herbivores* (HUR-bi-voarz). They are heterotrophs and are the first *food-consumers* of the ecosystem. The energy synthesized and stored by phytoplankton is transferred to the protoplasm of the herbivores. The *carnivores* (KAHR-ni-voarz), or flesh-eating animals, are sometimes divided into two groups: the *first-level carnivores,* which consume and use the energy of the herbivores; and the *second-level carnivores,* which prey on the first-level carnivores. The copepods, ostracods and other minute crustaceans in a lake are herbivores, while the minnows and the young fish that eat these crustaceans are carnivores. Since young fishes may eat small minnows and herbivorous crustaceans, you can see that young fishes may be carnivores of either the first or second level, depending on the availability of food.

The *scavengers* feed on dead organisms. They are important in the cycling of chemicals and in the transfer of energy to the animals in the ecosystem that feed on them. These "garbage collectors" are represented in the lake by the crayfish and some snails. Many fishes are partial scavengers. The bacteria and yeasts are the *decomposers* in a lake. They break the tissues and excretions of organisms into simpler substances through the process of decay. Other bacteria present in the mud bottom of the lake and in the soil convert the simpler substances left by the decomposers into nitrogen compounds, which are, in turn, used by the plants. The bacteria that do this are called *transformers*. Together, then, the decomposers and transformers return the nitrogen, phosphates, and other substances to the soil or water so that the plants can begin the cycle again. If decomposers did not exist, matter would not be available for reuse in the ecosystem.

Food chains in an ecosystem

The energy from the sun's radiation is converted into stored energy by the green plants. As the plants are consumed, much of the stored energy is released, but some of it is stored in the bodies of herbivores. When a carnivore consumes a herbivore, some of the stored energy is used, and some is stored in the carnivore. Energy passes from the carnivores to the scavengers, decomposers, and transformers. It is important to remember, however, that not all the energy stored by a herbivore is stored by the carnivore. Much of it is used in vital processes like metabolism, locomotion, and reproduction. Only the excess energy is converted or stored.

Let us examine the lake again for examples of energy transfer. Energy passes from the algae to the copepod to the minnow to the sunfish to the bass. If there is no predator to eat the bass, it is the top carnivore and may furnish the crayfish with food when it dies. The transfer of the sun's energy to a specific herbivore to a first-level carnivore to a second-level carnivore to a scavenger is called a *food chain*. Food chains are sometimes long and complex. The chain in the lake could be extended by pointing out that a bullfrog may eat the crayfish and a raccoon may eat the bullfrog. Other food chains may also operate within the lake. The frog might eat a crayfish or a minnow, and in turn be swallowed by a snake. If the conditions are right, the bass could conceivably consume the snake. Now it appears that the biotic relationships are complicated by the integration of the food chains.

If you were to list every organism in the lake and draw arrows to indicate which organisms are used for food by others, your diagram would represent all the biotic relationships involving energy transfer. Since there are so many possibilities, this diagram would actually resemble a web more than a chain. For this reason, food chains are sometimes called *food webs*.

Ecological pyramids

Food chains in an ecosystem are often represented as a pyramid, with the food-producers forming the base and the top car-

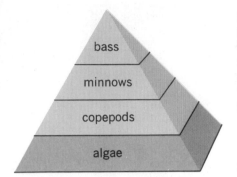

pond water and dissolved minerals

49-5 A pyramid based on the number of individuals in a pond.

49-6 A pyramid based on numbers (top) and one based on mass. The latter reflects truer energy relationships.

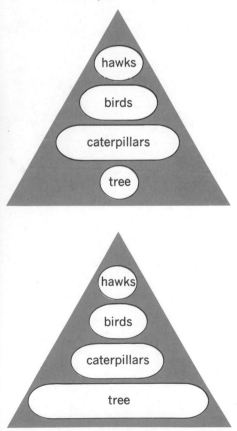

nivore at the apex (Figure 49-5). Such a *food pyramid* can be made to represent an actual food chain by counting the numbers of individuals involved in each step of the chain. One such count of a bluegrass field revealed the following average per acre: 5,842,424 producers to 707,624 herbivorous invertebrates to 354,904 ants, spiders, and predatory beetles to 3 birds and moles. Such population-density studies can help us to understand the energy relationships existing in the ecosystem, and even to predict future changes.

Another way of studying food chains is to determine the mass in each step of the chain. This might produce a clearer picture of the biotic relationships, resulting in a more gradually sloping pyramid. Let's consider a sample pyramid involving a willow tree, caterpillars feeding on it, birds feeding on the caterpillars, and hawks feeding on the birds. In this pyramid, one tree could support a number of caterpillars. Drawn on the basis of numbers, our pyramid would look like Figure 49-6 (top). However, if we were to represent this food chain by the masses of organisms, our pyramid would somewhat resemble Figure 49-6 (bottom).

Pyramids based on numbers of individuals or on mass indicate a condition found in the ecosystem at any particular moment. Remember, however, that conditions within the ecosystem are not static, they are always changing. The numbers of individuals and mass of a group of organisms depends not so much on the extent of their food supply at any given time, but more on how quickly the food supply is being replenished.

By having a knowledge of food chains and pyramids, ecologists may be able to predict future events. Returning to our lake ecosystem, let us suppose that conditions in the spring are favorable and an algal bloom results. The increase in the phytoplankton of the lake would be favorable to the organisms feeding on it. Many minnows, then, would be able to survive on the abundant food, whereas in less favorable years, many would have died. These favorable conditions would suggest a larger gross mass for the bass population in the next year.

A knowledge of food pyramids may aid us in solving a human social problem. Can a population explosion such as we have in the world today continue indefinitely without causing a critical food shortage? In crowded regions like the Far East, people must live largely on vegetable diets. They cannot afford to use the food substances necessary to raise herbivores for meat, because the plant level of the pyramid supplies far more food from the available soil. Remember that each level of a food chain uses up some of the energy originally obtained from the sun. More food energy is available closer to the base level of the pyramid. By finding more ways to get nearer to the base of the pyramid, it may be possible to accommodate larger human populations.

Special nutritional relationships

Most of the animals we have discussed in our study of food chains are *bulk-feeders*. That is, they consume tissues in bulk as whole organisms or parts of them. Most animals, including man, are bulk-feeders.

Another group of heterotrophs absorbs nutrients from dead tissues or products of organisms. These are the *saprophytes*. In this group are the bacteria that decompose plant and animal bodies; the molds that live on bread, fruit, leather, and other organic materials; the yeasts that ferment sugars; the fungi that live on dead trees; and many other organisms. While some saprophytes are destructive, others are very useful.

In another nutritional relationship, the individuals live in direct association with each other. This relationship is called *symbiosis,* which means "living together." Biologists usually further divide symbiosis into three different kinds: (1) parasitism; (2) mutualism; and (3) commensalism. In *parasitism,* the parasite lives in or on another organism, called a host. The parasite benefits from this association, while the host is harmed. Tapeworms and lampreys are examples of animal parasites. Other parasites include disease-causing bacteria, mildews, rusts, and smuts.

It is to the advantage of a parasite to keep its host alive. Death of the host results in loss of the parasite's habitat and food supply. Neither ticks, fleas, mosquitoes, nor the fungus that causes athlete's foot kills the host. They take only enough nourishment to sustain themselves, to grow, and to reproduce. The biologically successful parasite is often a degenerate organism. However, it may possess structures that free-living relatives lack. For instance, tapeworms have special hooks that hold them to the intestinal wall to prevent their being swept away. They also have thick skins or cuticles to protect them from the corrosive action of the host's digestive juices. Many parasites have complicated methods of dispersal, requiring two or more hosts before reaching maturity.

Every free-living organism appears to have its parasites, and many parasites have parasites of their own. In terms of numbers of individuals, there are more parasites than there are free-living organisms.

In *mutualism,* another form of symbiosis, two different kinds of organisms live together to the advantage of each. In some instances, the two organisms become so dependent on each other that neither can live alone. Termites, for example, can chew and ingest the cellulose in wood, but they cannot digest it. The cellulose is digested by protists living in the termite's digestive tract. The termite is provided with a means of digestion, and the protist is given a place to live. The association of an alga and a fungus in lichens is another example of mutualism.

49-7 A hermit crab with sea anemones living attached to its shell. What do biologists call such a relationship? (Walter Dawn)

The nonphotosynthetic fungus provides moisture and support for the alga, which, in turn, synthesizes food for the fungus and itself. Another well-known example of mutualism is the relationship between the flowers that supply nectar and the insects that, in seeking it, pollinate them. Many mammals are associated in a mutualistic relationship with specific birds that pick off and eat ticks.

In *commensalism,* one of the partners is benefited, while the other is neither benefited nor harmed. A good example of a commensal relationship is that existing between the remora and the shark. The remora is a small fish with a suction pad on top of its head. It attaches this pad to the lower side of a shark and feeds on scraps of the shark's food. The remora thus benefits, although the shark does not. But the shark is not harmed, either. The term *commensalism* means, literally, "common table," or "messmates."

Passive protection

Individual animals have some form of protection from bulk-food eaters. Claws, teeth, spines, stingers, pincers, and the ability to run fast are all used in defense. But many animals are able to survive because they can hide in their surroundings. Similarly, a predator uses concealment in hunting prey. Many animals would perish, either as prey or predator, if removed from their usual environment. A green katydid is nearly impossible to find among the green leaves of a lakeside tree, but if it should fall into the lake, it would be immediately vulnerable to birds, fishes, other insects, or frogs. The blending of an animal with its surroundings is a kind of camouflage. Many animals owe their existence to this protection.

Animal camouflage involves several principles. Sometimes the animal is colored or marked like its surroundings. We call such camouflage *protective coloration.* The orange background and black stripes of the tiger blend almost perfectly with the grasses and shadows of its environment. A covey of quail crouched in a thicket goes unnoticed until the birds become frightened and fly into the air. The common tree frog has irregular markings of brown and ashy gray that blend with the bark of a tree. As you can see, protective coloration is a common adaptation for survival. Some animals blend so perfectly with their surroundings that only the most careful observer can discover them.

You may recall from Chapter 34 that fishes illustrate a slightly different principle, called *countershading.* In having darker colors on their upper side and light colors on their lower side, fishes blend well into their surroundings.

Still another principle of animal camouflage is illustrated when an animal resembles something in its environment. This is called *protective resemblance.* Several kinds of butterflies resemble brown leaves when their wings are folded. The walking stick, a relative of the grasshopper, actually looks like a stick with legs.

Mimicry is another type of protective resemblance. In mimicry, however, the animal looks like another animal rather than a part of its environment. Several kinds of defenseless flies resemble stinging insects. This gives them nearly the same protection stingers have. Another example of mimicry is found in two butterflies, the viceroy and the monarch (Figure 49-8). The monarch is the common orange-and-black butterfly seen around milkweed plants. The viceroy looks almost exactly like the monarch. The monarch has such an unpleasant taste that it is avoided by birds. The more palatable viceroy escapes because it looks so much like the unpalatable monarch.

In studying animal camouflage, we must keep in mind the principle of cause and effect. The animal does not blend with its surroundings *in order to live*. Rather, it survives *because* it blends with its surroundings. Today, we see the result of many years of survival by those animals best adapted to their surroundings. Through slight variations in form and color, certain individuals resembled their surroundings more than others. They had a better chance to survive and produce more of their kind. It has taken many thousands of years for the animals we have just mentioned to develop. In the process, countless millions perished because they were not very well adapted to their environment.

49-8 Mimicry. Note the resemblance between the viceroy butterfly and the foul-tasting monarch butterfly. (Walter Dawn)

Adaptive behavior

Any set of behavior patterns that tends to insure the well-being of an animal can be considered *adaptive behavior*. Many biologists are currently devoting much time to the study of animal behavior. When a particular activity of an organism is natural or untaught, it is said to be instinctive. Migration of birds, complicated courtship behavior, caring for young, building nests, and spinning webs are all instincts. The term *instinct* is gradually being replaced by *behavior patterns*. *Ethology* is the study of these fixed patterns. Behavior pattern during courtship has been well studied in the stickleback, for example. A physiological change in the male causes his change of color, and he establishes a territory and builds a nest. The appearance of a female whose abdomen is distended with eggs causes the male to swim in a zigzag path. In turn, the female swims toward the male. The male leads her to the nest. She follows, and he indicates the entrance by pointing out the direction with his head. When the female enters the nest, he prods her with his nose, and she spawns. Next, the male chases her from the nest and spreads milt on the nest to fertilize the eggs. Then he repairs the nest, adjusts the eggs, and cares for them. This complicated chain of events is a fixed pattern, and each of the responses must occur;

if one does not, the behavior will not continue. Behavior is being carried out, then, as a series of fixed actions. Each action brings about a response that, in turn, causes another response.

By analyzing the sequence of events and responses in an animal's activity, the biologist has learned much about animal behavior. The hypotheses of the ethologist can be tested by predicting behavioral responses to various stimuli and then observing the animal in the laboratory and in the field to see if it behaves in accordance with the biologist's prediction.

SUMMARY

All living things are interrelated with others and with the physical environment. Energy, originally from the sun, is transferred from one organism to another, and chemicals are cycled in the ecosystem. Soil, temperature, water, light, atmospheric conditions, and earth changes determine plant growth.

The producers, consumers, scavengers, and decomposers all have an important function in the transfer of energy and the cycling of inorganic compounds in a biotic community. On the basis of the manner in which organic nutrients are received from the environment, heterotrophs can be classified as symbionts, saprophytes, or bulk feeders.

BIOLOGICALLY SPEAKING

habitat	topography	symbiosis
niche	producer	parasitism
limiting factor	consumer	mutualism
tolerance	scavenger	commensalism
poikilothermic	decomposer	saprophyte
homoiothermic	transformer	protective coloration
dormant period	food chain	countershading
aquatic	food web	protective resemblance
marine	food pyramid	mimicry
terrestrial	bulk-feeder	

QUESTIONS FOR REVIEW

1. Explain the importance of plankton.
2. How does the tolerance of an organism for an environmental factor limit its distribution?
3. How does soil control the distribution of plants and animals?
4. Name several important controlling physical factors of the environment.
5. What is meant by the terms *cold-blooded* and *warm-blooded*?
6. How do deciduous plants adjust to the freezing temperatures of winter?
7. How are environments classified according to the available water?
8. In what ways are plants adapted to live in environments of varying water content?
9. Although some organisms can live in total darkness, most are dependent on light. Explain.
10. In what environments is oxygen a limiting factor for organisms?
11. How does topography affect the distribution of living things?

12. Starting with the food-producers, name and define the various types of organisms in a food chain.
13. What is meant by a fixed pattern of behavior? Give an example.

APPLYING PRINCIPLES AND CONCEPTS

1. Choose some common organisms in your environment and discuss possible limiting factors for each.
2. Discuss the importance of research on the limiting factors of specific organisms.
3. Does man's tolerance to a wide variety of conditions enable him to extend his habitat? What other factors are involved?
4. Discuss the survival value of the constant internal temperature condition found in some animals.
5. Give some possible reason why evergreens do not freeze and die when covered with snow during the freezing temperatures of winter.
6. Discuss the adaptations of plants growing in areas of reduced light.
7. Make a diagram of a food web that exists in your local area.
8. Cite evidence of fixed patterns of behavior in the animals in your region. How can you be sure that these are fixed patterns?

RELATED READING

Books

BERRILL, N. J. *The Life of the Ocean*.
  McGraw-Hill Book Company, New York. 1967. The ecology of ocean life, with descriptions of a variety of plant and animal life found in the sea.
DAVID, STEPHEN, and JAMES LOCKIE, *Nature's Way*.
  McGraw-Hill Book Company, New York. 1969. A very readable book on representative organisms and how they are able to fit into their ecological niche.
DUDDINGTON, CHARLES, L., *Flora of the Sea*.
  Thomas Y. Crowell Company, New York. 1965. A comprehensive study of algae; suitable for high school use in a study of food chains.
FARB, PETER, and the Editors of *Life* Magazine *The Forest*.
  Time-Life Books (Time, Inc.), New York. 1962. Superb color photographs and clear text convey the concept of the forest as a community of living things bound to their physical environment.
GOETZ, DELIA, *Swamps*.
  William Monroe and Company, Inc., New York. 1961. Explains how swamps are formed, where they are located, and the various forms of life found in them.
KAYALER, LUCY, *Dangerous Air*.
  The John Day Co., Inc., New York. 1967. The frightening story of what air pollution is doing to life on earth as environments are forced to change.
MENNINGER, EDWARD ARNOLD, *Fantastic Trees*.
  The Viking Press, Inc., New York. 1967. Shows how most of such trees owe their eccentricities to environmental influences and the struggle to survive, while some defy explanation.
NIERING, WILLIAM A., *The Life of the Marsh: The North American Wetlands*.
  McGraw-Hill Book Company, New York. 1967. Origins of and changes in fresh and salt-water marshes and bogs, and the webs of organisms living therein.

CHAPTER FIFTY

PERIODIC CHANGES IN THE ENVIRONMENT

Alternating periods of activity

In the springtime, people living in rural areas may waken to the sounds of hundreds of birds announcing the arrival of a new day. At the same time, snails, slugs, and sowbugs are making their way on the ground to protected spots where they will remain throughout the daylight hours. Some distance away, a meadow mouse narrowly escapes a diving hawk that has spotted it while soaring in the early morning sky above. If all goes well for the mouse, it will be safe for the remainder of the day. This was not the first escape for the little four-footed creature. Three hours earlier, in the dark of night, it had narrowly escaped the sharp talons of a barn owl. Each morning, similar events occur wherever animals exist.

An organism that is active during the day is said to be *diurnal* (die-URN'l), while one that is active at night is *nocturnal* (nock-TURN'l). Each diurnal organism occupies a niche that may also be occupied by a nocturnal organism. The diurnal hawk's predatory niche, for example, is occupied by the owl during the night. Alternating periods of activity are called *periodicity,* and when the periodicity is regular, it is said to be *rhythmic.* Thus, the early bird getting its worm every morning shows rhythmic behavior. Although biologists have observed and studied rhythmic behavior in plants and animals for many years, there are still many challenging unsolved questions. Some animals have been described as having "internal clocks." What causes migrations of certain insects, birds, or mammals, and what causes hibernating animals to wake up?

Where rhythms have been observed and studied, light seems to be an important factor. It is not too surprising, then, to find rhythms based on daily, seasonal, and annual light variations. Rhythms that coincide with changes in the phases of the moon have also been studied. However, as you will see, factors other than light are important in the control of regular activity of organisms.

In the nocturnal white-footed mouse, daily behavior is not a simple matter of activity in the dark and rest in the daylight. When the animal is kept in constant darkness, its waking and sleeping periods continue as they would if it were exposed to normal periods of daylight and darkness. Thus, the animal has the capacity to "remember" its rhythm and is not entirely dependent on the external stimulus of light.

Even plants raised in a darkened room under constant temperature have been shown to have regular sleep movements they would have if they were exposed to normal periods of daylight and darkness. The same has been shown for chickens, lizards, and *Drosophila*. Although different species may show slight variations in the length of active and resting periods, most seem to be on a twenty-two-hour to twenty-six-hour cycle.

The daily rhythms

Similar environmental conditions are found in every forest, whether it be a lowland, a hilltop wood, or a lofty mountain forest. The tall trees provide areas of shade under which the shrub layer of the forest floor flourishes. In other areas, the sunny spaces favor meadows of various sizes where the grasses grow. Ferns cover the banks of streams. A summer morning may find the birds, chipmunks, and ground squirrels searching for food as the deer browse in the meadows. Before noon, however, the activity of the forest diminishes.

As evening approaches, many of the diurnal animals renew their activities for a short time. Dragonflies and bats dart over streams and ponds searching for both diurnal and nocturnal insects available in this transition period. The sounds of the cicadas and birds gradually fade as the chirping crickets begin their activity. The foxes, raccoons, skunks, owls, and mountain lions begin their nocturnal search for food.

Although respiration in plants also occurs during the daylight hours, the photosynthetic manufacture of sugars proceeds rapidly during those hours. In the darkness, when photosynthesis ceases, the manufactured sugars continue to be transported downward to storage tissues, and only respiration occurs. When plants bloom, the flowers may open and close at regular times of the day. The petals of the poppy flower, for example, open in the morning and close at night for several days. Many cacti bloom only at night and depend on nocturnal insects for pollination.

The desert environment, with its extreme temperature changes, leads to a sharper division between day and night activities. In the early morning hours, birds feed on insects and seeds, and jack rabbits come out of their burrows to look for food. The snakes, hawks, vultures, and ground squirrels are awake and active. Since the temperature may reach 170°F on the desert surface, noontime finds most of the desert creatures in the shade at the bases of cacti, sagebrush, or creosote bushes. Many of

50-1 Plants living here are adapted to the seasonal changes. The animals have adaptations for survival in the various conditions or they move away for the harsh winter season. (Walter Dawn)

50-2 The flowers of the night-blooming cereus, a cactus, open only in the evening or at night. (Fritz Henle, Photo Researchers)

50-3 The cardinal is a permanent resident in the midwest, eastern, and southeastern states. (Thase Daniel, Bruce Coleman, Inc.)

them take to their cooler burrows, where the moisture in the air may be more than twice that of the atmosphere.

In the early evening, many of these creatures again come out for a brief period. During the night, the Gila monster and the rattlesnake prey upon the many small nocturnal desert mammals. The bobcat, coyote, fox, and owl are some of the larger nocturnal desert carnivores.

In contrast to the desert are the polar regions, where very little activity takes place during the cold nights. During the brief summer, most organisms are active throughout the prolonged daylight.

Farther from the poles, however, are the vast regions of treeless plains, where the ground remains frozen most of the year. For brief periods, a shallow layer of earth at the surface thaws enough to allow lichens and some grasses to grow there. In the north, amid this plant life, a few warm-blooded animals such as the caribou, musk ox, Arctic hare, fox, lemming, and ptarmigan spend the daylight hours obtaining food.

In equatorial regions, the lengths of the day and night are nearly equal and the temperature is less variable. Light is, therefore, a more important factor in determining the periods of activity of equatorial organisms. In the equatorial forests, nocturnal and diurnal animals are both very numerous.

Day-night rhythms are also found in the oceans, where periodic vertical migrations have been studied. Vast numbers of copepods and shrimp are found at night near the surface, where they are able to feed on plankton. During the day, these herbivores sink to a lower level and may be found three hundred feet below the surface. Small carnivorous fishes follow the daily excursions of these plankton-feeders.

Seasonal community changes

Animals meet the problems of seasonal temperature changes in several ways. When the cold winter brings snow and freezing temperatures, an organism must be able to adjust or move away. Otherwise, it will die. The first freezing nights of autumn take a heavy toll of insect life. Although the mourning cloak butterflies find winter protection in a hollow tree or crevice, most adult insects have completed their life cycles before the beginning of autumn, and are killed by the first frosts. Species survival through the winter is insured by other means. Many of the moths spend winter as pupae within silk-insulated cocoons. Grasshoppers, crickets, and cicadas lay eggs in the ground or in the bark of trees. Stoneflies, mayflies, and dragonflies spend the winter as nymphs in the water, sheltered beneath the frozen surface of a pond or stream.

The familiar honeybee finds protection in numbers. During the winter months, the bees feed on the honey that had been stored in the spring and summer. This supplies them with energy as they remain active in their hives during the winter months. On a very cold winter day the temperature within the hive may be as much as 75°F higher than that outside.

Animals like the eastern cottontail rabbit, the white-tailed deer, the cardinal, and the bluejay are permanent residents of their regions. During extremely cold weather, they find protection in woods and thickets; but when snow is on the ground, food-getting becomes a serious problem.

Ground squirrels, chipmunks, woodchucks, and many reptiles and amphibians undergo true *hibernation* during cold weather. The rate of body metabolism drops greatly. Heart action and respiration decrease, and the animal loses consciousness. Greatly reduced activity lowers energy requirements to the minimum. An animal undergoing true hibernation cannot resume activity until the temperature of the environment increases and the body processes speed up.

Some animals that remain in a region during periods of unfavorable climatic conditions may enter a state of dormancy. That is, they reduce their life activities to the minimum necessary for survival. The bear finds a hollow log, a cave, or some other protected location and lives on stored fat during his *winter sleep*. Although the activities of the body slow down, a normal body temperature is maintained, and the bear can be awakened. It may even leave its shelter on a mild winter day. The skunk, raccoon, and opossum undergo a similar winter sleep.

During hot weather, many animals enter a period of dormancy sometimes called *summer hibernation*. The biological term for this is *estivation*. A frog may estivate in the cool mud at the bottom of a pond. The box turtle often escapes the heat by burying itself in a pile of leaves. The period of estivation may be several days or several weeks. The gopher tortoise of the southeastern states finds protection deep in a burrow in the ground. The California ground squirrel sleeps in its burrow, thus reducing both its food and water needs.

Many animals migrate to warmer regions when winter comes. We usually associate *migration* with birds, but some mammals also show the tendency to migrate. These seasonal journeys may cover thousands of miles. Some mammals make migratory journeys because of seasonal changes in the food supply. Others migrate to a more favorable climate regardless of food supply. Still others make seasonal journeys to regions where they can produce their young under the most favorable conditions.

The bighorn sheep spends its summers in the high meadows near the summits of the Rocky Mountains. As winter approaches, it moves down into the protection of the forests on the mountain slopes. Through the summer months, herds of Olympic elk browse in the high altitude of the mountains on the Olympic Peninsula. During the winter, these herds move to the more protected mountain valleys and nearby plains. With the coming of

50-4 Hibernation. Many animals, like this jumping mouse in his underground nest, pass the winter in a dormant state. (Allan Roberts)

spring, the herds move back up the slopes in long, single-file processions.

Among the most remarkable migrations is that of the fur seal. During the winter, females, young males, and pups roam the waters of the Pacific Ocean to as far south as California. The older males winter in the cold waters near Alaska and the Aleutian Islands. With the approach of the breeding season in the spring, the males migrate to the Pribilof Islands, north of the Aleutians. The males arrive several weeks before the females and battle for a territory. The females and young seals start their long journey of three thousand miles or more to the Pribilofs in the spring and arrive in June. A herd of fifty or more females gathers around each male. Pups from the past year's breeding are born almost immediately, and within a week, breeding occurs again. Then, the seals migrate southward.

A butterfly migration

The monarch, or milkweed, butterfly, one of the strongest insect fliers, makes a remarkable seasonal journey. In the latter part of the summer, these butterflies gather by the thousands in northern Canada and begin a long flight southward. Some of them travel to the Gulf states to spend the winter, but their habits there have not as yet been studied. Others travel a southern route along the Pacific Coast. At some time between the middle of October and the first of November, tens of thousands of these insects arrive on the Monterey Peninsula in the small town of Pacific Grove, California. Here they seek shelter in a specific grove of pines. The monarchs hang by their legs from the branches and needles in such large numbers that the trees appear to be solid brown. They stay there in a state of semihibernation until the winter is over. On warm sunny days throughout the fall and winter, many will be seen flying about local gardens gathering nectar.

In March, the monarchs fly out over Monterey Bay to begin their northward flight. As they fly north, they lay eggs on milkweed plants. It is unlikely that any of the travelers ever reach their northern home, because many die after the eggs are laid. However, after the eggs hatch, the larvae pupate, the young butterflies emerge, and the trip northward is continued. These new butterflies also lay eggs on milkweed as they progress.

By late summer, the monarchs begin gathering for their southward journey to the same locality and the same trees in which their ancestors of two generations before spent the winter. Although we attribute the monarch's behavior to instinct, the factors causing the migrations and the ability of these butterflies to find their way are not well understood.

Bird migration

The control of migration has been a most interesting problem for the biologists. Many birds fly long distances in the spring,

50-5 These monarch butterflies spend the winter in a pine grove in California after migrating from Canada. (D. Morton, National Audubon Society)

then nest and raise their young in a new home. They return to warmer climates in the fall. Migration may be prompted by food needs, climatic changes, or breeding habits. It is difficult to determine why some species leave abundant food and warmth in the tropics to migrate to breeding grounds in the Far North. Much more easily explained is the southward migration of insect-eaters when cold weather kills their prey, and the southward flight of water birds before the ponds and lakes freeze over. The tendency of fruit- and seed-eaters to follow their food supplies is also logical.

Some birds make their migratory flights at night and some during the day, depending on the species. Perhaps you are familiar with the night flights of geese during the spring and autumn, when they become confused by the lights of a city and circle about, honking noisily. The daylight flights of thousands of red-winged blackbirds and grackles are familiar sights during spring and fall.

While many birds migrate slowly, feeding along the way and averaging only twenty to thirty miles a day, others are marvels of speed and endurance. The ruddy turnstone travels each autumn from Alaska to Hawaii in a single flight, and the golden plover travels from Canada to South America, a distance of more than eight thousand miles, as you can see from the map in Figure 50-6.

Migratory routes

We do not know much about the instinct that governs the time and route of migration. Any given species follows the same route year after year and can be expected to arrive at a certain point within a few weeks of the same time each season, depending on the weather. Certain species travel northward along one route and return by an entirely different route, as if they wanted to vary the scenery. How do they know the way? Keen sight may help, but not over water or through dark nights and fogs. Even old birds that have made the flight before may remember the way. This cannot account for the unescorted flights of young birds. Biologists now believe that the sun guides birds on their daylight flights. There is also good indication that birds can allow for variations in position of the sun in different seasons and in different latitudes. This finding does not imply that birds have a true understanding of their position in relation to the sun.

What about the migratory flights of birds at night? Recent studies of night flights reveal even more amazing possibilities. Birds seem to be directed by the positions of the stars. Investigators in Europe have used caged birds in a planetarium in which star patterns of different seasons and various places in the earth can be projected on a dome representing the heavens. They have found that certain birds, including warblers, make a definite response to the positions of stars during the normal season for migration. Under a fall star pattern, they face toward the winter

50-6 The migratory route of the golden plover. Flying more than 8,000 miles in a single migratory flight, it breeds in northern Canada during the summer, and in the fall flies to South America, where it spends the winter.

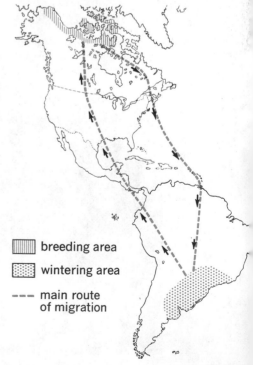

||||| breeding area

▨ wintering area

--- main route of migration

migratory home in their cages. Under a spring pattern, they face toward the summer home.

Seasonal bird study

Observing migration adds much to the study of birds as a hobby, because different species arrive and can be sighted as the seasons change. Each locality has *permanent residents* that remain the year around. Certain species may be present in the winter only, moving farther north with the coming of spring. These species are called *winter residents*. The *summer residents* spend summers in a given locality and migrate southward in the fall. Many species are found only at certain times in the spring and fall. These are the *migratory birds* that are passing through a given locality on their journey between wintering areas farther south and breeding areas farther north.

Lunar rhythms

There are many folk tales about planting certain crops with various phases of the moon. Although these are interesting stories, there are no experimental data, as yet, to support them. The greatest effect of the moon is seen along the coasts, because the moon is mainly responsible for ocean tides. Tidal extremes vary in different parts of the earth, depending on the shape of the coastline. In the Bay of Fundy, located between Maine and Nova Scotia, fifty-foot tides have been measured. On the coast of France, the Bay of St.-Malo has tides of thirty-nine feet. Although most regions do not have such spectacular tides, the rhythmic rise and fall of the water level affects the lives of the organisms living within this zone of tidal influence.

Many marine biologists spend their time studying the ways in which plants and animals are able to adjust to the changing

50-7 A tidal pool at low tide (left), and later the same area partially immersed by the incoming tide. (Walter Dawn)

conditions of such an environment. The greatest problem for these organisms is the danger of drying out at low tide. The small fishes and crustaceans find protection in the waters of the tide-pools that are left as the water recedes. Many limpets and snails resist drying out by clamping down tightly on the rocks. Mussels and barnacles preserve moisture by closing their shells tightly. Some sessile sponges and tunicates live under rock ledges protected by seaweed, which covers them during low tide.

Drying is only one of the problems the intertidal organisms face. In both summer and winter, tides cause wide variations in temperature. Fresh water from rains alters the salt content of tidepools. Thus, the organisms inhabiting the intertidal zone must adjust to a great variety of changes.

On certain nights from March to August, cars are parked bumper to bumper on the coast highway of California. Thousands of beach fires, added to the light of the moon, create a spectacular scene. The people are waiting for the turning tide, when the beach will be left shimmering with thousands of fish. These fish, called *grunion,* are caught and roasted over the fires. The grunion's behavior offers a precise example of lunar periodicity. Exactly at the turning of the tide on the second, third, and fourth of the nights during which the highest tides occur, pairs of these fish swim up the beach with the breaking waves. The female digs into the sand and deposits eggs about three inches below the surface. The male fertilizes the eggs, and on the next wave, the pair slips back into the sea. The eggs remain in the sand until the next unusually high tides, about ten days later, when they are washed out of the sand. The eggs hatch as soon as they are immersed, and the tiny fish swim away. Several marine annelid worms have also been observed to swarm and breed at definite phases of the moon.

50-8 What factors synchronize the spawning of these grunion? (Walter E. Harvey, National Audubon Society)

Annual rhythms

Many of the reproductive cycles of plants and animals are associated with seasonal changes and occur in a yearly rhythm. The female bear has her young during the winter in the protection of her den. Birds nest in the spring, thereby insuring full growth of the young before winter. Wildflowers bloom in the spring and produce seeds for the development of the next generation. Deciduous trees lose their leaves in the fall. These are all familiar expressions of annual cycles. You will be able to name many more.

What is the value of periodicity?

As you now know, an organism must be able to adjust to environmental changes if it is to survive. You have seen many ways in which various organisms adjust to such changes. The establishment of a synchronized rhythm for any entire population is of value for the survival of the species. The behavior of swarming grunion insures the aggregation of males and females necessary for the fertilization of the eggs. Male and female gametes of the oyster are shed into the surrounding water at the same time. Even though a female oyster may produce 114,-000,000 eggs, her efforts will be futile if a neighboring male does not release sperms at the same time. The search for factors causing synchronized behavior in populations occupies the time and efforts of many biologists.

Changing biotic communities

We have seen some of the daily and seasonal changes occurring in various communities. Plants and animals are continually on the move, but since the changes in plant populations of an area usually occur slowly, they are not always easy to identify. As plants change, animals find new homes. This changing of communities is called *succession.* If an area within an ecosystem were completely cleared of living things, the natural events that bring about the changes in its populations would be more apparent. The gradualness of changes results in a relatively stable community in equilibrium with local conditions.

Winds, fires, volcanic activity, and other events in nature, as well as man's clearing the land, may destroy the organisms living within a natural area. Then, if the area is left alone, succession starts. Eventually, a permanent community will reclaim the region. This process, from beginning to climax, may take as long as one hundred years.

Succession can occur even in a jar of water. If a culture medium is made by boiling hay and then exposing the culture medium to the air, it will soon be teeming with bacteria. Since the bacteria are the first organisms to enter the area, they are called *pioneers.* If a few drops of pond water containing several kinds of protists are added to the bacterial culture, the protists will multiply at varying rates. The flagellate population thrives on the bacteria, but as their numbers increase, their food becomes scarce, and they begin to dwindle. The disappearance of the flagellates is also speeded up by an increase in numbers of *Colpoda,* a ciliate resembling the paramecium. As the *Colpoda* feed on the flagellates, the medium becomes less acid, and the *Colpoda* population is replaced by the increasing numbers of paramecia and amebas, which can adjust to a more alkaline environment. As the amebas begin to increase and consume more ciliates, the ciliate population declines. This succession may take several months, and at the end of this stage, the organisms will all die unless more nutrients are added. If we were to add a few cells of green algae to the jar, they would multiply, and the addition of a few drops of pond water might allow some rotifers and crustaceans to develop. At this stage, there are a few of each of the organisms in the water, and as long as the plants receive enough light to

Pioneer stage—
bacteria

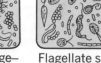

Flagellate stage

Colpoda stage

Paramecium—
ameba stage

Climax stage—balanced

50-9 Succession in a jar of water.

manufacture food, the numbers of each kind will remain about the same. This stable, or balanced, system may remain for months or even longer. The balance will, then, be the climax condition of the ecosystem in the jar.

Natural succession in a forest

Whenever a tree falls in a forest, succession begins. This type of succession is much more complex than that occurring in a hay infusion and lasts for a great many years. Whenever rocks are put in a bay for a breakwater, or wooden or steel piles are sunk for a pier, ecological succession occurs. Wherever succession has been studied, the sequence of organisms and their time of appearance can be predicted with surprising accuracy.

To examine a more complex series in a succession, we shall start with a section of bare soil in an open area. This area might be a region devastated by fire, one cut for trees and not reseeded or an abandoned agricultural field. We shall locate it in a broadleaved forest in the eastern United States, where beech and sugar-maple forests once grew over much of the land. First, the seeds of grasses and other open-field plants that may be dormant in the soil or carried in by animals or winds find the environment satisfactory. A meadow is produced by these pioneers, which may dominate the region for several years. Next, the seeds of elms, cottonwoods, and shrubs find their way into the meadow, marking the beginning of a forest. The larger plants shade the shorter grasses and field plants. Thus, the environment of the once open field is changed into an open, low woods. This second stage may soon become too shady even for the seedlings of the pioneer trees and shrubs, so the area will again begin to change.

The third stage may be the arrival of seeds of oaks, ashes, and other trees whose seedlings can grow well in a shady environment. These trees grow among the elms and cottonwoods and gradually become the dominant vegetation. Finally, a dense forest begins to form. The ground becomes moist and fertile, and beech and maple seedlings outdistance all other species in the competition for a place in which to live in the forest. Eventually they crowd out most of the other trees. Since the beech and maple trees assume final dominance in the region, we call them the *climax species*. If such a succession had occurred on a ridge, the climax species might have been an oak and hickory forest. Short grasses are the climax plants in the Great Plains.

Succession in ponds and lakes

Observation of ponds and lakes is excellent for the study of succession. As we learned earlier, the cattails and water lilies around the edges hold soil around their roots and build the soil up over a period of many years. When this occurs, the pond actually grows smaller. Through a close examination of the or-

50-10 Schematic diagram of succession in a pond (top to bottom): pioneer, open-pond stage; submerged vegetation stage; cattail stage; sedge meadow stage; climax forest stage.

50-11 You can see the grasses growing along the margin of this lake. With time, this shallow lake will be filled in, and climax vegetation will be growing where the lake once was. Compare this picture to the diagram in Figure 50-10. (A. Towle)

ganisms that grow from the pond's edge outward to the terrestrial climax plants of the region, we can predict the succession of plants and animals that will gradually obliterate the pond. The size of the pond or lake, the way in which its water is supplied, and its location also determine the extent to which succession fills it in. In addition, these conditions determine the time at which succession will occur and whether it will take a few years or hundreds of years.

In August, 1959, an earthquake occurred in the Madison River Canyon of Gallatin National Forest in Montana. A new lake was formed, and a large portion of the land on the mountainside was cleared when tons of earth, rocks, and trees fell into the canyon. Biologists are currently studying the changes occurring in the new lake and clearings. A knowledge of ecological succession enables us to predict changes, but phenomena such as this earthquake provide us with opportunities to prove our theories. For this reason, the United States Forest Service has set aside 37,800 acres of the earthquake areas as a preserve for public enjoyment and scientific study.

The drama in nature goes on endlessly. Conditions in any given area are never permanent or static. Each society of plants that occupies the area alters the environment, and often makes it unsuitable for other plants of the same kind but favorable for other kinds of plants that move in. Over a period of time, these changes in plant population prepare the way for the climax vegetation. Since the animal population depends on the plant population, the kinds and numbers of animals change as plant succession occurs.

SUMMARY

Periodic changes in communities are brought about by regular variations in light, temperature, and climatic conditions. These physical factors may act as stimuli or inhibitors, controlling the behavior of organisms. Rhythmic behavior sometimes continues, however, even when an animal has been placed in a laboratory under constant conditions. Therefore many biologists say that some animals have an "internal clock."

In cleared areas an orderly progression of living things occurs. The pioneers make the environment right for other organisms, which in turn create conditions allowing different plants and animals to move in. Gradually a balance is achieved and the climax vegetation dominates the flora. Ecological succession can be studied in an ecosystem as small as a jar of water or as large as a forest.

BIOLOGICALLY SPEAKING

diurnal	hibernation	succession
nocturnal	estivation	pioneers
periodicity	migration	climax species
rhythmic		

QUESTIONS FOR REVIEW

1. The owl and the hawk are both predators. Do they compete with one another? Explain.
2. What evidence is available to indicate that daily rhythms are not merely a matter of light and dark stimuli?
3. What factors make the equatorial environment favorable for growth of many organisms?
4. In what ways do animals adjust to seasonal changes?
5. In what ways are hibernation and estivation similar? How are they different?
6. What explanation has been given to account for the migrations of the monarch butterfly?
7. What environments are directly affected by lunar rhythms?
8. How does the grunion's behavior demonstrate lunar periodicity?
9. Of what value is periodicity to survival of some species?
10. By what methods can succession be studied?

APPLYING PRINCIPLES AND CONCEPTS

1. Discuss why man's activities are not entirely governed by external rhythms.
2. Discuss the senses that are well developed in nocturnal animals.
3. Discuss the ways in which a desert and an arctic treeless plain are similar.
4. Review and discuss possible causes for migration in birds.
5. Name and identify the permanent residents, winter residents, and summer residents of the bird population in your area.
6. What climax communities exist in the area in which you live? Identify any stages leading toward this climax.

RELATED READING

Books

CARTHY, J. D., *Animals and Their Ways.*
  Natural History Press (distr., Doubleday & Co., Inc.), Garden City, N.Y. 1965. A new look at nature in a highly recommended book.
FARB, PETER, and the Editors of *Life* Magazine, *Ecology.*
  Time-Life Books (Time, Inc.), New York. 1963. Diagrams and text explain the delicate balance that exists in nature, and the damage that man inflicts on the earth everywhere.
MCCORMICK, JACK, *The Life of the Forest.*
  McGraw-Hill Book Company, New York. 1966. About the interwoven lives of plants and animals in a forest, produced with the cooperation of the U.S. Department of Interior.

Article

MATHEWS, SAMUEL W., *"The Night the Mountain Moved."*
  National Geographic Magazine, March, 1960. How Montana's historic earthquake (partially in Yellowstone Park) wrenched the Rockies, created a lake, and touched off new geysers in the park.

CHAPTER FIFTY-ONE
BIOGEOGRAPHY

The distribution of plants and animals

Have you ever seen or felt cobwebs floating in the breeze of a spring or fall day? The young of many spiders spin silken threads that, when they are caught in air currents, can support the weight of the tiny creatures. Why do spiders travel in this way? Where are they going?

Perhaps you have had to pick out burrs and stickers from your socks after a hike. Cats, dogs, and other animals that roam are often covered with such seeds or fruits of weeds and other plants. As you learned in a previous chapter, fruits and seeds often have very intricate devices for getting from one place to another. A milkweed seed, for example, may travel through the air for miles on its fluffy parachute. Sailors have seen tiny airborne spiders more than two hundred miles from the nearest land. Spiders have also been found in the air at an altitude of ten thousand feet.

Methods of dispersal are of value to a particular species because they allow members of the species to go to new environments. The baby spider and the milkweed seed are likely to float to areas where there are not so many of their own kind. Of course, if a spider lands in the ocean, or if a milkweed seed lands in a lake or on a granite cliff, each will die. However, dispersal has much to do with producing the succession about which you learned in Chapter 50.

Biogeography is the study of the distribution of plants and animals throughout the various regions of the earth. Natural methods of dispersal may extend the range of a species, or human beings may deliberately or accidentally serve as dispersal agents. For example, the Scotch broom and the French broom are shrubs that have been imported because of their delicate, almost lace-like green foliage and bright yellow flowers. In the Coast Ranges and the Sierra Nevada foothills of California, the seeds produced by these garden plants have found a favorable environment. The plants have become so numerous that whole hillsides are sometimes a vivid yellow during the blooming season. The plants have "escaped" from cultivated gardens and have become locally established almost as thoroughly as if they were native.

Most people associate the pineapple with Hawaii. Actually, the pineapple is native to South America, but was imported to the Hawaiian Islands, where its growth and production have played a large part in the economy of this state. The introduction of European starlings, English sparrows, and Japanese beetles into North America is another example of dispersal by man.

Dispersal of plant seeds is a passive process. They must ride on currents of water or air. Spores, seeds, or fruits may stick to the fur of animals or to the feet of birds. Birds or mammals may eat some berries, which pass through the digestive tract undigested and then germinate after they have been voided.

Some animal forms are also dispersed passively. Eggs and larvae may float in water or air. Driftwood may carry mussels or shipworms to new locations. Logs may carry rodents for long distances. Drifting ice may carry a polar bear for hundreds of miles. Many animals seek new habitats, however, by actively swimming, flying, walking, running, or hopping.

Man actively extends his living area. He fills in tidelands with rocks and then soil. Entire communities, airports, shopping

51-1 This is a map of the world showing the distribution of vegetation and use of the land. What factors might be responsible for this uneven distribution?

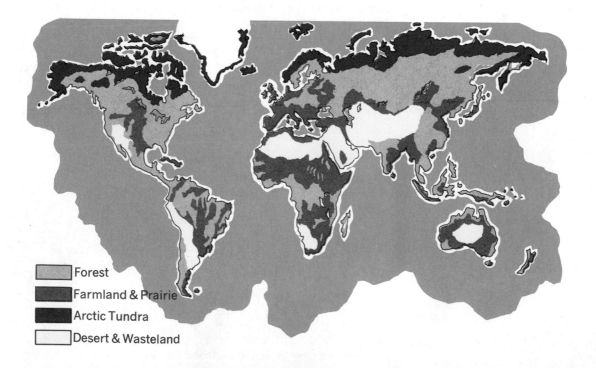

Forest

Farmland & Prairie

Arctic Tundra

Desert & Wasteland

centers, and industrial sites may occupy areas that were once covered with water. New designs and building materials make it possible to build houses on hillsides once considered far too steep for human habitation. By irrigation, man makes human life possible in deserts. Man's experiments in space may become a method of dispersal for man. Even now missiles being sent to the moon are sterilized in order to prevent establishment of microorganisms on our nearest space neighbor.

Barriers to dispersal

A frog could be transported across a large fresh-water lake on a log. It could jump into the water periodically to keep its skin moist. But if the log were floating in the ocean, the salt water would cause the death of the frog. Salt water, then, can act as a barrier to the dispersal of frogs. High mountains, deserts, lakes, rivers, and soil conditions are other *geographical barriers* across which many plants and animals cannot pass. Continents are barriers to marine organisms. For shallow-water marine forms, deep water is a barrier. Many marine organisms are limited in their dispersal by the lowering of salt concentration where rivers flow into the sea. Similarly, many brackish-water organisms are limited to estuaries, where fresh water and sea water mix.

Lack of food may act as a *biotic barrier,* keeping animals from moving into new areas. A zebra from Africa might find a suitable habitat in Eurasia, but it would not find enough food to sustain itself while attempting to cross the Sahara Desert. For deer, squirrels, and other forest animals, a desert would be a climatic as well as a biotic barrier.

Climatic barriers prevent the spread of many organisms. Although many mammals would be physically able to climb over a mountain range, the weather conditions in the mountains might keep them from doing so. A desert, with its dry, hot climate by day and near-freezing temperatures at night, will stop many transients.

Major climatic zones of North America

Temperature ranges in the various regions of North America divide it into *climatic zones.* Northern Canada, part of Alaska, Greenland, and other land masses of the polar regions lie in the area of the polar climate. We commonly speak of these areas as the *arctic region.* Similar climatic conditions prevail above the timber line on high mountains. Most of Canada and most of the United States lie in the area of mid-latitude climate. This is often called the *temperate region.* Florida is in a *semitropical region.* As you move southward through Mexico, the semitropics gradually become the *tropical region.*

Just as the climatic regions of the earth vary between the North or South poles and the equator, so do the principal kinds of living things. A journey from the Far North, for example,

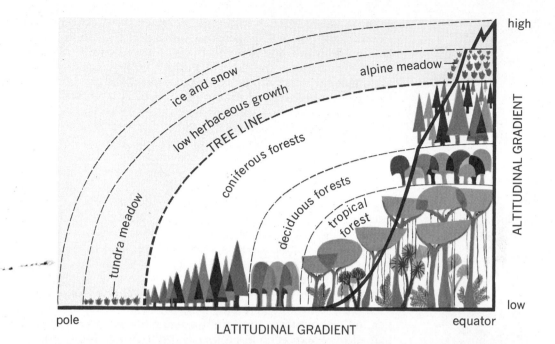

ice and snow

low herbaceous growth

TREE LINE

coniferous forests

alpine meadow

deciduous forests

tropical forest

tundra meadow

ALTITUDINAL GRADIENT

low

pole

LATITUDINAL GRADIENT

equator

would begin with a region of ice and snow that would gradually give way to low herbaceous vegetation. Still farther south, we find the large coniferous forest belts. The division between the low vegetation and the coniferous forest, called the *tree line,* is usually distinct. Next come the deciduous forests. Depending on the rainfall of the area in which we are traveling, the deciduous forest may give way to prairie grassland and then to a desert or a tropical forest.

In a few places, all of the climatic zones can be found within a small area. Figure 51-2 shows that the broad zones from polar to tropical are duplicated on a high mountain at the equator. Thus, you can see how climatic barriers can be produced by the topography of the land as well as by a change in latitude.

51-2 Vertical climatic regions of the earth are similar to horizontal climatic regions. Life zones on a high mountain can be compared to those found while traveling from the equator to either pole.

Biomes—regions with climax vegetation

The coniferous, deciduous, and tropical forests found in various climatic zones are made up of the climax species that result as living things interact with the climate and succession occurs. A large geographical region identified mainly by its climax vegetation is called a *biome.* Now, let us distinguish among the earth's biomes.

The Poles

Perhaps you think of the North Pole and the South Pole as large areas permanently covered with ice and snow, always cold, and with days and nights lasting several months. You are only partly correct, however, because there are several differ-

51-3 Mt. Erebus is the only active volcano on the Antarctic continent. It is 12,140 feet high. (A. Towle)

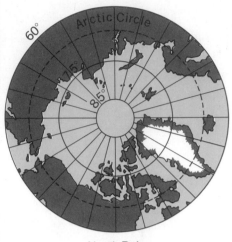

North Pole

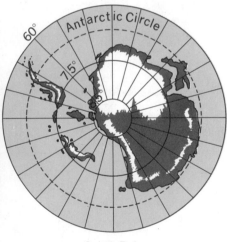

South Pole

51-4 Although the Southern Hemisphere is called the water hemisphere, its pole is covered by the large Antarctic continent. The Northern Hemisphere is called the land hemisphere, but its pole is located on a sheet of ice overlying the Arctic Ocean.

ences in the distribution of life at the poles. First, if you examine Figure 51-4 you will notice that Antarctica is, indeed, a large land mass. It is a continent almost as large as the United States and Canada combined. The Antarctic land is covered by ice to an average depth of one mile, and winds over its surface have been measured at more than 200 miles per hour. There is no month in the year when the mean temperature in the Antarctic goes above freezing. Because of these low temperatures, water is in the form of ice, and therefore, the Antarctic is virtually a desert. Over 90% of the world's ice is stored in Antarctica. However, it is believed that this southern continent was not always the frozen desert it is today. Scientists have found fossil leaves and coal deposits indicating a tropical climate once existed here. Presently there are no terrestrial mammals living in the Antarctic. Three flowering plants exist on the tip of the peninsula and the lichens and mosses are rare. The animals of this southern continent consist of penguins, a few visiting birds, mites, a wingless fly, and a rare insect living at the tip of the peninsula. About 50% of these small invertebrates are parasitic.

Of the five and one-half million square miles comprising Antarctica, only about three thousand square miles are ever bare of ice and snow. There is no interchange of plants and animals and the nearest tree is about 700 miles north of the farthest northward extension of the Antarctic peninsula. Man visits the vast area of the Antarctic for scientific studies, and recently a few tourists have set foot on the continent.

The Arctic offers a surprising contrast. Scientists have identified more than 100 species of flowering plants, many varieties of mosses, lichens, insects, birds, and mammals. Also within the Arctic live a million people, mostly North America's Eskimos and reindeer herdsmen of Europe and Siberia.

Several factors explain the differences found in the polar regions. The Arctic is largely covered by an ice-sheet seldom more than 15 feet thick, and the stored heat from the ocean below has a moderating effect on the temperature. The Arctic averages 35°F warmer than the frigid Antarctic. Thus, nine-tenths of the Arctic's lands lose their ice covering during the summer months. Temperatures may rise as high as 78°F and as the upper layers of tundra thaw, water is available to the many flowering plants.

The tundra

A large zone encircling the Arctic Ocean of the Northern Hemisphere is a biome known as the *tundra*. Since there is no large land mass at a corresponding latitude of the Southern Hemisphere, the area of southern tundra is very small compared to that of the North. The climate of the tundra is extremely cold, and the ground is permanently frozen a few feet below the surface. During the continuous daylight of summer, the surface thaw produces saturated bogs, many streams, and ponds.

Mosses and lichens form the prominent perennial vegetation, although some dwarf birches, alders, willows, and conifers may be found. The annual plants have a rapid growing season, and many produce large, brilliant flowers even though there are periods of freezing temperatures (Figure 51-5). Most of the birds are summer migrants, but the ptarmigan is a permanent resident. Many of the inhabitants of the tundra, such as the Arctic hare, lemmings, Arctic foxes, and polar bears, have white coats that act as protective coloration. In the summer, insects are very numerous, and the eggs they produce are resistant to freezing. Herds of caribou visit the tundra to graze on the moss and lichens.

The coniferous forest

Another biome occurs just south of the tundra in Europe, Asia, and North America. As with southern tundra, there is no large corresponding zone in the Southern Hemisphere, because there are no large land masses in these latitudes. Sometimes this large coniferous forest is called the *taiga* (TIE-guh). In this region, the growing season may be as long as six months, although the winter temperatures may be as severe as they are in the tundra. The tree line marking the transition from the tundra to the taiga may be quite noticeable. At this point, the most obvious tree is the spruce. In the taiga, soil is shallow because of glacial scraping years ago.

Farther south, the broad coniferous belt covers much of Canada, where alders, birches, and junipers may be found in groves. Here, where fire has destroyed large areas and where succession has occurred, the pioneer grasses are followed by aspens and birches. These are eventually replaced by the spruces, pines, and firs that form the climax community. Magnificent

51-5 Some plants of the tundra have large, showy flowers. Annual plants grow and reproduce during the brief season when the surface layer of ground is thawed. (Dr. William Steere; Walter Dawn)

51-6 A scene in a coniferous forest of the Northwest. (A. Towle)

51-7 Deciduous forests undergo seasonal change. During what season was each photograph taken? (Russ Kinne; Herbert Weihrich; John King)

stands of pine, spruce, and redwoods grow along the coastal ranges of Washington, Oregon, and California. Here, giant trees reach a height of two hundred feet or more. Rainfall may be as heavy as eighty inches per year, and fog blankets the area.

The prominent permanent residents of the coniferous forests are many. Moose are plentiful in areas that have not been excessively hunted or in which they are protected. Black bears roam these forests. Martins, wolverines, and lynxes may be found. Squirrels, voles, chipmunks, rabbits, and mice may be preyed upon by the bobcat, fox, and wolf. Beavers and porcupines are found in many of the coniferous forests. During the summer months, many birds breed in these forests, but in the fall, they migrate south. The numerous insects and other invertebrates found in the coniferous biome during the summer lie dormant during the cold winter months.

The deciduous forest

In areas of the temperate zone, the growing season may last for six months or more. Rainfall averages around forty inches per year, and where the soil is suitable, large *deciduous forests* occur. The eastern United States, England, central Europe, and parts of China and Siberia are or once were covered with large stands of deciduous trees. Although a similar zone occurs in South America, it is limited in size by inadequate rainfall.

Local conditions of soil, drainage, and variations in climate throughout the temperate zone provide conditions necessary for different climax communities. In the United States, the beech-maple forests are found in the north central regions, while the oak-hickory forests are common in the western and southern regions. Although most of the native chestnut trees were destroyed by blight, oak-chestnut forests formerly covered much of the Appalachian Mountain chain. Other common deciduous trees found in the temperate zone are the sycamore, elm, poplar, willow, and cottonwood. Although each of these specific forests has its typical animal species, there are several animals that are found in all deciduous forests. Deer are the common herbivores. Foxes, martins, raccoons, and squirrels inhabit the area. Wolves may wander between the taiga and the deciduous forests. Woodpeckers and other tree-nesting birds are plentiful.

The deciduous forest undergoes great seasonal change. In late spring and summer, the trees are green, and shrubs growing in their shelter produce beautiful blooms. During the fall, many areas are turned into artistic splendor by the multicolored leaves of trees preparing for winter. In the winter, the bare branches contrast with the white snow.

The deciduous forest biome region is probably one of the most important of the world. It is in this region that some of the world's largest cities have been built and man has achieved some of his greatest cultural and technological development.

The grasslands

In vast areas where rainfall is between ten and thirty inches each year, *grasslands* occur. The variable rainfall is not enough to support large trees but is sufficient for many species of grass. These natural pastures have been used by huge herds of grazing animals. However, improper use of the land by man has caused erosion of the topsoil, thus turning thousands of acres of grassland into bare wasteland. When the grasses are destroyed, essential topsoil is worn away by water and wind.

In North America, the grasslands occur in the Great Plains and tall-grass prairies. The tropical grasslands of Africa, with populations of giraffe, zebra, antelope, ostrich, and lion, form a familiar picture to most of us. A *savannah* is a grassland with scattered trees. In the West, the oak-grass savannah is familiar. South America also has large areas of savannah. In Australia, the grasslands and savannahs support cattle and sheep for food, grazing kangaroos, and burrowing animals. Wild dogs are found very often and are prominent predators.

In America, great herds of bison and antelope once grazed on the grasses of the plains. Burrowing mammals such as hares, prairie dogs, ground squirrels, and pocket gophers are still abundant and form an important link in the food chain, for they are eaten by weasels, snakes, and hawks. Locusts and grasshoppers are important members of the insect population.

The desert biome

It may surprise you to learn that there are hot and cold *deserts*. Death Valley, where the creosote bush is the climax vegetation, is a representative hot desert. In the typical cold desert, which occurs in the temperate zone, the sagebrush is the dominant shrub (Figure 51-9). Cold deserts are found in several of our northwestern states. In both hot and cold deserts, the plants are xerophytic. That is, they are adapted for living in areas where the rainfall may be only ten inches per year. The leaves are small, with thick, leathery outer layers that help conserve the plant's water. Other desert plants, like the cacti, have no typical leaves at all. Spines are nonfunctioning leaf vestiges.

Desert animals also have special adaptations. The reptiles, some insects, and birds excrete nitrogenous wastes in the form of uric acid. Uric acid can be excreted in an almost dry form, thus conserving valuable water. Mammals, however, cannot do this. They excrete nitrogenous waste in the form of urea dissolved in water. This means that water is lost from the mammal's body. Most desert mammals burrow during the day to conserve moisture. Some rodents are able to live on the small amount of water in the seeds and fruits they eat. Others get water from the tissues of cacti or other water-storing plants.

Desert herbs, grasses, and flowering plants burst forth in growth and color in a surprisingly short time after a rain. Many

51-8 When scattered trees are present, a grassland area, such as that above in Africa, is called a savannah. (Louise Boker, National Audubon Society)

51-9 The cold desert occurs in northern latitudes. Note the sagebrush, characteristic of these areas. (Inez and George Hollis, Photo Researchers)

of these plants are able to complete their cycle of growth, flowering, and seed production within a few weeks. Some of the larger perennial plants have very long tap roots. Most desert plants can hold water in the angles between the leaves or in spongy tissue.

The rain forest

Although each *rain forest* may have its particular flora and fauna, the conditions and ecological niches are similar in all rain forests. In areas of abundant water supply and a long growing season, life flourishes. There is a *temperate rain forest* on the northwest Pacific Coast in the Olympic Peninsula of Washington. The *tropical rain forests* are found on and near the equator, including most of Central America, northern South America, central Africa, southern Asia, the East Indies, the South Pacific Islands, and northeastern Australia. The seasonal temperature variation is usually less than the temperature variation between the day and night.

The numbers of plants in a rain forest produce a dense growth. Shorter trees grow beneath tall trees. The canopy produced may be so dense that few plants can grow on the ground. Even so, many smaller plants have become adapted to life in the tropical rain forest. Some have evolved long vines that make it possible for them to have roots in the moist ground and leaves high up toward the daylight. Other small plants grow high among the trees and thereby receive sufficient light for photosynthesis. Most tropical rain forest plants have very large leaves, because the conservation of water is not a problem for them. The critical

51-10 A tropical rain forest. (Shostal)

factors are finding a place to grow and obtaining sufficient light for photosynthesis.

Epiphytes (EP-i-FITES) are plants that attach themselves to trees, sometimes one hundred feet or more above the ground. They have thick, porous roots adapted to catching and holding rainfall. Many epiphytes also have leaves arranged so that they catch water, insects, falling leaves, and other debris. As the insects decompose, essential minerals are released for the epiphyte's use. Many species of orchids, bromeliads, mosses, ferns, and lichens, are epiphytes. Even though epiphytes do not take nourishment from the plants upon which they grow, they may cause minor injury by shading the leaves of the supporting plant or causing limbs to break from their weight. Spanish moss, too, is an epiphyte.

Although animal life in the rain forest is plentiful, it may not be obvious to the casual observer in the daytime because so much of it exists high in the trees. Except for an occasional bird-call or chatter from monkeys, the day in a rain forest is relatively quiet. But toward evening, everything seems to come alive. Ants, beetles, termites, and other insects are numerous, supplying many animals with food. The crickets and tree frogs begin singing, and the birds make more noise as they search for food. The tree-dwelling monkeys chatter excitedly and howl for the last time before settling down for the night. The nocturnal carnivorous cats, such as the jaguar in South America, the leopard in Africa, and the tiger in Asia, hunt for monkeys, deer, and other animals.

Many people confuse jungle growth with a rain forest. A typical rain forest is climax vegetation. Jungle is extremely dense ground growth that occurs along the edges of rivers or on land that was once cleared by man or by some natural event like a flood or fire. If left alone, most jungle eventually becomes rain forest. A jungle is, therefore, a kind of immature rain forest.

The marine biome

To those of you who live near the seashore, this section may be familiar. To those of you who have never seen a shoreline or looked into a tidepool, this section will provide you with a better understanding of the life in the oceans. The oceans cover more than two thirds of the earth's surface and support a great bulk of living material.

As you already know, the changing tides produce a rhythmic rise and fall of the water. Near the shore, the area periodically covered and uncovered by water is called the *intertidal zone*. This is exposed at low tide but covered at high tide. Around large land masses, the continental shelf slopes gradually to a depth of about 600 feet. At this point it drops sharply to six thousand or more feet. In some areas, trenches as deep as thirty thousand feet or more occur.

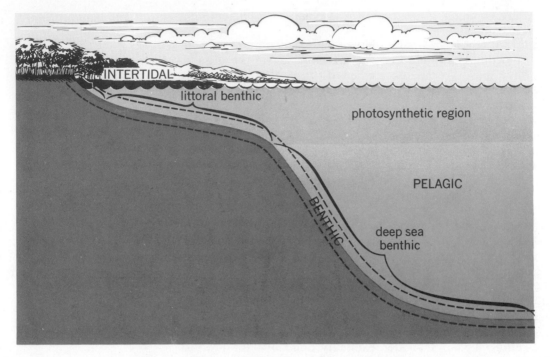

51-11 The ocean may be divided into several zones, each having characteristics that determine the kinds of organisms that are able to live therein.

The *marine biome* can be divided into the bottom, or *benthic zone,* and the ocean, or *pelagic zone* (Figure 51-11). The end of the shoreline comprising the continental shelf divides the benthic zone into the *littoral benthic zone* and the *deep sea benthic zone.* Some light can penetrate the waters above the continental shelf and the pelagic zone to a depth of about six hundred feet. This region is the most productive of the marine biome. It is in this region that photosynthesis occurs in microscopic suspended algae as well as in the large drifting algae.

Although plants in the ocean play an important role in energy production, they do not have as great an influence on the environment as do the plants on land. Compared to the terrestrial environment, the ocean provides a stable condition for life. Except in the intertidal zones, estuaries, and changing currents, the temperature and salinity remain fairly constant.

The basic food of the pelagic zone is the plankton. This is composed of diatoms, dinoflagellates, unicellular algae, protozoans, and the larval forms of many animals. Many copepods, small shrimp, small jellyfish, and worms are considered to be part of the plankton, even though they swim poorly. The copepods feed on the microscopic diatoms and algae and, in turn, provide the major food for the largest animal of all, the whale. The food chains of the ocean involve many carnivorous fishes, squids, and sharks.

Animals that live beneath the photosynthetic zone depend on sinking plankton, dead animals, and the swimming organisms that pass between the levels for food. Many of the forms that constantly inhabit the deeper dark regions of the ocean are very unusual in appearance, often having luminescent organs.

Recent studies have revealed that many animals live on the deep-sea bottom. For their nutrition, deep-sea scavengers depend on the descent of dead animals from above. Bacteria living in the soft ooze on the bottom break up complex organic molecules of dead organisms that have settled there. However, mineral exchange occurs in the ocean, as it does on the land. Currents cause upwellings of the deeper waters, and the minerals and compounds essential to life are again brought to the surface, where they can be synthesized into organic compounds by the phytoplankton and larger algae. The process of upwelling brings colder waters, rich in sedimentary materials, to the surface. Upwelling is a very conspicuous phenomenon near the coasts of Morocco, southwest Africa, California, and Peru. Since the rising waters bring nutrients to the photosynthetic zone, you can see why these areas are very rich in phytoplankton. Associated with the abundance of food-producers are the many consumers. These are very popular fishing areas.

The intertidal zone corresponds to the tropical rain forest on land. Here, in spite of exposure to drying conditions and greater variations in temperature and salinity, life is abundant. Space for growing is at a premium. Algae and many tiny colonial animals attach to the shells of snails, limpets, and kelp, as well as to rocks. Epiphytic algae are found on most kelps. Snails, periwinkles, and barnacles can be found high on rocks, where they are exposed to the air for long periods of time. They conserve water by clamping down tightly when the tide is out. Then, when covered by the tide, they browse on the algae. Other herbivores present in the intertidal zone are shrimp, small fishes, and copepods. Clams, mussels, oysters, and sponges filter many forms of microscopic life from the water. The starfish, sea anemone, larger fishes, octopus, and squid are carnivores. Sea urchins "graze" on algae. Various worms, crabs, and hermit crabs are familiar scavengers of the intertidal zone. Other worms and bacteria play an important role in breaking down waste materials and dead organisms for the recycling of essential elements.

The fresh-water biome

The *fresh-water biome* includes bodies of standing water such as lakes, ponds, and swamps, as well as bodies of moving water such as springs, streams, and rivers. Many of the plankton organisms of the ponds and lakes do not survive under conditions of running water. The strength of the current and the type of bottom are factors that determine the kinds of life able to inhabit a stream. A bottom of shifting, sandy soil greatly limits the life in a stream, but in streams with a sluggish current and a sandy bottom, many burrowing forms are found among rooted vegetation. Stony streams provide a habitat for actively swimming or clinging animals. The larvae of caddis flies, mayflies, dragonflies, and dobsonflies are common inhabitants of stony brooks and streams, where they constitute part of the trout's diet. Algae grow attached to the rocks, stumps, or stones. The plankton organisms of lakes have already been discussed. These floating organisms are found in streams only as occasional transients.

Water temperature is important to the organisms living in an aquatic environment. To raise the temperature of water requires a large amount of heat. Also, when water evaporates, a large amount of heat is absorbed from both the water and the atmosphere. These properties of water are well known to all who enjoy diving into a pool on a hot summer day. When one comes out of the pool, he may find even a warm breeze chilling as the water evaporates.

Another property of water that is very important to the inhabitants of a lake is that it becomes less dense, or lighter, as it freezes. Therefore, when water freezes, it floats. A covering layer of ice on the surface of the water prevents the lake from freezing solid, but it interferes with light penetration and oxygen exchange at the surface. In many lakes, living things can, and often do, suffocate during the cold months of the winter season.

Through osmosis, water is continually passing into fresh-water organisms. Thus, as you have learned in earlier chapters, fresh-water organisms must be able to excrete large quantities of water in order to survive. This problem is solved in protozoans by their contractile vacuoles. The efficient kidneys of the fresh-water fishes excrete the excess water that enters through the gill membranes. The problem in salt-water fishes is just the reverse. They live in water of a higher salt concentration than that in their own body fluids and are in danger of dehydration. This problem is solved in marine fishes with the excretion of salts by the gill membranes.

Another aspect of the distribution of organisms

It is fundamental that we understand why an organism is suited for living where it does, but biologists carry their questions further. They also may ask, How did this animal get here? Why are these animals similar in many ways to another animal living a thousand miles away? These questions plagued Darwin when he visited the Galápagos Islands on the H.M.S. *Beagle* in 1835. As a naturalist, Darwin was trained to observe, compare, contrast, and relate. The notes and observations he made were to play an important role in his theory of evolution and thus in the history of biology.

Although Darwin noticed similarities between the finches on the islands and those in Ecuador, he also noticed major differ-

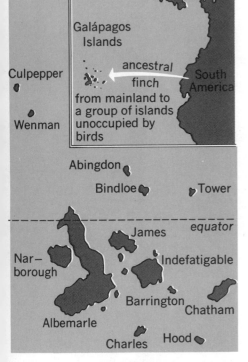

51-12 The Galápagos Islands. These islands were thrust up from the ocean floor and have never been connected to the mainland. Newly emerged, they offered a new environment for any organism that reached them.

ences in the shapes of their beaks and the specialized methods by which they obtained their food.

The Galápagos Islands are located about six hundred miles west of Ecuador (Figure 51-12). These islands were thrust up from the ocean floor and were never connected to the mainland. At the time of their formation, then, they offered a new environment for any organism reaching their shores. The six hundred miles of ocean proved an effective barrier against terrestrial organisms and most land birds. However, tropical storms may have provided a means for some flying organisms, seeds, spores, small arthropods, and drought-resistant eggs to reach the Galápagos. Natural rafts composed of uprooted trees and debris that had been washed out to sea from tropical rivers may have carried small mammals to distant shores. Undoubtedly, many such excursions have been made to the Galápagos, but most organisms must have died while en route to the unpopulated land from exposure, lack of food, or injury while landing in the wave-swept surf.

Once plants were established, the arriving animals were free to explore the several available niches. As you have already seen, competition is very keen among organisms occupying the same niche, but here on islands far from the mainland, a new arrival was free to explore many possibilities. Perhaps the first finches to survive on the islands were groundfeeders who found the grass seeds to be a plentiful source of food. Some of these finches may next have immigrated to the surrounding islands. Since the islands are separated by reasonably broad expanses of water, cross-breeding of the various island populations could not easily have occurred with any regularity.

All the small land birds on the Galápagos Islands are believed to be descendants of a type of small finch from the mainland. However, there are now several distinct species. Some of the birds are still ground finches, some feed primarily on cactus, and many are tree-dwellers. One even fills the role of a woodpecker. The woodpecker finch lacks the long tongue of the mainland woodpecker. Instead, it uses a cactus spine as a tool to probe cracks and holes to drive insects out for food. Another finch pecks at the tail feathers of larger birds and drinks the blood that flows.

The bills of these birds show variations that have developed from specialization. Some birds have bills that are thin like the warblers', and others have heavy bills that are well-suited for seed-cracking (Figure 51-13). How might the different species of birds have arisen from a common ancestor?

51-13 In studying the distribution of the finches on the islands of the Galápagos, Darwin looked for variation and adaptation. The finches he found offered an excellent example of adaptive radiation. Of the thirteen species found by Darwin, shown below are these: (a) *vegetarian tree finch,* a plant eater; (b) *large ground finch,* primarily a seed eater; (c) *tool-using,* or woodpecker, *finch,* an insect eater; (d) *large cactus ground finch,* primarily a plant eater; (e) *warbler finch,* an insect eater; (f) *small ground finch,* a plant and insect eater.

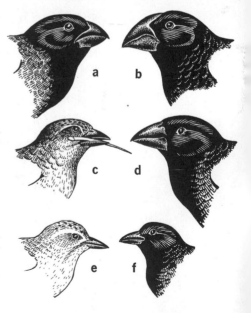

SUMMARY

The distribution of plants and animals depends on organic, climatic, and geographical factors. Dispersal involves both a method of travel and a favorable environment once a new location is reached. You are aware of the special adaptations that allow organisms to live under various conditions.

Large geographical regions, called biomes, may be characterized by climax vegetation. The major biomes are: the poles, tundra, coniferous forest, deciduous forest, grassland, desert, rain forest, marine, and fresh water. Although the poles may be grouped into the polar biome, different life-forms are found in the Arctic and the Antarctic. The Antarctic is a large isolated continent. Its temperature is frigid and its short summer months do not allow the exposure of much of its land. The North Pole, however, is located on an ice sheet, and the warmer oceanic waters below create a warmer climate. The land masses extending into the Arctic support many living things. More than a million people live in the Arctic.

The Galápagos Islands are of special interest to the biologist as they are believed to have been separated from the mainland for a long period of time. The Galápagos finches are used to illustrate theories of speciation and adaptive radiation.

BIOLOGICALLY SPEAKING

biogeography	biome	intertidal zone
geographical barrier	deciduous forest	marine biome
biotic barrier	grassland	benthic zone
climatic barrier	savannah	pelagic zone
climatic zone	desert	fresh-water biome
tree line	rain forest	

QUESTIONS FOR REVIEW

1. How do methods of dispersal limit the distribution of plants and animals?
2. Name the major barriers to dispersal and give an example of each.
3. Name and define the major climatic zones of North America.
4. What two factors are important in determining the flora and fauna within a climatic zone?
5. What reasons can you give for the great difference in the abundance of flora and fauna in the Arctic and the Antarctic?
6. Where are the coniferous forests in the United States?
7. Where are the deciduous forests in the United States?
8. Where are the grasslands in the United States?
9. What are the differences between hot and cold deserts?
10. What adaptations to desert climates are found in the mammals? in the birds? in the reptiles?
11. What environmental factors are found in a rain forest?
12. In the rain forests, the plants compete with one another for space in which to grow. How are the epiphytes adapted for living in such forests?
13. What extremes of environmental conditions occur in the intertidal zone?
14. What furnishes the food for organisms living in total darkness in the depths of the ocean?
15. What prevents the organisms in a lake from being frozen solid in the winter?
16. How has the study of the distribution of animals been used to support current evolutionary theory?

APPLYING PRINCIPLES AND CONCEPTS

1. What are some food chains found in coniferous forests?
2. Give some examples of organisms finding their way to a favorable new environment in spite of barriers.
3. Give evidence in support of the following statement: The same climatic regions are observed as one moves up a high mountain on the equator as one observes as he moves from the equator to either of the poles.
4. If you were an entomologist and found an unusual insect destroying the pineapple plants in Hawaii, what might be a good plan to follow in searching for a biotic method of control?
5. What environmental factors differ in a coniferous and a deciduous forest?
6. Discuss the adaptations of organisms in the intertidal zone.
7. How does excretion differ in salt-water and fresh-water organisms?

RELATED READING

Books

ALLEN, DURWARD L., *The Life of the Prairies and the Plains.*
McGraw-Hill Book Company, New York. 1967. Describes with vivid text and marvelous illustrations and photographs ecological relationships that exist on our great plains.

BEEBE, B. F., *Yucatan Monkey.*
David McKay Co., Inc., New York. 1967. The authentic story of the small spider monkey of the Yucatan Jungle.

BROOKS, MAURICE, *The Life of the Mountains.*
McGraw-Hill Book Company, New York. 1967. An excellent presentation on the ecology of the flora and fauna of our eastern and western areas: well-illustrated.

ERRINGTON, PAUL, *Of Predation and Life.*
Iowa State University Press, Ames, Iowa. 1967. Record of the author's experiences over half a century with wild animals of the marshes, woodlands, and other habitats of north-central U.S.

HYLANDER, CLARENCE J., *Wildlife Community: From the Tundra to the Tropics in North America.*
Houghton Mifflin Co., Boston. 1966. Environmental areas of North America as they control the various forms of wildlife found in them.

LEOPOLD, A. S., *The Desert.*
Time-Life Books (Time, Inc.), New York. 1962. Portrays the many moods of the desert and the odd ways in which plants and animals adjust to living in it.

PEDERSEN, ALWIN, *Polar Animals.*
Taplinger Publishing Co., Inc., New York. 1967. First-hand results of observations made during six years of studying and photographing animals in northeast Greenland.

PRUITT, WILLIAM O., JR., *Animals of the North.*
Harper & Row Publishers, New York. 1966. The drama of taiga life described by a naturalist-biologist as he sets forth the life cycles and interdependence of vegetation, animals, and man in the far north.

SANDERSON, IVAN, *Ivan Sanderson's Book of the Great Jungles.*
Julian Messner, New York. 1965. A bizarre, exotic world brought alive in this natural history book.

CHAPTER FIFTY-TWO
THE LAND WE LIVE IN

Environmental problems and responsibilities

When the pioneers pushed westward through the North American wilderness, they found abundant natural resources at every turn. Stands of hardwood trees extending from the Eastern coast to the Middle West composed the largest deciduous forest in the world. Beyond the forest were the prairies and plains where native grasses had occupied the land for thousands of years. Rich, fertile land with topsoil that had accumulated for centuries waited only the clearing of trees and plowing of grasses to provide fields for agriculture.

Wildlife abounded everywhere. Inland waters teemed with fish, waterfowl, and aquatic mammals. Large and small game roamed the forests and grasslands. No doubt, the early settlers thought the natural resources they found were inexhaustible.

For more than a century, it was man against nature. Waste and destruction of the natural environment continued with little or no thought of future consequences. Finally, a few decades ago, the public began to realize the penalties that must be paid. Much of the soil was gone—washed or blown away. Land that had supported forests lay in abandoned fields. Wildlife was rapidly disappearing. A wealth of natural resources had been exploited. None too soon, corrective conservation measures were started in an effort to halt the destruction of the natural environment that could lead only to disaster.

In recent years, new environmental problems arose. These problems do not involve a direct conquest of nature. They concern man's life in congested cities. They involve the artificial environment science, technology, and industry have provided. This environment includes office buildings, high rise apartments, housing projects, apartment complexes, residential communities, shopping centers, power plants, refineries, sprawling industries, and congested streets and highways. Thoughtlessness and indifference in the urban environment can be measured in terms of pollution of our rivers, lakes, and oceans and the air we breathe.

Who would have thought even a few years ago that concerned people would be asking how much longer man can survive?

Every factor of our environment is threatened today. It is appropriate that we begin with the soil, the source of the food that is basic to survival.

The loss of soil fertility

A century ago, land was still available to anyone who would go West and claim it. The soil contained a rich store of minerals accumulated through many centuries of the growth and decay of native vegetation. Fields of corn, wheat, cotton, and other crop plants rapidly replaced the native vegetation. Fields were planted year after year with little thought of **depletion** of soil minerals. When loss of soil fertility began to show in reduced crop yields, fields were abandoned for new and more profitable areas.

Continuous farming and removal of plants from a field results in another form of soil loss, the **loss of organic matter.** In a natural environment, native plants die and decompose each season. This adds humus to the topsoil. However, when crop plants are harvested and removed from the fields, little organic matter is left in the soil. Soil organisms die out and many of the processes necessary for maintaining soil fertility cease.

The scientific farmer prevents exhaustion of soil minerals by the application of *fertilizers* containing nitrates, phosphates, and potash (potassium compounds). He also practices *crop rotation* in each growing area. Many rotation plans follow a three-year cycle. The first crop might be corn, followed by wheat or oats, then by grass or clover. Clover, alfalfa, cowpeas, lespedeza, and other legumes are important in a rotation cycle because they support nitrogen-fixing bacteria on their roots. As we mentioned in the discussion of the nitrogen cycle, these bacteria form nitrates from atmospheric nitrogen. Valuable organic matter is added when these crops are plowed into the soil.

52-1 These two plots show the difference in corn growth that can result from soil treatment. The plot on the right was treated with a fertilizer containing a high nitrogen content; the other was not. (A. L. Lang)

52-2 Clover is used in crop rotation because it supports nitrogen-fixing bacteria on its roots. (Grant Heilman)

Problems of soil erosion

Of all forms of soil loss, *erosion* is the most advanced and destructive. Erosion is the loss of soil itself by the action of water and wind. Precious topsoil from millions of acres of our most productive land now lies in riverbeds and ocean bottoms. Some has been blown thousands of miles by the wind in violent dust storms. Some erosion has always occurred. Before any land was cultivated, however, soil formation generally kept pace with erosion. But when the land was stripped of its natural vegetation and poor farming methods exposed it to the forces of water and wind, a far more serious accelerated erosion began.

Water erosion and its control

In their eagerness to make every inch of land pay, many farmers have cultivated river bottoms, hillsides, and all other available locations rather than increasing yields from land in more favorable locations.

When surface water stands on exposed soil during seasonal floods, a thin layer of soil is dissolved. The receding water removes the soil in an action known as *sheet erosion.* In some cases, sheet erosion can be prevented by draining fields. Bottomlands subject to seasonal flooding should be returned to natural vegetation. This is important to flood control as well as soil conservation.

In rolling and hilly sections, where the rain falls on exposed soil, the flowing water dissolves soil and forms tiny channels, or rills. This is the beginning of *rill erosion.* Each time water flows down the slope, it follows the same rills. These deepen and widen and lead to *gully erosion.* If the gully is not checked, it may become a gulch or a canyon.

One solution to water erosion on a slope or a hillside is *contour farming.* The land is plowed across the slope of a hill rather than up and down. Each furrow serves as a small dam to check the flow of water. If this simple practice had been followed long ago, our lands today would be richer and our rivers deeper and clearer.

52-3 This hillside shows serious rill erosion. (Grant Heilman)

52-4 Soil conservation methods. Strip cropping, terracing, and gully control. (*Top, bottom right,* Grant Heilman; *bottom left,* courtesy of Pan Am)

Strip cropping frequently combines two important soil conservation measures. Broad strips are cultivated on the contour of a slope for growing *row crops,* such as corn, cotton, potatoes, or beans. These strips alternate with strips in which *cover crops* such as wheat, oats, alfalfa, clover, or grass are grown. The cover crops completely cover the surface of the soil and hold it securely. When water runs from a strip of row crop, it is checked as it enters a strip of cover crop. Crop rotation is often practiced in planting the strips.

Terracing is used to check the flow of water on cultivated steeply sloping land. A long slope is broken into a series of short ones by forming a series of banks. A terracing grader is used to form flat strips on the contour of the slope. Each strip is divided from another by a bank. Soil on the banks is held by plantings or by rocks. A drainage ditch at the base of each bank conducts water around the slope.

Gully control requires additional measures. One of these is to plant the slopes of the gully with trees, grasses, or other plants to act as soil binders and prevent further widening. Deepening can be prevented by building a series of dams across the gully. The dams slow the flow of water and settling soil gradually fills the gully.

The problem of wind erosion

Wind erosion is a serious form of soil loss, especially in the prairie and plains regions where strong winds frequently sweep across the treeless expanses. Originally, native grasses and other plants anchored the soil with their shallow, spreading root systems. However, much of this land is very fertile and the climate is ideal for growing cereal crops. Extensive areas have been plowed for agriculture.

The nation learned a costly lesson in soil conservation in the Southwest during the 1930's. During several late summer and early fall droughts, strong, hot winds blew much of the exposed topsoil away in gigantic clouds of dust. During these great dust storms, fine particles of soil filled the atmosphere as far away as the East Coast and several hundred miles over the Atlantic Ocean.

How can dust storms such as these be prevented in future years? Wind erosion is an especially difficult problem since large areas are often involved. Any local wind erosion control measure can be wiped out in a single dust storm. Consequently these projects must be undertaken on a very large scale and with the aid of state and national conservation agencies.

One such measure is the planting of windbreaks or *shelterbelts* along the margins of fields. When land is cultivated, furrows should be *plowed at right angles to the prevailing wind.* When wind blows across the furrows rather than down them, each furrow helps to stop the movement of soil. In sections where *irrigation* is possible, diversion of water into the fields during dry

52-5 A dust storm. What consequences can result from storms of this type? (Grant Heilman)

52-6 Planting rows of trees as windbreaks helps prevent serious erosion of the soil caused by unchecked wind. (Walter Dawn)

periods will check wind erosion, because moist soil does not blow away. Every inch of soil not cultivated regularly must be anchored firmly by the roots of grasses and other *soil-binding plants*.

It is imperative that dust storms be prevented in future years. Scientists have warned of possible tragic consequences. Atmospheric dust absorbs or reflects much of the sunlight before it reaches the earth. This reduction in sunlight over a period of time could lower the temperature of the earth's surface and cause another ice age.

It is interesting to know, too, that dust storms are not limited to our planet. In November, 1971, the spacecraft Mariner 9 went into orbit around the Planet Mars. Photographs from the spacecraft revealed a giant dust storm that blocked a view of the planet's surface for several weeks. The cause of this great storm is, of course, not known. However, such a storm was not occurring when photographs were received from an earlier Mariner showing the surface of Mars through a clear atmosphere.

Soil problems and water problems—a vicious circle

Disastrous flood and droughts, the two extremes in water problems, are inevitable results of the misuse of soil and its plant cover. Rains that should soak into the ground run off the surface of eroded land. Streams flood with muddy water from nearby fields. Then, there is too much water, but later in the season, there may be a severe water shortage. During floods, water washes soil away and during droughts, the wind blows it away. Soil erosion, floods, and droughts—these three disasters form a vicious circle.

The lowering of the water table is causing alarm in many parts of the country. This is due, in part, to the loss of runoff water during rainy periods. Increased demands for ground water in large cities is another cause. Numerous wells must supply much of the water for domestic use. Industries also require large amounts of water, much of which comes from wells. More and more, cities are drawing heavily on the supply of ground water as a source of cold water for air conditioning systems. This water is discharged into sewer systems and flows to rivers and lakes as surface water.

Water conservation methods

The water problem in America can never be solved until we have succeeded in reducing the amount of runoff water during heavy rains. Water conservation must begin in hilly and mountainous regions, called *watersheds*. Restoration of forests and other plant cover in these regions is imperative. Vegetation breaks the fall of the rain and accumulated humus and leaf litter soak up water like a sponge. Water stored in watersheds feeds rivers and streams as it flows in springs and rivulets.

Since rainfall is uneven and seasonal in most sections of the country, rivers will always have high and low water stages. However, a high water stage need not be a disastrous flood. As rivers rise, water should spread into flood plains, sloughs, and backwaters. As the high waters recede, these natural reservoirs should feed water back into the channel. Thus, the ground and natural surface reservoirs receive excess water during rainy periods and maintain the water supply during dry periods.

As another means of conserving water and preventing floods, large dams have been constructed in many parts of the country. These dams and the large reservoirs they form supply water for domestic and industrial use in many large cities. Many are sites of hydroelectric plants that supply electricity to large areas of the nation. By raising the water level, dams have made rivers navigable for long distances. In the West, water from reservoirs is used for irrigation and has made many semiarid regions ideal for crop production. The deep, clear lakes formed by large dams are ideal places for fishing, boating, and other water sports.

52-7   A flood may cause great damage in a residential area. (UPI)

Water pollution

Floods, droughts, and lowered water tables are only part of the water problems we face today. Hardly a stream, river, or lake is not polluted. Many are dangerously polluted. We have even polluted the oceans. Water pollution is a matter of great national concern. It is, also, a national disgrace. It is incredible that a nation such as ours with its advanced standards of health and high living standards would allow its waterways to become flowing sewers and death traps for aquatic life.

There are many sources of water pollution. We can discuss only a few of the major causes. One of the major problems is sewage and garbage disposal. During the settlement of our nation, many towns and villages were established on rivers and on the shores of large lakes. Navigation was one reason for this. Waste disposal was another. For many years, communities poured untreated sewage, garbage, and other refuse into nearby rivers, lakes, and oceans. As long as the towns were small and widely separated, bacteria decomposed organic refuse in the water and prevented high levels of pollution. Such polluting materials that are decomposed by natural processes are said to be *biodegradable*. Today, it is a different story. Towns have become large cities with dense populations. The load of sewage and other refuse poured into rivers and lakes far exceeds the rate of natural decomposition.

The pollution problem is complicated further by home disposal appliances that grind garbage and wash it into sewers. This is a much more convenient and sanitary method of eliminating garbage than collection and hauling to disposal plants as far as homes are concerned, but it adds to the problem of sewage disposal. In many areas, cannery wastes add still further to the bulk of organic matter poured into streams.

52-8   These fish died as a result of water pollution in this stream. (A. Devaney, Inc.)

52-9 Industrial waste is a major source of pollution in our environment. (A. Devaney, Inc.)

52-10 The oil polluting this beach destroys its beauty and wildlife. (Josef Muench)

Detergents, especially those with a high phosphate content, add to the pollution problem. Phosphates stimulate the growth of algae in streams and lakes. Overpopulation of algae upsets the natural balance in aquatic environments. A serious pollution problem results when such a large bulk of organic material decomposes in the water at the end of a growing season. Government regulations now restrict the content of detergents, especially phosphates.

Industrial wastes present a special water pollution problem. Power plants, steel mills, paper mills, refineries, automobile factories, and other industries often pour chemical wastes directly into streams. These include cyanide, acids, alkalis, mercury compounds, salts, solvents, and other toxic compounds. Industrial wastes present a special pollution problem since they are *nonbiodegradable* and are not decomposed by bacterial action.

Oil and other petroleum products from refineries, drilling and pumping operations, ship yards, and oil spills have destroyed wildlife and made water unfit for use in many areas. Often, there is a heavy toll of ducks and other swimming birds as well as shore birds when their feathers become saturated with oil. This may cause them to die of exposure or drown when they lose buoyancy because of the oil-soaked feathers. When oil slicks are washed to the shore, they foul the beach and cause heavy loss of shore and tidepool organisms.

Thermal pollution is the addition of heat to a river, lake, bay, or even to an area of an ocean. Power plants and other industries use water as a coolant. When the returning hot water is piped to a stream or a lake, an area of the water environment is made uninhabitable for many aquatic plants and animals.

Effects of water pollution

Destruction of aquatic life is an immediate result of water pollution. The presence of sewage, garbage, and other organic wastes in water results in rapid increase in the bacterial population to massive numbers. Bacterial decomposition of organic matter is an oxygen-consuming process. This chemical activity lowers the content of dissolved oxygen in the water to the point that many fish and other aquatic animals die of suffocation. This is especially true of the more desirable game fish with high oxygen requirements.

Chemical wastes poison aquatic plants and animals. Many of our major rivers that once supported large populations of fish are now little more than giant drains. The Great Lakes are examples of water pollution on a large scale. All of the lakes are polluted with sewage, garbage, and industrial wastes from the many large cities along their shores. The problem is probably most acute in Lake Erie. Recent surveys indicate that the many large cities and industries along its shores have been dumping up to 1 billion gallons of sewage and 9.5 billion gallons of water containing chemical wastes into the Lake each day. Fish have died out in one-third of the Lake and the death zone will increase if pollution continues.

Water pollution is also a human health hazard. Water polluted with sewage harbors bacteria and viruses that cause typhoid, dysentery, hepatitis, cholera, and other infectious diseases. Purification of polluted water for domestic use is a problem. Many cities must draw on distant water supplies because of heavy pollution in local waters.

Cases of typhoid, hepatitis, and other infectious diseases have been traced to the eating of clams, oysters, and other sea foods taken from polluted waters. It has been found, that certain fresh-water and salt water food fish consume and absorb harmful chemicals and concentrate them in their bodies. Such chemicals include mercury compounds and radioactive materials present in the water. This constitutes a serious health hazard, especially for local populations depending on the fish for a food supply.

Many recreational areas have been spoiled by pollution. For example, many of the cities on the Great Lakes provided beaches for swimming and other water recreation only a few years ago. Today, most of these beaches are closed and posted with signs warning of pollution. Many resort areas have suffered loss of business because of waters that are no longer fit for recreation.

Cleaning up our waters

Water pollution is a serious problem, but not a hopeless one. The public, generally, is aroused and willing to support corrective measures. This must begin with elimination of the sources of pollution as rapidly as possible. This will allow a slow cleaning of waters and reestablishment of plant and animal populations.

The Clean Waters Act, proposed by President Nixon, authorizes $4 billion over a four-year period as the Federal Government's share of matching funds for the construction of municipal waste treatment plants. Large industries are spending vast amounts of money to combat water pollution. The restoration of clean waters is a costly process and will require many years. However, with public support and cooperation it can be done.

Air pollution

It has been estimated that the smokestacks of industry, the chimneys of homes and apartments, the exhausts of automobiles, incinerators, jet airplanes, and other sources of pollution are pouring more than 70 billion tons of airborne wastes into the atmosphere each year. This is a staggering figure. Atmospheric pollution is an enormous problem.

We cannot escape atmospheric pollution. This is the air we must breathe. While we measure the effects of air pollution in terms of human health and discomfort, remember that plants and animals require this same air. We have no way of knowing how much wildlife has been destroyed by our carelessness and indifference in pollution of the atmosphere.

52-11   Factories such as these are sources of air pollution. (Robert Perron)

52-12   Smog over the Los Angeles Civic Center. (UPI)

Sources of air pollutants

We include all foreign materials introduced into the atmosphere as *pollutants*. Certain of these pollutants are droplets of liquids or small particles of solid materials. We refer to these as *particulates*. We see them in the air as smoke, dust, or haze. Tiny suspended droplets of pollutant materials are also called *aerosols*. We often speak of fine particles of carbon in the atmosphere as *soot*.

Various *gases,* mostly products of combustion, constitute a major problem in atmospheric pollution. Of these, *sulfur dioxide* (SO_2) is one of the most deadly. The principal source of sulfur dioxide is combustion of coal and oil. It has been estimated that the industrial smokestacks and home and apartment chimneys pour 100,000 tons of sulfur dioxide into the atmosphere each day. Sulfur dioxide is a heavy gas with a choking odor. It may combine with water in the atmosphere to form sulfuric acid (H_2SO_4). Both sulfur dioxide and sulfuric acid are damaging to man and to all other living things.

Automobile exhausts are the major source of another air pollutant, *nitric oxide* (NO). When this gas combines with oxygen in the atmosphere, a far more deadly gas, *nitrogen dioxide* (NO_2) is formed. Sunlight accelerates the process, referred to as a *photochemical reaction*. Nitrogen dioxide is irritating to the lungs and, in high concentrations, can be fatal. It combines with water to form *nitric acid* (HNO_3), that is damaging to humans as well as to plants and animals.

Carbon monoxide (CO) is a product of incomplete combustion of fuels such as coal, charcoal, wood, oil, and gasoline. The major source of this gas as an air pollutant is automobile exhausts. You are familiar with the deadly effect of this gas. Imagine the possible effects of 180,000 tons of carbon monoxide coming from the exhaust pipes of the nation's automobiles each day. This is more than 65 million tons per year!

What is smog?

As the name implies, *smog* is a combination of smoke and fog. When you think of air pollution, smog probably comes to your mind. No doubt, you have seen pictures of smog hanging low over a city or a valley.

There are several kinds of smog. In industrial communities, it usually consists of smoke containing sulfur dioxide from coal and oil combustion mingled with fog. Blast furnaces, power plants, and factories are often responsible for this type of smog.

Another type of smog is formed from gases pouring from automobile exhaust pipes. This smog contains nitrogen dioxide, hydrocarbons, carbon monoxide, and other harmful gases. The toxic substances in smog irritate the eyes and damage the lungs. They are also damaging to plants and animals.

What is a temperature inversion?

Under normal atmospheric conditions, warm air close to the earth rises and cools. Air pollutants rise with it and are dispersed through the upper atmosphere. The air is cleansed each time it rains and gases dissolve and fall to the earth. This is a normal occurrence.

From time to time, an atmospheric phenomenon known as *temperature inversion* may occur. A layer of cool air moves into an area below a layer of warm air. The warm air acts as a lid on the atmosphere and prevents the cool air from rising. Temperature inversions are most common in the fall of the year, due to atmospheric conditions.

In an area of high air pollution, a temperature inversion can be a disaster. Smog containing noxious gases is trapped in the lower atmosphere and may remain over an area for hours, days, or even weeks. Damage to health, serious illness, and death may be widespread. We have no way of knowing the extent of the damage to wildlife during such a crisis.

Reducing air pollution

It is impossible to eliminate air pollution entirely in an industrial nation such as ours. However, we can and must reduce it to a level well below the danger point. How can this be accomplished? Let's start at home. Do you burn trash in outdoor incinerators or open piles? Indoor incinerators are more efficient. Trash pick up and municipal disposal is even better. Do you burn leaves in the fall? These can be put in a compost pile or put in bags for collection if you live in a city. These are thoughtful things you can do.

We look to industry to do its part. Many industries are leading the battle against air pollution. Factories are installing anti-smoke devices. Many industries are converting from coal and oil to gas. Power plants are installing precipitators, cyclones, filters, and other devices to reduce or eliminate air pollution. Automobile manufacturers have reduced the compression in engines to burn low test gasoline and reduce the lead and other products of combustion in higher octane gasoline. Millions of dollars are being spent for research on smaller internal combustion engines and the possibility of steam power or electric power in automobiles of the future.

Pollution by radiation

Air can also be polluted with particles of radioactive materials. Some of these particles are carried great distances by air currents in the atmosphere and may settle to the earth as *fallout*.

It is not known how long some radioactive particles can remain in the upper atmosphere before they return to earth. *Strontium-90* is produced in certain nuclear explosions in the atmosphere. When this radioactive material settles on the earth, it is taken up

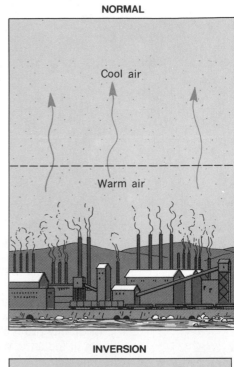

NORMAL

Cool air

Warm air

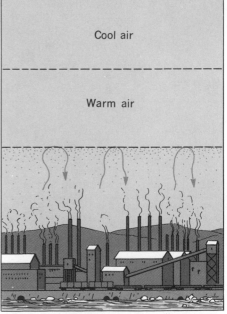

INVERSION

Cool air

Warm air

52-13 Cities with large quantities of pollutants in the atmosphere are often affected by a temperature inversion.

from the soil by plants. If these plants are eaten by animals, strontium-90 may be deposited in the flesh, milk, or bones. If the animals are consumed by humans, the strontium-90 is deposited in the bones in the same manner as calcium is deposited. Once a certain level of concentration is reached, radiations from the strontium-90 can destroy tissues, cause cancer, and even cause death.

Fortunately, radioactive materials are closely supervised and dangers are avoided. Nations have agreed not to contaminate the atmosphere with tests of atomic devices in the atmosphere. However, if there is atomic warfare in any part of the world, pollution of the atmosphere with radioactive materials could be a major disaster.

SUMMARY

It took us less than two centuries to carve a great nation out of a wilderness, but at what a price! We have led the world in science, technology, agriculture, and industry. Many generations of Americans have prospered, but what about future generations?

Many people blame the farmer for soil problems and blame industry for water and air pollution. But what about *you*? Are you adding to the problem? We all want the best quality produce from the farm at the lowest possible price. We all want the products of industry. Whether as producers or consumers, we are all involved in the environmental problems we are facing today.

A nation that can send astronauts to the moon, produce color television, and conquer polio can certainly restore its soil, clean its water, and purify its air. It can and it will, with the help of every person.

BIOLOGICALLY SPEAKING

subsoil	gully erosion	biodegradable
humus	contour farming	nonbiodegradable
peat	strip cropping	particulate
topsoil	terracing	aerosol
soil flora	wind erosion	photochemical reaction
depletion	shelterbelt	smog
crop rotation	irrigation	temperature inversion
sheet erosion	watershed	fallout
rill erosion	pollution	

QUESTIONS FOR REVIEW

1. Describe the composition of topsoil and subsoil.
2. What are soil flora?
3. How does overproduction lead to mineral depletion?
4. Describe three types of water erosion.
5. Why are cover crops and row crops alternated in strip cropping?
6. In what kind of situation would terracing be practiced?
7. Outline several methods of preventing wind erosion.
8. What is a watershed?

9. List several sources of water pollution.
10. Distinguish between biodegradable and nonbiodegradable water pollutants and give an example of each.
11. What is thermal pollution?
12. Explain how the decomposition of sewage and other organic materials in water kills fish and other aquatic animals.
13. What are particulates in the atmosphere?
14. List several toxic gases associated with air pollution.
15. What is smog?
16. Identify two sources of pollutants that form smog.
17. Describe the atmospheric conditions that cause a temperature inversion.
18. In what respect is strontium-90 a hazardous form of radioactive fallout?

APPLYING PRINCIPLES AND CONCEPTS

1. Discuss the principle of crop rotation as a soil conservation measure.
2. Discuss the combination of strip cropping, contour farming, and crop rotation as a conservation measure in a hilly agricultural area.
3. Discuss several damaging results of water pollution.
4. Discuss the chemical steps to nitric acid in smog from automobile exhausts.
5. Discuss various methods for reducing air pollution.

RELATED READING

Books

AYLESWORTH, THOMAS G., *This Vital Air, This Vital Water.*
Rand McNally and Company, Chicago. 1968. Presents the facts about air, water, and noise pollution throughout the world in clear, straightforward language.
CARR, DONALD E., *Death of the Sweet Waters.*
W. W. Norton and Co., Inc., New York. 1966. The entire story of the acute shortage of usable fresh water, and the lack of intelligent water management.
PRINGLE, LAURENCE, *The Only Earth We Have.*
The Macmillan Company, New York. 1969. Describes the dangers to the earth from air and water pollution, undisposed and disposable wastes, pesticides and from the disruption of plant-animal communities.
TALLEY, NAOMI, *To Save the Soil.*
The Dial Press, Inc., New York. 1965. Clearly outlines the nature and causes of soil erosion and traces the history of soil conservation in the U.S.
VANDERSAL, WILLIAM R., *The Land Renewed.*
Henry Z. Walck, Inc., New York. 1968. A well illustrated book describing how our land can be developed, restored, and improved for ourselves and for future generations.
WISE, WILLIAM, *Killer Smog.*
Rand McNally and Co., Chicago. 1968. Vividly describes the four days of the London smog and warns how such a pollution tragedy awaits any of our major cities if nothing is done to prevent it.

CHAPTER FIFTY-THREE
FOREST AND WILDLIFE RESOURCES

Forest and wildlife conservation

The environmental problems we face today are not limited to the soil, water, and atmosphere. For many years, we have wasted our forests and destroyed much of our wildlife. Part of this destruction was the inevitable result of agricultural development and the growth of cities. However, much of it was the result of greed, carelessness, and indifference. A *total* conservation program must include not only improvement of the physical environment but restoration of our forests and wildlife as well.

Some forest facts

The original forests of America covered nearly half our land—a total of more than 822 million acres. Forests occupied much of the eastern and western parts of our country. Prairies, plains, and arid lands covered much of the large central and southwestern area.

The two great forest belts of the East and West were, in turn, divided into distinct types of forests, as shown in Figure 53-1. The type of forest in each belt was determined by such environmental factors as temperature, rainfall, soil, and topography.

The *central hardwood forest* occupies much of eastern and central United States and extends to the prairies in the Midwest. Beech, maples, oaks, hickories, ashes, and black walnut are among the valuable timber trees of this forest. The *northern forest* lies in the region of the Great Lakes and the northeastern states and extends west to Minnesota. This forest also occupies much of Canada. Species of pine, spruce, and balsam fir mingle with birch, aspen, maples, linden, and other hardwoods in the northern forest. A mixed forest of conifers and hardwood trees extends down the Appalachian Mountains to Tennessee and North Carolina. Several species of pine mingle with hardwoods in the *southern forest* occupying the Southeastern and Gulf Coastal states.

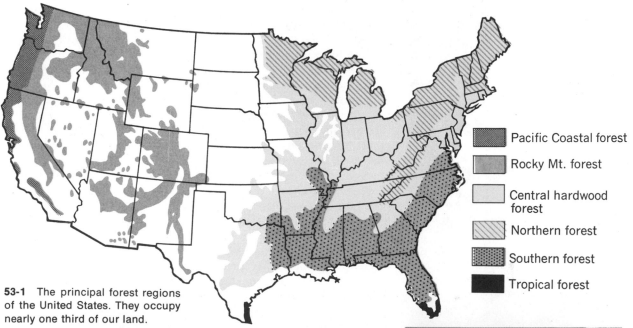

Pacific Coastal forest

Rocky Mt. forest

Central hardwood forest

Northern forest

Southern forest

Tropical forest

53-1 The principal forest regions of the United States. They occupy nearly one third of our land.

The *Rocky Mountain forest* covers the mountain slopes to the timberline. Various species of pine, spruce, fir, and larch compose much of this extensive forest. Perhaps the most magnificent of all forests is the *Pacific Coastal forest,* extending from California to Washington and far into Canada. Here, the towering Douglas fir, coast redwood, sugar pine, white fir, and western cedar form dense and valuable stands of timber.

Forests—their use and misuse

For more than a century, the greatest drain on forests has been the demand for lumber. In earlier days, hardwoods were used for construction as well as for furniture and other articles. In more recent years, conifers have supplied most of the construction lumber. The wood of conifers is ideal for construction because it is soft and easily worked.

The forests supply an enormous amount of *pulpwood,* used in making newsprint, high-quality book paper, stationery, packaging paper, and other paper products. Pulpwood comes from the coniferous forests of the United States and Canada. Hardwoods yield valuable *distillation products,* including wood alcohol, oxalic acid, charcoal, and lampblack. Beech, maple, and birch are commonly used in hardwood distillation. The Southern states lead the nation in the production of *pine products,* including turpentine, rosin, and pine tar. Other products of the forest include maple sugar and *tannic acid,* extracted from the bark of hemlock, chestnut oak, and tan-bark oak.

53-2 What differences can you see between the coniferous forest (above) and the deciduous forests? (Charlie Ott, National Audubon Society; A. Towle)

53-3 These logs of pulpwood will be used in the manufacturing of a variety of paper products. (Katherine Young)

53-4 Block-cutting removes an entire stand of timber at once. Stands of timber remaining protect the land and the new seedlings. (courtesy of Weyerhaeuser)

Timber and forest products represent only a part of the value of forests, however. We have already discussed the importance of forests in *regulating the water supply* and preventing floods, especially in watersheds. They also *prevent soil loss* by serving as soil binders. When you consider the amount of water a large tree absorbs from the ground and transpires from its leaves to the atmosphere each day, you can understand how forests *affect the climate* and precipitation over large areas.

It was inevitable that large areas of the original forests would be cleared for agriculture. This was especially true of the central hardwood forest that grew in deep, fertile soil. However, this necessary clearing is only part of the story of forest destruction. Much of the destruction was useless waste. During lumbering operations, entire forests were leveled and left an expanse of brush and stumps. Often, more than half of the timber removed was wasted. Forests were leveled in hilly regions unsuited to agriculture and the land was left to the ravages of erosion. It is tragic to look back and realize that our forests could have supplied all of the timber needed for the building of our nation without being seriously depleted if they had been used wisely.

Forest management

Fortunately, the condition of our forests was realized before it was too late. As early as 1905, officials in Washington, among them President Theodore Roosevelt, became alarmed about the condition of our forests. Accordingly, Congress created the United States Forest Service under the Department of Agriculture. Vast tracts of timber, especially in the West, were set aside as national forests.

The United States Forest Service has built a splendid record through its years of activity. State and local conservation departments have looked to this agency for guidance and assistance in carrying out forest conservation programs. A staff of foresters and biologists conducts continuous research in an effort to discover better conservation methods, controls for forest diseases and insect enemies, more efficient lumbering practices, and new uses for timber products. Owners and managers of private forest lands work closely with these agencies.

National, state, and many privately owned forests are now managed on a *sustained yield* basis. In managed forests, trees are grown as crops to be harvested at regular intervals. All of the trees are valuable timber species. Trees that are crowded, crooked, damaged, diseased, or of no timber value are removed in *improvement cutting*. As timber trees mature, they are removed in *selective cutting*. The forest is a constant source of timber, yet is never cut extensively. Mature trees are left to provide seeds for the future forest. Every foot of space produces good timber and every tree is a nearly perfect specimen.

Block-cutting is another kind of lumbering used in planted stands of timber in which all the trees are about the same age. In such lumbering, a complete block of trees is removed after which the block is replanted or reseeded. Stands of timber left around the block provide protection of the exposed land and natural reseeding. When small trees are established, another block can be cut. This method is used in stands of Douglas fir. A modified method, called *strip cutting* is often used in harvesting spruce.

Reforestation is a lengthy and costly process, but is a vital part of the conservation program. Large areas, useless for agriculture, have been cleared unwisely in past years. These regions, as well as heavily lumbered and burned-out areas should be returned to forests as rapidly as possible.

The various forest states, the United States Forest Service, and many private lumber and paper companies maintain large nurseries where seedlings of timber species are grown. The owners of nearly 48 million acres of forest land in 45 states are growing trees as a crop in the Tree Farm Program. Pine is being used in the reforestation of many hardwood forest regions because it matures rapidly and yields valuable construction lumber.

Fire, a major enemy of forests

Fire is a major tragedy in forests. Aside from the danger to human life, a fire destroys standing timber and consumes the seeds and young trees of the future forest. A large fire may even burn into the rich humus of the forest floor. The toll of animal life in a large forest fire cannot even be estimated. Disaster continues as rains pour over the blackened earth and debris and wash much of the remaining humus into streams. It may be 50 years or more before the scars of a fire are hidden in a new forest.

The causes of forest fires are shocking. According to a recent report of the United States Forest Service, the following are causes listed in the order of frequency:

Incendiarists (those who set fires deliberately).
Debris burners (those who let brush fires get out of control).
Smokers (those who throw lighted cigarettes, cigars, and matches from automobiles).
Lightning
Campers (those who leave live coals in campfires).
Railroads
Lumbering
Miscellaneous

Notice that the first three causes of forest fires are the result of deliberate destruction or carelessness. All of these fires could have been avoided. Lightning, the only natural cause, is fourth in frequency.

National and state forests are under the watchful eyes of rangers in fire towers during the danger season. Firefighters,

53-5 Fire, whether caused by lightning or the carelessness of man, destroys plant and animal life and has other, far-reaching effects. (Sally Kaicher; Sonja Bullaty; both National Audubon Society)

equipped with trucks, water tanks, chemical extinguishers, and specially equipped airplanes are standing by to fight a fire as soon as it is reported. However, full public cooperation is essential in combating this major enemy of the forest.

Wildlife conservation

The term *wildlife* includes all native animals. However, we shall limit the discussion of wildlife conservation to problems relating to fish, bird and mammal populations.

Wildlife conservation problems are directly related to other environmental problems. The decline in much of our wildlife population is the direct result of destruction of habitats. This was not an inevitable result of the development of our nation. Cities and farms can prosper without the wholesale destruction of wildlife we have witnessed in past years.

Fish conservation

There is no possibility of restoring a thriving population of desirable fishes to polluted waters. The first step in fish conservation must be the removal of sewage, garbage, chemical wastes, and other pollutants from our rivers and lakes. A second phase is restoration of aquatic environments in which fishes can thrive and multiply.

The productive stream has rapids and ripples, shallows and depths, and undercut banks. Larger rivers should be fed from productive backwaters. These are ideal spawning areas for many fish species. A productive environment must support both large and small fish as well as frogs, crayfish, aquatic insects, tiny crustaceans, protozoans, and other animals as parts of a complicated society and food chain. The aquatic environment must also support a normal population of algae and water plants.

Dams across rivers interfere with fish migration unless *fish ladders* are provided. A fish ladder is a channel around a dam through which fish can travel upstream. The fast flow of water is broken by a series of staggered plates projecting into the water from the sides. In other fish ladders, a long slope is broken into a series of steps like terraces which can be leaped by fish traveling upstream. Fish ladders are especially important in salmon streams.

With the decline in fish populations in many of our natural waters, numerous artificial lakes and ponds have been stocked with game fish. These include *farm ponds* constructed in the low corners of fields with earthen dams. Numerous small lakes have been formed by the removal of large quantities of soil, sand, and gravel for building up grades and overpasses for super highways. Many of these are stocked with bass, crappies, bluegills, and other game fish. Great numbers of artificial lakes have been formed by building dams and backing up water in valleys. Lakes of this type are ideal for fishing and boating.

53-6 This fire tower permits observation of many square miles of forest for signs of fire. (L. A. Forestry Commission, AFI)

53-7 Fish ladders are built around large hydro-electric dams to enable fish to travel upstream during their spawning season. (U.S. Department of Interior)

While artificial bodies of water may satisfy the fisherman, they can never replace the rivers, streams, and natural lakes as environments for fishes and other aquatic animal life.

Conservation of our valuable allies—the birds

The advancing tide of civilization has made dangerous inroads into the populations of many songbirds and game birds. Much of this destruction was useless and avoidable.

The cutting of forests, clearing of underbrush, and burning of fields have destroyed large areas that served as bird habitats. Unnecessary drainage of marshes and lowering of the water in ponds and lakes have deprived water birds and wading birds of both food and nesting sites.

In former years, many thousands of birds were slaughtered for flesh or feathers. Fortunately, such market hunting is forbidden by both state and federal laws. But conservation measures came too late to save some species that were once common but are now extinct.

The extinction of the passenger pigeon is a story familiar to most people. In the early 1800's, Audubon described flocks of passenger pigeons so large they darkened the sky. For many years, these birds were slaughtered in their roosting areas, gathered in sacks to be sold for a few cents, or fed to hogs. An introduced epidemic disease may have eliminated the last survivors. Certainly no thought was given to the conservation of this species until it was too late. The last survivor of what is estimated to have been over two billion passenger pigeons died in the Cincinnati Zoo in 1914.

Endangered bird species

Whether a bird species can survive in a changing environment depends on several factors. Reproductive habits are important, as are adjustments to different food supplies. Some bird species have actually increased with civilization. Others are holding their own. Several, however, are endangered and may soon become extinct unless rigid conservation measures to save them are successful. These include the ivory-billed woodpecker, which may now be extinct, the prairie chicken, the California condor, numbering about 60, the American (Southern) bald eagle, reduced to less than 1,500, the Everglades kite of which about 60 remain, the osprey, brown pelican, and the whooping crane.

It is true that large numbers of plant and animal species have flourished and died out through the ages. Perhaps the extinction of some species in our time is a normal occurrence. Yet biologists hate to see this happen, especially if we are partially responsible, if something can be done about it. Something *is* being done to save the whooping crane.

In spite of close watching and protection, native whooping cranes have not been able to increase their numbers to any extent

53-8 Some endangered birds. (William Finley, Allan Cruickshank, National Audubon Society; George H. Harrison from Grant Heilman)

for many years. In 1966, there were only 43 wild whooping cranes. Seven birds in captivity raised the total known number to 50. In 1964, an artificial propagation program was started in Patuxent, Maryland. With an agreement between the Bureau of Sport Fisheries and Wildlife and the Canadian Wildlife Service, eggs were collected from wild nests for incubation and rearing of the young at the research center.

The whooping crane normally lays two eggs and often rears only one of the chicks that hatch. It was found that collection of one egg from several nests did not reduce the wild population. The collected eggs are shipped by jet airplane to the research center in special rubber-lined cases warmed with hot water bottles. After hatching in incubators, the chicks are given specially prepared diets. Captive birds will be returned to the wild flocks. By 1971, the total number of whooping cranes had increased to 21 in captivity and 59 in the wild — a total of 80 birds, or an increase of 60 percent. These figures are very encouraging. Perhaps other endangered bird species can be saved by similar methods.

Problems relating to game birds

Migratory ducks and geese are favorite *game birds* for hunters. However, the decline in certain species of these birds has been due only in part to hunting. A major factor has been the draining of marshes which supply both food, such as wild rice, and nesting sites. Many have died in oil slicks and chemically poisoned waters. In the western prairies, large numbers of ducks have died of alkali poisoning in ponds and lakes when the water is low and salts become concentrated. In some regions, thousands of ducks and geese have died of botulism when they consumed the deadly toxin in probing for food in the mud of ponds and lakes when the water is low.

Upland game birds, including the quail, partridge, grouse, wild turkey, and pheasant are widely hunted by sportsmen. These birds can survive regulated hunting only if adequate food and cover are provided, especially during the winter months. Many farmers allow trees, shrubs, and tall grasses to grow along their fence rows as cover. A few rows of grain left at the margin of fields provide winter feed for these and other birds.

The insecticide dilemma

One of the more recent problems seriously affecting our wildlife populations, especially fish and birds, is the indiscriminate use of *insecticides*. It is nearly impossible to grow agricultural crops, fruit, garden vegetables, flowers, or lawns without the use of chemical poisons to protect the plants from insect attack. We also use them widely to kill mosquitoes, flies, roaches, and other household pests. In our chemical warfare on insects, we have dusted, sprayed, and filled the atmosphere with insect-killing aerosols. Unfortunately, the killing effects of insecticides have gone far beyond the harmful insects we intended to destroy. Valuable insects, such as bees and mantids are destroyed in the process. Insecticides penetrate the soil and wash into streams with ground water where they poison fish and other aquatic animals.

Many insecticides contain mercury, lead, and other heavy metals. These are retained in the tissues of animals that consume them and are built up in concentration from animal to animal in a food chain. A top carnivore may contain high concentrations of these lethal substances. Large numbers of birds, frogs, snakes, and other insect-eating animals are, undoubtedly, poisoned in this manner. Several species of predatory fish have been found to contain dangerous concentrations of mercury compounds and other poisons in their tissues. Fortunately, recent laws prohibit the sale of insecticides containing mercury and other heavy metals that are cumulative in animal tissues.

DDT—a special problem

Near the close of World War II, a newly developed insecticide, DDT (dichloro-diphenyl-trichloroethane) was used effectively in Northern Italy to kill body lice and control an epidemic of typhus. DDT became popular immediately. It was used as a powder and a spray to kill mosquitoes, household pests, garden pests, and agricultural pests.

DDT has a rapid paralyzing effect on insects. It is, also, absorbed and retained in their exoskeletons. Since it breaks down very slowly, its effects are long lasting. This was once thought to be a benefit, but this very quality has caused great concern. When DDT is sprayed into the atmosphere, it settles on the ground and is washed into streams and carried to the oceans. Traces of DDT have been found in ocean water as far away as the Antarctic. In an aquatic environment, it clings to algae and other aquatic plants that are consumed by herbivorous animals. As in the case of other insecticides, concentrations of DDT increase from one animal to another in the food chain and may reach lethal concentrations in a top level carnivore.

Bird populations are suffering from the effects of DDT in still another way. When concentrations build up, especially in insect-eating and predatory birds, the reproductive system and lime-producing glands are affected. This causes the bird to lay thin-shelled eggs that are easily broken in the nest. This has been observed in the nests of eagles and the California brown pelican.

As DDT and other dangerous insecticides are banned from the market, the question is how we can continue to control insect pests. Undoubtedly, certain insecticides must be used. However, they must be used carefully and must be of a type that will not be a danger to wildlife.

53-9 The beaver is a valuable fur-bearing mammal. (Leonard Lee Rue III, National Audubon Society)

53-10 The American bison, once on the brink of extinction, has slowly increased in number only because it has been protected on large game preserves. (Leonard Lee Rue III, National Audubon Society)

The ideal answer to the insect problem is *biological control*. By conserving such insects as the praying mantis and ladybird beetle, we are using insects to fight insects. Fish destroy great numbers of aquatic insects. Frogs, toads, snakes, and other small animals are also valuable in insect control. However, birds are our best allies. Yet, we are killing these animals with chemical poisons. This is the insecticide dilemma.

Conservation of smaller mammals

The first drain on the smaller mammals, especially the valuable fur-bearers, occurred early in our history. Long before the pioneers began their journey westward, trappers had explored the wilderness in search of fortunes in furs. The prize, especially in the Pacific Northwest, was the beaver. "Empire builder" is an appropriate name for this valuable fur-bearer which played a major role in the settlement of this vast wilderness country. The pelts of beaver, mink, otter, martin, muskrat, fox, raccoon, and other fur-bearing mammals brought an annual revenue of over $100 million to the early trappers.

Today, rigid laws protect these valuable mammals. Much of the supply of pelts for the fur garment industry is coming from fur farms and ranches, thus reducing the demand for natural skins.

Protected areas, such as parks and residential communities have surprisingly large mammal populations. These include squirrels, raccoons, opossums, groundhogs, chipmunks, and other smaller mammals. Residents are often not aware of these mammals because they are active at night. However, animals killed on city streets and highways are evidence of their numbers. The loss of animals on highways is a problem for which there seems to be no solution.

Protection of larger mammals

The slaughter of the plains bison is a story of senseless waste. A century ago, herds of bison numbering many thousands thundered across the Great Plains. Then came the buffalo hunters. Bison were slaughtered by the hundreds of thousands for pure sport or to feed the crews of workmen building railroads through the West. At one time, these herds were reduced to a few hundred stragglers and it appeared that the bison might vanish entirely.

The white-tailed deer and black bear vanished from most of the eastern and central states many years ago. Elk, mule deer, and antelope declined rapidly in the West.

Today, due to conservation measures, the situation is greatly improved. White-tailed deer populations have been reestablished in the forest areas of many eastern and central states. Herds of bison, elk, mule deer, and antelope are increasing in the West. The problem, now, is limiting the population of large mammals to the available ranges and food supply. This requires scientific game management.

National parks and national forests

The twenty-nine national parks with their six million acres are to be preserved forever. They are part of the game preserves and recreational areas dedicated to those who seek undisturbed and unspoiled natural beauty.

In addition, there are 149 national forests covering more than 181 million acres. The national forests supply timber for about 12 percent of the nation's forest products. They provide homes for wildlife and ranges for herds of big game mammals. During the summer months, when drought strikes the Great Plains, national forests are opened for supervised cattle and sheep grazing. In addition, they provide recreational areas for campers and sportsmen who fish in their lakes and streams.

SUMMARY

The forests and wildlife of America belong to you. What you do to preserve them will be your contribution to future generations. If you live on a farm or ranch, think of the birds and animals along your fence rows or in the brush or woodland you have not cultivated. If you live in the city, protect the squirrels in the park, the rabbits in the vacant lot, and the birds that come to your yard. If you fish and hunt, respect the laws and the limits. Be sure your campfire is really out. The fewer signs of human presence you leave behind you in a natural area, the more you are doing to maintain the beauty of America. Generations to come will appreciate your thoughtfulness.

BIOLOGICALLY SPEAKING

sustained yield	wildlife	upland game bird
block cutting	fish ladder	insecticide
strip cutting	farm pond	biological control
reforestation	game bird	

QUESTIONS FOR REVIEW

1. List the major forest belts and regions in the United States.
2. List various forest products and identify each with a forest region.
3. Distinguish between improvement cutting and selective cutting in a forest sustained yield program.
4. What is block-cutting?
5. List seven causes of forest fires and indicate whether each cause is deliberate, the result of carelessness, accidental, or natural.
6. Explain the purpose of a fish ladder.
7. List eight endangered bird species.
8. Describe the work at the research center in Patuxent, Maryland to increase the population of whooping cranes.
9. List several reasons for decline in the populations of certain wild ducks and geese.

10. Describe several far-reaching effects of the use of DDT in insect control.
11. What is meant by biological control of insects?

APPLYING PRINCIPLES AND CONCEPTS

1. Discuss the management of a forest on a sustained yield basis.
2. Discuss the role of sportsmen in a wildlife conservation program.
3. Discuss possible reasons why some bird species are now endangered while others are maintaining their numbers or even increasing.
4. In what respect are many insecticides a biological dilemma?
5. Discuss various management problems in the restoration of populations of large mammals, such as deer, elk, and bison.

RELATED READING

Books

BARKER, WILL, *Freshwater Friends and Foes.* Acropolis Books, Washington, D.C. 1967. A conservation-minded book on how pollution affects wildlife.

GRAHAM, FRANK JR., *Since Silent Spring.* Houghton Mifflin Co., Boston. 1970. Describes the background of Rachel Carson's *Silent Spring* and armed with her warnings, goes on to tell how the pesticide problem stands today.

GRAHAM, ADA, AND FRANK GRAHAM JR., *Wildlife Rescue.* Cowles Book Company, Inc., New York. 1970. Tells about four special people who have devoted their lives to the rescue of wild animals whose existence has been threatened by "progress."

LAYCOCK, GEORGE, *America's Endangered Wildlife.* W. W. Norton and Co., New York. 1969. A provocative, well illustrated book in which the author describes a long list of threatened animals, how and where they live, why they are endangered, and what is being done to save them.

LAYCOCK, GEORGE, *Wild Refuge.* The Natural History Press, Garden City, New York. 1969. Takes the reader on a tour of the more important wildlife refuges as he reads of conservationists' efforts to help preserve America's vanishing wildlife.

McCoy, J. J., *Saving Our Wildlife.* Crowell-Collier Press, New York. 1970. Tells of the destruction of much North American wildlife and describes the attempts that have been made to save the wildlife resources we have now.

PINNEY, ROY, *Wildlife in Danger.* Duell, Sloan, and Pearce, New York. 1966. Describes why animals, including birds are extinct or in danger of vanishing.

STOUTENBERG, ADRIEN, *A Vanishing Thunder.* Doubleday and Co., Inc., Garden City, N.Y. 1967. A dramatic account of vanishing birdlife in America, based on naturalists' records.

abdomen in arthropods, the body region posterior to the thorax.

abiogenesis (or *spontaneous generation*), a disproved belief that certain nonliving or dead materials can be transformed into living organisms.

abscission layer the two rows of cells near the base of a leaf petiole that are involved in the natural falling of the leaf.

absorption the process by which water and dissolved substances pass into cells.

acetylcholine a chemical substance released at the motor end plate, causing muscle contraction.

actin slender filaments arranged in bundles in the composition of myofibrils of muscles.

active absorption water intake through the mechanism of osmosis.

active transport requiring the expenditure of energy, the passage of a substance through a cell membrane.

adaptation an adjustment to conditions in an environment.

adaptive behavior a set of behavior patterns that tends to insure the well-being of an animal.

adaptive radiation a branching out of a population through variation and adaptation to occupy many environments.

addiction the body's need for a drug that results from the use of the drug.

adductor muscles those in bivalves that control the opening and closing of the valves.

adenoid a mass of lymph tissue that grows from the back wall of the nasopharynx.

ADP (adenosine diphosphate) a low-energy compound found in cells that functions in energy storage and transfer.

adrenal glands two ductless glands located one above each kidney.

adsorption the process by which a thin film of one substance adheres to the surface of another substance.

adventitious root one that develops from the node of a stem or from a leaf.

aerobic requiring free atmospheric oxygen for normal activity.

aerosols tiny suspended droplets of pollutants in the air.

agglutinin an immune substance in the blood that causes specific substances, including bacteria, to clump.

agglutinogen a protein substance on a blood cell's surface that is responsible for blood types.

air bladder a thin-walled, elliptical sac found in fish that allows the animal to maintain a level in the water.

air sacs in insects, the enlarged spaces in which the tracheae terminate; in birds, cavities extending from the lungs; in man, thin-walled divisions of the lungs.

alimentary canal those organs that compose the food tube in animals and man.

allantois an extraembryonic membrane that in birds and reptiles aids in respiration and excretion.

allele one of a pair of genes responsible for contrasting traits.

alternation of generations a type of life cycle in which the asexual reproductive stage alternates with the sexual reproductive stage.

altricial birds usually laying fewer than 6 eggs.

alveoli microscopic protrusions in the lungs in which the exchange of gases takes place.

amebocytes amebalike cells in sponges that function in circulation and excretion.

amino acids substances from which organisms build protein; the end products of protein digestion.

ammonification the release of ammonia from decaying protein by means of bacterial action.

amnion the innermost fetal membrane, forming the sac that encloses the fetus.

amnionic fluid secreted by the amnion and filling the cavity in which the embryo lies.

amniote egg an egg laid on land by a reptile or bird, having an amnion, other membranes, and a shell.

anaerobic deriving oxygen for life activity from chemical changes and, in some organisms, being unable to live actively in free oxygen.

analogous organs those that are similar in function.

anaphase a stage of mitosis during which chromosomes migrate to opposite poles.

annual a plant that lives for one season only.

annulus the ring on the stipe of a mushroom that marks the point where the rim of the cap and the stipe were joined.

antenna a large "feeler" in insects and certain other animals.

antennule a small "feeler" in the crayfish and certain other animals.

anther that part of the stamen which bears pollen grains.

antheridium a sperm-producing structure found in some plants.

anthocyanin a red, blue, or purple pigment dissolved in cell sap.

antibiotic a germ-killing substance produced by a bacterium, mold, or other fungus.

antibody an immune substance in the blood and body fluids.

anti-codon the triplet nucleotide base code in transfer RNA which is the opposite of the messenger RNA codon.

antigen a substance, usually a protein, which when introduced into the body stimulates the formation of antibodies.

antipodals three nuclei found in the embryo sac at the end farthest from the micropyle.

antitoxin a substance in the blood that counteracts a specific toxin.

antivenin a serum used against snakebite.

anus the opening at the posterior end of the digestive tube or alimentary canal.

aorta the great artery leading from the heart to the body.

apical cell the terminal, or tip, cell of a growing plant.

appendage an outgrowth of the body of an animal, such as a leg, fin, or antenna.
appendicular skeleton skeletal members forming the limbs of a vertebrate.
aquatic living in water.
aqueous humor the watery fluid filling the cavity between the cornea and lens of the eye.
arachnoid mater the middle of the three membranes of the brain and spinal cord.
archegonium an egg-producing structure found in some plants.
arteriole a tiny artery that eventually branches to become capillaries.
artery a large, muscular vessel that carries blood away from the heart.
ascus In Ascomycetes, the saclike structure that contains the spores.
asexual reproduction reproduction without eggs and sperms.
assimilate to incorporate digested and absorbed molecules into the makeup of an organism.
association fibers nerve processes connecting different parts of the cerebral cortex.
asters the fibrils that form and radiate from the centriole like rays from a star during cell division.
ATP (adenosine triphosphate) a high-energy compound found in cells that functions in energy storage and transfer.
atrioventricular node the structure in the heart that relays the beat to the muscles of the lower heart.
atrioventricular valves the heart valves located between the atria and ventricles.
atrium a thin-walled upper chamber of the heart that receives venous blood.
auditory nerve the nerve leading from the inner ear to the brain.
autonomic nervous system a division of the nervous system that regulates the vital internal organs.
autosome any paired chromosome other than the sex chromosomes.
autotrophs organisms capable of organizing organic molecules from inorganic molecules.
auxin a plant hormone that regulates growth.
axil the angle between a leaf stalk and a stem.
axon a nerve process that carries an impulse away from the nerve body.
bacillus a rod-shaped bacterium.
bacteria a group of microscopic, one-celled protists.
bacteriolysin an antibody found in the blood that causes a specific kind of bacteria to dissolve.
bacteriophage one of several kinds of viruses that can destroy bacteria.
barrier anything that prevents the spread of organisms to a new environment.
basidium a club-shaped structure found in the club fungus which bears the spores.
belly a term used to refer to the body of a striated muscle.
benthic zone bottom part of the marine biome.
biennial a plant that lives two seasons.
bile a brownish-green emulsifying fluid secreted by the liver and stored in the gall bladder.
binary fission the division of cells into two approximately equal parts.
binomial nomenclature the system that gives an organism a name composed of two parts.
biodegradable polluting materials decomposed by natural processes.
biogenesis the biological principle that life arises from life.
biogeography the study of the distribution of plants and animals throughout the earth.
biome a large geographical region identified mainly by its climax vegetation.
biosphere the area in which life is possible on our planet.
biosynthesis the organization of organic molecules by living organisms.
biotic community all the living organisms in an ecosystem.
bivalve a mollusk possessing a shell of two valves hinged together.
blade the thin, green portion of a leaf, usually strengthened by veins.
blastocoel the space between the ectoderm and endoderm in the early stages of development.
blastula an early stage in the development of an embryo, in which cells have divided to produce a hollow sphere.
bony layer the hard region of a bone between the periosteum and the marrow.
book lungs air-filled respiratory sacs in spiders.
Bowman's capsule the cup-shaped structure forming one end of the tubule and surrounding a glomerulus in the nephron of a kidney.
brain stem an enlargement at the base of the brain where it connects with the spinal cord.
branchial arteries those that lead to and from the gills of the fish.
breathing the mechanical process of getting air into and out of the body.
bronchial tube a subdivision of a bronchus within a lung.
bronchiole one of numerous subdivisions of the bronchial tubes within a lung.
bronchus a division of the lower end of the trachea, leading to a lung.
Brownian movement an oscillating or bouncing movement seen in colloids or bacteria and due to random movement of molecules.
bud an undeveloped shoot of a plant, often covered by scales.
bud scale a small, leaflike structure covering the growing point of a twig.
budding the uniting of a bud with a stock; also, a form of asexual reproduction in yeast and hydra.
bulb a form of underground stem composed largely of thick scale leaves.
bulbus arteriosus a muscular, bulblike structure attached to the ventricle of the fish heart.
bulk feeders animals that consume food as whole or parts of organisms.
bursa a fluid-filled sac forming a cushion in the knee and shoulder joints.
Calorie (food) used to measure food energy, the amount of heat required to raise the temperature of 1,000 grams (1 kilogram), or 1 liter, of water one centigrade degree.

calyx   the sepals, collectively.
cambium   a ring of meristematic cells in roots and stems that forms secondary xylem and phloem.
canine   (or *cuspid*) teeth for tearing, enlarged in certain mammals.
capillarity   a force causing water to move upward through a tube with a small diameter.
capillary   small, thin-walled vessel through which exchanges occur between blood and tissue fluid.
capsule   a thick slime layer surrounding some bacteria; also, the spore case of mosses.
carapace   the hard-shield covering on the back of an animal such as a crab, lobster, and turtle.
cardiac muscle   muscle composing the heart wall.
carnivore   a meat-eating organism.
carotene   a reddish-orange pigment in chloroplasts.
carotid arch   anterior branch of the truncus arteriosus in the frog.
cartilage   a strong, pliable supporting tissue in vertebrates.
catalyst   a substance that accelerates a chemical reaction without itself being altered chemically.
cell   a unit of structure and function of an organism.
cell theory   a belief that the cell is the unit of structure and function of life and that cells come only from preexisting cells by reproduction.
cell wall   the outer, nonliving cellulose wall secreted around plant cells.
cementum   the covering of the root of a tooth.
central nervous system   the brain and spinal cord and the nerves arising from each.
central neuron   a nerve cell in the brain or spinal cord that connects a motor and a sensory nerve.
centriole   a cytoplasmic body lying just outside the nucleus in animal cells.
centromere   a single granule that, during cell division, attaches a pair of chromatids after replication.
centrosome   a small, dense area of cytoplasm appearing outside the nucleus during division of animal cells.
cephalization   concentration of nervous tissue at the anterior end of an organism.
cephalothorax   a body region in crustaceans and certain other animals, consisting of the fused head and the thorax.
cerebellum   the brain region between the cerebrum and medulla, concerned with equilibrium and muscular co-ordination.
cerebrospinal fluid   a clear fluid in the brain ventricles and surrounding the spinal cord.
cerebrum   the largest region of the human brain, considered to be the seat of emotions, intelligence, and voluntary nervous activities; also present in all other vertebrates.
cervix   the neck of the uterus.
chelicera   first appendages of the spider, serving as poison fangs.
cheliped   a claw foot used in food-getting and for protection in some arthropods.
chemosynthesis   the organization of organic compounds by organisms by means of energy from inorganic chemical reactions instead of energy from light.
chemotropism   the response of an organ or an organism to chemicals.
chitin   a material present in the exoskeleton of insects and other arthropods.
chlorenchyma   chlorophyll-containing parenchyma cells.
chlorophyll   green pigments essential to food manufacture in plants.
chloroplast   a cell plastid containing chlorophyll.
cholinesterase   a chemical substance released at the motor end plate that neutralizes acetylcholine.
chorion   a membrane that forms early during development and attaches, in mammals, to the uterine wall. In birds and reptiles, this membrane is found under the shell.
choroid layer   the second and innermost layer of the eyeball.
chromatid   during cell division, each part of a double-stranded chromosome, joined by a centromere, after replication.
chromatin material   fine strands in the nucleoplasm believed to be forms of the chromosomes.
chromatophores   structures containing pigments in the skin of fishes, frogs, and other animals.
chromoplasts   pigment-containing bodies, other than chloroplasts, in certain plant cells.
chromosomal aberration   an alteration in the structure or number of a chromosome.
chromosome   a rod-shaped gene-bearing body in the cell nucleus, composed of DNA joined to protein molecules.
chrysalis   the hard covering of the pupa of a butterfly.
chyme   partly digested, acidic food as it leaves the stomach.
cilia   tiny, hairlike projections of cytoplasm.
ciliary muscles   those that control the shape of the lens.
cleavage   the rapid series of divisions that the fertilized egg undergoes.
cleavage furrow   an indentation that appears during the telophase of dividing animal cells.
climax plant   a species that assumes final prominence in a region.
clitellum   a swelling on the earthworm when involved in reproduction.
cloaca   a chamber below the large intestine in certain vertebrates into which waste is emptied.
coccus   a sphere-shaped bacterium.
cochlea   the hearing apparatus of the inner ear.
cocoon   a silken case containing the pupa of a moth.
codon   one or more groups of base triplets of messenger RNA that code a specific amino acid.
coelom   the space between the mesodermal layers that forms the body cavity of an animal.
coenzyme   a nonprotein molecule that works with an enzyme in catalyzing a reaction.
cohesion   the clinging together of molecules, as in a column of liquid.
coleoptile   a protective sheath encasing the primary leaf of the oat and other grasses.

collar cells   flagellated cells in sponges that set up water currents.

colloid   a gelatinous substance, such as protoplasm or egg albumen, in which one or more solids are dispersed through a liquid.

colon   the large intestine.

commensalism   one organism living in or on another, with only one of the two benefiting.

companion cells   long, narrow, nucleated cells bordering sieve tubes in phloem tissue.

compound eye   an eye composed of numerous lenses and containing separate nerve endings.

conditioned reaction   an acquired behavior pattern in which a particular response regularly follows a specific stimulus.

conidia   a name given to spores of molds.

conifer   a cone-bearing gymnosperm.

conjugation   a primitive form of sexual reproduction in *Spirogyra* and certain other algae and fungi in which the content of two cells unites; also, an exchange of nuclear substance in the paramecium.

connective tissue   a type of tissue that lies between groups of nerve, gland, and muscle cells.

conservation   the preservation and wise use of natural resources.

consumers   the heterotrophs in an environment.

contractile vacuole   a large cavity in protozoans associated with the discharge of water from the cell and the regulating of osmotic pressure.

contraction (muscle)   the shortening of muscle fibers.

control   in testing a hypothesis, experiments are usually conducted in duplicate, since the control lacks the experimental variable.

conus arteriosus   a large vessel lying against the front side of the frog's heart.

convergent evolution   the type in which organisms of entirely different origin evolve in a manner that results in certain similarities.

convolution   one of many irregular, rounded ridges on the surface of the brain.

cork   a tissue formed by the cork cambium that replaces the epidermis in woody stems and roots.

cork cambium   a layer of cells in the outer bark of a woody stem that produces cork tissue.

corm   a shortened underground stem in which the leaves are reduced to thin scales.

cornea   a transparent bulge of the sclerotic layer of the eye in front of the iris, through which light rays pass.

corolla   the petals of a flower, collectively.

corpus luteum   refers to the follicle after the ovum is discharged.

cortex   in roots and stems, a storage tissue; in organs such as the kidney and brain, the outer region.

cotyledon   a seed leaf present in the embryo plant that serves as a food reservoir.

countershading   a form of protective coloration in which darker colors on the upper side of the animal fade into lighter colors on the lower side.

Cowper's gland   located near the upper end of the male urethra. It secretes a fluid which is added to the sperms.

cranial nerves   the twelve pairs of nerves communicating directly with the human brain.

cristae   membranous infolded partitions in mitochondria.

crop   an organ of the alimentary canal of the earthworm, bird, and certain other animals that serves for food storage.

crossing over   the exchange of segments of two chromosomes, and the genes in the segments, when the two chromosomes twist around each other during synapse in meiosis.

cuticle   a waxy, transparent layer covering the upper epidermis of certain leaves; also, the outer covering of an earthworm.

cyclosis   flowing, or streaming, of cell cytoplasm.

cyst   in lower animals and plants, a spore with a capsule covering constituting a resting stage in some algae; a resting stage of a protozoan or lower animal, enclosed by a protective wall.

cytolysis   the swelling and bursting of cells when put into a hypotonic medium.

cytoplasm   the protoplasmic materials lying outside the nucleus and inside the cell membrane.

cytoplasmic matrix   clear-appearing portion of the cytoplasm that suspends various visible bodies.

daughter cells   newly formed cells resulting from the division of a previously existing cell, called a mother cell. The two daughter cells receive identical nuclear materials.

decay   the reduction of the substances of a plant or animal body to simple compounds.

deciduous   woody plants that shed their leaves seasonally.

decomposers   organisms that break down the tissues and excretions of other organisms into simpler substances through the process of decay.

deficiency disease   a condition resulting from the lack of one or more vitamins.

dehydration   loss of water from body tissues.

dendrite   a branching nerve process that carries an impulse toward the nerve cell body.

denitrification   the process carried on by denitrifying bacteria in breaking down ammonia, nitrites, and nitrates and liberating free nitrogen.

dentine   a substance that is relatively softer than enamel, forming the bulk of a tooth.

deoxygenation   the process during which oxygen is removed from the blood or tissues.

depressant   a drug having an anesthetic effect on the nervous system.

dermis   the thick, active layer of tissue lying beneath the epidermis.

diaphragm   a muscular partition separating the thoracic cavity from the abdominal cavity.

diastole   part of the cycle of the heart during which the ventricles relax and receive blood.

dicotyledon   a seed plant with two seed leaves, or cotyledons.

diffusion the spreading out of molecules in a given space from a region of greater concentration to one of lesser concentration.

digestion the process during which foods are chemically simplified and made soluble for use.

dihybrid an offspring having genes for two contrasting characters.

diploid term used to indicate a cell that contains (or an organism whose cells contain) a full set of homologous pairs of chromosomes.

diurnal active during the day.

division of labor specialization of cell functions resulting in interdependence.

division plate a wall of cellulose that forms across the dividing plant cell, forming a common boundary between daughter cells.

DNA (deoxyribonucleic acid) a supermolecule consisting of alternating units of nucleotides, composed of deoxyribose sugar, phosphates, and nitrogen bases.

dominance principle first observed by Mendel, that one gene may prevent the expression of an allele.

dormancy a period of inactivity.

double cross genetic process in which four pure-line parents are mixed in two crosses.

double fertilization in a spermatophyte one sperm fertilizes the egg and one unites with the polar nuclei to form the endosperm nucleus.

ductus arteriosus a connection between the pulmonary artery and the aorta during fetal life.

duodenum the region of the small intestine immediately following the stomach.

dura mater the outer of the three membranes of the brain and spinal cord.

ecology the study of the relationship of living things to their surroundings.

ecosystem a unit of the biosphere in which living and nonliving things interact, and in which materials are used over and over again.

ectoderm the outer layer of cells of a simple animal body; in vertebrates, the layer of cells from which the skin and nervous system develop.

ectoplasm the outer layer of thin, clear cytoplasm, as in the ameba.

egestion elimination of insoluble, nondigested particles from a cell.

egg a female reproductive cell.

elongation region the area behind the embryonic region of a root or stem in which cells grow in length.

embryo a developing organism.

embryo sac the tissue in a plant ovule that contains the egg, the antipodals, the polar nuclei, and the synergids.

embryonic membrane one of the delicate coverings of a developing organism.

embryonic region the area near the tip of a root or stem in which cells are formed by division.

enamel the hard covering of the crown of a tooth.

endocrine gland a ductless gland that secretes hormones directly into the blood.

endoderm the inner layer of cells of a simple animal body; in vertebrates, the layer of cells from which the linings of the digestive system, liver, lungs, and so on develop.

endodermis a single layer of cells located at the inner edge of the cortex of a root.

endoplasm the inner portion of cytoplasm, as in the ameba.

endoplasmic reticulum a complex system of parallel membranes that extend from the plasma membrane to the nuclear membrane of a cell and that function as a system of canals.

endoskeleton internal framework of vertebrates composed of bone and/or cartilage.

endosperm the tissue in some seeds containing stored food.

endosperm nucleus the body formed by the fusion of a sperm with the polar nuclei during the double fertilization of a spermatophyte.

endospore a resting stage of some bacteria that is resistant to adverse environmental conditions.

endotoxins insoluble poisons formed by certain bacteria within the cells.

environment surroundings of an organism; all the external forces that influence the expression of an organism's heredity.

enzyme an organic catalyst.

epicotyl in a seed, the part of the embryo plant that lies above the attachment of the cotyledons and from which the stem and leaves will develop.

epidermis the outer tissue of a young root or stem, a leaf, and other plant parts.

epiphyte a plant that grows on a tree or other plant.

epithelial tissue that composing the covering of various body organs.

epiglottis a cartilaginous flap at the upper end of the trachea.

erosion the loss of soil by the action of water or wind.

esophagus the food tube, or gullet, that connects the mouth and the stomach.

estivation a period of summer inactivity in certain animals.

ethology the study of behavior patterns in animals.

etiolated a condition of pale yellow leaves and stems of plants when they are grown in the dark.

Eustachian tube a tube connecting the pharynx with the middle ear.

evaporation movement of water in the form of water vapor from the earth to the atmosphere.

evolution the slow process of change by which organisms have acquired their distinguishing characteristics.

excretion the process by which metabolic waste materials are removed from living cells or from the body.

excurrent pore in sponges, the osculum.

excurrent siphon the structure in the clam through which water passes out of the body.

exoskeleton the hard outer covering or skeleton of certain animals, especially arthropods.

exotoxin a soluble toxin excreted by certain bacteria and absorbed by the tissues of the host.
experimental factor (also called *single variable*) the one condition involved in testing a hypothesis, and differing from the control in an experiment.
expiration the discharge of air from the lungs.
extensor a muscle that straightens a joint.
extraembryonic membrane one that functions during development of reptiles, birds, and mammals, but that does not become part of the embryo.
eyespot the sensory structure in the euglena and planarian that is believed to perceive light and dark.
Fallopian tube oviduct in the mammal.
fallout radioactive particles that settle on the Earth from the atmosphere.
fang a hollow tooth of a poisonous snake, through which venom is ejected.
feces intestinal solid waste material.
feedback an operating mechanism in the body that produces a delicate check-and-balance system.
fermentation glucose oxidation that is anaerobic and in which lactic acid or alcohol is formed.
fertilization the union of sperm and egg.
fetus mammalian embryo after the main body features are apparent.
fiber an elongated, thick-walled strengthening tissue in a woody plant.
fibrin a substance formed during blood clotting by the union of thrombin and fibrinogen.
fibrinogen a blood protein present in the plasma, involved in clotting.
fibrovascular bundle in higher plants, a strand containing xylem and phloem.
filament a stalk of a stamen, bearing the anther at its tip; in algae, a threadlike group of cells.
fission asexual reproduction of unicellular organisms by division into two equal daughter cells.
flaccid limp, due to loss of turgor pressure.
flagellate an organism bearing one or more whiplike appendages, or flagella.
flagellum a whiplike projection of cytoplasm used in locomotion by certain simple organisms and by the sperms of many multicellular organisms.
flexor a muscle that bends a joint.
follicle an indentation in the skin from which hair grows; also, a mass of ovarian cells that produces an ovum.
food any substance absorbed into the cells of the body that yields material for energy, growth, and repair of tissue and regulation of the life processes, without harming the organism.
food chain the transfer of the sun's energy from producers to consumers as organisms feed on one another.
food pyramid a quantitative representation of a food chain, with the food producers forming the base and the top carnivore at the apex.
food web complex food chains existing within an ecosystem.
forebrain that part of the brain composed of the cerebrum.
fossil the imprint or preserved remains of an organism that once lived.
fovea a small, sensitive spot on the retina of the eye where cones are specially abundant.
fragmentation a process whereby pieces of an organism may break off and regenerate into whole organisms; an asexual method of reproduction.
fresh-water biome a body of standing water or a river or stream in which life can exist.
fungus a protist lacking chlorophyll and therefore deriving nourishment from an organic source.
gall bladder a sac in which bile from the liver is stored and concentrated.
gamete a male or female reproductive cell, or germ cell.
gametophyte the stage that produces gametes in an organism having alternation of generations.
gamma globulin a blood protein sometimes used to give temporary immunity to diseases.
ganglion a mass of nerve cells lying outside the central nervous system.
gastric caecum a pouchlike extension from the stomach of a grasshopper.
gastric fluid glandular secretions of the stomach.
gastrovascular cavity the central cavity of Coelenterata.
gastrula a stage in embryonic development during which the primary germ layers are formed.
gemmule a coated cell mass produced by the parent sponge and capable of growing into an adult sponge.
gene that portion of a DNA molecule that is genetically active and capable of replication and mutation.
gene frequency the extent to which a gene occurs in a population.
gene linkage the assemblage of genes in a linear arrangement on a chromosome.
gene pool all the genes present in a given population.
generative nucleus the nucleus in a pollen grain that divides to form two sperm nuclei.
genetic code the sequential arrangement of the bases in the DNA molecule, which controls traits of an organism.
genetics the science of heredity.
genotype the hereditary makeup of an organism.
geotropism the response of plants to gravity.
germination growth of a seed when favorable conditions occur.
gestation period the period between fertilization and the birth of a mammal.
gibberellins growth regulating substances promoting cell elongation in plants.
gill an organ modified for absorbing dissolved oxygen from water; in mushrooms, a platelike structure bearing the reproductive hyphae and spores.
gill arch a cartilaginous structure in fishes to which the gill filaments are attached.
gill filament one of many threadlike projections forming the gills in fishes.
gill raker fingerlike projections of the gill arches in fishes.

gill slits in chondrichthyes, openings to the gills; in chordates, openings in the throat that appear during embryological development.

girdling removal of a ring of bark down to the cambium layer and thus causing death of the tree.

gizzard an organ in the digestive system of the earthworm and birds modified for grinding food.

glomerulus the knob of capillaries in a Bowman's capsule.

glottis the upper opening of the trachea in land vertebrates.

glycolysis (lactic acid fermentation) the conversion of pyruvic acid to lactic acid by an organism when ample oxygen is not available; that is, under anaerobic conditions.

Golgi apparatus small groups of parallel membranes in the cytoplasm near the nucleus.

gonads the male and female reproductive organs in which the gametes are produced.

grafting the union of the cambium layers of two woody stems, one of the stock and the other of the scion.

grana disklike bodies in chloroplasts.

green gland an excretory organ of crustaceans.

guanine crystals excretory products causing color and found in the scales, skin, eyes, and air bladders of fishes.

guard cell one of the two epidermal cells surrounding a stoma.

gullet the passageway to a food vacuole in paramecia; the food tube or esophagus.

guttation the forcing of water from the leaves of plants, usually when the stomata are closed.

habitat place where an organism lives.

halophyte a plant living in soil with a high salt content, often above that of seawater.

haploid a term used to indicate a cell, such as a gamete, that contains only one chromosome of each homologous pair; also, an organism having cells of this type.

hard palate forms the roof of the mouth in the chewing area.

Haversian canals numerous channels penetrating the bony layer of a bone.

hemoglobin an iron-containing protein compound giving red corpuscles their color.

hemotoxin a poison that destroys red blood cells and breaks down the walls of small blood vessels.

Henle's loop a widened loop of a kidney tubule that enters the cortex.

hepatic portal vein a vessel carrying blood to the liver before the blood returns to the heart.

herbaceous an annual stem with little woody tissue.

herbivores plant-eating animals.

heredity the transmission of traits from parents to offspring.

hermaphroditic having the organs of both sexes.

heterocysts in *Nostoc*, empty cells with thick walls that enable the filaments to break into pieces.

heterogametes male and female gametes that are unlike in appearance and structure.

heterotrophs organisms that are unable to synthesize organic molecules from inorganic molecules; that is, nutritionally dependent on other organisms or their products.

heterozygous refers to an organism in which the paired genes for a particular trait are different.

hibernate to spend the winter months in an inactive condition.

hilum the scar on a seed where it was attached to the ovary wall.

hindbrain that part of the brain that is composed of the cerebellum, the pons, and the medulla.

holdfast the special cell at the base of certain algae that anchors them to the substrata.

homeostasis a steady state that an organism maintains by self-regulating adjustments.

homoiothermic warm-blooded, as applied to birds and mammals.

homologous organs those similar in origin and structure but not necessarily in function.

homologue a single chromosome of a homologous pair.

homozygous refers to an organism in which the paired genes for a particular trait are identical.

hormone the chemical secretion of a ductless gland producing a definite physiological effect.

host in a parasitic relationship, the organism from which the parasite derives its food supply.

humus organic matter in the soil formed by the decomposition of plant and animal remains.

hybrid an offspring from a cross between parents differing in one or more traits.

hybridization (or *outbreeding*) the crossing of different strains, varieties, or species to establish new genetic characteristics.

hydrolysis the chemical breakdown of a substance by combination with water.

hydrophytes plants that grow in water or partially submerged in water in very wet surroundings.

hydrotropism the response of roots to water.

hypertonic solution a solution containing a higher concentration of solutes and a lower concentration of water molecules than another solution.

hypha a threadlike filament of the vegetative body of a fungus.

hypocotyl that part of a plant embryo from whose lower end the root develops.

hypothesis a scientific idea or working theory.

hypotonic solution a solution containing a lower concentration of solutes and a higher concentration of water molecules than another solution.

ileum the third and longest region of the small intestine.

immune therapy the assistance and stimulation of the natural body defenses in preventing disease.

immunity the ability of the body to resist a disease by natural or artificial means.

incisor one of the cutting teeth in the front of both jaws in mammals.

incomplete dominance a blend of two traits, resulting from a cross of these characteristics.

incubation the providing of ideal conditions for growth and development.

incurrent pore one of many holes in the sponge through which water passes into the animal.

incurrent siphon   the structure in a clam through which water passes into the body.

indehiscent   one of a class of dry fruits that do not open to discharge seeds at maturity.

independent assortment, law of   a law based on Mendel's hypothesis that the separation of gene pairs on a given pair of chromosomes, and the distribution of the genes to gametes during meiosis, are entirely independent of the distribution of other gene pairs on other pairs of chromosomes.

individual characteristics   traits that are inherited but that make an organism different.

innate behavior   inborn behavior.

inoculation   the voluntary addition of bacteria, viruses, or other microorganisms to a culture medium.

insertion   the attachment of a muscle at its movable end.

inspiration   the intake of air into the lungs.

instinct   a natural urge, or drive, not depending on experience or intelligence.

intelligent behavior   activities of an organism involving problem-solving, judgment, and decision.

interdependence   the dependence of cells on other cells for complete functioning, or of organisms on the activities of other organisms.

interface   the molecular boundary of a colloid, such as cytoplasm.

interferon   a cellular chemical defense against a virus.

internode   the space between two nodes.

interphase   the period of growth of a cell that occurs between mitotic divisions.

intertidal zone   the area of the beach that is periodically covered and uncovered by water.

intestinal fluid   a digestive secretion of the intestinal glands.

invertebrate   an animal lacking a backbone.

involuntary muscle   one that cannot be controlled at will, like smooth muscle.

iris   the muscular, colored portion of the eye, behind the cornea and surrounding the pupil.

irritability   the ability to respond to a stimulus.

islets of Langerhans   groups of cells in the pancreas that secrete insulin.

isogametes   male and female gametes that are structurally similar.

isolation   the confinement of a population to a certain location because of barriers.

isotonic solution   a solution containing the same concentration of solutes and the same concentration of water molecules as another solution.

isotopes   different forms of an element resulting from varying numbers of neutrons.

Jacobson's organs   tiny pits containing nerve endings sensitive to odors and located close to the front of the roof of the mouth of a snake.

jejunum   a section of the small intestine lying between the duodenum and the ileum.

kidney   an excretory organ that excretes urine.

kinetic energy   energy that is actively expressed.

kinin (also *cytokinin*)   a plant hormone, but one not transported out of the cell.

labium   the lower mouthpart, or "lip," of an insect.

labrum   the two-lobed upper mouthpart of an insect.

lacteal   a lymph vessel that absorbs the end products of fat digestion from the intestinal wall.

larva   an immature stage in the life of an animal.

larynx   the voice box; also called the "Adam's apple."

lateral bud   a bud that develops at a point other than at the end of a stem.

lateral line   a row of pitted scales along each side of the fish, functioning as a sense organ.

layering   propagation by stimulating the growth of roots on a stem, as in burying a stem in the ground and then cutting it when roots form.

leaf scar   a mark on a twig left at the point of attachment of a leaf stalk in a previous season.

leaf trace   the continuous passageway formed from bundles of conducting tissues branching from the vascular region of the stem, then passing through the petiole and into the leaf.

leaflet   a division of a compound leaf.

leafy stem   a term used to denote one of the stages of a moss plant.

lenticel   a small pore in the epidermis or bark of a young stem through which gases are exchanged.

lethal gene   one that bears a characteristic that is usually fatal to the organism.

leucoplast   a colorless plastid serving as a food reservoir in certain plant cells.

ligament   a tough strand of connective tissue that holds bones together at a joint.

limiting factor   any factor that is essential to organisms and for which there is competition.

line breeding   (or *inbreeding*)   the crossing of closely related strains of plants or animals to preserve certain genetic traits.

liver   the largest gland in the human body, associated with digestion and sugar metabolism.

lung   an organ for air breathing and external respiration in higher animals.

lymph   the clear, liquid part of blood that enters the tissue spaces and lymph vessels.

lysis   the dissolution or destruction of cells.

lysogenic phage   a virus that invades a bacterial cell without causing immediate destruction. It is passed along to daughter bacteria and becomes destructive at a later time.

lysosome   a spherical body within the cytoplasm of most cells.

lysozyme   an enzyme that dissolves the cell walls of many bacteria.

lytic cycle   the cycle of a virulent phage resulting in destruction of a bacterial cell.

macronucleus   the large nucleus of the paramecium and certain other protozoans.

Malpighian tubules   long excretory tubules attached to the junction of the stomach and intestine of the grasshopper, and that collect nitrogenous wastes from the blood.

mammary glands   those found in female mammals that secrete milk.

mandible   a strong, cutting mouthpart of arthropods; a jaw, as in the beak of a bird or the jawbone structure of a mammal.

mantle   a thin membrane covering the visceral hump of a mollusk; in some, it secretes a shell.

marine   inhabitants of salt water.

marsupial   a pouched mammal.

mass selection   the picking of ideal plants or animals from a large number to serve as parents for further breeding.

matrix   a gelatinous secretion of cells of *Nostoc* and certain other blue-green algae.

maturation region   the area of a root or stem where embryonic cells differentiate into tissues.

maxilla   a mouthpart of an arthropod; the upper jaw of vertebrates.

maxillary teeth   small, conical teeth projecting from the upper jawbone of the frog.

maxilliped   a "jaw foot," or first thoracic appendage, of the crayfish and other arthropods.

medulla   in the kidney, the inner portion containing pyramids that, in turn, contain numerous tubules; in the adrenal gland, the inner portion that secretes epinephrine.

medulla oblongata   the enlargement at the upper end of the spinal cord, at the base of the brain.

medusa   the bell-shaped, free-swimming form in the jellyfish.

megaspore mother cells   diploid cells in the plant ovary that divide twice, forming four haploid megaspores.

megaspores   four cells formed from the megaspore mother cell, three of which disintegrate and one of which develops into the embryo sac.

meiosis   the type of cell division in which during oogenesis and spermatogenesis there is reduction of chromosomes to the haploid number.

meninges   the three membranes covering the brain and spinal cord.

menstruation   the periodic breakdown and discharge of uterine tissues that occur in the absence of fertilization.

meristematic tissue   small, actively dividing cells that produce growth in plants.

mesentery   a folded membrane that connects to the intestines and the dorsal body wall.

mesoderm   the middle layer of cells in an embryo.

mesoglea   a jellylike material between the two cell layers composing the body of a coelenterate.

mesophyll   photosynthetic tissue composed of chlorenchyma cells and located between the upper and lower epidermis of a leaf.

mesophytes   plants that occupy neither extremely wet nor extremely dry surroundings.

mesothorax   the middle portion of the thorax of an insect, bearing the second pair of legs and the first pair of wings.

messenger RNA   the type of RNA that is thought to receive a code for a specific protein from the DNA in the nucleus and to act as a template for protein synthesis on the ribosome.

metabolism   the sum of the chemical processes of the body.

metamorphosis   a marked change in structure of an animal during its growth.

metaphase   the stage of mitosis in which the chromosomes line up at the equator.

metathorax   the posterior portion of the thorax of an insect, bearing the third pair of legs and the second pair of wings.

micronucleus   a small nucleus found in the paramecium and certain other protozoans.

micropyle   the opening in the ovule wall through which the pollen tube enters.

microspore mother cells   diploid cells in the anther that divide twice, forming four haploid microspores.

microspores   four cells, formed from the microspore mother cell, that develop into pollen grains.

midbrain   that part of the brain composed of nerve fibers connecting the forebrain to the hindbrain.

middle lamella   a thin plate, composed largely of pectin, forming the middle portion of a wall between two adjacent plant cells.

midrib   the large central vein of a pinnately veined leaf.

migration   seasonal movement of animals from one place to another.

milt   sperm-containing fluid of fishes.

mimicry   a form of protective coloration in which an animal closely resembles another kind of animal or an object in its environment.

mitochondria   rod-shaped bodies in the cytoplasm known to be centers of cellular respiration.

mitosis   the division of chromosomes preceding the division of cytoplasm.

molar   a large tooth for grinding, highly developed in herbivores.

molting   shedding of the outer layer of exoskeleton of arthropods, or of a scale layer of reptiles.

monocotyledon   a flowering plant that develops a single seed leaf, or cotyledon.

monoecious   bearing staminate and pistillate flowers on different parts of the same plant.

monohybrid   an offspring from a cross between parents differing in one trait.

monosomy   the presence of a single homologous chromosome in all body cells.

monotreme   an egg-laying mammal.

mosaic   a pattern of leaf arrangement for greatest light exposure.

mother cell   a cell that has undergone growth and is ready to divide.

motor end plate   the terminus of the axon of a motor nerve in a muscle.

motor neuron   one that carries impulses from the brain or spinal cord to a muscle or gland.

motor unit   the nerve cell and the individual muscle fibers it stimulates to contract.

mucous membrane   a form of epithelial tissue that lines the body openings and digestive tract and secretes mucus.

mucus   a slimy secretion of mucous glands.

multiple alleles   one of two or more pairs of genes that act together to produce a specific trait.

muscle tissue   cells that are specialized to contract and cause movement.

mutation   a change in genetic makeup resulting in a new characteristic that can be inherited.

mutualism   a form of symbiosis in which two organisms live together to the advantage of both.

mycelium   the vegetative body of molds and other fungi, composed of hyphae.

mycoplasmas   a group of viruslike organisms.

mycorrhiza   a fungus that lives in a symbiotic relationship with the roots of trees and other plants.

myofibrils   fine, parallel threads arranged in groups to form a muscle fiber.

myosin   a form of thick filament that together with actin filaments composes a myofibril.

narcotics   a group of drugs that have a pronounced effect on the nervous system and that are addictive with continued use.

nare   nostril.

nastic movement   turgor movement in plants, such as the daily opening and closing of flowers.

natural selection   the result of survival in the struggle for existence among organisms possessing those characteristics that give them an advantage.

nematocyst   a stinging cell in coelenterates.

nephridia   the excretory structures in worms, mollusks, and certain arthropods.

nephron   one of the numerous excretory structures in the kidney, including Bowman's capsule, the glomerulus, and the tubules.

nerve cord   part of the central nervous system in chordates, lying above the notochord and extending down the dorsal side of the body.

nerve impulse   an electrochemical stimulus causing change in a nerve fiber.

nervous tissue   specialized cells for transmitting impulses for coordination, perception, or automatic body functions.

neuron   a nerve cell body and its processes.

neurotoxin   a poison that affects the parts of the nervous system that control breathing and heart action.

niche   the particular role played by organisms of a species in relation to those of other species in a community. No two species occupy the same niche.

nictitating membrane   a thin, transparent covering, or lid, associated with the eyes of certain vertebrates, such as birds; a third eyelid.

nitrification   the action of a group of soil bacteria on ammonia, producing nitrates.

nitrogen cycle   a series of chemical reactions in which nitrogen compounds change form.

nitrogen fixation   the process by which certain bacteria in soil or on the roots of leguminous plants convert free nitrogen into nitrogen compounds that the plants can use.

nocturnal   an organism active at night.

node   a growing region of a stem from which leaves, branches, or flowers develop.

nondisjunction   the failure of homologous chromosomes to segregate during meiosis.

"nonsense" codon   genetically inactive nucleotide bases at the end of a strand of DNA.

notochord   a rod of cartilage running longitudinally along the dorsal side of lower chordates and always present in the early embryological stages of vertebrates.

nuclear membrane   a living membrane surrounding the nucleus.

nucleolus   a small, spherical body within the nucleus.

nucleoplasm   the dense, gelatinous living content of the nucleus.

nucleotide   a unit composed of a ribose or deoxyribose sugar, a phosphate, and an organic base. Many such units make up an RNA or DNA molecule.

nucleus   the part of the cell that contains chromosomes; also, the central mass of an atom, containing protons and neutrons.

olfactory lobe   the region of the brain that registers smell.

olfactory nerve   the nerve leading from the olfactory receptor endings to the olfactory lobe.

oogenesis   the process of the development of female reproductive cells whereby the diploid chromosome number is reduced to the haploid.

oogonium   an egg-producing cell in certain thallophytes.

ootid   a cell that matures into an egg.

operculum   the gill cover in fishes.

opsonins   antibodies that prepare bacteria for ingestion by phagocytic cells.

optic lobe   the region of the brain that registers sight.

optic nerve   the nerve leading from the retina of the eye to the optic lobe of the brain.

oral groove   a deep cavity along one side of the paramecium and similar protozoans.

organ   different tissues grouped together to perform a function or functions.

organelles   specialized structures present in the cell.

organic variation   differences that occur among individual organisms within a species.

organism   a complete and entire living thing.

origin   the attachment of a muscle at its immovable end.

osculum   an opening in the central cavity of sponges through which water leaves the animal.

osmosis   the diffusion of water through a semipermeable membrane from a region of greater concentration of water to a region of lesser concentration.

osmotic system   the separation of two different solutions by a selectively permeable membrane.

ossification   the process by which cartilage cells are replaced by bone cells, resulting in a hardening of the body framework as the organism grows.

ovary the basal part of the pistil containing the ovules; a female reproductive organ.
oviduct a tube in a female through which eggs travel from an ovary.
oviparous producing offspring from eggs hatched outside the body, as, birds and most fish.
ovipositor an egg-laying organ in insects.
ovoviviparous bringing forth the young alive after they have developed without placental connection.
ovule a structure in the ovary of a flower that can become a seed when the egg is fertilized.
pancreas a gland located near the stomach and duodenum that is both endocrine and digestive.
pancreatic fluid a digestive secretion of the pancreas.
papilla a small elevation on the skin of an animal; projections on the tongue containing taste buds.
parasite an organism that takes nourishment from a living host.
parasympathetic nervous system a division of the autonomic nervous system.
parathyroid one of the four small ductless glands embedded in the thyroid.
parenchyma the thin-walled, soft tissue in plants forming cortex and pith.
parotid gland one of the salivary glands in the side of the face in front of and below the ear.
parthenogenesis the development of an egg without fertilization.
passive absorption water intake through a root by the forces of transpiration and cohesion.
passive transport the movement of molecules by their own energy during diffusion.
pasteurization the process of killing and/or retarding the growth of bacteria in milk and alcoholic beverages by heating to a selected temperature so that the flavor is retained.
pathogenic disease-causing.
pectoral girdle the framework of bones by which the forelimbs of vertebrates are supported.
pedicel the stalk that supports a single flower.
pedipalps the second pair of head appendages in spiders.
pelagic zone the deep part of the marine biome.
pellicle a thickened membrane surrounding the cell of a paramecium.
pelvic girdle the framework of bones by which the hind limbs of vertebrates are supported.
pelvis the hip girdle; in man consisting of the ilium, ischium, and pubis bones; also, the central portion of a kidney.
perennials plants that grow through more than two growing seasons.
pericardium the membrane around the heart.
pericycle the tissue in roots from which secondary roots arise.
periderm the corky layer forming the outer edge of a root after secondary thickening.
periodicity alternating periods of activity.
periodontal membrane the fibrous structure that anchors the root of the tooth in the jaw socket.
periosteum the tough membrane covering the outside of a bone.
peripheral nervous system the nerves communicating with the central nervous system and other parts of the body.
permeable membrane one that allows substances to pass through it.
petal one of the colored parts of the flower. (In some flowers the sepals are also colored.)
petiole the stalk of a leaf.
phage a bacteriophage, or virus, that reproduces in a bacterium.
phagocytic cells those that engulf bacteria and digest them by means of enzymes.
pharynx the muscular throat cavity, extending up over the soft palate and to the nasal cavity.
phenotype the outward appearance of an organism as the result of gene action.
phloem the tissue in leaves, stems, and roots that conducts dissolved food substances.
photochemical reaction a chemical reaction accelerated by sunlight.
photolysis the splitting of water molecules by light energy in the photo phase of photosynthesis.
photoperiodism the dependence of some plants on the relation between the length of light and the length of darkness in a given day.
photoreceptor an organ that is sensitive to light.
photosynthesis the process by which certain living plant cells combine carbon dioxide and water in the presence of chlorophyll and light energy, to form carbohydrates and release oxygen.
phototropism the response of plants to light.
phycocyanin the bluish pigment found in blue-green algae.
phycoerythrin a red pigment dissolved in the cell sap of certain red algae.
pia mater the inner of the three membranes of the brain and spinal cord.
pinocytosis the engulfing of large particles in pockets in a cell membrane.
pistil the part of a flower bearing the ovary at its base.
pith a storage tissue of roots and stems consisting of thin-walled parenchyma cells.
pituitary gland a ductless gland composed of two lobes, located beneath the cerebrum.
placenta a large, thin membrane in the uterus, in the area of the chorionic villi, that transports substances between the mother and developing young by means of the umbilical cord.
plankton minute floating organisms suspended near the surface in a body of water that serve as food for larger animals.
planula a ciliated swimming larva of a jellyfish.
plaques holes in a colony of bacteria resulting from destruction of cells by a bacteriophage.
plasma the liquid portion of blood tissue.
plasma membrane a thin, living membrane located at the outer edge of the cytoplasm.
plasmodium name given to the vegetative body of a typical slime mold.
plasmolysis the collapse of cell protoplasm due to loss of water.

plastids living bodies in the cytoplasm of plant cells.

plastron the lower shell of the turtle.

platelet the smallest of the solid components in the blood, releasing thromboplastin in clotting.

pleural membrane one of two membranes surrounding each lung.

plexus a mass of nerve cell bodies.

poikilothermic cold-blooded, in reference to certain animals.

polar nuclei the two nuclei in the embryo sac in flowers that fuse with one of the sperm nuclei to form the endosperm nucleus.

pollen the microgametophyte produced in the anther of a spermatophyte.

pollen sacs structures in the anther containing pollen grains.

pollen tube the tube formed by a pollen grain when it grows down the style of a pistil.

pollination the transfer of pollen from anther to stigma.

pollution the addition of impurities.

polyp one of the stages in the life cycle of coelenterates.

polyploidy the condition in which cells contain more than twice the haploid number of chromosomes.

pons a part of the hindbrain located in the brain stem.

population a group of individuals of any one kind of organism in a given ecosystem.

portal circulation an extensive system of veins that lead from the stomach, pancreas, small intestine, and colon, then unite and enter the liver.

potential energy energy that has not been actively released as, for example, chemical energy.

precipitin a blood antibody that causes bacteria to settle out in lymph nodes and other body filters.

precocial birds laying many eggs, often 12–20, and starting incubation when the last egg is laid.

predator any animal that preys on other animals.

premolars large teeth for grinding.

primary germ layers the ectoderm, endoderm, and mesoderm.

primary oocyte the structure in the female gonads that divides to form the secondary oocyte and first polar body.

primary root the first root of the plant to issue from the seed.

primary tissues the first tissues developing from the meristematic region.

proboscis a tubular mouthpart in certain insects; also, the trunk of an elephant.

proglottid a segment of a tapeworm's body.

prolegs in caterpillars, the extra pairs of fleshy legs at the end of the abdomen.

propagation the multiplication of plants by vegetative parts.

prophase the stage of mitosis in which chromosomes shorten and appear distinctly double and the nuclear membrane disappears.

prostate a gland located near the upper end of the urethra in the male.

prostomium a kind of upper lip in the earthworm.

protein synthesis a universal phase of cell anabolism whereby protein molecules are built up from amino acid molecules.

prothallus the tiny, heart-shaped gametophyte that develops from the spore of the fern.

prothorax the first segment of an insect's thorax, to which are attached the head and first pair of legs.

protonema a filamentous gametophyte structure produced by a spore in mosses.

protoplasm organized complex system of substances found in living organisms.

proventriculus first portion of the stomach of a bird.

pseudocoel a cavity or false coelom that is not lined with specialized covering cells.

pseudopodium a "false foot" of the ameba or amebalike cells.

psychedelic drugs a class of drugs, including LSD, that affect the mind by changing the perception of all the senses, the interpretation of space and time, and the rate and content of thought.

puberty the age at which the secondary sex characteristics appear.

pulmocutaneous arches posterior branches of the truncus arteriosus in the frog from which arteries to the lungs and skin arise.

pulmonary pertaining to the lungs.

pulse regular expansion of the artery walls caused by the beating of the heart.

Punnett square a grid system used in computing possible combinations of genes resulting from random fertilization.

pupa the stage in an insect having complete metamorphosis that follows the larva stage.

pupil the opening in the front of the eyeball, the size of which is controlled by the iris.

pyloric caeca pouches extending from the upper end of the intestine in fishes.

pyloric valve a sphincter valve regulating the passing of substances from the stomach to the duodenum.

pyramid a conical projection in the medulla of the kidney.

pyrenoid a small protein body that serves as a center for starch formation in *Spirogyra*.

quadrate bone the bone in the snake's skull to which the lower jaw is attached.

quarantine isolation of plants or animals to prevent the spread of infection.

radicle embryonic root in a seed.

radioactive refers to an element that spontaneously gives off radiations.

radioautography a method of labeling cells with an isotope, and then exposing a photographic plate to show the location of the radioactive material.

reaction time the elapsed time between the moment when a stimulus is received and that in which a response occurs.

receptacle the end of the flower stalk bearing the reproductive structures.

receptor a cell, or group of cells, that receives a stimulus.

recessive refers to a gene or character that is masked when a dominant allele is present.

rectum the posterior portion of the large intestine, above the anus.

red corpuscles (or *erythrocytes*) disk-shaped blood cells containing hemoglobin.

red marrow found in flat bones and the ends of long bones; forms red corpuscles and certain white corpuscles.

reduction division the reduction of chromosomes during meiosis from the diploid number to the haploid number.

reflex action a nervous reaction in which a stimulus causes the passage of a sensory nerve impulse to the spinal cord or brain, from which, involuntarily, a motor impulse is transmitted to a muscle or gland.

regeneration the ability of organisms to form new parts.

renal relating to the kidneys.

replication self-duplication, or the process whereby a DNA molecule makes an exact duplicate of itself.

reproduction the process through which organisms produce offspring.

resolution the production under a microscope of a visible image in which details can be seen.

respiration the exchange of oxygen and carbon dioxide between cells and their surroundings, accompanied by oxidation and energy release.

response the reaction to a stimulus.

reticuloendothelial system fixed phagocytic cells in small blood vessels of the liver, spleen, and other organs.

retina the inner layer of the eyeball, formed from the expanded end of the optic nerve.

Rh factor any one of six or more protein substances found on the surface of the red blood cells of most people.

rhizoid a rootlike structure that carries on absorption.

rhizome horizontal underground stem, often enlarged for storage, that can carry on vegetative propagation.

rhythmic regular periodicity in organisms.

ribosomes tiny, dense granules attached to the endoplasmic reticulum and lying between its folds. They contain RNA and protein-synthesizing enzymes.

rickettsiae a group of organisms that cause disease, and midway between the viruses and bacteria in size.

rind the outer region of a corn stem containing sclerenchyma cells.

RNA (ribonucleic acid) a nucleic acid in which the sugar is ribose. A product of DNA, it serves in controlling certain cell activities, including protein synthesis.

rod a cell of the retina of the eye that receives impulses from light rays and that is sensitive to shades but not to colors.

root cap a tissue at the tip of a root that projects the tissues behind it.

root hair a projection of an epidermal cell of a young root.

root pressure that which is built up in roots due to water intake and the resulting turgor.

rostrum a protective area that is an extension of the carapace in crustaceans.

ruminant a cud-chewing ungulate; for example, a cow.

saliva a fluid secreted into the mouth by the salivary glands.

salivary gland a group of secretory cells producing saliva.

saprophyte an organism that lives on dead or nonliving organic matter.

scales the epidermal plates forming the outer covering in fishes and reptiles.

scavengers animals that feed on dead organisms.

scion the portion of a twig grafted onto a rooted stock.

sclerenchyma a plant-strengthening tissue, including fibers, mechanical tissue, and stone cells.

sclerotic layer the outer layer of the wall of the eyeball.

scolex knob-shaped head with hooks or suckers, as seen on some parasitic flatworms.

scrotum the pouch outside the body that contains the testes.

scutes the broad scales on the lower side of the snake's body.

secondary oocyte a cell that results from reduction division and develops into the ootid.

secondary root a branch root developing from the pericycle of another root.

secondary tissues those produced by the vascular cambium, a lateral secondary meristematic tissue, of a root or stem.

secretion formation of essential chemical substances by cells.

sedative an agent that depresses body activities.

seed a complete embryo plant surrounded by an endosperm and protected by seed coats.

segregation, law of Mendel's first law, based on his third hypothesis, stating that a pair of factors (genes) is segregated, or separated, during the formation of gametes (spores in lower plants).

selectively permeable membrane one that lets substances pass through more readily than other substances, but changes its permeability for a substance continually.

semen fertilizing fluid consisting of sperms and fluids from the seminal vesicle, prostate gland, and Cowper's gland.

semicircular canals the three curved passages in the inner ear that are associated with balance.

semilunar valves cup-shaped valves at the base of the aorta and the pulmonary artery that prevent backflow into the heart ventricles.

seminal receptacles structures that receive sperm cells in certain animals.

seminal vesicles structures that store sperm cells in certain animals.

seminiferous tubules a mass of coiled tubes in which the sperms are formed within the testes.

sensory neurons those that carry impulses from a receptor to the spinal cord or brain.

sepal the outermost part of a flower, usually green and not involved in the reproductive process.

septum a wall separating two cavities or masses of tissues; as, the nasal or heart septum.

serum   a substance (usually an extract of blood containing antibodies) used in treating disease and to produce immediate passive immunity.
serum albumin   a blood protein necessary for absorption.
serum globulin   a blood protein that contains antibodies.
sessile   a leaf lacking a petiole.
setae   bristles on the earthworm used in locomotion.
sex chromosomes   the two kinds of chromosomes (X and Y) that determine the sex of an offspring.
sex-influenced character   a characteristic that is dominant in one sex, recessive in the other.
sex-limited character   a characteristic that develops only in the presence of sex hormones.
sex-linked character   a recessive characteristic carried on the X type of sex chromosome.
sexual reproduction   that involving the union of a female gamete, or egg, and a male gamete, or sperm.
shell membrane   a double lining around the albumen and beneath the shell of a bird's egg.
shock   a body condition resulting from the lowering of fluid volume in the bloodstream.
shoot apex   a conical mass of meristematic tissue deep within the bud scales.
sieve plate   in the starfish, the opening of the water-vascular system to the outside.
sieve tube   a conducting tube of the phloem.
simple eye   a small, photosensitive organ of many lower animals.
sinoatrial node   a mass of tissue in the right atrium of the heart in which the beat originates.
sinus   a space or cavity.
sinus venosus   a thin-walled sac, formed by an enlargement of the cardinal vein of the fish and the venae cavae of the frog, that lies at the entrance to the heart.
skeletal muscle   that which is striated and voluntary.
slime layer   that which surrounds a bacterium.
small intestine   the digestive tube, about twenty-three feet long in man, that begins at the pylorus.
smog   a combination of the pollutant smoke with fog.
smooth muscle   that which is involuntary and that is found lining the walls of the intestine, stomach, and arteries.
soft palate   forms the roof of the mouth behind the chewing area.
soil flora   bacteria, molds, and other fungi that live in the soil.
solar plexus   the large nerve ganglion of the sympathetic nervous system, located in the abdomen.
solute   the dissolved substance in a solution.
solution   a homogeneous mixture of two or more substances.
solvent   the dissolving component of a solution.
sori   small clusters of sporangia that appear on fern leaves when mature.
spawn   to discharge gametes directly into water, as fish do; a mass of eggs discharged by a fish or other aquatic animal.
speciation   the development of a species.
sperm   a male reproductive cell.
spermatid   a structure formed from a secondary spermatocyte that matures into a sperm.
spermatocyte   (primary) a structure formed by meiosis from a spermatogonial cell; (secondary) a structure formed by division of a primary spermatocyte.
spermatogenesis   the process of the development of male reproductive cells whereby the diploid chromosome number is reduced to the haploid.
sphincter muscle   a ring of smooth muscle that closes or contracts an opening or a tube.
spicule   the material forming the skeleton of certain sponges.
spinal cord   the main dorsal nerve of the central nervous system in vertebrates, extending down the back from the medulla.
spinal nerves   large nerves connecting the spinal cord with various parts of the body.
spindle   the numerous fine threads formed between the poles of the nucleus during mitosis.
spinnerets   structures on the tip of a spider's abdomen containing numerous silk tubes.
spiracles   external openings of the insect's tracheal tubes on the thorax and abdomen.
spirillum   a spiral-shaped bacterium.
spirochete   a group of spiral-shaped one-celled organisms resembling both protozoans and bacteria.
spongin   fibers composing the skeleton of certain sponges.
spongocoel   the central cavity in sponges.
spongy mesophyll   loosely constructed leaf tissue containing many spaces.
spontaneous generation   a disproved belief that certain nonliving or dead materials could be transformed into living organisms.
sporangiophore   in molds, an ascending hypha bearing sporangia.
sporangium   a structure that produces spores.
spore   an asexual reproductive cell.
sporophyte   the stage that produces spores in an organism having alternation of generations.
stamen   the male reproductive part of the flower bearing an anther at its tip.
statocyst   the balancing organ of the crayfish.
sternum   breastbone.
stigma   the part of the pistil that receives pollen grains.
stimulant   an agent that increases or elevates body activity.
stimulus   a factor or environmental change capable of producing activity in protoplasm.
stipe   the stalk portion of a fruiting body of a mushroom.

stipule a leaflike or scalelike structure at the base of many leaf petioles.

stock the plant on which a scion has been grafted; a line of descent.

stolon a transverse hypha of a mold; also, a horizontal, creeping underground stem.

stomach an organ that receives ingested food, prepares it for digestion, and begins protein digestion.

stomata pores regulating the passage of air and water vapor to and from the leaf.

style the stalk of the pistil.

sublingual gland one of the pair of salivary glands lying under the tongue.

submaxillary gland one of the pair of salivary glands lying in the angle of the lower jaw.

substrate a layer or substance an organism takes root in or adheres to.

succession the changing plant and animal populations of a given area.

successive osmosis the cell-to-cell diffusion of water.

succulent leaf a thick, fleshy leaf.

suspension a mixture formed by particles that are larger than ions or molecules.

sweepstakes dispersal the movement of organisms into new areas despite strong barriers.

swimmerets appendages on the abdomen of a crustacean.

symbiosis the relationship in which two organisms live together for the advantage of each.

sympathetic nervous system a division of the autonomic nervous system.

synapse the space between nerve endings.

synapsis the coming together of homologous pairs of chromosomes during meiosis.

synergid one of two structures formed on either side of the egg in the embryo sac of flowers.

syngamy the fusion of gametes during fertilization.

synovial fluid a secretion of cartilage that lubricates a joint.

systemic relating to the body.

system a group of organs forming a functional unit.

systole part of the cycle of the heart during which the ventricles contract and force blood into the arteries.

systolic blood pressure arterial pressure produced when the ventricles contract.

taproot the main root of a plant, often serving as a food reservoir.

taste buds flask-shaped structures in the tongue containing nerve endings that are stimulated by flavors.

teliospore a two-celled, black winter spore of wheat rust.

telophase the last stage of mitosis, during which two daughter cells are formed.

telson the posterior segment of the abdomen of certain crustaceans.

temperate phage a phage that injects its DNA into a bacterial cell without causing the production of phage particles.

temperature inversion an atmospheric phenomenon in which a layer of cold air moves under a lid of warm air.

template a pattern, such as a specific sequence of bases in the messenger RNA molecule that acts as a pattern for the synthesis of a protein molecule.

tendon a strong band of connective tissue in which the fleshy portion of a muscle terminates.

tendril a part of a plant modified for climbing.

tentacle a long appendage, or "feeler," of certain invertebrates.

terrestrial land-living.

testa the outer seed coat.

testes the male reproductive organs of higher animals.

tetrad a group of four cells.

tetraploid a term used to indicate that a cell has four sets of homologous chromosomes.

thallus a plant body, such as an alga, that lacks differentiation into stems, leaves, and roots and that does not grow from an apical point.

theory a hypothesis that is continually supported by experimental evidence.

thermal pollution the addition of heat to a body of water.

thigmotropism a response of an organ or an organism to an object; for example, the wrapping of the tendrils of a climbing plant around an object.

thoracic duct a vessel carrying lymph and emptying into the left subclavian vein.

thorax the middle region of the body of an insect between the head and abdomen; the chest region of mammals.

thrombin a substance formed in blood clotting as a result of the reaction of prothrombin, thromboplastin, and calcium.

thromboplastin a substance essential to blood clotting formed by disintegration of blood platelets.

thymus one of the ductless glands, situated in the upper chest cavity.

thyroid the ductless gland, located in the neck on either side of the larynx, that regulates metabolism.

tissue a group of cells that are similar in structure and function.

tissue fluid that which bathes the cells of the body; called lymph when contained in vessels.

tolerance an organism's ability to withstand an environmental condition.

tone (muscle) the condition in which flexor and extensor muscles oppose each other, resulting in a continuous state of slight contraction.

tonsil a mass of lymphatic tissue in the throat of higher animals.

topography the physical features of the earth.

toxin a poisonous substance produced by bacteria and other organisms that acts in the body or on foods.

toxin-antitoxin a mixture of diptheria antitoxin and toxin, formerly used to develop active immunity.

toxoid toxin weakened by mixing with formaldehyde or salt solution, used extensively to develop immunity to diphtheria, scarlet fever, and tetanus.

trachea an air tube in insects and spiders; the windpipe in air-breathing vertebrates.

tracheids thick-walled conducting tubes that strengthen woody tissue.

tracheophyte a plant that has vessels for the conduction of fluids.

transfer RNA a form of RNA thought to deliver amino acids to the template formed by messenger RNA on the ribosomes.

transformation in pneumococcus, the change from a noncapsulated to a capsulated form, brought about by the transfer of DNA.

transformers bacteria that change the simpler substances left by the decomposers into nitrogen compounds that are used by plants.

translocation the movement of water, dissolved foods, and other substances in plants.

transpiration the loss of water from plants.

transpiration-cohesion theory an explanation of water translocation based on the loss of water through leaves and an attraction of water molecules for one another.

transpiration pull one of the forces involved in the rise of water in a stem. As cells lose water to the atmosphere, water enters them from adjacent cells, resulting in upward movement of water.

trichocysts sensitive protoplasmic threads in the paramecium, concerned with protection.

trisomy the presence of three homologous chromosomes in all body cells.

trochophore a larval form of mollusks.

tropism an involuntary growth response of an organism to a stimulus.

truncus arteriosus a branch of the conus arteriosus in the frog.

tube feet movable suction discs on the rays of most echinoderms.

tube nucleus one of the three nuclei present in a pollen tube.

tuber an enlarged tip of a rhizome swollen with stored food.

tubule a tiny collecting tube extending from a Bowman's capsule of a kidney.

turbinate one of three layers of bones in the nasal passages.

turgor the stiffness of plant cells due to the presence of water.

tympanic membrane the eardrum.

tympanum a membrane in certain arthropods, serving a vibratory function.

umbilical cord found in female mammals, leading from the placenta to the embryo.

unit character, principle of Mendel's concept that the various hereditary characteristics are controlled by factors (genes), and that these factors occur in pairs.

urea a nitrogenous waste substance found chiefly in the urine of mammals, but also formed in the liver from broken-down proteins.

urediospore a one-celled, red summer spore of wheat rust.

ureter a tube leading from a kidney to the bladder or cloaca.

urethra the tube leading from the urinary bladder to an external opening of the body.

uric acid a nitrogenous waste product of cell activity.

urinary bladder the sac at the base of the ureters that stores urine.

urine the liquid waste filtered from the blood in the kidney and voided by the bladder.

uropod a flipper, or developed swimmeret, at the posterior end of the crayfish.

uterus the organ in which young mammals are nourished until they are ready for birth.

uvula the extension of the soft palate.

vaccination method of producing immunity by inoculating with a vaccine.

vaccine a substance used to produce immunity.

vacuolar membrane a membrane surrounding a vacuole in a cell and regulating the movement of materials in and out of the vacuole.

vacuole one of the spaces scattered through the cytoplasm of a cell and containing fluid.

vagina cavity of the female immediately outside and surrounding the cervix of the uterus.

vagus nerve the principal nerve of the parasympathetic nervous system.

valve one of the shells of a mollusk; also, a fold or flap of tissue controlling the direction of blood flow, as in the heart or veins of some organisms.

vasa deferentia ducts that transport sperms from the testes.

vasa efferentia tiny tubes in the reproductive system of the frog through which sperm pass.

vascular bundles strands of phloem and xylem found in the roots, stems, and leaves of higher plants.

vascular cylinder the innermost region of a root, containing the xylem and phloem.

vascular rays sheets or ribbons of parenchyma cells radiating from the pith through the xylem of a woody stem.

vascular tissue fluid-conducting tissues characteristic of the tracheophytes.

vector (or *arthropod carrier*) an insect or other arthropod that carries infectious organisms.

veins strengthening and conducting structures in leaves; vessels carrying blood to the heart.

vena cava a large collecting vein found in many vertebrates.

venation the arrangement of veins through the leaf blade.

venom the poison secreted by glands of poisonous snakes or other animals.

ventricle a muscular chamber of the heart; also, a space in the brain.

venules small branches of veins.

vermiform appendix a fingerlike outgrowth of the intestinal caecum in man.

vertebra a bone of the spinal column of vertebrates.

vertebrate an animal with a backbone.

vessel a large, tubular cell of xylem through which water and minerals are conducted.

vestigial organs those that are poorly developed and not functioning.

villi   microscopic projections of the wall of the small intestine that increase the absorbing surface.

virulence   the potency or the ability of an organism to cause disease.

virulent phage   a bacteriophage that produces a lytic cycle of destruction.

viruses   particles that are noncellular and have no nucleus, no cytoplasm, and no surrounding membrane. They may reproduce in living tissue.

visual purple   the chemical in the rods of the eye necessary for their functioning in reduced light.

vitamin   an organic substance that, though not a food, is essential for normal body activity.

vitreous humor   a transparent substance that fills the interior of the eyeball.

viviparous   refers to animals that bear their young alive, and that nourish them before birth by means of a placenta.

vocal cords   those structures within the larynx that vibrate to produce sound.

voluntary muscle   a striated muscle that can be controlled at will.

vomerine teeth   scalelike teeth in the roof of the frog's mouth that aid in holding prey.

water cycle   the continuous movement of water from the atmosphere to the earth and from the earth to the atmosphere.

watershed   a hilly region, usually exending over a large area, that conducts surface water to streams.

water table   the level at which water is standing underground.

water-vascular system   the circulatory system of certain echinoderms.

white corpuscles   colorless, nucleated blood cells.

withdrawal symptoms   nervous reactions and hallucinations resulting from the lack of a drug to which the victim is addicted.

X chromosome   a sex chromosome present singly in human males and as a pair in females.

xanthophyll   a yellow pigment, one of the chlorophyll pigments found in certain chromoplasts.

xerophyte   a plant that requires very little water to live.

xylem   the woody tissue of a root or stem that conducts water and dissolved minerals upward.

Y chromosome   a sex chromosome found only in males.

yellow marrow   fills the central cavity of a long bone and is primarily composed of fat cells.

yolk   the part of the bird's egg from which the egg cell obtains its nourishment.

yolk sac   an extraembryonic membrane providing food for the embryo.

zoospores   in *Ulothrix* and certain other algae, the flagellated cells that leave the mother cell and later develop into new organisms.

zygospore   the dormant form of some protists when the zygote forms a thick protective wall.

zygote   a fusion body formed when two gametes unite.

Page references for illustrations are printed in **boldface**.

The American Time Capsule